Intermediate Algebra

SIXTH EDITION

M.A. Munem

C. West

Macomb College

 KENDALL/HUNT PUBLISHING COMPANY
4050 Westmark Drive   Dubuque, Iowa 52002

Book Team

Chairman and Chief Executive Officer   Mark C. Falb
Senior Vice President, College Division   Thomas W. Gantz
Director of National Book Program   Paul B. Carty
Editorial Development Manager   Georgia Botsford
Developmental Editor   Angela Willenbring
Vice President, Production and Manufacturing   Alfred C. Grisanti
Assistant Vice President, Production Services   Christina E. O'Brien
Prepress Editor   Ellen Kaune
Managing Editor, College Field   David L. Tart
Associate Managing Editor, College Field   W. Ray Wood
Acquisitions Editor, College Field   David Mattaliano

Preface

This sixth edition aims to do exactly what its title suggests: to prepare students to work actively at *solving problems, studying examples, and investigating troublesome points in algebra* through *graphical, numerical, verbal* and *analytic representations.* To accomplish these goals we continue to focus on a number of features that help students *explore, learn,* and conceptually *understand* the topics in intermediate algebra.

Improvements in this edition evolved out of the generous response from users of the previous editions as well as a talented team of reviewers and colleagues. In addition, we incorporated other improvements based on a market survey of instructors, course outlines, and college catalogues. These improvements were fundamental to this book's growth and effectiveness in the classroom.

Style

The textbook was written for student comprehension. Great care was taken to write a book that is mathematically correct and accessible to students. Short sentences and clear explanations are used. To make the book relevant to students, special practical applications and models throughout the book were used.

Organization

In this edition, we continue to incorporate the best aspects of reform in a meaningful yet easy-to-use manner. We reordered the sequence of topics traditionally found in intermediate algebra books. Here are some of the organizational changes.

1. **Review Material:** The review material from beginning algebra was placed separately in Chapter R in the beginning of the book to help students with varied backgrounds in algebra.
2. **Principal Changes:** The material in Chapters 4 and 5 on polynomial expressions and functions, as well as, rational expressions and functions has been extensively revised and reorganized. This allows the embedding of graphs, zeros of functions, intercepts, and solving equation graphically in these chapters.
3. **Increased Coverage:** Increased discussion of radicals, quadratic functions as well as exponential and logarithmic functions (in Chapters 6, 7, and 8) encourage students to make connections among these topics. Integrated reminders were also used throughout the book to help students improve their study skills and serve to emphasize the importance of these concepts throughout the chapters.
4. **Functions:** Functions were introduced in Chapter 2, and they were used throughout the book. Increased emphasis has been placed on the use of functions that describe interesting real-world situations.

5. **Systems of Equations:** Systems of linear equations were introduced in Chapter 3 to provide students with problem-solving tools. This will give the student the opportunity to translate problem situations into systems of linear equations. It also provides a useful reinforcement to translating problems into linear equations.

6. **Graphing:** Graphing was introduced in Chapter 1 of this edition and was integrated throughout the book. Sketching graphs by hand and equation recognition are fundamental parts of learning mathematics at this level. *Interpreting graphs, exploring new ideas from graphs,* and *extracting information from graphs* were of primary importance in this edition. In addition, the text contains numerous graphs, models, and illustrations to visually clarify and reinforce concepts.

Special Features

The following pedagogical features influenced some of the decisions we made in preparing this edition.

1. **Four-Color Design:** New to this edition was a fundamental four-color design. For easy reference, important definitions, properties and procedures were highlighted in color screened. Artwork was clarified with the use of multiple colors. It also allowed for easy identification of important features.

2. **Real-World Chapter Openers:** Each chapter opener focuses on a real-world application together with a colored photograph describing the application. Also there is a reference showing where this application is worked out in the chapter.

3. **Margin Notes:** Helpful hints were strategically placed in the margins for the purpose of reinforcing important concepts. They were highlighted with a different color as a quick reference.

4. **Section Objectives:** In this edition, we continued to open each section with a list of skills that the student should learn in that section.

5. **Revised Examples:** This edition continues to make algebraic skills attainable by providing numerous illustrative examples that are presented one step at a time. All examples, whether they are drill or applied, are revised and many new examples are added. They have a title so that students can see the purpose of each example. Each section includes solving applied problems from various disciplines, thus showing students the current application of these ideas. We focused on adding examples at the end of each section that involve applications.

6. **Revised Problem Sets:** All problems sets have been reorganized and rewritten. They were carefully graded in level of difficulty. They have been checked and rechecked for accuracy by professors who class-tested this edition of the book. The exercises in each problem set have been grouped into three categories:

   (a) The first category was designed to provide the practice needed to *master the concepts* in the book. They were generally modeled after the examples and follow the order of presentation of the book.

   (b) The second category provided a broad range of applications and models that require students to *apply the concepts*. Special emphasis has been put into the design of these problems.

   (c) The third category challenged students to *develop and extend the concepts*. These problems encourage *critical thinking*, and provide an excellent opportunity to stimulate class discussions that foster *collaborative learning* among students.

   The problems in the last two categories offer the opportunity for writing assignments such as a report or a class project. Also new to this edition are many modeling data problems that ask students to find and interpret mathematical models from the real-life data.

1. **Review Problems and Chapter Tests:** As with previous editions, each chapter concludes with a review problem set. Also, a chapter test follows the comprehensive collection of chapter review problems. Chapter tests focus on the problems in the chapter so that students can determine if they are ready to take an actual class test.

2. **Applications and Models:** All applications, models, and data have been thoroughly updated in this edition so as to link the student's own experience to these real-world situations. For this reason, throughout the book, we have placed emphasis on applied problems drawn from a variety of disciplines. We also believe that we offer an approach to mathematical modeling that will capture the student's attention. Based on our classroom experiences, we have selected certain types of problems and organized their solutions through a sequence of questions that help students through the difficult thought processes of developing mathematical modes.

3. **Calculator Exploration:** The use of a grapher (a term used for a **graphing calculator**) is *optional,* although it is likely that some students and instructors will want to make use of this device. To assist these students and instructors optional graphing activities that used graphers were included throughout the book. All of the optional grapher's material can be omitted without loss of continuity. Scientific calculators will be needed to go through the material of this book.

Supplements

The sixth edition is accompanied by the following supplemental material.

- **For the Instructor:** An instructor's test manual includes six different test forms for each chapter of the textbook. These tests have been written at graded levels of difficulty, with three of the tests for each chapter made up of multiple choice questions, and three of standard problem-solving questions. Also available to the instructor are answers to all even numbered problems in the book. Both the test bank and even answers are available in hard copy or CD.

- **For the Student:** A student solution manual offers worked-out solutions to all odd numbered problems in the book as well as all problems in the chapter tests. This manual is packaged with the text in the form of a CD.

Acknowledgments

In preparing this edition, we have drawn from our own experience in teaching intermediate algebra as well as the feedback provided by our friends, colleagues, and students. We offer them all our grateful thanks. At the risk of omitting some, they include the following:

William Coppage, Wright State University

Linda Exley, DeKalb Community College – Clarkston

Merle Friel, Humboldt State University

Gene Hall, DeKalb Community College – Clarkston

Gus Pekara, Oklahoma City Community College

Stuart Thomas, Oregon State University

Roger Willig, Montgomery County Community College

Jim Wolfe, Portland Community College

Special thanks are due to our many colleagues at Macomb College who have taught from previous editions. We are especially grateful to our colleagues

Victoria Ackerman

Steve Fasbinder

Marion Rappa

who worked on the book in various stages of development. Their advice and comments were very helpful.

A special note of appreciation is also due to our colleagues Christina Guastella and Robert Eckert who proofread the page proofs and checked the accuracy of the problems. In order to guarantee the accuracy of the answers provided, all the problems and examples were checked by an independent problem solver, Julie Berens.

Finally, we were very grateful to the staff of Kendall/Hunt for their cooperation and assistance throughout the project. We would also like to express our sincere gratitude to the staff of Carlisle Publisher Services who carried out the details of production in a dedicated and caring way.

M. A. Munem
C. West

Contents

Review of Basic Concepts

Chapter Contents

The Norman style of architecture is characterized by massive construction, carved decorations, and rectangular windows surmounted by semicircular arches. The photograph below shows a window from St. Mary's Church in Detroit, MI, built in 1835.

The area of such a window is given by the formula

$$A = 2ry + \frac{1}{2}\pi r^2$$

where r is the radius of the semicircle and y is the length of the rectangle. Find the total area of a Norman window with width 6 feet and length 8 feet. This is discussed in example 2 of section R6 on page 34.

The study of Intermediate Algebra depends on successful completion of a Beginning Algebra course. This review material is intended to highlight particular topics that are important, and to establish a common language. Review is the key word in this introduction.

Objectives

1. Introduce Notation
2. Use Exponents
3. Use Order of Operations

R1. Notation and the Order of Operations

Introducing Notation

Algebra begins with a systematic study of the addition, subtraction, multiplication, and division operations which serve as a basis for all arithmetic calculations. In order to achieve generality, letters of the alphabet such as a, b, c, x, and y are used to represent numbers. A letter that represents an arbitrary number is called a **variable.** The four arithmetic operations are described in symbols and words in Table 1, where a and b represent any two numbers.

TABLE 1

Operation	Operation in Symbols	Operation in Words
Addition	$a + b$	sum of a and b
Subtraction	$a - b$	difference of a and b
Multiplication	ab, $a \cdot b$, $a(b)$, $(a)b$	product of a and b
Division	a/b, $a \div b$, $\dfrac{a}{b}$, $b \neq 0$	quotient of a and b

We can use the relationship between division and multiplication to clarify division involving zero. Suppose we try to divide 5 by 0. Let us say the answer is a number represented by the letter c. That is,

$$c = 5 \div 0$$

this means that

$$5 = c \cdot 0.$$

This is impossible, since $c \cdot 0 = 0$.

In algebra, we say that 5/0 is undefined. However, $0 \div 5 = 0$ since $5 \cdot 0 = 0$.

In general,

Note that 0/0 is indeterminate

$$\frac{a}{0} \text{ is undefined and } \frac{0}{a} = 0, \text{ where } a \neq 0.$$

Table 2 shows six additional symbols that are used for comparing numbers and expressions.

TABLE 2

Comparison in Symbols	Interpretation in Words	Example
$a = b$	a is equal to b	$14 = 10 + 4$
$a \neq b$	a is not equal to b	$3 \neq 5$
$a \leq b$	a is less than or equal to b	$3 \leq 6$
$a < b$	a is less than b	$-3 < 6$
$a \geq b$	a is greater than or equal to b	$5 \geq -5$
$a > b$	a is greater than b	$3 > -2$
$a > 0$	a is positive	$7 > 0$
$a < 0$	a is negative	$-2 < 0$
$a \geq 0$	a is non-negative	$0 \geq 0$

EXAMPLE 1   **Translating Words into Symbols and Vice Versa**

Write the following statement in symbols.

(a) The sum of 4 and twice y is greater than 5.

(b) Express $4 + 3y > 10$ in words.

Solution   (a) $4 + 2y > 5$

(b) The sum of 4 and the product of 3 and y is greater than 10.

Using Exponents

Algebraic notation is designed to clarify ideas and simplify calculations by allowing us to write expressions compactly and efficiently. For instance, the product $3 \cdot 3$ can be written as 3^2, read "3 squared". The product $3 \cdot 3 \cdot 3 \cdot 3$ can be written as 3^4, read "3 to the fourth power". The expression 3^4 is called an **exponential form**, whereas $3 \cdot 3 \cdot 3 \cdot 3$ is called an **expanded** or **factored form** of 3^4 and each number is called a **factor** of 3^4. Since $3 \cdot 3 \cdot 3 \cdot 3 = 81$, we say that the **value** of 3^4 is 81. The use of exponents provides a compact notation for products. Thus,

$$x \cdot x = x^2, \quad x \cdot x \cdot x = x^3, \quad \text{and} \quad x \cdot x \cdot x \cdot x = x^4.$$

In general, if n is a positive integer, then

Note that $x^1 = x$.

$$x^n = \overbrace{x \cdot x \cdot x \cdot \ldots \cdot x}^{n \text{ factors}}$$

In using the **exponential notation** x^n, we refer to x as the **base** and n as the **exponent** or **power**.

EXAMPLE 2   **Using Exponential Notation**

(a) Find the value of the expression 5^4.

(b) Write the expression $a \cdot b \cdot b \cdot b \cdot c \cdot c$ in exponential notation.

Solution   (a) $5^4 = 5 \cdot 5 \cdot 5 \cdot 5 = 625$ 5 is the base and 4 is the power

(b) We have 1 factor of a, 4 factors of b, and 2 factors of c. Hence the exponent of a is 1, of b is 4, and of c is 2.

$$a \cdot b \cdot b \cdot b \cdot b \cdot c \cdot c = ab^4c^2$$

Using the Order of Operations

Consider the problem of evaluating the expression

$$2 + 3 \cdot 5$$

If we add 2 and 3 first, then multiply by 5, we get 25. If we take the product of 3 and 5 first, then add 2, we get 17. In order to avoid having two different values for the same expression, the following steps describe the order in which the operations should be performed.

The Order of Operations

A fraction bar is a grouping symbol which groups the numerator and the denominator.

1. **Grouping Symbols:** Perform operations inside grouping symbols. Grouping symbols include **parentheses ( )**, **brackets [ ]**, **braces { }**, and **fraction bar rules.** Work from the innermost grouping symbols to the outermost.
2. **Powers:** Find the value of any powers indicated by exponents.
3. **Multiply (Divide):** Perform all multiplication and division from left to right.
4. **Addition (Subtraction):** Perform all additions and subtractions from left to right.

EXAMPLE 3   **Using the Order of Operations**

Find the value of each expression.

(a) $3 \cdot 2^3 + \dfrac{10}{5} - 3^2$ (b) $5[4^3 + 3(6^2 - 3 \cdot 2)]$ (c) $\dfrac{5^2 + 2 \cdot 5}{3 + 10 \div 5}$

Solution   (a) $3 \cdot 2^3 + \dfrac{10}{5} - 3^2$ Given

$= 3 \cdot 8 + \dfrac{10}{5} - 9$ Powers

$= 24 + 2 - 9 = 17$ Multiplication, division, and then addition, subtraction

(b) $5[4^3 + 3(6^2 - 3 \cdot 2)]$ Given

$= 5[64 + 3(36 - 6)]$ Powers and multiplication

$= 5[64 + 3(30)]$ Innermost parentheses

$= 5[64 + 90]$ Multiplication inside parentheses

$= 5[154] = 770$ Add and multiply

(c) $\dfrac{5^2 + 2 \cdot 5}{3 + 10 \div 5}$ Use of a fraction bar

$= (25 + 10) \div (3 + 2)$ Simplify numerator and denominator separately

$= 35 \div 5 = 7$ Divide

An expression formed from any combination of numbers and variables by using the four operations in Table 1 as well as raising to powers or taking roots is called an **algebraic expression.** If an algebraic expression consists of parts connected by plus or minus signs, each of the parts together with the sign preceding it is called a **term.** For example,

$$4x^2 + 3x - \frac{2}{x}$$

is an algebraic expression with terms:

$$4x^2, \; +3x \text{ and } -\frac{2}{x}.$$

EXAMPLE 4 **Evaluating an Expression for Specific Values**

Evaluate $\dfrac{5x + y^2}{3}$ for $x = 12$ and $y = -3$.

Solution   Using parentheses to substitute 12 for x and -3 for y in the expression yields

$$\frac{5x + y^2}{3} = \frac{5(12) + (-3)^2}{3} \qquad \text{Power and multiplication}$$

$$= \frac{60 + 9}{3} \qquad \text{Fraction bar is the grouping symbol}$$

$$= \frac{69}{3} = 23$$

EXAMPLE 5 **Translating a Situation into Symbols**

The cruise control of a car is set at y miles per hour (mi/hr), then the brakes are applied to drop the speed by 30 miles per hour. To pass a truck, the speed is doubled.

(a) Write an algebraic expression that describes the speed when passing the truck.

(b) Find the speed while passing the truck if the cruise control is initially set at:
   (i) 55 miles per hour.
   (ii) 65 miles per hour.

Solution   (a) Let y equal the initial speed in miles per hour. Then $y - 30$ is the speed dropped by 30 miles per hour, and $2(y - 30)$ is the speed doubled while passing the truck.

(b) We substitute the different initial speeds into this expression.
   (i) If the initial speed is 55 mi/hr, then the speed passing the truck is
$2(y - 30) = 2(55 - 30) = 2(25) = 50$ mi/hr.
   (ii) If the initial speed is 65 mi/hr, then the speed passing the truck is
$2(y - 30) = 2(65 - 30) = 2(35) = 70$ mi/hr.

◈ PROBLEM SET R1.

In Problems 1–4, write each statement in symbols.

1. (a) The sum of 4 and 5 equals 9.
   (b) The sum of 5 and 4 equals 9.
   (c) The sum of 4 and 5 equals the sum of 5 and 4.

2. (a) The difference of 12 and 2 equals 10.
   (b) The difference of 2 and 12 does not equal 10.
   (c) The difference of 12 and 2 does not equal the difference of 2 and 12.

3. The product of x and 5 is greater than the quotient of x and 5.

4. The difference of twice x and 7 is less than or equal to 21.

In Problems 5–8, express each algebraic statement in words.

5. (a) $5x = 20$

   (b) $5x - 4 = 20$

6. (a) $\dfrac{x}{3} = 7$

   (b) $\dfrac{x}{3} + 2 = 7$

7. (a) $x + 2 < 10$

   (b) $10 > x + 2$

8. (a) $3x \le 15$

   (b) $15 \ge 3x$

In Problems 9 and 10, write each expression in expanded form and find its value.

9. (a) 6^2

   (b) 3^6

10. (a) 11^2

    (b) 2^{11}

In Problems 11 and 12, write each expression in exponential form.

11. (a) $2 \cdot 2 \cdot x \cdot y \cdot y \cdot y$

    (b) $2 \cdot 2 \cdot 3 \cdot 3 \cdot y \cdot y \cdot z$

12. (a) $3 \cdot 3 \cdot 3 \cdot u \cdot u \cdot u$

    (b) $3 \cdot 3 \cdot 3 \cdot 4 \cdot u \cdot v \cdot v$

In Problems 13–20, apply the order of operations to find the value of each expression.

13. (a) $6 + 12 - 2 - 8$

    (b) $6 + (12 - (2 + 8))$

14. (a) $2 \cdot \frac{3}{3} \cdot 10$

    (b) $2 \cdot 3 \div 3 \cdot 10$

15. (a) $3(2 + 3(7 - 4))$

    (b) $3 \cdot 2 + 9(7 - 4)$

16. (a) $8(3 + (20 - 9)) + 6$

    (b) $24 + 160 - 72 + 6$

17. $3^4 - 2^4$

18. $(3 - 2)^4 + 7^3$

19. $\dfrac{(9 + 6)^2}{5}$

20. $12 + 3 \cdot 4^2 - 5$

In Problems 21 and 22, find the value of each expression.

21. (a) $\dfrac{24}{6} + 2$

    (b) $\dfrac{24}{6 + 2}$

22. (a) $\dfrac{10}{5} + 5$

    (b) $\dfrac{10}{5 + 5}$

In Problems 23–28, evaluate each expression by substituting the given values for the variables.

23. $7 + 7x + 7y$ for $x = 2$, $y = 3$

24. $\dfrac{x + y}{x - y}$ for $x = 5$, $y = 3$

25. (a) $10(x - y) - 4(x - y)$ for $x = 20$, $y = 10$

    (b) $6(x - y)$ for $x = 20$, $y = 10$

26. (a) $\dfrac{4x + 4y}{4x}$ for $x = 10$, $y = 10$

    (b) $1 + \dfrac{y}{x}$ for $x = 10$, $y = 10$

27. (a) $(x - y)^2$ for $x = 5$, $y = 2$

    (b) $x^2 - 2xy + y^2$ for $x = 5$, $y = 2$

28. (a) $(x + 1)^3$ for $x = 4$

    (b) $x^3 + 3x^2 + 3x + 1$ for $x = 4$

In Problems 29–32, write each sentence in symbols.

29. (a) The product of 5 and the sum of 3 and 4 equals 35.

    (b) The sum of the product of 5 and 3 and the product of 5 and 4 equals 35.

30. (a) The product of 3 and the difference of 6 and 4 equals 6.

    (b) The difference of the product of 3 and 6 and the product of 3 and 4 equals 6.

31. (a) One number is 10 times 20 more than another.

    (b) One number is 20 more than 10 times another.

32. (a) One number is the square of the sum of two other numbers.

    (b) One number is the sum of the squares of two other numbers.

In Problems 33 and 34, express each algebraic statement in words.

33. (a) $3y - 1 \ne 5$

    (b) $3(y - 1) = 5$

34. (a) $\dfrac{y}{4} + 3 = 7$

    (b) $\dfrac{y + 3}{4} \ne 7$

In Problems 35–38, insert parentheses so that each statement is true.

35. (a) $4 + 5 \cdot 2 + 3 = 21$

    (b) $4 + 5 \cdot 2 + 3 = 45$

    (c) $4 + 5 \cdot 2 + 3 = 29$

36. (a) $12 \div 3 + 1 + 2 = 7$

    (b) $12 \div 3 + 1 + 2 = 5$

    (c) $12 \div 3 + 1 + 2 = 2$

37. (a) $4 + 3 \cdot 8 - 1 + 6 = 31$

    (b) $4 + 3 \cdot 8 - 1 + 6 = 33$

    (c) $4 + 3 \cdot 8 - 1 + 6 = 7$

38. (a) $4 \cdot 6 + 1 - 5 + 4 \cdot 3 = 1$

    (b) $4 \cdot 6 + 1 - 5 + 4 \cdot 3 = 35$

    (c) $4 \cdot 6 + 1 - 5 + 4 \cdot 3 = 20$

In Problems 39–44, find the value of each expression, if possible. (Hint: Recall that division by zero is undefined.)

39. (a) $\dfrac{7(4-4)}{3(5+3)}$

(b) $\dfrac{3(5+3)}{7(4-4)}$

40. (a) $\dfrac{9+3\cdot 2}{24-8\cdot 3}$

(b) $\dfrac{24-8\cdot 3}{9+3\cdot 2}$

41. $\dfrac{25^2+18\div 2\cdot 3+8\cdot 5}{2\cdot 2}$

42. $\dfrac{3^2(3+2)+3(3+2)^2}{5\cdot 2-2(3-1)}$

43. $\dfrac{(5+7)^2-(3\cdot 4)^2}{5^2+7^2-3\cdot 4^2}$

44. $\dfrac{5^2+7^2-3\cdot 4^2}{(5+7)^2-(3\cdot 4)^2}$

In Problems 45 and 46, write the calculator key strokes necessary to find the value of each expression on your calculator.

45. (a) $3+4\div 5$

(b) $(3+4)\div 5$

(c) $4+\dfrac{5(3-2)}{10}$

46. (a) $\dfrac{1}{3+4}$

(b) $\dfrac{1}{3}+\dfrac{1}{4}$

(c) $(4+17)(18-3)$

47. House Value: A house that originally cost $95,000 doubled in value over several years. Due to a recession, it then decreased in value by $y.

(a) Write an expression for the value of the house after the recession.

(b) Find its value if $y = \$5000$ and $y = \$7000$.

48. Sunset: A week ago the sun set at 6:55 PM. Each day for the next five days the sun sets x minutes earlier.

(a) Write an expression for the time the sun sets after the five days.

(b) At what time would it set if $x = 4$ minutes?

49. Depreciation: A computer for an office costs $w and will be used for five years, at which time it will have a salvage value of $300. If the company uses straight line depreciation (the difference of cost and salvage value divided by useful life),

(a) Write an expression for the yearly depreciation amount.

(b) If the computer costs $4100, what is the yearly depreciation amount?

R2. Properties of Real Numbers

Objectives

1. Introduce Sets
2. Create a Number Line
3. Categorize Numbers in Sets
4. Introduce the Properties of Real Numbers

In this section, we review the sets of numbers and their properties.

Introducing the Notation of Sets

From experience you know that a family is a well-defined collection of people. A person in a family is called a member of the family. Last names distinguish families like Abel, Bonnucci, Carter, Donic, etc. To generalize this idea, a **set** is a well-defined collection of objects or numbers. An object in a set is called an **element** of the set. Capital letters, such as $A, B, C,$ and $D,$ are often used as labels for sets. Braces, {}, are used to enclose the elements of a set which are separated by commas. Thus, we write

$$A = \{1, 2, 3, 4, 5\}$$

to represent the set A whose elements are 1, 2, 3, 4, 5. The symbol $\in$ is used to mean that an element belongs to a set. The notation

$$3 \in A$$

is read "3 is an element of the set A". The symbol $\notin$ indicates that an element does not belong to a set.

If a set contains no elements, we call it the **empty set** or the **null set** and is denoted by {} or the symbol $\varnothing$. For example, the set of all numbers that are both odd and even is the empty set.

The set $\{\varnothing\}$ does not represent the empty set since this set contains one element, the symbol $\varnothing$.

The set $A = \{1,3,5,7,11,13\}$ contains a finite number of elements and is considered a **finite set.** The set of odd counting numbers

$$B = \{1,3,5,7,\ldots\}$$

contains an infinite number of elements and is considered an **infinite set.** The three dots (ellipsis) indicate that the list of elements continues without ending.

Consider the sets $A = \{1,2,3\}$ and $B = \{1,2,3,5\}$. Every element of set A is also an element of set B. We say, "A is a **subset** of B" and write

$$A \subseteq B$$

For example, the subsets of the set $\{1,2,3\}$ are $\varnothing$, $\{1\}$, $\{2\}$, $\{3\}$, $\{1,2\}$, $\{1,3\}$, $\{2,3\}$, and $\{1,2,3\}$.

Sometimes it is easier to describe the elements of a set rather than listing them. In such cases, we may use a new notation called **set-builder notation** that includes conditions that define the set. Using this notation, we describe the set A that consists of all numbers greater than 3 as

$$A = \{x \mid x > 3\}$$

which is read "the set of all numbers x such that x is greater than 3".

Creating a Number Line

Sets of numbers and relations among such sets can be visualized by the use of a *number line* or *coordinate axis.* A **number line** is constructed by drawing a line, fixing a point called the **origin,** and associating that point with the number 0. We choose an arbitrary *unit length* and mark off ticks of this length to the right and left of the origin. The unit distance between consecutive tick marks is called the **scale.** The *positive numbers* extend to the right of the origin and *negative numbers* to the left (Figure 1).

Figure 1

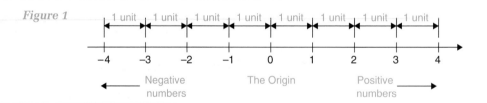

The point associated with a number on a number line is called its **graph,** and the number is called the **coordinate** of that point. For example, in Figure 2 the coordinates of points A, B, and C on the number line are the -3, 2, and 5, respectively.

Figure 2

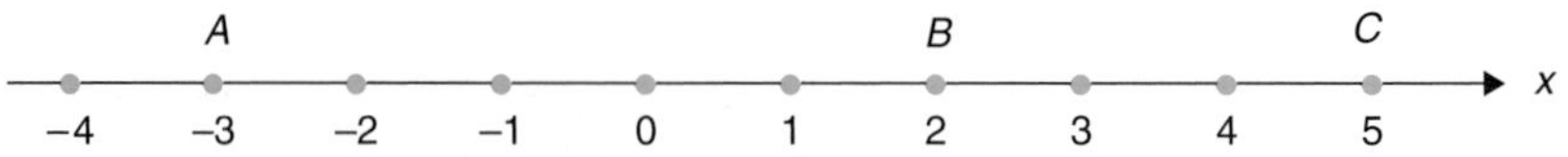

In this textbook, sometimes we identify the number line by placing x at the end of the arrow. Thus, the coordinates of points A, B, and C may also be referred to as $x = -3$, $x = 2$, and $x = 5$, respectively.

Two numbers on a number line that are the same distance from the origin 0, but on opposite sides of 0 are called **negatives** or **opposites** of each other. For example, Figure 3 shows that the negative (or opposite) of 4 is -4, also the negative

Figure 3

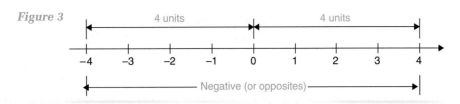

(or opposite) of -4 is written as $-(-4)$ or 4. This example illustrates the fact that if a is any number, then

$$-(-a) = a.$$

Categorizing Numbers in Sets

In algebra we encounter many different sets of numbers.

1. The set of **natural numbers,** N, is also called the **counting numbers** or **positive integers,**

$$N = \{1, 2, 3, 4. . .\}.$$

This set is illustrated in Figure 4.

Figure 4

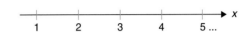

The set of non-negative integers is called the set of **whole numbers** *{0, 1, 2, 3, 4. . .}*

2. The set of **integers,** I, consist of the counting numbers, their negatives, and zero. Figure 5 shows the graph of the integers on a number line.

Figure 5

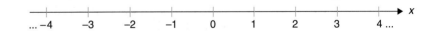

3. The **rational numbers,** Q, consist of all numbers that can be written in the form a/b, in which a and b are integers and $b \neq 0$. Note that because the denominator of the rational number a/b may equal 1, then every integer is considered a rational number. Some examples of rational numbers are

$$\frac{1}{2}, \frac{4}{3}, \frac{20}{1} = 20, \frac{0}{3} = 0, \text{ and } \frac{-7}{2}.$$

In set-builder notation, the rational numbers are

$$Q = \left\{\frac{a}{b} \middle| a, b \in I \text{ and } b \neq 0\right\}.$$

Every rational number can also be written in decimal form. To express a rational number as a decimal, divide the numerator by the denominator. When a rational number is written in decimal form, it can be shown that the decimal either terminates or is a repeating decimal. For example, the decimal forms of

$$\frac{2}{5} = 2 \div 5 = 0.4 \text{ and } \frac{3}{4} = 3 \div 4 = 0.75$$

terminate, whereas the decimal forms of

Be careful when using a calculator to change a fraction to its equivalent decimal form. The repeating pattern may not be visible if the repetition is more than 5 or 6 digits.

$$\frac{2}{3} = 0.6666... \text{ and } \frac{1}{7} = 0.142857142857...$$

are repeating and nonterminating. These decimals can be designated with an over bar to indicate that portion of the decimals that repeats. Thus,

$$\frac{2}{3} = 0.6666... = 0.\overline{6} \text{ and } \frac{1}{7} = 0.142857142857... = 0.\overline{142857}.$$

EXAMPLE 1

Expressing Fractions as Decimals

Express each rational number as a decimal.

(a) $\dfrac{3}{8}$ (b) $\dfrac{-10}{3}$

Solution

(a) $\dfrac{3}{8} = 3 \div 8 = 0.375$

(b) $\dfrac{-10}{3} = -10 \div 3 = -3.333... = -3.\overline{3}$

Every terminating decimal can be changed to a fraction where the denominator is a power of 10. For example,

$$0.7 = \frac{7}{10^1} = \frac{7}{10} \text{ and } 1.023 = \frac{1023}{10^3} = \frac{1023}{1000}.$$

We will see later in this book that every decimal with a repeating pattern can also be written as a fraction.

4. The **irrational numbers,** *J,* have decimal expansions that neither terminate nor repeat digits. Examples of irrational numbers with their decimal forms are:

$$\sqrt{2} = 1.4142135... \text{ and } \pi = 3.1415926...$$

To emphasize that a numerical value is only an approximation, we often use the symbol ≈ and write

$$\sqrt{2} \approx 1.41 \text{ and } \pi \approx 3.14.$$

Most calculators have a round off command. Check your manual to locate this command.

5. The **real numbers,** *R,* is the set of rational and irrational numbers. Thus, the real numbers are numbers which can be written as terminating or repeating decimals (rational numbers), or nonrepeating nonterminating decimals (irrational numbers). Figure 6 shows the relationships among these sets.

Figure 6

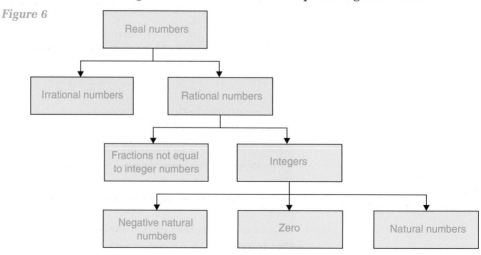

Introducing the Properties of Real Numbers

Now we consider basic properties of real numbers that serve as a foundation for algebra. The letters a, b, and c in Table 1 represent any real number.

TABLE 1

Property	Symbolic Form	Example
Commutative Property for Addition	$a + b = b + a$	$3 + 4 = 4 + 3$
Commutative Property for Multiplication	$a \cdot b = b \cdot a$	$3 \cdot 4 = 4 \cdot 3$
Associative Property for Addition	$a + (b + c) = (a + b) + c$	$7 + (3 + 9) = (7 + 3) + 9$
Associative Property for Multiplication	$a(b \cdot c) = (a \cdot b) \cdot c$	$3 \cdot (5 \cdot 2) = (3 \cdot 5) \cdot 2$
Distributive Property	$a \cdot (b + c) = a \cdot b + a \cdot c$	$6 \cdot (5 + 7) = 6 \cdot 5 + 6 \cdot 7$
Identity for Addition is 0	$a + 0 = 0 + a = a$	$7 + 0 = 0 + 7 = 7$
Identity for Multiplication is 1	$a \cdot 1 = 1 \cdot a = a$	$8 \cdot 1 = 1 \cdot 8 = 8$
Additive Inverse of a is $-a$	$a + (-a) = 0$	$4 + (-4) = 0$
Multiplicative Inverse of a is $1/a$	$a \cdot \dfrac{1}{a} = 1, a \neq 0$	$5 \cdot \dfrac{1}{5} = 1$

EXAMPLE 2

Applying Properties of Real Numbers

State the property that justifies each statement.

(a) $5 \cdot (1 \cdot 2) = (5 \cdot 1) \cdot 2$

(b) $5 \cdot (1 + 2) = (1 + 2) \cdot 5$

(c) $5 \cdot (1 + 2) = 5 \cdot 1 + 5 \cdot 2$

(d) $5 \cdot (1 + 2) = 5 \cdot (2 + 1)$

Solution

(a) The grouping is changed so this is an example of the associative property for multiplication.

(b) The order of multiplication is changed and this is an example of the commutative property for multiplication.

(c) The parentheses are removed and this is an example of the distributive property.

(d) The order of addition is changed and this is an example of the commutative property for addition.

It is possible for the additive inverse to be positive. If $a = -5$, then $-a = -(-5) = 5$

Many calculators have inverse keys. The additive inverse key is $\boxed{+/-}$. It changes the sign of the displayed number. The multiplicative inverse key is $\boxed{1/x}$ or $\boxed{x^{-1}}$. It changes the displayed number to its multiplicative inverse by taking its reciprocal.

We now look at relationships in the set of real numbers. Table 2 lists the basic properties of equality.

TABLE 2

Property	Description	Example
Reflexive	$a = a$	$2 = 2$
Symmetric	If $a = b$, then $b = a$	If $5 = 2 + 3$, then $2 + 3 = 5$
Transitive	If $a = b$ and $b = c$, then $a = c$	If $7 = 14 \div 2$ and $14 \div 2 = 1 + 6$, then $7 = 1 + 6$
Substitution	If $a = b$, then a can be substituted for b in any statement involving b without affecting the truth of the statement.	If $3 + 1 = 4$, then $4 + 2 = (3 + 1) + 2$

EXAMPLE 3   **Identifying Properties of Equality**

State the property of equality that justifies each statement.

(a) $\sqrt{9} = 3$, then $3 = \sqrt{9}$

(b) If $x + 1 = 4$ and $x = a$, then $a + 1 = 4$

(c) If $a = x$ and $x = 4$, then $a = 4$

Solution   (a) Symmetric property because the numbers are reversed.

(b) Substitution property because the value of x is substituted into the first equation.

(c) Transitive property, the value of a is found by passing through the value of x ▣

 PROBLEM SET R2.

In Problems 1 and 2, list all the possible subsets of each set.

1. (a) $\{0,1\}$

(b) $\{a,b,c\}$

2. (a) $\{1/2\}$

(b) $\{1/2, 1.3/2, 2\}$

In Problems 3 and 4, use set-builder notation to describe each set.

3. (a) A is the set of all numbers less than or equal to zero.

(b) B is the set of all negative numbers.

(c) C is the set of all non-negative numbers.

4. (a) A is the set of all numbers greater than zero.

(b) B is the set of all positive numbers.

(c) C is the set of all non-positive numbers.

In Problems 5 and 6, describe each set using set notation.

5. (a) A is the set of integers within 5 units of the origin on a real number line.

(b) B is the set of natural numbers within one unit of 10 on a number line.

(c) C is the set of all real numbers, except 5.

(d) D is the set of integers, except 5.

6. (a) K is the set of even integers.

(b) L is the set of odd integers.

(c) M is the set of numbers greater than 4 but not greater than 7.

(d) N is the set of all numbers between 0 and 1, excluding 0, but including 1.

In Problems 7 and 8, represent the elements of each set on a number line.

7. (a) $A = \left\{ \dfrac{22}{7}, \pi, \dfrac{223}{71} \right\}$

  (b) $B = \{ 1.4141, \sqrt{2}, 1.4142 \}$

8. (a) $C = \left\{ -\sqrt{3}, -\sqrt{4}, -\sqrt{5} \right\}$

  (b) $D = \left\{ \dfrac{1}{3}, 0.3, 0.\overline{3} \right\}$

In Problems 9–12, express each rational number as a decimal. If it repeats, use a bar over the repeating digits.

9. (a) $\dfrac{9}{10}$

  (b) $\dfrac{9}{100}$

  (c) $\dfrac{9}{1000}$

10. (a) $\dfrac{1}{3}$

  (b) $\dfrac{2}{3}$

  (c) $\dfrac{4}{3}$

11. (a) $\dfrac{2}{9}$

  (b) $\dfrac{5}{9}$

  (c) $\dfrac{7}{9}$

12. (a) $\dfrac{1}{11}$

  (b) $\dfrac{3}{11}$

  (c) $\dfrac{12}{11}$

In Problems 13–16, identify all the subsets of the real numbers to which each of the following numbers belongs.

13. (a) -51

  (b) 51

14. (a) $\dfrac{2}{10}$

  (b) $\dfrac{10}{2}$

15. (a) $\sqrt{120}$

  (b) $\sqrt{121}$

16. (a) $3.03030303\ldots$

  (b) $3.030030003\ldots$

In Problems 17–20, use a calculator and round each number as indicated.

17. To five decimal places:

  (a) π

  (b) $\dfrac{22}{7}$

  (c) $\dfrac{512}{163}$

18. To five decimal places:

  (a) $(\sqrt{2})^2$

  (b) $(1.414)^2$

  (c) $(1.4142)^2$

19. To five decimal places:

  (a) $(1.732)^2$

  (b) $(\sqrt{3})^2$

  (c) $(1.733)^2$

20. To five decimal places:

  (a) $(3.1622)^2$

  (b) $(\sqrt{10})^2$

  (c) $(3.1623)^2$

In Problems 21 and 22, find the negative (or opposite) of a, then graph a and its negative on a number line.

21. (a) $a = \dfrac{3}{5}$

  (b) $a = -\dfrac{2}{3}$

22. (a) $a = -0.7$

  (b) $a = \dfrac{2}{3}$

In Problems 23–26, state the property or properties that justify each statement.

23. (a) $3(5 + 1) = 3(5) + 3(1)$

  (b) $3(5 + 1) = (5 + 1)3$

  (c) $3(5 + 1) = 3(1 + 5)$

24. (a) $2 + 3 \cdot 5 = 3 \cdot 5 + 2$

  (b) $(2 + 3) \cdot 5 = 2 \cdot 5 + 3 \cdot 5$

  (c) $2 + 3 \cdot 5 = 5 \cdot 3 + 2$

25. (a) $4 + (3 + 7) = (3 + 7) + 4$

  (b) $4 + (3 + 7) = (4 + 3) + 7$

  (c) $4 + (3 + 7) = 4 + (7 + 3)$

26. (a) $7 \cdot (8 \cdot 9) = (7 \cdot 8) \cdot 9$

  (b) $7 \cdot (8 \cdot 9) = 7 \cdot (9 \cdot 8)$

  (c) $7 \cdot (8 \cdot 9) = (8 \cdot 9) \cdot 7$

In Problems 27–30, state the property of equality that justifies each statement.

27. If $4 = 2x - 1$, then $2x - 1 = 4$

28. If $4 = 2x - 1$ and $2x - 1 = y$, then $4 = y$

29. If $2x + 1 = 3$ and $x - 1 = 3$, then $2x + 1 = x - 1$

30. If $x = 3$, then $2x + 1 = 2(3) + 1$

In Problems 31 and 32, plot each of the following on a number line,

31. $\dfrac{1}{2}, 0, -\sqrt{4}, \sqrt{2}, -\sqrt{10}$

32. $\dfrac{1}{4}, \dfrac{3}{8}, -\sqrt{8}, \pi, -\sqrt{15}$

In Problems 33 and 34, rewrite each expression using only the stated property.

33. (a) $(19 + 82) \cdot 7 = \underline{\quad}$, distributive property

  (b) $(19 + 82) \cdot 7 = \underline{\quad}$, commutative property for addition

  (c) $(19 + 82) \cdot 7 = \underline{\quad}$, commutative property for multiplication

34. (a) $3x + 4x = \underline{\quad}$, distributive property

  (b) $5 \cdot \dfrac{1}{8} + 3 \cdot \dfrac{1}{8} = \underline{\quad}$, distributive property

  (c) $2\sqrt{7} + 5\sqrt{7} = \underline{\quad}$, distributive property

In Problems 35 and 36, list the elements of each set that are (a) natural numbers N, (b) integers I, (c) rational numbers Q, (d) irrational numbers J, and (e) real numbers R.

35. $\left\{-\sqrt{9}, -\sqrt{5}, 0, \dfrac{25}{5}, \dfrac{25}{4}\right\}$

36. $\left\{-\dfrac{100}{5}, -\dfrac{100}{12}, 0, 4.\overline{25}, 4.76776777677776...\right\}$

37. Tipping: Calculate a 15% tip on a restaurant bill of $24.
  (a) Multiply $(0.15)(24.00)$.
  (b) Add a 10% tip to half of a 10% tip.
  (c) Identify a, b, and c in $(a + b)c = ac + bc$ to help explain why the distributive property guarantees that a 15% tip equals a 10% tip plus a 5% tip.
  $(0.15)(24.00) = (0.10)(24) + (0.05)(24)$

38. Give an example to show that the commutative and associative properties do not hold for subtraction.

39. Give an example to show that the commutative and associative properties do not hold for division.

40. Determine the number of subsets that exist for a set containing the following number of elements.
  (a) 0   (b) 1   (c) 2   (d) 3   (e) 4

41. Which of the following indicates the number of subsets of a set containing n elements, 2^n or $2n$? (Hint: See Problem 40)

42. (a) Explain why the associative property of multiplication allows us to write a triple product such as $12 \cdot 33 \cdot 42$ without parentheses.
  (b) Explain why the associative property of addition allows us to write a triple sum such as $12 + 30 + 45$ without parentheses.

In Problems 43–46, state the property that justifies each step.

43. $-2(-3) + 2(-3) = (-2 + 2)(-3)$ (a)_____
  $= 0(-3)$ (b)_____
  $= 0$ Definition of $0(-3)$

44. $5x + 7y + 15x = 5x + 15x + 7y$ (a)_____
  $= (5 + 15)x + 7y$ (b)_____
  $= 20x + 7y$ Definition of $5 + 15$

45. $(x + y) - y = x + (y - y)$ (a)_____
  $= x + [1 \cdot y + 1 \cdot (-y)]$ (b)_____
  $= x + 1[y + (-y)]$ (c)_____
  $= x + 1 \cdot 0$ (d)_____
  $= x + 0$ Definition of $0 \cdot y$
  $= x$ (e)_____

46. $x + (-x) = 1 \cdot x + 1(-x)$ (a)_____
  $= 1[x + (-x)]$ (b)_____
  $= 1 \cdot 0$ (c)_____
  $= 0$ Definition of 1

R3. Operations with Integers

Objectives

1. Introduce Absolute Value
2. Add and Subtract Integers
3. Multiply and Divide Integers

In this section we extend the rules for addition, subtraction, multiplication, and division of positive numbers to all integers.

Introducing Absolute Value

Although we can use the number line to add integers, it is often convenient to give a precise description of the addition process. For this purpose we consider the concept of absolute value. Figure 1 shows that points P and Q are 5 units from the origin.

Figure 1

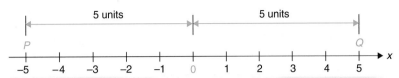

Geometrically, the *absolute value* of a number is the distance between the number and zero on the number line, regardless of direction. Since -5 and 5 each is 5 units from the origin, 0, they have the same absolute value, which we denote by two vertical bars. That is, $|-5|$ and $|5|$. The statements

$$|-5| = 5 \text{ and } |5| = 5$$

are read as "the absolute value of -5 is 5" and "the absolute value of 5 is 5", respectively. In general, we define the absolute value of any real number x as follows:

Definition of Absolute Value

If x is a number, then $|x|$ is the **absolute value** of x, where

$$|x| = \begin{cases} x, \text{ if } x \geq 0, x \text{ is non–negative} \\ -x, \text{ if } x < 0, x \text{ is negative} \end{cases}$$

For example according to this definition,

$$|3| = 3, |0| = 0, \text{ and } |-2| = -(-2) = 2$$

EXAMPLE 1   **Evaluating Absolute Value**

Find the value of each expression.

(a) $|26|$ (b) $|-26|$ (c) $-|-26|$

Solution   We use the definition of absolute value.

(a) When $x > 0$, $|x| = x$, here $26 > 0$, so $|26| = 26$

(b) When $x < 0$, $|x| = -x$, here $-26 < 0$, so $|-26| = -(-26) = 26$

(c) From part (b) we know that $|-26| = 26$. Thus, $-|-26| = -26$

Adding and Subtracting Integers

The justification for the addition rule is seen by drawing arrows on the number line to indicate a change in position. An arrowhead to the right indicates a positive move and an arrowhead to the left a negative move. The length of the arrow corresponds to the size of the move. For example, the sum $2 + 3$ can be interpreted on a number line as follows: start at the origin, 0, and move 2 units to the right. Then move 3 more units to the right to find the sum 5. Thus, $2 + 3 = 5$ (Figure 2).

Figure 2

Similarly, to show the sum $3 + 2 = 5$, we start by moving 3 units to the right followed by moving 2 more units to the right to find the sum 5. We can also use the number line to add two negative numbers. For instance, the sum $(-2) + (-3)$ is interpreted as follows: start at the origin, move 2 units to the left, then move 3 more units to the left to the find the sum -5 (Figure 3). That is $(-2) + (-3) = -5$.

Figure 3

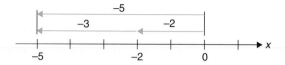

In general, when we add two numbers with like signs, we add their absolute values and keep their common sign. For example, $(+2) + (+3) = +(|+2| + |+3|) = +5$ and

$$(-2) + (-3) = -(|-2| + |-3|) = -5$$

We can also use a number line to illustrate adding two numbers with different signs. We represent the sum $-3 + 8$ on a number line as follows: Start at the origin and move 3 units to the left, then move 8 units to the right to find the sum 5 (Figure 4). Thus $-3 + 8 = 5$.

Figure 4

The following rules generalize the addition of integers.

Rules For Adding Signed Numbers

1. If the signs of the numbers are the **same,** we add their absolute value and keep their common sign.
2. If the signs of the numbers are **different,** we find the **difference** of their absolute values and repeat the sign of the number with the largest absolute value.

EXAMPLE 2

Adding Integers

Find each sum.

(a) $(-15) + (-72)$ (b) $10 + (-230)$ (c) $-14 + 17 + 25$

Solution

(a) Since both numbers are negative, their sum is negative. That is

$$(-15) + (-72) = -(|-15| + |-72|) = -87$$

(b) The numbers have different signs and the negative number has the larger absolute value. That is,

$$10 + (-230) = -(|-230| - |10|) = -220.$$

(c) Working from left to right, we add -14 and 17 first, then add 25 to obtain:

$$-14 + 17 + 25 = 3 + 25 = 28$$

Now we look at the operation of subtraction. The **difference,** $a - b$, is defined as the sum of a and the additive inverse of b. In symbols,

$$a - b \text{ means } a + (-b)$$

Thus, we can rewrite a difference as a sum by changing the sign of the number being subtracted; then apply the rules for addition. For example,

$$-7 - 3 = (-7) + (-3) = -10 \text{ and } 12 - (-5) = 12 + (+5) = 17$$

EXAMPLE 3   **Subtracting Integers**

Find each difference.

(a) $-99 - 12$ (b) $60 - (-50)$ (c) $(-37) - (-12)$

Solution   We rewrite each subtraction as an addition, then add.

(a) $-99 - 12 = -99 + (-12) = -111$

(b) $60 - (-50) = 60 + 50 = 110$

(c) $-37 - (-12) = -37 + 12 = -25$

Multiplying and Dividing Integers

Multiplication is defined as repeated addition. For example, $4 \cdot 2$ means adding two four times, that is,

$$4 \cdot 2 = 2 + 2 + 2 + 2 = 8$$

Also, this product can be described as adding four twice, that is,

$$2 \cdot 4 = 4 + 4 = 8.$$

The same approach applies to the product of a positive and a negative number. For instance, we interpret $4(-2)$ as

$$(-2) + (-2) + (-2) + (-2) = -8.$$

Figure 5 illustrates this product on a number line.

Figure 5

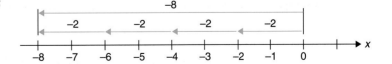

This example illustrates the general statement:

The product of a positive number and a negative number is negative.

Now consider the product of two negative numbers. To explore this multiplication, we will calculate $-2(-4 + 4)$ in two different ways. Adding the numbers inside the parentheses first yields

$$-2(-4 + 4) = -2(0) = 0$$

Applying the distributive property yields

$$-2(-4 + 4) = -2(-4) + (-2)4$$

We know $(-2)4 = -8$ and $-2(-4 + 4) = 0$, so by substitution we have

$$0 = -2(-4) + (-8)$$

The additive inverse of -8 is $+8$. Thus, the product of -2 and -4 must be the positive number 8.

In general, we have the following rules for multiplying the numbers a and b, where a and b are positive numbers.

Rules for Multiplying Signed Numbers

1. If the signs of the numbers are the **same,** the product is **positive.** That is, $a \cdot b = ab$ and $(-a)(-b) = ab$
2. If the signs of the numbers are different, the product is negative. That is, $a(-b) = -ab$ and $(-a)b = -ab$.

For example, $3 \cdot 5 = 15$, $(-3)(-5) = 15$ and $3(-5) = -15$.
When multiplying more than two integers, we note the following pattern.

$$(-1)(-2) = +2$$
$$(-1)(-2)(-3) = -6$$
$$(-1)(-2)(-3)(-4) = +24$$
$$(-1)(-2)(-3)(-4)(-5) = -120$$

The number of positive factors has no effect on the sign of the product.

As the pattern suggests, the product of integers is positive for an even number of negative factors and negative for an odd number of negative factors.

EXAMPLE 4   **Multiplying More Than Two Integers**

Determine whether each product is positive or negative. Then find each product.

(a) $3(-2)(-7)(6)(-2)$ (b) $-2(-2)(3)(-1)(-4)$

Solution   (a) Since three of the factors are negative, the product is negative. So that,

$$3(-2)(-7)(6)(-2) = -504$$

(b) Since four of the factors are negative, the product is positive. That is,

$$-2(-2)(3)(-1)(-4) = +48$$

Multiplication and division of numbers are related in the following way:

$$a \div b = c \text{ means } a = bc$$

The result of a division, *c,* is called a **quotient.** For example, $6 \div 2 = 3$ because $6 = 2(3)$ and 3 is called the quotient.

Two numbers are called **multiplicative inverses** or **reciprocals** of each other, if their product is 1. For example 5 and $\frac{1}{5}$ are reciprocals because $5 \cdot \frac{1}{5} = 1$, as are $\frac{-3}{7}$ and $\frac{-7}{3}$ because $\left(\frac{-3}{7}\right)\left(\frac{-7}{3}\right) = 1.$

Using the concept of a multiplicative inverse, or reciprocal, division may be defined in terms of multiplication as follows:

Division of Numbers

If a and b are numbers, with $b \neq 0$, then

$$a \div b = a \cdot \frac{1}{b}$$

This definition says that to divide a by b, multiply a by the multiplicative inverse of b. For example, $12 \div 3$ can be written as:

$$12 \div 3 = 12 \cdot \left(\frac{1}{3}\right) = 4$$

Quotients of non-zero numbers follow the same rules as multiplication, that is:

1. If two numbers have the **same signs,** their quotient is **positive.**
2. If two numbers have **different signs,** their quotient is **negative.**

EXAMPLE 5   **Dividing Integers**

Perform each division.

(a) $45 \div (-15)$ (b) $-45 \div 15$ (c) $-45 \div (-15)$

Solution

(a) $45 \div (-15) = 45 \cdot \dfrac{1}{(-15)} = -3$ Multiply 45 by the reciprocal of -15

(b) $-45 \div 15 = (-45) \cdot \dfrac{1}{15} = -3$ Multiply -45 by the reciprocal of 15

(c) $-45 \div (-15) = (-45) \cdot \dfrac{1}{(-15)} = 3$ Multiply -45 by the reciprocal of -15 ▦

EXAMPLE 6   **Applying the Order of Operations to Integers**

Find the value of the expression.

$$(-3)^2 + 7(-6 + 4) \div (-2)$$

Solution

$(-3)^2 \neq -3^2$. The base of $(-3)^2$ is -3, whereas the base of -3^2 is 3.

The order of operations applies to all real numbers, so that:

$$
\begin{aligned}
(-3)^2 + 7(-6 + 4) \div (-2) &= (-3)^2 + 7(-2) \div (-2) &&\text{Parentheses} \\
&= 9 + 7(-2) \div (-2) &&\text{Powers} \\
&= 9 + (-14) \div (-2) &&\text{Multiplication} \\
&= 9 + 7 = 16 &&\text{Division, addition} \quad ▦
\end{aligned}
$$

PROBLEM SET R3.

In Problems 1 and 2, find the value of each expression.

1. (a) $|-25|$
   (b) $|25|$
   (c) $-|-25|$

2. (a) $-|100|$
   (b) $|-100|$
   (c) $-|-100|$

In Problems 3 and 4, find each sum by using a number line.

3. (a) $-3 + 1$
   (b) $-3 + 1 - 1$
   (c) $-3 + 1 - 2$

4. (a) $2 + (-4) - 1$
   (b) $4 + (-2) + 3$
   (c) $2 + (-5) - 3$

In Problems 5 and 6, translate each phrase into symbols.

5. (a) 5 minus negative 3
   (b) Negative 3 minus 1

6. (a) Negative 2 plus negative 4
   (b) Negative 7 minus negative 3

In Problems 7–10, find each sum.

7. (a) $15 + (-5)$
   (b) $-15 + 5$

8. (a) $18 + (-18)$
   (b) $7 - 12 + 5$

9. (a) $-8 + 25 + (-15)$
   (b) $-8 + (-15) + 25$

10. (a) $-1.4 + 1.2 - 3$
   (b) $-6.3 + (-3.6) + 2$

In Problems 11–14, find each difference.

11. (a) $-7 - 11$
   (b) $-7 - (-11)$

12. (a) $-111 - (-111)$
   (b) $-1212 - (1212)$

13. (a) $-503 - (-483)$
   (b) $-483 - (-503)$

14. (a) $-1.6 - (-1.6)$
   (b) $-4.3 - (-1.5)$

In Problems 15 and 16, find each product.

15. (a) $-100(-50)$
   (b) $100(50)$
   (c) $-100(50)$
   (d) $100(-50)$

16. (a) $24(18)$
   (b) $-24(18)$
   (c) $-24(-18)$
   (d) $24(-18)$

In Problems 17 and 18, determine whether each product is positive or negative. Then find the product.

17. (a) $-3(-5)(-15)$
   (b) $-3(-5)(-15)(-20)$

18. (a) $-2(7)(-10)(14)$
   (b) $-2(7)(-10)(-14)$

In Problems 19 and 20, find each quotient.

19. (a) $-30 \div (-15)$
   (b) $30 \div (-15)$
   (c) $-30 \div 15$

20. (a) $-288 \div 144$
   (b) $288 \div (-144)$
   (c) $(-288) \div (-144)$

In Problems 21–30, find the value of each expression.

21. (a) $-2[9 + (-11)]$
   (b) $-8[[-7) - (-3)]$

22. (a) $6[-8 - (-5)]$
   (b) $7[-4 + (-3)]$

23. (a) $-5(-15 \div 3)$
   (b) $-4[-35 \div (-7)]$

24. (a) $[28 - (-7)] \div (-5)$
   (b) $[-48 - (-12)] \div (-2)$

25. $-15 \div 3 + 2 \cdot (-5)$

26. $11(-16) \div (-4) - (-3)$

27. $0(-7)$

28. $0 \div (-2)$

29. $-3 \div 0$

30. $2(10) \div [(-3) + 3]$

In Problems 31–38, find the value of each expression.

31. (a) $(-5)^2$
   (b) -5^2

32. (a) $(-1)^6$
   (b) -1^6

33. (a) $(-4)^3$
   (b) -4^3

34. (a) $\left(-\dfrac{1}{2}\right)^4$
   (b) $-\left(\dfrac{1}{2}\right)^4$

35. $5(-2)^3(-3)^3 \div (-8)$

36. $5^2 \cdot (-3)2(-2)^3 \div -10$

37. $-2[(-3) - (-2)^3] + (4 - 6)$

38. $3 + [(-4) + 2]^3 \cdot (-8)$

In Problems 39–44, find the value of each expression given the indicated values of x and y.

39. For $x = -1$ and $y = -2$,
   (a) x^2y
   (b) xy^2

40. For $x = -3$ and $y = 5$,
   (a) $2xy$
   (b) $-2xy$

41. For $x = -5$ and $y = -1$,
   (a) $(x + y)^2$
   (b) $x^2 + 2xy + y^2$

42. For $x = 1$ and $y = -2$,
   (a) $(x - y)^2$
   (b) $x^2 - 2xy + y^2$

43. For $x = -3$ and $y = -4$,
   (a) $\dfrac{x^2 + y^2}{x + y}$
   (b) $\dfrac{x^2 - y^2}{x + y}$

44. For $x = -2$ and $y = -3$,

(a) $\dfrac{x^3 - y^3}{x - y}$ (b) $\dfrac{x^3 + y^3}{x + y}$

45. Temperature: One night the thermometer outside read $18°F$ and in the morning it read $-2°F$.

(a) Write an expression for the difference between the night and morning temperatures.

(b) Find the value of the expression.

46. Finance: On Monday you open a savings account with $270. You deposit $100 on Tuesday, withdraw $40 on Wednesday, and withdraw $65 on Thursday.

(a) Write an expression that shows the activity in your account.

(b) Find the value of the expression.

47. Sports: A football team starts with the ball on its own 20 yard line. In a series of three plays, the team loses 4 yards, gains 7 yards, and then loses 5 yards.

(a) Write an expression that shows the team's field position.

(b) Find the value of the expression.

48. Tides: Suppose that on a typical day high tide was 6 feet and low tide was -2 feet.

(a) Write an expression for the difference between the high and low tides.

(b) Find the value of the expression.

Objectives

1. Reduce and Simplify Rational Numbers
2. Determine Ratio and Proportion
3. Multiply and Divide Rational Numbers
4. Add and Subtract Rational Numbers

R4. Operations with Rational Numbers

In this section we review operations with rational numbers. Recall that a rational number can be expressed as

$$\frac{a}{b}, \text{ where } b \neq 0$$

with numerator a and denominator b. If $b = 0$, then $\dfrac{a}{b}$ is undefined.

Reducing and Simplifying Rational Numbers

Consider the rational number 15/25. By writing the numerator and denominator of this number in factored form, and dividing out (canceling) the common factors, we have:

$$\frac{15}{25} = \frac{3 \cdot 5}{5 \cdot 5} = \frac{3}{5} \cdot 1 = \frac{3}{5}$$

Since the numerator and denominator of 3/5 has no common factors other than 1 (or -1), we say that the rational number 3/5 is **simplified** or **reduced to lowest terms.** In general, to *simplify* a rational number, or *reduce to lowest terms,* we use the following principle.

Fundamental Principle of Rational Numbers

For any rational number $\dfrac{a}{b}$ and any number $k \neq 0$

$$\frac{a \cdot k}{b \cdot k} = \frac{a}{b}$$

Thus, the fundamental principle says that multiplying or dividing the numerator and denominator of a rational number by the same non-zero number produces an equivalent rational number. Two rational numbers are said to be **equivalent** if they simplify to the same number. For example,

$$\frac{24}{28} = \frac{6 \cdot 4}{7 \cdot 4} = \frac{6}{7}$$

Thus, the rational numbers 24/28 and 6/7 are equivalent.
It should be noticed that for the rational number 4/7.

$$\frac{4}{7} = \frac{-4}{-7} = -\frac{-4}{7} = -\frac{4}{-7} \text{ and } -\frac{4}{7} = \frac{-4}{7} = \frac{4}{-7} = -\frac{-4}{-7}.$$

EXAMPLE 1   **Reducing Rational Numbers**

Reduce each rational number to lowest terms, if possible.

(a) $\dfrac{16}{-48}$ (b) $\dfrac{-26}{-91}$ (c) $\dfrac{35}{26}$

Solution   By factoring the numerator and denominator of each rational number, we have,

(a) $\dfrac{16}{-48} = \dfrac{1 \cdot 16}{-3 \cdot 16} = \dfrac{1}{-3} \cdot \dfrac{16}{16} = \dfrac{1}{-3} \cdot 1 = -\dfrac{1}{3}$

(b) $\dfrac{-26}{-91} = \dfrac{26}{91} = \dfrac{2 \cdot 13}{7 \cdot 13} = \dfrac{2}{7} \cdot \dfrac{13}{13} = \dfrac{2}{7} \cdot 1 = \dfrac{2}{7}$

(c) There are no common factors in 35 and 26, so 35/26 is already simplified.

Determining Ratios and Proportions

A ratio a/b is a quotient of two quantities, a and b. The ratio a/b can also be written as $a : b$. For example, if there are 2618 students and 154 teachers in a school, then the ratio of students to teachers is given by:

$$\frac{2618}{154} = \frac{17 \cdot 154}{1 \cdot 154} = \frac{17}{1}$$

Therefore, there are 17 students per one teacher.
   A very important use of ratio is *percent*. By definition, a **percent** is a ratio whose denominator is 100. *Percent means parts per 100 or parts of 100* and it is denoted by the symbol %. Rational numbers or decimals are often expressed as percents. For instance

$$\frac{3}{100} \text{ or } 0.03 \text{ means } 3\%$$

Thus, 3% of 60 is 0.03(60) = 1.8. Also, if we ask what percent of 400 is 36, then we write

$$\frac{36}{400} = \frac{9}{100} = 0.09 = 9\%$$

EXAMPLE 2   **Using Percents**

(a) What is 70% of 300?   (b) What percent of 2000 is 40?

Solution   (a) $70\% \times 300 = 0.70 \times 300 = 210$

(b) $\dfrac{40}{2000} = 0.02 = 0.02 \times 100\% = 2\%$

As you can see, ratios are another name for rational numbers. Thus, two ratios are *equivalent* if they simplify to the same number. When two ratios are equal, we call it a *proportion.* More formally, we say that:

> Two ratios $\dfrac{a}{b}$, and $\dfrac{c}{d}$, are **proportional** if $\dfrac{a}{b} = \dfrac{c}{d}$.

Notice that when the two ratios $\dfrac{2}{3}$ and $\dfrac{10}{15}$ are proportional, the "cross products" are equal. That is, $\dfrac{2}{3} = \dfrac{10}{15}$ if and only if $2(15) = 3(10)$.

In general we write:

> $\dfrac{a}{b} = \dfrac{c}{d}$ if and only if $ad = bc$

EXAMPLE 3   **Rewriting as a Ratio**

Rewrite each statement as a ratio of equal values.

(a) 33 dimes to 22 nickels   (b) 12 meters to 30 centimeters

Solution   (a) $\dfrac{\text{value of 33 dimes}}{\text{value of 22 nickels}} = \dfrac{33 \cdot 10}{22 \cdot 5} = \dfrac{3 \cdot \cancel{11} \cdot \cancel{5} \cdot 2}{2 \cdot \cancel{11} \cdot \cancel{5}} = \dfrac{3}{1}$

(b) $\dfrac{\text{12 meters as centimeters}}{\text{30 centimeters}} = \dfrac{12 \cdot 100}{30} = \dfrac{2 \cdot \cancel{6} \cdot \cancel{5} \cdot 20}{\cancel{5} \cdot \cancel{6}} = \dfrac{40}{1}$

Multiplying and Dividing Rational Numbers

The multiplication of two rational numbers is performed by finding the product of their numerators divided by the product of their denominators.

Definition of Multiplication

> If a/b and c/d are rational numbers with $b \neq 0$ and $d \neq 0$, then
>
> $$\frac{a}{b} \cdot \frac{c}{d} = \frac{a \cdot c}{b \cdot d}$$

When two rational numbers containing common factors are to be multiplied, we may multiply the rational numbers then reduce the product to lowest terms. Another method is to factor the numerator and denominator, divide each by the common factors, then multiply.

EXAMPLE 4 **Multiplying Rational Numbers**

Find the product of $-\dfrac{22}{65} \cdot \dfrac{15}{11}$.

(a) Multiply the rational numbers, then reduce the product to lowest terms.

(b) Factor the numerator and denominator, divide by the common factors, then multiply.

Solution (a) Multiply and then reduce

$$-\frac{22}{65} \cdot \frac{15}{11} = -\frac{22 \cdot 15}{65 \cdot 11} = -\frac{330}{715} = -\frac{6}{13} \cdot \frac{55}{55} = -\frac{6}{13}$$

(b) Factor, divide by common factors, then multiply

$$-\frac{22}{65} \cdot \frac{15}{11} = -\frac{2 \cdot \cancel{11}}{5 \cdot 13} \cdot \frac{3 \cdot 5}{\cancel{11}} = -\frac{6}{13}$$

To divide two rational numbers, find the reciprocal of the divisor (second rational number) and multiply.

Definition of Division

> If a/b and c/d are rational numbers with $b \neq 0$, $c \neq 0$, and $d \neq 0$, then
>
> $$\frac{a}{b} \div \frac{c}{d} = \frac{a}{b} \cdot \frac{d}{c} = \frac{a \cdot d}{b \cdot c}$$

EXAMPLE 5 **Dividing Rational Numbers**

Perform each division and reduce the results to lowest terms.

(a) $\dfrac{5}{8} \div \dfrac{3}{4}$ (b) $\dfrac{39}{95} \div \dfrac{-26}{35}$

Solution (a) $\dfrac{5}{8} \div \dfrac{3}{4} = \dfrac{5}{8} \cdot \dfrac{4}{3}$ Division is multiplication by the reciprocal.

$$= \frac{5}{2 \cdot \cancel{4}} \cdot \frac{\cancel{4}}{3} = \frac{5}{6}$$ Factor and divide out common factor.

(b) $\dfrac{39}{95} \div \dfrac{-26}{35} = \dfrac{39}{95} \cdot \dfrac{35}{-26} = -\dfrac{3 \cdot \cancel{13} \cdot \cancel{5} \cdot 7}{\cancel{5} \cdot 19 \cdot 2 \cdot \cancel{13}} = -\dfrac{21}{38}$

Adding and Subtracting Rational Numbers

To add or subtract rational numbers having the same denominator, we add (or subtract) the numerators and repeat the denominator. In symbols:

$$\frac{a}{b} + \frac{c}{b} = \frac{a + c}{b} \quad \text{and} \quad \frac{a}{b} - \frac{c}{b} = \frac{a - c}{b}$$

EXAMPLE 6

Adding and Subtracting Rational Numbers with the Same Denominator

Combine the following rational numbers.

(a) $\dfrac{2}{7} + \dfrac{3}{7}$ (b) $\dfrac{7}{9} - \dfrac{5}{9}$ (c) $\dfrac{17}{20} + \dfrac{19}{20} - \dfrac{3}{20}$

Solution

Some calculators can perform the arithmetic of rational numbers displaying answers in rational number form.

(a) $\dfrac{2}{7} + \dfrac{3}{7} = \dfrac{2+3}{7} = \dfrac{5}{7}$

(b) $\dfrac{7}{9} - \dfrac{5}{9} = \dfrac{7-5}{9} = \dfrac{2}{9}$

(c) $\dfrac{17}{20} + \dfrac{19}{20} - \dfrac{3}{20} = \dfrac{17+19-3}{20} = \dfrac{33}{20}$

To add or subtract rational numbers having different denominators, we use *the fundamental principle of rational numbers* in reverse to change the form of each rational number so that they have a common denominator, then add (or subtract) as before.

Definition of Addition and Subtraction

$$\frac{a}{b} + \frac{c}{d} = \frac{ad}{bd} + \frac{cb}{db} = \frac{ad+bc}{bd} \quad \text{and} \quad \frac{a}{b} - \frac{c}{d} = \frac{ad}{bd} - \frac{cb}{db} = \frac{ad-bc}{bd}$$

In algebra, an improper rational number such as 29/24 is preferred over the equivalent mixed number 1 5/24.

For example, to add 3/8 and 5/6, we have

$$\frac{3}{8} + \frac{5}{6} = \frac{3 \cdot 6}{8 \cdot 6} + \frac{5 \cdot 8}{8 \cdot 6} = \frac{3 \cdot 6 + 5 \cdot 8}{8 \cdot 6} = \frac{18 + 40}{48} = \frac{58}{48} = \frac{29}{24}$$

This process of using the product of the denominators as a common denominator often creates denominators that are larger than necessary. The problem generally will be less complicated if the *least common denominator,* LCD, is used. The **least common denominator** is the least common multiple of all the denominators, that is, it is the smallest positive integer that is a multiple of the denominators.

If the LCD is not obvious, then we find it as follows:

1. Factor each denominator into a product of *prime numbers.* (A **prime number** is a positive integer whose only factors are itself and 1.)
2. Form the product of all the different prime factors, raising each prime factor to the highest power that occurs in any of the denominators.
3. This product is the LCD.

For example, in the preceding example, 3/8 + 5/6, the prime factors of 8 are $2 \cdot 2 \cdot 2 = 2^3$ and the prime factors of 6 are $2 \cdot 3$. The LCD of the rational numbers is $2^3 \cdot 3 = 24$. We use the fundamental principle of rational numbers to build each rational number to an equivalent rational number with the denominator of 24 as follows.

$$\frac{3}{8} + \frac{5}{6} = \frac{3}{8} \cdot \frac{3}{3} + \frac{5}{6} \cdot \frac{4}{4} = \frac{9}{24} + \frac{20}{24} = \frac{29}{24}$$

EXAMPLE 7   **Combining Rational Numbers with Different Denominators**

Combine and simplify.

(a) $\dfrac{1}{12} + \dfrac{3}{40}$ (b) $\dfrac{11}{18} - \dfrac{7}{75}$

Solution   (a) To find the LCD, factor each denominator:

$$12 = 2^2 \cdot 3, \ 40 = 2^3 \cdot 5.$$

The LCD is the product of every different factor to its highest power,

$$2^3 \cdot 3 \cdot 5 = 120.$$

Convert each rational number into an equal rational number with 120 as the denominator and then add.

$$\frac{1}{12} + \frac{3}{40} = \frac{1}{12} \cdot \frac{10}{10} + \frac{3}{40} \cdot \frac{3}{3} = \frac{10}{120} + \frac{9}{120} = \frac{19}{120}$$

(b) The LCD is $2 \cdot 3^2 \cdot 5^2 = 450$ since $18 = 2 \cdot 3^2$ and $75 = 3 \cdot 5^2$.

This can be checked with a calculator using decimal approximations.

$$\frac{11}{18} - \frac{7}{75} = \frac{11}{18} \cdot \frac{25}{25} - \frac{7}{75} \cdot \frac{6}{6} = \frac{275}{450} - \frac{42}{450} = \frac{233}{450}$$

PROBLEM SET R4.

In Problems 1–6, reduce each rational number to lowest terms.

1. (a) $\dfrac{15}{45}$
   (b) $\dfrac{270}{420}$

2. (a) $\dfrac{36}{90}$
   (b) $\dfrac{48}{120}$

3. (a) $-\dfrac{65}{95}$
   (b) $-\dfrac{150}{510}$

4. (a) $\dfrac{-75}{45}$
   (b) $\dfrac{-210}{150}$

5. (a) $\dfrac{-12}{-18}$
   (b) $\dfrac{-15}{-25}$

6. (a) $\dfrac{-270}{-450}$
   (b) $\dfrac{-200}{-150}$

In Problems 7–10, rewrite each rational number as a percent.

7. (a) $\dfrac{4}{5}$
   (b) $\dfrac{7}{20}$

8. (a) $\dfrac{3}{50}$
   (b) $\dfrac{11}{25}$

9. (a) $\dfrac{3}{8}$
   (b) $\dfrac{7}{8}$

10. (a) $\dfrac{2}{3}$
    (b) $\dfrac{7}{16}$

In Problems 11–16, change each percent to a rational number.

11. (a) 7%
    (b) 25%

12. (a) $9\dfrac{1}{2}\%$
    (b) $\dfrac{3}{4}\%$

13. (a) $\dfrac{1}{2}\%$
    (b) $\dfrac{5}{8}\%$

14. (a) 35%
    (b) 75%

15. (a) 3.5%
    (b) 10.6%

16. (a) 0.2%
    (b) 6.4%

17. (a) What is 15% of 200?
    (b) What is 4.2% of 200?

18. (a) What is 17% of 300?
    (b) What is 8.5% of 300?

19. What percent of 400 is 80?

20. What percent of 156 is 26?

In Problems 21–24, rewrite each statement as a ratio.

21. (a) 15 inches to 25 inches
    (b) 12 miles to 18 miles

22. (a) 40 feet to 65 feet
    (b) 18 ounces to 24 ounces

23. (a) 75 seconds to 2 minutes

(b) 8 inches to 3 feet

24. (a) 4 feet to 4 yards

(b) 5 gallons to 12 quarts

In Problems 25–30, find each product by

(a) multiplying the rational numbers, then reducing the product to lowest terms, and

(b) dividing the common factors, then multiplying.

25. $\dfrac{5}{5} \cdot \dfrac{7}{5}$

26. $-\dfrac{11}{7}\left(\dfrac{7}{11}\right)$

27. $\dfrac{-11}{24} \cdot \dfrac{24}{33}$

28. $\dfrac{-40}{21}\left(\dfrac{-7}{20}\right)$

29. $\dfrac{11}{100} \cdot \dfrac{24}{132}$

30. $\dfrac{-100}{90} \cdot \dfrac{45}{25}$

In Problems 31 and 32, divide the rational numbers and express the answers reduced to lowest terms.

31. (a) $\dfrac{21}{20} \div \dfrac{9}{15}$

(b) $\dfrac{110}{9} \div \dfrac{121}{9}$

32. (a) $\dfrac{-5}{18} \div \dfrac{-25}{12}$

(b) $\dfrac{-3}{20} \div \dfrac{9}{100}$

In Problems 33–36, combine the rational numbers and reduce the answers to lowest terms.

33. (a) $\dfrac{2}{5} + \dfrac{1}{5}$

(b) $\dfrac{7}{12} + \dfrac{5}{12}$

34. (a) $\dfrac{5}{11} - \dfrac{3}{11}$

(b) $\dfrac{7}{15} + \dfrac{-3}{15}$

35. (a) $\dfrac{-7}{4} - \dfrac{4}{3} + \dfrac{3}{4}$

(b) $-\dfrac{21}{5} - \dfrac{16}{25} + \dfrac{7}{50}$

36. (a) $\dfrac{2}{5} - \dfrac{3}{8} + \dfrac{7}{20}$

(b) $\dfrac{5}{6} - \dfrac{2}{9} + \dfrac{4}{3}$

37. Sports: A football team won 10 of its 16 games played with no ties.

(a) What is the ratio of wins to games played?

(b) What percent of the games played did the team win?

(c) What is the ratio of wins to losses?

38. Gas Mileage: A midsize car travels 252 miles on 9 gallons of gasoline. What is the ratio of miles to gallons?

39. Marketing: The price of a microwave oven went from $99 to $115. Find the ratio of the increase in price to the original price.

40. Marketing: The price of a computer went from $1800 to $1100. Find the ratio of the decrease in price to the original price.

41. Environment: Environmentalists estimate that paper takes up about half of our landfill space. If newspapers are 2/5 of the paper in the landfills, what fraction of landfill space is taken up by newspapers?

42. College Enrollment: At a college, 3/5 of the students take a mathematics course. Of those, 1/3 take algebra. What fraction of the students take algebra?

43. Land Distribution: It is estimated that there are 2,300,000,000 acres of land in America. Native Americans own 1/50 of the land. How many acres of land are owned by Native Americans?

44. Activities: Suppose that a person sleeps 1/3 of the day and works 5/12 of the day. What fraction of the day is left for other activities?

R5. Properties of Positive Exponents

Objectives

1. Introduction to the Properties of Exponents

2. Solve Applied Problems

Note the difference between $(-5)^4$ and -5^4.

Recall from section R1 that the exponential expression x^n can be written as

$$x^n = \overbrace{x \cdot x \cdot x \cdot \ldots \cdot x}^{n\text{-times}}$$

where the right side of the equation is called the *expanded form.* For example,

(a) $2^7 = 2 \cdot 2 \cdot 2 \cdot 2 \cdot 2 \cdot 2 \cdot 2 = 128$

(b) $(-2)^7 = (-2) \cdot (-2) \cdot (-2) \cdot (-2) \cdot (-2) \cdot (-2) \cdot (-2) = -128$

(c) $(-5)^4 = (-5) \cdot (-5) \cdot (-5) \cdot (-5) = 625$

(d) $-5^4 = -5 \cdot 5 \cdot 5 \cdot 5 = -625$

Introducing the Properties of Exponents

To multiply exponential expressions with the same base such as $a^2 \cdot a^3$, we can write the product as follows:

$$a^2 \cdot a^3 = (a \cdot a)(a \cdot a \cdot a) = a^5$$

Notice that the exponent of the product, a^5, is the sum of the exponents of the factors a^2 and a^3. That is,

$$a^2 \cdot a^3 = a^{2+3} = a^5$$

Expressions such as $x^4 y^2$ cannot be simplified because the bases are different.

In other words, when we multiply exponential expressions with the *same base* we repeat the base and add the exponents. In general, we have the following property.

Product Property of Exponents

If m and n are positive integers, then $a^n \cdot a^m = a^{n+m}$

Now we consider the meaning of $(a^2)^3$—observe that $(a^2)^3$ means a^2 raised to the third power, that is,

$$(a^2)^3 = a^2 \cdot a^2 \cdot a^2 = (a \cdot a) \cdot (a \cdot a) \cdot (a \cdot a) = a^{2 \cdot 3} = a^6$$

That is, to raise an exponential expression to a power, repeat the same base and multiply the exponents. In general, we have the following property.

Power Property for Exponents

If m and n are positive integers, then $(a^m)^n = a^{m \cdot n}$

EXAMPLE 1

Illustrating the Properties of Exponents

Use the properties of exponents to simplify each expression.

(a) $x^{10} \cdot x^3$ (b) $(y^3)^4$ (c) $(-t^5)^6$

Solution

(a) $x^{10} \cdot x^3 = x^{10+3} = x^{13}$ Product Property

(b) $(y^3)^4 = y^{3 \cdot 4} = y^{12}$ Power Property

(c) $(-t^5)^6 = (-t^5)(-t^5)(-t^5)(-t^5)(-t^5)(-t^5) = t^{30}$ Expanded Form

We can write the power of a **product** involving two bases such as $(a^2 b^3)^4$ in expanded form as follows:

$$(a^2 b^3)^4 = (a^2 b^3)(a^2 b^3)(a^2 b^3)(a^2 b^3)$$
$$= a^2 a^2 a^2 a^2 b^3 b^3 b^3 b^3$$
$$= a^{2+2+2+2} b^{3+3+3+3}$$
$$= a^8 b^{12}$$

If we multiply the exponents, the result is the same.

$$(a^2b^3)^4 = (a^2)^4(b^3)^4 = a^8b^{12}$$

In general, we have:

Power of a Product Property

If n is a positive integer, then $(ab)^n = a^nb^n$

We can also write the power of a **quotient** $(a/b)^3$ involving two different bases in expanded form as follows:

$$\left(\frac{a}{b}\right)^3 = \frac{a}{b} \cdot \frac{a}{b} \cdot \frac{a}{b} = \frac{a^3}{b^3}$$

Notice that if we raise the numerator and denominator to the third power, the result is the same.

$$\left(\frac{a}{b}\right)^3 = \frac{(a)^3}{(b)^3} = \frac{a^3}{b^3}$$

In general, we have the following property

Power of a Quotient Property

If n is a positive integer, then $\left(\dfrac{a}{b}\right)^n = \dfrac{a^n}{b^n}, b \neq 0$

EXAMPLE 2   **Power Property of Exponents**

Use the properties of exponents to simplify each expression.

(a) $(-x^3y)^4$ (b) $\left(\dfrac{x^2}{y}\right)^5$ (c) $\left(\dfrac{2x^5}{3^2y^2}\right)^3$

Solution

(a) $(-x^3y)^4 = (-x^3)^4y^4 = x^{12}y^4$ Power of a Product

(b) $\left(\dfrac{x^2}{y}\right)^5 = \dfrac{(x^2)^5}{y^5} = \dfrac{x^{10}}{y^5}$ Power of a Quotient

(c) $\left(\dfrac{2x^5}{3^2y^2}\right)^3 = \dfrac{(2x^5)^3}{(3^2y^2)^3}$ Power of a Quotient

$= \dfrac{2^3(x^5)^3}{(3^2)^3(y^2)^3} = \dfrac{8x^{15}}{729y^6}$ Power of a Product

To develop a property for dividing exponential expressions, consider the following quotient for $a \neq 0$:

$$\frac{a^6}{a^4} = \frac{a \cdot a \cdot a \cdot a \cdot a \cdot a}{a \cdot a \cdot a \cdot a} = \left(\frac{a}{a}\right) \cdot \left(\frac{a}{a}\right) \cdot \left(\frac{a}{a}\right) \cdot \left(\frac{a}{a}\right) \cdot a \cdot a = 1 \cdot 1 \cdot 1 \cdot 1 \cdot a \cdot a = a^2$$

Here the exponent of the quotient, a^2, is the difference of the exponents of a^6 and a^4.

$$\frac{a^6}{a^4} = a^{6-4} = a^2$$

In general, we have the following property:

Quotient Property for Exponents

If m and n are positive integers where $m > n$, and $a \neq 0$, then $\dfrac{a^m}{a^n} = a^{m-n}$.

EXAMPLE 3   **Illustrating the Quotient Property**

(a) $\dfrac{x^5}{x^3}$ (b) $\dfrac{(-x)^5}{(-x)^3}$ (c) $\dfrac{-x^6}{(-x)^2}$ (d) $\dfrac{(2x)^5}{(3x)^2}$

Solution

(a) $\dfrac{x^5}{x^3} = x^{5-3} = x^2$

(b) $\dfrac{(-x)^5}{(-x)^3} = (-x)^{5-3} = (-x)^2 = (-x)(-x) = x^2$

In part d of this example, we simplify first because of the different bases, 2x and 3x, and then use the quotient property.

(c) $\dfrac{-x^6}{(-x)^2} = \dfrac{-x \cdot x \cdot x \cdot x \cdot x \cdot x}{(-x)(-x)} = \dfrac{-x \cdot x \cdot x \cdot x \cdot x \cdot x}{x \cdot x} = -\dfrac{x^6}{x^2} = -x^{6-2} = -x^4$

(d) $\dfrac{(2x)^5}{(3x)^2} = \dfrac{2^5 x^5}{3^2 x^2} = \dfrac{32}{9} x^{5-2} = \dfrac{32}{9} x^3$

The following is a summary of all these properties.

Properties of Exponents

If a and b are real numbers and m and n are positive integers, then

$a^m \cdot a^n = a^{m+n}$ — Product of Exponents

$\dfrac{a^m}{a^n} = a^{m-n}, a \neq 0, m > n$ — Quotient of Exponents

$(a^m)^n = a^{m \cdot n}$ — Power Property

$(ab)^n = a^n b^n$ — Power of a Product

$\left(\dfrac{a}{b}\right)^n = \dfrac{a^n}{b^n}, b \neq 0$ — Power of a Quotient

Solving Applied Problems

Many applications require the use of very large numbers. For example, the speed of light is approximately

300,000,000 meters per second.

Scientific notation expresses such numbers, using the form

$$s \times 10^n$$

where s is a number between 1 and 10, including 1 ($1 \leq s < 10$), and n is an integer. We write 300,000,000 in scientific notation as

$$3.0 \times 10^8$$

Notice in 3.0×10^8 that 3 is multiplied by $10^8 = 100,000,000$ (100 million), moving the decimal 8 places to the right.

Most calculators can express numbers in scientific notation by using the command $\boxed{\text{EE}}$ or $\boxed{\text{EXP}}$ (check your manual if you do not find these commands). To enter 3.0×10^8 on a calculator, enter 3.0, then $\boxed{\text{EE}}$ or $\boxed{\text{EXP}}$ and the exponent 8. Any operations that the calculator can perform can be done in scientific notation. If the result of a calculation has more digits than the calculator can display, the answer is automatically written in scientific notation. For example, we have seen that using the laws of exponents, the product $2^{15} \cdot 2^{25} = 2^{15+25} = 2^{40}$. On the calculator $2^{15} = 32,768$ and $2^{25} = 33,554,432$, but $2^{40} = 1.09951162778E12$. The result is written in scientific notation.

EXAMPLE 4   **Using Scientific Notation**

A light-year is the distance light travels in one year. Find this distance in meters (m). Express the answer in scientific notation.

Solution   The speed of light is approximately 300,000,000 meters per second. A light-year is the distance that light travels in one year. So a light-year is given by:

$$\frac{300,000,000 \text{ m}}{1 \text{ sec}} \cdot \frac{3600 \text{ sec}}{1 \text{ hr}} \cdot \frac{24 \text{ hr}}{1 \text{ day}} \cdot 365.25 \text{ days} =$$

9,467,280,000,000,000 m or 9.46728×10^{15} m

Therefore, light travels 9.46728×10^{15} meters in one year.

EXAMPLE 5   **Multiplying and Dividing Expressions in Scientific Notation**

Use the properties of exponents to simplify each expression. Write each answer in scientific notation.

(a) $(4.0 \times 10^7)(2.0 \times 10^5)$ (b) $\dfrac{3.5 \times 10^7}{5 \times 10^4}$

Solution   (a) $(4.0 \times 10^7)(2.0 \times 10^5) = (4.0 \times 2.0)(10^7 \times 10^5)$

$$= 8.0 \times 10^{7+5}$$

$$= 8.0 \times 10^{12}$$

(b) $\dfrac{3.5 \times 10^7}{5 \times 10^4} = \left(\dfrac{3.5}{5}\right) \times \left(\dfrac{10^7}{10^4}\right) = 0.7 \times 10^{7-4}$

$$= 0.7 \times 10^3$$

$$= 7 \times 10^2$$

PROBLEM SET R5.

In Problems 1–8, simplify each expression.

1. (a) $3^2 \cdot 3^3$
   (b) $x^2 \cdot x^3$

2. (a) $2^4 \cdot 2^2$
   (b) $y^4 \cdot y^2$

3. (a) $(-4)^5 \cdot (-4)^7$
   (b) $-x^5 \cdot (-x)^7$

4. (a) $(-10)^6 \cdot (-10)^4$
   (b) $-y^6 \cdot (-y)^4$

5. $y^3 \cdot y^4 \cdot y^5$

6. $(-y)^9 \cdot (-y)^3 \cdot (-y)$

7. (a) $\dfrac{13^5}{13^2}$
   (b) $\dfrac{x^{50}}{x^{20}}$

8. (a) $\dfrac{14^9}{14^6}$
   (b) $\dfrac{x^{90}}{x^{60}}$

In Problems 9 and 10, simplify each expression.

9. (a) $\dfrac{-x^{25}}{x^{20}}$
   (b) $\dfrac{(-x)^{25}}{(-x)^{20}}$

10. (a) $\dfrac{-x^{13}}{x^{10}}$
    (b) $\dfrac{(-x)^{13}}{(-x)^{10}}$

In Problems 11–14, simplify each expression by removing the parentheses.

11. (a) $(12^2)^3$
    (b) $(x^{20})^{30}$

12. (a) $(13^2)^4$
    (b) $(x^{20})^{40}$

13. (a) $(y^{13})^5$
    (b) $(-y^5)^{13}$

14. (a) $(-y^{20})^3$
    (b) $(-y^3)^{20}$

In Problems 15–20, simplify each expression.

15. (a) $-4\left(\dfrac{x^3 y^5}{x^2 y^3}\right)^2$
    (b) $-\left(\dfrac{4x^3 y^6}{x^2 y^3}\right)^2$

16. (a) $\dfrac{3}{2}\left(\dfrac{a^5 b^3}{a^2 b^6}\right)^3$
    (b) $-\left(\dfrac{3a^5 b^3}{2a^2 b^6}\right)^3$

17. $(x^2 y^3)(xy^2)^2$

18. $(-x^2 y^3)(3xy^4)^2$

19. $\dfrac{x^{12}(y^3)^3}{(x^2 y)^5}$

20. $\dfrac{x^6 y^4}{(xy^2)^2}$

In Problems 21 and 22, rewrite each number in scientific notation.

21. (a) 385,000
    (b) 385 million
    (c) 38,000

22. (a) 47,000
    (b) 47 million
    (c) 47 billion

In Problems 23–26, use the properties of exponents to simplify each expression. Write each answer in scientific notation.

23. (a) $(2 \times 10^5)(4.8 \times 10^9)$
    (b) $\dfrac{4.8 \times 10^{15}}{2 \times 10^9}$

24. (a) $(3.4 \times 10^7)(2 \times 10^4)$
    (b) $\dfrac{3.4 \times 10^7}{2 \times 10^4}$

25. $\dfrac{(9.75 \times 10^8) \cdot (1.5 \times 10^3)}{(7.5 \times 10^2) \cdot (2.5 \times 10^4)}$

26. $\dfrac{(1.8 \times 10^5) \cdot (2.4 \times 10^{19})}{(3.6 \times 10^7) \cdot (4.8 \times 10^{11})}$

In Problems 27–30, simplify each expression.

27. $\dfrac{a^2}{b^2} + \dfrac{b^2}{a^2}$

28. $\dfrac{t^3}{u^3} + \dfrac{u^3}{t^3}$

29. $-\left(\dfrac{3a^5 b^{13}}{2a^2 b^6}\right)^3$

30. $\left(\dfrac{-4t^{12} s^3}{8t^3 s^2}\right)^5$

In Problems 31 and 32, rewrite each statement so that all numbers are expressed in scientific notation.

31. (a) The diameter of the star Betelgeuse is approximately 358,400,000 kilometers.
    (b) There are approximately 300,000,000 inches in 5000 miles.

32. (a) Oil reserves in the United States are about 35,000,000,000 barrels.
    (b) The number of shares traded on the New York Stock Exchange in one day is 1,250,000,000.

In Problems 33–37, express each answer in scientific notation.

33. **Electronics:** A computer can do 86 million operations in 4.3 seconds. How long, in nanoseconds, does it take this computer to do one operation? (1 billion nanoseconds = 1 second)

34. **Astronomy:** In late 1993 shortly after it was repaired, the Hubble Space Telescope took a dramatic photo of M100, a spiral galaxy far from earth. Scientists estimate that the closest stars in M100 are 35 million light-years from us, whereas the most distant stars are 80 million light-years from us. How many meters away from us are the closest and farthest stars? (Hint: see example 4)

35. **Physics:** The mass of the Earth is 6×10^{21} tons. The mass of the sun is about 300,000 times the mass of the Earth. Use scientific notation to find how many tons is the mass of the sun.

36. **Chemistry:** What is the number of atoms in 15 grams of hydrogen if one gram of hydrogen

contains 602,300,000,000,000,000,000,000 atoms? (Round to two decimal places.)

37. **Astronomy:** The mass of the sun is approximately 1.97×10^{29} kilograms, and our galaxy (the Milky Way) is estimated to have a total mass of 1.5×10^{11} suns. The mass of the known universe is at least 10^{11} times the mass of the galaxy. Calculate the mass of the known universe.

Objectives

1. Introduce Dimensional Analysis
2. Apply Formulas from Geometry
3. Apply the Pythagorean Theorem
4. Apply Formulas from Business
5. Apply Formulas from Science

R6. Formulas

Many applications in mathematics and other sciences require the use of *formulas* for their solutions. A **formula** is an equation that expresses a relationship between several quantities. In this section, we shall use formulas from geometry, business and finance, physical sciences, and medicine. As we work with formulas, it is often necessary to find a specific value when other quantities are known. To expedite this concept, we extend our work to include *dimensional analysis.*

Introducing Dimensional Analysis

In applications from mathematics and other sciences, we refer to the units of measurement as **dimensions** and the study of these units is called **dimensional analysis.** In working with dimensional analysis, all the properties of algebra apply and we use *conversion factors* to allow us to change units of measure. For example, we write 3 feet = 1 yard as

$$\frac{3 \text{ feet}}{1 \text{ yard}}$$

and we read this conversion as *3 feet per yard.* Other examples of conversion factors include the following:

$$\frac{60 \text{ minutes}}{1 \text{ hour}} \text{ which reads "60 minutes per hour"}$$

$$\frac{24 \text{ hours}}{1 \text{ day}} \text{ which reads "24 hours per day"}$$

EXAMPLE 1 **Using Dimensional Analysis**

According to the July 13, 2004, edition of *USA Today* a British pound was worth 1.8643 US dollars and $1 US was worth 108.23 Japanese yen. Find the conversion of 1 British pound to Japanese yen.

Solution $$\frac{1.8643 \text{ US dollar}}{1 \text{ British pound}} \cdot \frac{108.23 \text{ yen}}{1 \text{ US dollar}} = \frac{201.77 \text{ yen}}{1 \text{ British pound}}$$

Applying Formulas from Geometry

There are several formulas in geometry that we use often. Table 1 shows formulas that give the area A and the perimeter P (or circumference C) of some plane figures.

TABLE 1

Name	Area (A)	Perimeter (P) or Circumference (C)	Figures
Square	$A = s^2$	$P = 4s$	
Rectangle	$A = lw$	$P = 2l + 2w$	
Circle	$A = \pi r^2$	$C = 2\pi r$	
Triangle	$A = \dfrac{1}{2} ch$	$P = a + b + c$	

EXAMPLE 2   **Finding the Area of a Norman Window**

The Norman style of architecture is characterized by massive construction, carved decorations, and rectangular windows surmounted by semicircular arches. Suppose a window in St. Mary's Church has dimensions 6 feet wide by 8 feet long. Find the total area of the window.

Solution   Referring to Table 1 and Figure 1, the area of the rectangle is given by:

$$A = lw = (6 \text{ ft})(8 \text{ ft}) = 48 \text{ ft}^2$$

Figure 1

and the area of a semicircle is

$$A = \frac{1}{2} \pi r^2$$

We will use 3.14 as an approximation for π and half the width of the rectangle is the radius of the semicircle.

$$A = \frac{1}{2} (3.14)(3 \text{ ft})^2$$

$$= 14.13 \text{ ft}^2$$

Therefore, the area of the window is $48 + 14.13 = 62.13$ square feet.

Throughout this book, we will use some names and prefixes for units used in the solutions of applied problems and related arts. These names and prefixes are listed in Table 2.

Common formulas for volumes and surface areas are given in Table 3.

TABLE 2

Name	Prefix
inch(es)	in
square inch	in²
foot or feet	ft
square feet	ft²
yard(s)	yd
square yards	yd²
centimeter(s)	cm
square centimeter	cm²
meter(s)	m
cubic meter	m³
feet per second	ft/sec
miles per hour	mi/hr

TABLE 3

Name	Volume (V)	Surface Area (S)	Figure
Cube	$V = s^3$	$S = 6s^2$	
Rectangular Solid	$V = lwh$	$S = 2lh + 2hw + 2lw$	
Circular Cylinder	$V = \pi r^2 h$	$S = 2\pi r^2 + 2\pi rh$	
Sphere	$V = \dfrac{4}{3}\pi r^3$	$S = 4\pi r^2$	

EXAMPLE 3

Volume and Surface Area of a Balloon

A spherical balloon is being inflated with helium. At one instant, the radius of the balloon is 60 inches.

(a) Find the volume of the balloon in cubic feet.

(b) Find the surface area of the balloon in square feet. (Use $\pi \approx 3.14$)

Solution

First we convert the radius to feet, so that

$$60 \text{ inches} = (60 \text{ in.}) \, \frac{1 \text{ foot}}{12 \text{ inches}} = 5\text{ft}.$$

(a) Substituting 5 ft for r in the volume formula, we have

$$V = \frac{4}{3}\pi r^3$$

$$= \frac{4}{3}(3.14)(5 \text{ ft})^3 = 523.33 \text{ ft}^3$$

Therefore, the volume of the balloon is 523.33 cubic feet.

(b) Now substituting 5 ft for r in the surface area formula, we have

$$S = 4\pi r^2$$

$$= 4(3.14)(5\text{ft})^2 = 314 \text{ ft}^2$$

Therefore, the surface area of the balloon is 314 ft².

Applying the Pythagorean Theorem

Figure 2

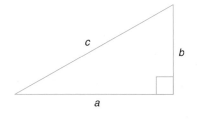

The Pythagorean Theorem serves as a guideline for solving certain types of problems involving right triangles. The theorem relates the lengths of the sides of a right triangle. The side opposite the triangle's right angle is called the **hypotenuse.** The other two sides are called the **legs** of the triangle. Figure 2 shows a right triangle with legs a and b and hypotenuse c.

The Pythagorean Theorem

> The sum of the square of the lengths of the legs of a right triangle (Figure 2) is equal to the square of the length of the hypotenuse.
>   That is,
>
> $$a^2 + b^2 = c^2.$$

It should be noted that if $a^2 + b^2 = c^2$, then the triangle is a right triangle. This is commonly referred to as the *Converse of the Pythagorean Theorem.*

EXAMPLE 4   **Using the Pythagorean Theorem**

To prepare for an outdoor rock concert, the stage crew placed two large speaker towers on stage in such a way that the distance from the control booth to the first speaker tower is 25 feet, and the distance from the booth to the second tower is 60 feet. Assume the triangle formed is a right triangle with the distance between the towers as the hypotenuse (Figure 3). How far apart are the towers?

Figure 3

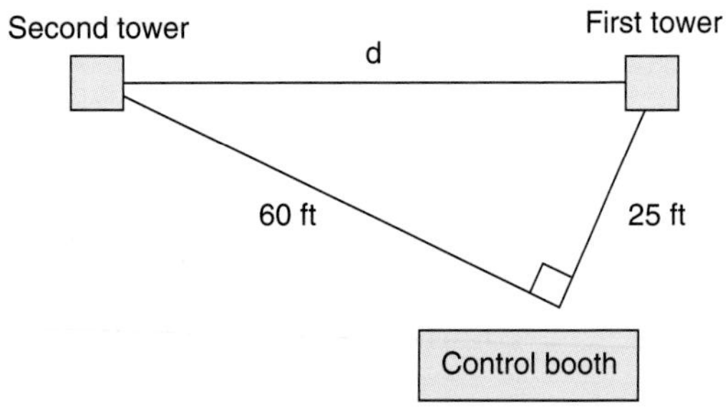

Solution   Figure 2 shows the given lengths of the legs of the right triangle. Use the Pythagorean Theorem to find the length d of the hypotenuse as follows.

$$d^2 = (60 \text{ ft})^2 + (25 \text{ ft})^2$$
$$= 3600 \text{ ft}^2 + 625 \text{ ft}^2$$
$$= 4225 \text{ ft}^2$$

so $d = \sqrt{4225 \text{ ft}^2} = 65$ ft.

Therefore, the distance between the speaker towers is 65 feet.

Applying Formulas from Business

Many types of investments and loans use formulas. For instance, the basic formula for computing simple interest, I, on an investment or loan is given by:

$$I = Prt$$

where P is the principal, r is the interest rate, and t is time. The total amount, T, of money accumulated when P dollars earn at an interest rate of r for t years is given by

$$T = P + Prt.$$

EXAMPLE 5   **Computing Interest**

A deposit of $4,000 was left in an account that pays 7.5% simple interest.

(a) Find the amount of interest after 3 years.

(b) How much money is accumulated in this account after 3 years?

Solution   (a) Substituting $4,000 for P, 0.075 per year for r, and 3 years for t yields

$$I = Prt$$

$$= (\$4,000)\,\frac{0.075}{1\ \text{year}}\,3\ \text{years} = \$900$$

Therefore, the amount of interest earned is $900.

(b) The total amount accumulated in the account after 3 years is given by:

$$T = P + I = P + Prt$$

$$= \$4000 + \$900 = \$4900.$$

In the next example, we use the formula

$$A = P(1 + r)^t$$

to compute the amount A of accumulated principal and interest in a savings account if P is saved at an annual compounded interest rate r for t years. Compounded interest earns interest on both accumulated interest and principal.

EXAMPLE 6   **Calculating Compound Interest in a Savings Account**

Compute the amount of money in an Individual Retirement Account (IRA) if $2000 is compounded annually at an interest rate of 6% for 30 years.

Solution   Substituting 2000 for P, 0.06 for r, and 30 for t yields

$$A = P(1 + r)^t$$

$$= 2000(1 + 0.06)^{30}$$

$$= 2000(1.06)^{30} = 11,486.98.$$

The total amount accumulated is approximately $11,486.98.

Applying Formulas from Science

There are several formulas in science that we use quite often. They will be used periodically throughout this textbook. For example, if you travel at an average rate of 55 miles per hour for 4 hours, then the distance traveled is found by multiplying the rate by the time. Thus, the distance d is given by

$$d = \frac{55\ \text{miles}}{1\ \text{hour}}\,(4\ \text{hours}) = 220\ \text{miles}.$$

This is stated as **distance equals rate times time.** In symbols, we use the formula

$$d = rt.$$

EXAMPLE 7   **Computing Distance**

A motorist drove 3 hours on the Ohio Turnpike at 65 miles per hour and 2 hours on the Pennsylvania Turnpike at 55 miles per hour. What is the total distance traveled?

Solution   The distance traveled on the Ohio Turnpike is given by

$$d = rt = \frac{65\ \text{miles}}{1\ \text{hour}}\,(3\ \text{hours}) = 195\ \text{miles}.$$

The distance traveled on the Pennsylvania Turnpike is given by

$$d = rt = \frac{55 \text{ miles}}{1 \text{ hour}} \, (2 \text{ hours}) = 110 \text{ miles}.$$

The total distance traveled is 195 mi + 110 mi = 305 mi.

The formulas $C = \dfrac{5}{9}(F - 32)$ and $F = \dfrac{9}{5}C + 32$ show the conversion from Fahrenheit to Celsius and from Celsius to Fahrenheit respectively, where C is the temperature reading on the Celsius scale and F is the temperature reading on the Fahrenheit scale.

EXAMPLE 8   **Converting Temperatures**

On a given day the temperature reading on the Canadian side of Niagara Falls is 25°C. What is the corresponding temperature on the US side?

Solution   In Canada, the temperature reading is in Celsius while in the US the temperature is read in Fahrenheit. Using the formula, $F = \dfrac{9}{5}C + 32$

We have
$$F = \frac{9}{5}(25) + 32$$
$$= 45° + 32° = 77°.$$

Therefore, the temperature reading on the US side of Niagara Falls is 77°F.

 PROBLEM SET R6.

In Problems 1–4, use the formula for the area of a rectangle $A = lw$ and dimensional analysis to complete the following conversions. Note that 100 centimeters = 1 meter and 10 millimeters = 1 centimeter.

Item	Square Meters	Square Centimeters	Square Millimeters
1. Area of a sheet of paper		588	
2. Area of a desktop	1.5		
3. Area of a dollar bill		100	
4. Area of a postage stamp			480

In Problems 5 and 6, use dimensional analysis to complete each conversion (5280 feet = 1 mile).

5. 2.7 miles = _____ feet

6. 144 square inches = _____ square miles

In Problems 7–18, find the shaded area of the region drawn in the given figure. (Use 3.14 ≈ π and round to two decimal places.) Also find the perimeter of each region in problems 7–14.

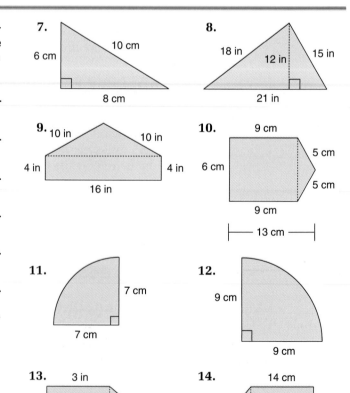

15.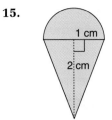

16.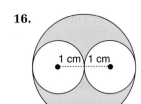

17.

18.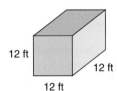

In Problems 19–26, find (a) the volume and (b) the surface area of each solid figure. Round to two decimal places.

19.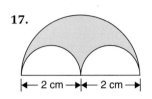

20.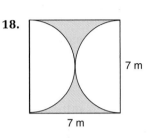

21.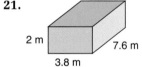

22.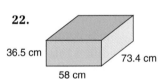

23.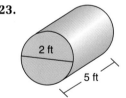

24.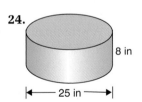

25.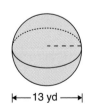

26.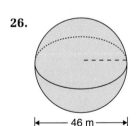

In Problems 27–32, use the Converse of the Pythagorean Theorem to determine if the triangle whose sides are *a, b,* and *c* units is a right triangle.

27. $a = 5, b = 12, c = 13$

28. $a = 8, b = 15, c = 17$

29. $a = 12, b = 20, c = 24$

30. $a = 12, b = 30, c = 32$

31. $a = 12, b = 35, c = 37$

32. $a = 31, b = 19, c = 37$

In Problems 33–36, use the formula $I = Prt$ to complete the following table.

	Interest I (in dollars)	Principal P (in dollars)	Interest Rate r	Time t (in years)
33.		2000	6%	3
34.		1345	4%	4
35.	312	1560		5
36.	1250.20	8930		2

In Problems 37–40, use the formula $d = rt$ to complete the following table.

	Distance d	Rate r	Time t
37.		550 miles/hour	4 hours
38.	2200 kilometers	110 kilometers/hour	
39.	1000 feet		3.2 seconds
40.	660 miles	55 miles/hour	

In Problems 41–44, use the conversion formulas for temperature:

41. (a) Convert 25°C to Fahrenheit
    (b) Convert 23°F to Celsius

42. (a) Convert 53°C to Fahrenheit
    (b) Convert 51.8°F to Celsius

43. (a) Convert 17°C to Fahrenheit
    (b) Convert 14°F to Celsius

44. (a) Convert 2°C to Fahrenheit
    (b) Convert 5°F to Celsius

45. Carpeting: Suppose that we wish to carpet a rectangular room whose length is 14 feet and width 12 feet, with a carpet costing $12.95 per square yard and a pad costing $2.85 per square yard.
    (a) What is the cost of the carpet?
    (b) What is the cost of the pad?
    (c) What is the total cost of the carpet and the pad?

46. Geometry: A cement slab is 16 feet long, 12 feet wide, and 6 inches thick.
    (a) How many cubic yards of concrete does the slab contain?
    (b) If "ready mix" concrete sells for $56 per cubic yard, what is the cost of the concrete?

47. Geometry: Figure 4 shows three tanks in the forms of a cube, a cylinder, and a sphere with the given dimensions.

Figure 4

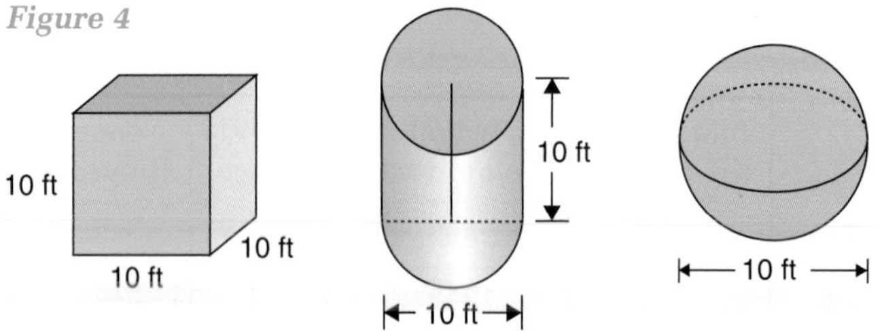

10 ft

10 ft
10 ft

10 ft

←— 10 ft —→

←— 10 ft —→

(a) Which of the three tanks has the largest surface area?

(b) Which of the three tanks has the largest volume?

48. Geometry: A tank in the shape of a right circular cylinder with a radius of 10 feet and a height of 14 feet is to be painted. According to the label on the can of paint, a gallon covers 400 square feet.

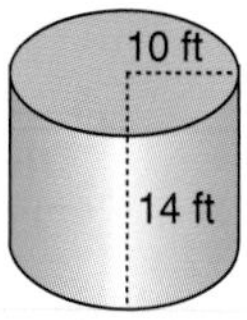

10 ft

14 ft

(a) How many gallons of paint will be needed to paint the tank?

(b) How many gallons of water will the tank hold? (1 cubic foot = 7.48 gallons)

49. Geometry: An orange grove, in the form of a square with sides 0.3 miles long, is sold for $1,036,800.

(a) What is the area of the grove in acres if one acre = 43,560 square feet and one mile = 5280 feet?

(b) What is the cost per acre of this grove?

50. Simple Interest: If $6000 is borrowed from a bank at an interest rate of 6.5% for a period of 8 months, how much interest is owed?

51. Travel Time: Determine the time for the space shuttle to travel a distance of 5950 miles if its average speed is 27,200 kilometers per hour. (Use 1 mile = 1.6 kilometers.)

52. Boston Marathon: Find the average speed of the man holding the record time of 2 hours and 8 minutes in the Boston Marathon. The length of the course is 26 miles and 385 yards. (Use 1 mile = 1760 yards.)

53. Doubling Principal: Compute the amount of money in a savings account (use $T = P + Prt$):

(a) A principal of $1000 is earning at a rate of 8% for 5 years.

(b) A principal of $2000 is earning at a rate of 8% for 5 years.

54. Doubling Interest Rate: Compute the amount of money in a savings account (use $T = P + Prt$):

(a) A principal of $5000 is earning at a rate of 4.75% for 6 years.

(b) A principal of $5000 is earning at a rate of 9.5% for 6 years.

55. Doubling Principal: Compute the amount of money in a savings account (use $A = P(1 + r)^t$):

(a) A principal of $1000 is earning at a rate of 8% compounded annually for 5 years.

(b) A principal of $2000 is earning at a rate of 8% compounded annually for 5 years.

56. Doubling Interest Rate: Compute the amount of money in a savings account (use $A = P(1 + r)^t$):

(a) A principal of $5000 is earning at a rate of 4.75% compounded annually for 6 years.

(b) A principal of $5000 is earning at a rate of 9.5% compounded annually for 6 years.

Cumulative Review of Basic Concepts

1. Write each statement in symbols.

(a) Five added to the sum of −5 and 3.

(b) The product of −2 and −6 decreased by 3.

(c) The quotient of −30 and 5 decreased by 2.

2. Write an equivalent sentence in words.

(a) $2(x + 1) < 5$

(b) $3x + 7 > 5x$

(c) $\dfrac{x}{2} - 1 = 4$

(d) $5 - 8 \neq 7$

3. (a) Write 3^5 in expanded form.

(b) Write the expression $3 \cdot 3 \cdot 3 \cdot x \cdot x \cdot x \cdot x$ in exponential form.

4. (a) Apply the order of operations to find the value of the expression.

$$(15 + 5 \cdot 3^2 - 6) \div (24 - 6 \cdot 3)$$

(b) Find the value of the expression $x^3 - 3xy^2 + y^3$

if $x = 2$ and $y = 3$.

5. State the properties that justify each statement.

(a) $4(5y) = (4 \cdot 5)y$

(b) $7(1) = 7$

(c) $(3 + x) + 7 = (x + 3) + 7$

(d) $4 + (-4) = 0$

(e) $7 + 0 = 7$

(f) $4(x - 2) = 4x - 8$

(g) $(3 + 2) + x = (3 + x) + 2$

(h) $(2x - 3)4 = 8x - 12$

6. Consider the set of real numbers
$\{-7, -1.35, 0, 3/4, 1, \pi, \sqrt{2}, 5\}$.

(a) Which of the numbers in this set are not integers?

(b) Which numbers are rational numbers?

(c) Which numbers are irrational numbers?

7. Express each rational number as a repeating decimal.

(a) $\dfrac{3}{9}$ (b) $\dfrac{5}{18}$ (c) $\dfrac{2}{7}$ (d) $\dfrac{2}{11}$ (e) $\dfrac{-7}{100}$

(f) Use a calculator to round off each number to four decimal places.

8. Find the value of each expression.

(a) $|16| + |-4|$

(b) $|35| - |-20|$

(c) $-16 + (-30)$

(d) $-15 - (-7)$

(e) $-18 + 9 - 4$

(f) $(-8)(-7)$

(g) $(-1)(-2)(-1)(-5)$

(h) $(-2)(-4)(-1)(-6)$

(i) $(-63) \div (-9)$

(j) $(36) \div (-4)$

(k) $(-5)^2 + (-1)^3$

(l) $5(4 - 6)^2 - 4(3 - 7)^2$

(m) $\dfrac{3^2 + 4^2}{(3 - 4)^2}$

(n) $\dfrac{5^2 - 7^2}{(5 - 7)^2}$

9. Find the value of each expression.

(a) $(2x + y)^2$ for $x = -1$ and $y = 3$.

(b) $\dfrac{x^2 - y^2}{x + y}$ for $x = -2$ and $y = 1$.

10. (a) Reduce each rational number to lowest terms.
$$\frac{525}{735}, \frac{-385}{-455}, \text{ and } \frac{-279}{310}$$

(b) Rewrite each rational number as a percent.
$$\frac{3}{5}, \frac{5}{16}, \text{ and } \frac{5}{8}$$

(c) Change each percent to a rational number.
$$4\%, \frac{3}{8}\%, \text{ and } 21\%$$

(d) What is 12% of 340?

(e) What is the ratio of 35 seconds to 2 minutes?

11. Perform each operation and simplify.

(a) $\dfrac{5}{7} \cdot \dfrac{7}{10}$

(b) $-\dfrac{11}{13} \cdot \dfrac{13}{11}$

(c) $\dfrac{21}{10} \div \dfrac{7}{15}$

(d) $\dfrac{-7}{18} \div \dfrac{-14}{12}$

(e) $\dfrac{5}{12} + \dfrac{-3}{8}$

(f) $\dfrac{3}{20} + \dfrac{7}{30}$

(g) $\dfrac{9}{16} - \dfrac{-7}{12}$

(h) $\dfrac{1}{30} - \dfrac{-9}{40}$

(i) $\dfrac{1}{2} - \dfrac{1}{3} + \dfrac{1}{4}$

(j) $\dfrac{1}{8} - \dfrac{1}{4} + \dfrac{1}{5}$

12. Simplify each expression.

(a) $(-3)^4 \cdot (-3)^2$

(b) $(x^3)^5$

(c) $\dfrac{x^{10}}{x^4}$

(d) $\dfrac{(-x)^{12}}{(-x)^5}$

(e) $\left(\dfrac{x^2 y^3}{xy}\right)^3$

(f) $\dfrac{x^7 y^4}{(xy^2)^2}$

(g) $(x^3 y^2)^2 (x^4 y^5)^3$

(h) $(x^2 y^4)^3 (xy^2)^2$

13. (a) Write the number 730,000,000 in scientific notation.

(b) Write 2.35×10^5 in expanded form.

(c) Simplify $\dfrac{(2.4 \times 10^7) \cdot (3.5 \times 10^4)}{(1.2 \times 10^5) \cdot (0.7 \times 10^2)}$.

14. Use the Pythagorean Theorem to determine whether or not the triangle with sides a, b, and c is a right triangle.

(a) $a = 8$, $b = 15$, $c = 17$

(b) $a = 5$, $b = 10$, $c = 13$

15. (a) Convert 11°C to Fahrenheit.

(b) Convert 46.4°F to Celsius.

16. Use the formula $I = Prt$ to find I if:

(a) $P = 1500$, $r = 3\%$, $t = 2$

(b) $P = 1200$, $r = 3.7\%$, $t = 4$

17. A gambler begins his day in a casino with $100 and wins $70. After an hour, he doubled his money. The next hour he loses $123.

(a) Write an expression with these numbers to describe the situation.

(b) Find the value of the expression. That is, how much money does he have in the end?

18. Describe an everyday occurrence that exhibits the commutative property.

19. **Markdown:** A dress that usually sells for $300 is marked down 25%. What is the sale price of the dress? Use the formula:

sale price = regular price − markdown.

20. **Markdown:** Suppose the regular price for a calculator is $90. If the calculator's price is marked down 20%, what is the markdown?

21. **Markup:** A carpet dealer marks up his merchandise 25% on dealer's cost. If the carpet costs the dealer $600, how much will the dealer sell the carpet? Use the formula:

$$\text{retail price} = \text{cost} + \text{markup}.$$

22. **Markdown:** The regular price of a special computer monitor is $750. If it is reduced by $150, by what percent is the price reduced?

23. **Grades:** In an algebra class, 3/5 of the students are females. Of the female students, 1/3 earn a B in the course, what fraction of the students in the class earn a B in the course and are female?

24. **Distance:** Two cars cross an intersection. One is traveling due west at a rate of 60 miles per hour. The other is traveling due south at a rate of 50 miles per hour. How far are the two cars from each other after three hours? Round off the answer to two decimal places.

25. **Distance:** A man is standing on a dock pulling a rope which is 10 feet above water. The rope is 45 feet long and attached to his boat. How far away from the dock is the boat? Round off to one decimal place.

26. **Right Triangle:** A 16-foot ladder is leaning against a vertical wall. How far above the ground is the top of the ladder when the bottom of the ladder is 10 feet from the wall? Round off to one decimal place.

27. **Area:** Figure 5 shows the design of a flower bed in the shape of a right triangle and two semicircles. Find the total area of the flower bed. Round off the answer to two decimal places.

Figure 5

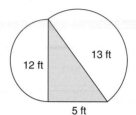

28. **Compound Interest:** Compute the amount of money in an IRA account if $3000 is compounded annually at an interest rate of 7% for 25 years.

Linear Equations and Inequalities

In New York harbor stands a monument to liberty commonly called the Statue of Liberty. The torch of this great statue is 305 feet above the ground. If the height of one-half of the pedestal is 74 feet shorter than the height of the statue, find the height of the pedestal. In this chapter, we will develop methods to solve such a problem. This is solved in example 4 of section 1.3 on page 60.

305 ft

In this chapter we apply the algebraic skills developed in the preceding chapter to solve equations and inequalities. Both equations and inequalities are used to describe events from science and business as well as from social sciences and everyday life. Mathematics is quickly becoming the language not only of science but of much of human endeavor.

Objectives

1. Solve Linear Equations
2. Solve Linear Equations Involving Fractions
3. Classify Identities and Inconsistent Equations

1.1 Solving Linear Equations in One Variable

An **equation** is a statement that two mathematical expressions are equal. Substituting a particular number for the variable produces an equation that is either true or false. If a true statement results from such a substitution, we say the number **satisfies** the equation and this number is called a **solution** or a **root** of the equation. For instance, 3 is a solution of the equation $4x + 7 = 19$, because the substitution of 3 for x makes the equation a true statement. That is,

$$4(3) + 7 = 12 + 7 = 19$$

But -4 is not a solution, because -4 makes the equation false.

$$4(-4) + 7 = -16 + 7 = -9 \neq 19$$

Two or more equations are said to be **equivalent** if they have *exactly* the same solution. For example, the following equations are equivalent because they all have the same solution, namely, 3.

Equation	Substitute $x = 3$
$4x + 7 = 19$	$4(3) + 7 = 12 + 7 = 19$
$4x = 12$	$4(3) = 12$
$x = 3$	$x = 3$

We can change an equation

$$P = Q$$

into an equivalent one by performing any of the following operations.

Property	Let $P = Q$, then	Illustration Let $x = 3$
1. Addition (Subtraction) Property Add (subtract) the same quantity to both sides of the equation.	(a) $P + R = Q + R$ (b) $P - R = Q - R$	$x + 7 = 3 + 7$ $x - 7 = 3 - 7$
2. Multiplication (Division) Property Multiply (divide) both sides of the equation by the same non zero quantity.	(a) $PR = QR$ (b) $\dfrac{P}{R} = \dfrac{Q}{R}, R \neq 0$	$5x = (5)(3)$ $\dfrac{x}{6} = \dfrac{3}{6}$
3. Symmetric Property Interchange the two sides of the equation.	$Q = P$	$3 = x$

To **solve** an equation means to find all of its solutions. The usual method for solving an equation is to write a sequence of equations, starting with the given one, in which each equation is equivalent to the previous one. The last equation should express the solution directly.

Solving Linear Equations

Each of the following equations,

$$5x - 3 = 7, \quad 5 + 8(x + 2) = 23 - 2(2x - 5), \quad \text{and} \quad \frac{x}{7} - \frac{x - 1}{2} = \frac{11}{4}$$

contains one variable, and each variable has power one. They are examples of **linear equations** or **first degree equations.** In this section, we consider linear equations in one variable. It should be noted these linear equations have only *one* solution; that is, only one value of the variable and no other value will make the original equation true. For instance, we solve the linear equation $5x - 3 = 7$ as follows:

$$5x - 3 = 7 \quad \text{Start with the given equation}$$
$$5x - 3 + 3 = 7 + 3 \quad \text{Use the Addition Property}$$
$$5x = 10$$
$$\frac{5x}{5} = \frac{10}{5} \quad \text{Use the Division Property}$$
$$x = 2 \quad \text{An equivalent equation with an obvious solution.}$$

Check:

x	Left Side	Right Side	True/False
2	$5(2) - 3 = 10 - 3 = 7$	7	True

In general, we may solve linear equations in one variable by using the following strategy.

Strategy For Solving Linear Equations

1. **COMBINE**
   (a) Multiply each side of the equation by the least common denominator (LCD), if fractions exist in the equation.
   (b) Remove all parentheses or grouping symbols on each side of the equation.
   (c) Combine like terms on each side of the equation.
2. **REARRANGE** Apply the addition (or subtraction) property to collect all terms containing the variable for which we are solving on one side of the equation and all other terms on the other side.
3. **ISOLATE** Divide each side of the equation by the coefficient of the variable.
4. **CHECK** Substitute the solution into the original equation to verify that the equation is true.

EXAMPLE 1

Solving a Linear Equation

Solve each equation.

(a) $5 + 4x = 33$ (b) $-4t + 7 = -3t + 5$

Solution

(a)
$$5 + 4x = 33$$
$$5 + 4x - 5 = 33 - 5$$
$$4x = 28$$
$$\frac{4x}{4} = \frac{28}{4}$$
$$x = 7$$

(b)
$$-4t + 7 = -3t + 5 \quad \text{Given}$$
$$-4t + 7 - 7 = -3t + 5 - 7 \quad \text{Rearrange}$$
$$-4t = -3t - 2 \quad \text{Rearrange}$$
$$-4t + 3t = -2$$
$$-t = -2 \quad \text{Isolate}$$
$$-1(-t) = -1(-2)$$
$$t = 2$$

Check: (a)

x	Left Side	Right Side	True/False
7	$5 + 4(7) = 5 + 28 = 33$	33	True

Check: (b)

x	Left Side	Right Side	True/False
2	$-4(2) + 7 = -8 + 7 = -1$	$-3(2) + 5 = -6 + 5 = -1$	True

EXAMPLE 2

Using the Distributive Property to Solve an Equation

Solve the equation $5 + 8(x + 2) = 23 - 2(2x - 5)$.

Solution

We begin by using the distributive property and combine like terms.

$5 + 8(x + 2) = 23 - 2(2x - 5)$	Given
$5 + 8x + 16 = 23 - 4x + 10$	Combine, use the distributive property
$8x + 21 = -4x + 33$	Combine like terms
$8x + 4x + 21 - 21 = -4x + 4x + 33 - 21$	Rearrange, add 4x and −21 to each side
$12x = 12$	
$x = 1$	Isolate, divide each side by 12

Check:

x	Left Side	Right Side	True/False
1	$5 + 8(1 + 2) = 5 + 8(3)$ $= 5 + 24 = 29$	$23 - 2(2(1) - 5) = 23 - 2(2 - 5)$ $= 23 - 2(-3) = 23 + 6 = 29$	True

EXAMPLE 3

Solving an Equation Involving Decimals

Solve the equation $1.3(2 - 5.2y) = 0.2 - (1.76y - 3.4)$.

Solution

Here we can use a calculator for some computations.

$1.3(2 - 5.2y) = 0.2 - (1.76y - 3.4)$	Given
$2.6 - 6.76y = 0.2 - 1.76y + 3.4$	Combine, use the distributive property
$-6.76y + 1.76y = 0.2 + 3.4 - 2.6$	Rearrange, add 1.76y and −2.6 to each side
$-5y = 1.00$	
$y = -0.2$	Isolate, divide each side by −5

If an equation involves decimals, the solution should be expressed in decimals.

A check will show that -0.2 is the solution. We recommend that the reader always check the answer.

Solving Linear Equations Involving Fractions

We indicated in step 1 of our strategy that when an equation contains fractions, we multiply each side of that equation by the least common denominator (LCD) to clear the fractions. This process produces an equivalent equation with integer coefficients which is easier to solve.

EXAMPLE 4   **Solving Equations Involving Fractions**

Solve each equation:

(a) $\dfrac{2x}{3} + \dfrac{1}{2} = \dfrac{5}{6}$ (b) $\dfrac{y}{4} + \dfrac{y-2}{3} = \dfrac{5}{3}$

Solution   (a)

$$\frac{2}{3}x + \frac{1}{2} = \frac{5}{6} \qquad \text{Given}$$

$$6\left(\frac{2}{3}x + \frac{1}{2}\right) = 6\left(\frac{5}{6}\right) \qquad \text{Combine, multiply each side by 6, the LCD}$$

$$6\left(\frac{2}{3}x\right) + 6\left(\frac{1}{2}\right) = 6\left(\frac{5}{6}\right) \qquad \text{Combine, use the distributive property}$$

$$4x + 3 = 5$$

$$4x = 2 \qquad \text{Rearrange, subtract 3 from each side}$$

$$x = \frac{2}{4} = \frac{1}{2} \qquad \text{Isolate, divide each side by 4}$$

Therefore, the solution is $\dfrac{1}{2}$. A check will show the answer is correct.

(b)

$$\frac{y}{4} + \frac{y-2}{3} = \frac{5}{3} \qquad \text{Given}$$

$$12\left(\frac{y}{4} + \frac{y-2}{3}\right) = 12\left(\frac{5}{3}\right) \qquad \text{Combine, multiply by the LCD, 12}$$

$$12\left(\frac{y}{4}\right) + 12\left(\frac{y-2}{3}\right) = 20 \qquad \text{Combine, use the distributive property}$$

$$3y + 4(y-2) = 20 \qquad \text{Combine, use the distributive property}$$

$$3y + 4y - 8 = 20 \qquad \text{Combine}$$

$$7y - 8 = 20 \qquad \text{Combine, add like terms}$$

$$7y = 28 \qquad \text{Rearrange}$$

$$y = 4 \qquad \text{Isolate}$$

Check:

y	Left Side	Right Side	True/False
4	$\dfrac{4}{4} + \dfrac{4-2}{3} = \dfrac{4}{4} + \dfrac{2}{3}$	$\dfrac{5}{3}$	True
	$\dfrac{12}{12} + \dfrac{8}{12} = \dfrac{20}{12} = \dfrac{5}{3}$		

Classifying Identities and Inconsistent Equations

Each of the equations considered in Examples 1–4 were true for only one value of the variable. These linear equations are called **conditional equations.** There are two other possibilities, identities, and inconsistent equations. An equation that is true for all real numbers is called an **identity.** For instance,

$$7(x + 2) = 7x + 14 \text{ and } 5x - 2x = 3x$$

are examples of an *identity.*

An equation that has no solution is called an **inconsistent equation.** The equation

$$4(5x - 1) = 20x - 1$$

is *inconsistent* as can be seen by simplifying each side.

$$20x - 4 = 20x - 1$$

There are no values of x which would make this equation true. To determine whether an equation is an identity, conditional equation, or an inconsistent equation, solve as we have been doing. The equivalent equation will be true for all values of the variable, only one value of the variable, or no values of the variable.

EXAMPLE 5   **Classifying Equations**

Determine which of the given equations is inconsistent and which is an identity.

(a) $3(4x + 5) = 4 + 12x$ (b) $2(x + 5) = 5x + 10 - 3x$

Solution

(a) $3(4x + 5) = 4 + 12x$ Given

$\quad\quad 12x + 15 = 4 + 12x$ Combine

$\quad\quad\quad\quad\quad 15 = 4$ Rearrange, subtract $12x$ from each side

Since $15 = 4$ is a false statement, the equation has no solution and it is called *inconsistent.*

(b) $2(x + 5) = 5x + 10 - 3x$ Given

$\quad\quad 2x + 10 = 5x + 10 - 3x$ Combine

$\quad\quad 2x + 10 = 2x + 10$ Combine

Remember that 0 can be a solution to an equation. This is not the same as "no solution".

Notice that any number we substitute for x in the original equation leads to a true statement. Therefore, the solution consists of all real numbers and the equation is called an *identity.*

PROBLEM SET 1.1

Mastering the Concepts

In Problems 1–22, solve each equation and check the results.

1. (a) $2x - 5 = 11$
    (b) $2x + 7 = 7$

2. (a) $3w - 8 = 13$
    (b) $4 - 3w = 16$

3. (a) $19 - 15x = -131$
    (b) $84 = 15x - 51$

4. (a) $104 - 12t = 152$
    (b) $12t + 108 = 144$

5. (a) $18 - x = 2 + 3x$
   (b) $12 - 3m = 4m + 26$
6. (a) $2 - 5y = -3y + 6$
   (b) $2x - 14 = 7 - x$
7. (a) $2u + 1 = 5u + 19$
   (b) $2x - 3x = 7 - 15x$
8. (a) $7y - 18 = 3y - 10$
   (b) $7 + 8x - 12 = 2x - 8 + 5x$
9. $3y - 2y + 7 = 12 - 4y$
10. $2t - 9t + 3 = 6 - 5t$
11. $1 - 2(5 - 2y) = 26 - 3y$
12. $7t - 3(9 - 5t) = -5t$
13. $8(5x + 1) - 36 = -3(x - 5)$
14. $2(1 + 2y) = 3(2y - 4)$
15. $2(x + 5) - (3 - x) = 16$
16. $7(x - 3) = 4(x + 5) - 47$
17. $6(c - 10) + 3(2c - 7) = -45$
18. $5(x - 3) - (x - 1) = -14$
19. $3(x - 2) + 5(x + 1) = -45$
20. $4(x + 4) - 5(2 - x) = 3(6x - 2)$
21. $2[3x - (x - 3)] = 3(x - 3)$
22. $16 - 2[4 - 3(1 - x)] = 14 - 5x$

In Problems 23–36, solve each equation by multiplying each side by the least common denominator.

23. $\dfrac{t}{6} - \dfrac{t}{7} = 5$

24. $\dfrac{y}{2} - \dfrac{y}{3} = 4$

25. $\dfrac{4x}{3} - 1 = \dfrac{5x}{6}$

26. $\dfrac{x}{4} + \dfrac{2x}{3} = \dfrac{33}{12}$

27. $\dfrac{5x - 15}{7} - \dfrac{2x}{3} = \dfrac{2}{3}$

28. $\dfrac{3y}{5} - \dfrac{13}{15} = \dfrac{y + 2}{12}$

29. $\dfrac{t + 5}{7} + \dfrac{t - 3}{4} = \dfrac{5}{14}$

30. $\dfrac{w - 2}{3} + \dfrac{w + 1}{4} = -1$

31. $\dfrac{4x + 1}{10} = \dfrac{5x + 2}{4} - \dfrac{5}{4}$

32. $\dfrac{4x - 1}{10} - \dfrac{5x + 2}{4} = -4$

33. $\dfrac{y + 9}{4} - \dfrac{6y - 9}{14} = 2$

34. $\dfrac{8t + 10}{5} - \dfrac{6t + 1}{4} = \dfrac{3}{20}$

35. $\dfrac{3x - 2}{3} + \dfrac{x - 3}{2} = \dfrac{5}{6}$

36. $\dfrac{3u - 6}{4} - \dfrac{u + 6}{6} + \dfrac{2u}{3} = 5$

In Problems 37–44, solve each equation involving decimals.

37. $0.7x + 3 = 0.5x + 2$
38. $1.5y + 4 = 1.2y - 2$
39. $x - 0.1x - 4.5 = 0.9$
40. $0.6x - 2.5 = 0.03x + 8.9$
41. $0.02(y - 100) = 62 - 0.06y$
42. $0.05(t - 100) = 0.2t - 35$
43. $0.5x - 0.1x - 0.2x = 0.05$
44. $0.75x - 0.5x = 0.625x + 0.75$

In Problems 45–48, determine whether each equation is an identity, a conditional equation, or an inconsistent equation.

45. (a) $6(x + 2) = 6x + 12$
    (b) $4(x + 1) - 4 = 4x$
46. (a) $5(x + 3) = 5x + 3$
    (b) $5(x + 1) - 5 = 0$
47. $4(w - 2) + 2 = 3w - 7 + w$
48. $16 - 2[4 - 3(1 - x)] = 14 - 6x$

Applying the Concepts

49. **Car Rental:** A car rental company charges for its midsize models $75 a week plus $0.15 for each mile driven for that week. The total charges C (in dollars) for a week is given by the equation

$$C = 75 + 0.15n$$

where n is the number of miles driven for a week.
    (a) What are the total charges if a midsize rented car is driven 400 miles in a week?
    (b) If the total charge for a week is $165, how many miles are driven for that week?

50. **Telephone Charges:** The cost of a long distance telephone call between two cities during business hours is $0.28 for the first minute and $0.15 for each additional minute or a fraction of a minute. The total cost C (in dollars) of long distance calls is given by the equation

$$C = 0.28 + 0.15(t - 1)$$

where t is the time for a call in minutes and $t > 1$.
    (a) What is the total charge for a phone call that lasted 41 minutes?
    (b) What is the length of a long distance phone call if the total cost is $3.28?

51. **Real Estate:** A real estate agent gets a commission of 7% of the sale price of a house. The net price N (in dollars) of a house after the commission is paid is given by the equation

$$N = p - 0.07p$$

where p (in dollars) is the sale price of the house.

(a) What is the net price of a house if it is sold for $140,000?

(b) If a couple wants to sell their house and have $148,800 after they pay the commission, what should be the selling price of the house?

52. Coin Problem: A vending machine accepts nickels, dimes, and quarters. When the coin box is emptied, the total value of the coins is found to be $24.15. Suppose that the box contains n nickels, $n + 5$ dimes, and $n/2$ quarters, where n satisfies the equation

$$0.05n + 0.10(n + 5) + 0.25(n/2) = 24.15$$

Solve the equation for n, then find the number of coins of each kind in the box.

Developing and Extending the Concepts

In Problems 53 and 54, solve each equation as follows:

(a) Multiply each side of the equation by a power of 10 that will eliminate the decimals.

(b) Solve using decimals.

(c) Explain if you prefer method (a) or (b) and why.

53. $0.2(2x + 6.1) + 0.4(2.1x + 3) = 0.56$ -1.5

54. $3y - 3(1.9y - 4.1) - 0.5 = 2(1.6y)$

In Problems 55–58, solve each equation by multiplying each side by the least common denominator.

55. $\dfrac{2y + 3}{3} + \dfrac{3y + 4}{6} = \dfrac{y - 4}{9}$

56. $\dfrac{x - 2}{2} + \dfrac{x + 3}{4} = \dfrac{6x + 6}{6}$

57. $\dfrac{2t - 1}{4} + \dfrac{3t + 4}{3} = \dfrac{57 - 4t}{12}$

58. $\dfrac{y - 2}{4} - \dfrac{y + 5}{6} = \dfrac{y - 2}{9}$

59. From geometry, the sum of the interior angles of a triangle is 180°. Use Figure 1 with the given measurements to find t.

Figure 1

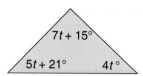

60. From geometry, two angles that are on opposite sides of the intersection of two lines are called *vertical angles.* It can be shown that vertical angles have the same measure.

(a) Use Figure 2a to find x and the angles.

Figure 2a

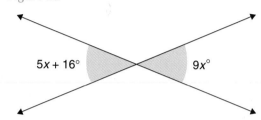

(b) Use Figure 2b to find y and the angles.

Figure 2b

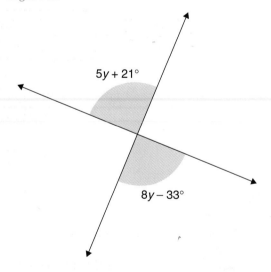

Objectives

1. Solve Literal Equations

2. Solve Formulas

3. Use Formulas to Solve Applied Problems

1.2 Solving Literal Equations and Formulas

Thus far we have worked with equations that contain one variable, represented by a letter. To analyze a relationship between more than one variable or letter, we often use **literal equations.** In these equations, some of these letters represent variables, others represent constants. In mathematics, common examples of literal equations that express relationships between two or more quantities are called **formulas.** For example:

$$I = Prt, \quad F = \frac{9}{5}C + 32, \quad \text{and} \quad ax + by = c$$

are formulas for simple interest, the conversion from Celsius to Fahrenheit, and the standard form of a linear equation. These formulas describe real-life situations in mathematical terms.

Solving Literal Equations

We will work with literal equations and formulas that produce linear equations. We use the equation solving strategy illustrated in section 1.1 to solve them. That is, we bring all terms containing the variable for which we wish to solve to one side of the equation, and all other terms to the opposite side. Then, we divide each side by the coefficient of that variable. As always, we must be careful not to divide by zero.

EXAMPLE 1 **Solving a Literal Equation**

Solve each equation for the indicated variable:

(a) $4x - a = x + 8a$ for x (b) $2x + 5y - 6 = 0$ for y

Solution

(a) $\quad 4x - a = x + 8a$ Given

$\quad\quad 4x - a - x = 8a$ Rearrange, add $-x$ to each side

$\quad\quad\quad 4x - x = 8a + a$ Rearrange, add a to each side

$\quad\quad\quad\quad 3x = 9a$ Combine

$\quad\quad\quad\quad\quad x = 3a$ Isolate, divide by 3

Check:

x	Left Side	Right Side	True/False
$3a$	$4(3a) - a = 12a - a = 11a$	$(3a) + 8a = 11a$	True

(b) $2x + 5y - 6 = 0$ for y Given

$\quad\quad\quad 5y = -2x + 6$ Rearrange, add $-2x + 6$ to both sides

$$y = \frac{-2x + 6}{5}$$ Isolate

The check is left to the reader.

EXAMPLE 2 **Solving a Literal Equation Involving Fractions**

Solve the equation $\dfrac{y - 3x}{b} = 2 - \dfrac{x}{b}$ for y, $b \neq 0$

Solution

$$\frac{y - 3x}{b} = 2 - \frac{x}{b}$$ Given

$$b\left(\frac{y - 3x}{b}\right) = b\left(2 - \frac{x}{b}\right)$$ Multiply each side by b, the LCD

$$y - 3x = 2b - x$$ Combine

$$y = 2b + 2x$$ Rearrange

A check will show the answer is correct.

Solving Formulas

Many applications from different sciences require the use of *formulas* for their solutions. For instance,

$$P = 2L + 2w, \quad d = rt, \text{ and } A = P(1 + rt)$$

are formulas for the perimeter of a rectangle, the distance in terms of rate and time, and the accumulated amount of money at simple interest rate. To solve a formula for a specific variable or letter, we follow the same strategies as we do for solving equations.

EXAMPLE 3

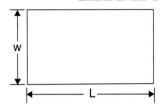

Solving a Formula

The formula, $P = 2L + 2w$, represents the perimeter (distance around the rectangle) of a rectangle where L is the length and w is the width of the rectangle. Solve this formula for w.

Solution

2 is not a factor of the numerator and so the fraction cannot be reduced.

$P = 2L + 2w$	Given
$P - 2L = 2w$	Rearrange
$\dfrac{P - 2L}{2} = w$	Isolate
$w = \dfrac{P - 2L}{2}$	Interchange sides

EXAMPLE 4

Solving a Formula

The formula $A = P(1 + rt)$ gives the accumulated sum of money in dollars, when a principal of P dollars is invested at an interest rate r for t years. Solve the formula for t.

Solution

$A = P(1 + rt)$	Given
$P(1 + rt) = A$	Interchange the two sides
$P + Prt = A$	Combine
$Prt = A - P$	Rearrange
$t = \dfrac{A - P}{Pr}$	Isolate

Using Formulas to Solve Applied Problems

The process of describing a real-world situation in mathematical terms is called **mathematical modeling.** This process may involve constructing an equation or collecting data. Once a model is established, it can be used to analyze the situation and make predictions. Some mathematical models are very accurate because they can be described by formulas. Others are too complicated to be precisely modeled. For instance, if a volume is being sought in a real-world situation, we write a *formula* for the volume and replace quantities such as length, width, and height by letters or values given in the situation. These quantities should be represented by the same units of measurement such as feet, meters, or miles. Recall from section R5 that *dimensional analysis* is needed to convert one system of measurement into another.

EXAMPLE 5   **Modeling by Using a Formula for Area**

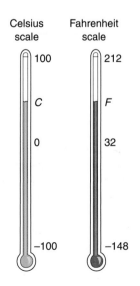

The public tennis courts in a city each measure 40 feet wide and 90 feet long.

(a) What is the surface area of each court?

(b) How many gallons of paint are needed for each court to be repainted? Assume one gallon of paint covers 400 square feet of surface.

Solution   (a) The area of a rectangle is length times width, $A = Lw$

$$A = (40 \text{ ft})(90 \text{ ft}) = 3600 \text{ ft}^2$$

(b) If one can of paint covers 400 ft², then

$$3600 \text{ft}^2 \cdot \frac{1 \text{ can}}{400 \text{ ft}^2} = 9 \text{ cans of paint are needed.}$$

EXAMPLE 6   **Modeling by Using a Formula from Chemistry**

The formula $F = \frac{9}{5}C + 32$ expresses temperature F in degrees Fahrenheit in terms of the temperature C in degrees Celsius.

(a) Solve the formula for C.

(b) Complete the following table and then use the table to predict when the temperature is the same in both scales.

°F	−40	32	212
°C			

Solution

$$F = \frac{9}{5}C + 32 \qquad \text{Given}$$

$$F - 32 = \frac{9}{5}C \qquad \text{Rearrange}$$

$$\frac{5}{9}(F - 32) = C \qquad \text{Isolate}$$

$$C = \frac{5}{9}(F - 32) \qquad \text{Interchange}$$

(b) If we replace F by −40, 32, and 212 in the above result, we obtain the corresponding values for C. Notice the table shows the temperature reading is the same at −40 on both scales.

°F	−40	32	212
°C	−40	0	100

Celsius scale

100

C

0

−100

Fahrenheit scale

212

F

32

−148

PROBLEM SET 1.2

Mastering the Concepts

In Problems 1–14, solve each equation for the indicated variable and check the results.

1. $6x + 7c = 37c$ for x
2. $4u - 19a = 5u$ for u
3. $ad + b = c$ for d
4. $ad - b = c$ for d
5. $12z - 4b = 6z - 7b$ for z
6. $13f + 6g = 8f - 9g$ for g
7. $4x - 3a - (10x + 7a) = 0$ for x
8. $27t - 4b - (15t - 6b) = 0$ for t
9. $5(4r - 3c) - 2(7r - 9c) = 0$ for r
10. $3(a - 2b) + 4(b + a) = 5$ for a
11. $5(mu - 2d) - 3(mu - 4d) = 9d$ for u
12. $9t + 7h - (11h - 13t) = 6t$ for t
13. $8(y - 2b) - 3(5y + 11b) = 0$ for y
14. $a(y - a) = -ab + 2b(y - b)$ for y

In Problems 15–24, solve each equation for y.

15. $\dfrac{x}{3} + \dfrac{y}{4} = 1$

16. $\dfrac{x}{7} - \dfrac{y}{4} = 1$

17. $\dfrac{x}{9} + \dfrac{y}{11} = 1$

18. $-\dfrac{x}{9} + \dfrac{y}{5} = 1$

19. $\dfrac{3x}{7} + \dfrac{2y}{9} = 1$

20. $\dfrac{7x}{3} - \dfrac{3y}{2} = 1$

21. $x = 0.01y - 0.03$

22. $8.7x = 3.9y + 9.6$

23. $\dfrac{x - 2y}{5} = \dfrac{3(y + 2)}{4} + 3$

24. $\dfrac{x + 3y}{5} = \dfrac{x - y}{2} + \dfrac{1}{2}$

In Problems 25–36, solve each formula for the indicated variable.

25. $A = P + Prt$, for P (Business)
26. $IR + Ir = E$, for r (Physics)
27. $\dfrac{T}{C} = \dfrac{x}{y}$ for C (Ecology)
28. $S(1 - r) = a$, for r (Mathematics)
29. $P = c - 30\%c$, for c (Business)
30. $BS = F + BV$, for B (Business)
31. $y = mx + b$, for x (Geometry)
32. $a = x - \dfrac{y_1}{y_2}$ for y_1 (Calculus)
33. $C = 2\pi r$, for r (Geometry)
34. $V = \pi r^2 h$, for h (Geometry)
35. $P = 2l + 2w$, for l (Geometry)
36. $A = 2\pi r^2 + 2\pi rh$, for h (Geometry)

In Problems 37 and 38, solve each formula for the indicated variable.

37. $V = \dfrac{Bh}{3}$ (volume of a pyramid)
    (a) For base B
    (b) For height h

38. $V = \dfrac{\pi r^2 h}{3}$ (volume of a cone)
    (a) For height h
    (b) For radius r

In Problems 39–42, solve each formula for the indicated variable.

39. $v = -32t + v_o$ (velocity in free-fall)
    (a) For initial velocity v_o
    (b) For time t

40. $A = \left(\dfrac{b_1 + b_2}{2}\right)h$ (area of a trapezoid)
    (a) For one of the bases b_1
    (b) For height h

41. $1 = \dfrac{x}{a} + \dfrac{y}{b}$ (*intercept* form of a line)
    (a) For a, the x intercept
    (b) For b, the y intercept

42. $y - y_1 = m(x - x_1)$ (point-slope form of a line)
    (a) For slope m
    (b) For coordinate x_1

Applying the Concepts

43. **Circumference:** A runner jogged 0.1 mile in one lap around a circular track in a field house. What is the greatest distance measured straight across the track in feet? (Use $C = \pi d$, $\pi \sim 3.14$, and 5280 feet = 1 mile, round to one decimal place.)

44. **Area:** Suppose we wish to carpet the room shown in Figure 1. If the carpet, pad, and

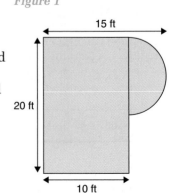

Figure 1

labor cost a total of $18 per square yard, what is the cost?

45. Area: You want to make a circular table cloth with a ruffle along the bottom. The table cloth is to hang 30 inches over the edge of a round table top (Figure 2).

Figure 2

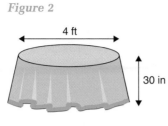

4 ft

30 in

(a) Find the length of the ruffle. (Use $C = 2\pi r$)
(b) If the ruffle costs $2.39 per yard, how much will it cost?

46. Volume of a Coal Pile: A pile of coal is in the shape of a cone. Its volume is given by the formula

$$V = \frac{\pi r^2 h}{3},$$

where r is the radius of the circular base and h is the height of the pile (Figure 3).

Figure 3

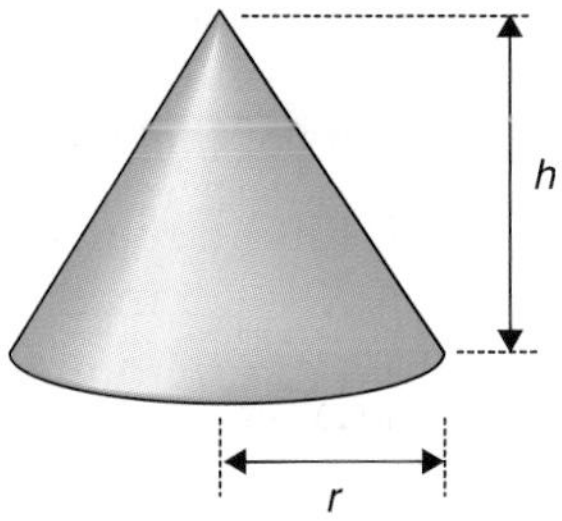

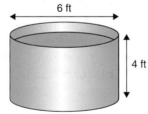

h

r

(a) Solve the formula for h.
(b) What is the height of the pile if its volume is 1000 cubic meters and its radius is 8.7 meters.

47. Hot Tub Volume: A hot tub is in the shape of a right circular cylinder Figure 4. It measures 6 feet in diameter and 4 feet in depth.

Figure 4

6 ft

4 ft

(a) Find the volume V of the tub using $V = \pi r^2 h$ where r is the radius and h the depth.
(b) How long would it take to fill the hot tub if the rate of water flowing into the tub is 1.2 cubic feet per minute?
(c) Solve the formula for h.
(d) How deep is the water if there is 84.78 cubic feet of water in the tub?

48. Depth of Aquarium: Suppose you have two fish aquariums. One measures 4 feet long and 18 inches wide with a depth of 2 feet (Figure 5). If you pour all the water into the other aquarium which is 7 feet long and 1.5 feet wide, what is the depth of water in the second aquarium? (Use $V = lwh$)

Figure 5

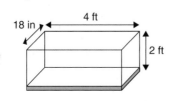

18 in

4 ft

2 ft

49. Tax Credit: A certain state offers to give a tax credit for installing solar energy panels. The credit is 15% of the cost of the solar panels over $1000 and is modeled by the equation

$$T = 0.15(C - 1000)$$

where T is the tax credit and C is the cost of the panels and $C > 1000$.

(a) What is the tax credit for panels costing $1760 ($C = 1760$)?
(b) Solve this equation for C in terms of T.
(c) Use part (b) to determine the cost of panels if a tax credit of $252 is desired.

50. Depreciation: For tax purposes, companies use the depreciation formula

$$D = \frac{C - S}{n}$$

where D (in dollars) is the yearly depreciation of an item, C is its original cost, S is its salvage value, and n is its useful life.

(a) Solve the formula for S.
(b) What is the salvage value of a copy machine if its original cost was $20,000 and its yearly depreciation is $2000 over a life of 10 years?

51. Volume: The external liquid hydrogen tank of the space shuttle Endeavor is fabricated in the shape of a right circular cylinder as in Figure 6. The volume V of a tank in this shape is given by the formula

Figure 6

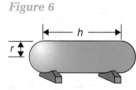

r h

$$V = \pi r^2 \left(h + \frac{4r}{3} \right).$$

(a) Solve the formula for h.
(b) Find the height h of the tank in meters if its total volume is 1174 cubic meters and the radius of each hemisphere is 4.2 meters. Round off the answer to one decimal place.

52. Storage: A cistern is to be fabricated in the shape of a right circular cylinder with a hemisphere at the bottom (Figure 7). The volume V of the cistern is given by the formula

Figure 7

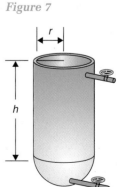

r

h

$$V = \pi r^2 h + \frac{2}{3}\pi r^3$$

(a) Solve this formula for h.
(b) Find h if r is 3 feet and V is 810 cubic feet.

Developing and Extending the Concepts

In Problems 53–56, solve each equation for the indicated variable.

53. $(ax + 7)(ax - 3) = ax(ax + 1)$, for x

54. $A = 2lw + 2lh + 2wh$ for l

55. $\dfrac{a^2y + 8}{b} = \dfrac{a^2y - 10}{3b}$ for y

56. $\dfrac{b - x}{2} = \dfrac{2 + b}{4} - \dfrac{3x}{5}$ for x

57. Polygons: The number of diagonals d of a polygon is one half the product of the number of sides n and the number of sides minus 3.
   (a) Write the formula for finding the number of diagonals d.
   (b) Find the number of diagonals for a six-sided polygon.

58. Chain Length: The approximate length L of a chain joining two sprockets (toothed wheels) of a bicycle of radii r and R (Figure 8) is given by the formula

Figure 8

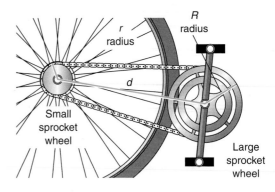

$$L = 2d + \frac{1}{2}(2\pi r) + \frac{1}{2}(2\pi R)$$

where d is the distance between the center of the two sprockets. Consider a bicycle having a chain of length 64 inches and a distance between the center of the sprockets of 24 inches.

   (a) Find r, rounded to one decimal place, if $R = 3.5$ inches.
   (b) Find R, rounded to one decimal place, if $r = 1.5$ inches.

Objectives

1. Introduce a Problem Solving Format
2. Solve a Variety of Applied Problems

1.3 Applied Problems

In this section, we introduce a strategy for solving applied problems. These techniques were described in (the Hungarian mathematician) George Polya's book, *How to Solve It.* The method focused on the problem solving process rather than just obtaining the correct answer.

Introducing a Problem Solving Format

One reason algebra is so important is that it can be used to solve problems in many disciplines, including business, engineering, economics, social sciences, and the physical sciences. Often the most challenging step in solving a word problem or application is translating the situation described in the problem into algebraic form. The complete result, including clearly stated assumptions and the appropriate formula, is a mathematical model. Polya developed the following three step process for solving applied problems.

Problem Solving Process

Step 1. Make a plan
Step 2. Write the solution
Step 3. Look back

Step 1. Make a Plan
         The following list is very useful in applying this plan.
         (a) Identify the *given.* (What information or assumptions are known.)
         (b) Identify the *goal.* (What questions need to be answered.)
         (c) Draw a diagram whenever possible.
         (d) Arrange and organize the given information in the form of a table or a chart whenever possible.

(e) Select one of the unknown quantities and label it with a symbol or a variable.

(f) Write algebraic relationships among the quantities involved, then combine these relationships into a single equation.

Step 2. Write the Solution

(a) Solve the resulting equation in part (f) of step 1.

(b) Justify each step of the solution.

Step 3. Look Back

Interpret the solution in terms of the original problem to determine whether or not the answer is reasonable.

EXAMPLE 1

Solving a Number Problem

Find three consecutive odd integers whose sum is 63.

Solution

Make a Plan

The *given* is that there are three odd integers and they are consecutive. This means that these integers follow one another on the number line; that is, they are in order. Since they are consecutive odd integers, they differ by 2. The *goal* is to find the three integers. We can guess a few numbers so as to better understand the situation. Consider the following table.

First Odd Integer	Second Odd Integer	Third Odd Integer	Sum of the Integers is 63
5	$5 + 2 = 7$	$7 + 2 = 9$	$5 + 7 + 9 = 21$
11	$11 + 2 = 13$	$13 + 2 = 15$	$11 + 13 + 15 = 39$

Observe the pattern in the numbers and translate it into algebra:

Let x = first odd integer, then

$x + 2$ = second consecutive odd integer, and

$(x + 2) + 2 = x + 4$ = third consecutive odd integer.

The following diagram is a visual description of the *given*.

First odd	+	Second odd integer	+	Third odd integer	=	Sum of the integers
x	+	$x + 2$	+	$x + 4$	=	63

Translating the symbols into an equation, we have

$$x + (x + 2) + (x + 4) = 63$$

Write the Solution

Now we solve the equation

$x + (x + 2) + (x + 4) = 63$	Given
$3x + 6 = 63$	Combine
$3x = 57$	Rearrange
$x = 19$	Isolate
$x + 2 = 19 + 2 = 21$	
$x + 4 = 19 + 4 = 23$	

Therefore, the three integers are 19, 21, and 23.

Look Back

These three integers, 19, 21, and 23, are consecutive and odd. Their sum is 63. So our answers, 19, 21, and 23, are reasonable.

As you try to work applied problems, you will notice the frequent occurrence of certain words in the statements of these problems. Table 1 presents some common English phrases used in word problems together with their symbolic translations, where x and y represent variables.

TABLE 1

English Phrases	Algebraic Expressions
The sum of x and y	$x + y$
The difference of x and y	$x - y$
The product of x and y	xy
The quotient of x and y	$\dfrac{x}{y}$
The square of the difference of x and y	$(x - y)^2$
Twice the sum of x and y	$2(x + y)$
The sum of twice x and y	$2x + y$
The sum of two consecutive integers	$x + (x + 1)$

Solving a Variety of Applied Problems

Our equation-solving skills enable us to solve a wide range of applications.

In many business problems such as, rates discounts, and interest, we use *percents* (%) which means *by the hundred.* To calculate with percents, we express them as decimals.

EXAMPLE 2   **Solving a Geometry Problem**

The most popular sport in many countries of the world is soccer which is played on a rectangular grass field. The length of a soccer field is 20 meters more than twice the width. The field's perimeter is 310 meters. What are the dimensions of the field?

Solution   **Make a Plan**

The key words in this problem are *rectangular* and *perimeter.* The two unknowns are the length, L, and width, w, of the rectangular field. The perimeter P of a rectangle is found using the formula

$$P = 2L + 2w.$$

First, let us guess some numbers to better understand the problem.

Width, Guess a Number	Length is 20 Meters More Than Twice the Width	Perimeter $P = 2L + 2w$	Condition $P = 310$
10 m	$20 + 2(10) = 40$	$2(10) + 2(40) = 100$	$100 \neq 310$
30 m	$20 + 2(30) = 80$	$2(30) + 2(80) = 220$	Too small
50 m	$20 + 2(50) = 120$	$2(50) + 2(120) = 340$	Too big
40 m	$20 + 2(40) = 100$	$2(40) + 2(100) = 280$	Too small
45 m	$20 + 2(45) = 110$	$2(45) + 2(110) = 310$	Correct!

This table uses the same construction as an electronic spreadsheet. It can be used with any type of problem. The principle is to guess a reasonable value for the answer and work it through the wording of the problem. Does it satisfy the conditions? If yes, you found the answer; if not guess again. This process can be very long and cumbersome. Once the pattern is recognized, translate it into algebra.

Let w = the width of the field in meters,

$L = 2w + 20$ = the length of the field in meters.

The following diagram is a visual description of the *given*.

Two times the length of the rectangle $2(2w + 20)$	+ +	Two times the width of the rectangle $2w$	= =	Perimeter of the rectangle 310

Translating the problem into symbols, we obtain the equation

$$2(2w + 20) + 2w = 310$$

Write the Solution

$2(2w + 20) + 2w = 310$	Given
$4w + 40 + 2w = 310$	Combine
$6w + 40 = 310$	Combine
$6w = 270$	Rearrange
$w = 45$	Isolate
$2w + 20 = 110$	

The width of the soccer field is 45 meters and the length is 110 meters.

Look Back

We can check this result in the words of the problem. The perimeter is

$$P = 2L + 2w = 2(110) + 2(45) = 220 + 90 = 310.$$

Also, the length of 110 meters is 20 meters more than twice the width of 45 meters.

EXAMPLE 3

Cutting a Length of Wire

An electrician needs to cut a 17 foot wire into three parts. If the longest piece must be 3 times as long as the shortest piece, and the middle piece is 2 feet longer than the shortest piece, how long should each piece be?

Solution

Make a Plan

The *given* is a length of wire which is 17 feet long and is being cut into three pieces. Our *goal* is to determine the length of each piece. We assign a variable to the length of the shortest piece.

Let x = the length of the shortest piece in feet.

$3x$ = the length of the longest piece in feet.

$x + 2$ = the length of the middle piece in feet.

The following diagram is a visual description of the *given*.

Length of the shortest piece	+	Length of the middle piece	+	Length of the longest piece	=	Overall length
x	+	$x + 2$	+	$3x$	=	17

Translating the problem into symbols, we have the equation

$$x + x + 2 + 3x = 17.$$

Write the Solution

$x + x + 2 + 3x = 17$	Given equation
$5x + 2 = 17$	Combine
$5x = 15$	Rearrange
$x = 3$	Isolate
$x + 2 = 5$	
$3x = 9$	

Look Back

The lengths are 3 feet, 5 feet, and 9 feet. The sum of the length of the three pieces is $3 + 5 + 9 = 17$ feet. The longest piece is three times the length of the shortest piece and the middle piece is 2 feet longer than the shortest piece.

EXAMPLE 4

Solving a Length Problem

The torch of the Statue of Liberty is 305 feet above the ground. If half the height of the pedestal is 74 feet shorter than the height of the statue, how tall is the pedestal?

305 ft

Solution

Make a Plan

The total height of the Statue of Liberty Monument is 305 feet which is the sum of the height of the statue plus the height of the pedestal. Our goal is to find the height of the pedestal. We will make a table of numbers to understand the wording of this problem.

If the Height of Pedestal Is	Height of Statue	Height of Whole Monument
100	$\frac{1}{2}(100) + 74 = 124$	100 + 124 = 224 too small
200	$\frac{1}{2}(200) + 74 = 174$	200 + 174 = 374 too big

The table gives us the pattern for the algebra and a suggestion as to the size of the pedestal, between 100 and 200 feet. We write the algebra.

$$\text{Let } p = \text{height of the pedestal}$$

$$\frac{1}{2}p + 74 = \text{height of the statue}$$

$$p + \frac{1}{2}p + 74 = 305 = \text{whole height}$$

Write the Solution

$$p + \frac{1}{2}p + 74 = 305$$

$$\frac{3}{2}p + 74 = 305 \qquad \text{Combine}$$

$$\frac{3}{2}p = 231 \qquad \text{Rearrange}$$

$$p = 154 \qquad \text{Isolate}$$

Looking Back

The pedestal is 154 feet and the statue is $\frac{1}{2}(154) + 74 = 151$. The whole monument is 154 + 151 = 305 feet.

EXAMPLE 5   **Solving a Rate Problem**

Detroit

New York

San Francisco

On a recent trip across the country, a jetliner flew from San Francisco to Detroit at an average speed of 500 miles per hour. It then continued to New York at an average speed of 550 miles per hour. If the entire trip covered 3075 miles and the total flying time was 6 hours, what was the distance of each leg of the trip?

Solution   **Make a Plan**

Motion problems are based on the distance formula $d = rt$ where d is the distance, r is the average rate, and t is time. This trip consists of two legs, San Francisco to Detroit and Detroit to New York.

Let t = time (in hours) traveled on the first leg,

and $6 - t$ = time (in hours) traveled on the second leg.

For each leg of the trip, we write expressions to represent rate, time, and distance. This information is summarized in the following table.

	Rate r (miles per hour)	Time t (hours)	Distance $d = rt$ (miles)
First leg of the trip	500	t	$500t$
Second leg of the trip	550	$6 - t$	$550(6 - t)$

The following diagram is a visual description of the *given*.

Distance from San Francisco to Detroit	+	Distance from Detroit to New York	=	Total distance
$500t$	+	$550(6 - t)$	=	3075

Since the total distance traveled is 3075 miles, we obtain the equation:

$$500t + 550(6 - t) = 3075$$

Write the Solution

$500t + 550(6 - t) = 3075$	Given
$500t + 3300 - 550t = 3075$	Combine
$-50t = -225$	Rearrange
$t = 4.5$	Isolate

Substituting 4.5 for t, yields

$$500t = 500(4.5) = 2250 \text{ mi.}$$
$$550(6 - t) = 550(6 - 4.5) = 825 \text{ mi.}$$

Therefore, the first leg was 2250 miles long; the second was 825 miles.

Look Back

As a check, notice that the time spent traveling from San Francisco to Detroit is 4.5 hours. To find this distance, substitute the values of r and t in the equation $d = rt$, so that we have

$$d = 500(4.5) = 2250 \text{ miles.}$$

The time flying from Detroit to New York is $6 - 4.5 = 1.5$ hrs. Thus the distance is

$$d = 550(1.5) = 825 \text{ miles.}$$

Total distance traveled is $2250 + 825 = 3075$ miles.

All these examples are *analogous;* they are solved using the same mathematical techniques. An important problem solving strategy is recognizing when a technique used to solve one problem can be used to solve another seemingly different, but analogous problem.

PROBLEM SET 1.3

Mastering the Concepts

1. **Number Problem:** Find three consecutive even integers whose sum is 60.

2. **Number Problem:** (a) Find a number such that three more than twice the number is 57.
   (b) Find a number such that twice the sum of three and the number is 56.

3. **Number Problem:** Find three consecutive even integers such that twice the sum of the first and third is twelve more than twice the second.

4. **Age Problem:** A mother is five times as old as her son. If twice the age of the son is 3 years less than half of the age of the mother, how old are they?

5. **Number Problem:** Find three consecutive even integers such that twice the sum of the first and third is four more than three times the second.

6. **Home Address:** Four houses on one side of a street have addresses that are consecutive odd numbers. Find each address if the sum of these numbers is 23,672.

7. **Swimming Pool Dimensions:** The length of a rectangular swimming pool is 2 feet more than twice its width (Figure 1). The perimeter of the pool is 94 feet. What are the length and width of the pool?

Figure 1

8. **Rectangular Dimensions:** A carpet layer wishes to determine the cost of carpeting a rectangular room whose perimeter is 52 feet. Suppose that the length of the room is 4 feet more than its width. Find
   (a) the length and width of the room;
   (b) the cost of the carpet if it costs $18.95 per square yard.

9. **Plot Dimensions:** A rectangular plot of farmland is bounded on one side by a river and the other three sides by a single-strand fence 797 meters long. What is the length and the width of the plot if its length is 1 meter more than twice its width. No fence is used on the river side (Figure 2).

Figure 2

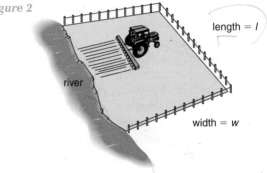

10. **Playground Dimensions:** A recreation department plans to build a rectangular playground enclosed by 1000 meters of fence. If the length of the playground is 4 meters less than three times the width, find the length and the width of the playground (Figure 3).

Figure 3

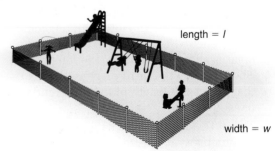

11. **Computer Monitor Dimensions:** The length of a rectangular computer monitor is 6 inches less than twice its width. The perimeter of the face of the monitor is 36 inches. Find the length and width of the monitor.

12. **Frame Dimensions:** The molding of a picture frame (the frame's perimeter) is 2.78 meters. The length of the finished frame is 0.1 meter more

than twice the width (Figure 4). What are the dimensions of the frame?

Figure 4

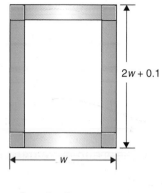

$2w + 0.1$

w

13. **Plumbing:** A plumber wishes to cut a 26 foot length of pipe into two pieces. The longer piece needs to be 7 feet less than twice the length of the shorter piece. Where should the plumber cut the pipe?

14. **Carpentry:** A carpenter needs two pieces of lumber which together are 21 feet long. The longer piece needs to be 1 foot longer than three times the length of the shorter piece. Find the length of each piece.

15. **Rope Cutting Problem:** A scout leader wishes to cut a rope for tents into three pieces whose lengths are consecutive odd integers. If the length of the first and the third piece is 58 feet, find the length of the middle piece.

16. **Cable Cutting Problem:** A cable installer has 105 meters of cable which is to be cut into three pieces. The second piece is 5 meters shorter than the first piece, and the length of the third piece is twice the length of the second. Find the length of each piece.

17. **Wire Cutting Problem:** A wire 180 centimeters long is to be cut into three pieces. The length of one piece is three times the length of the second. The length of the third piece is 12 centimeters shorter than twice the length of the second. Find the length of each piece.

18. **Fabric Cutting Problem:** A fabric store worker has a 57 yard bolt of cloth which is to be cut into three lengths. The length of the second piece is 5 yards shorter than twice the length of the first. The length of the third piece is 2 yards longer than twice the length of the first. Find the length of each piece.

19. **Travel Distance:** A train travels through the mountains at an average speed of 35 miles per hour. It then continues on flat land at an average speed of 85 miles per hour. If the entire trip covers 1200 miles and each leg of the trip takes the same number of hours, how many miles is each leg?

20. **Travel Time:** Last weekend two college students were returning to school 200 miles away. During a blizzard, they were only able to average 25 miles per hour. For the rest of their trip, they averaged 55 miles per hour. If the entire trip took 4 hours, how long were they driving in the blizzard?

21. **Travel Distance:** Two marathon runners leave the same point at the same time and travel in opposite directions. One runs at a rate that is 0.75 miles per hour faster than the other. After 3 hours they are 44.25 miles apart. How far did each runner travel?

22. **Travel Time:** On the first part of a 27.5 mile trip, the average speed was 48 miles per hour. Later, due to traffic, the average speed was reduced to 35 miles per hour. If one spent five times as long on the first part as on the second part, what is the total time of the trip?

23. **Average Speed:** A snowmobile follows a trail to a nearby camp in 10 minutes. A skier, averaging 16 miles per hour slower than the snowmobile, starts from the same point on the trail and following the snowmobile's track covers the same distance in 50 minutes. What was the average speed of the snowmobile?

24. **Average Speed:** A helicopter travels 30 miles per hour faster than a speeding car. The car had a half hour head start. If the helicopter overtakes the car in 1.5 hours, what is the speed of each?

25. **Average Speed:** An Amtrak train leaves Las Vegas heading west towards Los Angeles. At the same time, 272 miles away a bus leaves Los Angeles traveling east towards Las Vegas. If the train averages 31 miles per hour faster than the bus

Figure 5

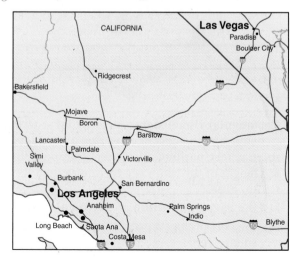

and they pass each other in 2 hours, what is the average speed of the train and the bus? (Figure 5)

26. **Travel Time:** Two cars start at the same interstate entrance and travel in the same direction at average speeds of 72 and 65 miles per hour respectively. How long will it take before the two cars are 17.5 miles apart?

27. **Travel Distance:** Two planes leave the same airport at the same time and fly in opposite directions. After 2.5 hours, they are 2425 miles apart. If one flies 70 miles per hour faster than the other, what are their speeds?

28. **Travel Distance:** A trucker travels from one city to another at the rate of 55 miles per hour, and returns on the same route at the rate of 45 miles per hour. Find the total distance traveled for the round trip if the total driving time is 5 hours.

Developing and Extending the Concepts

29. **Cash Register Change:** A fitness center usually needs three times as many $5 bills as $10 bills to transact its daily business. If the cash register contains $750 in fives and tens at the beginning of a business day, how many of each bill does the register contain?

30. **Coin Problem:** A vending machine accepts nickels, dimes, and quarters. When the coins are emptied, the total value of the coins is found to be $24.15. Find the number of coins of each kind in the box, if there are twice as many nickels as quarters and 5 more dimes than nickels.

31. **Age Difference:** A man is 25 years older than his son. Fourteen years from now the man will be twice as old as his son will be. What are the present ages of the man and his son?

32. **Clothing Sale:** A clothing store has discounted the price of a sweater by 20%. If the sale price of the sweater is $49.95, what was the original price?

33. **Insurance Discount:** Some insurance companies reward students with safe driving records and good grades by discounting their annual premiums. If a student's annual premium is discounted by 15% to $874.65, what is the original annual premium?

34. **Computing Salaries:** An employee's new annual salary is $36,516, which includes a 5% pay raise and a 2.4% cost-of-living allowance. What was the employee's original salary?

35. **Tire Sale:** A tire was sold for $58.50 which included a 6% sales tax and an 11% federal tax. What was the price of the tire before the taxes?

36. **Consumer Cost:** A customer has $70 to spend in a grocery store. She spends $28.20 for canned goods. The remaining amount is needed to buy a total of 9 pounds of steak and chicken. If the cost per pound for steak and chicken is $6.95 and $2.80 respectively, how many pounds of each meat can she afford to buy?

37. **Tipping:** For groups of four or more diners, many restaurants automatically add a 15% gratuity as well as a tax of 8% to the original bill. If the bill (including gratuity and tax) for four people came to $193, how much was the tip?

38. **Credit Card Charge:** A customer carries an average daily balance on her credit card of $12.70 per day. The customer is offered a card from Bank A with no annual fee and 18% finance charge. Bank B offers a card with 15% finance charge and $39 annual fee.

   (a) Which card is more economical for the customer?

   (b) What average daily balance would give the same total yearly charges?

39. **Aerobic:** A runner burns three times as many calories running as walking the same distance. Suppose a runner burns a total of 765 calories running 3 kilometers and walking 8 kilometers. At this rate, how many calories are burned by running 1 kilometer?

40. **Operating Budget:** Suppose 60% of the salary budget for a college goes to faculty salaries, 25% to administrative salaries, 10% to support staff salaries, and the remaining $1,820,000 to maintenance salaries. How much money goes to each group?

41. **Work Force:** When a new factory opened recently, twice as many men as women applied for work. Five percent of all the applicants are hired. If 3% of the men who applied were hired, what percentage of the women who applied were hired?

42. **Stock Prices:** Two different stocks were sold for $3000 each. On one stock the gain was 10% of its original value, on the other the loss was 10% of its original value. Determine the total gain or loss from both stocks.

43. **Automobile Maintenance:** The mileage on a car's odometer reads 6253.6. The next service is scheduled to take place at about 10,000 miles. If the car is driven 170 miles per week, in how many weeks will it need to be serviced?

44. **Bank Charges:** A bank charges its customers a general fee of $6 per month plus an additional fee of 10 cents per check written from a checking account. Over the past nine months, a customer was charged $62 in service fees. How many checks did the customer write during that period?

45. Geology: While some climbers began a hike up Mt. St. Helens, the temperature at the base was 20°C. The temperature dropped 5°C with every 1000 meters of elevation gained. When the climbers reached the top of their ascent, the temperature was 7.5°C. How high up had they climbed?

46. Car Rental: For one of its compact cars, a car rental company charges customers $33 per day. In addition, the company charges 25 cents per mile for each mile driven over 400 miles. A customer rented a compact car for four days and paid $186.50. How many miles did the customer drive?

47. Car Racing: After 125 miles of the Indianapolis 500 car race, Car A has a lead of 0.31 miles over Car B. If Car A averages 185 miles per hour and Car B averages 192 miles per hour, how long will it take Car B to overtake Car A?

48. Energy Saving: By installing a $50 thermostat that automatically reduces the temperature setting while at work and at night, a condominium owner hopes to cut the annual heating bill by 15%. If the monthly heating bill on average is $85 before the new thermostat, how long will it take to recover the cost of the thermostat?

49. Factory Outlet Sales: A retail factory outlet store sells running shoes for $45 a pair. At this price,

the profit on each pair of shoes is 25% of the cost to the retailer. What is the retailer's profit on each pair of shoes?

50. Metal Alloys: An archaeologist discovers a crown weighing 800 grams in the tomb of an ancient Egyptian king. Evidence indicates that the crown is made of a mixture of gold and silver. Gold weighs 19.3 grams per cubic centimeter, silver weighs 10.5 grams per cubic centimeter. It is found that the crown weighs 16.2 grams per cubic centimeter. How many grams of gold does the crown contain?

Figure 6

51. Geometry: A ladder rests against a vertical wall. The angle A made by the ladder and the ground is 6° less than 7 times the angle B made by the ladder and the wall (Figure 6). What is the measure of angle B?

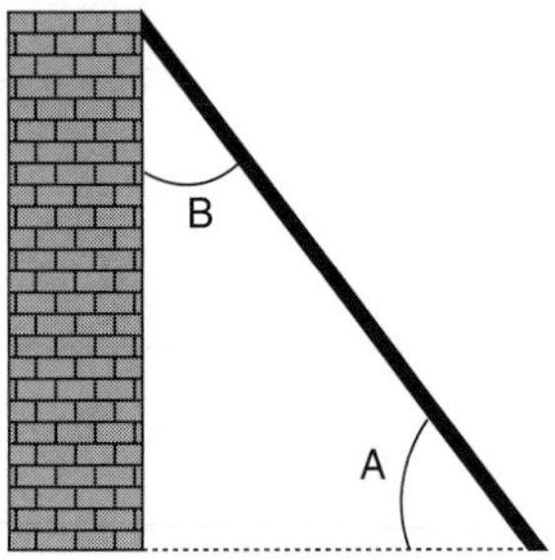

52. Package Design: An open box is to be constructed by removing equal squares, each of length x inches, from the corners of a piece of cardboard and then folding up the sides (Figure 7). If the cardboard measures 12 inches by 16 inches, find the volume of the box if its length is twice its width.

Figure 7

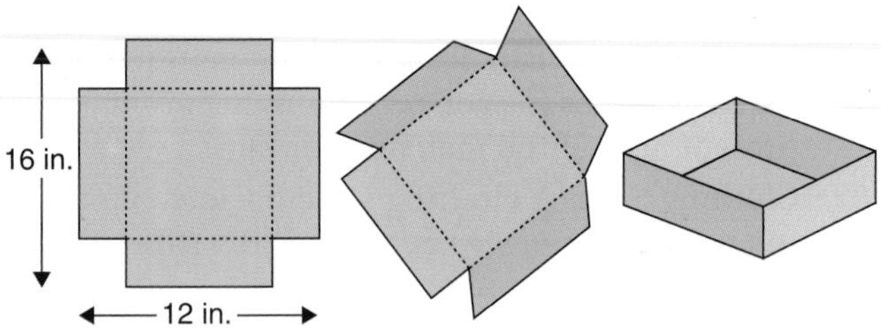

16 in.

12 in.

Objectives

1. Solve Mixture Problems
2. Solve Work Problems

1.4 Additional Applied Problems

In this section, we introduce two additional applied problems involving mixture and work. We will continue to use the Polya problem solving strategy.

Solving Mixture Problems

Mixture problems occur in everyday living as the first example shows.

EXAMPLE 1   **Solving a Nut Mixture Problem**

Suppose you want to make a 3 pound almond and peanut mixture for a party. You can spend $9.00; almonds cost $4.24 per pound and peanuts cost $1.76 per pound. How many pounds of almonds and peanuts should you buy?

Solution   **Make a Plan**

We will arrange this information in a table and take a few guesses for the weight of the almonds. Since we need a total of 3 pounds of nuts, the sum of the weights must always be 3.

Weight of Almonds	Weight of Peanuts	Cost of Almonds	Cost of Peanuts	Total Cost
1 lb.	2 lbs.	$4.24(1) = $4.24	$1.76(2) = $3.52	$4.24 + $3.52 = $7.76
2 lbs.	1 lb.	$4.24(2) = $8.48	$1.76(1) = $1.76	$8.48 + $1.76 = $10.24

The condition of the expense has not been satisfied, but we have seen a pattern. Now we will use variables and a diagram.

Let x = number of pounds of almonds

$3 - x$ = number of pounds of peanuts

Weight of Almonds	Weight of Peanuts	Cost of Almonds	Cost of Peanuts	Total Cost
x lb.	$3 - x$ lb.	4.24x$	$1.76(3 - x)$	4.24x$ + $1.76(3 - x)$

Write the Solution

The equation comes from the condition that we spend $9.00 and then we solve the equation.

$$\$4.24x + \$1.76(3 - x) = \$9.00$$
$$4.24x + 1.76(3) - 1.76x = 9.00 \qquad \text{Combine}$$
$$2.48x + 5.28 = 9$$
$$2.48x = 3.72 \qquad \text{Rearrange}$$
$$x = 1.5 \qquad \text{Isolate}$$
$$3 - x = 1.5$$

We should mix 1.5 pounds of each nut.

Look Back

If we mix 1.5 pounds each of almonds and peanuts, we will have 3 pounds of nuts. The cost of the almonds is $4.24(1.5) = $6.36 and the cost of peanuts is $1.76(1.5) = $2.64.

The total cost is $6.36 + $2.64 = $9.00.

Recall that *percents* (%) mean *by the hundred*. To calculate with percents, we express them as decimals.

EXAMPLE 2   **Solving an Investment Problem**

A financial planner advised some customers to invest part of their money in a bond fund that paid 4% simple interest and the rest in a real estate fund that earned 13% simple interest. If a customer has $24,000 to invest and wants a return of $2031 the first year, how much money should be invested in each fund?

Solution   **Make a Plan**

The formula for simple interest is $I = Prt$ where P is the principal, r is the rate, and t is the time in years. The *given* is a total principal of $24,000, two rates of interest earning a total of $2031, and time is one year.

Let x = amount (in dollars) invested at 13%

$24,000 - x$ = amount (in dollars) invested at 4%

We organize the information in the following table.

	Principal (dollars)	Rate (% per year)	Time (year)	Interest (dollars)
Real Estate Fund	x	0.13	1	$0.13x$
Bond Fund	$24,000 - x$	0.04	1	$0.04(24,000 - x)$

We know that the sum of interest earned by both investments is a total of $2031 for the year. The following diagram is a visual description of this *given*.

Interest from real estate fund	+	Interest from bond fund	=	Total interest earned
$0.13x$	+	$0.04(24,000 - x)$	=	2031

Write the Solution

$$0.13x + 0.04(24,000 - x) = 2031 \qquad \text{Given}$$
$$0.13x + 960 - 0.04x = 2031 \qquad \text{Combine}$$
$$0.09x = 1071 \qquad \text{Rearrange}$$
$$x = 11,900 \qquad \text{Isolate}$$
$$24,000 - x = 24,000 - 11,900 = 12,100$$

Thus, $11,900 was invested in the real estate fund and $12,100 was invested in the bond fund.

Look Back

To check this solution, we calculate the earnings from each investment.

$$\$11,900 \text{ at } 13\% \text{ yields } \$11,900(0.13) = \$1547$$
$$\$12,100 \text{ at } 4\% \text{ yields } \$12,100(0.04) = \$484$$

The total amount of interest is $2031. The total amount invested is $11,900 + $12,100 = $24,000.

Solving Work Problems

In the following work problems, we are assuming that work is done at a constant rate either by a machine or a person. The rate at which work is done multiplied by the time worked is the part of the task completed. For example, if a copier works at the rate of 10 pages per minute and runs for 5 minutes, it has copied 50 pages. This comes from multiplying rate times time:

$$\frac{10 \text{ pages}}{1 \text{ minute}} \cdot 5 \text{ minutes} = 50 \text{ pages}$$

Further, we assume that if two or more people (or machines) are working together, their rates do not change compared to working alone.

EXAMPLE 3

Solving a Work Problem

Suppose you can cut the grass in your yard in 2 hours and 15 minutes and your brother can cut the same area in 1 hour and 30 minutes. If you work together, how long will it take to cut the grass?

Solution

15 minutes =
$$\frac{15}{60} = \frac{1}{4} = 0.25$$
and 30 minutes =
$$\frac{30}{60} = \frac{1}{2} = 0.5$$

Make a Plan

The whole job is 1 and is composed of your efforts and your brother's contribution. Your rate is $\dfrac{1 \text{ job}}{2.25 \text{ hours}}$, and your brother's rate is $\dfrac{1 \text{ job}}{1.5 \text{ hours}}$.

Let t = number of hours working together.

This information is organized in the table.

Worker	Rate	Time	Part of Job
You	$\dfrac{1 \text{ job}}{2.25 \text{ hours}}$	t hours	$\dfrac{t}{2.25}$
Brother	$\dfrac{1 \text{ job}}{1.5 \text{ hours}}$	t hours	$\dfrac{t}{1.5}$

The equation for this situation reflects the fact that the sum of the work done by each is 1. The equation is:

$$\frac{t}{2.25} + \frac{t}{1.5} = 1$$

Write the Solution

$$\frac{t}{2.25} + \frac{t}{1.5} = 1 \qquad \text{Given}$$

$$1.5t + 2.25t = 3.375 \qquad \text{Combine, multiply by the LCD, } 1.5(2.25)$$

$$3.75t = 3.375 \qquad \text{Combine}$$

$$t = 0.9 \qquad \text{Isolate}$$

which is 0.9 hours $\dfrac{60 \text{ minutes}}{1 \text{ hour}} = 54$ *minutes.*

Look Back

If you work for 0.9 hours at the rate of $\dfrac{1\text{ job}}{2.25\text{ hours}}$, then 0.4 of the job is done. Your

brother works for 0.9 hours at the rate of $\dfrac{1\text{ job}}{1.5\text{ hours}}$, then 0.6 of the job is done.

Together, $0.4 + 0.6 = 1$, the whole job is finished.

EXAMPLE 4   **Solving a Work Problem**

A newspaper uses two printing presses to make 50,000 copies of the morning paper. The first press can produce 4000 copies per hour. Together they take 5 hours to produce the morning edition. What is the hourly rate of the second press?

Solution   **Make a Plan**

Let t = hourly rate of the second press

The given information is organized in the following table.

Press	Rate	Time	Part of Job
First	$\dfrac{4000\text{ copies}}{1\text{ hour}}$	5 hours	4000(5) copies
Second	$\dfrac{t\text{ copies}}{1\text{ hour}}$	5 hours	$5t$ copies

The equation for this situation is

$$4000(5) + 5t = 50{,}000$$

Write the Solution

$4000(5) + 5t = 50{,}000$	Given
$20{,}000 + 5t = 50{,}000$	Combine
$5t = 30{,}000$	Rearrange
$t = 6{,}000$	Isolate

Look Back

The first press makes $4000(5) = 20{,}000$ copies.
The second press makes $6000(5) = 30{,}000$ copies.
Together the morning edition of 50,000 is produced.

◆◆PROBLEM SET 1.4

Mastering the Concepts

1. **Investments:** Suppose an investor earned $2820 at the end of one year from a $32,000 investment in mutual funds. The investor bought two types of funds. One investment fund earned 9% annual interest, and the other earned 8.5% annual interest. How much was invested in each type of fund?

2. **Investments:** $18,000 is invested in two types of bonds for one year. Part of the money is invested in a high-risk, high-growth fund that earned 13% annual interest. The rest is invested in a safe fund that earned 5% annual interest. If $1497.92 in interest is earned, how much was invested at each rate?

3. **Investments:** An inheritance is divided into two investments. One investment pays 7% annual interest whereas the second investment, which is twice as large as the first, pays 10% annual interest. If the combined annual income from both investments is $4050, how much money is the total inheritance?

4. **Assets of a Retirement Fund:** A portfolio manager purchased 10,000 shares in mutual stock funds and mutual bonds funds valued at $146,000 for a retirement fund. At the time of the purchase, the stock funds sold for $11 per share, while the bond funds sold for $17 per share. How many shares of each did the retirement fund purchase?

5. **Investments:** Suppose one investment in a money market fund earned 4.25% annual interest; the other investment in a stock fund earned 7.75% annual interest. Together the two investments totaled $3450. If a total of $230 for the year is earned, how much was invested in each fund? Round to the nearest dollar.

6. **Auto Loan:** A customer purchased a car and financed $16,000. The customer borrowed part of the money from a bank charging 10% annual interest, and the rest from a credit union at 8% annual interest. If the total interest for the year was $1390, how much was borrowed from the bank and how much was borrowed from the credit union?

7. **Juice Mixture:** A fruit juice company mixes pineapple juice that sells for $5.50 per gallon with 100 gallons of orange juice that sells for $3.00 per gallon. How much pineapple juice is used to make a pineapple-orange juice drink selling for $3.50 per gallon?

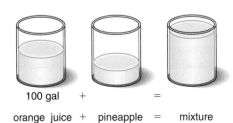

100 gal + =

orange juice + pineapple = mixture

8. **Milk Mixture:** How many liters of whole milk containing 3.5% butterfat must be mixed with 3 liters of skim milk (containing no butterfat, 0%

butterfat) to obtain a mixture containing 2% butterfat?

9. **Meat Mixture:** A meat market manager mixes hamburger having 30% fat content with hamburger that has 10% fat content in order to obtain 400 pounds of hamburger with 25% fat content. How much hamburger of each type should the manager use?

10. **Antifreeze Mixture:** A car radiator has a capacity of 8 quarts of coolant that is 20% antifreeze. A mechanic needs to drain part of the radiator fluid and replace it with 100% antifreeze to bring the coolant in the system to 60% antifreeze. How much of the original coolant should be drained and replaced by 100% antifreeze to achieve the desired mix?

11. **Gravel Driveway:** A farmer wishes to make a gravel driveway. The mixture needed is two parts pea gravel that costs $9.00 per ton and one part sand that costs $8.00 per ton. The gravel company gave a bid of $390. How much gravel and sand are they going to deliver?

12. **Seed Mixture:** A nursery has one kind of grass seed selling at 75 cents per pound and another kind of grass seed selling at $1.10 per pound. How many pounds of each kind should be mixed to produce 50 pounds of mixture of seed that will sell for 90 cents per pound?

13. **Painting:** Jack can paint a house in 4 days. Maven can paint the same house in 5 days. How long would it take both of them, painting together, to paint the house?

14. **Mowing:** Mike can mow a lawn in 3 hours. Paul can mow the same lawn in 2 hours. How long would it take both of them working together to mow the lawn?

15. **Filling a Pond:** A pond can be filled with runoff water in 6 hours. It can be filled with well water in 9 hours. How long will it take to fill the pond using both runoff and well water?

16. **Sanding:** Tom can sand a racquetball court in 4 hours. His friend Eric can sand the same court in 6 hours. How long will it take them working together to sand the court?

17. **Copying:** A marketing company uses two photocopiers to make 30,000 copies for advertising. One copier can produce 3000 copies per hour. What is the hourly rate of the second copier if together they take 4 hours to produce the advertising?

18. **Mailing:** A post office needs to distribute 21,000 pieces of mail. An experienced clerk can sort 2000 pieces of mail in one hour. Together, with a new clerk, they take 7 hours to do the job. What is the hourly rate of the new clerk?

19. **Gleaning:** Lori and Jason are gleaning (or collecting) avocados from a grove. Jason can glean 350 avocados per hour. Lori can glean 400 avocados per hour. If they start at the same time, how long will it take both of them to collect a total of 10,500 avocados?

20. **Recycling:** Sam and Linda need to sort 39,000 pieces of recyclable cans and bottles. Sam can sort 3000 pieces per hour. Together Sam and Linda take 6 hours to sort all the pieces. What is Linda's hourly rate?

Developing and Extending the Concepts

21. **Student Loans:** A college student has two loans totaling $15,000. One of the loans is at 3% annual simple interest and the other at 5% annual simple interest. After one year, the student owes $578.48 in interest. What is the amount of each loan?

22. **Mixture:** A farmer has two brands of fertilizer containing nitrogen and water. One brand is 8% nitrogen, the other is 16% nitrogen. The farmer needs to combine the two types of solution into one tank to make 200 gallons of a solution that is 13% nitrogen. How much of each brand should be used?

23. **Mixture:** Pat bought two types of sweet tea: one with 12.5% sugar and the other with 20% sugar. How many milliliters of each type should be mixed to have 960 milliliters of sweet tea with a 15% sugar concentration?

24. **Plowing:** A farmer started plowing his 40 acre field at a rate of 2 acres per hour. His tractor broke down and he borrowed his neighbor's tractor to finish the job at the rate of 2.4 acres per hour. If the whole job took 18 hours, how long did each tractor work?

25. **Geometry:** The sum of the measures of the interior angles of a triangle is 180°. Is it possible that the three angles can have the following measurement? If yes, find the measurements of the angles. If no, explain why.
    (a) Three consecutive integers.
    (b) Three consecutive even integers.
    (c) Three consecutive odd integers.

26. **Biology:** A biologist wishes to determine the volume of blood in the circulatory system of an animal. She injects 6 milliliters of a 9% solution of a biologically inert chemical and waits until it is thoroughly mixed with the animal's blood. Then she withdraws a small sample of blood and determines that 0.03% of the sample consists of the biologically inert chemical. Find the original volume of blood in the circulatory system.

27. **Basketball Court:** The perimeter of a rectangular basketball court is 80 meters. Its length is 2 meters shorter than twice its width. Find the dimensions of the court.

28. **Fencing Cost:** The length of a rectangular yard is 10 meters less than 4 times its width. The total cost for a fence to enclose the entire yard is $2706. Find the dimensions of the yard if the fence costs $8.20 per meter.

29. **Lunch Meal:** A post office worker has $10 and wishes to have lunch. If the state sales tax is 6% and the tip is 15%, what is the maximum price of the lunch that can be afforded?

30. **Printing:** Suppose a press can print 50 fliers per minute, and another press can print 90 fliers per minute. How long will it take for both presses running together to print 5150 fliers if the faster printer starts 5 minutes after the slower one and they both continue until the job is completed?

Objectives

1. Introduce Interval Notation
2. Solve Linear Inequalities
3. Solve Compound Inequalities
4. Solve Applied Problems

1.5 Solving Linear Inequalities

So far we have used linear equations to model certain types of applied problems. In section R1 we introduced the following inequality statements:

> greater than	< less than
≥ greater than or equal to	≤ less than or equal to

Recall that if point a lies to the left of point b on a number line (Figure 1), we say that b is *greater than a* (or equivalently, that a is *less than b*) and we write $b > a$ (or $a < b$). In other words, $b > a$ (or $a < b$) means that $b - a$ is positive.

The symbols $<$, $>$, $\leq$, and $\geq$ are called **inequality signs** and the expressions on the left and right of these signs are called the **sides** or **members** of the inequality. The inequalities $b > a$ or $a < b$ are said to be **strict** because they do not allow the possibility of equality. However, the inequalities $b \geq a$ and $a \leq b$ which allow the possibilities of equality are said to be **nonstrict**. The inequality $x \leq a$ is true when x is replaced by a or any number less than a: we say that these replacements **satisfy** the inequality. The set of all points on a real number line that satisfy an inequality is called the **graph** of the inequality. For example,

$$x > 2, \ -3 < x < 4, \text{ and } x \leq -1$$

have solutions that can be graphed on a real number line. Some relationships among quantities can only be described by inequalities. For instance, for a business to make a profit, the revenue, R, must be greater than the cost, C. A profit will result if $R > C$. In particular, if

$$C = 200 + 15x \text{ and } R = 3x,$$

where x is the number of items, then the number of items that must be sold for the business to have a profit is determined by solving the inequality

$$3x - (200 - 15x) > 0.$$

To solve such an inequality, we need to introduce interval notation.

Introducing Interval Notation

Certain sets of real numbers, called *intervals*, have an important role to play in the study of inequalities. The basic types of intervals are summarized in Table 1. Note that a *parenthesis* is used to graph inequalities with symbols $<$ or $>$, and a *bracket* is used to graph inequalities with symbols $\leq$ or $\geq$. In Table 1, we let a and b be real numbers such that $a < b$.

The special symbols $+\infty$ and $-\infty$, called **positive infinity** and **negative infinity,** are used to indicate that the interval extends indefinitely to the right or to the left.

Figure 1

TABLE 1

Interval Type	Inequality Notation	Interval Notation	Graph
Closed Interval:			
Numbers between a and b, Inclusive	$a \leq x \leq b$	$[a,b]$	
Open Interval:			
Numbers between a and b	$a < x < b$	(a,b)	
Numbers Greater than a	$x > a$	(a,∞)	
Numbers Less than b	$x < b$	$(-\infty,b)$	
Half-Open Interval:			
Numbers Greater than or Equal to a and Less than b	$a \leq x < b$	$[a,b)$	
Numbers Greater than a and Less than or Equal to b	$a < x \leq b$	$(a,b]$	
Numbers Greater than or Equal to a	$x \geq a$	$[a,\infty)$	
Numbers Less than or Equal to b	$x \leq b$	$(-\infty,b]$	

EXAMPLE 1   **Graphing Inequalities**

Graph each inequality and express each using interval notation.

(a) $-3 \leq x \leq 5$ (b) $x \geq 4$ (c) $x < -\dfrac{3}{2}$

Solution

(a) The graph of this inequality is the set of all real numbers on the number line that are greater than or equal to -3 and less than or equal to 5 (Figure 2a). The interval notation is $[-3,5]$.

(b) The graph of $x \geq 4$ is the set of all real numbers on the number line that are greater than or equal to 4 (Figure 2b). The interval notation is $[4,\infty)$.

(c) The graph of $x < -\dfrac{3}{2}$ is the set of all real numbers on the number line that are less than $-\dfrac{3}{2}$ (Figure 2c). The interval notation is $\left(-\infty, -\dfrac{3}{2}\right)$.

Figure 2

     (a)   (b)   (c)

Solving Linear Inequalities

Inequalities such as

$$x - 3 < 5, \ y + 2 \geq 7, \text{ and } 2t + 3 < 11 - 2t$$

are examples of linear inequalities in one variable. We **solve** an inequality for a variable by finding all values of the variable for which the inequality is true. Such values are called **solutions.** The set of all solutions of an inequality is called its **solution set.** Two inequalities are said to be **equivalent** if they have exactly the same solution set. To solve a linear inequality, we proceed in much the same way as in solving linear equations. Use the following *properties of inequalities* to complete a solution.

TABLE 2

Properties of Inequalities	Equivalent Inequality	Illustration
1. Addition (Subtraction) Property Add (subtract) any real number to each side of an inequality.	If $a < b$, then $a + c < b + c$ $a - c < b - c$	Since $5 < 13$, then $5 + 7 < 13 + 7$ $5 - 7 < 13 - 7$
2. Multiplication (Division) Property Multiplying (dividing) each side of an inequality by a *positive* number yields an equivalent inequality.	If $a < b$, then $ac < bc, c > 0$ $\dfrac{a}{c} < \dfrac{b}{c}, c > 0$	$5 < 13$, then $5(7) < (13)(7)$ $\dfrac{5}{7} < \dfrac{13}{7}$
Multiplying (dividing) each side of an inequality by a *negative* number yields an equivalent inequality with the order of the inequality reversed.	If $a < b$, then $ac > bc, c < 0$ $\dfrac{a}{c} > \dfrac{b}{c}, c < 0$	$5 < 13$, then $5(-7) > (13)(-7)$ $\dfrac{5}{-7} > \dfrac{13}{-7}$

Note that if $-3x > 0$, then $x < 0$, because the product of two negative numbers is positive. Remember when you multiply or divide by a negative number, you must reverse the inequality symbol.

Each of these properties is true if the symbols $<$ and $>$ are replaced by $\leq$ and $\geq$, respectively. In addition a, b, and c can be either real numbers or algebraic expressions. Consider the effect of adding (subtracting) and multiplying (dividing) each side of the inequality $4 < 6$ by a positive number (see Table 3) or a negative number (see Table 4).

TABLE 3

	Add 2	Subtract 2	Multiply by 2	Divide by 2
Given	$4 < 6$	$4 < 6$	$4 < 6$	$4 < 6$
Result	$6 < 8$	$2 < 4$	$8 < 12$	$2 < 3$

TABLE 4

	Add -2	Subtract -2	Multiply by -2	Divide by -2
Given	$4 < 6$	$4 < 6$	$4 < 6$	$4 < 6$
Result	$2 < 4$	$6 < 8$	$-8 > -12$	$-2 > -3$

Notice in Table 4, when we multiply or divide each side of the inequality $4 < 6$ by -2, the inequality symbol is reversed.

EXAMPLE 2   **Solving a Linear Inequality**

Solve the inequality $2x + 3 < 11 - 2x$, graph and write the solution in interval notation.

Solution

$x < 2$ is equivalent to $2 > x$. Traditionally, we rewrite inequalities so that the variable is on the left side.

$2x + 3 < 11 - 2x$	Given
$4x + 3 < 11$	Rearrange
$4x < 8$	Rearrange
$x < 2$	Isolate

The solution set of the inequality consists of all real numbers less than 2. In interval notation, $(-\infty, 2)$ is the solution set. The graph of this solution set is in Figure 3.

To check, choose a value in the solution set; the inequality is true. Choose a value not in the solution set; the inequality is false. The endpoint of the interval should also be checked.

Figure 3

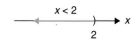

x	Left Side	Right Side	True/False
1	$2(1) + 3 = 5$	$11 - 2(1) = 9$	$5 < 9$, True
2	$2(2) + 3 = 7$	$11 - 2(2) = 7$	$7 < 7$, False
3	$2(3) + 3 = 9$	$11 - 2(3) = 5$	$9 < 5$, False

EXAMPLE 3   **Solving a Linear Inequality**

Solve the inequality $4(x - 2) \geq 3(x - 2) - 4$, graph the solution and write the solution in interval notation.

Solution

$4(x - 2) \geq 3(x - 2) - 4$	Given
$4x - 8 \geq 3x - 6 - 4$	Distribute
$4x - 8 \geq 3x - 10$	Combine
$x - 8 \geq -10$	Rearrange
$x \geq -2$	

The solution set of the inequality consists of all real numbers greater than or equal to -2. In interval notation, $[-2, \infty)$ is the solution set. The graph of this solution set is in Figure 4.

We check by picking a value in the solution set outside the solution set, and the endpoint of the solution set.

Figure 4

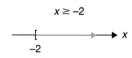

x	Left Side	Right Side	True/False
-3	$4(-3 - 2) = -20$	$3(-3 - 2) - 4 = -19$	$-20 \geq -19$, False
-2	$4(-2 - 2) = -16$	$3(-2 - 2) - 4 = -16$	$-16 \geq -16$, True
0	$4(0 - 2) = -8$	$3(0 - 2) - 4 = -10$	$-8 \geq -10$, True

Solving Linear Inequalities

Inequalities such as

$$x - 3 < 5, \quad y + 2 \geq 7, \quad \text{and} \quad 2t + 3 < 11 - 2t$$

are examples of linear inequalities in one variable. We **solve** an inequality for a variable by finding all values of the variable for which the inequality is true. Such values are called **solutions.** The set of all solutions of an inequality is called its **solution set.** Two inequalities are said to be **equivalent** if they have exactly the same solution set. To solve a linear inequality, we proceed in much the same way as in solving linear equations. Use the following *properties of inequalities* to complete a solution.

TABLE 2

Properties of Inequalities	Equivalent Inequality	Illustration
1. Addition (Subtraction) Property Add (subtract) any real number to each side of an inequality.	If $a < b$, then $a + c < b + c$ $a - c < b - c$	Since $5 < 13$, then $5 + 7 < 13 + 7$ $5 - 7 < 13 - 7$
2. Multiplication (Division) Property Multiplying (dividing) each side of an inequality by a *positive* number yields an equivalent inequality.	If $a < b$, then $ac < bc, c > 0$ $\dfrac{a}{c} < \dfrac{b}{c}, c > 0$	$5 < 13$, then $5(7) < (13)(7)$ $\dfrac{5}{7} < \dfrac{13}{7}$
Multiplying (dividing) each side of an inequality by a *negative* number yields an equivalent inequality with the order of the inequality reversed.	If $a < b$, then $ac > bc, c < 0$ $\dfrac{a}{c} > \dfrac{b}{c}, c < 0$	$5 < 13$, then $5(-7) > (13)(-7)$ $\dfrac{5}{-7} > \dfrac{13}{-7}$

Note that if $-3x > 0$, then $x < 0$, because the product of two negative numbers is positive. Remember when you multiply or divide by a negative number, you must reverse the inequality symbol.

Each of these properties is true if the symbols $<$ and $>$ are replaced by $\leq$ and $\geq$, respectively. In addition a, b, and c can be either real numbers or algebraic expressions. Consider the effect of adding (subtracting) and multiplying (dividing) each side of the inequality $4 < 6$ by a positive number (see Table 3) or a negative number (see Table 4).

TABLE 3

	Add 2	Subtract 2	Multiply by 2	Divide by 2
Given	$4 < 6$	$4 < 6$	$4 < 6$	$4 < 6$
Result	$6 < 8$	$2 < 4$	$8 < 12$	$2 < 3$

TABLE 4

	Add -2	Subtract -2	Multiply by -2	Divide by -2
Given	$4 < 6$	$4 < 6$	$4 < 6$	$4 < 6$
Result	$2 < 4$	$6 < 8$	$-8 > -12$	$-2 > -3$

Notice in Table 4, when we multiply or divide each side of the inequality $4 < 6$ by -2, the inequality symbol is reversed.

EXAMPLE 2

Solving a Linear Inequality

Solve the inequality $2x + 3 < 11 - 2x$, graph and write the solution in interval notation.

Solution

$x < 2$ is equivalent to $2 > x$. Traditionally, we rewrite inequalities so that the variable is on the left side.

$2x + 3 < 11 - 2x$	Given
$4x + 3 < 11$	Rearrange
$4x < 8$	Rearrange
$x < 2$	Isolate

The solution set of the inequality consists of all real numbers less than 2. In interval notation, $(-\infty, 2)$ is the solution set. The graph of this solution set is in Figure 3.

To check, choose a value in the solution set; the inequality is true. Choose a value not in the solution set; the inequality is false. The endpoint of the interval should also be checked.

Figure 3

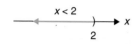

x	Left Side	Right Side	True/False
1	$2(1) + 3 = 5$	$11 - 2(1) = 9$	$5 < 9$, True
2	$2(2) + 3 = 7$	$11 - 2(2) = 7$	$7 < 7$, False
3	$2(3) + 3 = 9$	$11 - 2(3) = 5$	$9 < 5$, False

EXAMPLE 3

Solving a Linear Inequality

Solve the inequality $4(x - 2) \geq 3(x - 2) - 4$, graph the solution and write the solution in interval notation.

Solution

$4(x - 2) \geq 3(x - 2) - 4$	Given
$4x - 8 \geq 3x - 6 - 4$	Distribute
$4x - 8 \geq 3x - 10$	Combine
$x - 8 \geq -10$	Rearrange
$x \geq -2$	

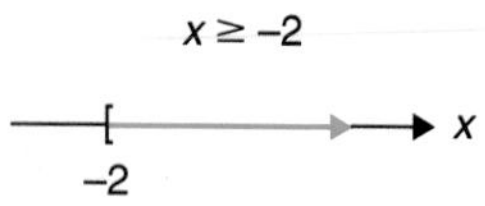

The solution set of the inequality consists of all real numbers greater than or equal to -2. In interval notation, $[-2, \infty)$ is the solution set. The graph of this solution set is in Figure 4.

We check by picking a value in the solution set outside the solution set, and the endpoint of the solution set.

Figure 4

$x \geq -2$

$\xleftarrow{\qquad}[\xrightarrow{\qquad\qquad}} x$
$\quad -2$

x	Left Side	Right Side	True/False
-3	$4(-3 - 2) = -20$	$3(-3 - 2) - 4 = -19$	$-20 \geq -19$, False
-2	$4(-2 - 2) = -16$	$3(-2 - 2) - 4 = -16$	$-16 \geq -16$, True
0	$4(0 - 2) = -8$	$3(0 - 2) - 4 = -10$	$-8 \geq -10$, True

EXAMPLE 4

Solving a Linear Inequality

Solve the inequality $\dfrac{2-x}{2} \le -\dfrac{x+1}{5}$, graph the solution and write in interval notation.

Solution

$$\dfrac{2-x}{2} \le -\dfrac{x+1}{5} \qquad \text{Given}$$

$$10\left(\dfrac{2-x}{2}\right) \le 10\left(-\dfrac{x+1}{5}\right) \qquad \text{Multiply}$$

$$5(2-x) \le -2(x+1) \qquad \text{Combine}$$

$$10 - 5x \le -2x - 2 \qquad \text{Distribute}$$

$$10 - 3x \le -2 \qquad \text{Rearrange}$$

$$-3x \le -12 \qquad \text{Rearrange}$$

$$x \ge 4 \qquad \text{Isolate}$$

Dividing by −3 reverses the inequality.

Figure 5

$x \ge 4$

The solution set of the inequality consists of all real numbers greater than or equal to 4. In interval notation, $[4,\infty)$ is the solution set. The graph of this solution set is in Figure 5.

We pick three points to check.

x	Left Side	Right Side	True/False
3	$\dfrac{2-3}{2} = -\dfrac{1}{2}$	$-\dfrac{3+1}{5} = -\dfrac{4}{5}$	$-\dfrac{1}{2} \le -\dfrac{4}{5}$, False
4	$\dfrac{2-4}{2} = -1$	$-\dfrac{4+1}{5} = -1$	$-1 \le -1$, True
5	$\dfrac{2-5}{2} = -\dfrac{3}{2}$	$-\dfrac{5+1}{5} = -\dfrac{6}{5}$	$-\dfrac{3}{2} \le -\dfrac{6}{5}$, True

Solving Compound Inequalities

Two inequalities connected with the words "and" or "or" are called **compound inequalities.** Examples of these inequalities are:

$$-1 \le x \text{ and } x \le 2, \quad x \le -5 \text{ or } x \ge 1$$

The solution set of a compound inequality with the word "and" is the set of real numbers *common* to the solution sets of each inequality. For example, the solution set of

$$-1 \le x \text{ and } x \le 2$$

consists of all real numbers x that are greater than or equal to -1 *and* less than or equal to 2. That is, all real numbers between -1 and 2, including the end points. This is also written as

$$-1 \le x \le 2 \text{ or as an interval } [-1,2].$$

Note that a compound inequality containing the word "or" may not be compressed as the "and" compound inequality is compressed.

Graphing these inequalities provides a visual idea of the solutions. Figure 6a shows the graph of the inequalities $-1 \le x$ and $x \le 2$.

The solution set of a compound inequality containing the word "or" is the set of real numbers in the solution set of either or both of the two inequalities. For example, the solution set of

$$x \le -5 \text{ or } x \ge 1, \text{ as an interval } (-\infty,-5] \text{ or } [1,\infty)$$

consists of all real numbers x that are less than or equal to -5, or all real numbers that are greater than or equal to 1 (Figure 6b).

Figure 6

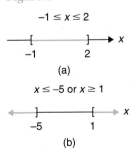

$-1 \le x \le 2$

$-1 \quad 2$

(a)

$x \le -5$ or $x \ge 1$

$-5 \quad 1$

(b)

EXAMPLE 5

Solving Compound Inequalities

Solve each compound inequality, express the solution set in interval notation, and graph.

(a) $-3 < 2x + 1 \leq 5$ (b) $5x - 2 < -7$ or $5x - 2 > 8$

Solution

(a) Our goal is to isolate the variable x in the middle.

$$-3 < 2x + 1 \leq 5 \qquad \text{Given}$$
$$-3 - 1 < 2x + 1 - 1 \leq 5 - 1 \qquad \text{Rearrange}$$
$$-4 < 2x \leq 4 \qquad \text{Combine}$$
$$-2 < x \leq 2 \qquad \text{Isolate}$$

Therefore, the solution set consists of all real numbers greater than -2 and less than or equal to 2. In interval notation, the solution set is $(-2, 2]$ and the graph is drawn in Figure 7a.

(b) We solve each inequality individually.

$$5x - 2 < -7 \quad \text{or} \quad 5x - 2 > 8 \qquad \text{Given}$$
$$5x - 2 + 2 < -7 + 2 \quad \text{or} \quad 5x - 2 + 2 > 8 + 2 \qquad \text{Rearrange}$$
$$5x < -5 \quad \text{or} \quad 5x > 10 \qquad \text{Combine}$$
$$x < -1 \quad \text{or} \quad x > 2 \qquad \text{Isolate}$$

The solution set consists of all real numbers less than -1 or greater than 2. In interval notation, the solution set is $(-\infty, -1)$ or $(2, \infty)$ and the graph is drawn in Figure 7b.

Figure 7

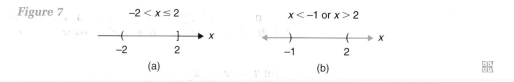

(a)   (b)

Solving Applied Problems

The strategy for solving word problems in section 1.3 applies equally to word problems involving linear inequalities. Table 5 illustrates how inequality statements are translated into symbols.

TABLE 5

Word Statement	Algebraic Statement
x is at least 20	$x \geq 20$
x is at most 15	$x \leq 15$
x is no more than 20	$x \leq 20$
x is not less than 7	$x \geq 7$

EXAMPLE 6

Solving a Cost Problem

A mathematics department wishes to purchase a number of graphing calculators. Each calculator costs $60, plus 8% sales tax. In addition, a $50 delivery charge for the entire shipment must be paid. If the department has allocated at most $3200 for the purchase, how many calculators can be ordered?

Solution   **Make a Plan**

The cost of each calculator is $60 + 0.08($60) = $64.80.

Let x = the number of graphing calculators, then

$64.80x = total cost of the graphing calculators.

To solve this problem we write an inequality that requires the cost of the calculators, $64.80x, plus the delivery charge, $50, to be less than or equal to $3200. The diagram visualizes the *given*.

Total cost of calculators $64.80x	+ +	Delivery charge $50	≤ ≤	Total purchase price $3200

Write the Solution

We solve the inequality as follows:

$$64.80x + \$50 \le \$3200 \qquad \text{Given}$$
$$64.80x \le 3150 \qquad \text{Rearrange}$$
$$x \le 48.6 \qquad \text{Isolate}$$

Enough money was allocated to order 48 or fewer calculators.

Look Back

We do not round up to 49 since it is not less than 48.6: $64.80(49) + $50 = $3225.20. Such a purchase would be over the budget amount of $3200.

EXAMPLE 7   **Temperature Range**

The average daily range of temperature for a city during a late summer month varies from at least 72°F to no more than 98°F.

(a) Describe the temperature range for one day.

(b) If the average temperature decreases 2°F per day, describe the temperature range after one week.

Solution   **Make a Plan**

Let T represent the temperature during a summer month. We write a compound inequality stating that the temperature is to be between 72°F and 98°F inclusive. The temperature drops 2°F each day for a week. We decrease the temperatures by 7(2°F).

Write the Solution

The average range of temperature is $72° \le T \le 98°$. Since the temperature decreases 2°F per day for a week, the temperature range after one week is

$$72° - 7(2°) \le T \le 98° - 7(2°)$$
$$72° - 14° \le T \le 98° - 14° \qquad \text{Combine}$$
$$58° \le T \le 84° \qquad \text{Combine}$$

Look Back

By decreasing the temperature 2°F each day for seven days, we see that 72°F drops to 58°F and 98°F to 84°F.

PROBLEM SET 1.5

Mastering the Concepts

In Problems 1–8, graph each inequality and express each using interval notation.

1. (a) $-1 < x \leq 1$
    (b) $-1 \leq x < 1$

2. (a) $-1 \leq x \leq 1$
    (b) $-1 < x < 1$

3. (a) $x \geq -1$
    (b) $x < -1$

4. (a) $x < 1$
    (b) $x \leq 1$

5. (a) $x \leq 2$ or $x \geq 4$
    (b) $x \geq 2$ or $x \leq 4$

6. (a) $x \geq 4$ or $x \leq 5$
    (b) $x \leq 4$ or $x \geq 5$

7. (a) $x \leq -10$ or $x \geq -10$
    (b) $x \leq -10$ and $x \geq -10$

8. (a) $x < 0$ or $x > 0$
    (b) $x < 0$ and $x > 0$

In Problems 9 and 10, state the property of inequalities that justifies each statement.

9. (a) If $x < 4$ then $2x < x + 4$
    (b) If $x < 4$ then $2x < 8$
    (c) If $y \leq -2$ then $y + 2 \leq 0$
    (d) If $y \leq -2$ then $-y \geq 2$

10. (a) If $y \geq y - 1$ then $2y \geq 2y - 2$
    (b) If $y \geq y - 1$ then $-2y \leq 2 - 2y$
    (c) If $x - 1 \leq 4$ then $x \leq 5$
    (d) If $a \leq x$ then $a - 2 \leq x - 2$

In Problems 11–40, solve each inequality and express the solution using interval notation. Also graph the solution on a number line.

11. $5 + 3x \geq 8$

12. $-5x + 2 > 12$

13. $-3x + 4 < 14 + 2x$

14. $x + 6 \leq 4 - 3x$

15. $8 - 9x \leq -x$

16. $4 - x \geq 3x$

17. $2(3x - 2) < 7 + 4x - 1$

18. $6 - (7 - x) \geq 2(x - 3)$

19. $-(x - 1) > 2(x + 1/2)$

20. $4(3 - x) \geq 2(x - 6)$

21. $5x \geq -3(x - 2)$

22. $-3(x + 7) \geq -4(2x - 1)$

23. $-9(x - 3) - 8(4 - x) \geq -2x$

24. $-4(x - 1) \geq 2(x + 1) - 4$

25. $3(x - 5) + 10 < 2(x + 4)$

26. $9(x + 2) < -6(4 - x) + 18$

27. $3(x + 2) - 2 \geq -(x + 5) + x$

28. $4(x + 4) > -2(x - 3) + 1$

29. $\dfrac{1}{9}(3x - 2) < \dfrac{1}{3}(1 - 4x)$

30. $\dfrac{1}{6}(2x - 7) < \dfrac{1}{2}(x + 1)$

31. $\dfrac{x}{3} - 25 < \dfrac{x}{4} - 2x$

32. $\dfrac{3x}{2} \geq -6 - \dfrac{x}{2}$

33. $\dfrac{x}{2} - \dfrac{x}{3} \leq 4$

34. $\dfrac{x}{6} + 1 \geq \dfrac{x}{3}$

35. $\dfrac{4x - 2}{2} > \dfrac{3x + 6}{3}$

36. $\dfrac{x + 4}{2} < \dfrac{2x - 3}{3}$

37. $\dfrac{x + 4}{-2} \leq \dfrac{x - 3}{5}$

38. $\dfrac{4x + 17}{7} > x + 2$

39. $\dfrac{5x + 1}{3} > \dfrac{3x + 5}{4}$

40. $\dfrac{2x + 4}{-5} \leq \dfrac{3x - 3}{-3}$

In Problems 41–52, solve each compound inequality, express the solution in interval notation, and graph the solution on a number line.

41. $3 \leq x + 1 \leq 5$

42. $-2 \leq x - 3 \leq 3$

43. $-9 < 3x < 6$

44. $-4 \leq 2x \leq 8$

45. $-3 \leq 4x - 1 \leq 5$

46. $1 < 8 - 3x < 12$

47. $4x - 5 < -3$ or $4x - 5 > 3$

48. $5x - 4 > 1$ or $5x - 4 < -1$

49. $-5 - 4x > 3$ or $-5 - 4x < -3$

50. $4x - 7 \geq 3$ or $4x - 7 \leq -3$

51. $2x - 1 < -7$ or $2x - 1 > 7$

52. $4x + 3 < -5$ or $4x + 3 > 5$

Applying the Concepts

53. Sales Commission: A sales representative for a tennis club earns a base salary of $500 per week plus a commission of $15 for every new membership sold. How many memberships must be sold in a week in order to earn at least $725 for that week?

54. Weekly Income: A newspaper carrier earns $7 per week plus $0.18 for each newspaper delivered to a home. How many newspapers should be delivered in order for the carrier to earn more than $30 per week?

55. Telephone Charges: Suppose the cost for an international telephone call is $3.50 for the first minute and $1.85 for each additional minute (or part of a minute). If you do not want your total cost to exceed $30, how long can you talk?

56. Car Rental: A vacationer has two choices of car rental agencies. The first agency charges a flat fee of $44.95 per day with unlimited miles. The second charges $19.95 per day plus $0.19 per mile.
(a) Assuming a seven day rental, what range of miles driven would make the flat fee rental more economical?
(b) What range of miles driven per day would make the second rental more economical?
(c) What range of miles driven in three days would make the flat fee rental more economical?

57. Test Averages: A student scored 65, 80, and 74 on the first three tests during the term. What does the student need to score on the fourth test to ensure an average score that is above 75?

58. Diet Program: A clinic advertises that a person can reduce their weight by at least 1.5 pounds per week by exercising and special dieting. At the beginning of a diet program, a person weighs 195 pounds. What is the maximum number of weeks before this person's weight will be reduced to 170 pounds?

59. Energy Consumption: The average American home uses at least 90 but no more than 120 kilowatt hours of electricity per month.
(a) Use an inequality to express the average range of kilowatt hours per day, assuming that one month is 30 days.
(b) Use an inequality to express the average range of kilowatt hours per week and per year.

60. Investment: Suppose you invest $7000 in a mutual fund that pays simple interest for one year. If you earn at most $924 at the end of 18 months, what is the maximum annual interest rate for this investment?

61. Geometry: The length of a rectangle is 17 centimeters less than three times the width. Find, to the nearest positive integer, the width of the rectangle if the perimeter is between 222 and 238 centimeters.

62. Travel Times: A traveler flying between New York and Paris can choose among several planes with different average speeds. Suppose the slowest plane averages 515 miles per hour for the trip, and the fastest averages 975 miles per hour.
(a) Find the range in travel times for the 3645 mile trip.
(b) Assume the jet stream (high-altitude winds) blows west to east at a velocity of 50 miles per hour. Find the range of this trip traveling with the jet stream.

Developing and Extending the Concepts

In Problems 63 and 64, solve each inequality, write it in interval notation, and graph the solution set.

63. $2x - 3 \leq 3x + 1 \leq 4x - 5$

64. $2x - 1 \leq 3x + 7 \leq x + 9$

65. What is wrong with the following argument? Suppose that $0 < x < 1$, then

$$x^2 < x$$
$$x^2 - 1 < x - 1$$
$$(x - 1)(x + 1) < x - 1$$
$$x + 1 < 1$$
$$x < 0$$

But we said that $0 < x < 1$, what do you think?

66. Suppose that $\dfrac{b}{a} > 1$. Is $a - b$ positive or negative if:
(a) a and b are positive numbers,
(b) a and b are negative numbers?

Objectives

1. Solve Equations Involving Absolute Value
2. Solve Inequalities Involving Absolute Value
3. Solve Applied Problems

1.6 Solving Equations and Inequalities Involving Absolute Value

In Section R3, we learned that the absolute value of a number can be described geometrically as the distance between the number and zero on a number line. As a result, equations and inequalities involving absolute values of algebraic expressions can also be solved geometrically and algebraically. Here, we consider both approaches. First we will review the absolute value concept. If a is a point on the number line, then the distance between a and 0 is represented by $|a|$ and called the absolute value of a regardless of whether the point a is to the left (Figure 1a) or to the right of 0 (Figure 1b).

Figure 1

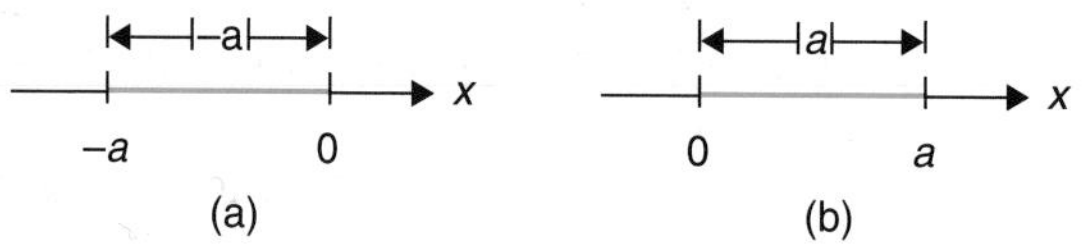

(a)   (b)

Algebraically, absolute values are written as:

$$|x| = \begin{cases} x \text{ if } x > 0 \\ 0 \text{ if } x = 0 \\ -x \text{ if } x < 0 \end{cases}$$

Absolute values can be used to find the distance between numbers on the number line. The distance between -4 and -7 is represented on the number line as 3 units regardless if we move from -7 to -4 or from -4 to -7.

It is written as

$$|-4 - (-7)| = |-4 + 7| = |3| = 3$$

or

$$|-7 - (-4)| = |-7 + 4| = |-3| = 3.$$

More generally,

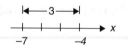

$|a - b|$ is the number of units of distance between the point with coordinate a and the point with coordinate b.

This statement is true no matter which point is to the left of the other. That is, point a can be to the left (Figure 2a) or to the right of point b (Figure 2b).

Also $|a - b| = |b - a|$

Figure 2

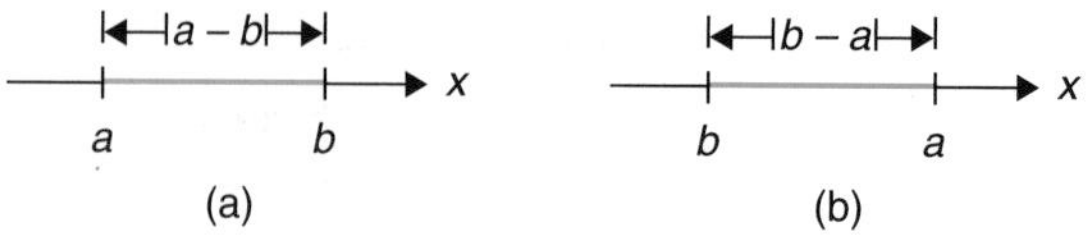

(a)   (b)

Solving Equations Involving Absolute Value

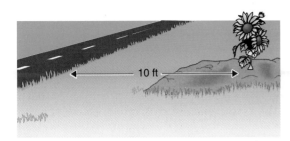

10 ft

Along a narrow strip of garden 10 feet from the street is a flower. If we want to plant another flower exactly 4 feet away from the original plant, where should we place it? We can place a plant 4 feet to the left (6 feet from the street) or 4 feet to the right of the original plant (14 feet from the street). If we let x represent the distance of the new plant from the street, then the distance between the plants is 4. In symbols, it is expressed by the **absolute value equation:**

$$|x - 10| = 4$$

The solutions 6 and 14 of the equation were found by understanding the situation, not the algebra. Now, let us find a method to solve any absolute value equation.

An equation containing absolute value such as the following:

$$|x| = 3, \quad |x - 2| = 5, \quad \text{and} \quad |4x - 1| = 7,$$

is called an **absolute value equation.** We can solve an absolute value equation such as $|x - 10| = 4$ geometrically by interpreting the equation to mean those points $x - 10$ on the number line which are exactly 4 units away from zero. There are two such points: -4 which is exactly four units to the left of zero, and $+4$ which is exactly four units to the right of zero (Figure 3).

Figure 3

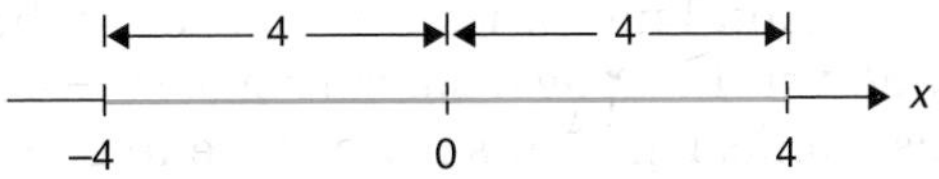

$$
\begin{array}{c}
\mid\!\!\leftarrow\!\!-4\!\!-\!\!\rightarrow\!\!\mid\!\!\leftarrow\!\!-4\!\!-\!\!\rightarrow\!\!\mid \\
\end{array}
$$

-4	0	4	→ x

So we have two possibilities:

$$x - 10 = -4 \text{ or } x - 10 = 4$$

$$x = 6 \text{ or } \qquad x = 14$$

Therefore, the solutions of the equation $|x - 10| = 4$ are 6 and 14.

The strategy for algebraically solving an absolute value equation is to rewrite the equation as two equivalent linear equations, then solve the resulting linear equations.

In general, we solve an absolute value equation as follows:

Let u be a variable expression, and a be a real number.

1. If $a > 0$, then $|u| = a$ is equivalent to $u = -a$ or $u = a$ (Figure 4).
2. If $a = 0$, then $|u| = a$ is equivalent to $u = 0$.
3. If $a < 0$, then $|u| = a$ has no solution.

Figure 4

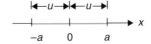

$$
\mid\!\!\leftarrow\!u\!\rightarrow\!\!\mid\!\!\leftarrow\!u\!\rightarrow\!\!\mid
$$

$-a$	0	a	→ x

EXAMPLE 1   **Solving Absolute Value Equations**

Solve each equation.

(a) $|-x - 3| = 6$ (b) $|4x - 1| = 7$ (c) $|5x - 1| = -1$

Solution   (a) Using $u = -x - 3$ and $a = 6$ (Figure 5) in the equation $|u| = a$,

Figure 5

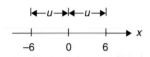

$$-x - 3 = -6 \quad \text{or} \quad -x - 3 = 6$$
$$-x = -3 \quad \text{or} \qquad -x = 9$$
$$x = 3 \quad \text{or} \qquad x = -9$$

Check   The two solutions are 3 and -9. Check these in the original equation.

x	Left Side	Right Side	True/False
3	$|-3 - 3| = |-6| = 6$	6	True
-9	$|-(-9) - 3| = |9 - 3| = |6| = 6$	6	True

Figure 6

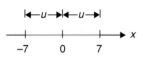

(b) Using $u = 4x - 1$ and $a = 7$ (Figure 6) in the equation $|u| = a$,

$$4x - 1 = -7 \quad \text{or} \quad 4x - 1 = 7$$
$$4x = -6 \qquad\qquad 4x = 8$$
$$x = -\frac{3}{2} \quad \text{or} \qquad x = 2$$

The solutions are $-\dfrac{3}{2}$ and 2. Check these solutions in the original equation.

(c) This equation $|5x - 1| = -1$ has *no solutions* because there is no value of x which would make $|5x - 1|$ negative.

Solving Inequalities Involving Absolute Value

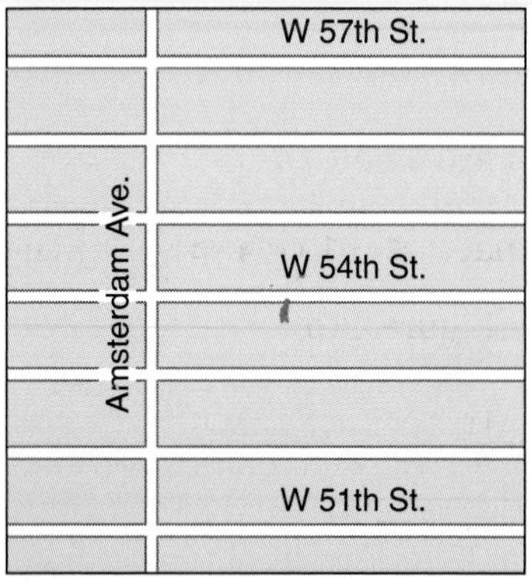

In New York City, a plaintiff got a personal protection order against a defendant which required the defendant to live more than three blocks away. If the plaintiff lives on 54th street, then the distance between the residences must be more than three blocks. From the map, we see that one possibility is that the defendant may live north of 57th street or south of 51st street. This situation can be described in one of the following *absolute value inequalities* where the distances are not equal to but greater than or less than:

$$|x - 54| > 3.$$

They can also be solved by using geometric interpretation on a real number line or by an algebraic method.

For example, $|x| < 3$ requires that the value of x is to be less than 3 units from zero. That is, x is between -3 and 3 which is written as

$$-3 < x < 3.$$

In interval notation the solution set is $(-3, 3)$ (Figure 7a). Similarly, $|x| > 3$ requires that the value of x is more than 3 units from zero. That is, $x < -3$ or $x > 3$. In interval notation the solution set is $(-\infty, -3)$ or $(3, \infty)$ (Figure 7b).

Figure 7

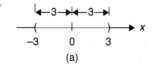
(a)

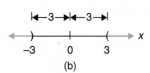
(b)

More generally, we have the following results:

Solving an Absolute Value Inequality

Figure 8(a)

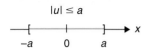

Let u be a linear expression, and let a be a real number.

1. If $a \geq 0$, rewrite the **absolute value inequality**
   (a) $|u| \leq a$ as $-a \leq u \leq a$ (Figure 8a)
   (b) $|u| \geq a$ as $u \leq -a$ or $u \geq a$ (Figure 8b)
2. Solve the resulting linear inequalities.

Figure 8(b)

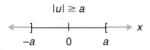

Statements comparable to these can be made for strict inequalities (those involving $<$ or $>$).

EXAMPLE 2

Solving Absolute Value Inequalities

Solve each inequality and graph the solution set on a number line.

(a) $|x - 3| < 7$ (b) $|7x - 2| \geq 9$

Solution

(a) $|x - 3| < 7$ Use $|u| < a$ where $u = x - 3$ and $a = 7$ (Figure 9)

Figure 9

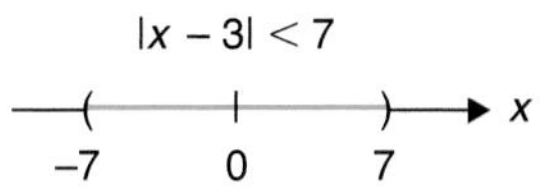

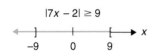

$$-7 < x - 3 < 7 \qquad \text{Equivalent inequalities}$$
$$-4 < x < 10 \qquad \text{Add 3 to all three parts.}$$

Therefore, the solution set consists of all real numbers that are greater than -4 and less than 10; that is, between -4 and 10. In interval notation, the solution set is $(-4,10)$ (Figure 11a).

Figure 10

$|7x - 2| \geq 9$

(b) $|7x - 2| \geq 9$ Use $|u| \geq a$ for $u = 7x - 2$ and $a = 9$ (Figure 10)

$$7x - 2 \leq -9 \text{ or } 7x - 2 \geq 9 \qquad \text{Equivalent inequalities}$$
$$7x \leq -7 \qquad\quad 7x \geq 11 \qquad \text{Rearrange}$$
$$x \leq -1 \text{ or } \qquad x \geq \frac{11}{7} \qquad \text{Isolate}$$

The solution set consists of all real numbers x such that $x \leq -1$ or $x \geq \frac{11}{7}$. In interval notation, the solution set is $(-\infty,-1]$ or $[\frac{11}{7}, \infty)$ (Figure 11b).

Figure 11

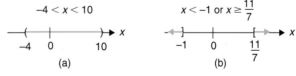

(a)   (b)

EXAMPLE 3

Solving Absolute Value Inequalities

Solve each absolute value inequality.

(a) $|x - 3| < 0$ (b) $|x - 3| \geq 0$

Solution

(a) There is no solution of $|x - 3| < 0$ because there are no values of x which would make the absolute value less than zero.

(b) Since the absolute value of every number is always greater than or equal to zero, the solution of $|x - 3| \geq 0$ is the set of all real numbers, R.

Solving Applied Problems

Equations and inequalities containing absolute values are extremely useful in solving applied problems especially those involving data analysis. For instance, many specifications usually allow a certain amount of variation so that measured dimensions (such as thickness, distance, and margins of error) differ from exact dimensions by acceptable margins, known as **tolerances.** The next example provides a way to represent such variations.

EXAMPLE 4   **Solving a Manufacturing Problem**

A manufacturer of helicopter blades determines that the manufactured thickness x of a blade may differ from the design thickness of 17.48 millimeters by no more than 0.12 millimeters (the tolerance is 0.12 millimeters).

(a) Write an absolute value inequality that expresses this relationship.

(b) Solve the inequality to find the acceptable range of thicknesses.

Solution   (a) Let x = the manufactured thickness of the blade.
$|x - 17.48|$ = the difference between the manufactured thickness and design thickness.
   The tolerance of 0.12 millimeters can be expressed using the absolute value inequality: $|x - 17.48| \le 0.12$.

(b) $|x - 17.48| \le 0.12$ Given

$\qquad -0.12 \le x - 17.48 \le 0.12$ Equivalent inequality

$\qquad 17.36 \le x \le 17.60$ Rearrange

Figure 11

Therefore, the manufactured thickness of the blade must be between 17.36 millimeters and 17.60 millimeters (Figure 11).
   We can check this result by substitution.

Thickness	Right Side	Tolerance	True/False
17	$|17 - 17.48| = 0.48$	0.12	$0.48 \le 0.12$, False
17.4	$|17.4 - 17.48| = 0.08$	0.12	$0.08 \le 0.12$, True
17.68	$|17.68 - 17.48| = 0.2$	0.12	$0.2 \le 0.12$, False

PROBLEM SET 1.6

Mastering the Concepts

In Problems 1 and 2, write each expression without absolute value signs and simplify.

1. (a) $|(-4) - (-2)|$
   (b) $|5 - \pi|$
   (c) $|\sqrt{5} - 2|$

2. (a) $|(-2) - (-4)|$
   (b) $|\sqrt{3} - 3|$
   (c) $|3 - \sqrt{7}|$

In Problems 3 and 4, find the distance between points A and B.

3. (a) $A = -5, B = 4$
   (b) $A = -10, B = -8$
   (c) $A = 6, B = 3$

4. (a) $A = -6, B = -7$
   (b) $A = -8, B = 4$
   (c) $A = 4, B = -2$

In Problems 5–24, solve each equation algebraically.

5. (a) $|x - 5| = 6$
   (b) $|5 - x| = 6$

6. (a) $|y - 3| = 4$
   (b) $|3 - y| = 4$

7. (a) $|x - 2| = 1$
   (b) $|x + 2| = 1$

8. (a) $|x - 4| = 2$
   (b) $|x + 4| = 2$

9. (a) $|y - 5| = 0$
   (b) $|y - 5| = -2$

10. (a) $|x - 7| = -1$
   (b) $|x + 7| = 0$

11. (a) $|3x| = 0$
   (b) $|-3x| = 0$

12. (a) $|-7x| = 14$
   (b) $|-7x| = -14$

13. (a) $|x| + 2 = 8$
   (b) $|x + 2| = 8$

14. (a) $|x| - 4 = 4$
   (b) $|x - 4| = 4$

15. $-4|5t| = 2$

16. $8|2x| = 4$

17. $|3t + 2| = 0$

18. $|5 - 4x| = 0$

19. $|3(x + 7)| = 12$

20. $|2(x - 10)| = 1$

21. $5 = \left| \dfrac{3x}{6} - \dfrac{9}{6} \right|$

22. $2 = \left| \dfrac{4x}{5} + \dfrac{4}{5} \right|$

23. $\left| \dfrac{5x}{2} + \dfrac{10}{2} \right| = \dfrac{5}{2}$

24. $\left| \dfrac{-5x}{2} - \dfrac{10}{2} \right| = \dfrac{5}{2}$

In Problems 25–48, solve each inequality algebraically. Write the solution set in interval notation and graph it on a number line.

25. (a) $|x - 4| \geq 5$
   (b) $|x - 4| < 5$

26. (a) $|x - 2| \geq 3$
   (b) $|x - 2| < 3$

27. (a) $|x + 14| \geq 25$
   (b) $|x + 14| < 25$

28. (a) $|x + 12| \geq 30$
   (b) $|x + 12| < 30$

29. $|3x| < 15$

30. $|6x| \geq 1$

31. (a) $|2x - 1| \leq 3$
   (b) $|2x - 1| \geq 3$

32. (a) $|8x + 3| \geq 3$
   (b) $|8x + 3| < 3$

33. (a) $|5x + 1| \geq -2$
   (b) $|5x + 1| \leq -2$

34. (a) $|2x - 12| \geq 4$
   (b) $|2x - 12| < 4$

35. $\left| \dfrac{x}{2} \right| > \dfrac{1}{2}$

36. $\left| \dfrac{1}{3}x \right| \leq 2$

37. $|2x - 1| - 2 < 3$

38. $|2x + 7| - 2 \geq 9$

39. $|3x + 1| + 2 \geq 7$

40. $|5x - 3| - 4 \leq 3$

41. $\left| \dfrac{2}{3}x - 1 \right| \leq \dfrac{1}{3}$

42. $\left| \dfrac{x}{2} + 2 \right| > \dfrac{5}{2}$

43. $\left| \dfrac{1}{3} - \dfrac{x}{6} \right| \geq \dfrac{1}{2}$

44. $\left| \dfrac{x}{7} + \dfrac{3}{7} \right| 1$

45. (a) $|x + 7| < 0$
   (b) $|x + 7| \leq 0$

46. (a) $2|x + 7| < 0$
   (b) $2|x + 7| \leq 0$

47. (a) $|x + 6| > 0$
   (b) $|x + 6| \geq 0$

48. (a) $3|x + 6| > 0$
   (b) $3|x + 6| \geq 0$

Applying the Concepts

49. Body Temperature: During and after surgery, research has determined that a patient's temperature T (in degrees Fahrenheit) must not differ from 98.6°F by no more than 2.3°F. Write an absolute value inequality to express this situation, then solve the inequality.

50. Manufacturing: Manufacturers of compact discs (CDs) permit only very small tolerances. The measured diameter, x, of a CD may differ from the exact diameter of 117 mm by no more than 0.01 mm. Write an absolute value inequality to express this situation, then solve the inequality.

51. Electric Motor Design: In the design of an electric motor, it was determined that the measured voltage needed to run the motor may differ from the exact voltage of 220 by no more than 25 volts. Write an absolute value inequality that expresses this situation, then solve the inequality to find the lower and upper limits for voltage for which the motor will run.

52. Heating and Cooling: A thermostat is set at 68°F so that the heating and cooling system will activate when the room temperature is more than 3.5°F from that setting. Write an absolute value inequality to express this relationship and solve the inequality.

53. Car Production: An automobile manufacturer predicted that the number of cars to be produced by the automobile industry for a given year must differ from the expected 4,000,000 demand for cars by no more than 500,000 cars. Use an absolute value inequality to describe this situation, then solve this inequality.

54. Stock Brokerage: A stockbroker uses a computer program to be alerted on the price change of stocks. To trigger the alert, the current price x must differ from $16.25 by more than $0.75. Write an absolute value inequality to express this situation, then solve the inequality.

55. Medicine: In administering medicine, a physician has determined that the dosage x cc for the patient must differ from the prescription of 2.5 cc. by no more than 0.2 cc. Write an absolute value inequality that describes this situation, then solve this inequality to determine the lower and upper limits of the dose of medicine to be given to the patient.

56. Grading: At the beginning of the semester, a mathematics professor informed the class that in order to receive a grade of C in this algebra course, a student's score x must differ from 71.5% by no more than 8%. Write an absolute value inequality that describes this situation, then solve the inequality.

Developing and Extending the Concepts

In Problems 57 and 58, write an absolute value inequality that represents each statement. Then solve the inequality.

57. (a) The distance between x and 3 is 8.
   (b) The distance between x and 3 is almost 8.
58. (a) The distance between x and -5 is 4.
   (b) The distance between x and -5 is at least 4.

In Problems 59 and 60, write an absolute value inequality whose solution is shown in the graph.

59a. x
   -5 5

59b. x
   -4 4

60a. x
   -2 2

60b. x
   -3 3

 CHAPTER 1 REVIEW PROBLEM SET

In Problems 1–22, solve each equation.

1. $3(x + 5) - 6x = 6$
2. $6x + 2(x - 3) + 5 = -25$
3. $5[3t - (2 - t)] = 3(2t - 1)$
4. $3[(2x - 5) - (x + 1)] = 3$
5. $0.13(x + 300) = 61 - 0.09x$
6. $0.4(t + 200) = 89 - 0.5t$
7. $0.09(15 - y) + 0.04y = 0.07(25)$
8. $0.05(300 - y) - 0.1y + 105 = 0$
9. $\dfrac{3x}{5} - \dfrac{2x}{3} = 1$
10. $\dfrac{t}{3} + \dfrac{8}{5} = \dfrac{8 - t}{5}$
11. $\dfrac{2x + 1}{4} - \dfrac{3x - 4}{5} = \dfrac{1}{2}$
12. $\dfrac{x - 3}{10} + \dfrac{x - 1}{6} = \dfrac{3}{5}$
13. $\dfrac{2t + 6}{3} + \dfrac{2(t - 3)}{5} = 4$
14. $\dfrac{y - 3}{4} + \dfrac{y - 5}{6} = \dfrac{2y - 1}{3}$
15. $|3 - 2x| = 1$
16. $|7t + 4| = 3$
17. $|4x - 2| + 1 = 3$
18. $|2y + 3| - 4 = 1$
19. $|4 - y| = |y - 2|$
20. $|2x - 1| = |2x - 5|$
21. $\left|\dfrac{x - 3}{4}\right| + 4 = 4$
22. $\left|\dfrac{5t - 3}{2}\right| + 2 = 6$

In Problems 23–30, solve each equation for the indicated variable.

23. $y = -2x + 5$, for x
24. $8x - 12b = 4(2x - 3b)$, for x
25. $A = \dfrac{1}{2} h(a + b)$, for b
26. $S = \dfrac{n}{2}(f + 1)$, for f
27. $S = 2\pi r^2 + 2\pi rh$, for h
28. $A = 2hw + 2lw + lh$, for l
29. $\dfrac{1}{a} + \dfrac{1}{b} = \dfrac{1}{c}$, for b
30. $y = \dfrac{3 - x}{x - 1}$, for x

In Problems 31–48, solve each inequality. Graph the solution set and give the solution in interval notation.

31. $8x + 14 \leq 5x + 20$
32. $2(x - 3) \geq 3x + 4$
33. $3(x + 2) > 11 - 2(2 - x)$
34. $3(x - 4) + x < 2(6 + 2x)$

35. $2 \leq x + 5 \leq 8$

36. $2 \leq 3x - 4 \leq 8$

37. $2x - 1 < 5$ and $2x - 1 > 3$

38. $x < -1$ and $3x > x - 6$

39. $|2x - 3| \leq 7$

40. $|5 - 6x| \leq 29$

41. $|x + 2| - 5 < -3$

42. $|3x - 1| + 2 < -3$

43. $\left| \dfrac{x - 4}{2} \right| - 1 \geq 7$

44. $\left| \dfrac{5x - 5}{6} \right| - 2 \geq 3$

45. $\left| \dfrac{x + 2}{2} \right| - 1 > 5$

46. $\left| \dfrac{2x - 3}{5} + 7 \right| + 6 < 8$

47. $\left| \dfrac{3x + 4}{5} \right| - \dfrac{2}{5} \leq 1$

48. $\left| 2\left(\dfrac{3 - x}{5} \right) \right| - \dfrac{3}{5} < \dfrac{6}{5}$

49. For what values of x will each inequality be true?

(a) $|3x + 2| \leq 0$

(b) $|3x + 2| \geq 0$

50. (a) Criticize the following solution

$$\frac{3}{x} \geq 5$$

$$\frac{3}{x} \cdot x \geq 5 \cdot 2x$$

$$3 \geq 10x$$

$$\frac{3}{10} \geq x$$

(b) Write the correct solution of $\dfrac{3}{x} \geq 5$.

51. Solve the inequality $-3|2x - 5| \geq -9$.

52. For what values of x will the expression

$\dfrac{7}{\sqrt{4 - 2x}}$ be a real number?

53. **Age Difference:** Milt is 15 years older than Linda. The sum of their ages is 85. Find Linda's age.

54. **Sales Tax:** The state of Michigan has a 6% sales tax on nonfood items. What is the original price of a computer if the total cost of the computer, including tax, is $1166?

55. **Geometry:** A rectangular pool has a perimeter of 68 feet. If the length of the pool is 4 feet more than twice its width, find the length and the width of the pool.

56. **Basketball:** In winning a basketball game, a professional basketball team made only 5 free throws (one free throw is worth one point). The remainder of their points came from two and three point baskets. If the number of baskets from the field (two or three point baskets) totaled 45, and if the team scored 110 points in the game, how many three point baskets did the team make in that game?

57. **Roofing:** Two ingredients are used to make flat roof top filter. The first ingredient contains 35% coal tar, and the second contains 60% coal tar. How much of each ingredient should be mixed to make 60 tons of roof top filter that contains 50% coal tar?

58. **Investments:** A financial planner has $80,000 to invest. Part of the money is invested in a stock paying 9% annual simple interest for 2 years, the balance is invested in a mutual fund paying 7% annual simple interest for 3 years. If the total interest earned is $15,300, how much money is placed in each investment?

59. **Aircraft:** An aircraft manufacturer determines that one of its aircraft has twice as many seats as another. Find the number of seats for each aircraft if the total number of seats is 336.

60. **Biology:** The international whaling commission determined that the average weight of a blue whale is about 5 tons more than three times the average weight of a humpback whale. Determine the average weight of each whale if the total of their average weights is 117 tons.

61. **Painting:** A painter can paint a racquetball court in 8 hours. His assistant needs an additional 2 hours to paint the same court working by himself. How long will it take both of them working together to paint the court?

62. **Boating:** The speed of a motor boat in still water is 24 miles per hour. If the boat travels 27 miles upstream in the same time it takes to travel 45 miles downstream, what is the speed of the current?

63. **Telephone Fees:** A certain telephone service charges $1.57 for the first 3 minutes of a cellular call and $0.01 for each additional second. How long may one speak if the charge is not to exceed $4.72?

64. **Test Scores:** A math student has scores of 65, 68, 78, and 75 on her tests. Find the range of scores needed on her final examination so that she can earn a C in the course. Assume that the final examination counts as two tests and a grade of C is earned if the final course average is between 69.5 and 79.4.

In Problems 65–68 solve each equation or inequality and show the solution on a number line.

65. $-6 \le \frac{2}{3}(4 - 2x) \le 8$

66. $|2x - 5| = 2x - 5$

67. $|2x - 5| = -(2x - 5)$

68. $|x - 2| \le |x + 3|$

 CHAPTER 1 PRACTICE TEST

1. Solve each equation.
   (a) $3x + 2(45 - x) + 5 = 10$
   (b) $4(2y - 3) + 55 = 5(4 - 3y)$
   (c) $0.05(x + 2) + 0.10x = 2$

2. Solve each equation. Check for extraneous solutions.
   (a) $\frac{x}{8} + \frac{2x}{3} = \frac{19}{24}$
   (b) $\frac{3}{x - 3} + 4 = \frac{x}{x - 3}$

3. Solve each equation.
   (a) $|3y| = 24$
   (b) $|2 - 3x| = 7$
   (c) $|3x + 5| = 0$

4. (a) Solve for y, $3a + 6y = 8y - 9y$
   (b) Solve the formula $A = Prt + P$ for P.

5. Graph each solution set on a number line.
   (a) $x \le -3$
   (b) $-6 \le x \le 0$
   (c) $x \le 5$ or $x > 8$

6. Solve each inequality. Express each answer in interval notation and show the solution on a number line.
   (a) $-4x \ge 12$
   (b) $15x < -45$
   (c) $-5 < 3x + 7 \le 10$
   (d) $\frac{5 - 3x}{2} \le \frac{2x + 1}{3}$
   (e) $|x| \le 3$
   (f) $|x - 4| \ge 1$

7. **Wholesale Pricing:** A carpet dealer marks up her merchandise by 21% over the wholesale cost. If a carpet is sold for $1,815, what is the wholesale price?

8. **Triangle:** The perimeter of a triangle is 17 inches. One side is twice as long as another, and the third is 5 inches long. What is the length of the shortest side?

9. **Current:** A sailboat travels 7 miles per hour in still water. The sailboat travels 20 miles upstream in the same time as it travels 50 miles downstream. What is the speed of the current?

10. **Numbers:** If 1 is added to five times a number, the result is greater than 9 subtracted from three times the number. Find all possible such numbers.

Linear Graphs and Functions

Chapter Contents

The concept of slope is used to determine the *grade* of a road or the *change of elevation*. When a 4% grade is described, it means that the incline changes 4 feet vertically for every 100 feet of horizontal change.

During a heart stress test, the technician changes the grade of the treadmill. Some questions of interest to us would be:

(a) What is the grade of the treadmill if its vertical rise is 0.6 feet and the horizontal length of the treadmill is 8 feet?

(b) What is the rise of a 10% grade on a 6-foot treadmill?

These questions are discussed in example 7 of section 2.3 of this chapter.

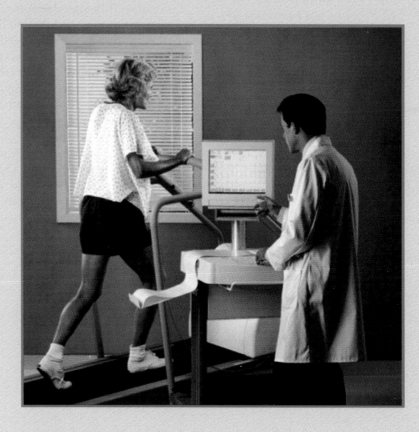

Objectives

1. Plot Points in the Plane
2. Graph by Point Plotting
3. Solve Linear Equations Graphically
4. Solve Applied Problems

2.1 Graphing Techniques

Many algebraic relationships can be classified if we show a visual picture that interprets them. The picture that shows the relationship between two variables is referred to as a graph. The mathematical development that allows us to construct a graph that describes a relationship is called the **Cartesian coordinate system.**

Plotting Points in the Plane

Earlier we have seen that a number line associates points on the line with real numbers. Extending this concept, we can associate points on the plane with ordered pairs of real numbers. One reference system used to represent this association is called a Cartesian, or rectangular coordinate system.

Figure 1 displays the *Cartesian,* or *rectangular coordinate system.* It consists of two perpendicular number lines, one horizontal and one vertical, called the coordinate axes. These axes intersect at a point called the **origin,** which is the zero point on each number line. The horizontal line is often called the *horizontal axis* or the *x axis.* The positive portion of the *x* axis is to the right of the origin and the negative portion is to the left. The vertical line is called the *vertical axis* or the *y axis,* with the positive portion above the origin and the negative portion below it. A plane endowed with a Cartesian coordinate system is called a **Cartesian plane or the *xy* plane.**

Any **ordered pair** of real numbers (a,b) can be represented as a point P in the coordinate system. The first number, a, called the **x coordinate** (or abscissa) of P, indicates the location of P to the right or left of the y axis. The second number, b, called the **y coordinate** (or ordinate) of P, indicates the location of P above or below the x axis. The coordinates of the origin are $(0,0)$. The point P with coordinates (a,b) is called the **graph** of the ordered pair (a,b). We *plot,* or *locate,* the position of (a,b) in the plane by placing a dot at the point P (Figure 1). The association between point P and ordered pair (a,b) seems so natural that in practice we write

$$P = (a,b)$$

For example if the x coordinate of a point is 3, and the y coordinate is 5, then the ordered pair representing that point is $(3,5)$ (Figure 2).

The coordinate axes divide the plane into four regions called **quadrants,** which are described and displayed in Figure 3. All points in the *xy* plane lie on one of the four quadrants, or on the coordinate axes.

Figure 1

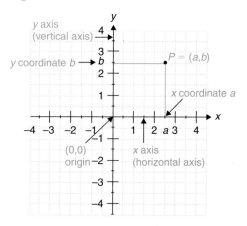

Figure 2

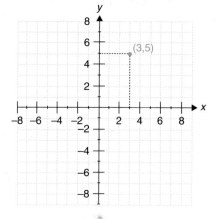

Figure 3

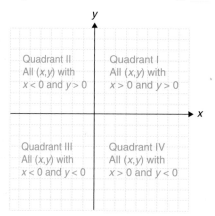

EXAMPLE 1   **Plotting Points**

Plot the following points on the same coordinate system. Also specify their quadrant.

(a) $A = (1,2)$ (b) $B = (2,1)$ (c) $C = (0,-5)$ (d) $D = (-4,5)$

(e) $E = (-2,-3)$ (f) $F = (0,0)$ (g) $G = (3,-4)$ (h) $H = (-5,0)$

Solution   Figure 4 shows the locations of all the points. Notice that the point $(1,2)$ and $(2,1)$ are different points as are $(0,-5)$ and $(-5,0)$. The points whose x coordinates are 0, C and F, are on the y axis and the points whose y coordinates are 0, F and H, are on the x axis. The point $F = (0,0)$ is called the origin and is on both axes. The quadrant to which each point belongs is in the chart.

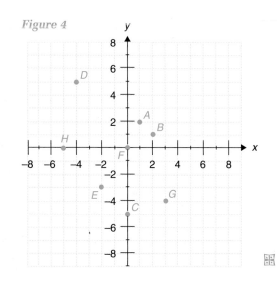

Figure 4

Point	Quadrant
$A = (1,2)$	I
$B = (2,1)$	I
$C = (0,-5)$	y axis
$D = (-4,5)$	II
$E = (-2,-3)$	III
$F = (0,0)$	Origin
$G = (3,-4)$	IV
$H = (-5,0)$	x axis

Graphing by Point Plotting

In Chapter 1, we solved equations involving one variable and represented their solutions on a number line. Now we extend this concept to equations involving two variables. The Cartesian coordinate system enables us to explore the connection between algebra and geometry; this connection is illustrated graphically by representing the solutions of equations involving two variables by ordered pairs of numbers. For example, a solution of the equation

$$y = 3x + 2$$

is given by $(-2,-4)$. If we substitute -2 for x and -4 for y in the above equation, the result is a true statement. However, if we substitute 1 for x and 4 for y, the result is a false statement and we say that $(1,4)$ is not a solution of the equation. It should be noticed that there are many other solutions to the above equation, in fact, there are an infinite number of solutions.

EXAMPLE 2

Verifying Solutions of an Equation

Determine which ordered pair is a solution of the equation $y = -3x + 5$.

(a) $(-2,11)$ (b) $(1,2)$ (c) $(2,1)$ (d) $(0,4)$

Solution

We substitute the first number for x and the second for y, so that

(a) $y = -3x + 5$	(b) $y = -3x + 5$	(c) $y = -3x + 5$	(d) $y = -3x + 5$
$11 = -3(-2) + 5$	$2 = -3(1) + 5$	$1 = -3(2) + 5$	$4 = -3(0) + 5$
$11 = 6 + 5$	$2 = -3 + 5$	$1 = -6 + 5$	$4 = 0 + 5$
$11 = 11$ True	$2 = 2$ True	$1 = -1$ False	$4 = 5$ False

The solutions are (a) $(-2,11)$ and (b) $(1,2)$.

A **graph** of an equation in two variables x and y is the set of points (x,y) whose co-ordinates satisfy the equation. One method of graphing equations is known as the *point plotting method.* In this method, we list some points that satisfy the equation in a table of values by choosing a few values of x and then find the corresponding values of y. We then draw a curve through the points and obtain the graph. The simplest curves in the plane are straight lines. Any equation in x and y whose graph is a straight line is called a **linear equation.** For instance

$$y = 2x - 4$$

is an example of a linear equation. The largest exponent that appears in either variable is 1. For this reason we call linear equations **first-degree equations.** The following equations

$$y = x, \quad y = 3x, \quad y = 2x - 6, \quad 3x - 2y = 6$$

are examples of linear equations or first-degree equations.

EXAMPLE 3

Graphing by Point Plotting

Graph $y = -x$ by point plotting.

Solution

We first find some ordered pairs that are solutions to the equation by selecting any values of x and finding the corresponding values of y. We will put these values in the form of a table, plot the points, and then draw the graph (Figure 5).

Figure 5

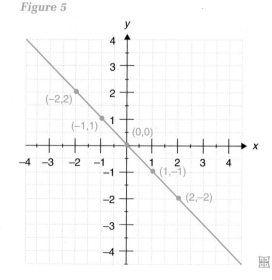

x	$-x = y$	(x,y)
-2	2	$(-2,2)$
-1	1	$(-1,1)$
0	0	$(0,0)$
1	-1	$(1,-1)$
2	-2	$(2,-2)$

EXAMPLE 4   **Graphing by Point Plotting**

Graph $y = 2/3\, x + 1$ by point plotting.

Solution   We select some values of x and find the corresponding values of y. Put these values in the form of a table, plot the points, and draw the graph (Figure 6).

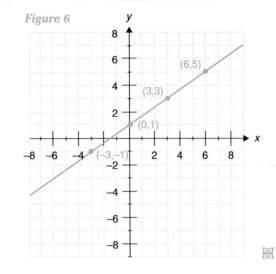

Figure 6

x	$y = \dfrac{2}{3}x = 1$	(x,y)
−3	$y = \dfrac{2}{3}(-3) + 1 = -2 + 1 = -1$	(−3,−1)
0	$y = \dfrac{2}{3}(0) + 1 = 0 + 1 = 1$	(0,1)
3	$y = \dfrac{2}{3}(3) + 1 = 2 + 1 = 3$	(3,3)
6	$y = \dfrac{2}{3}(6) + 1 = 4 + 1 = 5$	(6,5)

EXAMPLE 5   **Graphing by Point Plotting**

Graph $3x + y = 6$ by point plotting.

Solution   By solving the equation for y in terms of x it is easier to construct a table.

$$3x + y = 6 \quad \text{or} \quad y = -3x + 6$$

Now we make a table of values by choosing any three values for x and finding the corresponding values for y. Then we plot these points and graph the line (Figure 7).

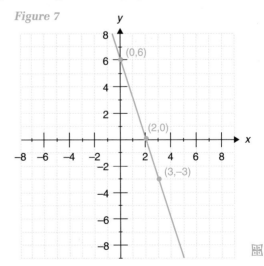

Figure 7

x	−3x + 6 = y	(x,y)
0	−3(0) + 6 = 6	(0,6)
2	−3(2) + 6 = 0	(2,0)
3	−3(3) + 6 = −3	(3,−3)

Solving Linear Equations Graphically

Earlier in Chapter 1, we solved linear equations in one variable algebraically such as

$$4x + 5 = 33.$$

Now we can graph this equation in the *xy* plane and solve it graphically. To solve this equation graphically, we first rewrite the equation in the equivalent form

$$4x + 5 - 33 = 0 \quad \text{or} \quad 4x - 28 = 0.$$

Then we graph the equation in the *xy* plane by point plotting the ordered pairs from the table below.

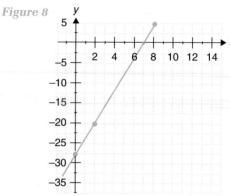

Figure 8

x	4x − 28 = y	(x,y)
0	4(0) − 28 = −28	(0,−28)
2	4(2) − 28 = −20	(2,−20)
8	4(8) − 28 = 4	(8,4)

Upon examining the graph of this equation (Figure 8), we notice that when $x = 7$, $y = 0$. By checking, the reader will verify that 7 is the solution of both the equation $4x - 28 = 0$ and the original equation $4x + 5 = 33$.

EXAMPLE 6   **Solving a Linear Equation Graphically**

Use a graph in the *xy* plane to solve the equation

$$-2x + 13 = 7.$$

Solution   We begin by writing an equivalent equation and setting it equal to zero.

$$0 = -2x + 6$$

Then, we graph the equation.

$$y = -2x + 6$$

by point plotting the ordered pairs from the table below

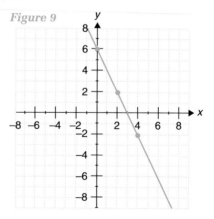

Figure 9

x	−2x + 6 = y	(x,y)
0	−2(0) + 6 = 6	(0,6)
2	−2(2) + 6 = 2	(2,2)
4	−2(4) + 6 = −2	(4,−2)

The solution of the equation

$$-2x + 13 = 7$$

is the value of *x* when $y = 0$ for the equation

$$y = -2x + 6.$$

Figure 9 shows the graph of $y = -2x + 6$. The line crosses the *x* axis when *x* is 3 and *y* is 0. Therefore, 3 is the solution of the equation.

Solving Applied Problems

If the point (2,5) is on the graph of an equation, it means that when 2 is substituted for x in the equation, the resulting value of y is 5. The converse is also true, that is, if 5 is substituted for y then the resulting value of x is 2. When reading a graph, we know the relationship from either view. Consider the following example.

EXAMPLE 7   **Finding a Temperature Conversion**

The relationship between Fahrenheit temperature F and Celsius temperature C is given by the model

$$F = 1.8C + 32.$$

(a) Sketch by point plotting the graph of this model using C as the horizontal axis and F as the vertical axis.

(b) Use the graph to determine the Fahrenheit temperature F when the Celsius temperature is 10°.

(c) Use the graph to find the Celsius temperature when the Fahrenheit reading is 5°.

Solution

Figure 10

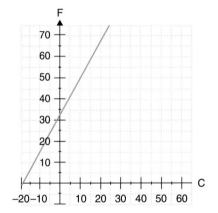

C	$1.8C + 32 = F$	(C,F)
-10	$1.8(-10) + 32 = 14$	$(-10,14)$
-5	$1.8(-5) + 32 = 23$	$(-5,23)$
5	$1.8(5) + 32 = 41$	$(5,41)$

(a) Figure 10 is the graph by point plotting of the equation $y = 1.8x + 32$.

(b) To find the Fahrenheit temperature equivalent to 10°C, we look for a point on the graph with first coordinate 10. The point is (10,50), so that the equivalent Fahrenheit reading is 50°.

(c) If the Fahrenheit reading is 5°, we are looking for a point on the graph whose second coordinate is 5. The point is $(-15,5)$; so the equivalent temperature to 5°F is -15°C.

EXAMPLE 8   **Determining Cable Television Charges**

A cable television company charges a $50 installation fee plus a $25 monthly service fee. The total charge C (in dollars) for its services is given by the linear model

$$C = 50 + 25t$$

where t is the number of months for the service.

(a) Sketch the graph of this equation by point plotting using t as the horizontal axis and C as the vertical axis.

(b) From the graph, find the total cost for 9 months of service.

(c) If the total cost of service is $175, for how many months is the service used?

Solution

(a) Figure 11 shows the graph that models this situation, obtained by plotting the ordered pairs in the table below.

(b) On the graph we are looking for a point with first coordinate 9. The second coordinate of this point is 275. Therefore, the total charges are $275.

(c) To find t when $C =$ $175, we are looking for a point on the graph whose second coordinate is 175. The point is (5,175). Therefore, it takes 5 months of service to spend $175.

Figure 11

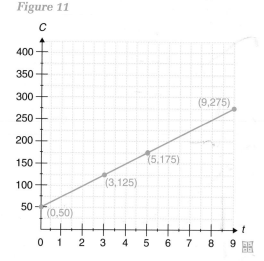

t	$C = 50 + 25t$	(t,C)
0	$C = 50 + 25(0) = 50$	(0,50)
3	$C = 50 + 25(3) = 125$	(3,125)
5	$C = 50 + 25(5) = 175$	(5,175)
9	$C = 50 + 25(9) = 275$	(9,275)

◆ PROBLEM SET 2.1

Mastering the Concepts

In Problems 1 and 2, plot the given set of points, then specify their quadrant location.

1. (3,4), (−2,4), (5,−1), (0,7), (−7,0), and (0,−6).

2. (3,−2), (−2,3), (3,0), (0,−3), (−3,0), and (0,3).

In Problems 3–8, find the coordinates of the point A as described.

3. A is located three units to the left of the y axis and four units above the x axis.

4. A is located seven units to the right of the y axis and two units below the x axis.

5. A is located on the negative x axis, eight units from the origin.

6. A is located on the positive y axis, five units from the origin.

7. The coordinates of A are equal and A is located in the third quadrant five units below the x axis.

8. The coordinates of A are equal and are opposite in sign. A is located in the fourth quadrant and 3 units below the x axis.

In Problems 9–12, verify if the given points satisfy the given equations.

9. $y = -4x + 9$
(a) (2,−1)
(b) (−2,17)

10. $y = 5x - 3$
(a) (−2,−13)
(b) (4,23)

11. $2y + 3x = 6$
(a) (3,0)
(b) (0,2)

12. $4x - 3y = 24$
(a) (0,−8)
(b) (6,0)

In Problems 13–30, graph each equation by point plotting.

13. $y = -2x + 2$

14. $y = 2x - 1$

15. $y = \dfrac{1}{2}x + 3$

16. $y = -\dfrac{1}{3}x + 2$

17. $3x + y = 2$

18. $-x + 2y = 4$

19. $3x - 2y = 6$

20. $2x - 3y = -6$

21. $2y - 6 = 0$

22. $3y + 4 = 0$

23. $y + 2 = 3(x - 1)$

24. $y - 1 = -2(x + 4)$

25. $\dfrac{x}{3} + \dfrac{y}{4} = 1$

26. $\dfrac{y}{2} - \dfrac{3x}{5} = 1$

27. $2x + 4 = 0$

28. $3x - 9 = 0$

29. $y = -x$

30. $y = 3x$

In Problems 31–38, solve each equation graphically.

31. $5x + 7 = 32$

32. $2x - 3 = 31$

33. $\dfrac{2x}{3} + \dfrac{1}{2} = -\dfrac{3}{2}$

34. $\dfrac{x}{3} + 1 = -\dfrac{2}{3}$

35. $-4x - 19 = 2x + 5$

36. $6x + 13 = 8x + 15$

37. $4x = 0$

38. $-6y = 12$

In Problems 39–42, match the graph to the equation.

39. $y = -3$

40. $x = -3$

41. $x = 4$

42. $y = 4$

(a)
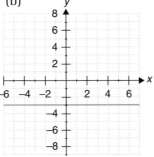

(b)

(c)
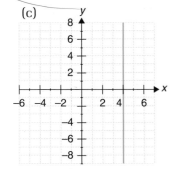

(d)
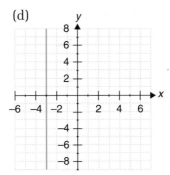

Applying the Concepts

43. Car Rental: A car rental company charges $80 per week plus 10 cents for each mile driven on its midsize cars. The total weekly cost C (in dollars) is given by the linear model

$$C = 80 + 0.10m$$

where m represents the weekly mileage driven.

(a) Sketch the graph of this equation using m as the horizontal axis and C as the vertical axis.

(b) What is the cost C when the car has been driven 350 miles in a week?

(c) How many miles are driven if the cost is $95?

44. Photo Development: A photo development center charges $2.00 for each roll of film

developed plus 20 cents for each print. The total charge C (in dollars) is given by the linear model

$$C = 2 + 0.20n$$

where n represents the number of prints developed.

(a) Sketch the graph of this equation using n as the horizontal axis and C as the vertical axis.

(b) What is the total cost for developing and printing 25 exposures?

(c) If the cost is $9, how many prints were made?

45. Cellular Phone Cost: A cellular phone company charges $50 for each phone plus $25 per month under a special sale plan. The total cost C (in dollars) of operating the phone is given by the linear model

$$C = 50 + 25t$$

where t represents the number of months of operation of a cellular phone.

(a) Sketch the graph of this equation using t as the horizontal axis and C as the vertical axis.

(b) How many months has the phone been in service if the total cost is $250?

(c) What is the cost of the service after 18 months?

46. Wages: A fitness center offered its manager a $2,000 monthly salary plus 5% of the net monthly income of the center. The manager's monthly salary S (in dollars) is given by the linear model

$$S = 2000 + 0.05n$$

where n represents the monthly net income of the center (in dollars).

(a) Graph this model with n as the horizontal axis and S as the vertical axis.

(b) If the manager's monthly income is $3500, what is the net income of the center?

(c) If the center's net income for a given month is $35,000, what will the manager's salary be for that month?

Developing and Extending the Concepts

In Problems 47–52, determine the quadrant(s) in which each point is located.

47. $(x,3)$, $x < 0$

48. $(-4,y)$, $y > 0$

49. (x,y), $x > 0$, $y < 0$

50. $(-x,y)$, $x < 0$, $y > 0$

51. (x,y), $xy > 0$

52. (x,y), $xy < 0$

In Problems 53–56, graph by point plotting and then find x if $y = 0$ and find y if $x = 0$.

53. $\dfrac{1}{3}x + \dfrac{3}{4}y = 9$

54. $\dfrac{x}{5} - \dfrac{y}{3} = 1$

55. $x = 2y + 20$

56. $\dfrac{x}{12} - \dfrac{y}{9} = 1$

Objectives

1. Use Intercepts to Graph Linear Equations
2. Graph Special Linear Equations in One Variable
3. Solve Equations and Inequalities Involving Absolute Values Graphically
4. Solve Applied Problems

2.2 Graphing Linear Equations Using Intercepts

An equation of the form

$$Ax + By = C$$

where A, B, and C are constant real numbers, and A and B are not both zero is called a **linear equation in standard form.** Examples are

$$3x + 2y = 6, \quad -x + 3y = 6, \quad \text{and} \quad 4x - 3y = 12.$$

Using Intercepts to Graph Linear Equations

At times it is easier to draw the graph of a linear equation in standard form by first determining where the graph crosses the x and y axes. Figure 1 shows the graph of the equation $3x + 2y = 6$ which crosses the x axis at the point (2,0). We call this point the **x intercept point,** and 2 is the **x intercept** of the graph. The graph also crosses the y axis at (0,3), which is called the **y intercept point,** and 3 is the **y intercept** of the graph.

Figure 1

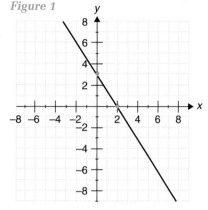

In general, Table 1 below explains how the x and y intercepts may be determined algebraically.

TABLE 1 Intercepts of the Graph of an Equation that Relates x and y

Terminology	Description	Illustration	Determination of Intercepts
x intercept	The x coordinate of the points where the graph intersects the x axis.		Set $y = 0$ and solve for x.
y intercept	The y coordinate of the points where the graph intersects the y axis.		Set $x = 0$ and solve for y.

EXAMPLE 1

Using Intercepts to Graph an Equation

Find the intercepts and sketch the graph of the equation

$$3x - 4y = 12.$$

Solution   The x intercept of the graph of the equation $3x - 4y = 12$ is obtained by setting $y = 0$ and solving for x.

$$3x - 4(0) = 12$$
$$3x = 12$$
$$x = 4$$

To find the y intercept, set $x = 0$ and solve for y.

$$3(0) - 4y = 12$$
$$-4y = 12$$
$$y = -3$$

Figure 2

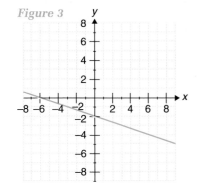

Thus the x and y intercepts of the graph are 4 and -3, respectively. Figure 2 shows the graph of the equation.

We must be careful when graphing linear equations using only two points like the intercept points. If one of the points is wrong, the graph will be wrong. It is always good to plot a third point as a check point to ensure the correctness of the graph.

EXAMPLE 2   **Using Intercepts to Graph an Equation**

Find the intercepts and sketch the graph of the equation

$$y = -\frac{1}{3}x - 2.$$

Solution   To find the y intercept, set $x = 0$ and solve for y, so that

$$y = -\frac{1}{3}(0) - 2 = -2$$

To find the x intercept, set $y = 0$ and solve for x, so that

Figure 3

$$0 = -\frac{1}{3}x - 2$$

$$3(0) = 3\left(-\frac{1}{3}x - 2\right) \qquad \text{Multiply by the common denominator, 3.}$$

$$0 = -x - 6$$

$$x = -6$$

Therefore, the x and y intercepts of the graph of the equation are -6 and -2, respectively. We also plot the check point $(3, -3)$, since this point is also on the graph of the equation. Figure 3 shows the graph of the equation.

Graphing Special Linear Equations in One Variable

Graphs of equations of the form

$$Ax + By = 0$$

where A and B are not both zero, have the same x and y intercept, namely (0,0). To graph such an equation, we use the origin as one point and select at least two other values of x with the corresponding values of y. The next example shows this method.

EXAMPLE 3   **Graphing a Special Linear Equation**

Sketch the graph of the equation

$$6x - 4y = 0.$$

Solution   If we set $x = 0$, we find that $y = 0$, so the graph contains the origin $(0,0)$. We will solve the equation for y and then use a table to find two other points on the graph. The graph of this equation is in Figure 4.

Figure 4

$$6x - 4y = 0$$
$$-4y = -6x$$
$$y = \frac{3}{2}x$$

x	$y = \dfrac{3}{2}x$	(x,y)
2	$y = \dfrac{3}{2}(2) = 3$	$(2,3)$
−4	$y = \dfrac{3}{2}(-4) = -6$	$(-4,-6)$

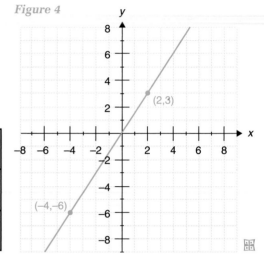

Other special linear equations of the form

$$x = h \quad \text{and} \quad y = k$$

where h and k are constants, are graphed by writing each equation as follows:

$$x = h, \quad 1x + 0y = h$$
$$y = k, \quad 0x + 1y = k.$$

These graphs will always be *vertical* and *horizontal* lines respectively, as the next example shows.

EXAMPLE 4   **Graphing Special Linear Equations**

Sketch the graph of each equation

(a) $y = -3$ (b) $x = 2$

Solution   (a) The equation $y = -3$ can be written in the form

$$y = 0x - 3.$$

Thus, for any value of x selected, y will be -3. Figure 5a shows the graph of the equation $y = -3$.

(b) The equation $x = 2$ can be written as

$$1x + 0y = 2.$$

Thus, for every value of y selected, x will have a value of 2. Figure 5b shows the graph of the equation.

Figure 5

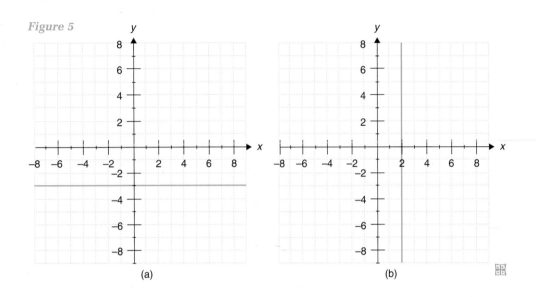

(a)   (b)

Note that the x axis is a horizontal line with equation $y = 0$. The y axis is a vertical line with equation $x = 0$. Table 2 illustrates the characteristics that apply to all other equations of horizontal and vertical lines. Assume $h \neq 0$ and $k \neq 0$.

TABLE 2

Equation	Graph	x Intercept	y Intercept
$y = k$	horizontal line	none	k
$x = h$	vertical line	h	none

Solving Equations and Inequalities Involving Absolute Values Graphically

Equations and inequalities in one variable involving absolute values such as

$$|x - 3| = 7$$
$$|x - 3| < 7$$

or

$$|x - 3| > 7$$

can also be solved graphically. The procedure is somewhat similar to the one in the previous section. This is illustrated in the following examples.

EXAMPLE 5   **Solving an Absolute Value Equation Graphically**

Use the graph of $y = |x - 5| - 3$ (Figure 6) to solve the equation

$$|x - 5| = 3.$$

Figure 6

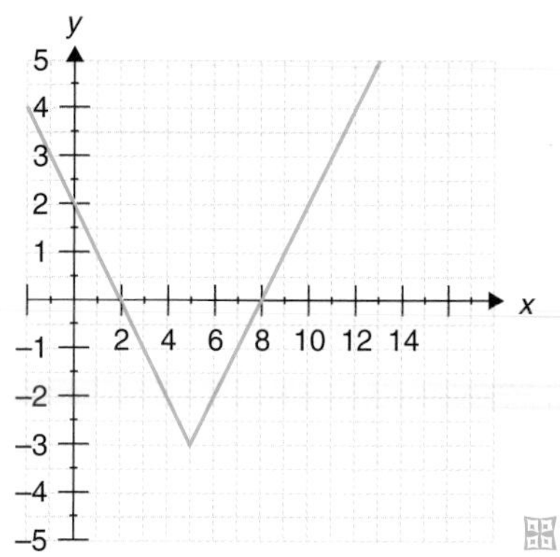

Solution   The equations $|x - 5| = 3$ and $|x - 5| - 3 = 0$ are equivalent equations. We are looking on the graph (Figure 6) for the point, or points, where the second coordinate is 0. In other words, we are looking for the x intercepts. This V-shaped graph crosses the x axis twice, at $(2,0)$ and at $(8,0)$. Therefore, the values of x which make y zero are 2 and 8.

EXAMPLE 6   **Solving an Inequality Graphically**

Use the graph of $y = |x - 5| - 3$ (Figure 6) to solve the inequality

$$|x - 5| < 3.$$

Solution   The inequalities $|x - 5| < 3$ and $|x - 5| - 3 < 0$ are equivalent inequalities. On the graph of

$$y = |x - 5| - 3$$

we observe the values of x which make $y < 0$ (Figure 6). That is, the x values of points on the graph which are below the x axis. Such values of x are between 2 and 8 or in the interval $(2,8)$. In the table below, we pick values of x less than 2, between 2 and 8, and greater than 8 to check our solution.

x	Left Side	Right Side	True/False
0	$|0 - 5| = 5$	3	$5 < 3$ False
3	$|3 - 5| = 2$	3	$2 < 3$ True
11	$|11 - 5| = 6$	3	$6 < 3$ False

Solving Applied Problems

In business, profit is the difference between revenue and cost. That is, the difference between the money collected or taken in and the money paid out in expenses to run the business. The following example shows the relationship between intercepts on a graph and profit.

EXAMPLE 7   **Graphing the Profit Line**

Suppose the profit of a company which makes boxes is modeled by the equation

$$P = 20x - 100, \quad 0 \le x \le 10$$

where x (in hundreds per day) is the number of boxes produced and P (in dollars) is the daily profit.

(a) Graph this model by finding the intercepts where x is the horizontal axis and P is the vertical axis.

(b) Interpret the meaning for the business of these intercepts.

Solution   (a) To find the intercepts, first let $x = 0$.

$$P = 20(0) - 100 = -100$$

Then let $P = 0$.

$$0 = 20x - 100$$
$$-20x = -100$$
$$x = 5$$

Thus the intercepts are $(0,-100)$ and $(5,0)$. The graph is in Figure 7.

Figure 7

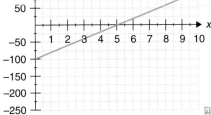
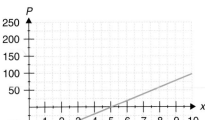

(b) The intercept $(0,-100)$ means that when 0 boxes are produced the profit is -100. That is, there is a loss of \$100. This is called the fixed over-head, even though nothing is pro-duced, there are still expenses to be paid. The second intercept $(5,0)$ is called the break even point because this is the level of production at which income equals expenses. All the bills are paid but nothing is left over.

PROBLEM SET 2.2

Mastering the Concepts

In Problems 1–20, find the x and y intercepts of each equation and sketch the graph.

1. $3x - 4y = 24$
2. $2y = 3x + 6$
3. $2x = 3y + 1$
4. $-6x + 3y = 12$
5. $2x - 7y = 21$
6. $4x + 10y = 20$
7. $-6x + 3y = 12$
8. $2y - x = 5$
9. $10y = -6x + 14$
10. $3y + x = 4$
11. $2y - x = 0$
12. $y - 2x = 1$
13. $8y + 24 = 0$
14. $-2y = -7$
15. $2x - 3 = 0$
16. $4y + 5 = 0$
17. $2y + 7 = 0$
18. $4x - 8 = 0$
19. $3x - 1 = 0$
20. $2x + 3 = 0$

In Problems 21–24, find the x and y intercepts of the graphs.

21.
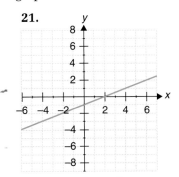

22.
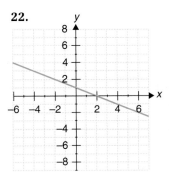

23.
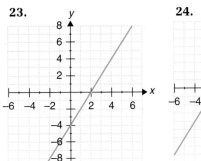

24.
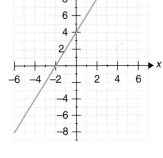

25.
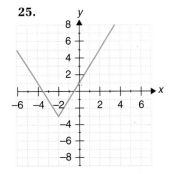

26.
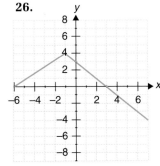

In Problems 27–30, sketch a graph of a line with the following properties.

27. x intercept of 4 and y intercept of -2
28. x intercept of -2 and y intercept of -5

29. x intercept of -1.5 and y intercept of -2.5

30. x intercept of 8.1 and y intercept of -1

In Problems 31–34 use the given graphs to find the solution of the equations and inequalities.

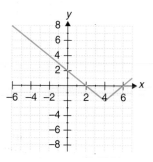

31. $y = |x - 4| - 2$
   (a) $|x - 4| = 2$
   (b) $|x - 4| < 2$

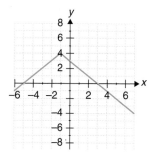

32. $y = -|x + 4| + 4$
   (a) $|x + 1| = 4$
   (b) $|x + 1| < 4$

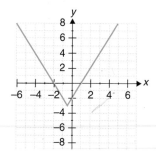

33. $y = |2x + 1| - 3$
   (a) $|2x + 1| = 3$
   (b) $|2x + 1| > 3$

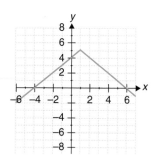

34. $y = -|x - 1| + 5$
   (a) $|x - 1| = 5$
   (b) $|x - 1| > 5$

Applying the Concepts

35. Profits: A service company's business plan includes a profit graph as seen in Figure 8. The horizontal axis represents the number of service calls made (x) per week and the vertical axis represents profit in thousands of dollars (P) per week.
   (a) When no service calls are made, what is the profit?

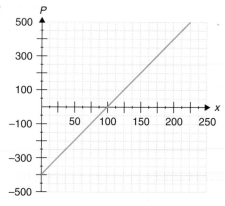

Figure 8

   (b) How many service calls must be made to have a profit of zero?

36. Profit: Suppose the same company in problem 35 has the profit graph in Figure 8.
   (a) When 75 service calls are made, what is the profit?
   (b) How many service calls must be made to have a profit of 100,000?

37. Tank: A water tank is being filled according to the model in the given graph. The horizontal axis (x) is the number of hours that the pump is running and the vertical axis is the depth of the water in feet (y).

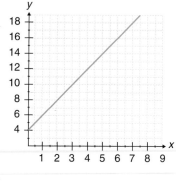

   (a) How much water is in the tank before the pump starts?
   (b) How deep is the water when the pump has been running for 3 hours?
   (c) How long will it take the pump to achieve a depth of 6 feet in the tank?

38. Tank: A water tower drains water according to the model in the given graph. The horizontal axis (x) is the number of hours that the drain is open and the vertical axis is the depth of the water in feet (y).

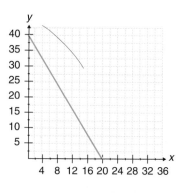

   (a) How deep is the water in the tank before the drain is opened?

(b) How deep is the water after 4 hours?

(c) How long will it take to empty the tank?

Developing and Extending the Concepts

39. Given the graph of $y = x^2 - 2x - 8$, find

(a) x intercepts

(b) y intercept

(c) Solve the inequality $x^2 - 2x - 8 < 0$.

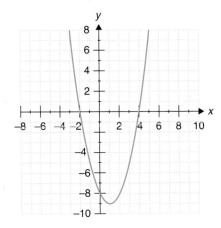

40. Given the graph of $y = 15 - 2x - x^2$, find

(a) x intercepts

(b) y intercept

(c) Solve the inequality $15 - 2x - x^2 > 0$.

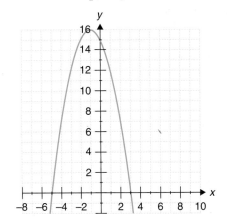

41. One of the forms of depreciation for income tax purposes is the **straight line depreciation method** in which the value V dollars of a property is expressed in terms of t years by a linear equation. The graph in Figure 9 shows the value of a color printer purchased by a small business which is being depreciated over a period of 6 years. Use the graph in Figure 9 to find:

(a) the original purchase price of the color printer.

(b) the depreciated value of the color printer after 6 years.

(c) After how many years is the value of the printer $3000?

(d) When is the value of the printer zero?

Figure 9

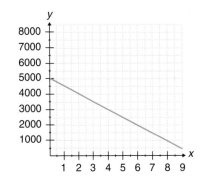

Objectives

1. Find the Slope of a Line

2. Determine Parallel and Perpendicular Lines

3. Solve Applied Problems

2.3 Graphing Linear Equations Using Slopes

In everyday life, we speak of a highway ramp that has a *grade,* the *pitch* of a roof, or the *steepness* of a ski slope. In each case, the measurement of the steepness of a line is an important concept in many areas of mathematics.

Finding the Slope of a Line

Informally, the idea of the slope is used when we refer to the steepness of a hill, the grade of a road, or the pitch of a roof. If a roof has a $10 - 12$ pitch, it means that

Figure 1

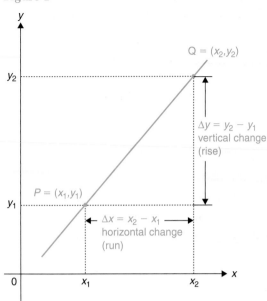

for every 12 feet of length, there is a rise of 10 feet in the roof. In mathematics, the word slope has a precise meaning. Suppose that a nonvertical line contains two different points $P = (x_1, y_1)$ and $Q = (x_2, y_2)$. As we move along the line from P to Q, the **horizontal change** (also called the **run**) between P and Q is

$$\Delta x = x_2 - x_1$$

and the **vertical change** (also called the **rise**) is

$$\Delta y = y_2 - y_1$$

(Figure 1) where the symbols Δx and Δy are read as "delta x" and "delta y" respectively. The ratio of the vertical change to the horizontal change is the **slope** of a line.

More formally, we have the following definition.

Definition of the Slope of a Line

If $P = (x_1, y_1)$ and $Q = (x_2, y_2)$ are two different points of a nonvertical line, the **slope m** of the line containing P and Q is given by

$$m = \frac{\text{change in } y}{\text{change in } x} = \frac{\Delta y}{\Delta x} = \frac{y_2 - y_1}{x_2 - x_1}.$$

EXAMPLE 1   **Finding the Slope of a Line**

Use the coordinates of the given points on the graph of each line in Figure 2 to determine its slope.

Figure 2

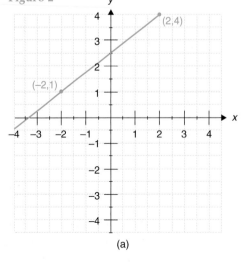

(a)

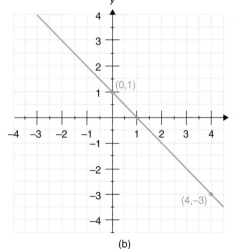

(b)

Solution (a) Using Figure 2a, the coordinates of the given points are $(-2,1)$ and $(2,4)$, so we choose $x_1 = -2$, $y_1 = 1$, $x_2 = 2$, and $y_2 = 4$, and the slope m is given by

$$m = \frac{\Delta y}{\Delta x} = \frac{y_2 - y_1}{x_2 - x_1} = \frac{4 - 1}{2 - (-2)} = \frac{3}{4}.$$

Notice, if we choose $x_1 = 2$, $y_1 = 4$, $x_2 = -2$, and $y_2 = 1$, the slope has the same value.

$$m = \frac{\Delta y}{\Delta x} = \frac{y_2 - y_1}{x_2 - x_1} = \frac{1 - 4}{-2-2} = \frac{-3}{-4} = \frac{3}{4}$$

Observe that the line rises from left to right.

(b) From Figure 2b, the coordinates of the given points are $(0,1)$ and $(4,-3)$, so we choose $x_1 = 0$, $y_1 = 1$, $x_2 = 4$ and $y_2 = -3$, and the slope m is given by

$$m = \frac{\Delta y}{\Delta x} = \frac{y_2 - y_1}{x_2 - x_1} = \frac{-3 - 1}{4 - 0} = -1.$$

Here the line falls from left to right.

The next example shows that a line is horizontal if $y_1 = y_2$ and it is vertical if $x_1 = x_2$.

EXAMPLE 2   **Finding Slopes of Horizontal and Vertical Lines**

Sketch the graph of each line containing the given points and find the slope.

(a) $(-1,4)$ and $(3,4)$ (b) $(2,-3)$ and $(2,2)$

Solution   Figure 3a shows the horizontal line containing the points $(-1,4)$ and $(3,4)$.

Figure 3b shows the vertical line containing the points $(2,-3)$ and $(2,2)$.

Figure 3

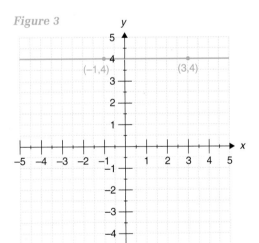

(a)

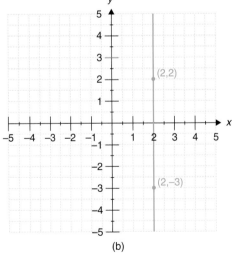

(b)

(a) The slope of the horizontal line in Figure 3a is given by

$$m = \frac{\Delta y}{\Delta x} = \frac{y_2 - y_1}{x_2 - x_1} = \frac{4 - 4}{3 - (-1)} = \frac{0}{4} = 0.$$

(b) The slope of the vertical line in Figure 3b is given by

$$m = \frac{\Delta y}{\Delta x} = \frac{y_2 - y_1}{x_2 - x_1} = \frac{2 - (-3)}{2 - 2} = \frac{5}{0} \text{ which is undefined.}$$

It is very important to be able to visualize the difference between positive and negative slopes. As Figure 4a shows, the lines that have positive slopes rise from left to right. Also, as Figure 4b shows, the lines that have negative slopes fall from left to right.

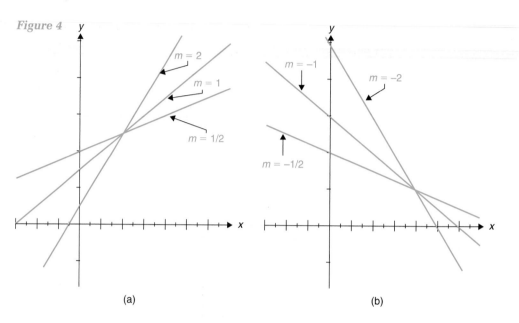

Figure 4

(a)   (b)

Table 1 describes the connection between slopes and directions of lines.

TABLE 1 Slope and Direction of a Line

Slope Values (m)	Line Direction	Sample Graph
Positive	Rises from left to right	$m > 0$
Negative	Falls from left to right	$m < 0$
Zero	Horizontal	$m = 0$
Not defined	Vertical	m undefined

EXAMPLE 3

Using Slope to Graph a Line

Sketch the line that contains the point $(-2,3)$ and has the given slope.

(a) $m = \dfrac{3}{5}$ (b) $m = -\dfrac{5}{3}$

Solution

(a) The slope is $m = \dfrac{\Delta y}{\Delta x} = \dfrac{3}{5}$.

If we start at the point $(-2,3)$ and move 3 units up and 5 units to the right, we find a second point $(3,6)$ on this line. We draw a line containing these two points (Figure 5a).

(b) The slope is $m = \dfrac{\Delta y}{\Delta x} = -\dfrac{5}{3} = \dfrac{-5}{3}$.

Once again, we start at the point $(-2,3)$ and move 5 units down and 3 units to the right, we find a second point $(1,-2)$ on this line. We draw a line containing the two points (Figure 5b).

Figure 5

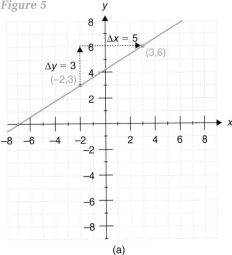

(a)   (b)

We can determine the slope of a line directly from its equation as the next example shows.

EXAMPLE 4

Finding the Slope of a Line Defined by an Equation

Find the slope of the line defined by the equation $2x - 3y = 12$.

Figure 6

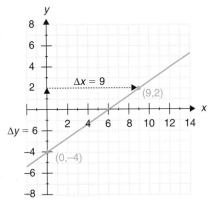

Solution

The slope of this line can be found by using any two points on the line. For instance,

if $x = 0$ then $y = -4$

and

if $x = 9$, then $y = 2$ (Figure 6).

Using the points $(0,-4)$ and $(9,2)$ we find

$$m = \frac{\Delta y}{\Delta x} = \frac{y_2 - y_1}{x_2 - x_1}$$

$$= \frac{2 - (-4)}{9 - 0} = \frac{6}{9} = \frac{2}{3}.$$

If we were to solve the equation in Example 4 for y we would have

$$y = \frac{2}{3}x - 4.$$

We observe that the coefficient of x is 2/3 which is the same as the slope of the line.

The constant, -4, of the equation is the y intercept.

In general, the equation of a line with slope m and y intercept b can be written in the form

$$y = mx + b.$$

This equation is called the **slope-intercept form** of an equation of a line.

EXAMPLE 5   **Using an Equation to Find the Slope**

Express the equation $2x - 3y + 6 = 0$ in slope-intercept form and determine the slope and y intercept. Sketch the graph.

Solution   We solve the equation for y in terms of x.

$$2x - 3y + 6 = 0$$

$$-3y = -2x - 6$$

$$y = \frac{-2x}{-3} - \frac{6}{-3}$$

$$y = \boxed{\frac{2}{3}} x + \boxed{2}$$

$$\underset{\boxed{\text{slope}}}{\swarrow} \qquad \underset{\boxed{y \text{ intercept}}}{\searrow}$$

Thus the slope is 2/3 and the y intercept is 2. Since the slope is 2/3, we start at the point $(0,2)$ and move 2 units up and 3 units to the right to find a second point $(3,4)$ (Figure 7).

Figure 7

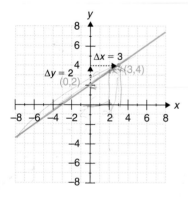

Determining Parallel and Perpendicular Lines

The concept of slope provides a tool for determining whether lines in the Cartesian plane are parallel or perpendicular. To explore this idea, consider the relationship between the graphs of the lines, each with slope 2, and defined by the equations

$$y = 2x \quad \text{and} \quad y = 2x + 4$$

Table 2 lists some solutions of both equations.

TABLE 2

x	−2	−1	0	1	2
$y = 2x$	−4	−2	0	2	4
$y = 2x + 4$	0	2	4	6	8

By plotting these points and drawing the two lines on the same coordinate system, we see that the graph of $y = 2x + 4$ can be obtained by "shifting" the graph of $y = 2x$ four units upward (Figure 8a). So the two lines are parallel.

Now consider the graph of the line L whose equation is given by $y = 2/3\, x + 1$ and whose slope is $m = 2/3$ (Figure 8b). If we rotate this line 90° in the counter clockwise direction, we get a new line L_1 perpendicular to L.

Figure 8

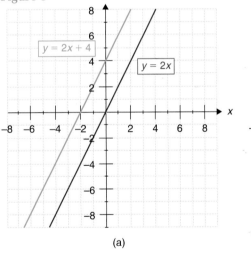

(a)   (b)

By looking at the grid, we see that we can select two points, P_1 and Q_1, on the new line L_1 whose coordinates are $(-4,7)$ and $(-2,4)$ respectively. Then the slope of the new line L_1 is given by

$$m_1 = \frac{4 - 7}{-2 - (-4)} = -\frac{3}{2}.$$

Notice that the rise and run of the new line L_1 are interchanged compared to L, that is, the slope m_1 is the negative reciprocal of m, so that

$$m \cdot m_1 = \left(\frac{2}{3}\right)\left(\frac{-3}{2}\right) = -1.$$

In general, we have the following properties:

Properties of Parallel and Perpendicular Lines

Suppose that L_1 and L_2 are two distinct nonvertical lines with slopes m_1 and m_2 respectively.

1. L_1 is parallel to L_2 if and only if $m_1 = m_2$.
2. L_1 is perpendicular to L_2 if and only if

$$m_1 = -\frac{1}{m_2} \quad \text{or} \quad m_1 \cdot m_2 = -1.$$

All vertical lines are parallel to each other and are perpendicular to horizontal lines.

EXAMPLE 6   **Determining if Two Lines Are Parallel or Perpendicular**

Determine whether L_1 and L_2 are parallel or perpendicular. Then sketch the graphs of the lines.

(a) L_1 contains the points $(-4,1)$ and $(0,-5)$ and L_2 contains the points $(-2,5)$ and $(2,-1)$

(b) L_1: $-3x + 2y - 3 = 0$ and L_2: $2x + 3y - 10 = 0$

Solution   The lines L_1 and L_2 that contain the points are graphed in Figure 9a. The slopes m_1 and m_2 of L_1 and L_2 are respectively given by

$$m_1 = \frac{\Delta y}{\Delta x} = \frac{y_2 - y_1}{x_2 - x_1} = \frac{-5 - 1}{0 - (-4)} = -\frac{6}{4} = -\frac{3}{2}$$

and

$$m_2 = \frac{\Delta y}{\Delta x} = -\frac{-1 - 5}{2 - (-2)} = -\frac{6}{4} = -\frac{3}{2}.$$

The two lines, L_1 and L_2, are parallel because their slopes are equal (Figure 9a).

(b) We write both equations in slope-intercept form and compare their slopes.

$$L_1: -3x + 2y - 3 = 0 \qquad\qquad L_2: 2x + 3y - 10 = 0$$

$$2y = 3x + 3 \qquad\qquad\qquad 3y = -2x + 10$$

$$y = \frac{3}{2}x + \frac{3}{2} \qquad\qquad\qquad y = \frac{-2}{3}x + \frac{10}{3}$$

Since $m_1 \cdot m_2 = \dfrac{3}{2} \cdot \dfrac{-2}{3} = -1$, the two lines are perpendicular (Figure 9b).

Solving Applied Problems

The concept of slope is used in many applications and models in which two quantities are changing at a fixed rate. For instance, numbers such as 2% and 4% are

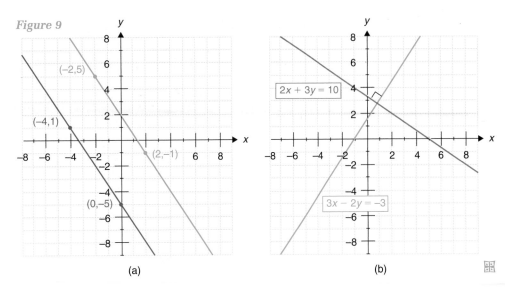

Figure 9

(a)

(b)

often used to refer to the **grade** of a road or the **change of elevation** on a hill or a mountain. That is the steepness of a road. When we describe a road as having a 4% grade, we mean that the road changes 4 feet vertically for every 100 feet of horizontal change. Fitness centers also use the grade of a treadmill to measure its effect on the heart rate of their members. Another interpretation of slope is **the rate of change of one quantity per unit change of another,** such as fuel efficiency in miles per one gallon or wages in dollars per one hour. Also, the **pitch** of the roof of a house is an application of slope. The ratio of the rise to run determines the pitch or steepness of a roof.

EXAMPLE 7

Finding a Change of Elevation of a Treadmill

During a heart stress test, the technician changes the grade of the treadmill.

(a) What is the grade of the treadmill if its vertical rise is 0.6 foot and the horizontal length of the treadmill is 8 feet?

(b) What is the rise of a 10% grade on a 6-foot treadmill?

Solution

(a) The grade of the treadmill is the slope as a percent.

$$m = \frac{\Delta y}{\Delta x} = \frac{y_2 - y_1}{x_2 - x_1} = \frac{0.6}{8} = 7.5\%$$

The grade is 7.5%, as shown in Figure 10.

(b) The rise on a 10% treadmill is the change in y in the slope formula

$$m = \frac{\Delta y}{\Delta x} = \frac{y_2 - y_1}{x_2 - x_1}$$

$$0.10 = \frac{\Delta y}{6} \quad \text{or} \quad \Delta y = 0.60$$

Therefore, a 0.60 foot elevation on a 6-foot treadmill would give a 10% grade.

Figure 10

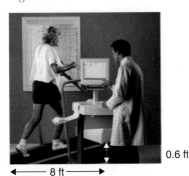

0.6 ft

8 ft

◈ PROBLEM SET 2.3

Mastering the Concepts

In Problems 1–4, find the coordinates of the indicated points on the graph of each line and then find its slope.

1.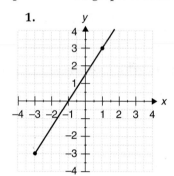

2.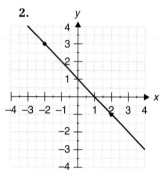

3.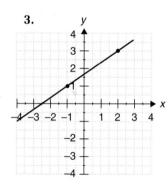

4.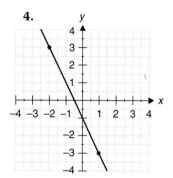

In Problems 5–22, sketch the graph of each line containing the given points, then find the slope of each line. Indicate whether the line containing the pair of points rises, falls, is horizontal, or is vertical.

5. $(3,-2)$ and $(5,3)$

6. $(3,4)$ and $(5,2)$

7. $(2,-1)$ and $(-3,4)$

8. $(1,-4)$ and $(2,3)$

9. $(1,5)$ and $(-2,3)$

10. $(0,0)$ and $(-2,5)$

11. $(0,0)$ and $(4,3)$

12. $(-2,-3)$ and $(5,1)$

13. $(-1,4)$ and $(3,4)$

14. $(-2,5)$ and $(3,5)$

15. $(3,2)$ and $(3,-4)$

16. $(2,-1)$ and $(2,4)$

17. $(0,1)$ and $(2,0)$

18. $(0,-2)$ and $(3,0)$

19. $(2.5,-3)$ and $(4.5,1)$

20. $(3.4,1)$ and $(3.4,5)$

21. $(-2,-4)$ and $(2,4)$

22. $(3.9,-1)$ and $(-4.1,3)$

In Problems 23–30, sketch the graph of the lines described in parts (a) and (b) on the same coordinate axis.

23. (a) $(0,2)$, $m = 2/3$
    (b) $(0,2)$, $m = -3/2$

24. (a) $(3,-2)$, $m = -2$
    (b) $(3,-2)$, $m = 1/2$

25. (a) $(-1,2)$, $m = 3$
    (b) $(-1,2)$, $m = -1/3$

26. (a) $(4,1)$, $m = 4/5$
    (b) $(4,1)$, $m = 5$

27. (a) $(-2,3)$, $m = 0$
    (b) $(-2,3)$, m is undefined

28. (a) $(2,-0.5)$, $m = 0$
    (b) $(2,-0.5)$, m is undefined

29. (a) $(0,-5)$, $m = 5/8$
    (b) $(0,-5)$, $m = -4$

30. (a) $(-5, 0)$, $m = 4/3$
    (b) $(-2,6)$, $m = -3$

In Problems 31–36, express each equation in slope-intercept form, then determine the slope and the y intercept. Also sketch the graph.

31. (a) $-3x - y - 6 = 0$
    (b) $x - y - 4 = 0$

32. (a) $2x + 3y - 12 = 0$
    (b) $x - 2y - 4 = 0$

33. (a) $x + y = 0$
    (b) $x - y = 0$

34. (a) $2x + y = 0$
    (b) $2x - y = 0$

35. (a) $y - 2 = 0$
    (b) $x + 2 = 0$

36. (a) $x - 3 = 0$
    (b) $y + 3 = 0$

In Problems 37–44, determine whether L_1 and L_2 are parallel, perpendicular, or neither. Sketch a graph of the lines.

37. (a) L_1 contains $(2,4)$ and $(3,8)$
    (b) L_2 contains $(5,1)$ and $(4,-3)$

38. (a) L_1 contains $(1,8)$ and $(-1,2)$
    (b) L_2 contains $(0,-2)$ and $(2,4)$

39. (a) L_1 contains $(2,4)$ and $(3,8)$
    (b) L_2 contains $(8,-2)$ and $(-4,1)$

40. (a) L_1 contains $(2,1)$ and $(6,3)$
    (b) L_2 contains $(2,3)$ and $(3,1)$

41. (a) $L_1: 3x - 2y = 7$
    (b) $L_2: 6x - 4y = 5$

42. (a) $L_1: x - 2y - 2 = 0$
    (b) $L_2: 2x - 4y + 7 = 0$

43. (a) $L_1: 2x - 3y = 10$
    (b) $L_2: 3x - 2y = 6$

44. (a) $L_1: 4y - 3x + 5 = 0$
    (b) $L_2: 4x + 3y + 2 = 0$

Applying the Concepts

45. Handicapped Access Ramp: To design a handicapped access ramp, the law states that for every vertical rise of 1 foot, a 12 foot run is required (Figure 11).

(a) What is the grade of the ramp?

(b) What is the horizontal distance from the foot of the ramp to a point 2.6 feet above the horizontal sidewalk in front of a building?

Figure 11

46. Grade of a Ramp: A ramp for water skiing is 14 feet high and 700 feet long, as in Figure 12.

Figure 12

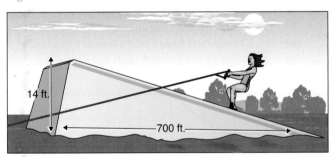

14 ft.

700 ft.

(a) What is the grade of the ramp?

(b) When the skier is 10 feet above the water, how much horizontal distance has he or she traversed?

47. Grade of a Treadmill: A cardiologist changes the grade of a treadmill to measure the heart rate of a patient.

(a) What is the grade of the treadmill if its vertical rise is 0.4 feet and the horizontal length of the treadmill is 8 feet?

(b) What is the rise of a 15% grade of a 6 foot treadmill?

48. Pitch of a Roof: The **span of a roof** is the horizontal distance between the outside extensions of the roof, while the *rise* is the vertical distance from the top of the roof to the center of the span (Figure 13). The *pitch* is the ratio of the rise to the span of a roof. Determine the rise of a house whose span is 28 feet with a pitch of (a) 1/4 (b) 1/2 (c) 4/7.

Figure 13

rise

run

49. Rate of Change: Suppose that the outside temperature rise on a certain day in January for a city in the Midwest is fairly linear from 6 A.M. to 2 P.M. If the temperature at 6 A.M. was $-10°F$, and by 2 P.M. was up to 15°F,

(a) What was the hourly rate of change?

(b) If the rate of increase of temperature is uniform during that day, estimate the temperature at noon.

50. Rate of Change: On a certain day in 1973, the cost of one gallon of unleaded gasoline in a city was 27 cents. On the same date in 1997, the cost of a gallon of unleaded gasoline was $1.15 for the same city. Assume the increase in cost of unleaded gasoline is linear over these years.

(a) What was the annual rate of change in the cost of one gallon of unleaded gasoline over this time period?

(b) If this rate of increase continued uniformly, predict the price of one gallon of unleaded gasoline on the same date in the year 2004.

Developing and Extending the Concepts

In Problems 51–53, if $P = (x_1, y_1)$ and $Q = (x_2, y_2)$ are two points in the xy plane, then the coordinates of the **midpoint** M of the line segment $\overline{PQ}$ are given by

$$M = \left(\frac{x_1 + x_2}{2}, \frac{y_1 + y_2}{2} \right)$$

The *midpoint* of a line segment is the point located halfway between the two endpoints of the line segment. For example, the coordinates of the midpoint of the line segment containing (1,4) and (5,8) are given by

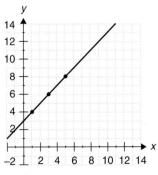

$$M = \left(\frac{1 + 5}{2}, \frac{4 + 8}{2} \right) = (3,6).$$

Find the coordinates of the midpoint M of the line segment $\overline{PQ}$.

51. (a) $P = (5,6)$ and $Q = (-7,8)$
    (b) $P = (-4,7)$ and $Q = (-3,0)$
52. (a) $P = (-2,3)$ and $Q = (4,-2)$
    (b) $P = (2,-5)$ and $Q = (-1,-3)$
53. $P = (8.32, 7.46)$ and $Q = (3.72, 6.38)$
54. **Map Coordinates:** If the map coordinates of two towns are $(39.14, 61.78)$ and $(49.07, 32.15)$, what are the coordinates of a point halfway between them?
55. Many applications can be modeled by linear equations like

$$y = kx$$

which expresses a **direct variation** between the variables x and y. In this case we say that y is **directly proportional** to x, or y **varies directly as x.** The number k is called the **constant of variation.**

(a) If y is directly proportional to x, and $y = 3$ when $x = 6$, find y when $x = 12$.
(b) If you were to graph the linear equation in part (a), what role does the value of k play in the graph?

56. **Wages:** Linda's total weekly wage T (in dollars) varies directly as the number of hours h that she works. Suppose that her weekly pay for 40 hours of work is $340.
(a) How much does she earn per hour?
(b) Sketch the graph of the linear equation in part (a) where T is the vertical axis and h is the horizontal axis.
(c) Interpret the slope.

57. **Relative Weight:** The weight W of an object on earth is directly proportional to the weight w of the same object on the moon. A 180 pound astronaut would weigh 30 pounds on the moon.
(a) Write an equation that relates W to w.
(b) Sketch the graph of the equation in part (a) where W is the vertical axis and w is the horizontal axis.
(c) Interpret the slope.

58. **Money Exchange:** The number of US dollars d is directly proportional to the number of Japanese yen y. On June 30, 2004, $100 US dollars bought 110.2 yen.
(a) Find the constant of proportionality k and the equation which will convert US dollars to Japanese yen.

(b) Sketch the graph of the equation in part (a). Let u represent US dollars on the horizontal axis and y represent Japanese yen on the vertical axis.
(c) Interpret the slope.
(d) At this rate of exchange, how many US dollars can be bought for 551 yen?

59. **Physics:** In physics, **Hook's Law** states that the distance d (in centimeters) that a spring will stretch is directly proportional to the mass m (in kilograms) of an object hanging from the spring.

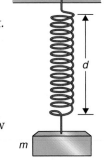

(a) Suppose that a 4 kilogram ball stretches a spring 50 centimeters. Find the constant of proportionality k.
(b) Sketch the graph of the line in part (a) where d is the vertical axis and m is the horizontal axis.
(c) Interpret the slope.
(d) Use the graph to predict how far the spring will stretch with a 6 kilogram ball.

60. **Average Speed:** On a trip from Chicago, for 4 hours a motorist kept track of the distance traveled every half hour. She passed the given mile markers at the following times.

Distance in Miles	30	55	85	125	160	195	230	265
Time in Hours	0.5	1.0	1.5	2.0	2.5	3.0	3.5	4.0

(a) Graph the data for the distance d in terms of time t where t is the horizontal axis and d is the vertical axis.
(b) Find and interpret the slope of each segment in part (a).

61. **Stock Prices:** The graph in Figure 14 shows a stock's closing price for consecutive trading days.
(a) What is the slope of each line segment?
(b) On what days did the price increase?
(c) On what days did the price remain constant?
(d) On what days did the price decrease?
(e) On what days did the price drop the fastest?

Figure 14

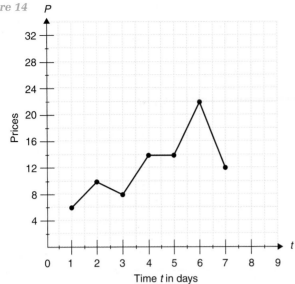

Time *t* in days

62. Use the concept of slope to show that the points $A = (-4,-1)$, $B = (0,2)$, $C = (2,2)$, and $D = (-2,-1)$ are the vertices of a parallelogram (a four sided figure with opposite sides parallel).

63. Use the concept of slope to determine whether the triangle with vertices *A*, *B*, and *C* is a right triangle.
  (a) $A = (-4,2)$, $B = (1,4)$, $C = (3,-1)$
  (b) $A = (8,5)$, $B = (1,-2)$, $C = (-3,2)$

64. Determine the value of *k* so that the given conditions are satisfied.
  (a) The lines $x - ky + 6 = 0$ and $2x + 4y - 5 = 0$ are parallel.
  (b) The lines $x - ky + 6 = 0$ and $2x + y = 4$ are perpendicular.

Objectives

1. Use the Point-Slope Form of a Line
2. Use the Slope-Intercept Form of a Line
3. Find Equations of Horizontal and Vertical Lines
4. Solve Applied Problems

2.4 Finding Equations of Lines

So far we have used linear equations to sketch graphs of lines. In this section, we reverse the process; we use information about the graphs of lines to find their equations. In the process, we examine different forms of linear equations so that we can choose the form that is most appropriate for a given situation.

Using the Point-Slope Form of a Line

Suppose we are given a nonvertical line *L* with slope *m* and a fixed point

$$P_1 = (x_1, y_1)$$

on *L* (Figure 1). If $P = (x,y)$ is any other point on the line *L*, then the slope *m* of the line is given by

Figure 1

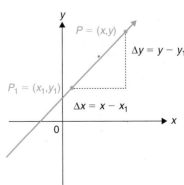

$$m = \frac{y - y_1}{x - x_1}.$$

Multiplying each side of the equation by $x - x_1$, we get the equation

$$m (x - x_1) = y - y_1$$

or

$$y - y_1 = m (x - x_1).$$

Because the latter equation displays the coordinates of the point (x_1,y_1) on the line L and the slope of L, it is called the *point-slope equation*.

Point-Slope Equation

> The **point-slope equation** of a line that contains the point (x_1,y_1) and has a slope of m, where m is defined, is given by
>
> $$y - y_1 = m(x - x_1).$$

EXAMPLE 1   **Finding an Equation of a Line in Point-Slope Form**

Find an equation of the line that has a slope of 3/2 and contains the point $(-1,3)$:

(a) in point-slope form

(b) in the standard form $Ax + By = C$.

Solution   (a) Using point-slope equation, we have

$$y - y_1 = m(x - x_1) \qquad \text{Point-slope equation}$$

$$y - 3 = \frac{3}{2}(x - (-1)) \qquad \text{Substitute } -1 = x_1, 3 = y_1 \text{ and } m = \frac{3}{2}$$

$$y - 3 = \frac{3}{2}(x + 1) \qquad \text{Simplify}$$

(b) We express the equation in the standard form as follows

$$2(y - 3) = 3(x + 1) \qquad \text{Multiply each side by 2 (the LCD)}$$

$$2y - 6 = 3x + 3$$

$$-3x + 2y = 6 + 3$$

$$-3x + 2y = 9$$

or $\qquad 3x - 2y = -9 \qquad$ Standard form of the equation

EXAMPLE 2   **Finding the Equation of a Line Given Two Points**

Find an equation of the line that contains the points $(-1,5)$ and $(2,20)$.

Solution   We first calculate the slope using $x_1 = -1$, $y_1 = 5$, $x_2 = 2$, and $y_2 = 20$.

$$m = \frac{\Delta y}{\Delta x} = \frac{y_2 - y_1}{x_2 - x_1}$$

$$= \frac{20 - 5}{2 - (-1)} = \frac{15}{3} = 5$$

Notice that if we use the other point, we will get the same equation.
$$y - 20 = 5(x - 2)$$
$$y - 20 = 5x - 10$$
$$y = 5x + 10$$

Substituting $m = 5$, $x_1 = -1$, and $y_1 = 5$ into $y - y_1 = m(x - x_1)$ yields

$$y - 5 = 5(x - (-1))$$

$$y - 5 = 5x + 5$$

$$y = 5x + 10$$

Figure 2

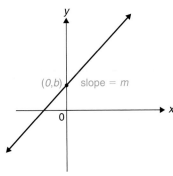

Using the Slope-Intercept Form of a Line

If the slope m and the y intercept b are given (Figure 2), we can write the equation of the line by substituting 0 for x_1 and b for y_1 in the point-slope form and simplifying.

$$y - b = m(x - 0)$$
$$y - b = mx$$
$$y = mx + b$$

Because the latter equation displays the slope m and the y intercept b, it is called the **slope-intercept form** of the equation of a line.

EXAMPLE 3 **Finding an Equation of a Line in Slope-Intercept Form**

Find an equation of each line with the given information.

(a) The line has a slope 4 and y intercept 3.

(b) The line has a slope 4 and contains the point (1,7).

Solution (a) Substituting $m = 4$ and $b = 3$ into the equation

$$y = mx + b$$

yields the equation

$$y = 4x + 3.$$

(b) To find the y intercept b, we substitute $m = 4$ and the coordinates of the point (1,7) into the equation

$$y = mx + b$$

so that

$$7 = 4(1) + b$$

or

$$b = 3.$$

Now we substitute 4 for m and 3 for b into the equation

$$y = mx + b$$

to obtain

$$y = 4x + 3.$$

Finding Equations of Horizontal and Vertical Lines

In section 2.3 we learned that:

1. A **horizontal line** has slope $m = 0$ with y intercept k and an equation

$$y = k$$

for a constant k.

2. A **vertical line** has an undefined slope with x intercept h and an equation

$$x = h$$

for a constant h.

EXAMPLE 4   **Finding an Equation of a Horizontal and Vertical Line**

Find an equation of a line that contains (3,4) if the line is:

(a) horizontal and (b) vertical (Figure 3).

Figure 3

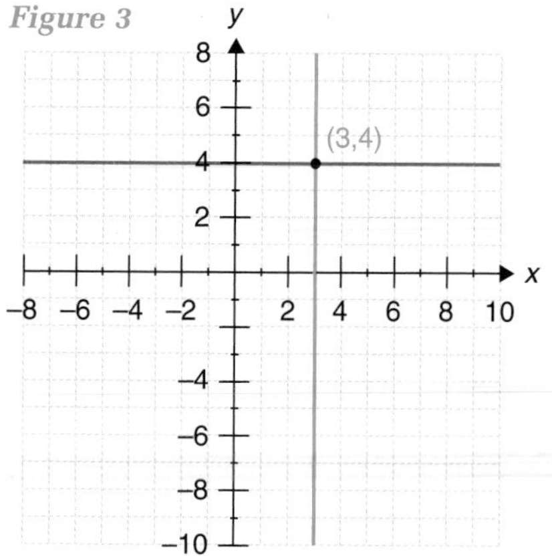

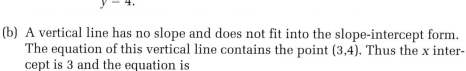

Solution   (a) A horizontal line has a slope of 0 and this line contains the point (3,4). Thus, the *y* intercept is 4. We substitute into the slope-intercept form to get

$$y = 0x + 4$$
$$y = 4.$$

(b) A vertical line has no slope and does not fit into the slope-intercept form. The equation of this vertical line contains the point (3,4). Thus the *x* intercept is 3 and the equation is

$$x = 3.$$

EXAMPLE 5   **Finding the Equations of Parallel and Perpendicular Lines**

Find a slope-intercept equation and sketch the graph of the line *L*, containing the point (2,7) that is:

(a) parallel to the line $-5x + 4y = 8$;

(b) perpendicular to the line $-5x + 4y = 8$.

Solution   First we express the equation $-5x + 4y = 8$ in slope-intercept form as follows:

$$4y = 5x + 8 \text{ or } y = \frac{5}{4}x + 2$$

so the slope is $\frac{5}{4}$.

(a) The slope of a line parallel to the given line is also $\frac{5}{4}$. By letting

$(x_1, y_1) = (2,7)$ and $m = \frac{5}{4}$, we have

$$y - y_1 = m(x - x_1)$$

$$y - 7 = \frac{5}{4}(x - 2)$$

$$y - 7 = \frac{5}{4}x - \frac{10}{4}$$

$$y = \frac{5}{4}x - \frac{5}{2} + 7$$

$$y = \frac{5}{4}x + \frac{9}{2} \text{ (Figure 4a).}$$

Figure 4a

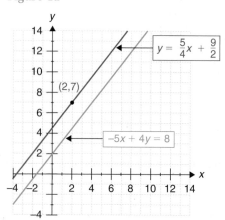

(b) The slope of a line perpendicular to the given line is the negative reciprocal, $-4/5$ (remember that $-4/5 = -4/5$). Using the point-slope form with

Figure 4b

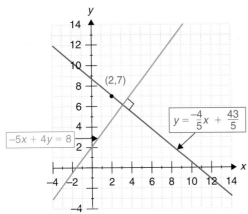

$(x_1,y_1) = (2,7)$ and $m = \dfrac{-4}{5}$, we have

$$y - y_1 = m(x - x_1)$$

$$y - 7 = \frac{-4}{5}(x - 2)$$

$$y - 7 = \frac{-4}{5}x + \frac{8}{5}$$

$$y = \frac{-4}{5}x + \frac{8}{5} + 7$$

$$y = \frac{-4}{5}x + \frac{43}{5} \text{ (Figure 4b).}$$

Solving Applied Problems

Many applied problems can be modeled by linear equations. Their graphs may be helpful in obtaining information about the relationships stated in the problems.

EXAMPLE 6   **Solving a Straight Line Depreciation Problem**

The graph in Figure 5 shows the book value R of a typical van which is depreciated by a small business firm.

(a) Express the book value R of the van as a linear equation in terms of a number of years t after the van was purchased.

(b) Find and interpret the R-intercept of the line.

(c) Find and interpret the slope of the line.

(d) Does the model make sense if the van is sold after 8 years? Explain.

Solution   (a) Using Figure 5, the point $(1, 17{,}500)$ represents the book value of the van 1 year after its purchase, while the point $(4, 10{,}000)$ represents its book value after 4 years.

Figure 5

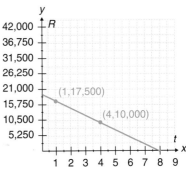

To find the slope of the line, we substitute $t_1 = 1$, $R_1 = 17{,}500$, $t_2 = 4$, $R_2 = 10{,}000$ into the slope formula to obtain

$$m = \frac{\Delta R}{\Delta t} = \frac{R_2 - R_1}{t_2 - t_1}$$

$$= \frac{17{,}500 - 10{,}000}{1 - 4} = \frac{7{,}500}{-3} = -2{,}500.$$

Substituting $m = -2{,}500$ and $(t_1,R_1) = (1, 17{,}500)$ into the slope-point form, we have

$$R - R_1 = m(t - t_1)$$

$$R - 17{,}500 = -2{,}500(t - 1)$$

$$R - 17{,}500 = -2{,}500t + 2{,}500$$

$$R = -2{,}500t + 20{,}000.$$

(b) The R intercept, \$20,000, represents the original purchase price of the van.

(c) The slope of the line, $-2{,}500$, represents the rate of decrease in book value. This means that for each elapsed year, there is a reduction of \$2,500 in the book value of the van; this is called **straight line depreciation.**

(d) When $R = 0$, then

$$0 = -2{,}500t + 20{,}000$$

$$2{,}500t = 20{,}000$$

$$t = 8.$$

This means the van will be fully depreciated (its book value is zero) in 8 years. While its value might reach zero for accounting purposes, the van probably retains a **salvage value** for old vans.

EXAMPLE 7   **Finding Temperature Conversions**

The freezing temperature of water is 0° on the Celsius scale and 32° on the Fahrenheit scale. The boiling temperature of water is 100° Celsius and 212° Fahrenheit. Assume that the relationship between the temperature scales is linear.

(a) Determine an equation relating the temperature C on the Celsius scale to the temperature F on the Fahrenheit scale.

(b) Graph the equation of the line.

(c) Find and interpret the intercepts of the line.

(d) Find and interpret the slope of the line.

Solution   (a) Using F for Fahrenheit and C for Celsius, the data points are $(F_1, C_1) = (32, 0)$ and $(F_2, C_2) = (212, 100)$. Using the slope formula

$$m = \frac{C_2 - C_1}{F_2 - F_1}$$

$$= \frac{100 - 0}{212 - 32} = \frac{100}{180} = \frac{5}{9}.$$

The equation of the line is

$$C - 0 = \frac{5}{9}(F - 32)$$

or

$$C = \frac{5}{9}(F - 32).$$

(b) Figure 6 shows the graph of the line.

(c) The F intercept is 32 and the C intercept is $-\dfrac{160}{9}$. This means when $F = 0°$,

$$C = \frac{5}{9}(0 - 32) = -\frac{160°}{9}$$

Figure 6

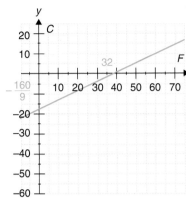

and when $C = 0°$,

$$0 = \frac{5}{9}(F - 32)$$

$$0 = F - 32 \qquad \text{Multiply both sides by } \frac{9}{5}$$

$$F = 32°.$$

(d) The slope of the line is $\frac{5}{9}$. This means that for a 5° change in Celsius, there is a 9° change in Fahrenheit measure.

PROBLEM SET 2.4

Mastering the Concepts

In Problems 1–8, write an equation of the line in the slope-intercept form for the given graph.

1.

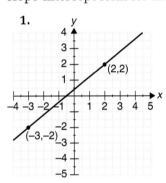

2.

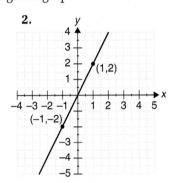

3.

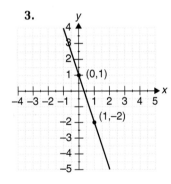

4.

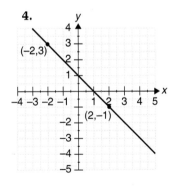

5.

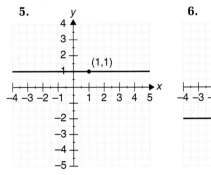

6.

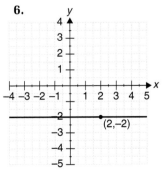

7.

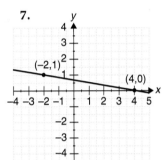

8.

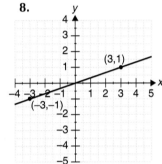

In Problems 9–16, write an equation of the line in slope-intercept form, $y = mx + b$, satisfying the following conditions.

9. $m = 2$, $b = -3$ **10.** $m = 3$, $b = 1$

11. $m = -2/3$, $b = 2$ **12.** $m = -1/2$, $b = 3$

13. $m = 3$ and contains $(0,2)$

14. $m = -4$ and contains $(0,-2)$

15. $m = -5/8$ and contains $(0,5)$

16. $m = -3/4$ and contains $(0,-5)$

In Problems 17–32, write an equation of the line satisfying the given conditions in (a) point-slope form, $y - y_1 = m(x - x_1)$ and (b) slope-intercept form, $y = mx + b$.

17. $m = 5$ and contains $(-1,2)$

18. $m = -3$ and contains $(7,3)$

19. $m = -3/7$ and contains $(5,-1)$

20. $m = -3/4$ and contains $(-4,6)$

21. $m = 3/8$ and contains $(0,0)$

22. $m = 4/5$ and contains $(-1,-2)$

23. $m = 0$ and contains $(3,-4)$

24. $m = 0$ and contains $(-4,2)$

25. $m = 5/3$ and contains $(-1,-1)$
26. $m = -5/11$ and contains $(1,3)$
27. contains $(2,3)$ and $(1,7)$
28. contains $(-2,3)$ and $(-4,2)$
29. contains $(4,-2)$ and $(-4,-5)$
30. contains $(3,-5)$ and $(3,4)$
31. contains $(-5,-8)$ and $(-1,-9)$
32. contains $(-1,-4)$ and $(-3,5)$

In Problems 33–46, write an equation of the line satisfying the given conditions in standard form,

$$Ax + By = C.$$

33. $m = 2$ and contains $(-3,1)$
34. $m = -3/4$ and contains $(5,-2)$
35. $m = -2/3$ and contains $(4,0)$
36. $m = 1$ and contains $(4,7)$
37. $m = 4/5$ and contains $(3,-4)$
38. $m = 5/3$ and contains $(3,0)$
39. $m = -4/3$ and contains $(-8,2)$
40. $m = -5/6$ and contains $(-2,3)$
41. contains $(4,5)$ and $(5,-3)$
42. contains $(-1,2/3)$ and $(1/2,-1)$
43. contains $(-4,0.3)$ and $(3,-3.4)$
44. contains $(5,-1/2)$ and $(3/2,-3/4)$
45. m is undefined and contains $(-1,8)$
46. m is undefined and contains $(5,-4)$

In Problems 47–52, write an equation of the line satisfying the given conditions in (a) slope-intercept form, $y = mx + b$ and (b) standard form, $Ax + By = C$.

47. contains $(3,-1)$; parallel to $x + 2y + 7 = 0$
48. contains $(-2,1)$; parallel to $5x - 7y - 8 = 0$
49. contains $(1,5)$; perpendicular to $2x + 3y - 1 = 0$
50. contains $(-3,2)$; perpendicular to $x + 2y - 5 = 0$
51. has y intercept 3; parallel to $x - 2y = 5$
52. has y intercept 5; perpendicular to $2x - 3y = 6$

Applying the Concepts

In Problems 53–59, use the given grids to sketch each graph.

53. Telephone Charges: A telephone company charges $1.80 for a 12 minute long distance domestic call, and $2.80 for a 20 minute call to the same number at the same time of day. Let the relationship between the cost C (in dollars) and the time t (in minutes) for a call be linear.

  (a) Write the information as two ordered pairs in the form (t,C).
  (b) Find an equation of the line that expresses C in terms of t and contains the data points.

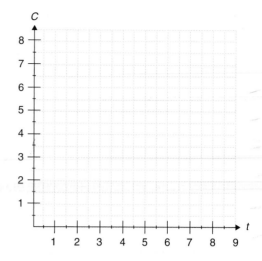

  (c) Sketch the graph of the line using t as the horizontal axis and C as the vertical axis.
  (d) Use the equation to predict the cost of a 40 minute call.
  (e) Use the equation to predict the time in minutes for a call that costs $6.15.
  (f) Are the points in part (d) and (e) on the graph?

54. Temperature Changes: An airline chart shows that the temperature T in degrees Fahrenheit at an altitude of 5,000 feet is 42°. At an altitude of 15,000 feet, the temperature is 5°. Assume the relationship between the air temperature T (in degrees Fahrenheit) and the altitude h (in feet above sea level) is linear for

$$0 \le h \le 20,000.$$

  (a) Write the information as two ordered pairs in the form (h,T).
  (b) Find an equation of the line that expresses T in term of h and contains the data points.
  (c) Sketch the graph of the line using h as the horizontal axis and T as the vertical axis.
  (d) Use the equation to predict the air temperature at an altitude of 20,000 feet.

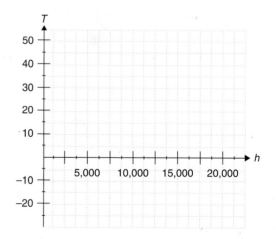

(e) Use the equation to predict the altitude at which the air temperature is 0°.

(f) Are the points in part (d) and (e) on the graph?

55. Cable Television Charges: A cable television company charges a $30 installation fee and $35 per month for cable service.

(a) Write a linear equation that determines the total cost C (in dollars) for t months of cable television service.

(b) Sketch the graph of the equation in part (a) using t as the horizontal axis and C as the vertical axis.

(c) Use the equation to find the total cost for six months of service.

(d) Use the equation to find the number of months of cable television service if the total cost is $170.

(e) Are the values in parts (c) and (d) on the graph?

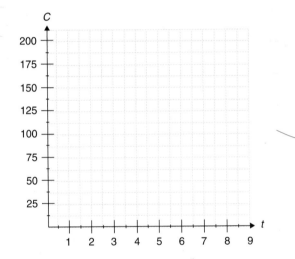

56. Pollution: In 2005, tests showed that water in a lake was polluted with 7 milligrams of mercury compounds per 1,000 liters of water. To clean up the lake, it was determined that the pollution level must drop at the rate of 0.75 milligrams of mercury compound per 1,000 liters of water per year.

(a) Write a linear equation that expresses the pollution level L (in milligrams per 1,000 liters) in terms of t years.

(b) Sketch the graph of the equation that represents this model if $t = 0$ corresponds to 2005 and $0 \le t \le 9$. Use t as the horizontal axis and L as the vertical axis.

(c) Use the graph in part (b) to predict when the lake will have 4 milligrams of mercury pollution.

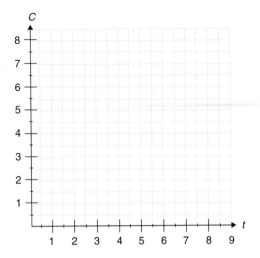

57. Straight-Line Depreciation A business firm is depreciating the value of a photocopier originally purchased for $7,500 over a period of 5 years. Suppose that the *trade-in* or *salvage value* of the photocopier at the end of 5 years is $1,100.

(a) Write a linear equation that expresses the value V (in dollars) of the photocopier in terms of time t (in years).

(b) Sketch the graph of the equation that represents this model for $0 \le t \le 5$, using t as the horizontal axis and V as the vertical axis.

(c) Use the equation of this model to determine the value of the copier after 2 years.

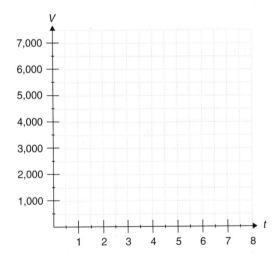

58. Banking: A bank charges its customers $4.00 per month as a maintenance fee for the first 12 checks and $0.30 per for each additional check written on its checking account

(a) Write a linear equation that expresses the total monthly cost C (in dollars) in terms of the number of checks n written for more than 12 checks.

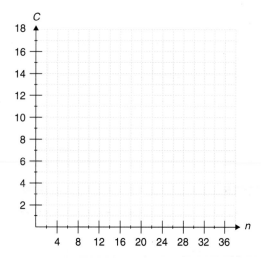

(b) Sketch the graph of the line using the cost as the vertical axis and the number of checks over 12 as the horizontal axis.

(c) What is the C intercept of the graph? Interpret its meaning.

(d) Use the graph to predict the number of checks over 12 written for a month if the total monthly charges are $10.

59. **Aerobic Exercise:** Physiologists have determined that the maximum safe pulse rate for a person doing aerobic exercise is 220 minus the person's age. The exercise is considered effective when the person's pulse rate is between 60% and 80% of their maximum pulse rate. Assume that the relationship between the pulse rate P and the age t (in years) is linear.

(a) Write two linear equations which describe (i) the 60% pulse rate and (ii) the 80% pulse rate in terms of the age t.

(b) Graph these two models using P as the vertical axis and t as the horizontal axis.

(c) Use the models in part (a) to determine the interval of an effective pulse rate for a 40-year-old person.

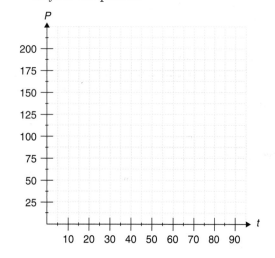

(d) Use the models in part (a) to determine a person's age whose pulse rate is between 102 and 136.

60. **Rental Occupancy:** A real estate company manages an apartment complex with 160 units. When the rent for each unit is $500 per month, all the apartments are occupied. However, when the rent for each unit is increased to $525 per month, 15 units become vacant. Assume that the relationship between the monthly rent R (in dollars) and the number n of occupied units is linear.

(a) Write an equation that expresses the monthly rent R in terms of the number n of occupied units for $50 \leq n \leq 160$.

(b) Use the model in part (a) to determine the monthly rent per unit if 120 units are occupied.

(c) Use the model in part (a) to determine how many units will be occupied if the monthly rent per unit is $600.

(d) Does the model make sense if the monthly rent per unit is $475? Why?

Developing and Extending the Concepts

61. (a) Suppose that a line L has nonzero x and y intercepts of a and b respectively. Show that an equation of L can be written in the **intercept form**

$$\frac{x}{a} + \frac{y}{b} = 1.$$

(b) Use the results of part (a) to find an equation of the line in intercept form where
   (i) x intercept is 5 and y intercept is 6
   (ii) x intercept is -2 and y intercept is 7.

62. Find a value of k so that each of the following conditions hold.

(a) The line $3x + ky + 2 = 0$ is parallel to the line $6x - 5y + 3 = 0$.

(b) The line $y = (2 - k)x + 2$ is perpendicular to the line $3x - y - 1 = 0$.

63. Let $P_1 = (-1, -3)$ and $P_2 = (5, 7)$ be two points.

(a) Find the midpoint M of the line segment $\overline{P_1 P_2}$.

(b) Find an equation in the point slope form of the perpendicular bisector of the segment.

Objectives

1. Define and Describe Functions
2. Evaluate Functions
3. Graph a Function
4. Use the Vertical Line Test
5. Find the Domain and Range of a Function
6. Solve Applied Problems

2.5 Introducing Functions

Historically the term "function" was first used by Gottfried Leibniz (1646–1716) in the seventeenth century to denote the association of quantities with a curve. In the eighteenth century, many of the concepts and the notation were further developed by the Swiss mathematician Leonhard Euler (1707–1783). This section discusses functions from both an algebraic and a graphic point of view. We will extend this discussion to different types of functions throughout this book.

Defining and Describing Functions

To understand the concept of functions, consider the following situations in which one quantity corresponds to another.

1. To each person there corresponds a birth date, that is, the month, day, and year born.
2. To each registered car there corresponds a unique license number.
3. To each circle there corresponds a unique area.

There are three characteristics common to all functions—an *input set,* an *output set,* and a *rule* that determines the association between the members of the two sets. The following table lists these characteristics for the above three examples.

TABLE 1

Set of Inputs	Set of Outputs	Rule (established by)
1. persons	birthdays	aging
2. cars	license numbers	licensing procedure
3. circles	areas	rule for area

These distinctions are central to the following definition.

Definition of a Function

> A **function** is a rule of correspondence between two sets D and R so that for each *input* value in D there exists one and only one *output* value in R.

In this definition, all the input elements in D are called the **domain** of the function and all the output elements in R are called the **range** of the function. The rule of a function produces exactly one output in the range for each input in the domain. It is acceptable that different inputs may produce the same output.

EXAMPLE 1

Determining Whether a Correspondence is a Function

Determine which correspondence represents a function. Find the domain and range of the function.

(a) A child corresponds with their biological mother.

(b) A doctor corresponds to their patients.

Solution   Figure 1 shows the correspondences.

Figure 1

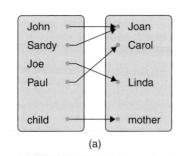
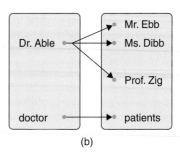

(a)   (b)

(a) This correspondence is a function because every child has one and only one biological mother. The domain is all the people born and the range is all the mothers.

(b) This correspondence is not a function because a doctor has more than one patient. Although this correspondence is not a function, its domain is the set of all doctors and its range is the set of all patients.

Using the correspondence in example 1, we can make the following observations about a function.

1. Each input in the domain must be matched with one output in the range. In example 1a, everyone has only one mother.

2. Two or more inputs of the domain may be matched with the same output in the range. In example 1a, John and Sandy have the same mother, Joan; John and Sandy are called siblings.

3. No input of the domain is matched with two different outputs in the range. In example 1a, note that people do not have two biological mothers.

Evaluating Functions

Functions may be defined by an equation or a formula involving two variables, by a table of data, or by a graph. For instance, the equation

$$y = 7x - 3$$

represents y as a function of the variable x. We call x the domain element, the input, or the **independent variable,** and y the range element, the output, or the **dependent variable.** When the function is defined by an equation, calculations may be performed in order to determine which element of the range corresponds with each element of the domain. For example,

$$\text{for } x = -2, \quad y = 7(-2) - 3 = -14 - 3 = -17$$
$$\text{for } x = 1, \quad y = 7(1) - 3 \quad = 7 - 3 \quad = 4$$
$$\text{for } x = 3, \quad y = 7(3) - 3 \quad = 21 - 3 \quad = 18$$

The inputs are values of x substituted into the equation to obtain the resulting output values of y. For a more concise notation, if we call the function f, we can use x to represent arbitrary inputs and $f(x)$ to represent the corresponding output. We read $f(x)$ as "f of x" or "the **value** of f at x". In this notation, the function given by $y = 7x - 3$ is written as $f(x) = 7x - 3$ and the above calculations would be written as

$$f(-2) = 7(-2) - 3 = -17$$
$$f(1) = 7(1) - 3 \quad = 4$$
$$f(3) = 7(3) - 3 \quad = 18.$$

The process of finding the value of $f(x)$ for a given value of x is called **evaluating a function.** Although we often use f as a convenient symbol to represent a function, other letters of the alphabet can be used to designate functions such as g, h, and k. Also letters other than x may be used to represent the independent variable. For example,

$$f(r) = 7r - 3, \quad g(t) = 7t - 3, \quad \text{and} \quad h(s) = 7s - 3$$

all define the same function. *Note that a function f and a function value f(x) are not the same: f is a rule of correspondence, but f(x) is an output value of f.*

EXAMPLE 2   **Using Functional Notation**

Suppose that $f(x) = 5x + 7$, find the indicated value

(a) $f(1)$ (b) $f(0)$ (c) $f(-4)$

Solution   To find the outputs we replace x by the inputs 1, 0, and -4 and simplify

(a) $f(1) = 5(1) + 7 \quad = 5 + 7 \quad = 12$

(b) $f(0) = 5(0) + 7 \quad = 0 + 7 \quad = 7$

(c) $f(-4) = 5(-4) + 7 = -20 + 7 = -13$

In many applied problems the input can be a letter or an algebraic expression as the next example shows.

EXAMPLE 3   **Evaluating a Function**

Let $f(x) = -3x + 4$. Find each value and simplify the result.

(a) $f(c)$ (b) $f(5b)$ (c) $f(t + 2)$ (d) $\dfrac{f(t + 2) - f(t)}{2}$

Solution   To find an output for a given input, we replace x in each case by the given letter or expression and simplify. Thus

(a) $f(c) = -3c + 4$

(b) $f(5b) = -3(5b) + 4 = -15b + 4$

(c) $f(t + 2) = -3(t + 2) + 4 = -3t - 6 + 4 = -3t - 2$

(d) $\dfrac{f(t + 2) - f(t)}{2} = \dfrac{[-3(t + 2) + 4] - [-3t + 4]}{2}$

$$= \frac{-3t - 6 + 4 + 3t - 4}{2}$$

$$= \frac{-6}{2} = -3$$

Note that the value of this quotient is the *slope* of the line

$$y = -3x + 4.$$

Graphing a Function

Functions can be described by tables and graphs. Such descriptions provide insight into many real world relationships. For example, the graph in Figure 2

Figure 2

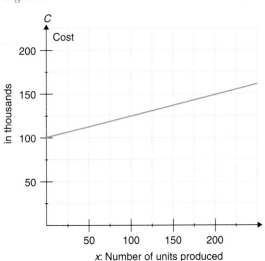

x: Number of units produced

shows the total cost (in thousands of dollars) for a manufacturing company that produces x units of its product. This is the graph of a function because each input (units produced) corresponds to exactly one output (cost). The horizontal scale represents the number of units produced, the domain elements, and the vertical scale, the range elements, represents the total cost C (in thousands of dollars).

From the graph we observe that \$100,000 was spent initially ($x = 0$), and \$150,000 was spent when 200 units were produced. Notice that as the number of units produced increases, the total cost also increases.

Quite often, businesses depend on equations that "closely fit" data-defined functions in order to predict future trends. For example, the linear function

$$f(x) = 0.25x + 100$$

approximates the data obtained from the graph of the function in Figure 2. To predict the cost (in thousands of dollars) spent to produce 230 units, we replace x by 230 and obtain

$$f(230) = 0.25(230) + 100 = 157.50.$$

Figure 3

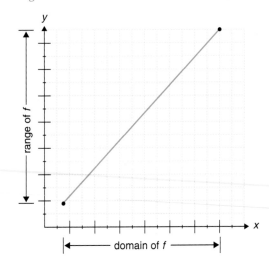

domain of f

Therefore, we predict \$157.50 or \$157,500 will be spent to produce 230 units.

Before we can analyze the graph of a function in the manner just illustrated in Figure 2, we must be able to sketch its graph. By definition, the **graph** of a function f is the set of all points $(x, f(x))$ for which x is in the domain of the function f. Graphically, *the domain of a function represents the set of all points on the horizontal axis (x axis) and the range represents the set of all points on the vertical axis (y axis).* The domain and range of a function are found from its graph (Figure 3). We *graph* a function in the same way we graph an equation. We substitute representative numbers from its domain into the equation that defines the function. So we obtain points $(x, f(x))$ that lie on the graph, then we plot these points and connect them with a line or a curve.

The standard form of a linear equation

$$Ax + By = C$$

can be expressed in the form of a function by solving for y in terms of x

$$By = -Ax + C$$

$$y = -\frac{A}{B}x + \frac{C}{B}$$

where $m = -\dfrac{A}{B}$ and $b = \dfrac{C}{B}$.

The function form of the latter linear equation is given by

$$f(x) = mx + b$$

and is called a **linear function.** Its graph is a straight line.

EXAMPLE 4   **Writing an Equation in the Form $f(x) = mx + b$**

Write the equation $4x - y = -2$ in the form

$$f(x) = mx + b.$$

Then sketch the graph.

Solution   We solve for y in terms of x to obtain

$$y = 4x + 2.$$

The associated linear function is given by

$$f(x) = 4x + 2.$$

Figure 4

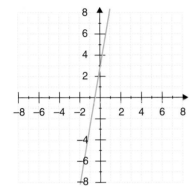

The graph of f is the same as the graph of the equation $y = 4x + 2$ which can be obtained by plotting the y intercept, 2, and then using the slope, 4, to find another point on the line. The line could also be graphed by finding and plotting the x and y intercepts, $-\frac{1}{2}$ and 2, respectively (Figure 4).

Using the Vertical Line Test

An examination of the graph of an equation can tell us whether or not the equation defines a function. Consider the graphs in Figure 5. In Figure 5a, every choice of input, x value, will correspond to one and only one output, y value; so, the graph represents a function. In Figure 5b, a single input value, say $x = 2$, corresponds to two output values, 2 and -2, and so the graph cannot represent a function. In this case, the points $(2,2)$ and $(2,-2)$ are positioned on a vertical line (Figure 5b). This leads to the following test.

Figure 5

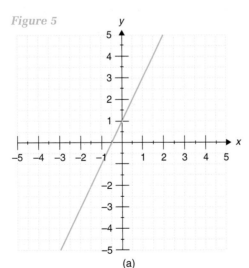

(a)

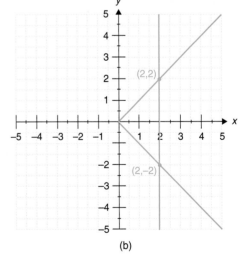

(b)

The Vertical Line Test

If a vertical line moving across a graph intersects the graph only one point at a time for each value in the domain, then the curve represents the graph of a function.

Finding the Domain and Range of a Function

By carefully examining the graph of a function, we may be able to determine its domain and range. It should be noted that when no endpoints are shown on the graph, we assume that the graph continues indefinitely. Also, unless otherwise stated, we assume that the domain of a function f is taken to be the set consisting of every real number for which the rule of f produces a real number.

EXAMPLE 5   **Using Graphs to Determine the Domain and Range**

Use the vertical line test to determine if the graph in Figure 6 is a function and then determine the domain and range of the function.

Figure 6

Solution   Figure 6 suggests a vertical line that crosses the graph from left to right, intersects the graph one point at a time. Thus the graph represents a function. The graph appears to start with the end point on the left at $x = -2$ and the lowest point at $y = 0$. It extends indefinitely to the right and upward. Hence, the graph suggests that the domain of the function consists of all real numbers such that $x \geq -2$ or $[-2, \infty)$ and the range consists of real numbers $y \geq 0$ or $[0, \infty)$.

Solving Applied Problems

Functions can be used in many applications and mathematical models.

EXAMPLE 6   **Determining the Cost of a Product**

A company was established in 2000, and every year since its founding the cost per unit produced has increased at a constant rate of $0.35. In 2003 the cost per unit was $10.60, $t = 0$ represents 2000.

(a) Find a function that expresses the cost C (in dollars) per unit as a function of time t (in years).

(b) Sketch the graph of the function.

(c) Predict the cost per unit in the year 2014 assuming the rate of increase remains the same.

(d) In what year will the cost per unit be $10.95?   *Figure 7*

Solution   (a) Since the rate of increase is constant, the function is linear and of the form $C(t) = mt + b$. Here the slope is 0.35 and $C(3) = 10.60$. Thus we have,

$$C(3) = 10.60 = 0.35(3) + b$$

$$b = 10.60 - 1.05 \text{ and } b = 9.55$$

The function is $C(t) = 0.35t + 9.55$

(b) Figure 7 displays the sketch of the graph.

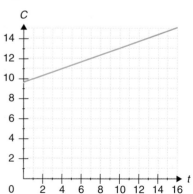

(c) In the year 2014, $t = 14$ and the cost per unit is

$$C(14) = 0.35(14) + 9.55 = \$14.45$$

(d) We are given $C(t) = 10.95 = 0.35t + 9.55$

$$1.40 = 0.35t$$

$$4 = t$$

So 4 years after 2000, that is 2004, the cost per unit was $10.95.

EXAMPLE 7   **Using Direct Variation**

In photography, the time T (in seconds) to make an enlargement of a photo negative is directly proportional to the area A (in square inches) of the enlargement. Suppose it takes 30 seconds to make a 6 by 8 enlargement.

(a) Find a linear function T in terms of A to model this situation.

(b) Use part (a) to find the time T it takes to make a 12 by 16 enlargement.

Solution   (a) Since T is directly proportional to A, we write the equation

$$T(A) = kA$$

where k is the constant of variation. For a given value of area, $A = 6(8) = 48$ square inches, there is exactly one value for $T = 30$ seconds, and so

$$30 = k(48)$$

$$k = \frac{30}{48} = \frac{5}{8}$$

The expression $T(A) = \frac{5}{8} A$ describes a linear function.

(b) Substituting $12(16) = 192$ for A, we have

$$T(192) = \frac{5}{8} (192) = 120$$

Therefore, it takes 120 seconds to make a 12 by 16 enlargement.

◈ **PROBLEM SET 2.5**

Mastering the Concepts

In Problems 1–4, determine which of the given correspondences represents a function. Find the domain and range of each function.

1. (a)

   (b)

2. (a)
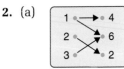

   (b)

3. (a) Each US citizen corresponds to their social security number.
   (b) Each person's age corresponds to their height.

4. (a) Each person's height corresponds to their age.
   (b) Each VCR in a store corresponds to its price.

In problems 5–14, find the indicated value of each given function.

5. $f(x) = 3x$
   (a) $f(-2)$
   (b) $f(0)$
   (c) $f(2)$

6. $g(x) = -3x$
   (a) $g(-3)$
   (b) $g(0)$
   (c) $g(3)$

7. $f(x) = x - 2$
   (a) $f(-2)$
   (b) $f(0)$
   (c) $f(2)$

8. $g(x) = 2 - x$
   (a) $g(-3)$
   (b) $g(0)$
   (c) $g(3)$

9. $f(x) = -4x + 3$
   (a) $f(0)$
   (b) $f(3/4)$
   (c) $f(-1/2)$

10. $g(x) = 5x + 2$
    (a) $g(-2/5)$
    (b) $g(0)$
    (c) $g(3/5)$

11. $h(x) = \dfrac{3x + 1}{5}$
    (a) $h(0)$
    (b) $h(-1/3)$
    (c) $h(2)$

12. $F(x) = \dfrac{-1 - 3x}{5}$
    (a) $F(0)$
    (b) $F(-1/3)$
    (c) $F(1/3)$

13. $f(x) = 6(2x - 3) + 4$
    (a) $f(0)$ (b) $f(7/6)$ (c) $f(3)$

14. $g(x) = 3(-2x + 1) - 5$
    (a) $g(0)$ (b) $g(-1)$ (c) $g(2)$

In Problems 15–18, find each value and simplify the results.

15. $f(x) = 7x + 3$
    (a) $f(t)$
    (b) $f(t + 2)$
    (c) $f(t) + 2$
    (d) $f(t + 2) - f(t)$

16. $g(x) = 3 - 7x$
    (a) $g(m)$
    (b) $g(m - 2)$
    (c) $g(m) - g(2)$
    (d) $g(m - 2) - g(2)$

17. $g(x) = 5 - 4x$
    (a) $g(2t)$
    (b) $g(t + h)$
    (c) $g(t) + h$
    (d) $g(t + h) - g(2t)$

18. $h(x) = 4x - 5$
    (a) $h(3p)$
    (b) $h(3p + 1)$
    (c) $h(3p) + 1$
    (d) $h(3p + 1) - h(3p)$

In Problems 19–24, find and simplify the value of the expression

$$\frac{f(t + h) - f(t)}{h}.$$

This expression is called the **difference quotient** of f (see example 3d on page 131). Compare this value to the slope of the line.

19. $f(x) = 3x + 11$
20. $f(x) = -7 + 4x$
21. $f(x) = 4 - 9x$
22. $f(x) = 5 - 13x$
23. $f(x) = \dfrac{-5}{2}x - 1$
24. $f(x) = \dfrac{3x}{7} + 4$

In Problems 25–32, sketch the graph of each function.

25. (a) $f(x) = x$
    (b) $g(x) = -x$

26. (a) $f(x) = \dfrac{x}{2}$
    (b) $g(x) = \dfrac{-x}{2}$

27. $f(x) = 3x - 4$

28. $g(x) = 4 - 3x$

29. $f(x) = \dfrac{-x}{4} + 1$

30. $g(x) = \dfrac{1}{2}x - 2$

31. $g(x) = -2$

32. $f(x) = 0$

33. Figure 8 shows the distance d (in miles) traveled in time t (in hours) over a period of 3.5 hours.
    (a) What is the domain?
    (b) What is the range?

Figure 8

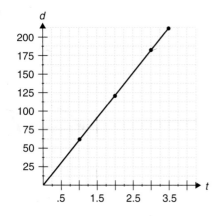

(c) Estimate the distance traveled after 1.5 hours.

(d) Estimate the time it took to travel 180 miles.

34. Figure 9 shows the relationship between the number of units *N* sold of a product and the price *p* (in dollars) of that product.

(a) Is this the graph of a function?

(b) What is the domain?

(c) What is the range?

(d) Estimate the number of units sold if the price is $4.50 per unit.

(e) Estimate the price per unit if 25 units were sold.

Figure 9

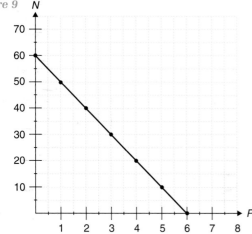

In Problems 35 and 36, use the vertical line test to determine whether or not each graph represents a function.

35. (a)   (b)

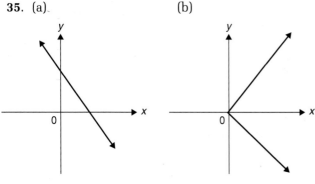

36. (a)   (b)

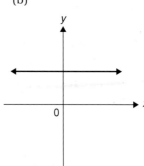

In Problems 37 and 38, use the graph of the given function to determine its domain and its range.

37. (a)   (b)

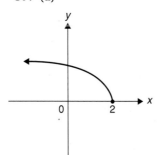

38. (a)   (b)

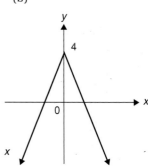

Applying the Concepts

39. **Car Rental:** The cost *C* (in dollars) for renting a midsize car for one day is shown in Figure 10 where *m* is the number of miles driven for that day.

(a) Find a function that models the cost *C* (in dollars) in terms of the number of miles *m* driven.

(b) Use the model to estimate the cost of driving 130 miles for that day.

(c) Use the model to determine the number of miles driven if the total cost for that day is $26.

(d) Use the model to predict the total cost if 300 miles were driven for that day.

Figure 10

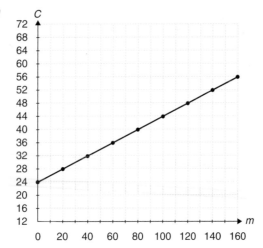

40. **Car Prices:** In 2000, the average price paid for a certain new car model was $16,800. The average price increased in a linear pattern to $21,080 in 2004.
    (a) Find a function that models the average price P paid (in dollars) in terms of the time t (in years after 2000).
    (b) Use the data points to graph the function.
    (c) Use the equation to predict the average price of that same model in the year 2008.
    (d) Use the equation to predict the year in which the average price of this model is $23,220.
    (e) Compare these values to points on the graph.

41. **Aerobics:** An aerobic dancer burns 350 calories per hour. The number of calories burned, C, is directly proportional to the number of hours danced, t.
    (a) Express C as a function of t to model the situation.
    (b) Sketch the graph of C.
    (c) Use the graph to determine the number of calories burned for an aerobic dancer who danced for 2.5 hours.

42. **Total Cost:** In 2004, the sales tax on a nonfood item in the state of Michigan was 6%.
    (a) Express the total cost C (in dollars) of a nonfood item as a function of the price p (in dollars) of the item.
    (b) Sketch the graph of $C = f(p)$ by using p as the horizontal axis and C as the vertical axis.
    (c) Use the equation to determine the total cost of an item that was purchased for $175.
    (d) Use the equation to predict the sale price before tax if its total cost is $200.
    (e) Compare the points on the graph to the values in parts (c) and (d).

43. **Tuition Cost:** The yearly cost C (in dollars) of tuition and other required fees for attending a certain university can be modeled by the linear function

    $$C(t) = 1,500t + 8,500, \ 0 \le t \le 10$$

    where t is the number of years with $t = 0$ representing 1997.
    (a) Use this model to approximate the total yearly cost of attending this university in the year 2007.
    (b) Use this model to predict in what year the total yearly cost of attending the university will be $13,000.
    (c) Does this model make sense for predicting the total yearly cost after 30 years? Explain.

44. **Ideal Weight:** The ideal weight W (in pounds) of a man is estimated by multiplying his height h (in inches) by 5 and subtracting 190.
    (a) Express W as a linear function of h to model this situation.
    (b) Find the ideal weight of a man who is 6 feet tall.
    (c) Sketch the graph of this model.
    (d) What restrictions should be put on h to make this model reasonable?

45. **Ideal Shoe Size:** A manufacturer of modern scientifically designed walking shoes estimates the ideal shoe size S for a man is obtained by multiplying his foot length l (in inches) by 3 and subtracting 22.
    (a) Express S as a linear function of l to model this situation.
    (b) Find the shoe size of a man if the length of his foot is 10.5 inches.
    (c) Sketch the graph of this model.
    (d) What restrictions should be put on l to make this model reasonable?

46. **Ecology:** An ecologist investigating the effect of air pollution on plant life determines that the percentage, p (in percent), of diseased trees and shrubs at a distance of x (in kilometers) from an industrial city is given by the linear model

    $$p(x) = 32 - 0.06x, \ 50 \le x \le 500.$$

    (a) Find $p(100)$, $p(200)$, and $p(400)$.
    (b) Sketch the graph of p.
    (c) Use the equation to determine the percent p of diseased trees at a distance of 300 kilometers.

Developing and Extending the Concepts

In Problems 47–50, let $f(x) = 3x - 2$ and $g(x) = 2x + 7$. Write an expression for each function and simplify the results.

47. (a) $f(x) + g(x)$ (b) $f(x) - g(x)$

48. (a) $2f(x) + 3g(x)$ (b) $2f(x) - 3g(x)$

49. (a) $f(x) \cdot g(x)$ (b) $x \cdot f(x)$

50. (a) $\dfrac{f(x)}{g(x)}$ (b) $\dfrac{g(x)}{f(x)}$

51. Find an equation of a linear function such that $f(3) = 7$ and $f(4) = 9$.

52. Let $f(x) = 5x + 1$
  (a) $f(s + t)$
  (b) $f(s) + f(t)$
  (c) Is $f(s + t) = f(s) + f(t)$?
  (d) $f(at)$
  (e) $a \cdot f(t)$
  (f) Is $f(at) = a \cdot f(t)$?

53. Sketch the graph of a line that does not represent a function.

54. Suppose that $y = 2$ regardless of the value of x.
  (a) Is y a function of x? Explain.
  (b) Is x a function of y? Explain.

55. The following data were collected by measuring the weight y (in pounds) of a man on the moon and his weight x (in pounds) on Earth.

weight x (lbs.) on Earth	120	132	156	162
weight y (lbs.) on the moon	20	22	26	27

(a) Plot the data points on a graph using x as the horizontal axis and y as the vertical axis.

(b) Analyze the graph to determine a linear model for y as a function of x to fit the data.

(c) Find and interpret the slope of the line.

(d) Use the model to predict the weight of a person on Earth if his weight on the moon is 32 pounds.

56. Let $f(x) = |2x - 1| + 1$ and $g(x) = -|1 - 2x| - 1$.

(a) Algebraically find (i) $f\left(\dfrac{1}{2}\right)$ (ii) $g\left(\dfrac{1}{2}\right)$
   (iii) $f(0)$ (iv) $g(0)$

(b) Use the graphs of f and g in Figure 11 to estimate the domains and ranges of f and g.

Figure 11

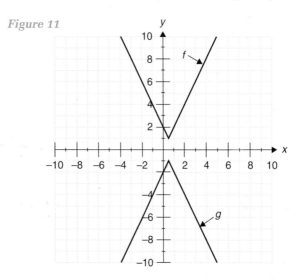

2.6 Graphing Linear Inequalities

Objectives

1. Graph Linear Inequalities in Two Variables
2. Graph Compound Inequalities
3. Solve Applied Problems

Graphing Linear Inequalities in Two Variables

Recall from section 2.1 that the graph of a linear equation in two variables is a straight line. Here we graph linear inequalities in two variables that can be written in the *standard form* $Ax + By < C$ where A, B, and C are real numbers and A and B are not both zero. The symbol $<$ may be replaced by $\geq$, $\leq$, or $>$. Examples of linear inequalities are:

$$3x - y < 6, \quad 5x - y > 7, \quad 0x + y \leq 1, \quad \text{and} \quad x + 0y \geq 2.$$

A **solution** of a linear inequality in two variables is an ordered pair of numbers (x,y) that make the inequality true. All the ordered pairs that make the inequality true are called the **solution set**. The **graph** of an inequality in two variables represents its solution set. To graph a linear inequality in x and y, we

Figure 1

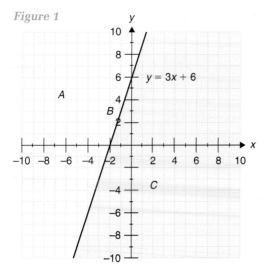

first graph the **boundary line** of the related equation. This line divides the xy plane into two regions called **half planes,** one above the line and one below it (assuming this line is not vertical). To determine which half-plane satisfies the inequality, we test a point in either region.

For example, to solve the inequality $y \leq 3x + 6$, first, we graph the boundary line $y = 3x + 6$ (Figure 1). This line divides the xy plane into an upper half-plane A and a lower half-plane C. The set of all points on the line is represented by B. The point $(0,0)$ is in the half-plane C. It makes the inequality true

$$0 \leq 3(0) + 6.$$

The solution of the inequality $y \leq 3x + 6$ consists of all points in the half-plane C and all points on the boundary line B because the inequality is *less than or equal to.* The following step procedure may be used to graph linear inequalities in two variables.

Graphing Linear Inequalities in Two Variables

Step 1. Sketch the graph of the boundary line obtained by replacing the inequality sign with an equal sign and graph this related equation. If the inequality has the symbol $<$ or $>$, draw a dashed line. A dashed line indicates that the boundary line is not part of the solution set. If the inequality has the symbols $\leq$ or $\geq$, draw a solid line to indicate that the points on the line are solutions of the inequality.

Step 2. Determine which half-plane corresponds to the inequality. To do this, select any convenient **test point** (x,y) not on the boundary line. Substitute the coordinates of this test point into the original inequality.
   (a) If the point makes the original inequality true, shade the half-plane containing the test point.
   (b) If the point makes the original inequality false, shade the other half-plane.

EXAMPLE 1   **Graphing a Linear Inequality**

Sketch the graph of the inequality $y < x + 2$.

Solution   **Step 1.** The boundary line for this inequality is the graph of the related equation $y = x + 2$. We draw the graph of a dashed boundary line because the inequality contains the symbol $<$ (which is the strict inequality, less than but not equal to).

Step 2. To determine which half-plane is the solution, we select the test point $(1,2)$ not on the boundary line.

Figure 2

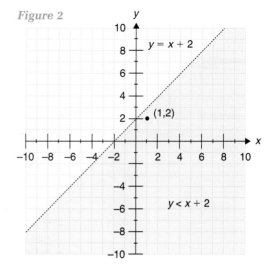

Replacing x by 1 and y by 2 in the original inequality leads to the following statement

$$y < x + 2$$

$$2 < 1 + 2$$

$$2 < 3 \text{ which is true.}$$

Because (1,2) satisfies the inequality, so does every point on the same side of the boundary line as (1,2). Therefore, we shade the lower half-plane that contains the point (1,2) (Figure 2).

EXAMPLE 2

Graphing a Linear Inequality

Sketch the graph of the inequality $2x - y \leq 1$.

Figure 3

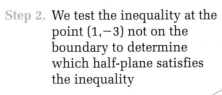

Solution

Step 1. Because the inequality contains the symbol $\leq$ (meaning less than or equal to), we graph the related equation as a solid boundary line $2x - y = 1$.

Step 2. We test the inequality at the point $(1,-3)$ not on the boundary to determine which half-plane satisfies the inequality

$$2x - y \leq 1.$$

When $x = 1$ and $y = -3$, we have

$$2(1) - (-3) \leq 1$$

$$5 \leq 1 \text{ False}$$

This point does not satisfy the inequality, so the correct half-plane to shade is the upper half-plane that does not contain the point $(1,-3)$ (Figure 3).

EXAMPLE 3

Graphing Special Linear Inequalities

Sketch the graph of each inequality

(a) $y \geq 3$ (b) $x < -1$

Solution

(a) First we draw the boundary line as a solid horizontal line $y = 3$ since the symbol $\geq$ is used. Next we choose $(-2,1)$ as a test point and find that $1 \geq 3$ is false. Therefore, we shade the upper half-plane that does not contain the point $(-2,1)$ (Figure 4a).

(b) We begin by drawing the boundary line as the dashed line $x = -1$, since equality is not included. Then we choose $(0,3)$ as a test point and find that $0 < -1$ is false. Therefore, we shade the half-plane to the left of the vertical boundary line which does not contain the point $(0,3)$ (Figure 4b).

Figure 4

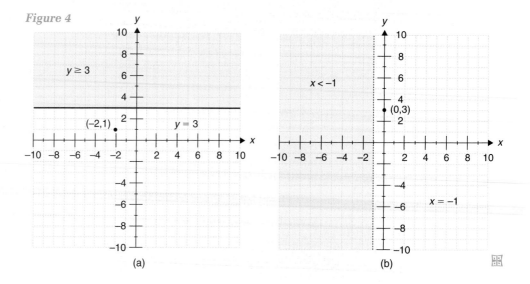

(a)   (b)

Graphing Compound Inequalities

In section 1.5 we solved compound inequalities containing one variable. Linear inequalities containing two variables can also be connected with the words "and" or "or" to form compound inequalities.

EXAMPLE 4   **Graphing a Compound Inequality**

Sketch the graph of the compound inequality

$$x \geq 2 \text{ and } y \geq 4x + 3.$$

Solution   Figure 5a shows the graph of $x \geq 2$ to be the region to the right of and including the solid line $x = 2$. Figure 5b shows the region above and including the solid boundary line $y \geq 4x + 3$. Because the inequalities are connected with the word "and", the graph of the compound inequality is the area common to both or the intersection of the two graphs (Figure 5c).

Figure 5

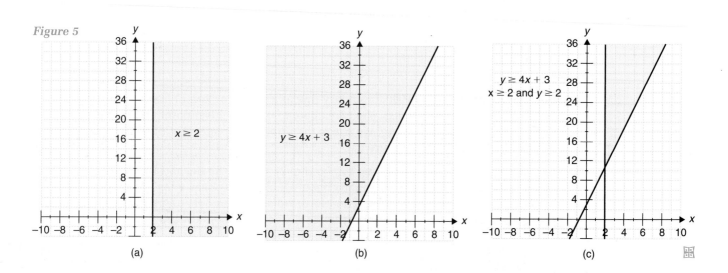

(a)   (b)   (c)

EXAMPLE 5   **Graphing a Compound Inequality**

Sketch the graph of the compound inequality

$$y \geq x - 2 \text{ or } y \leq 1.$$

Solution   Figure 6a shows the graph of $y \geq x - 2$ as the region above and including the solid boundary line $y = x - 2$. Figure 6b is the graph of $y \leq 1$ as the region below and including the boundary line $y = 1$. This compound inequality uses the connector "or" so that the graph is all the points in one graph or the other as shown in Figure 6c.

Figure 6

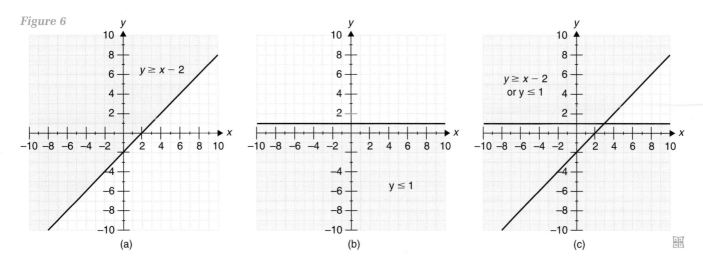

(a)   (b)   (c)

EXAMPLE 6   **Identifying a Region by Using Linear Inequalities**

Identify the region that satisfies all the conditions

$$\begin{cases} x + 2y \leq 8 \\ x \geq 0, y \geq 0. \end{cases}$$

Solution   The graph of the inequality $x + 2y \leq 8$ consists of all the points in the plane on or below the solid line $x + 2y = 8$ (Figure 7a). The graph of $x \geq 0$ consists of all points on or to the right of the y axis, while the graph of $y \geq 0$ includes all points in the plane on or above the x axis (Figure 7b). Therefore, the graph of these inequalities is the region common to all three. That is, the shaded region in the first quadrant which includes part of the coordinate axes (Figure 7c).

Figure 7

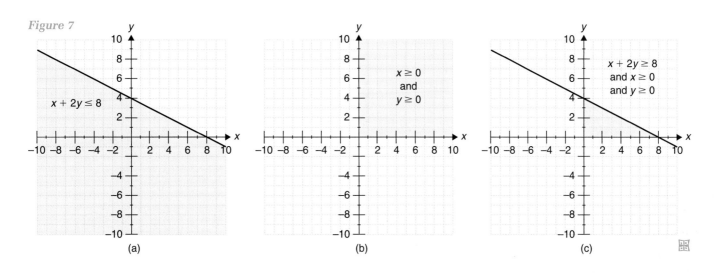

(a)   (b)   (c)

Solving Applied Problems

Linear inequalities in two variables are used in a variety of applications and models as the next example shows.

EXAMPLE 7 **Solving a Nutrition Problem**

A nutritionist determines that an apple contains 100 calories and a pear contains 120 calories. Suppose that a person eats an apple and a pear for a mid-morning snack several times a week. The person wishes to keep the total number of calories per week for the snack to be at most 1,180. If x and y represent the number of apples and pears consumed per week respectively, write a linear inequality to model this situation and graph it.

Solution Since x and y represent the number of apples and pears consumed per week, the total number of calories consumed per week for the snack is modeled by the following linear inequality

$$100x + 120y \leq 1,180$$

with the restriction that $x \geq 0$ and $y \geq 0$. Figure 8 shows the shaded region which describes the solution set of this inequality.

Figure 8

◈ PROBLEM SET 2.6

Mastering the Concepts

In Problems 1 and 2, match the shaded region with the appropriate inequality.

(i) $y > 2x - 3$ (ii) $x + y > 4$ (iii) $10y \geq -x + 5$ (iv) $y \geq x$ (v) $y < 3$ (vi) $x \geq 2$

1. (a)

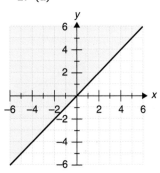

2. (a)

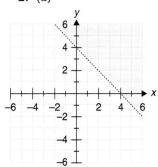

1. (b)

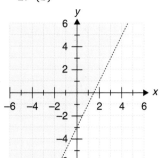

2. (b)

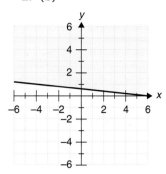

1. (c)

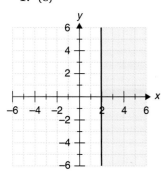

2. (c)

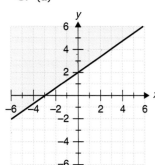

In Problems 5 and 6, the boundary line for the graph of the given inequality is shown. Shade the correct half-plane that indicates the solution of the inequality.

5. (a) $y \le 2x - 3$

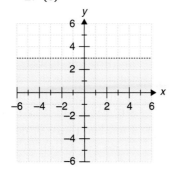

6. (a) $-2x + y < 6$

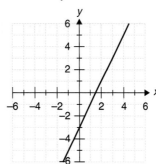

In Problems 3 and 4, write an inequality whose solution set is the given graph.

3. (a)

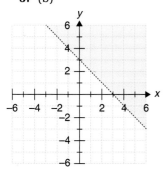

4. (a)

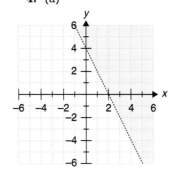

5. (b) $y \ge -3x$

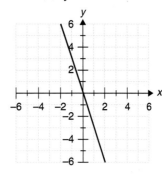

6. (b) $y \ge 2$

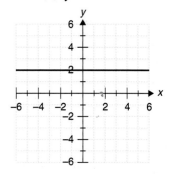

In Problems 7–20, sketch the graph of the given inequality.

7. $y \le 2x + 5$	**8.** $y \ge 3x + 4$
9. $y > 3x$	**10.** $y \le -4x$
11. $y > -2x + 3$	**12.** $2y > 3x + 7$
13. $2x + y \le 3$	**14.** $5y \ge 7$
15. $3y \le -4$	**16.** $y \le -1$
17. $x \ge 2$	**18.** $2x - 6 < 0$
19. $x \le 2$	**20.** $x < -2$

3. (b)

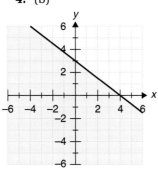

4. (b)

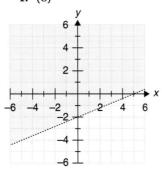

In Problems 21 and 22, write a compound inequality whose solution is described in the given graph.

3. (c)

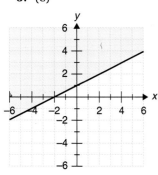

4. (c)

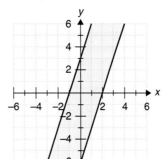

21. (a)

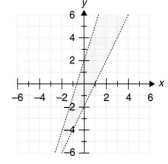

22. (a)

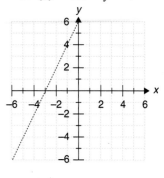

21. (b)

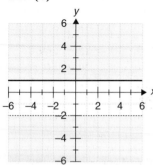

22. (b)

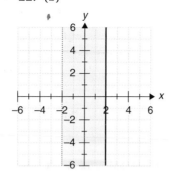

21. (c)

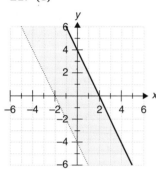

22. (c)

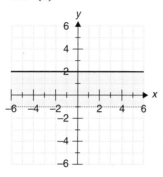

In Problems 23–32, sketch the graph of each compound inequality.

23. $y \geq -2$ and $x \leq 1$

24. $y \leq 2$ or $x \geq -1$

25. $y \leq -3$ or $x \geq 2$

26. $y \geq -2$ and $x \leq 1$

27. $2x + y < 3$ and $x + y < 1$

28. $-3x + y > 2$ and $x - 2y \leq 4$

29. $2x + y < 4$ or $x - y < 2$

30. $y \geq 3$ or $y < -2x + 1$

31. $\begin{cases} 2x + y \leq 10 \\ x \geq 0 \quad y \geq 0 \end{cases}$

32. $\begin{cases} 2x + 3y \leq 18 \\ x \geq 0 \quad y \geq 0 \end{cases}$

Applying the Concepts

33. Inventory Levels: An electronics store sells two brands of VCRs, A and B. Customer demand indicates that it is necessary to stock at least twice as many units of brand A as of brand B. Due to limitation of space, there is room for no more than 100 units in the store. Write a linear inequality that describes all possibilities for stocking the two brands and produce its graph.

34. Investments: A retirement fund invests in two mutual funds, X and Y. The X fund pays 5% annual simple interest and the Y fund pays 6% annual simple interest. The total interest from

both investments for one year must be at least $10,600. Write a linear inequality that describes this situation and produce its graph.

35. Ticket Prices: For an upcoming event in an auditorium, tickets are priced at $8 and $5. The total sales for that event must be at least $3,000. Write a linear inequality that describes this situation and produce its graph.

36. Fuel Economy: A certain make of car gets 16 miles per gallon in city driving and 24 miles per gallon in highway driving. The car is driven at most 380 miles on a full tank of gasoline. Write a linear inequality that describes this situation and produce its graph.

Developing and Extending the Concepts

37. Figure 9 shows the graph of the equation $|x + y| = 2$. Use the graph to describe the region which indicates the solution of each inequality.

(a) $|x + y| \leq 2$ (b) $|x + y| \geq 2$

Figure 9

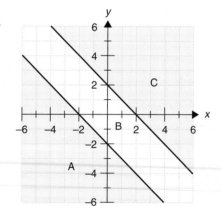

38. Figure 10 shows the graph of the equation $|y| = 1$. Use the graph to describe the region which indicates the solution of each inequality.

(a) $|y| \geq 1$ (b) $|y| \leq 1$

Figure 10

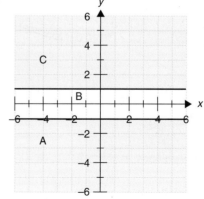

In Problems 39–42, sketch the graph of each absolute value inequality.

39. $|x - y| < 0$
41. $|y - x| \geq 0$

40. $|y - 3x| < 0$
42. $|3x - y| \leq 0$

In Problems 43 and 44, write inequalities that describe the indicated quadrant.

43. (a) Quadrant II
    (b) Quadrants I and IV
44. (a) Quadrant IV
    (b) Quadrants II and III

In Problems 45 and 46, sketch the graph of each inequality that fits the description.

45. A fast food restaurant has 20 Level I and 15 Level II employees. For a 7 hour day, each Level I employee is paid $50 and each Level II employee is paid $40. The daily payroll of the restaurant must never exceed $1,400.

46. A pen company manufactures two types of pens, A and B, which it sells for $2.75 and $1.50 respectively. Their total production is at most 15,000 pens per week. In order to remain profitable, their gross receipts must be at least $20,000.

CHAPTER 2 REVIEW PROBLEM SET

In Problems 1–6, sketch the graph of each linear equation by point plotting.

1. (a) $y = x - 3$
    (b) $y = x + 3$

2. (a) $y = \frac{3}{4}x$
    (b) $y = -\frac{3}{4}x$

3. (a) $y = 2x - 3$
    (b) $y = -2x + 3$
5. $2x - 3y + 12 = 0$

4. (a) $y = -3x + 1$
    (b) $y = -3x - 1$
6. $3x - 2y - 12 = 0$

In Problems 7–10, find the x and y intercepts of each equation and sketch the graph.

7. $2x - 5y = 10$
9. $4x + 12 = 0$

8. $-3x + 6y + 12 = 0$
10. $3y - 9 = 0$

In Problems 11–16, sketch the graph of each line containing the given points, then find the slope of each line segment. Indicate whether each line rises, falls, is horizontal, or is vertical.

11. $(2,-2)$ and $(7,3)$
13. $(-3,5)$ and $(1,3)$
15. $(3,0)$ and $(-5,0)$

12. $(7,4)$ and $(11,12)$
14. $(-1,4)$ and $(2,1)$
16. $(3,4)$ and $(3,5)$

In Problems 17 and 18, given a point and the slope, sketch the graph of each pair of lines on the same coordinate axis.

17. (a) $(1,2)$ and $m = \frac{3}{4}$
    (b) $(1,2)$ and $m = -\frac{4}{3}$

18. (a) $(4,-5)$ and $m = \frac{4}{7}$
    (b) $(4,-5)$ and $m = -\frac{7}{4}$

In Problems 19 and 20, express each equation in slope-intercept form, then determine the slope and the y intercept. Also sketch the graph.

19. (a) $3x - 4y - 8 = 0$
    (b) $3x + 4y + 8 = 0$
20. (a) $5x + 7y - 11 = 0$
    (b) $-7x + 5y - 11 = 0$

In Problems 21–24, determine whether L_1 and L_2 are parallel, perpendicular, or neither. Sketch a graph of the lines.

21. L_1 contains $(1,3)$ and $(-1,-1)$
    L_2 contains $(2,1)$ and $(-3,-9)$
22. L_1 contains $(4,6)$ and $(-2,-3)$
    L_2 contains $(3,0)$ and $(6,-2)$
23. $L_1: 3x + 2y - 6 = 0$ and $L_2: 2x - 3y - 6 = 0$
24. $L_1: x - 2y - 5 = 0$ and $L_2: 2x + y - 5 = 0$

In Problems 25–28, find an equation of each line L in slope-intercept form. Sketch the graph.

25. (a) $m = 3$ and contains $(1,1)$
    (b) $m = -2$ and contains $(-3,-2)$
26. (a) $m = -2$ and contains $(1,5)$
    (b) contains $(4,1)$ and $(2,3)$
27. (a) $m = 0$ and contains $(-3,4)$
    (b) m is undefined and contains $(5,3)$
28. (a) $m = 0$ and $b = -2$
    (b) m is undefined and contains $(1,7)$

In Problems 29–32, find an equation of the line L in point slope form and in general form. Sketch the graph.

29. L is parallel to the line $3x - 2y + 5 = 0$ and contains $(2,3)$.

30. L is parallel to the line $2x - 5y - 7 = 0$ and contains $(-3,4)$.

31. L is perpendicular to the line $2y + 3x + 6 = 0$ and contains $(-7,5)$.

32. L contains the point $(-1,2)$ and is perpendicular to the line segment containing $(3,4)$ and $(-5,6)$.

In Problems 33–38, sketch the graph of each inequality.

33. (a) $y \leq -2$
   (b) $x \geq 3$

34. (a) $x \leq 2$
   (b) $y \geq 3$

35. $y < -2x + 4$

36. $y > 3x - 7$

37. $y - 3x \geq 2$

38. $4x - 2y \geq 3$

In Problems 39 and 40, sketch the graph of each compound inequality.

39. (a) $y \geq -1$ and $x \leq 2$
   (b) $x - 2y < 6$ and $x + y > 2$

40. (a) $2x + y \leq 6$ and $x + y \leq 2$
   (b) $x \geq 2$ and $y \leq -3$

In Problems 41–44, find the indicated value of each given function.

41. $f(x) = 2x + 3$
   (a) $f(3)$
   (b) $f(t + 1)$
   (c) $f(-2)$
   (d) $f(t + 1) - f(t)$

42. $g(x) = |x + 5|$
   (a) $g(-8)$
   (b) $g(-5) - g(0)$
   (c) $\sqrt{g(-14)}$
   (d) $g(2) - g(-2)$

43. $f(x) = \sqrt{4 - x^2}$
   (a) $f(0)$
   (b) $\dfrac{1}{f(0)}$
   (c) $f(-2)$
   (d) $f(-1)$

44. $h(x) = \dfrac{x}{|x|}$
   (a) $h(-2)$
   (b) $h(4)$
   (c) $h(-3)$
   (d) $h(5)$

In Problems 45 and 46, find and simplify the value of the expression

$$\frac{f(t + h) - f(t)}{h}.$$

45. $f(x) = 3 - 5x$

46. $f(x) = \frac{2}{3}x + 5$

In problems 47–50, sketch the graph of each function.

47. (a) $f(x) = 3x$
   (b) $g(x) = -3x$

48. (a) $f(x) = -\dfrac{x}{3}$
   (b) $g(x) = \dfrac{x}{3}$

49. $f(x) = 3 - 4x$

50. $g(x) = \dfrac{x}{4} + 3$

In Problems 51 and 52, determine whether or not the curve is the graph of a function. Use the graph of the functions to determine the domain and range of each function.

51. (a)

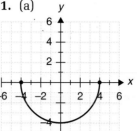

(b)

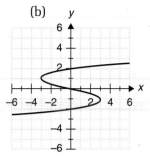

52. (a)

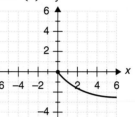

(b)

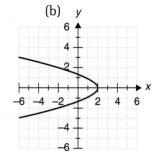

53. Gasoline Consumption: The amount of gasoline A (in gallons) in a fuel tank of a sport utility vehicle can be related to the number of miles m driven by the model.

$$A(m) = 20 - 0.05m, \ 0 < m \leq 400$$

(a) Sketch the graph of this function.
(b) Find and interpret the intercepts of the graph.
(c) How much gasoline is left in the tank after driving 200 miles?
(d) If the gas gauge registers one-fourth of the tank (20 gallon capacity), how far has this vehicle been driven?

54. Diet Program: The daily calorie allowance C to an ideal weight w (in pounds) of a young man is modeled by the function

$$C(w) = 18w - 100, \ 170 < w \leq 190.$$

(a) Sketch the graph of this model.
(b) Use the graph to determine the calorie intake for a man whose ideal weight is 180 pounds.
(c) Compute the slope of this line and indicate the units.
(d) What is the significance of the slope in terms of this problem?

55. Printing Cost: A printing company charges $20 for printing 300 business cards and $30 for printing 500 business cards. Assume the relationship between the cost C (in dollars) and the number n of cards is a linear function.

(a) Write an equation that expresses C in terms of n.

(b) Sketch the graph of this function.

(c) Use the graph to determine the cost of printing 750 business cards.

56. **Appreciation:** A home is purchased for $225,000 in 1998 and is expected to double in value by the year 2010.

(a) Find a linear function that expresses the cost C (in dollars) of the home after t years.

(b) Sketch the graph of the function.

(c) Use the equation to determine the value of the home after 8 years.

57. **Recycling:** Suppose that a recycling center recycled 500 tons of newspaper in the month of January and 530 tons of newspaper in the month of April. Suppose the increase was at a uniform rate. Let t denote the number of tons of newspaper recycled in the nth month (where $n = 1$ represents January). Assume the relationship between t and n is linear, where

$$1 \leq n \leq 12.$$

(a) Find a function that expresses t in terms of n.

(b) Find and interpret the slope of the line.

(c) In how many months does the amount of paper equal 570 tons?

58. **Hockey Statistics:** Suppose that a professional hockey player scores 12 goals in the first 10 games of the season and assume this pace continues throughout the 82 games of the season.

(a) If the relationship is linear, express the number of goals, N, in terms of the number, t, of games played.

(b) What is the significance of the slope in this problem?

(c) Sketch the graph of this function.

(d) Use this model to predict the approximate number of goals a player will score for the entire season.

In Problems 59–62, use the fact that the *tangent line* to the graph of $y = x^2 + 1$ at the point (a,b) on the graph has a slope of $2a$.

(a) Find an equation of the tangent line to the graph of $y = x^2 + 1$ at each of the given points.

(b) Graph $y = x^2 + 1$ and each tangent line to see whether or not the line appears to be tangent to the curve at the given point.

59. $(1,2)$

60. $(-1,2)$

61. $(-0.5,1.25)$

62. $(0.5,1.25)$

CHAPTER 2 PRACTICE TEST

1. Use point plotting to sketch the graph of each linear equation.
   (a) $y = 4x$ (b) $y = -2x + 3$
   (c) $x + 3y - 6 = 0$

2. Find the x and y intercepts of each linear equation. Also sketch the graph.
   (a) $3x - 2y - 6 = 0$ (b) $5x - 15 = 0$
   (c) $5y + 5 = 0$

3. Sketch the graph of each line L with the given information. Indicate whether the line rises or falls, find the slope.
   (a) L contains $(-3,2)$ and $(-1,7)$
   (b) $m = -\frac{2}{3}$ and contains $(2,5)$

4. Express the equation $5x - 3y - 15 = 0$ in slope-intercept form. Find the slope and the y intercept. Sketch the graph.

5. Determine whether the given lines are parallel or perpendicular or neither.
   L_1: contains $(8,-6)$ and $(-4,3)$
   L_2: contains $(3,3)$ and $(9,11)$

6. Find an equation of a line L in slope-intercept form and in general form. Also graph the line.
   (a) L contains $(0,5)$ and $(7,-2)$.
   (b) L contains $(-4,-1)$ and $m = -\frac{4}{3}$.
   (c) L contains $(1,-3)$ and is perpendicular to the line $7x - 3y - 4 = 0$.

7. Sketch the graph of each inequality.
   (a) $y \leq -x + 3$ (b) $y \geq 2x + 5$
   (c) $2x > 4$

8. Sketch the graph of the inequalities $x \leq 3$ and $y \geq 3$.

9. Let $g(x) = \sqrt{16 - x^2}$, find each value.
   (a) $g(0)$ (b) $g(4)$
   (c) $[g(2)]^2$ (d) $g(\sqrt{7})$

10. Find the domain and range of the graph in Figure 11.

Figure 11

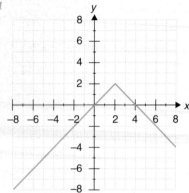

11. In business, the total cost (*y* in dollars) of producing *x* items of a product is given by an equation

$$y = mx + b$$

where *b* is the fixed cost and *m* is the variable cost. Suppose it cost $4,500 to start a business selling ice cream cones. Each cone costs $0.45 to produce.

(a) Write a cost function for this business.

(b) What is the cost *y* (in dollars) of producing 500 ice cream cones based on this model?

(c) How many ice cream cones can be produced if the total cost is $5,355?

12. The graph in Figure 12 shows the value of a photocopier over the first 5 years of ownership. Suppose that the trade in or salvage value of the photocopier is $1,000 after 5 years.

(a) Write a linear function that expresses the value *V* (in dollars) of the photocopier in terms of *t* (years).

(b) What is the initial value of the photocopier?

(c) What is the annual depreciation of the photocopier in each of the first 5 years?

Figure 12

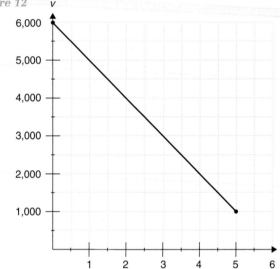

Systems of Linear Equations and Inequalities

In aviation the term *jet stream* is used to refer to a constant wind at high altitudes. Suppose a jetliner flies 2500 miles from Los Angeles to New York with the jet stream (with the wind) in 3.75 hours and makes the return trip from New York to Los Angeles against the jet stream (against the wind) in 4.4 hours. To find the speed of the jetliner in still air and the speed of the wind, we use a system of linear equations. See example 3 in section 3.4.

Up to now, we have solved equations and inequalities in one variable. In this chapter, we turn our attention to a set of two or more equations or inequalities containing two or more variables, called **systems.** Examples are:

$$\begin{cases} x + y = -7 \\ 3x - 4y = -35 \end{cases} \quad \begin{cases} 3x - y + z = 5 \\ x + 2y - 2z = -3 \\ 2x + 3y + z = 4 \end{cases} \text{ and } \begin{cases} x + 4y \le 9 \\ 5x - 2y < 7 \end{cases}$$

These systems are commonly used in geometry, science, business, and economics to describe relationships, limitations, and constraints. To **solve** a system of linear equations in two or more variables means to find values of the variables that make all equations in the system true.

Objectives

1. Solve a Linear System by Graphing
2. Explore the Geometry–Algebra Connection
3. Solve Applied Problems

3.1 Solving Linear Systems in Two Variables Graphically

Suppose we are to find the dimensions of a rectangular basketball court whose perimeter is 80 meters, and whose length is 2 meters shorter than twice its width. If we let x be the length of the court (in meters) and y be the width (in meters), we get the following system of linear equations:

$$\begin{cases} 2x + 2y = 80 \\ x = 2y - 2 \end{cases}.$$

Such a collection of equations is called a **system of linear equations in two variables.** The **solution** of this system is a set of ordered pairs that makes both equations true. Now we ask the questions: Is there one such ordered pair? Is there more than one pair? Is there no pair at all? To answer these questions, we examine the graphs of both equations on the same coordinate axis.

Solving a Linear System by Graphing

A common error is to solve a system for only one of the two variables and not both variables.

In this section, we use *graphs* of linear equations to understand why some systems have one solution, while others have an infinite number of solutions, and still others have no solution at all. The **graphical method** gives us a way of finding the numerical solution of a linear system of equations and also lends a visual meaning to the solution. In this case, the **points of intersection** of the two graphs of the linear equations in the system identify an ordered pair as the solution of the system. To locate the point of intersection in a graph drawn by hand requires judgment and estimation usually resulting in limited accuracy. However, as we shall see, these graphs can be produced easily and with great accuracy by using a grapher.

EXAMPLE 1   **Solving a System by Graphing**

Solve the system $\begin{cases} 2x + y = 0 \\ 3x + 4y = 5 \end{cases}$ graphically.

Solution   First we solve each equation for y in terms of x:

$$2x + y = 0 \qquad 3x + 4y = 5$$
$$y = -2x \qquad 4y = -3x + 5$$
$$y = \frac{-3}{4}x + \frac{5}{4}.$$

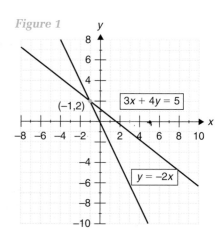

Figure 1

Now we graph both equations on the same coordinate system. As Figure 1 indicates, the two lines intersect at the point $(-1,2)$, and this point is the solution of the system.

Check

Equations	$x = -1$ $y = 2$	True/False
$2x + y = 0$	$2(-1) + 2 = 0$	True
$3x + 4y = 5$	$3(-1) + 4(2)$	True
	$= -3 + 8 = 5$	

Thus $(-1, 2)$ is the solution for the system because it makes both equations true.

EXAMPLE 2   **Solving a System by Graphing**

Solve the system $\begin{cases} 2x - y = 2 \\ -2x + y = 4 \end{cases}$ graphically.

Solution   Once again, we solve each equation for y:

$$2x - y = 2 \qquad\qquad -2x + y = 4$$
$$y = 2x - 2 \qquad\qquad y = 2x + 4$$

From the graph in Figure 2, the two lines appear to be parallel. If we examine both equations closely, we see that the slopes are the same. The y intercepts are -2 and 4. Thus, the lines are parallel and do not intersect; the system has no solution.

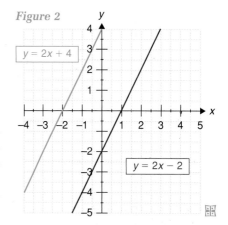

Figure 2

EXAMPLE 3   **Solving a System Graphically**

Solve the system $\begin{cases} 2x - y = 2 \\ -6x + 3y = -6 \end{cases}$ graphically.

Solution   We solve each equation for y in terms of x:

$$2x - y = 2 \qquad\qquad -6x + 3y = -6$$
$$y = 2x - 2 \qquad\qquad 3y = 6x - 6$$
$$\qquad\qquad\qquad y = 2x - 2$$

Figure 3 shows that both equations describe the same line. By examining the second equation, we see that it is the first equation multiplied by -3. Thus, the two equations are equivalent and have the same ordered pair solutions. So, the system has an infinite number of solutions.

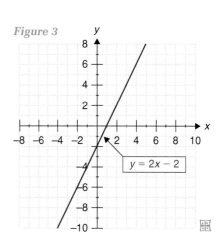

Figure 3

Exploring the Algebra – Geometry Connection

In graphing a system of two linear equations in two variables, x and y, we expect the following three possible outcomes.

1. The equations describe lines which intersect at one point. Such a system has one solution (x,y), the point of intersection (see example 1). It is called a **consistent** system and the equations are said to be **independent.**

2. The equations describe parallel lines. The graphs have no point of intersection (see example 2). Such a system has no solutions and is called an **inconsistent** system.

3. The equations describe the same line. The graphs coincide (see example 3). Such a system has an infinite number of solutions and it is called a **consistent** system. The equations are said to be **dependent.**

The three possibilities show the algebra – geometry connection and are summarized in Table 1.

TABLE 1

Geometry			
Algebra	The equations have different slopes when written in the form $y = mx + b$.	The equations have the same slopes but different y intercepts when written in the form $y = mx + b$.	The equations have the same slopes and the same y intercept when written in the form $y = mx + b$.
Type of Solution	Two lines intersect at one point. The system isconsistent and the equations are independent.	Two lines are parallel. The system is inconsistent.	Two lines coincide. The system is consistent and the equations are dependent.

EXAMPLE 4   **Solving Consistent and Inconsistent Systems**

Write each equation of the given system in the slope-intercept form. Then sketch the graphs of the equations in each system on the same coordinate axis. Indicate whether the system is consistent or inconsistent, and whether the equations are dependent or independent. Do not solve the system.

(a) $\begin{cases} x - 2y = 4 \\ -2x + 4y = -8 \end{cases}$ (b) $\begin{cases} 3x - 4y = -12 \\ \dfrac{1}{2}x = \dfrac{2}{3}y + 2 \end{cases}$ (c) $\begin{cases} 2x + 3y = 6 \\ -3x + 4y = 8 \end{cases}$

Solution   (a) $x - 2y = 4$ $-2x + 4y = -8$

$\qquad -2y = -x + 4$ $4y = 2x - 8$

$\qquad y = \dfrac{1}{2}x - 2$ $y = \dfrac{1}{2}x - 2$

These equations are equivalent because they have the same slope and the same y intercept. Figure 4a shows that the lines coincide. The system is consistent and the equations are dependent.

(b) $3x - 4y = -12$ $\qquad\qquad\qquad$ $\dfrac{1}{2}x = \dfrac{2}{3}y + 2$

$\qquad\qquad -4y = -3x - 12$ $\qquad\qquad$ $\dfrac{2}{3}y + 2 = \dfrac{1}{2}x$

$\qquad\qquad\qquad y = \dfrac{3}{4}x + 3$ $\qquad\qquad$ $\dfrac{2}{3}y = \dfrac{1}{2}x - 2$

$\qquad\qquad\qquad\qquad\qquad\qquad\qquad\qquad y = \dfrac{3}{4}x - 3$

The equations have the same slope but different y intercepts. Figure 4b shows that the lines are parallel. The system is inconsistent.

(c) $2x + 3y = 6$ $\qquad\qquad\qquad\qquad$ $-3x + 4y = 8$

$\qquad\qquad 3y = -2x + 6$ $\qquad\qquad\qquad$ $4y = 3x + 8$

$\qquad\qquad\quad y = \dfrac{-2}{3}x + 2$ $\qquad\qquad\qquad$ $y = \dfrac{3}{4}x + 2$

The equations have different slopes. Figure 4c shows that the lines intersect at one point. The system is consistent and the equations are independent.

Figure 4

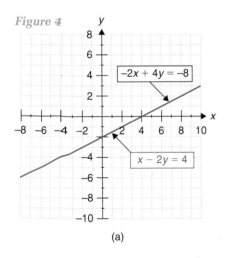

(a)

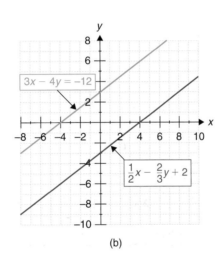

(b)

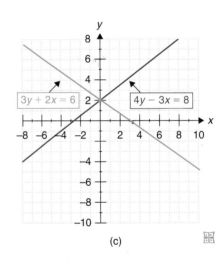

(c)

Solving Applied Problems

Some models from economics that arise in practical situations can be described by systems of linear equations. Suppose that p denotes the price per unit of a commodity, and q denotes the number of units of the commodity *demanded* in the marketplace at price p. The equation that relates q to p is called a **demand function.** If this equation is linear, then the graph is a line (Figure 5a). Linear demand functions usually have negative slopes; as prices drop more people want to buy the commodity. On the other hand, let p still denote price, but let q denote the number of units *supplied* in the marketplace at price p. The equation that relates q to p is called a **supply function.** If this equation is linear, then the graph is a line (Figure 5b). Linear supply functions usually have positive slopes; the more prices rise the more companies want to sell the commodity. If the demand and supply functions are graphed on the same coordinate system, the point where they intersect is called the **market equilibrium point** (Figure 5c). At this point, the quantity

Figure 5

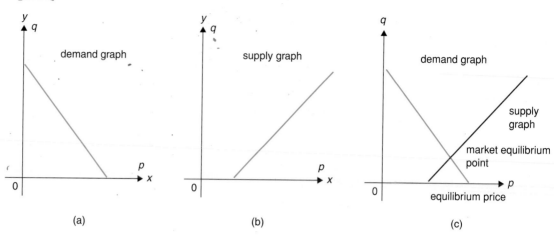

(a)   (b)   (c)

demanded will be equal to the quantity supplied. Under the usual interpretations of price, demand, and supply, only the portions of the demand and supply graphs that fall in the first quadrant are economically meaningful.

EXAMPLE 5   **Modeling Supply and Demand**

Suppose that the weekly demand D (in thousands) of a camera is related to the price of p dollars by the function:

$$D(p) = -\frac{1}{25}p + 7.4.$$

Also suppose that a manufacturer is willing to supply S (in thousands) of these cameras per week at a market price of p dollars per unit according the supply function:

$$S(p) = \frac{1}{30}p + 3.$$

(a) Sketch the graphs of both functions on the same coordinate system. Use p as the horizontal axis and q as the vertical axis.

(b) Find the market equilibrium point.

(c) Find the equilibrium price.

Figure 6

Solution

(a) First we graph both equations on the same coordinate system. Figure 6 shows the graphs of both equations.

(b) Using the graph in Figure 6, we see that the market equilibrium point is the point of intersection of the two graphs $(p,q) = (60,5)$.

(c) The equilibrium price is the p coordinate of the point of intersection, so that $p = 60$. This means that the demand is equal to the supply when the price of each camera is $60.

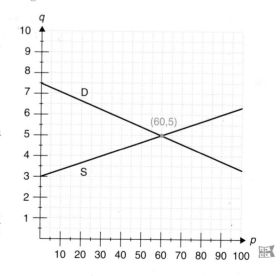

◆ **PROBLEM SET 3.1**

Mastering the Concepts

In Problems 1–12, solve each system graphically by sketching the graphs of the equations in the system on the same coordinate system. Then estimate the point of intersection (if any) of the two lines.

1. $\begin{cases} y = -x + 8 \\ y = 2x + 11 \end{cases}$ **2.** $\begin{cases} y = 2x \\ y = -3x - 5 \end{cases}$

3. $\begin{cases} x - 2y = 5 \\ 5x + y = 3 \end{cases}$ **4.** $\begin{cases} 2x = y + 3 \\ 4x - 2y = 5 \end{cases}$

5. $\begin{cases} y = 3x - 8 \\ 2x - y = 6 \end{cases}$ **6.** $\begin{cases} y = 2x - 3 \\ 4x - 2y = 6 \end{cases}$

7. $\begin{cases} x + 3y = 6 \\ 2x + 6y = 8 \end{cases}$ **8.** $\begin{cases} 2x - y = 6 \\ -4x + 2y = 10 \end{cases}$

9. $\begin{cases} 4x + 6y = 12 \\ 2x + 3y = -6 \end{cases}$ **10.** $\begin{cases} 2x + 5y = -9 \\ 3x - 4y = -2 \end{cases}$

11. $\begin{cases} 2x - 3y = -6 \\ -x + \dfrac{3}{2}y = 3 \end{cases}$ **12.** $\begin{cases} -2x + y = -8 \\ x - \dfrac{1}{2}y = 4 \end{cases}$

In Problems 13–18, the graphs of systems of linear equations are given. Use these graphs to determine the number of solutions of each system of equations. Do not solve the system.

13. 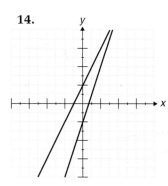   **14.**

15. 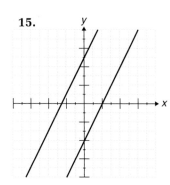   **16.**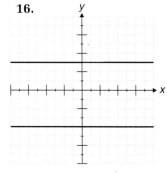

17. 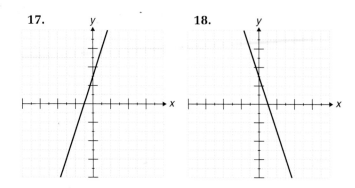   **18.**

In Problems 19–24, write the two equations in the given system in slope-intercept form. Sketch the graphs of the equations in each system on the same coordinate system. Then indicate whether the system is consistent and dependent, consistent and independent, or inconsistent.

19. $\begin{cases} x + y = 6 \\ x - y = 4 \end{cases}$ **20.** $\begin{cases} x - y = -9 \\ x + y = -1 \end{cases}$

21. $\begin{cases} 2x + 3y = 5 \\ 4x + 6y = 7 \end{cases}$ **22.** $\begin{cases} -x + 2y = 1 \\ 3x - 6y = 5 \end{cases}$

23. $\begin{cases} 2x + 3y = 5 \\ -6x - 9y = -30 \end{cases}$ **24.** $\begin{cases} 2x - y = 0 \\ 4x - 2y = 0 \end{cases}$

In Problems 25–30, write the two equations in the given system in slope-intercept form. Indicate the number of solutions of the system. Do not solve the system.

25. $\begin{cases} 1.5x - 2y + 1 = 0 \\ 4y - 3x + 7 = 0 \end{cases}$ **26.** $\begin{cases} -2y = 3 \\ -4x + 8y = 7 \end{cases}$

27. $\begin{cases} x + y = 0.5 \\ x - y = -3 \end{cases}$ **28.** $\begin{cases} 3y = 2x - 6 \\ 3y = 2x + 4 \end{cases}$

29. $\begin{cases} x - y = 6 \\ -x + y = -6 \end{cases}$ **30.** $\begin{cases} 2x - 3 = 2y \\ y = x - 15 \end{cases}$

Applying the Concepts

31. Modeling Supply and Demand: In a particular city, the monthly supply and demand for a new type of tennis racket of price p (in dollars) and q quantity, are modeled by the equations:

$$\begin{cases} S(p) = 38p - 1710 \\ D(p) = -46p + 4002 \end{cases} \quad \$45 \le p \le \$87.$$

(a) Sketch the graphs of the two equations on the same coordinate system. Let the horizontal axis be p and the vertical axis be q.

(b) Use the graphs of these models to estimate the equilibrium point. That is, at what price does the quantity supplied equal the quantity demanded?

32. **Modeling Supply and Demand:** In a particular area, the monthly supply and demand for a popular CD at price p (in dollars) is modeled by the equations:

$$\begin{cases} D(p) = -3p + 180 \\ S(p) = 4p + 110 \quad \$5 \le p \le \$15. \end{cases}$$

(a) Sketch the graphs of the two equations on the same coordinate system.

(b) Use the graphs of these models to estimate the equilibrium point. That is, at what price does the supply equal the demand?

Developing and Extending the Concepts

33. Describe the graphs of the two equations in a given system that has: (a) one solution, (b) no solution, and (c) an infinite number of solutions.

34. How do the graphs of the two equations in a given system differ?
   (a) If the system is consistent and independent.
   (b) If the system is consistent and dependent.
   (c) If the system is inconsistent.

35. Describe how you might write a pair of equations in a given system that has (a) no solution or (b) an infinite number of solutions.

36. Figure 7 shows the graphs of two linear equations. Which of the following ordered pairs could possibly be a solution for the system?
   (a) $(6,6)$ (b) $(-6,6)$ (c) $(-6,-6)$ (d) $(6,-6)$

Figure 7

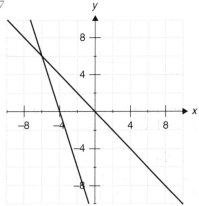

In Problems 37–40, determine the value of k so that the given system is classified as follows.

37. $\begin{cases} 2x - 3y = -6 \\ kx + 3y = 6 \end{cases}$ consistent and dependent.

38. $\begin{cases} 2x + ky = 10 \\ y = 5 - x \end{cases}$ consistent and dependent.

39. $\begin{cases} -x + 2y = 4 \\ x + ky = 0 \end{cases}$ inconsistent.

40. $\begin{cases} x - ky = -5 \\ 2x + y = 2 \end{cases}$ inconsistent.

In Problems 41 and 42, determine the restrictions on m and b so that the given system has the specified number of solutions.

41. $\begin{cases} y = m_1 x + b_1 \\ y = m_2 x + b_2 \end{cases}$ no solutions.

42. $\begin{cases} y = m_1 x + b_1 \\ y = m_2 x + b_2 \end{cases}$ one solution.

In Problems 43–46, graph each system and estimate the price p for which the supply and demand quantities are equal.

43. $\begin{cases} D(p) = 2100 - 100p \\ S(p) = -2400 + 400p \end{cases}$ $\$6 \le p \le \$12$

44. $\begin{cases} D(p) = 92 - 0.1p \\ S(p) = 47 + 0.05p \end{cases}$ $\$250 \le p \le \$400$

45. $\begin{cases} D(p) = 3774 - 14p \\ S(p) = -6850 + 90p \end{cases}$ $\$77 \le p \le \$110$

46. $\begin{cases} D(p) = 35 - 0.001p \\ S(p) = 11 + 0.005p \end{cases}$ $\$3500 \le p \le \$4300$

In Problems 47 and 48, translate the given information into a system of linear equations. Graph the system and estimate the solution of each system.

47. The sum of two numbers is -1 and their difference is 11. What are the two numbers?

48. A number is 4 less than a larger number. Two fifths of the larger number plus one third of the smaller is 6. What are the two numbers?

49. [C] **Data Analysis:** The following table shows the total employment in a certain industry from the years 1995 to 1997.

Year	1995	1996	1997
Employment	898	838	758

(a) Use the [REGRESSION] feature of a calculator to find the regression line or "best fit" line for these data points. Let $x = 0$ correspond to 1995.

(b) From statistics, it can be shown that the solution to the system:

$$\begin{cases} 3m + 3b = 2494 \\ 5m + 3b = 2354 \end{cases}$$

gives the slope, m, and y intercept, b, of the "best fit" line. Solve this system graphically and compare m and b to the line in part (a).

50. **Van Rental:** Two van rental companies A and B offer the following options for renting their vans. Company A will rent a 24 foot van for $40 plus $0.49 per mile. Company B will rent the same van for $66 plus $0.39 per mile.

(a) Suppose you are moving 200 miles round trip, which company is the cheapest to use?

(b) Suppose you are moving 350 miles round trip, which company is the cheapest to use?

(c) Write an equation for each company relating cost C (in dollars) as a function of miles, x.

(d) Write and solve graphically an algebraic system to determine the number of miles traveled when the cost of renting from A is equal to the cost of renting from B. Hint: scale the x axis from 0 to 400 miles with increments of 20 miles and the C axis from $40 to $250 with increments of $20.

Objectives

1. Solve Linear Systems by Substitution
2. Solve Linear Systems by Elimination
3. Examine Types of Systems Algebraically
4. Solve Applied Problems

3.2 Algebraic Solutions of Linear Systems in Two Variables

Although the graphing method provides an excellent means of approximating solutions to linear systems of equations in two variables, it does not always allow us to locate the exact solutions, especially if points of intersection are not located at integer values. For this reason, we use *algebraic methods* for solving these systems. We will see that such methods always give accurate results. In this section, we discuss two algebraic methods: the *substitution* and the *elimination methods.*

We use the first method to solve the system

$$\begin{cases} 2x + 4y = -4 \\ x - y = -5 \end{cases}$$

We find $x = y - 5$ from the second equation, and substitite $y - 5$ for x in the first equation to obtain

$$2(y - 5) + 4y = -4 \quad \text{or} \quad 2y - 10 + 4y = -4$$

$$6y - 10 = -4 \quad \text{or} \quad 6y = 6 \quad \text{or} \quad y = 1$$

so that $x = 1 - 5 = -4$

and the solution is $(-4, 1)$.

Solving Linear Systems by Substitution

The technique for solving a linear system of two equations in two variables by the **substitution method** is outlined in the following strategy.

Solving Linear Systems by Substitution

1. Choose one of the equations; solve it for one of the variables in terms of the other variable.
2. Substitute the expression from step 1 for the variable in the other equation to obtain an equation in one variable.
3. Solve the equation resulting from step 2.
4. Substitute the value from step 3 back into either of the original equations to find the value of the second variable.

EXAMPLE 1 **Solving a System by Substitution**

Use the substitution method to solve the system

$$\begin{cases} 3x + y = 7 \\ 2x - 3y = 1. \end{cases}$$

Solution

$3x + y = 7$	Choose first equation.
$y = -3x + 7$	Solve equation 1 for y in terms of x.
$2x - 3(-3x + 7) = 1$	Substitute $-3x + 7$ for y in equation 2.
$2x + 9x - 21 = 1$	
$11x = 22$	
$x = 2$	
$3(2) + y = 7$	Substitute 2 for x in equation 1.
$6 + y = 7$	
$y = 1$	

Thus, (2,1) is a solution of the system.

Check:

Equations	$x = 2$ $y = 1$	True/False
$3x + y = 7$	$3(2) + 1 = 7$	True
$2x - 3y = 1$	$2(2) - 3(1) = 1$	True

Therefore, (2,1) satisfies both equations.

EXAMPLE 2 **Solving a System by Substitution**

Use the substitution method to solve the system.

$$\begin{cases} \dfrac{x}{3} - \dfrac{y}{2} = \dfrac{8}{3} \\ -5x + 7y = -39 \end{cases}$$

Solution We begin by rewriting the first equation without fractions.

$6\left(\dfrac{x}{3} - \dfrac{y}{2}\right) = 6\left(\dfrac{8}{3}\right)$	Multiply by 6, the LCD.
$2x - 3y = 16$	
$2x = 3y + 16$	Solve for x.
$x = \dfrac{3}{2}y + 8$	
$-5x + 7y = -39$	Second equation.
$-5\left(\dfrac{3}{2}y + 8\right) + 7y = -39$	Substitute $\dfrac{3}{2}y + 8$ for x in equation 2.
$2\left[\dfrac{-15}{2}y - 40 + 7y = -39\right]$	

$$2\left[\frac{-15}{2}y - 40 + 7y\right] = 2(-39) \qquad \text{Multiply by 2, the LCD.}$$

$$-15y - 80 + 14y = -78$$

$$-y = 2$$

$$y = -2$$

Substituting -2 for y in equation 2, we have

$$-5x + 7(-2) = -39$$

$$-5x - 14 = -39$$

$$-5x = -25$$

$$x = 5$$

Therefore, the solution is $(5, -2)$. The reader is encouraged to check the solution.

Solving Linear Systems by Elimination

Another algebraic technique for solving linear systems of two equations in two variables is the **elimination method.** The strategy of this method is outlined in the following procedure.

Solving a System by Elimination

1. Write each equation in the form $Ax + By = C$.
2. Multiply the terms of one or both equations by appropriate numbers so that the coefficients of one of the variables are opposites.
3. Add the resulting equations to produce a single equation in one variable, and solve that equation for the variable.
4. Substitute the value obtained in step 3 into either equation in the original system and solve this equation for the second variable.

In using the elimination method, our goal is to transform the original system into an "equivalent" system so that the coefficients of one of the variables are opposites. **Equivalent systems** of equations are those systems that have exactly the same solutions. The following three operations produce equivalent systems.

1. Interchange two equations in the system.
2. Multiply one equation of the system by a non-zero constant.
3. Multiply another equation in the system by a non-zero constant and add it to a given equation in the system.

EXAMPLE 3   **Solving a System by Elimination**

Use the elimination method to solve the system.

$$\begin{cases} 3x - 2y = 4 \\ 2x + y = 5 \end{cases}$$

Solution   Our objective is to transform our original system into an equivalent system so that the coefficients of one of the variables are opposites. We then add the equations to eliminate this variable. Notice that each equation in the system is of the form $Ax + By = C$. If we multiply each side of equation 2 by 2, the coefficients of the y terms will be opposites.

$$\begin{cases} 3x - 2y = 4 \\ 2x + y = 5 \end{cases} \xrightarrow{\quad \text{no change} \quad} \begin{cases} 3x - 2y = 4 \\ 4x + 2y = 10 \end{cases}$$

Now we add the two equations in the resulting system to obtain an equation in one variable.

$$\begin{cases} 3x - 2y = 4 \\ 4x + 2y = 10 \end{cases}$$
$$7x \quad\quad = 14$$
$$x = 2$$

By substituting 2 for x into equation 1 in the original system, we have

$$3(2) - 2y = 4$$
$$6 - 2y = 4$$
$$-2y = -2$$
$$y = 1.$$

The solution is (2,1).

EXAMPLE 4   **Solving a System by Elimination**

Use the elimination method to solve the system.

$$\begin{cases} 0.3x + 0.2y = 0.8 \\ 0.2x - 0.3y = 1.4 \end{cases}$$

Solution   Each equation in the system is of the form $Ax + By = C$. We multiply each side of both equations by 10 to clear the decimals.

$$\begin{cases} 0.3x + 0.2y = 0.8 \\ 0.2x - 0.3y = 1.4 \end{cases} \xrightarrow{\quad \text{multiply each side by 10} \quad} \begin{cases} 3x + 2y = 8 \\ 2x - 3y = 14 \end{cases}$$

Next we make the coefficients of y opposites and eliminate y.

$$\begin{cases} 3x + 2y = 8 \\ 2x - 3y = 14 \end{cases} \xrightarrow{\quad \text{multiply each side by 3} \quad} \begin{cases} 9x + 6y = 24 \\ 4x - 6y = 28 \end{cases}$$

Now we add the two equations to obtain an equation in one variable.

$$\begin{cases} 9x + 6y = 24 \\ 4x - 6y = 28 \end{cases}$$
$$13x \quad\quad = 52$$
$$x = 4$$

By substituting 4 for x in equation 1 of the original system, we have

$$0.3(4) + 0.2y = 0.8$$
$$1.2 + 0.2y = 0.8$$
$$0.2y = -0.4$$
$$y = -2.$$

Thus the solution is $(4, -2)$. Notice that we could have substituted 4 for x in equation 1 in the first "equivalent system" to obtain

$$3(4) + 2y = 8$$
$$12 + 2y = 8$$
$$2y = -4$$
$$y = -2.$$

Thus, the same solution $(4, -2)$ is obtained.

Examining Types of Systems Algebraically

So far each example we have solved in this section produces exactly one solution. The next example illustrates systems with either no solution or an infinite number of solutions. These possibilities should always be kept in mind when solving a linear system, regardless of the method used.

EXAMPLE 5   **Solving Inconsistent and Dependent Systems**

Use the elimination method to solve each system.

(a) $\begin{cases} 2x - 3y = 1 \\ -4x + 6y = -2 \end{cases}$ (b) $\begin{cases} -2x + 4y = -4 \\ 3x - 6y = 0 \end{cases}$

Solution   (a) Each equation in the system is of the form $Ax + By = C$. The coefficients of x and y will be opposites if we perform the following multiplication.

$\begin{cases} 2x - 3y = 1 \\ -4x + 6y = -2 \end{cases}$ —— multiply each side by 2 ——→ —— no change ——→ $\begin{cases} 4x - 6y = 2 \\ -4x + 6y = -2 \end{cases}$

Adding, we get

$$\begin{cases} 4x - 6y = 2 \\ -4x + 6y = -2 \end{cases}$$
$$0 = 0.$$

We obtain the equation $0 = 0$ which is true for all values of x and y. This tells us that the system is *consistent* and the equations are *dependent*. Thus the system has an infinite number of solutions. If we write each equation in slope-intersect form, we have

$$y = \frac{2}{3}x - \frac{1}{3}.$$

We see that the graphs of these equations are identical (Figure 1a).

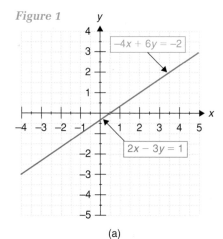

Figure 1

$-4x + 6y = -2$

$2x - 3y = 1$

(a)

Figure 1

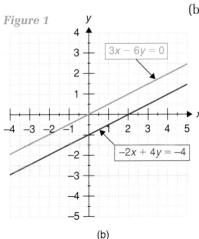

(b)

(b) Once again, each equation is of the form $Ax + By = C$. The coefficients of x and y will be opposites if we perform the following multiplication.

$$\begin{cases} -2x + 4y = -4 \quad\text{—— multiply each side by 3} \longrightarrow \\ 3x - 6y = 0 \quad\text{—— multiply each side by 2} \longrightarrow \end{cases} \begin{cases} -6x + 12y = -12 \\ 6x - 12y = 0 \end{cases}$$

Adding, we get

$$\begin{cases} -6x + 12y = -12 \\ \underline{6x - 12y = 0} \\ 0 = -12. \end{cases}$$
$$0 = -12.$$

There are no values of x and y for which $0 = -12$ is true, so the system is *inconsistent* and has no solutions. Figure 1b shows the graph that verifies our conclusion.

Solving Applied Problems

Linear systems arise naturally as models for real-world situations. In business, the point at which a company's costs equal its revenues is called the **break-even point.** If C represents the production cost (in dollars) of x units of a product, and R represents the revenue (in dollars) from the sale of x units of this product, then the break-even point is determined when $C = R$.

EXAMPLE 6   **Determining the Break-Even Point**

A fitness equipment manufacturer produces a special kind of exercise bicycle. The cost C (in dollars) of producing x bicycles is given by

$$C(x) = 220x + 180,000.$$

The revenue R (in dollars) from selling x bicycles is given by

$$R(x) = 310x.$$

The graph of the system is in Figure 2.

(a) Algebraically, find the number of units produced and sold in order to break even.

Figure 2

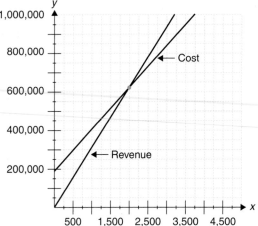

(b) Use the graph to indicate when a loss occurs.

(c) Use the graph to indicate when a profit occurs.

Solution   (a) The system of equations which models this situation is:

$$\begin{cases} C = 220x + 180,000 \\ R = 310x \end{cases}.$$

Since at the break-even point, $C = R$, we have the system:

$$\begin{cases} C = 220x + 180,000 \text{ — multiply each side by } -1 \rightarrow \\ C = 310x \text{\quad——— no change ———} \longrightarrow \end{cases} \begin{cases} -C = -220x - 180,000 \\ C = 310x \end{cases}$$

Adding gives us

$$\begin{cases} -C = -220x - 180{,}000 \\ \ \ C = \ \ \ 310x \end{cases}$$
$$\overline{\ \ \ 0 = \ \ \ \ 90x - 180{,}000}$$
$$90x = 180{,}000$$
$$x = 2{,}000.$$

Therefore, when 2000 units are produced and sold, the cost $620,000 of production equals the revenue $620,000.

(b) The graph of the system is in Figure 2. A loss occurs when cost exceeds revenue, that is, when the cost line is above the revenue line. Therefore, when fewer than 2,000 units are produced and sold, the company suffers a loss.

(c) A profit occurs when revenue exceeds cost, that is, when the revenue line is above the cost line. Therefore, when more than 2,000 units are produced and sold, the company enjoys a profit.

❖ PROBLEM SET 3.2

Mastering the Concepts

In Problems 1–18, use the substitution method to solve each system.

1. $\begin{cases} 2x - y = 5 \\ x + 3y = 13 \end{cases}$

2. $\begin{cases} 3x + y = 4 \\ 7x + 3y = 6 \end{cases}$

3. $\begin{cases} 5p - q = 13 \\ p + q = 1 \end{cases}$

4. $\begin{cases} 3r - 4t = 8 \\ 2r + 2t = 3 \end{cases}$

5. $\begin{cases} 2x - 3y = 4 \\ 3x + y = 1 \end{cases}$

6. $\begin{cases} 2x + 3y = 24 \\ 3x + 4y = -12 \end{cases}$

7. $\begin{cases} 3x - y = -2 \\ x + y = 6 \end{cases}$

8. $\begin{cases} 5x + 7y = 4 \\ -x + 6y = 14 \end{cases}$

9. $\begin{cases} 13a + 11b = 21 \\ 7a + 6b = -3 \end{cases}$

10. $\begin{cases} 2u - v = 3 \\ 3u + v = 22 \end{cases}$

11. $\begin{cases} 6x + y = -18 \\ -3x - y = 9 \end{cases}$

12. $\begin{cases} 0.2m - 0.3n = 0.1 \\ 0.3m + 0.2n = 2.1 \end{cases}$

13. $\begin{cases} 0.5x - 1.2y = 0.3 \\ 0.7x + 1.5y = 3.6 \end{cases}$

14. $\begin{cases} \dfrac{3}{4}x + 3y = 30 \\ -5x - y = 28 \end{cases}$

15. $\begin{cases} 5p - q = 13 \\ p + 3q = -\dfrac{5}{3} \end{cases}$

16. $\begin{cases} 5r + s = 3 \\ -r - 2s = -\dfrac{12}{5} \end{cases}$

17. $\begin{cases} \dfrac{1}{2}x + \dfrac{1}{6}y = 1 \\ \dfrac{2}{3}x - \dfrac{1}{2}y = -3 \end{cases}$

18. $\begin{cases} \dfrac{3}{4}x - \dfrac{1}{12}y = -\dfrac{5}{2} \\ \dfrac{3}{8}x - \dfrac{1}{6}y = -\dfrac{1}{2} \end{cases}$

In Problems 19–40, use the method of elimination to solve each system.

19. $\begin{cases} 2x - 3y = -39 \\ 4x + 3y = 3 \end{cases}$

20. $\begin{cases} 2x + 3y = 18 \\ -2x + 5y = -34 \end{cases}$

21. $\begin{cases} 7x - 3y = -70 \\ 7x + 2y = -70 \end{cases}$

22. $\begin{cases} 5x - 2y = -12 \\ 3x - 2y = -4 \end{cases}$

23. $\begin{cases} 2x - y = 1 \\ x + y = 2 \end{cases}$

24. $\begin{cases} 2x + y = 10 \\ 3x - y = 5 \end{cases}$

25. $\begin{cases} 2r + 4s = 2 \\ -r + s = 8 \end{cases}$

26. $\begin{cases} 7p + q = 5 \\ -2p + q = 3 \end{cases}$

27. $\begin{cases} 7p + q = 3 \\ 5p + q = 6 \end{cases}$

28. $\begin{cases} 3y + 2z = 5 \\ 2y + 3z = 1 \end{cases}$

29. $\begin{cases} 4u - 3v = 1 \\ 3u + 4v = 6 \end{cases}$

30. $\begin{cases} 13y + 5z = 2 \\ 6y + 2z = 2 \end{cases}$

31. $\begin{cases} u + 3v = 9 \\ u - v = 1 \end{cases}$

32. $\begin{cases} 5u + v = 14 \\ 2u + v = 5 \end{cases}$

33. $\begin{cases} 3x + 2y = 4 \\ 5x + 3y = 7 \end{cases}$

34. $\begin{cases} 2x - 7y = 5 \\ x + y = -8 \end{cases}$

35. $\begin{cases} -3x + y = 3 \\ 4x + 2y = 10 \end{cases}$

36. $\begin{cases} 5x - 2y = 35 \\ x + 4y = 25 \end{cases}$

37. $\begin{cases} 0.1x + 0.3y = 0.4 \\ 0.2x + 0.9y = 0.3 \end{cases}$

38. $\begin{cases} 0.5x + 0.25y = 25 \\ 0.35x + 0.1y = 25 \end{cases}$

39. $\begin{cases} \frac{1}{2}x + \frac{1}{3}y = 13 \\ \frac{1}{5}x + \frac{1}{8}y = 5 \end{cases}$

40. $\begin{cases} \frac{1}{3}x - \frac{1}{4}y = 2 \\ \frac{1}{4}x - \frac{1}{2}y = 7 \end{cases}$

In Problems 41–46, use any algebraic method to solve each system. Also determine whether each system is consistent and independent, consistent and dependent, or inconsistent. Confirm your results graphically.

41. $\begin{cases} 3x - 12y = 6 \\ 2x - 8y = 4 \end{cases}$

42. $\begin{cases} 2x + 6y = 8 \\ 3x + 9y = 12 \end{cases}$

43. $\begin{cases} 3x - 2y = 6 \\ -15x + 10y = 2 \end{cases}$

44. $\begin{cases} 4x - 9y = 12 \\ -8x + 18y = 10 \end{cases}$

45. $\begin{cases} \frac{1}{2}x + \frac{1}{3}y = 11 \\ \frac{1}{3}x - \frac{1}{5}y = 1 \end{cases}$

46. $\begin{cases} \frac{2}{3}x + \frac{3}{5}y = -17 \\ \frac{1}{2}x - \frac{1}{3}y = -1 \end{cases}$

Applying the Concepts

47. Break-Even Analysis: The cost C (in dollars) and the revenue R (in dollars) for producing and selling x units of a product are modeled by the equations:

$$\begin{cases} C = 8.5x + 75 \\ R = 10x \end{cases}$$

(a) Find the number of units that must be produced and sold in order to break even. That is, find the value of x for which $C = R$.

(b) Sketch the graphs of the equations that represent C and R.

(c) Use the graph in part (b) to indicate when a loss occurs.

(d) Use the graph in part (b) to indicate when a profit occurs.

48. Break-Even Analysis: Repeat problem 47 for the following cost and revenue models.

$$\begin{cases} C = 83x + 444 \\ R = 95x \end{cases}$$

49. Health Insurance: A family has a choice of two health insurance plans. The first option costs

$400 per month and covers 80% of their health care expenses. The second option will cost $350 per month and will cover 70% of their health expenses. The total cost to the family C_1 and C_2 (in dollars) of the two options after one year are modeled by the equations:

$$\begin{cases} \text{First option: } C_1 = 0.20x + 4800 \\ \text{Second option: } C_2 = 0.30x + 4200 \end{cases}$$

where x (in dollars) is the total health care expenses for one year. At what level of health care expenses will both options cost the same amount of money for one year? That is, find the value of x for which C_1 and C_2 are equal.

50. Car Rental: A car rental agency offers two options for renting its cars. The first option charges $20 for the first day and $18 for each additional day. The second option charges $26 for the first day and $15 for each additional day. The total costs C_1 and C_2 (in dollars) of the two options are given by the models:

$$\begin{cases} C_1 = 18x + 20 \\ C_2 = 15x + 26 \end{cases}$$

where x is the number of additional days after the first. After how many additional days will both options cost the same amount of money? That is, find the value of x for which $C_1 = C_2$.

Developing and Extending the Concepts

In Problems 51–54, use the given substitution for u and v and the elimination method to solve each system.

51. $\begin{cases} 2(x + 5) + 3(y - 2) = 10 \\ (x + 5) - 3(y - 2) = -4 \end{cases}$
Let $u = x + 5$, $v = y - 2$

52. $\begin{cases} 2(x - 3) - (y + 1) = -5 \\ 3(x - 3) + 2(y + 1) = 3 \end{cases}$
Let $u = x - 3$, $v = y + 1$

53. $\begin{cases} \frac{x + 2}{2} + \frac{y + 2}{3} = 11 \\ \frac{x + 2}{3} - \frac{y + 2}{5} = 1 \end{cases}$
Let $u = x + 2$, $v = y + 2$

54. $\begin{cases} \frac{-3(x - 2)}{7} + \frac{2(y + 4)}{5} = \frac{6}{5} \\ \frac{x - 2}{5} + \frac{7(y + 4)}{3} = 7 \end{cases}$
Let $u = x - 2$, $v = y + 4$

In Problems 55–58, use the substitution $u = 1/x$ and $v = 1/y$ and the method of elimination to solve each system.

55. $\begin{cases} \dfrac{2}{x} - \dfrac{1}{y} = 9 \\ \dfrac{5}{x} - \dfrac{3}{y} = 14 \end{cases}$

56. $\begin{cases} \dfrac{8}{x} + 11 = \dfrac{5}{y} \\ \dfrac{5}{x} - \dfrac{2}{y} = 1 \end{cases}$

57. $\begin{cases} \dfrac{3}{x} + \dfrac{2}{y} = 2 \\ \dfrac{1}{x} - \dfrac{1}{y} = 9 \end{cases}$

58. $\begin{cases} \dfrac{4}{x} + \dfrac{3}{y} = 1 \\ \dfrac{3}{x} + \dfrac{4}{y} = 6 \end{cases}$

In Problems 59 and 60, determine the values of a, b, and/or c so that the system:

59. (a) $\begin{cases} 4x - by = a \\ ax + 2y = b \end{cases}$ has (3, 2) as a solution

(b) $\begin{cases} 3x - cy = 5 \\ 3x - 4y = -2 \end{cases}$ is inconsistent

60. (a) $\begin{cases} 3x - 2y = 10 \\ cx + 6y = 5 \end{cases}$ is inconsistent

(b) $\begin{cases} x + cy = 1 \\ 3x + 18y = 3 \end{cases}$ is consistent and dependent

61. **Data Analysis:** A system of equations can be used to find an equation of a line that fits a pair of points. That is, if $y = mx + b$ contains the point (3,5), then m and b must satisfy the equation $3m + b = 5$. Find an equation of a line that fits the pair of points (3,5) and (7,8).

62. **Data Analysis:** As in problem 61, find the equation of a line that fits the pair of points $(-2,7)$ and $(4,-6)$.

Objectives

1. Solve Linear Systems with Three Variables by Elimination
2. Classify Linear Systems with Three Variables
3. Solve Applied Problems

3.3 Solving Linear Systems in More than Two Variables

Many practical situations often lead to more than two equations with more than two variables. In this section, we examine linear systems with three variables such as:

$$\begin{cases} 2x - 3y + 2z = -1 \\ 4x + 8y - 3z = -16. \\ x + 7y - 3z = -8 \end{cases}$$

A **solution** of a linear system of three equations in three variables is an ordered triple (a,b,c) that satisfies all three equations in the system. For instance, the ordered triple $(-3,1,4)$ is a solution to the above system, since it satisfies each of the equations.

$$\begin{cases} 2(-3) - 3(1) + 2(4) = -1 \text{ true} \\ 4(-3) + 8(1) - 3(4) = -16 \text{ true} \\ (-3) + 7(1) - 3(4) = -8 \text{ true} \end{cases}$$

Solving Linear Systems with Three Variables by Elimination

Like the linear systems of two equations in two variables, linear systems with three variables are **equivalent** if they have the same solution. Graphical methods for illustrating the solutions of linear systems of three equations in three variables are impractical because such graphs may only be drawn in a three-dimensional coordinate system. Also, the substitution method is cumbersome when solving systems with three variables. The elimination method is the most suitable technique to solve this kind of a system. Its use is outlined in the following strategy.

Solving a Linear System of Three Equations in Three Variables

1. Write each equation in the form $Ax + By + Cz = D$.
2. Select two equations in the system and eliminate one of the variables. The result is one equation in two variables.
3. Select one of the equations from step 2 and the third equation in the system. Eliminate the same variable from these equations. The result is another equation in the same two variables as step 2.
4. Solve this system of two equations in two variables.
5. Substitute the solution from step 4 into one of the original equations and solve for the third variable to complete the solution of the original system.
6. Check the solution in each of the original equations.

EXAMPLE 1

Solving a Linear System by Elimination

Use the elimination method to solve the system

$$\begin{cases} x + 2y - z = -3 \\ 2x + 5y + 2z = 10. \\ -3x - 2y + 4z = 13 \end{cases}$$

Solution

We look at the coefficients of the variables and choose to eliminate x. First we use equations 1 and 2, then use equations 1 and 3.

$$\begin{cases} x + 2y - z = -3 \\ 2x + 5y + 2z = 10 \end{cases} \quad\begin{matrix}\text{Multiply each side by } -2 \longrightarrow \\ \text{No change} \longrightarrow\end{matrix}\quad \begin{cases} -2x - 4y + 2z = 6 \\ \underline{2x + 5y + 2z = 10} \\ y + 4z = 16 \end{cases}$$

Add equations 1 and 3 to eliminate x.

$$\begin{cases} x + 2y - z = -3 \\ -3x - 2y + 4z = 13 \end{cases} \quad\begin{matrix}\text{Multiply each side by } 3 \longrightarrow \\ \text{No change} \longrightarrow\end{matrix}\quad \begin{cases} 3x + 6y - 3z = -9 \\ \underline{-3x - 2y + 4z = 13} \end{cases}$$
Add $ 4y + z = 4$

Now we have the resulting new system

$$\begin{cases} y + 4z = 16 \\ 4y + z = 4 \end{cases}.$$

We continue to use the elimination method to solve this system. We eliminate y.

$$\begin{cases} y + 4z = 16 \\ 4y + z = 4 \end{cases} \quad\begin{matrix}\text{Multiply each side by } -4 \longrightarrow \\ \text{No change} \longrightarrow\end{matrix}\quad \begin{cases} -4y - 16z = -64 \\ \underline{4y + z = 4} \end{cases}$$
Add $ -15z = -60$
$ z = 4$

Substitute 4 for z back into the second equation in the latter system in two variables to obtain

$$4y + 4 = 4$$
$$4y = 0$$
$$y = 0.$$

Now substitute 4 for z and 0 for y into any of the three original equations. We choose equation 1 to find x.

$$x + 2y - z = -3$$
$$x + 2(0) - 4 = -3$$
$$x = 1$$

Thus, the solution to the original system is (1,0,4). Once again we encourage the reader to check the answers.

Classifying Linear Systems with Three Variables

As with the solutions of linear systems with two variables, linear systems of three variables produce three possible solutions. Although these systems do not lend themselves to graphical solutions, it is of interest to describe some possible solutions. The graph of a linear equation in three variables is a "plane" in three dimensional space. Thus the solution of such a system is the intersection of three planes. These possibilities show the algebra – geometry connection and are summarized in Table 1.

TABLE 1

Geometry	The planes intersect at a point P. The system is *consistent* and the equations are *independent*.	All three planes do not have a common intersection. The system is *inconsistent*.	The planes intersect along a line. The system is *consistent* and the equations are *dependent*.
Algebra	The equations can be solved for x, y, and z.	The variables cancel each other, and the statement is false.	The variables cancel each other, and the statement is true.
Type of Solution	One solution is the ordered triple (a,b,c).	No real number solution.	An infinite number of ordered triples solutions.

EXAMPLE 2   **Solving an Inconsistent System**

Solve the system of equations by elimination.

$$\begin{cases} x + y - z = 9 \\ x + y - z = 7 \\ x - y + z = 3 \end{cases}$$

Solution

In a given linear system of three equations in three variables, if one step produces an identity while another step produces a false equation, then the system is inconsistent.

First, we eliminate y using the first two equations.

$$\begin{cases} x + y - z = 9 \\ x + y - z = 7 \end{cases} \quad \begin{array}{l} \text{Multiply each side by } -1 \\ \text{No change} \end{array} \longrightarrow \begin{cases} -x - y + z = -9 \\ \underline{x + y - z = 7} \end{cases}$$

$$\text{Add} \qquad\qquad\qquad\qquad\qquad\qquad\qquad\qquad\qquad 0 = -2$$

Since $0 = -2$ is false, the system is inconsistent and has no solution.

EXAMPLE 3   **Solving a Consistent System**

Solve the system of equations by elimination.

$$\begin{cases} x + 2y + 3z = 1 \\ 2x + 4y - z = 2 \\ 3x + 6y - 4z = 3 \end{cases}$$

Solution

First we eliminate z from the first two equations.

$$\begin{cases} x + 2y + 3z = 1 \\ 2x + 4y - z = 2 \end{cases} \quad \begin{array}{l} \text{No change} \\ \text{Multiply each side by } 3 \end{array} \longrightarrow \begin{cases} x + 2y + 3z = 1 \\ \underline{6x + 12y - 3z = 6} \end{cases}$$

$$\text{Add} \qquad\qquad\qquad\qquad\qquad\qquad\qquad\qquad 7x + 14y = 7$$

Next we eliminate z from the second and the third equations.

$$\begin{cases} 2x + 4y - z = 2 \\ 3x + 6y - 4z = 3 \end{cases} \quad \begin{array}{l} \text{Multiply each side by } -4 \\ \text{No change} \end{array} \longrightarrow \begin{cases} -8x - 16y + 4z = -8 \\ \underline{3x + 6y - 4z = 3} \end{cases}$$

$$\text{Add} \qquad\qquad\qquad\qquad\qquad\qquad\qquad -5x - 10y = -5$$

Now we solve the system

$$\begin{cases} 7x + 14y = 7 \\ -5x - 10y = -5 \end{cases} \quad \begin{array}{l} \text{divide each side by } 7 \\ \text{divide each side by } 5 \end{array} \longrightarrow \begin{cases} x + 2y = 1 \\ \underline{-x - 2y = -1} \end{cases}$$

$$\text{Add} \qquad\qquad\qquad\qquad\qquad\qquad\qquad 0 = 0$$

The identity $0 = 0$ indicates the system is consistent and the equations are dependent.

Solving Applied Problems

In many real-world situations, linear systems of three equations in three variables serve as mathematical models, as the next example shows.

EXAMPLE 4   **Finding the Operating Cost of a Car at Different Speeds**

A car manufacturer determined that the cost C (in cents per mile) of operating a car is given by the function

$$C(x) = ax^2 + bx + c$$

where x is the speed of the car (in miles per hour) and a, b, and c are constants.

(a) Write C in terms of x if the equation fits the data points (10,22), (20,20), and (50, 20).

(b) Use the function in part (a) to determine the cost of operating the car at 70 miles per hour.

Solution   (a) Since all the data points make the equation

$$C(x) = ax^2 + bx + c$$

true, we have

$$\begin{cases} 100a + 10b + c = 22 \\ 400a + 20b + c = 20 \\ 2500a + 50b + c = 20 \end{cases}$$

Solving this system for a, b and c, we get

$$a = 0.005, \; b = -0.35 \text{ and } c = 25.$$

So that the equation relating speed to cost reads

$$C(x) = 0.005x^2 - 0.35x + 25$$

(b) Here we replace x by 70 to obtain

$$C(70) = 0.005(70)^2 - 0.35(70) + 25$$

$$C(70) = 25$$

Therefore, if the car is driven at 70 miles per hour, it will cost 25 cents per mile to operate.

 PROBLEM SET 3.3

Mastering the Concepts

In Problems 1 and 2, determine whether or not the given ordered triple is a solution of the given system.

1. $\begin{cases} x + y + z = 6 \\ x + 2y + z = 9 \\ 2x + 2y - z = 6 \end{cases}$ (1,3,2)

2. $\begin{cases} -3x + 4y + z = -13 \\ 2x - 2y + 3z = -2 \\ 5x + 6y - 2z = 29 \end{cases}$ (2, −1, −3)

In Problems 3–16, use the method of elimination to solve each system.

3. $\begin{cases} x + y + 2z = 4 \\ x + y - 2z = 0 \\ x - y \quad\quad = 0 \end{cases}$

4. $\begin{cases} x - y + z = 2 \\ x - y - z = 4 \\ 2x - y + z = 0 \end{cases}$

5. $\begin{cases} x + y + \quad z = 8 \\ x - y + 2z = 6 \\ 2x + y + \quad z = 3 \end{cases}$

6. $\begin{cases} x + \quad y + \quad z = 6 \\ x + 2y + \quad z = 8 \\ x + \quad y + 2z = 9 \end{cases}$

7. $\begin{cases} x + y = 4 \\ 3x - y + 3z = 7 \\ 5x - 7y + 2z = -2 \end{cases}$

8. $\begin{cases} 2x + y + 3z = 30 \\ 4x - 3y + 6z = 50 \\ x + y - 2z = 30 \end{cases}$

22. $\begin{cases} 3u + v - w = 8 \\ 2u - v + 2w = 3 \\ u + 2v - 3w = 5 \end{cases}$

9. $\begin{cases} 2p + q - 3r = 9 \\ p - 2q + 4r = 5 \\ 3p + q - 2r = 15 \end{cases}$

10. $\begin{cases} u + 3v - w = -2 \\ 7u - 5v + 4w = 11 \\ 2u + v + 3w = 21 \end{cases}$

23. $\begin{cases} a - 2b + c = -5 \\ -3a + 6b - 3c = 15 \\ 2a - 4b + 2c = -10 \end{cases}$

11. $\begin{cases} 2r + 3s + t = 7 \\ r - 2s + 3t = -8 \\ 3r + s - t = -6 \end{cases}$

24. $\begin{cases} -2a - 4b + 6c = -8 \\ a + 2b - 3c = 4 \\ 4a + 8b - 12c = 16 \end{cases}$

12. $\begin{cases} 8s + 3t - 18u = -76 \\ 10s + 6t - 6u = -50 \\ 4s + 9t + 12u = 10 \end{cases}$

13. $\begin{cases} a - 5b + 4c = 8 \\ 3a + b - 2c = 4 \\ 9a - 3b + 6c = 6 \end{cases}$

25. $\begin{cases} \dfrac{p}{2} + \dfrac{q}{3} - \dfrac{r}{4} = -1 \\ \dfrac{p}{3} + \dfrac{r}{2} = 8 \\ \dfrac{2p}{3} + \dfrac{q}{3} - \dfrac{3r}{4} = -6 \end{cases}$

14. $\begin{cases} 2p + 3q - 2r = 3 \\ 8p + q + r = 2 \\ 2p + 2q + r = 1 \end{cases}$

15. $\begin{cases} 3x + y + 2z = 10 \\ -3x - 2y + 4z = -11 \\ 2y + z = 2 \end{cases}$

26. $\begin{cases} \dfrac{x}{6} + \dfrac{y}{3} + \dfrac{z}{2} = 1 \\ x - \dfrac{y}{2} + \dfrac{z}{2} = 1 \\ \dfrac{x}{6} - \dfrac{y}{2} + \dfrac{z}{3} = 0 \end{cases}$

16. $\begin{cases} 2x - y + 2z = 2 \\ -2x + 3y + 4z = 10 \\ -2x + y - z = 0 \end{cases}$

27. $\begin{cases} 0.2x + 0.1y + 0.1z = 0.6 \\ 0.3x + 0.2y + 0.2z = 1.0 \\ -0.1x + 0.3y + 0.1z = 0 \end{cases}$

In Problems 17–28, identify whether the given system is consistent and independent, consistent and dependent, or inconsistent. Describe the solution geometrically; refer to Table 1.

28. $\begin{cases} 0.5u + 1.5v - 0.5w = 2 \\ -1.5u - 2.5v + 0.5w = -4 \\ -0.5v + 1.5w = 7 \end{cases}$

17. $\begin{cases} p + q - r = 3 \\ p - 5q + r = 4 \\ -4p + 5q + r = 5 \end{cases}$

18. $\begin{cases} 2p + q - r = 4 \\ 4p - q + 2r = -2 \\ -2p + 2q - 3r = 6 \end{cases}$

19. $\begin{cases} x + y + z = 1 \\ 3x - 2y + 2z = 5 \\ 3x - 7y + z = 7 \end{cases}$

20. $\begin{cases} 3a - 5b - 2c = 9 \\ a - b + 2c = 3 \\ 2a - 3b = 1 \end{cases}$

21. $\begin{cases} 2p - q + 4r = -4 \\ 4p - 5q + 7r = 1 \\ -2p + 7q - 2r = 6 \end{cases}$

Applying the Concepts

29. Manufacturing: A study of a manufacturing company shows that the profit P (in thousands of dollars) is a function of the number n (in hundreds) of full time employees as shown by the following equation

$$P(n) = an^2 + bn + c, \ 0 \le n \le 100.$$

Find the values of a, b, and c such that this equation models the data (20,2000), (36,2512), and (70,200).

30. Height of a Rocket: A model rocket is launched and then it accelerates until the propellant burns out, after which it coasts upward to its highest point. The height h (in meters) above ground level t seconds after the burn out is modeled by the function

$$h(t) = at^2 + bt + c.$$

Find the values of a, b, and c such that this function models the data $(1,110)$, $(4,233)$, and $(7,68)$.

31. **Rainfall:** A monthly rainfall data set (x,y) for a city in a northwestern part of the United States fits the data $(2,5)$, $(4,3)$, and $(8,1)$, where x represents time in months, with $x = 1$ representing the month of January, and y represents the rainfall in inches. Find the values of a, b, and c such that the function

$$y(x) = ax^2 + bx + c$$

models these data. Round off the answers to two decimal places.

32. **Electric Circuits:** In the electric circuits shown in Figure 1, l_1, l_2, and I_3 represent the amount of current (in amperes) flowing across the 1-ohm, 2-ohm, and 3-ohms resistors. The system of linear equations used to find the currents l_1, l_2, and l_3 is given by

Figure 1

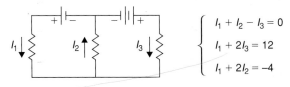

$$\begin{cases} l_1 + l_2 - l_3 = 0 \\ l_1 + 2l_3 = 12 \\ l_1 + 2l_2 = -4 \end{cases}$$

Solve the system for l_1, l_2, and l_3.

Developing and Extending the Concepts

33. Write a linear system of three equations in three variables that has $(3,2,-1)$ as a solution. How many possible systems can you have?

34. Use Table 1 to describe a way in which the graph of a linear system of three equations in three variables would indicate the system has:
    (a) one solution
    (b) no solution
    (c) an infinite number of solutions.

In Problems 35–40, extend the substitution method to solve the given linear system of three equations in three variables.

35. $\begin{cases} x + y = 5 \\ x + z = 1 \\ y + z = 2 \end{cases}$

36. $\begin{cases} 2x + 3y = 28 \\ 3y + 4z = 46 \\ 5x + 4z = 53 \end{cases}$

37. $\begin{cases} x + y + 2z = 11 \\ x - y + z = 3 \\ 2x + y + 3z = 17 \end{cases}$

38. $\begin{cases} x - 3y = -11 \\ 2y - 5z = 26 \\ 7x - 3z = -2 \end{cases}$

39. $\begin{cases} 2p - q + r = 8 \\ p + 2q + 3r = 9 \\ 4p + q - 2r = 1 \end{cases}$

40. $\begin{cases} s + 3t - u = 4 \\ 3s - 2t + 4u = 11 \\ 2s + t + 3u = 13 \end{cases}$

In Problems 41 and 42, solve each system by using the substitutions $u = 1/x$, $v = 1/y$ and $w = 1/z$.

41. $\begin{cases} \dfrac{3}{x} - \dfrac{4}{y} + \dfrac{6}{z} = 1 \\ \dfrac{9}{x} + \dfrac{8}{y} - \dfrac{12}{z} = 3 \\ \dfrac{9}{x} - \dfrac{4}{y} + \dfrac{12}{z} = 4 \end{cases}$

42. $\begin{cases} \dfrac{3}{x} + \dfrac{1}{y} - \dfrac{1}{z} = 5 \\ \dfrac{4}{x} - \dfrac{1}{y} + \dfrac{2}{z} = 13 \\ \dfrac{2}{x} + \dfrac{2}{y} + \dfrac{3}{z} = 22 \end{cases}$

43. Determine the value of a, b, and c so that

$$(a + b)x^2 + (c - 3a - 3b)x + 6a = 8x^2 - 17x + 30.$$

44. The fraction $1/18$ can be written as

$$\frac{1}{18} = \frac{x}{2} + \frac{y}{3} + \frac{z}{9}$$

where x, y, and z are solutions to the system

$$\begin{cases} 4x + 3y + 2z = -7 \\ -4x + y + 3z = -24 \\ 3x - 2y - z = 8 \end{cases}$$

(a) Solve the system.
(b) Verify that the sum of the fraction is $1/18$.

45. Solve each system.

(a) $\begin{cases} 1.5a - 0.2b + 4.5c = 2.74 \\ 5a - 4b - 8c = -1.7 \\ 0.3a + 5b - 8c = -4.91 \end{cases}$

(b) $\begin{cases} 2p - 3.8q + 2.1r = 3.26 \\ 3.7p - 0.2q + 0.05r = 0.41 \\ 1.5p + q - 0.2r = 0.05 \end{cases}$

46. Suppose that the point of intersection of the graphs of the equations

$$\begin{cases} ax - y + 2z = -7 \\ 2x + by + 3z = 4 \\ x + y + cz = -20 \end{cases}$$

is $(3,-1,2)$. Find a, b, and c.

Objectives

1. Solve Applied Problems by Using Linear Systems with Two Variables
2. Solve Applied Problems by Using Linear Systems with More Than Two Variables

3.4 Modeling with Linear Systems

Quite often it is desirable or even necessary to use two or more equations involving more than one variable to model a real-world situation. In such a case, we construct a model by writing a system of equations that describes and answers questions about the situation. In section 1.3, we solved applied problems by using one equation with one variable. In keeping with our emphasis on modeling, we present in this section a number of applied problems using linear systems of equations. We will rely on the three-step format that we first employed in solving a single equation with one variable.

Problem Solving
Step 1. Make a plan.
Step 2. Write the solution.
Step 3. Look back.

Solving Applied Problems by Using Linear Systems with Two Variables

If an applied problem involves two unknowns, we use two variables and so we need a system of two equations.

EXAMPLE 1   **Solving a Geometry Problem**

The perimeter of a flattened rectangular box of pizza is 56 inches. The length of one side of the box is 2 inches less than twice the width. What are the length and width of the pizza box?

Solution   **Make a Plan**

The two unknowns are the length and the width of the box. We use the formula for the perimeter P of a rectangle

$$P = 2l + 2w$$

Figure 1

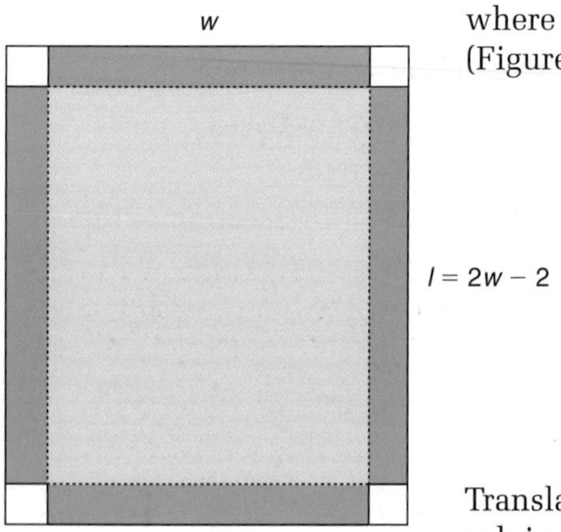

w

$l = 2w - 2$

where l is the length of the box in inches and w is the width of the box in inches (Figure 1). The following diagram is a visual description of the given.

Perimeter	=	56
$2l + 2w$	=	56

the length	=	2 less than twice the width
l	=	$2w - 2$

Translating the problem into symbols, then the length and width can be found by solving the system:

$$\begin{cases} 2l + 2w = 56 \\ l = 2w - 2 \end{cases}.$$

Write the Solution

We use substitution to solve the system. From equation 2, we substitute the expression for l into equation 1.

$$2(2w - 2) + 2w = 56$$
$$4w - 4 + 2w = 56$$
$$6w - 4 = 56$$
$$6w = 60$$
$$w = 10$$
$$\text{and } l = 2(10) - 2$$
$$l = 18$$

The width is 10 inches and the length is 18 inches.

Look Back

We check the results as follows:

Perimeter	Length is Two Less Than Twice the Width
$2l + 2w$	$2(10) - 2$
$= 2(18) + 2(10)$	$= 20 - 2$
$= 36 + 20 = 56$	$= 18$

EXAMPLE 2

Solving a Mixture Problem

A dietitian creates a nutritious breakfast by combining two types of grain. One type contains 10% protein and another contains 20% protein. How many grams of each type should be used to obtain a 100 gram mixture that contains 16% protein?

Solution

Make a Plan

The amount of protein is obtained by multiplying the quantity of grain by the percent of protein.

Let x = the number of grams of grain containing 10% protein.

y = the number of grams of grain containing 20% protein.

We organize the information in the following table.

	Amount of Grain	Percent of Protein	Total Amount of Protein
First Type	x	10%	$0.10x$
Second Type	y	20%	$0.20y$
Mixture	100	16%	$0.16(100)$

The quantity equation is

$$x + y = 100.$$

The protein equation is

$$0.10x + 0.20y = 0.16(100).$$

So we have the system of equations to solve

$$\begin{cases} x + y = 100 \\ 0.1x + 0.2y = 16 \end{cases}.$$

Write the Solution.

$$\begin{cases} x + y = 100 \\ 0.1x + 0.2y = 16 \end{cases} \quad \begin{array}{c} \text{——— no change ———} \rightarrow \\ \text{—— multiply each side by } -10 \longrightarrow \end{array} \quad \begin{cases} x + y = 100 \\ -1x - 2y = -160 \end{cases}$$

$$-y = -60$$
$$y = 60$$

Substitute this value into the original equation 1.

$$x + 60 = 100$$
$$x = 40$$

Therefore, the dietitian should mix 40 grams of the 10% grain with 60 grams of the 20% grain to get a mixture of 16% protein.

Look Back

Amount of Grain	Amount of Protein Mixed	Amount of Protein Required
40 + 60	0.10(40) + 0.20(60)	0.16(100)
= 100	= 4 + 12 = 16	= 16

EXAMPLE 3

Solving an Aviation Problem

A commercial jetliner flying with the wind takes 3.75 hours to fly the 2500 miles from Los Angeles to New York, but 4.4 hours to fly the same distance from New York to Los Angeles against the wind. Assuming that the jetliner travels at a constant speed and the wind blows at a constant rate, find the speed of the jetliner in still air and the wind speed.

Solution

Make a Plan

Here we choose a variable to represent the speed of the jetliner in still air and another variable to represent the wind speed. Using these variables, we can express the speed of the jetliner with and against the wind. We also use the equation $d = rt$ to write expressions for the distance traveled by the jetliner.

$$\text{Let } j = \text{speed of the jetliner in still air and}$$
$$w = \text{speed of the wind.}$$

The following table visualizes these results.

	Rate (in miles per hour)	Time (in hours)	Distance (in miles)
With the Wind (Los Angeles to New York)	$j + w$	3.75	$3.75(j + w)$
Against the Wind (New York to Los Angeles)	$j - w$	4.40	$4.40(j - w)$

Now we use a linear system of two equations to determine how the expressions for the distance are related.

$$\begin{cases} 3.75(j + w) = 2500 \\ 4.40(j - w) = 2500 \end{cases}$$

Write the Solution

We solve the system as follows.

$$\begin{cases} 3.75(j + w) = 2500 \\ 4.40(j - w) = 2500 \end{cases} \quad \begin{matrix} \text{divide by 3.75} \\ \text{divide by 4.40} \end{matrix} \longrightarrow \begin{cases} j + w = 666.67 \\ j - w = 568.18 \end{cases}$$

Add

$$2j = 1234.85$$

$$j = 617.43$$

To solve for w, substitute 617.43 in the equation

$$j + w = 666.67$$

$$617.43 + w = 666.67$$

$$w = 49.24.$$

Thus, the speed of the jetliner in still air is approximately 617.43 miles per hour, and the wind speed is approximately 49.24 miles per hour rounded to two decimal places.

Look Back

To check the solution, we calculate the distance traveling with the wind and against the wind.

Distance with Wind	Distance Against Wind
$3.75(617.43 + 49.24)$	$4.40(617.43 - 49.24)$
$= 2500.01$	$= 2500.04$

Solving Applied Problems Involving Systems with Three Variables

With our focal theme on solving applied problems, we now consider linear systems with three variables that are used in modeling real-world situations. The method of solving such systems is an extension to the method of solving linear systems with two variables.

EXAMPLE 4

Solving an Investment Problem

A financial planner advised a customer to invest part of her $10,000 in a safe money market fund that paid 5% simple interest, part in bonds that paid 7% simple interest, and the rest in a risky investment. If the risky investment earns 9% simple interest, the total earning from all the investments for one year is $656. However, if the risky investment pays nothing for the year, then the total earnings from first two investments is only $521. How much is invested in each fund?

Solution

Make a Plan

We have $10,000 to be divided into three investments: money market, bonds, and risky investments. To find the amount invested at each rate, we let

x = the amount (in dollars) invested in the money market at 5%

y = the amount (in dollars) invested in bonds at 7%

z = the amount (in dollars) invested in a risky investment at 9%.

Now, we use the formula

$$I = Prt$$

and analyze these investments in the following table.

Investment	Principal	Interest Rate	Interest
Money Market	x	5%	$0.05x$
Bonds	y	7%	$0.07y$
Risky	z	9%	$0.09z$

The three principals add to $10,000 and the sum of interests for one year is $656 and $521 depending on the return of the risky investment. The system of equations which represents this information is:

$$\begin{cases} x + y + z = 10000 \\ 0.05x + 0.07y + 0.09z = 656. \\ 0.05x + 0.07y = 521 \end{cases}$$

Write the Solution

We solve this system as follows:

$$\begin{cases} x + y + z = 10000 \\ 0.05x + 0.07y + 0.09z = 656 \\ 0.05x + 0.07y = 521 \end{cases}$$

— No Change $\longrightarrow$
— Multiply by 100 $\rightarrow$
— Multiply by 100 $\rightarrow$

$$\begin{cases} x + y + z = 10000 \\ 5x + 7y + 9z = 65600 \\ 5x + 7y = 52100 \end{cases}$$

We eliminate z from the first and second equation as follows:

$$\begin{cases} x + y + z = 10000 \\ 5x + 7y + 9z = 65600 \end{cases}$$

– Multiply each side by -9 $\rightarrow$
— No change $\longrightarrow$

$$\begin{array}{r} -9x - 9y - 9z = -90000 \\ 5x + 7y + 9z = 65600 \\ \hline \end{array}$$

Add

$$-4x - 2y = -24400$$

Now we have the following system of two equations in two variables

$$\begin{cases} -4x - 2y = -24400 \quad \text{—— Multiply each side by 5} \longrightarrow \\ 5x + 7y = 52100 \quad \text{—— Multiply each side by 4} \longrightarrow \end{cases} \begin{cases} -20x - 10y = -122000 \\ 20x + 28y = 208400 \end{cases}$$

$$\text{Add} \qquad\qquad\qquad\qquad\qquad\qquad\qquad\qquad\qquad 18y = 86400$$

$$y = 4800$$

Substitute this value for y in the equation.

$$5x + 7y = 52100$$

$$5x + 7(4800) = 52100$$

$$5x = 18500$$

$$x = 3700$$

Finally substitute these values into an original equation.

$$x + y + z = 10000$$

$$3700 + 4800 + z = 10000$$

$$z = 1500$$

Thus $3700 was invested in mutual funds, $4800 in bonds, and $1500 in the risky investment.

Look Back

The sum of the investments is $10,000 = $3700 + 4800 + 1500.
The money market fund earned 3700(0.05) = $185.
The bonds earned $4800(0.07) = $336.
The risky investment earned $1500(0.09) = $135.
If the risky investment gives a 9% return, then the total return is:

$$\$185 + 336 + 135 = \$656.$$

The earnings without the risky investment is $185 + 336 = $521.

EXAMPLE 5

Solving a Manufacturing Problem

A manufacturer of portable radios produces three models of radios, model A, model B, and model C. The manufacturer determines that the total time allocated for production, assembly, and testing are 638 hours, 253 hours, and 159 hours respectively. The table lists the information for each model.

	Model A	Model B	Model C	Total
Production Hours per Radio	1.4	1.6	1.8	638
Assembly Hours per Radio	0.5	0.6	0.8	253
Testing Hours per Radio	0.3	0.4	0.5	159

Use a linear system to determine how many of each model can be produced using all the allocated time.

Solution   **Make a Plan**

Let a = the number of model A radios produced

b = the number of model B radios produced

c = the number of model C radios produced.

Production requires $1.4a$ hours for model A, $1.6b$ for model B, and $1.8c$ for model C for a total of 638 production hours; we have

$$1.4a + 1.6b + 1.8c = 638.$$

Similarly, there are 253 hours available for assembly. We obtain

$$0.5a + 0.6b + 0.8c = 253.$$

The third equation considers that there are 159 hours available for testing, so we have

$$0.3a + 0.4b + 0.5c = 159.$$

The following system describes this model.

$$\begin{cases} 1.4a + 1.6b + 1.8c = 638 \\ 0.5a + 0.6b + 0.8c = 253 \\ 0.3a + 0.4b + 0.5c = 159 \end{cases}$$

Write the Solution

By solving the system, we obtain the solution

$$a = 150, b = 110, \text{ and } c = 140.$$

Therefore, the manufacturer should produce 150 model A, 110 model B, and 140 model C radios in order to use up the allotted time.

Look Back

By using the table, we have

$$\begin{cases} 1.4(150) + 1.6(110) + 1.8(140) = 638 \text{ true.} \\ 0.5(150) + 0.6(110) + 0.8(140) = 253 \text{ true.} \\ 0.3(150) + 0.4(110) + 0.5(140) = 159 \text{ true.} \end{cases}$$

 **PROBLEM SET 3.4**

Mastering the Concepts

1. **Carpeting:** A carpet layer determines that the perimeter of a rectangular wedding hall is 900 feet.
   (a) Find the length and the width of the hall if its length is 50 feet more than its width.
   (b) If the carpet costs $12.75 per square yard installed, how much does it cost to carpet the hall?

2. **Dimensions of a Pool:** The perimeter of a rectangular swimming pool is 192 feet. Find the length and the width of the pool if its width is one-third of its length.

3. **Designing a Frame:** An artist wishes to frame a picture with 84 inches of framing material. How wide will the frame be if its width is 75% of its length?

4. **Fencing:** A rancher has 40 meters of fencing to enclose a rectangular pen next to a barn, with the barn wall forming the length of the pen (Figure 2). Find the length and width of the pen if its length is twice its width.

Figure 2

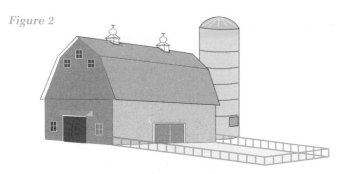

5. **Nutrition:** A dietitian arranges a special diet composed of two types of foods, A and B. Each ounce of food A contains 10 units of calcium and 4 units of iron; each ounce of food B contains 6 units of calcium and 4 units of iron. How many ounces of foods A and B should be used to obtain a food mix that contains 92 units of calcium and 44 units of iron?

6. **Blending Oil:** An oil company mixes two types of oil, low-sulfur and high-sulfur. The low-sulfur oil has a sulfur content of 2%, while the high-sulfur oil has a sulfur content of 7%. How many barrels of each type should be used to obtain a mixture of 10,000 barrels with a sulfur content of 4%?

7. **Aviation:** An airliner flying with the wind takes 3.5 hours to fly 1,875 miles, and 4 hours to fly the same distance against the wind. Assuming that the airliner travels at a constant speed and the wind blows at a constant rate, find the speed of the airliner in still air and the wind speed. Round off the answers to two decimal places.

8. **Water Speed:** A cruising riverboat travels 36 miles downstream (with the current) in 2 hours. It travels the same distance upstream (against the current) in 3 hours. Assuming the riverboat travels at a constant speed in still water and the rate of the current is constant, find the speed of the riverboat in still water and the rate of the current.

9. **Cycling:** Two bicycle riders start at the same point and travel in opposite directions. One travels 5 kilometers per hour faster than the other. In 3 hours they are 117 kilometers apart; how fast is each traveling?

10. **Fitness:** A student walks 4 miles from home to school each day. She jogs back from school to home. She makes the trip walking to school in one hour and takes half an hour to return. How fast does she walk and jog?

11. **Investment Portfolio:** An inheritance of $40,000 is invested in two municipal bonds which pay 6% and 7% simple annual interest. If the annual interest from both bonds is $2,550, how much is invested at each rate?

12. **Investment Portfolio:** An investment club invested $35,000 in two accounts. During one year, the first account earned 5% annual simple interest and the second account earned 6% annual simple interest. If the total interest from both investments for the year was $1880, how much was invested in each account?

13. **Investment Portfolio:** A total of $30,000 was invested in two mutual funds. One risky fund yields 10.5% annual simple interest; the other is a safe fund which earns 6% annual simple interest. At the end of the year, the combined interest from the two funds was $2340. How much was invested in each fund?

14. **Car Loan:** A new employee purchased a car and needed to borrow $16,000. He borrowed part of the money from the credit union at 9% annual simple interest and the remaining amount from his parents at 4% annual simple interest. If the total interest paid for one year was $1,190, how much was borrowed from the credit union?

15. **Coffee Mixtures:** A coffee shop mixes French roast coffee worth $9 per pound with kona coffee worth $4 per pound. How much of each type of coffee was used in order to obtain 20 pounds of a mixture worth $5 per pound?

16. **Window Cleaner Mixtures:** A window cleaner is made out of water and ammonia. Suppose that a window washer wishes to mix one brand of window cleaner that contains 5% ammonia with another brand that has 10% ammonia to form a 100 liters of a mixture that is 7.5% ammonia. How many liters of each brand should be used (Figure 3)?

Figure 3

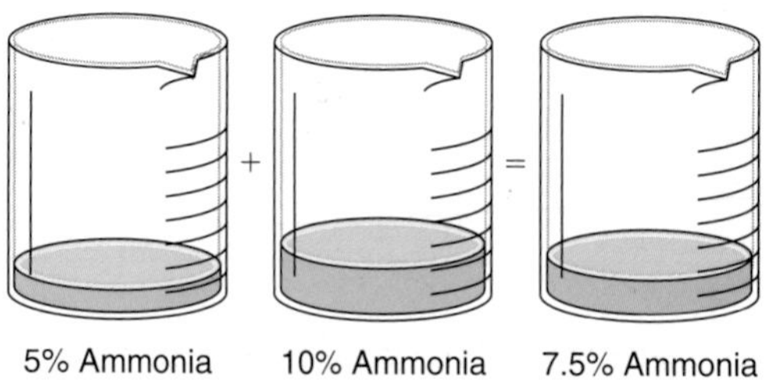

5% Ammonia   10% Ammonia   7.5% Ammonia

17. **Pharmaceutical Mixtures:** A pharmacist is to prepare a 50 fluid ounce solution that contains 54% glucose. The pharmacy has a 30% solution and a 90% solution in stock. How many ounces of each solution should be mixed to prepare the desired prescription?

18. **Antifreeze Mixtures:** An automobile radiator holds 16 liters of fluid. The radiator is filled with a mixture of 80% water and 20% antifreeze. How much of this mixture should be drained and replaced with pure antifreeze so that the final mixture will contain 40% antifreeze?

19. **Investment Portfolio:** A total of $20,000 is divided among three investments, a certificate of deposit paying 6% annual simple interest, stocks paying 7% annually, and municipal bonds paying 8% annually. The total return on the investments is $1390. The amount invested in the certificate of deposit plus the amount invested in municipal bonds is $11,000. Find the amount placed in each of the three investments.

20. **Geometry:** We know from geometry that the sum of the angles in any triangle is 180°. Suppose that the smallest angle *Figure 4* is 78° less than the largest one. Also suppose that the middle angle is three times as large as the smallest angle. What is the degree measure of each angle (Figure 4)?

21. **Ticket Sales:** A concert was held in a sports arena that seats 20,000 people. Tickets were sold at $20, $15, and $10. There were twice as many $15 seats as the $20 seats. If the concert was sold out and grossed $260,000, how many of each kind were sold?

22. **Nutrition:** A hospital dietitian prepares a meal consisting of chicken, potatoes, and peas. The dietitian determines that a patient's meal should contain 885 calories, 74 grams of protein, and 675 grams of sodium. Table 1 lists the compositions of these mixes.

TABLE 1

	Three Ounce Serving of Chicken	One Cup Serving of Potatoes	One Cup Serving of Peas	Total
Calories (cal)	140	160	125	885
Protein (gm)	27	4	8	74
Sodium (mg)	64	136	139	675

Use a linear system of three equations in three variables to determine how many servings of each should be used if all the allocation in Table 1 is used up.

23. **Manufacturing:** A manufacturer of color printers produces three models of color printers, Model A, Model B, and Model C. The manufacturer determines that the total time allocated for production, assembly, and testing are 100 hours, 100 hours, and 65 hours respectively. Table 2 lists the information for each model.

TABLE 2

One Printer	Model A	Model B	Model C	Total
Production hours	2	1	3	100
Assembly hours	2	3	2	100
Testing hours	1	1	2	65

Use a linear system to determine how many of each model can be produced if all the allocated time is used up.

24. **Airport Shuttle:** An airport shuttle service has three sizes of vans. The biggest size van holds 12 passengers, the middle size holds 10 passengers, and the smallest size holds 6 passengers. On a certain day, the manager has 15 vans available to accommodate 152 passengers. However, to save on fuel costs, the manager decides to use 2 more of the available large size vans than the small size ones. How many vans of each kind should be used?

Developing and Extending the Concepts

25. **Number Problem:** The difference between two numbers is 12. The sum of the larger number and twice the smaller is 75. Find the numbers.

26. **Puzzle:** A boy and his girlfriend are arguing over the difference in their ages. The boy says that he is 3 years older than her. The girl claims that 4 years ago the sum of their ages was 23. If they are both right, how old is each now?

27. **Coin Problem:** A vending machine box contains 45 coins worth $8.70 all in quarters and dimes. How many quarters and how many dimes were there in the box?

28. **Plumber Wages:** A plumber charges a fixed charge plus an hourly rate for service on a house call. The plumber charged $70 to repair a water tank that required 2 hours of labor and $100 to repair a water tank that took 3.5 hours. Find the plumber's fixed charge and hourly rate.

29. **Height of Eiffel Tower:** The height of the Eiffel Tower (with the television mast added to the

top) in Paris, France, is approximately 612 meters. An observer noticed that the difference in height between the tower and the television mast is approximately 262 meters. How tall is the tower (Figure 5)?

Figure 5

30. **Comparing Heights:** The sum of the heights of the tallest person and the shortest person in the world is approximately 3.30 meters. The difference in their heights is approximately 2.13 meters. How tall is each? [Source: *The Guinness Book of Records*]

31. **Membership Fees:** A health club offers two membership plans. An active membership costs an initial fee of $175 plus $4 per visit. A social membership has no initial fee, but costs $7 per visit. If a person decides to join for one year, how often would that person have to visit for the active membership to be the less expensive plan?

32. **Tips:** Together two waiters worked a total of 14 hours. The more experienced waiter averaged $15 per hour in tips, whereas the newer waiter averaged $10 per hour. The waiters contributed 10% of their tips into a pool distributed to other workers who bus the tables. After paying into the pool, the waiters found that they had a total of $153 in tips. How many hours did each waiter work? How much did each waiter take home in tips?

33. **Shoe Sales:** A sports clothing store sells both running and tennis shoes. The tennis shoes sell for $46 per pair and the running shoes sell for $79 per pair. During a one-day sale, 260 pairs of shoes were sold for a total of $14,435. How many of each type of shoes were sold?

34. **Hours of Employment:** A student holds two part-time jobs totaling 20 hours per week. The first job pays $5.50 per hour and the second job pays $6.50 per hour. How many hours did the student work at each job if the gross weekly income is $117?

35. **Savings Accounts:** A couple engaged to be married decided they wanted to have $15,000 in savings by their wedding date. To reach their goal, the bride-to-be must double her savings and the groom-to-be must triple his savings. If they still have a total of $11,000 to set aside, how much does each one currently have in savings?

36. **Earnings:** One sales representative earned a base salary of $200 per week plus 6% commission on all sales. A second representative earned a base salary of $120 per week plus 8% commission on all sales. At the end of a special promotional campaign, the two representatives had earned $27,000 and $29,400, respectively. How long was the promotional campaign and how much did each representative sell?

37. **Car Sales:** Two friends, living in different states, compared notes about the cars they had bought recently. One friend paid 6% sales tax whereas the other paid 8.5% tax on the car. Together, their cars cost a total of $36,350 and they paid a total of $2546 in sales tax. How much did each car cost?

38. **Number Problem:** The sum of three numbers is 225. The first number is 50 more than the second. The third is 25 more than the sum of the second and the first. Find the numbers.

39. **Consumer:** An employee goes to breakfast at a nearby coffee shop. One day, he spent $2.20 for a glass of juice, one bagel, and 2 cups of coffee excluding tax and tip. The second day he spent $2.35 for 2 bagels and 3 cups of coffee excluding tax and tip. The third day he spent $2.75 for a glass of juice, 2 bagels, and a cup of coffee excluding tax and tip. What was the price of each item?

40. **Landscaping:** Figure 6 shows three circular flower beds which are mutually tangent to each other and with centers at A, B, and C and radii, r_1, r_2, and r_3 respectively. The distances between the centers of the beds are as follows:

$$\overline{AB} = 7 \text{ feet}, \overline{BC} = 11 \text{ feet, and } \overline{AC} = 10 \text{ feet.}$$

Find the radii, r_1, r_2, and r_3, of the circular flower beds.

Figure 6

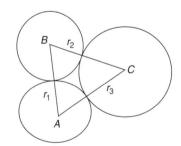

41. **Appliance Sales:** An appliance store sells dryers for $380, washing machines for $450, and microwave ovens for $140. On a certain day, the store sells 48 washers, dryers, and microwaves for total receipts of $12,630. If the number of microwaves sold for that day is twice the number of dryers, how many of each item was sold for that day?

42. Work Rate: Three painters *A, B,* and *C* can paint a court in a fitness center in 4 hours. When *A* and *C* work together, it takes 8 hours to paint the court. When *A* and *B* work together, it takes 6 hours to paint the court. How long would it take each of them to paint the court alone?

43. Fertilizer Mixture: A supplier of agriculture products has three types of fertilizer *A, B,* and *C* having nitrogen contents of 30%, 20%, and 15% respectively. The supplier plans to mix them, obtaining 600 pounds of fertilizer with a 25% nitrogen content. The mixture is to contain 100 pounds more of type *C* than of type *B*. How much of each type should be used?

44. Coin Problem: A vending machine which gives change must start each day with $12.50 in dimes and quarters. The number of dimes needs to be twenty more than the number of quarters. How many of each coin must be in the vending machine?

45. Coin Problem: A cashier begins with 4 times as many nickels as quarters in the register. If the coins have a value of $3.60, how many of each coin is in the register?

Objectives

1. Graph Linear Systems of Inequalities
2. Solve Applied Problems

3.5 Solving Systems of Linear Inequalities in Two Variables

In Section 2.5, we have learned how to graph linear inequalities of the form

$$3x - 4y \leq 12 \text{ or } x - 2y \geq 6.$$

Graphically, the solution set of each inequality is a line together with the region either below or above the line. Two inequalities considered together, such as

$$\begin{cases} 3x - 4y \leq 12 \\ x - 2y \geq 6 \end{cases}$$

is called a **system of two inequalities** in two variables.

Graphing Linear Systems of Inequalities

Just as in systems of linear equations in two variables, the **graph** of a system of linear inequalities is the intersection of the graphs of each inequality in the system. That is, the set of all points that satisfy each inequality in the system. In this section, we limit our study of solutions of systems of inequalities to graphical techniques.

EXAMPLE 1   **Solving a System of Inequalities**

Sketch the graph of the solution set of the system of inequalities.

$$\begin{cases} 1 \leq x \leq 5 \\ 2 \leq y \leq 4 \end{cases}$$

Solution   This system is equivalent to the system

$$\begin{cases} x \geq 1 \\ x \leq 5. \\ y \geq 2 \\ y \leq 4 \end{cases}$$

Figure 1

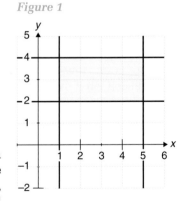

Figure 1 shows the solution set of the system which consists of the region to the left and including the line *x* = 5, to the right and including the line *x* = 1, above and including the line *y* = 2, and below and including the line *y* = 4.

EXAMPLE 2   **Solving a System of Inequalities**

Sketch the graph of the solution set for the system of inequalities.

$$\begin{cases} x + y > 3 \\ 3x - y \ge 6 \end{cases}$$

Solution   In slope-intercept form we have the inequalities

$$\begin{array}{c|c} x + y > 3 & 3x - y \ge 6 \\ y > -x + 3 & y \le 3x - 6. \end{array}$$

Figure 2a shows the graph of $y > -x + 3$.
Figure 2b shows the graph of $y \le 3x - 6$.
Figure 2c shows the region common to both graphs.
The solution set of the system consists of the points above

$$y = -x + 3$$

and below

$$y = 3x - 6,$$

including the points on the line

$$y = 3x - 6$$

from the point of intersection $(2.25, 0.75)$ and up.

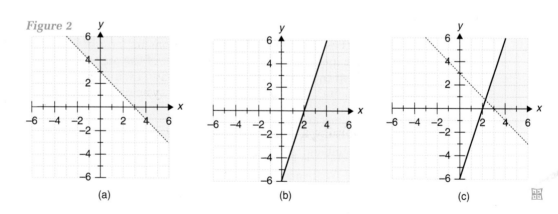

Figure 2

(a)   (b)   (c)

Systems of inequalities may include more than two inequalities, as seen in the next example.

EXAMPLE 3   **Solving a System of Inequalities**

Sketch the graph of the solution set for the system of inequalities.

$$\begin{cases} 2x - y \ge 4 \\ 2x + 3y \le 6 \\ x \ge 0 \\ y \ge 0 \end{cases}$$

Solution   Figure 3a shows the graph of $2x - y \geq 4$.

Figure 3b shows the graph of $2x + 3y \leq 6$.

Figure 3c shows the regions common to both graphs.

The solution set of the system consists of points below and including the line

$$2x - y = 4$$

and below and including the line

$$2x + 3y = 6.$$

The restrictions $x \geq 0$ and $y \geq 0$ limit the solution to quadrant I (Figure 3c).

Figure 3

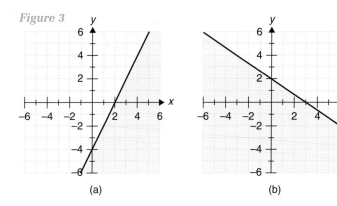

(a)   (b)

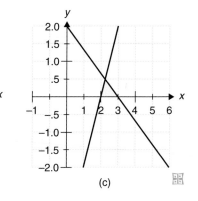

(c)

Solving Applied Problems

Many practical problems can be modeled by graphs of systems of inequalities.

EXAMPLE 4   **Solving a Relief Problem**

An airlift requires a total of at most 40 airplanes on a given day. Some of the planes are large and require a 4 person crew to fly. Other planes are small and require a 2 person crew. Both crews are selected from a pool of at most 120 crew members.

(a) Write a system of inequalities to describe this situation.

(b) Graph the solution set of this system.

Solution   **Make a Plan**

(a) First we find a system of two inequalities with two variables to describe this situation. One inequality will describe the number of planes in the airlift which is at most 40. The other will represent a different arrangement of crew members to total at most 120 people.

Let $x =$ number of large planes in the airlift

$y =$ number of small planes in the airlift.

We form the system of inequalities as follows:

$$\begin{cases} x + y \le 40 \\ 4x + 2y \le 120. \\ x \ge 0, y \ge 0 \end{cases}$$

Figure 4

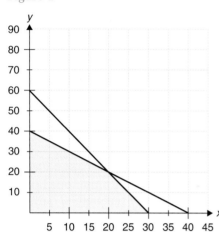

Write the Solution

We graph both inequalities on the same coordinate system. The solution set of this system is the set of all ordered pairs that satisfy each inequality. That is, the intersection of the solution set of the individual inequalities. The inequalities are graphed as follows:

Shade the region below and including the solid line $x + y = 40$.

Shade the region below and including the line $4x + 2y = 120$ or $2x + y = 60$.

The restrictions on the number of planes and crew members limits us to quadrant I. Therefore, the solution of the system of inequalities, possible airlift schedules, is the set of all points within the shaded region in Figure 4.

Look Back

All the points in the solution to this system of inequalities satisfy the conditions imposed in the problem. For example, (15,18) is a point in the shaded region. It would be possible to use 15 large airplanes and 18 small planes in the airlift. The number of planes being used, $15 + 18 = 33$, is less than 40. There would be enough crew members:

$$4(15) + 2(18) = 96 \text{ which is less than } 120.$$

Similarly, any point outside the shaded region would represent an impossible schedule. The point (21,24) is not in the shaded region; this would require more than 40 planes.

 PROBLEM SET 3.5

Mastering the Concepts

In Problems 1–6, graph the solution set of each system of inequalities.

1. $\begin{cases} x \ge 2 \\ y \le 3 \end{cases}$

2. $\begin{cases} x < -1 \\ y > 2 \end{cases}$

3. $\begin{cases} x \ge 0 \\ x + y < 4 \end{cases}$

4. $\begin{cases} y > 0 \\ x - y < 2 \end{cases}$

5. $\begin{cases} y \ge 0 \\ x - 2y \le 6 \end{cases}$

6. $\begin{cases} x \ge -1 \\ x + 3y \ge 6 \end{cases}$

In Problems 7–10, write a linear system of inequalities in two variables whose solution set is represented by the shaded region.

7.

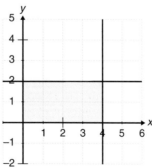

8.

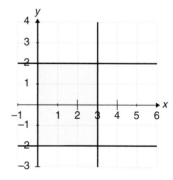

9.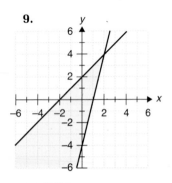

10.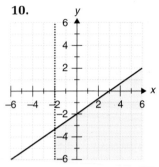

In Problems 11–14, sketch the graph of each linear system of inequalities.

11. $\begin{cases} x \geq 0, y \geq 0 \\ 3x + y \leq 6 \end{cases}$

12. $\begin{cases} x \geq 0, y \geq 0 \\ 2x - 3y \leq 12 \end{cases}$

13. $\begin{cases} x \geq 0, y \geq 0 \\ y \geq x, y \leq 4 \end{cases}$

14. $\begin{cases} x \geq 0, y \geq 0 \\ x \leq 4, y \leq 2 \end{cases}$

In Problems 15–30, sketch the graph of each system of linear inequalities.

15. $\begin{cases} x + y < 2 \\ 2x - y < -1 \end{cases}$

16. $\begin{cases} -x + y < 1 \\ -x + 3y \leq 5 \end{cases}$

17. $\begin{cases} x + y \leq 3 \\ x - y \geq 3 \end{cases}$

18. $\begin{cases} x + y > 5 \\ x - y < 9 \end{cases}$

19. $\begin{cases} x + 2y \leq 12 \\ x - y > 6 \end{cases}$

20. $\begin{cases} x + 2y > 12 \\ 3x - y < 1 \end{cases}$

21. $\begin{cases} x - y \geq 1 \\ x + y \leq 2 \end{cases}$

22. $\begin{cases} x - 2y \geq -2 \\ 2x + y \geq 2 \end{cases}$

23. $\begin{cases} 2x - 3y \leq -3 \\ 5x - 2y > 9 \end{cases}$

24. $\begin{cases} y \leq 2x + 4 \\ y \geq 3 - x \end{cases}$

25. $\begin{cases} x + y \leq 2 \\ -x + 3y \geq 4 \end{cases}$

26. $\begin{cases} 2x + y \geq 2 \\ x - 2y \leq 3 \end{cases}$

27. $\begin{cases} x \geq 0, y \geq 0 \\ 2x + y \leq 2 \\ x + 2y \leq 2 \end{cases}$

28. $\begin{cases} x \geq 0, y \geq 0 \\ x + y \geq 12 \\ 3x - y \geq 16 \end{cases}$

29. $\begin{cases} x \geq 0, y \geq 0 \\ x + y \geq 6 \\ 2x - 3y \geq 10 \end{cases}$

30. $\begin{cases} x \geq 0, y \geq 0 \\ x + y \leq 4 \\ 3x - y \leq 7 \end{cases}$

Applying the Concepts

31. Investments: An investment firm has at most $200,000 to invest in two stocks, A and B. Stock A sells at $56 a share and stock B at $86 a share; the total number of shares to be purchased of both stocks cannot exceed 3,000. Let x be the number of shares to be purchased of stock A and y be the number of shares to be purchased of stock B.

(a) Write a system of inequalities that indicates the restrictions on x and y.

(b) Graph the system showing the region of permissible values for x and y.

32. Sales Promotion: During a summer promotional campaign, a fitness center gave either one hour of free court time or a can of racquetball balls to new members. It cost the center $6 for each court and $3 for each can of balls. The center could accommodate at most 179 new members and budgeted at most $900 for the incentives.

(a) Write a system of inequalities that shows these restrictions.

(b) Graph the system showing the possible combination of incentives.

33. Comparing Salaries: The sum of the weekly salaries of a server and a busser is at most $1200. The server makes at least $200 more per week than the busser.

(a) Describe the distribution of salaries with a system of inequalities.

(b) Sketch the graph of the system.

(c) If the weekly salary of the server is $600, what is the most that the busser can make?

Developing and Extending the Concepts

In Problems 34–41, graph the system of inequalities.

34. $\begin{cases} |x + y| \leq 3 \\ x \geq 0, y \leq 0 \end{cases}$

35. $\begin{cases} |x - y| < 4 \\ x \leq 0, y \geq 0 \end{cases}$

36. $\begin{cases} |x| > |y| \\ x \leq 0, y \geq 0 \end{cases}$

37. $\begin{cases} |x| + |y| \leq 3 \\ x \geq 0, y < 0 \end{cases}$

38. $\begin{cases} 2x + 4y < 3 \\ -2x - y > 4 \end{cases}$

39. $\begin{cases} 3x - 5y \geq 15 \\ -2x + 6y \geq 12 \end{cases}$

40. $\begin{cases} 3x + y > 6 \\ x - y \geq 1 \end{cases}$

41. $\begin{cases} 2x - y \leq -2 \\ x - 2y \geq -2 \end{cases}$

CHAPTER 3 REVIEW PROBLEM SET

In Problems 1–4, solve each system graphically by sketching the graphs of the equations in the system on the same coordinate axis.

(a) Estimate the point of intersection (if any) of the two lines.

(b) Write each equation in slope-intercept form.

(c) Use the graph to indicate whether the system is consistent and independent, consistent and dependent, or inconsistent.

1. $\begin{cases} x + y = 5 \\ x - y = 1 \end{cases}$

2. $\begin{cases} x - 2y = 6 \\ 2x - 4y = 12 \end{cases}$

3. $\begin{cases} y = -2x + 6 \\ 4x + 2y = 8 \end{cases}$

4. $\begin{cases} 3x - 4y = 1 \\ 2x - y = -1 \end{cases}$

In Problems 5–8, use the substitution method to solve each system.

5. $\begin{cases} x + 2y = 3 \\ 2x + y = 5 \end{cases}$

6. $\begin{cases} 3x + 2y = 4 \\ 2x - 4y = 0 \end{cases}$

7. $\begin{cases} \frac{1}{2}x + \frac{1}{3}y = 8 \\ \frac{1}{4}x - \frac{1}{2}y = 0 \end{cases}$

8. $\begin{cases} 0.04x + 0.2y = 8 \\ 0.06x - 0.1y = 4 \end{cases}$

In Problems 9–12, use the elimination method to solve each system (if possible). Determine whether the system is consistent and independent, consistent and dependent, or inconsistent.

9. $\begin{cases} y = 4 - 2x \\ 2x - y = 4 \end{cases}$

10. $\begin{cases} 2x + 3y = 5 \\ 3x + 5y = 7 \end{cases}$

11. $\begin{cases} x - y = 1 \\ 3x - y = -1 \end{cases}$

12. $\begin{cases} 3x - 2y = 57 \\ 2x + 5y = -19 \end{cases}$

In Problems 13–24, use any method to solve each system (if possible).

13. $\begin{cases} x + y = \frac{9}{4} \\ 3x + 4y = 10 \end{cases}$

14. $\begin{cases} x - y = 1 \\ x + \frac{5}{3}y = \frac{4}{3} \end{cases}$

15. $\begin{cases} x + y + 2z = 11 \\ x - y + z = 3 \\ 2x + y + 3z = 17 \end{cases}$

16. $\begin{cases} x + 3y + z = -1 \\ 3x - 2y + 4z = -7 \\ 2x + y + z = 18 \end{cases}$

17. $\begin{cases} x - y + 2z = 1 \\ x - 4y + 3z = 1 \\ x + 2y - z = 1 \end{cases}$

18. $\begin{cases} x - y - z = 4 \\ -2x + 2y + 2z = -8 \\ 4x + 2y + 2z = 10 \end{cases}$

19. $\begin{cases} x + y + z = -1 \\ x + y + 2z = 0 \\ 3x + y + z = -5 \end{cases}$

20. $\begin{cases} 3x + y - z = 17 \\ 6x - 2y - 2z = 14 \\ 2x - y - 3z = 3 \end{cases}$

21. $\begin{cases} \frac{x}{2} + \frac{y}{3} - \frac{z}{4} = 2 \\ \frac{x}{2} + \frac{y}{2} = -1 \\ \frac{x}{2} + \frac{y}{3} + \frac{z}{4} = 1 \end{cases}$

22. $\begin{cases} \frac{p}{6} + \frac{q}{3} + \frac{r}{2} = 2 \\ p - \frac{q}{2} + \frac{r}{2} = -3 \\ \frac{p}{6} - \frac{q}{2} + \frac{r}{3} = -7 \end{cases}$

23. $\begin{cases} \frac{1}{x} + \frac{3}{y} - \frac{4}{z} = 0 \\ \frac{1}{x} - \frac{1}{y} + \frac{2}{z} = 4 \\ \frac{1}{x} + \frac{2}{y} - \frac{1}{z} = 4 \end{cases}$

24. $\begin{cases} \frac{2}{x} - \frac{1}{y} + \frac{5}{z} = 18 \\ \frac{2}{x} - \frac{3}{y} - \frac{4}{z} = -15 \\ \frac{1}{x} - \frac{4}{y} + \frac{3}{z} = 0 \end{cases}$

In Problems 25–32, sketch the graph of the solution set of each system of inequalities.

25. $\begin{cases} x \geq 2 \\ y \leq 2 \end{cases}$

26. $\begin{cases} x \geq 0 \\ y < 0 \end{cases}$

27. $\begin{cases} x \geq 1 \\ y \geq -1 \end{cases}$

28. $\begin{cases} x \geq y \\ y \geq x \end{cases}$ —

29. $\begin{cases} y - 3 \leq 0 \\ x + y > 2 \end{cases}$

30. $\begin{cases} y - 1 \leq 2 \\ x + 4 \geq 3 \end{cases}$

31. $\begin{cases} 2y - x < 4 \\ x + 2y \leq 6 \end{cases}$

32. $\begin{cases} 3y - x \geq -9 \\ x + 3y \geq -3 \end{cases}$

33. Geometry: The sum of the measure of two angles is 180°. If the difference between their angle measures is 32°, find the measure of each angle.

34. Discount: A publisher offered a college bookstore a 20% discount on hardback textbooks and a 10% discount on paperbacks. The regular price for the purchase of a shipment of books was $75,450, on which the total discount was $12,570. How much did the bookstore spend on each kind of book before the discount?

35. Merchandising: The total cost of a TV and a VCR is $890. If twice the cost of the VCR is $70 more than the cost of the TV, how much does the VCR cost?

36. Furniture Sale: A sofa and a matching chair were on sale for $1,990. If the sale price of the sofa was $485 less than twice the sale price of the chair, what was the sale price of the sofa?

37. Stock Portfolio: An investor owns stock in a Company A selling at $44 per share, and stock in a Company B selling at $72 per share. The investor owns 4 times as many shares of Company A as of Company B. If the investor's portfolio is worth $248,000, how many shares of each type does the investor own?

38. Fencing: A farmer has 720 feet of fencing to enclose a rectangular pasture. Because a river runs along the length of one side of the pasture, fencing will be needed on only three sides. Find the dimensions of the pasture if its length is 3 times its width.

39. Air Travel: Two airplanes leave the same airport at the same time and fly in opposite directions. Suppose that one airplane averages 120 miles per hour ground speed more than the other airplane. Find the ground speed of each airplane if they are 3000 miles apart in 3 hours.

40. Coin Collection: A coin collector has a collection of 174 coins consisting of dimes and nickels. If she thinks that each coin is worth 3 times its face value, then her collection is worth $40.35. How many of each type of coin does she have?

41. Inventory: A car dealer stocks minivans and sport utility vehicles and cannot hold more than 90 vehicles at one time. On the average, the manufacturer finances the dealership $15,000 on the purchase of a minivan and $18,000 on the purchase of a sport utility vehicle. How many vehicles of each type should the dealer stock if its total debt cannot exceed $1,446,000? Graph this region and interpret it.

42. Investment: A fund uses no more than $200,000 for two investments, one portion at 5% annual simple interest, and the rest at 6% annual simple interest. Suppose that the total interest from both investments for one year is at least $10,600. How much money should be put in each investment? Graph this region and interpret it.

In Problems 43–46, find the approximate point of intersection graphically.

43. $\begin{cases} x + y = \frac{9}{4} \\ 3x + 4y = 10 \end{cases}$

44. $\begin{cases} 3x - 6y = 3.9 \\ 0.2x - 0.4y = 0.5 \end{cases}$

45. $\begin{cases} 2x + 3y = 7 \\ 4x - 5y = 3 \end{cases}$

46. $\begin{cases} 4x + 6y = 3 \\ 7x - 3y = 12 \end{cases}$

CHAPTER 3 PRACTICE TEST

1. Solve each system graphically by sketching the graphs of the equations on the same coordinate axis. Then estimate the point of intersection (if any) of the two lines. Write the two equations in each system in slope intercept form.

(a) $\begin{cases} x - 2y = 1 \\ 2x + y = 7 \end{cases}$

(b) $\begin{cases} 2x + y = 1 \\ 4x + 2y = 3 \end{cases}$

(c) $\begin{cases} -x + 2y = 1 \\ 3x - 6y = -3 \end{cases}$

2. Indicate whether the systems in Problem 1 are consistent and dependent, consistent and independent, or inconsistent.

3. Use any method to solve the systems. Determine whether or not the systems are consistent.

(a) $\begin{cases} 2x - y = 3 \\ -3x + 2y = -7 \end{cases}$

(b) $\begin{cases} \frac{3}{5}x - \frac{y}{5} = 1 \\ 2x + y = 20 \end{cases}$

(c) $\begin{cases} x - y + z = 3 \\ x + y - z = 9 \\ x + y - z = 7 \end{cases}$

4. Determine whether or not $(-2,5,3)$ is a solution of the system

$\begin{cases} x + 2y - 3z = -1 \\ x + y - z = 0 \\ 3x - y + 3z = -2 \end{cases}$

5. Graph the solution set of each system of inequalities.

(a) $\begin{cases} x \geq 1 \\ y < 2 \end{cases}$

(b) $\begin{cases} x - y \leq 3 \\ x + 2y \leq 4 \end{cases}$

6. Linda paid for her $4.25 lunch with 55 coins. If all of the coins were dimes and nickels, then how many were there of each type?

7. An investor lost twice as much in a mutual fund as she did in a bond fund. If her losses totaled $12,690, then how much did she lose in each market?

Polynomial Expressions and Functions

Polynomial functions are another type of function which can accurately model events in real life. For example, they are used in business to model cost and profit, in physics to model motion, and in architecture to model design. This situation is explored in example 7 of section 4.2.

In this chapter, we discuss the algebra of polynomials: addition, subtraction, multiplication, and division. We introduce polynomial functions and examine some graphs of these functions. A most important tool in working with polynomials is factoring which is used in solving polynomial equations.

Objectives

1. Define Polynomials and Their Degrees
2. Evaluate Polynomial Functions
3. Explore Graphs of Polynomial Functions
4. Add and Subtract Polynomials
5. Solve Applied Problems

4.1 Introduction to Polynomial Functions

In this section we extend the properties of real numbers to special algebraic expressions known as polynomials.

Defining Polynomials and Their Degrees

Recall from Chapter 1 that the **terms** of an algebraic expression are the parts of the expression that are separated by plus or minus signs. A **polynomial** is an algebraic expression consisting of a finite number of terms in which all exponents of the variables are nonnegative integers and no variable appears in a denominator. For example,

The expression $4x$ is a one-term polynomial in the variable x.

The expression $3y - 5$ is a two-term polynomial in the variable y.

The expression $3t^2 + 4t + 7$ is a three-term polynomial in the variable t.

Note that the expressions:

$$\frac{3}{x^2}, \quad 4 - \frac{1}{y}, \quad \text{and} \quad \sqrt{2t^{-2} + 1}$$

are not polynomials because each expression has a variable in the denominator or a negative integer exponent. Expressions such as

$$5x^2 \text{ and } -12x^2$$

which differ only in their numerical coefficients are called **like terms** or **similar terms.** The expressions

$$4x^2 \text{ and } 3x^3$$

are not like terms because the exponents of x in each expression are different. The first term of a polynomial is called the **leading term.** Polynomials are identified by the number of their terms. Table 1 shows the description of some polynomials.

TABLE 1

Polynomial	Type	Examples
monomial	a polynomial of one term	$2x$, $3x^2$, $5x^3$, -14
binomial	a polynomial of two unlike terms	$2x + 1$, $3t^2 + t$, $2y^3 + 3y$
trinomial	a polynomial of three unlike terms	$x^2 + 5x + 2$, $3y^2 - 7y + 2$

Polynomials containing more than three terms are not given specific names. The term $5xy$ has two variable factors, x and y, and one numerical factor, 5, which is called the **numerical coefficient.** The numerical factors of the terms in a polynomial are called the **coefficients** of the polynomial. For example, the coefficients of the polynomial

$$-2x^5 - 3x^4 + 5x^3 - 4x^2 + 8$$

When a term is "missing", its coefficient is 0, 0 · x = 0

are -2, -3, 5, -4, 0, and 8. We usually list terms in order of *descending powers* with powers of x decreasing from 5 to 4 to 3 . . . to 0. The **leading coefficient** of a polynomial is the numerical factor of the highest degree term. The **degree of a non-zero term** in a polynomial is the sum of all the exponents of the variables in that term. The highest degree of all terms that appear with non zero coefficients in a polynomial is called the **degree of the polynomial.** For instance,

the degree of the monomial $5x^7$ is 7;

the degree of the binomial $5t^7 + 4t^8$ is 8

since the term $4t^8$ has the greatest degree, 8. A non-zero number such as 6 is called a **constant polynomial** with a zero degree. The number 0 is called a **zero polynomial** with no degree assigned to it. Table 2 contains specific illustrations of polynomials.

TABLE 2

Polynomial	Type	Degree	Leading Coefficient	Coefficients of Each Term
5	monomial	0	5	5
$2x + 7$	binomial	1	2	2, 7
$-3x^2 - 5x - 1$	trinomial	2	-3	$-3, -5, -1$
$x^4 - x^2 + x + 1$	polynomial	4	1	1, 0, -1, 1, 1

Table 3 describes polynomials according to their degrees.

TABLE 3

Polynomial	Description	Examples
Linear	a polynomial of degree 0 or 1	$3x + 2$, 5
Quadratic	a polynomial of degree 2	$4x^2 + x + 7$, $2y^2 - 7$
Cubic	a polynomial of degree 3	$-5x^3 + 2x^2 + 7x + 2$

EXAMPLE 1

Determining the Degree of a Polynomial

Find the degree of each polynomial and state the coefficient of each term including any missing terms.

(a) $4x^5 - 2x^3 + 1$ (b) $5x^3 - 2x^2 - 7x + 10$

Solution

We organize the results in a tabular form.

Polynomial	Degree of Polynomial	Coefficients
(a) $4x^5 - 2x^3 + 1$	5 from $4x^5$	4, 0, -2, 0, 0, 1
(b) $5x^3 - 2x^2 - 7x + 10$	3 from $5x^3$	5, -2, -7, 10

Evaluating Polynomial Functions

Polynomials that contain a single variable are often denoted by functional notation such as $P(x)$, read as "P of x" or $Q(x)$, read "Q of x". For example, polynomial expressions with functional notation such as

$$P(x) = 4x^5 - 2x^3 + 1 \quad \text{or} \quad Q(x) = 10 - 7x - 2x^2$$

are called **polynomial functions.** In each case, the expression used to describe each function is a polynomial. A polynomial function of the form

$$f(x) = 3x + 2$$

is called a **linear function.** A polynomial function of the form

$$f(x) = 3x^2 - 7x + 5$$

is called a **quadratic function.** These are functions because each value of x, the input element, returns one output element. The next example shows how to find the output value given an input value. This is called **evaluating** a polynomial at a given value of x.

EXAMPLE 2   **Evaluating a Polynomial Function**

Let $P(x) = 40 - 2x - 9x^2$; find each value.

(a) $P(-2)$ (b) $P(0)$ (c) $P(2)$ (d) $P(-u)$ (e) $P(2t)$

Solution   (a) Let $x = -2$, the value of the polynomial $40 - 2x - 9x^2$ is given by:

$$P(-2) = 40 - 2(-2) - 9(-2)^2 \qquad \text{Replace } x \text{ with } -2$$
$$= 40 + 4 - 9(4) \qquad \text{Order of operations}$$
$$= 40 + 4 - 36 = 8 \qquad \text{Simplify}$$

(b) When $x = 0$

$$P(0) = 40 - 2(0) - 9(0)^2 = 40 \qquad \text{Replace } x \text{ with } 0$$

(c) For $x = 2$

$$P(2) = 40 - 2(2) - 9(2)^2$$
$$= 40 - 4 - 9(4) \qquad \text{Order of operations}$$
$$= 40 - 4 - 36 = 0 \qquad \text{Simplify}$$

(d) If $x = -u$

$$P(-u) = 40 - 2(-u) - 9(-u)^2 \qquad \text{Substitute } -u \text{ for } x$$
$$= 40 + 2u - 9u^2$$

(e) Let $x = 2t$

$$P(2t) = 40 - 2(2t) - 9(2t)^2 \qquad \text{Substitute } 2t \text{ for } x$$
$$= 40 - 4t - 9(4t^2) \qquad \text{Order of operations}$$
$$= 40 - 4t - 36t^2$$

Exploring Graphs of Polynomial Functions

Figure 1 shows the graphs of the quadratic functions

$$P(x) = x^2 - 3x - 4 \quad \text{and} \quad Q(x) = -x^2 - x + 6.$$

Figure 1

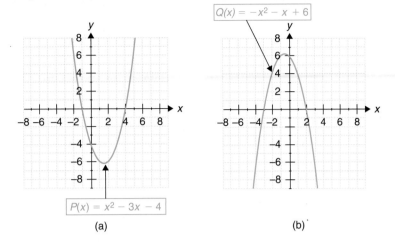

(a)

(b)

Observe that the graph of each function is a smooth continuous curve with no breaks or holes. Figure 1(a) shows that the function $P(x)$ has a y intercept at -4 and x intercepts at -1 and 4. Figure 1(b) shows that the function $Q(x)$ has a y intercept at 6 and x intercepts at -3 and 2.

Figure 2 shows the graphs of the cubic functions

$$R(x) = x^3 + 3x^2 - x - 3 \text{ and } T(x) = -x^3 + 4x.$$

Figure 2

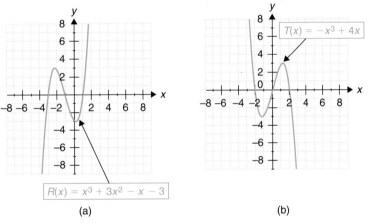

(a)

(b)

Figure 2(a), shows that the graph of $R(x)$ has a y intercept at -3 and the x intercepts are at -3, -1, and 1. In Figure 2(b), the graph of $T(x)$ shows that the y intercept is at the origin, 0, and the x intercepts are at -2, 0, and 2. The following table is a summary of all these important characteristics.

Function	y Intercept	x Intercepts
$P(x) = x^2 - 3x - 4$	-4	$-1, 4$
$Q(x) = -x^2 - x + 6$	6	$-3, 2$
$R(x) = x^3 + 3x^2 - x - 3$	-3	$-3, -1, 1$
$T(x) = -x^3 + 4x$	0	$-2, 0, 2$

Adding and Subtracting Polynomials

To find the sum of two or more polynomials, we group like terms together and then combine these like terms. For example, we add $3x^2$ and $4x^2$ as follows:

$$3x^2 + 4x^2 = (3 + 4)x^2 = 7x^2.$$

To find the difference of two polynomials, we change the signs of all the terms in the polynomial being subtracted then add the resulting like terms. For example, we subtract $-6y^2$ from $-4y^2$ as follows:

$$-4y^2 - (-6y^2) = -4y^2 + 6y^2 = (-4 + 6)y^2 = 2y^2.$$

In general, to *add* or *subtract* polynomials, remove parentheses (if present) and combine like terms. The procedure used for adding (or subtracting) polynomials is performed by extending the properties of numbers to polynomials.

If P, Q, and R are polynomials then:

	Property
1. $P + Q = Q + P$	Commutative Property
2. $P + (Q + R) = (P + Q) + R$	Associative Property
3. $P(Q + R) = PQ + PR$	Distributive Property
4. $(P + Q)R = PR + QR$	Distributive Property

Horizontal and vertical formats for adding and subtracting polynomials are illustrated in the next two examples.

EXAMPLE 3   **Using Horizontal Formats to Combine Polynomials**

Use a horizontal format to perform each operation and simplify.

(a) $(4y^3 + 7y - 3) + (y^3 - y + 2)$ (b) $(4y^3 + 7y - 3) - (y^3 - y + 2)$

Solution   Adding horizontally we have:

(a) $(4y^3 + 7y - 3) + (y^3 - y + 2)$ Given

$= 4y^3 + 7y - 3 + y^3 - y + 2$ Remove parentheses

$= (4y^3 + y^3) + (7y - y) + (2 - 3)$ Group like terms

$= (4 + 1)y^3 + (7 - 1)y - 1$ Distributive property

$= 5y^3 + 6y - 1$ Combine like terms

(b) We rewrite the subtraction problem as the addition of the additive inverse.

$(4y^3 + 7y - 3) - (y^3 - y + 2)$ Given

$= 4y^3 + 7y - 3 - y^3 + y - 2$ Interpret subtraction

$= (4y^3 - y^3) + (7y + y) + (-3 - 2)$ Group like terms

$= (4 - 1)y^3 + (7 + 1)y - 5$ Distributive property

$= 3y^3 + 8y - 5$ Combine like terms

To add or subtract polynomial functions, we add or subtract their corresponding polynomials. That is, if $P(x)$ and $Q(x)$ are functions whose domains intersect, then

$$(P + Q)(x) = P(x) + Q(x)$$

and

$$(P - Q)(x) = P(x) - Q(x).$$

EXAMPLE 4   **Combining Polynomial Functions**

Let $P(x) = 4x^2 - 5x + 7$ and $Q(x) = 2x^2 + 3x - 11$ be polynomial functions. Find each of the following.

(a) $(P + Q)(2)$ (b) $(P - Q)(-3)$ (c) $P(x) + Q(x)$ (d) $P(x) - Q(x)$

Solution   (a) Because $(P + Q)(2) = P(2) + Q(2)$, we will evaluate $P(2)$ and $Q(2)$ first.

$$P(2) = 4(2)^2 - 5(2) + 7 \qquad \text{Substitute 2 for } x$$

$$= 4(4) - 10 + 7 = 13 \qquad \text{Order of operations}$$

$$Q(2) = 2(2)^2 + 3(2) - 11 \qquad \text{Substitute 2 for } x$$

$$= 2(4) + 6 - 11 = 3 \qquad \text{Order of operations}$$

$$(P + Q)(2) = P(2) + Q(2) = 13 + 3 = 16$$

(b) Here, too, we will evaluate $P(-3)$ and $Q(-3)$ first.

$$P(-3) = 4(-3)^2 - 5(-3) + 7 \qquad \text{Substitute } -3 \text{ for } x$$

$$= 4(9) + 15 + 7 = 58$$

$$Q(-3) = 2(-3)^2 + 3(-3) - 11 \qquad \text{Substitute } -3 \text{ for } x$$

$$= 2(9) - 9 - 11 = -2$$

$$(P - Q)(-3) = P(-3) - Q(-3) = 58 - (-2) = 60$$

To make addition or subtraction of the functions P and Q easier, the solutions are often written with the like terms aligned vertically as shown below:

(c) For $P(x) + Q(x)$, we organize the terms as follows:

$$\begin{array}{r} 4x^2 - 5x + 7 \\ \underline{2x^2 + 3x - 11} \\ 6x^2 - 2x - 4 \end{array}$$

(d) For $P(x) - Q(x)$, we change the sign of $Q(x)$ and combine like terms

$$\begin{array}{r} 4x^2 - 5x + 7 \\ \underline{-2x^2 - 3x + 11} \\ 2x^2 - 8x + 18 \end{array}$$

Thus $P(x) + Q(x) = 6x^2 - 2x - 4$ and $P(x) - Q(x) = 2x^2 - 8x + 18$.

Solving Applied Problems

There are many applications involving polynomials. One commonly used in physics to model the free fall of an object is the polynomial function

$$h(t) = -16t^2 + v_0 t + s_0$$

where $h(t)$ is the height of an object (in feet) at time t (in seconds), v_0 is the initial velocity of the object (in feet per second) and s_0 is the initial height of the object (in feet).

EXAMPLE 5   **Finding the Height of an Object**

An object is thrown down from the top of a cliff 120 feet high, with an initial speed of 10 feet per second. Its height h after t seconds is modeled by the polynomial function:

$$h(t) = -16t^2 - 10t + 120, \, t \geq 0.$$

Find the height of the object after 2 seconds.

Figure 3

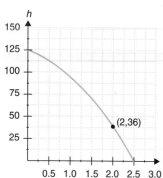

Solution

After 2 seconds, the height of the object is obtained by evaluating $h(2)$, so that,

$$h(2) = -16(2)^2 - 10(2) + 120$$

$$= -64 - 20 + 120 = 36.$$

Therefore, the height of the object is 36 feet.

Figure 3 shows the graphical description of this model. Notice the behavior of the graph that describes the height of the object at each point in the time of its fall. The point (2,36) is on the graph which corresponds to the height of the object after 2 seconds.

PROBLEM SET 4.1

Mastering the Concepts

In Problems 1–8, determine the degree of each polynomial and the coefficients of each term including the missing terms.

1. $-7x^2 + 3x - 9$
2. $x^3 + 5x^2 - 7x + 3$
3. $x^3 - 8x + 5$
4. $y^4 - 4y^2 + 11$
5. $u^4 + 5u^2 - 2$
6. $t^5 - t + 5$
7. $a^5 + 4a^2 - 3a - 1$
8. $8p^6 + 2p^4 - p - 4$

In Problems 9–16, arrange each polynomial in descending order, then find the leading term and the leading coefficient.

9. $x^2 - 4x^5 + x^3 - 11x + 2$
10. $y^3 + 3y^7 - 2y + 3$
11. $5 - 7x - 8x^4 + 4x^2$
12. $t^2 - 7 + 11t^3 + 2t^4$
13. $2x^3 - x + 9 - 10x^4$
14. $3x^2 - 4x^5 - x + 13$
15. $a^2 - 8 + 4a^4 + 7a$
16. $15 - 4x - 8x^4 + 7x^2$

In Problems 17–22, find each specified function value.

17. $P(x) = 6x^2 - 4x + 7$
    (a) $P(2)$ (b) $P(5)$
18. $Q(x) = -3x^2 + 2x + 13$
    (a) $Q(0)$ (b) $Q(4)$
19. $Q(t) = -5t^3 + 7t^2 - 1$
    (a) $Q(-2)$ (b) $Q(4)$

20. $P(u) = 8u^3 + 11u - 2$
    (a) $P(-3)$ (b) $P(2t)$

21. $P(x) = 2x^3 - 5x^2 + 2x + 3$
    (a) $P(0)$ (b) $P\left(-\frac{1}{2}t\right)$

22. $Q(x) = 8x^3 + 6x^2 - 4x + 1$
    (a) $Q\left(\frac{1}{2}\right)$ (b) $Q(-2y)$

In Problems 23–26, evaluate each polynomial for the given values of x.

23. $x^3 - 4x$
    (a) $x = -2$
    (b) $x = 0$
    (c) $x = 3$
    (d) $x = 4$

24. $-x^4 - 2x^2$
    (a) $x = -1$
    (b) $x = 0$
    (c) $x = 2$
    (d) $x = 3$

25. $3x^3 + 2x^2 - 3$
    (a) $x = -2$
    (b) $x = t$
    (c) $x = 2u$
    (d) $x = -3v$

26. $4x^5 + 6x + 3$
    (a) $x = -2$
    (b) $x = 3t$
    (c) $x = -2u$
    (d) $x = -\dfrac{v}{2}$

In Problems 27–30, use the given graph of a polynomial function to find the x and y intercepts of the graph. Also determine the domain of the function from its graph.

27.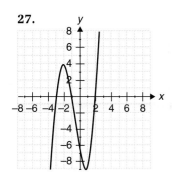

28.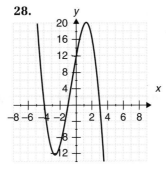

29.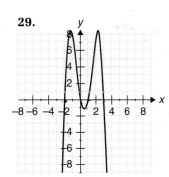

30.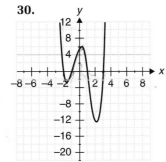

In Problems 31–38, perform each operation and simplify by using:

(a) the horizontal format   (b) the vertical format.

31. $(2x^2 - 3x + 2) + (5x^2 + 2x - 3)$

32. $(3x^2 + 2x + 1) - (5x^2 - 2x + 4)$

33. $(2t^2 + 7t + 8) - (t^2 - 3t + 5)$

34. $(6u^2 - 4u - 1) + (-u^2 - 2u + 2)$

35. $(4t^3 - 7t - 8) - (-2t^2 + 4t - 2)$

36. $(5y^3 - 3y^2 + 2y - 8) + (y - y^3 + 3y^2 - 1)$

37. $(x^6 - 2x^4 - 3x^2) - (x^5 - 2x^3 - 3x - 4)$

38. $(3x^4 - 4x^3 + 6x^2 + x - 1) + (4 - x + 2x^2 - 3x^3 - x^4)$

In Problems 39–42 find and simplify by any format.

(a) $P(-2) + Q(-2)$ (b) $P(-4) - Q(-4)$
(c) $P(x) + Q(x)$ (d) $P(x) - Q(x)$

39. $P(x) = 8x^2 + 3x - 7$, $Q(x) = -5x^2 + 2x - 1$

40. $P(x) = -2x^2 + 5x + 2$, $Q(x) = 3x^2 - 5x + 8$

41. $P(x) = -4x^3 + x^2 + 3$, $Q(x) = 3x^3 + x + 9$

42. $P(x) = 3x^4 - 2x^2 + 4$, $Q(x) = -x^4 + x^3 + 2$

Applying the Concepts

43. Free-Falling Object: A stone is dropped from the top of a building. Its height h (in feet) t seconds after being dropped is given by the formula:

$$h(t) = -16t^2 + 64.$$

Find the height of the stone at the specified times.

(a) $t = 0$ (b) $t = 1/2$ (c) $t = 3/4$

44. Horizontal-Moving Object: A particle is moving along a horizontal straight line according to the formula

$$s(t) = 4t^3 + 2t - 1$$

where $s(t)$ is the distance from a point 0 in meters and t is the time in seconds. Find the distance of the particle at the specified times.

(a) $t = 1$ (b) $t = 1/2$ (c) $t = 1.5$

45. Geometry: The length of a rectangle is 5 feet more than its width. Let x be the width of the rectangle in feet.

(a) Write a polynomial that represents the perimeter P of the rectangle and simplify the expression.

(b) If the width of the rectangle is 4 feet, find its perimeter.

46. Money: A newspaper vending machine accepts dimes and quarters only. Let x represent the number of dimes in the machine at the end of the day. Assume that the machine has six fewer quarters than dimes.

(a) Write an algebraic expression that represents the total value V of all coins in dollars.

(b) Find the total value of the coins in dollars when the machine has 100 dimes.

47. Profit: A furniture company that produces hand-crafted desks estimates that the weekly total cost C of manufacturing x desks is given by

$$C(x) = 6x^2 + 80x + 500.$$

(a) Find the polynomial for profit $P(x)$ if the total weekly revenue $R(x)$ obtained from selling x desks is given by $R(x) = 500x$. (Use $P(x) = R(x) - C(x)$)

(b) How much profit per week will the company make if it manufactures and sells 25, 35, and 50 desks weekly?

48. Manufacturing: A manufacturer produces x framed posters at a total daily cost $C(x)$ (in dollars) of:

$$C(x) = 5x + 50.$$

The total daily revenue $R(x)$ (in dollars) when x posters are sold is given by

$$R(x) = (48 - 2x)x.$$

(a) Find the polynomial that represents the daily profit $P(x)$ (in dollars).

(b) How much profit will the company make per day if it produces and sells 10 posters?

Developing and Extending the Concepts

In Problems 49–58, perform each operation and simplify using any format.

49. $(8x^2 + 3x - 7) + (-5x^2 + 2x + 1) + (2x^2 - 3x + 4)$

50. $(-2p^2 + 5p + 2) + (3p^2 - 2p + 3) + (p^2 - 3p - 7)$

51. $(t^3 - 2t^2 + 3t + 1) + (2t^3 + t^2 - 2t - 2) + (-t^3 + 3t^2 - 2t - 1)$

52. $(3y^4 - 4y^3 + y^2 - 2y + 3) + (7y^4 + 5y^3 + 2y^2 - y - 7) + (6y^3 - 6y + 5)$

53. $(2w^2 - 3w + 4) + (5w - 1 + w^2) - (w + 2w^2 - 6)$

54. $(-x^2 + 8x - 11) - (x^2 - 3x - 2) + (3x^2 - 10x + 3)$

55. $(-2x^2 + 5x + 2) + (3x^2 - 2x + 3) - (x^2 - 3x - 5)$

56. $(3x^3 - 2x^2 + 5x + 4) - (2x^3 + x^2 - 2x + 1) + (-2x^2 + 4x + 8)$

57. $(3y^4 - 4y^3 + y^2 - 2y + 4) + (7y^4 + 5y^3 + 2y^2 - y - 7) - (6y^3 - 6y + 5)$

58. $(t^2 + t + 12) - (t^2 - 2t + 5) + (t^2 + 3t - 4)$

59. Find the sum of the polynomial expressions and simplify.

$$\left(\frac{1}{2}x^2 - 2x^3 + \frac{1}{3}x - 3\right) + \left(\frac{1}{2}x^3 - \frac{3}{2}x^2 + \frac{1}{2}\right) + \left(\frac{3}{4} + 2x - \frac{5}{2}x^2 + \frac{1}{6}x^3\right)$$

60. Subtract the sum of $3x^3 - 2x^2 + 5$ and $3x^2 - x - 3$ from $2x^3 + 3x^2 + 7$.

61. Remove grouping symbols and combine terms
$$5x^2 - 3\{x - x[x + 4(x - 3)] - 5\}$$

62. Is the sum of two polynomials, each of degree 2, necessarily another polynomial of degree 2? Explain.

63. (a) Give an example of a polynomial of degree zero.
    (b) What is the degree of the product of a polynomial of degree zero and a polynomial of degree 2?

64. Two polynomials are equal if the coefficients of each like power terms are equal. Find the value of k so that
$$(3x^3 + 2x^2 + kx + 7) - (2x^3 + x^2 - 4x + 2) = x^3 + x^2 + 7x + 5.$$

Objectives

1. Multiply Monomials
2. Multiply Monomials by Polynomials
3. Multiply Polynomials
4. Multiply Binomials
5. Use Special Products
6. Solve Applied Problems

4.2 Multiplication of Polynomials

So far we have added and subtracted polynomials and polynomial functions. Just like numbers, polynomials such as $P(x)$ and $Q(x)$ can be multiplied. Their product is a polynomial $R(x)$ that gives the same values as $P(x) \cdot Q(x)$ for any value of x. We extend the properties for the multiplication of numbers to polynomials. If P, Q, and R are polynomials, then

	Property
1. $P \cdot Q = Q \cdot P$	Commutative Property
2. $P \cdot (Q \cdot R) = (P \cdot Q) \cdot R$	Associative Property
3. $P(Q + R) = PQ + PR$	Distributive Property
4. $(P + Q)R = PR + QR$	Distributive Property

In this section, we begin by multiplying monomials and consider **special products** for binomials, which will lead to faster ways to perform these multiplications. Then we multiply polynomials in general.

Multiplying Monomials

To multiply two monomials, we use the properties of exponents and the product rule along with the commutative and the associative properties.

EXAMPLE 1   **Multiplying Monomials**

Perform each multiplication and simplify.

(a) $(-8x^3)(-2x^2)$ (b) $(5x^2y)(-4x^3y^4)$

Solution   Using the properties of polynomials, we have:

(a) $(-8x^3)(-2x^2) = (-8)(-2)(x^3 \cdot x^2)$ Commutative property

$= 16x^{3+2} = 16x^5$ Product rule for exponents

(b) $(5x^2y)(-4x^3y^4) = 5(-4)(x^2 \cdot x^3)(y \cdot y^4)$ Commutative property

$= -20x^{2+3}\, y^{1+4} = -20x^5y^5$ Product rule for exponents ▦

Multiplying Monomials by Polynomials

We use the distributive property to multiply a monomial by a polynomial. Once the monomial factor is distributed, we can use the exponent rules to perform multiplication.

EXAMPLE 2   **Multiplying a Monomial by a Trinomial**

Find each product and simplify.

(a) $-2t^4(-3t^3 - 7t + 4)$ (b) $2x^2y(x^2 - 5xy - 3y^2)$

Solution   We use the distributive law to multiply.

(a) $-2t^4(-3t^3 - 7t + 4) = (-2t^4)(-3t^3) + (-2t^4)(-7t) + (-2t^4)(4)$

$= 6t^7 + 14t^5 - 8t^4$

(b) $2x^2y(x^2 - 5xy - 3y^2) = 2x^2y(x^2) + 2x^2y(-5xy) + 2x^2y(-3y^2)$

$= 2x^4y - 10x^3y^2 - 6x^2y^3$ ▦

Multiplying Polynomials

To multiply two polynomials, we multiply each term of one polynomial by each term of the second polynomial, then combine like terms. This can be accomplished by using the distributive property. Such multiplication can be carried out by employing either a horizontal or a vertical format.

EXAMPLE 3   **Multiplying a Binomial by a Trinomial**

Find the product of $(3t - 1)(2t^2 + 7t + 4)$.

Solution   First carry out the multiplication horizontally using the distributive property.

$(3t - 1)(2t^2 + 7t + 4)$

$= 3t(2t^2 + 7t + 4) - 1(2t^2 + 7t + 4)$ Distributive property

$= (3t)(2t^2) + (3t)(7t) + (3t)(4) - 1(2t^2) - 1(7t) - 1(4)$ Distributive property

$= 6t^3 + 21t^2 + 12t - 2t^2 - 7t - 4$ Multiply

$= 6t^3 + 19t^2 + 5t - 4$ Combine like terms

Using the alternate vertical format, we have:

$$\begin{array}{r} 2t^2 + 7t + 4 \\ (\times) \quad 3t - 1 \\ \hline 6t^3 + 21t^2 + 12t \\ (+) \quad -2t^2 - 7t - 4 \\ \hline 6t^3 + 19t^2 + 5t - 4 \end{array}$$

$\longleftarrow (2t^2 + 7t + 4)(3t)$

$\longleftarrow (2t^2 + 7t + 4)(-1)$

Like terms are written in the same column.

Just as we add, subtract, and multiply polynomials, it is possible to add, subtract, and multiply polynomial functions in order to obtain new polynomial functions. That is, if $f(x)$ and $g(x)$ define polynomial functions whose domains intersect, then

$$(f \cdot g)(x) = f(x) \cdot g(x).$$

EXAMPLE 4

Multiplying Polynomial Functions

Let $f(x) = 5x + 3$ and $g(x) = 2x^3 + 3x^2 - x + 2$. Find each product

(a) $f(2) \cdot g(2)$ (b) $f(x) \cdot g(x)$ (c) $(f \cdot g)(2)$

Solution

(a) First evaluate $f(2)$ and $g(2)$ and then find their product.

$$f(2) = 5(2) + 3 = 10 + 3 = 13 \qquad \text{Substitute 2 for } x$$
$$g(2) = 2(2)^3 + 3(2)^2 - (2) + 2 = 2(8) + 3(4) - 2 + 2 \qquad \text{Power}$$
$$= 16 + 12 - 2 + 2 = 28$$
$$f(2) \cdot g(2) = 13(28) = 364$$

(b) The product of the functions, $f(x) \cdot g(x)$, can also be performed by using either a horizontal or a vertical format. Using a horizontal format, we have:

$$f(x) \cdot g(x) = (5x + 3)(2x^3 + 3x^2 - x + 2)$$
$$= 5x(2x^3 + 3x^2 - x + 2) + 3(2x^3 + 3x^2 - x + 2)$$
$$= 10x^4 + 15x^3 - 5x^2 + 10x + 6x^3 + 9x^2 - 3x + 6$$
$$= 10x^4 + 21x^3 + 4x^2 + 7x + 6.$$

Using the vertical format, we have

$$\begin{array}{r} 2x^3 + 3x^2 - x + 2 \\ (\times) \quad 5x + 3 \\ \hline 10x^4 + 15x^3 - 5x^2 + 10x \\ (+) \quad 6x^3 + 9x^2 - 3x + 6 \\ \hline 10x^4 + 21x^3 + 4x^2 + 7x + 6. \end{array}$$

$\longleftarrow 5x(2x^3 + 3x^2 - x + 2)$

$\longleftarrow 3(2x^3 + 3x^2 - x + 2)$

(c) The value of $(f \cdot g)(2)$ is the product in part (b) evaluated at $x = 2$.

$$(f \cdot g)(2) = 10(2)^4 + 21(2)^3 + 4(2)^2 + 7(2) + 6$$
$$= 10(16) + 21(8) + 4(4) + 14 + 6 = 364$$

Notice that the two results in parts (a) and (c) are the same.

Multiplying Binomials

To multiply two binomials, once again, we can apply the distributive property. We use a horizontal format and try to discover a method to enable us to perform the multiplication efficiently. For example, to find the product $(x + 2)(x + 3)$, we multiply each term of $x + 2$ by every term of $x + 3$, so we write:

$$(x + 2)(x + 3) = x(x + 3) + 2(x + 3)$$
$$= x^2 + 3x + 2x + 6$$
$$= x^2 + 5x + 6.$$

The result of the preceding multiplication is a trinomial whose terms are determined as follows.

First term: $(x + 2)(x + 3) = x^2 + 5x + 6$
$(x)(x)$

Middle term: $(x + 2)(x + 3) = x^2 + 5x + 6$
$(3)(x)+(2)(x)$

Last term: $(x + 2)(x + 3) = x^2 + 5x + 6$
$(2)(3)$

A shortened process for multiplying the above binomials is called the **FOIL** method which is based on the distributive property. It provides a speedy process for organizing the multiplication of the two binomials. FOIL is an acronym which helps the reader to remember the pattern: the product of the **First** terms, plus the product of the **Outside** terms, plus the product of the **Inside** terms, plus the product of the **Last** terms. The four operations are outlined in Figure 1.

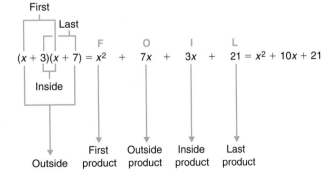

Figure 1

EXAMPLE 5   **Multiplying Binomials by FOIL**

Multiply the binomials by FOIL and simplify the results.

(a) $(x + 2)(2x - 3)$ (b) $(5x^2 - 3x)(2x^2 - 7x)$ (c) $(4x - 3y)(3x - 2y)$

Solution   (a) We distribute $x + 2$ over the sum $2x - 3$ using the FOIL method.

(i) F: $(x + 2)(2x - 3)$ $(2x)(x) = 2x^2$ product of the **First** terms

(ii) O: $(x + 2)(2x - 3)$ $(-3)(x) = -3x$ product of the **Outer** terms

(iii) I: $(x + 2)(2x - 3)$ $(2)(2x) = 4x$ product of the **Inner** terms

(iv) L: $(x + 2)(2x - 3)$ $(-3)(2) = -6.$ product of the **Last** terms

The Inner and Outer products can be combined.

Thus the product is obtained by combining like terms.

$$(x + 2)(2x - 3) = 2x^2 - 3x + 4x - 6$$
$$= 2x^2 + x - 6$$

(b) $(5x^2 - 3x)(2x^2 - 7x) = \overset{\text{F}}{(5x^2)(2x^2)} + \overset{\text{O}}{(5x^2)(-7x)} + \overset{\text{I}}{(-3x)(2x^2)} + \overset{\text{L}}{(-3x)(-7x)}$

$$= 10x^4 - 35x^3 - 6x^3 + 21x^2$$

$$= 10x^4 - 41x^3 + 21x^2$$

(c) $(4x - 3y)(3x - 2y) = \overset{\text{F}}{4x(3x)} + \overset{\text{O}}{4x(-2y)} + \overset{\text{I}}{(-3y)(3x)} + \overset{\text{L}}{(-3y)(-2y)}$

$$= 12x^2 - 8xy - 9xy + 6y^2$$

$$= 12x^2 - 17xy + 6y^2$$

Using Special Products

Using the FOIL method, we can determine a pattern for some special products that occur frequently in algebra. For instance, to find $(x + 5)^2$ using the FOIL method, we have

$$(x + 5)^2 = (x + 5)(x + 5) = \overset{\text{F}}{x(x)} + \overset{\text{O}}{5x} + \overset{\text{I}}{5x} + \overset{\text{L}}{5(5)}$$

$$= x^2 + 10x + 25$$

The result is a perfect square trinomial and the same pattern applies to other special products. It is recommended that you learn these special formulas.

Special Products of Polynomials

Assume That a and b Are Real Numbers or Algebraic Expressions		Result
1. $(a + b)^2 = a^2 + 2ab + b^2$	Squaring a binomial	Perfect square
2. $(a - b)^2 = a^2 - 2ab + b^2$	Squaring a binomial	Perfect square
3. $(a + b)(a - b) = a^2 - b^2$	Product of the sum and difference of two binomials	Difference of squares
4. $(a + b)(a^2 - ab + b^2) = a^3 + b^3$	Product which leads to the sum of two cubes	Sum of cubes
5. $(a - b)(a^2 + ab + b^2) = a^3 - b^3$	Product which leads to the difference of two cubes	Difference of cubes

EXAMPLE 6   **Multiplying by Using Special Products**

Find each product.

(a) $(3x + 2)^2$ (b) $(3x^2 - 4y^3)^2$ (c) $(y - 4)(y^2 + 4y + 16)$

(d) $(2m - 3)(2m + 3)$ (e) $(u + 2)(u^2 - 2u + 4)$

Solution   (a) $(3x + 2)^2$ is the square of a binomial where $a = 3x$ and $b = 2$. So we have

$(x - 5)^2$ is not the same as $x^2 + 25$.

$$(a + b)^2 = a^2 + 2ab + b^2$$
$$(3x + 2)^2 = (3x)^2 + 2(3x)(2) + 2^2$$
$$= 9x^2 + 12x + 4.$$

(b) $(3x^2 - 4y^3)^2$ is the square of a binomial where $a = 3x^2$ and $b = 4y^3$. So we have:

$$(a - b)^2 = a^2 - 2ab + b^2$$
$$(3x^2 - 4y^3)^2 = (3x^2)^2 - 2(3x^2)(4y^3) + (4y^3)^2$$
$$= 9x^4 - 24x^2y^3 + 16y^6.$$

(c) This is the form of the difference of cubes with $a = y$ and $b = 4$

$$(y - 4)(y^2 + 4y + 16) = y^3 - 64.$$

(d) This is a product of the sum and difference of two terms with $a = 2m$ and $b = 3$.

$$(2m - 3)(2m + 3) = (2m)^2 - (3)^2$$
$$= 4m^2 - 9.$$

(e) Using special product 4 with $a = u$ and $b = 2$, we have

$$(u + 2)(u^2 - 2u + 4) = u^3 + 2^3$$
$$= u^3 + 8.$$

Solving Applied Problems

The product of polynomial expressions can be used to solve different types of applied problems as the next example shows.

EXAMPLE 7   **Finding Revenue in Production**

A company manufacturing widgets is creating a revenue model for its product at different prices. It has determined that the cost per item and quantities sold depending on the price of the widgets are as follows.

The cost C (in dollars) when x is the selling price is given by the model

$$C(x) = -0.12x^2 + 1.2x + 2, \quad x > 0$$

and the quantity Q that are sold when x is the selling price is given by

$$Q(x) = -5x + 105, \quad x > 0.$$

The revenue including a 10% profit margin is the product:

$$R(x) = 1.10Q(x)C(x) \,, x > 0$$

(a) Find R as a function of x.

(b) Find R when the price, x, is $3, $6, and $9.

Solution   (a) The revenue function R is given by

$$R(x) = 1.10Q(x)C(x) = 1.10(-5x + 105)(-0.12x^2 + 1.2x + 2).$$

First we use a vertical format to multiply the polynomials:

$$
\begin{array}{r}
-0.12x^2 + 1.2x + 2 \\
-5x + 105 \\
\hline
0.6x^3 - 6x^2 - 10x \\
-12.6x^2 + 126x + 210 \\
\hline
0.6x^3 - 18.6x^2 + 116x + 210.
\end{array}
$$

Next we multiply by the constant 1.10.

$$R(x) = 1.10(0.6x^3 - 18.6x^2 + 116x + 210)$$

or

$$R(x) = 0.66x^3 - 20.46x^2 + 127.6x + 231$$

(b) We evaluate the revenue function at each price:

$$R(3) = 0.66(3)^3 - 20.46(3)^2 + 127.6(3) + 231 = \$447.48$$
$$R(6) = 0.66(6)^3 - 20.46(6)^2 + 127.6(6) + 231 = \$402.60$$
$$R(9) = 0.66(9)^3 - 20.46(9)^2 + 127.6(9) + 231 = \$203.28$$

Figure 2

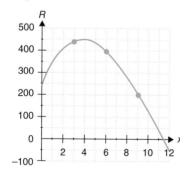

Figure 2 shows the graph of R. Notice how the points (3,447.48), (6,402.60), and (9,203.28) fit on the graph of R. The domain of this type of model is all positive real numbers.

PROBLEM SET 4.2

Mastering the Concepts

In Problems 1–10, perform each multiplication and simplify.

1. (a) $(2x^2)(3x^4)$
   (b) $(5x^2y)(7x^3y^2)$
2. (a) $(7m^4)(-8m^3)$
   (b) $(-3xy^4)(5x^4y^5)$
3. (a) $(3y)(5y^3)(-2y^3)$
   (b) $(2x^2y^3)(-3x^2yz^3)$
4. (a) $\left(\frac{2}{3}b\right)(15b^3)(-2b^2)$
   (b) $\left(\frac{4}{9}x^2y^3\right)\left(-\frac{3}{2}x^4y^2\right)$
5. (a) $-4y^2(3y-4)$
   (b) $5x^2(3x^4+2)$
6. (a) $2x^2(3x^2+7x+5)$
   (b) $3x^4(2x^3+7x-6)$
7. (a) $4x^2y^3(-2xy^4+7xy^2+3)$
   (b) $2xy^2(5x^3y^2-2xy+5)$
8. (a) $0.2x(10x^3y^2+6xy^3-20)$
   (b) $-0.3a^2b^2(9ab^3+6a^2b-1)$
9. (a) $0.3x^2y^3(9x^4y-4x^2y^3+5xy)$
   (b) $4.6m^3n(1.3m^4n^2-2.6mn^2+5n^4)$
10. (a) $-0.4c^2d^3(8c^3d^2+7cd^2-5d^4)$
    (b) $0.6x^3y^2(8x^2y-2.6xy^3+5.9y^5)$

In Problems 11–24, use the vertical format to find each product.

11. $(3x+2)(3x^2+5x-4)$
12. $(2u-3)(u^2-3u+2)$
13. $(3y+5)(y^2-5y+7)$
14. $(4t^2+7)(t^3+5t-1)$
15. $(2u^2-3)(2u^2-2u+4)$
16. $(x^2+3)(x^2-2x+5)$
17. $(3x-2)(3x^2-5x+4)$
18. $(x-2y)(x^2-3xy-y^2)$
19. $(m^2+3)(m^3+2m^2-3m+4)$
20. $(4-x^2)(x^4-2x^2-3x+4)$
21. $(y^2-5y+3)(y^2+2y+3)$
22. $(u^2-2u+5)(u^2+3u+1)$
23. $(x^2+2x-2)(x^3-3x^2+3)$
24. $(2t^2-5t+1)(t^3-t^2+t-3)$

In Problems 25 and 26, for the given functions find:

(a) $(f\cdot g)(2)$ (b) $(g\cdot f)(2)$
(c) $(f\cdot g)(x)$ (d) $(g\cdot f)(x)$

25. $f(x)=3x^2-5$ and $g(x)=-2x^2+3$
26. $f(x)=2x^2+3$ and $g(x)=-x^2+4x-1$

In Problems 27–38, perform each binomial multiplication.

27. (a) $(x+1)(x+2)$ 28. (a) $(x+3)(x+4)$
    (b) $(x-1)(x-2)$ (b) $(x-3)(x-4)$
29. (a) $(x+5)(x-3)$ 30. (a) $(x-6)(x+2)$
    (b) $(x-5)(x+3)$ (b) $(x+6)(x-2)$
31. (a) $(5x+4)(x-1)$
    (b) $(5x-4y)(x+y)$
32. (a) $(3x+2)(2x-4)$
    (b) $(3x-2y)(2x+4y)$
33. $(2x^2+3)(5x^2-1)$
34. $(x^2+y^3)(x^3-y)$
35. $(0.4x+3y)(2x-0.5y)$
36. $(3.2t-0.4s)(0.5t-3.2s)$
37. $\left(\frac{x}{3}+4y\right)\left(3x-\frac{y}{4}\right)$
38. $\left(\frac{x}{5}-3y\right)\left(\frac{x}{2}+5y\right)$

In Problems 39–48, use a special product to perform each multiplication.

39. (a) $(w-7)(w+7)$
    (b) $(7-w)(7+w)$
40. (a) $(2m-9)(2m+9)$
    (b) $(9+2m)(9-2m)$
41. (a) $(x+1)^2$
    (b) $(x-1)^2$
42. (a) $(x+4)^2$
    (b) $(x-4)^2$
43. (a) $(4y^2+5z)^2$
    (b) $(4y^2-5z)^2$
44. (a) $(9y+8z)^2$
    (b) $(9y-8z)^2$
45. (a) $(x+y)(x^2-xy+y^2)$
    (b) $(x-y)(x^2+xy+y^2)$

46. (a) $(x - 2y)(x^2 + 2xy + 4y^2)$

(b) $(x + 2y)(x^2 - 2xy + 4y^2)$

47. $\left(3x - \dfrac{1}{4}\right)\left(3x + \dfrac{1}{4}\right)$

48. $\left(7x - \dfrac{3}{7}\right)\left(7x + \dfrac{3}{7}\right)$

Applying the Concepts

49. Geometry: The length of a rectangle is 5 feet more than its width. Let x (in feet) be the length of the rectangle.

(a) Write a simplified expression that represents the area of the rectangle.

(b) Find the area of the rectangle if its length is 8 feet.

50. Flier's Design: Advertising fliers are to be printed on rectangular sheets of paper x centimeters long and $(2x - 5)$ centimeters wide. Suppose the margins at the top and bottom are each 3 centimeters and the margins at the sides are 2 centimeters (Figure 3).

Figure 3

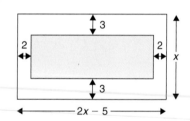

(a) Write a simplified expression that represents the area of the printed part of the flier.

(b) Find the area of the printed part of the flier if $x = 15$ centimeters.

51. Agriculture: The total yield from a crop is the number of trees times the yield per tree. An orange grower has determined that if 120 trees are planted per acre, each tree will yield approximately 60 boxes of oranges over the growing season. If the number of trees is increased, the yield per tree will be reduced due to crowding. For each additional tree planted over 120, the yield is expected to decrease by 2 boxes per tree. Let x represent the number of trees planted over 120 per acre.

(a) Write a simplified expression that represents the total yield in terms of x.

(b) Find the total yield if 125 trees are planted ($x = 5$).

52. Package Design: An open box is to be constructed by removing equal squares, each of length x centimeters from the corners of a piece of cardboard that measures 48 centimeters by 32 centimeters, and then folding up the sides (Figure 4).

Figure 4

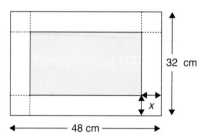

(a) Write an expression that represents the volume of the box.

(b) Simplify the resulting expression, then find the volume when $x = 6$ centimeters.

Developing and Extending the Concepts

In Problems 53–66, simplify each expression.

53. $-4x^2(5x - 10)^2$

54. $-(2x + 7)(x + 4)(x - 3)$

55. $(x - 4)(3x + 4)^2$

56. $(4 - 2x)^4$

57. $[(5x - 1) + y][(5x - 1) - y]$

58. $(2x + y)(2x - y)(4x^2 - y^2)$

59. $[(x + y) + 5]^2$

60. $[4 - (3x - 2y)]^2$

61. $[x + (3y + 4)][x - (3y + 4)]$

62. $(2x + 5y)^2(2x - 5y)^2$

63. $(x + 2y)^4$

64. $(2a - b)^4$

65. $[(y - 3) - (x + 1)]^2$

66. $[(y + 2) - (x + 4)]^2$

4.3 Division of Polynomials

In this section, we consider division of polynomials starting with division by monomials, then we look at division of polynomials containing more than one term using long division. Also, we introduce a technique for dividing a polynomial by a binomial of the form $x - c$, called *synthetic division*.

Dividing a Polynomial by a Monomial

We will use the law of exponents for division

$$\frac{x^m}{x^n} = x^{m-n}, \quad \text{and} \quad x^0 = 1.$$

We divide a polynomial by a monomial by subtracting the exponents when bases are the same, as the next example shows.

EXAMPLE 1   **Dividing a Polynomial by a Monomial**

Divide the following.

(a) $\dfrac{x^5}{x^2}$ (b) $\dfrac{7x^3y^4}{3xy^2}$ (c) $\dfrac{6x^3 - 3x^2 + 7x - 5}{3x}$.

Solution

(a) $\dfrac{x^5}{x^2} = x^{5-2}$ Quotient rule for exponents

$= x^3$

(b) $\dfrac{7x^3y^4}{3xy^2} = \dfrac{7}{3}x^{3-1}y^{4-2}$ Quotient rule for exponents

$= \dfrac{7}{3}x^2y^2$

(c) $\dfrac{6x^3 - 3x^2 + 7x - 5}{3x} = \dfrac{6x^3}{3x} - \dfrac{3x^2}{3x} + \dfrac{7x}{3x} - \dfrac{5}{3x}$ Divide each term by $3x$

$= 2x^2 - x + \dfrac{7}{3} - \dfrac{5}{3x}$ Simplify each quotient

In general, if P, Q and R, $R \neq 0$ are monomials, then

$$\frac{P + Q}{R} = \frac{P}{R} + \frac{Q}{R}$$

Dividing a Polynomial by a Binomial

Now we look at division of a polynomial by a polynomial containing more than one term such as

$$3x^2 - x - 10 \text{ divided by } x + 2.$$

When a divisor has more than one term, the steps used in arithmetic of **divide, multiply, subtract, bring down the next term,** form a repetitive procedure for polynomials also called *long division*.

EXAMPLE 2   **Dividing a Polynomial by a Binomial**

Divide $3x^2 - x - 10$ by $x + 2$.

Solution   The following steps have the same structure as that used in numerical long division.

$$x + 2 \overline{)3x^2 - x - 10}$$

Arrange the terms of $3x^2 - x - 10$ and $x + 2$ in descending powers of x.

$$\begin{array}{r} 3x \\ x + 2 \overline{)3x^2 - x - 10} \end{array}$$

Divide $3x^2$ by x, $\dfrac{3x^2}{x} = 3x$ align like terms.

$$\begin{array}{r} 3x \\ x + 2 \overline{)3x^2 - x - 10} \\ 3x^2 + 6x \end{array}$$

Multiply each term in $x + 2$ by $3x$ and align the terms of the product under like terms.

$$\begin{array}{r} 3x \\ x + 2 \overline{)3x^2 - x - 10} \\ \underline{-3x^2 - 6x} \\ -7x \end{array}$$

Subtract $3x^2 + 6x$ from $3x^2 - x$ by changing the sign of each term and adding.

$$\begin{array}{r} 3x \\ x + 2 \overline{)3x^2 - x - 10} \\ \underline{-3x^2 - 6x} \\ -7x - 10 \end{array}$$

Bring down -10 from the expression $3x^2 - x - 10$.

$$\begin{array}{r} 3x - 7 \\ x + 2 \overline{)3x^2 - x - 10} \\ \underline{-3x^2 - 6x} \\ -7x - 10 \end{array}$$

Dividing $-7x$ by x, $\dfrac{-7x}{x} = -7$ to find the second term of the quotient.

$$\begin{array}{r} 3x - 7 \\ x + 2 \overline{)3x^2 - x - 10} \\ \underline{-3x^2 - 6x} \\ -7x - 10 \\ -7x - 14 \end{array}$$

Multiply -7 by $x + 2$

$$\begin{array}{r} 3x - 7 \\ x + 2 \overline{)3x^2 - x - 10} \\ \underline{-3x^2 - 6x} \\ -7x - 10 \\ \underline{7x + 14} \\ 4 \end{array}$$

Subtract by changing signs in $-7x - 14$ and add. The remainder is 4.

The procedure illustrated in example 2 is called a *division algorithm*. The original expression may be written in the form

$$3x^2 - x - 10 = (x + 2) \cdot (3x - 7) + 4$$

Dividend   Divisor   Quotient   Remainder

In general, the division of a polynomial $P(x)$ by a divisor $D(x)$ is given by

$$\boxed{P(x) = D(x)Q(x) + R(x).}$$

The polynomial $Q(x)$ is called the **quotient,** $D(x)$ is called the **divisor,** and $R(x)$ is called the **remainder.** Note that $R(x)$ may be zero.

Now we summarize the general procedure for dividing a polynomial by another.

Long Division of Polynomials

1. **Arrange** both the divisor and the dividend in descending powers, noting any missing terms.
2. **Find** the first term of the quotient by dividing the first term of the dividend by the first term of the divisor.
3. **Multiply** the quotient obtained in step 2 by the entire divisor.
4. **Subtract** the product obtained in step 3 from the dividend, and bring down the next term.
5. **Repeat** the procedure in steps 2, 3, and 4 until the degree of the remainder is less than the degree of the divisor.

EXAMPLE 3   **Using Long Division to Divide Polynomials**

Divide $6x^3 + 7x^2 + 4x + 5$ by $2x + 1$.

Solution

$$2x + 1 \overline{)6x^3 + 7x^2 + 4x + 5}$$
Arrange in descending powers.

$$\begin{array}{r} 3x^2 \\ 2x + 1 \overline{)6x^3 + 7x^2 + 4x + 5} \end{array}$$
Divide $6x^3$ by $2x$ to get $3x^2$.

$$\begin{array}{r} 3x^2 \\ 2x + 1 \overline{)6x^3 + 7x^2 + 4x + 5} \\ 6x^3 + 3x^2 \end{array}$$
Multiply quotient, $3x^2$, by divisor, $2x + 1$, to get $6x^3 + 3x^2$.

$$\begin{array}{r} 3x^2 + 2x \\ 2x + 1 \overline{)6x^3 + 7x^2 + 4x + 5} \\ -6x^3 - 3x^2 \\ \hline 4x^2 + 4x \\ -4x^2 \pm 2x \\ \hline 2x \end{array}$$
Subtract the product from the dividend and bring down next term, $4x$.

$$\begin{array}{r} 3x^2 + 2x + 1 \\ 2x + 1 \overline{)6x^3 + 7x^2 + 4x + 5} \\ -6x^3 - 3x^2 \\ \hline 4x^2 + 4x \\ -4x^2 - 2x \\ \hline 2x + 5 \\ -2x - 1 \\ \hline 4 \end{array}$$
Repeat the process.

The remainder

The quotient is $3x^2 + 2x + 1$ and the remainder is 4. When there is a non-zero remainder, the dividend, the quotient, and the remainder can be written as:

$$\frac{6x^3 + 7x^2 + 4x + 5}{2x + 1} = \boxed{3x^2 + 2x + 1} + \frac{\boxed{4}}{\boxed{2x + 1}} \leftarrow \text{Remainder}$$

Quotient   Divisor

Check: (Quotient)(Divisor) + Remainder = $(2x + 1)(3x^2 + 2x + 1) + 4$

$$= 6x^3 + 7x^2 + 4x + 5 = \text{Dividend}$$

The quotient in example 3 involved division by a first degree binomial. The same process may be used for divisors of degrees greater than 1 as the next example shows.

EXAMPLE 4   **Dividing Polynomials**

Divide $-3w^3 + 2w^4 + 5w^2 + 2w + 7$ by $-w + w^2 + 1$.

Solution   First we arrange both polynomials in descending powers of w. The remaining steps are shown as follows:

Dividing Polynomials by Synthetic Division

When the divisor is of the form $x - c$, the long division can be carried out by a shortcut procedure called **synthetic division.** The key steps in the synthetic division process are summarized as follows.

Synthetic Division	Example
Step 1: Arrange the terms of the polynomial $P(x)$ in descending powers of x, and use the 0 coefficient for any missing powers of x.	$P(x) = 3x^3 - x^2 + 2x - 18$ divided by $D(x) = x - 2$
Step 2: Consider the divisor in the form $x - c$, $x - 2$. Write the value of c, 2, draw a vertical line, and then list the coefficients of $P(x)$.	2 $\mid$ 3 −1 2 −18
Step 3: Leave space below the row of coefficients, draw a horizontal line, and copy the leading coefficient of the dividend below the line.	2 $\mid$ 3 −1 2 −18 3

(continue)

Step 4: Multiply the value of c, 2, by the leading coefficient, 3, and place the product, 6, under the second coefficient, −1, and above the horizontal line. Add the product, 6, and the coefficient, −1, placing the result, 5, below the horizontal line.	2 | 3 −1 2 −18 6 3 5
Next multiply this sum, 5, by c, 2, and place the product under the next coefficient, 2. Add and place the sum, 12, under the horizontal line.	2 | 3 −1 2 −18 6 10 3 5 12
Continue in this way until the all of the coefficients have been multiplied and added. Isolate the very last sum, 6, with a short vertical line.	2 | 3 −1 2 −18 6 10 24 3 5 12 | 6
Step 5: In the last row, the numbers 3, 5, and 12 are the coefficients of the quotient Q(x) and the last number, 6, is the remainder R.	$Q(x) = 3x^2 + 5x + 12$ $R = 6$

Notice in the above illustration that $P(2) = 3(2)^3 - 2^2 + 2(2) - 18 = 6$, the remainder. When a polynomial $P(x)$ is divided by $x - c$, the reminder R is equal to $P(c)$, the value of $P(x)$ at c. This is true because if

$$P(x) = (x - c)Q(x), \text{ then}$$

$$P(c) = (c - c)Q(c) + R(c) = R(c)$$

so that the remainder $R = P(c)$

For example, we use the above result and synthetic division to find $p(2)$ when $p(x) = 2x^3 + 5x^2 - 7x + 4$ as follows:

$$
\begin{array}{r|rrr|r}
2 & 2 & 5 & -7 & 4 \\
  &   & 4 & 18 & 22 \\
\hline
  & 2 & 9 & 11 & 25 = R
\end{array}
$$

so that $p(2) = 25$.

EXAMPLE 5   **Using Synthetic Division to Evaluate a Function**

Let $P(x) = 3x^4 - 10x^3 + 14x - 3$, use synthetic division to find $P(-2)$.

Solution   We use synthetic division to divide $P(x)$ by $x - (-2)$. First we write the value of $c = -2$ and all the coefficients of the dividend in descending order, then we have the following display:

$$
\begin{array}{r|rrrr|r}
-2 & 3 & -10 & 0 & 14 & -3 \\
   &   & -6 & 32 & -64 & 100 \\
\hline
   & 3 & -16 & 32 & -50 & 97 = R
\end{array}
$$

Here we have $Q(x) = 3x^3 - 16x^2 + 32x - 50$, and $R = 97$.

By evaluating the polynomial of $x = -2$, it also yields 97, so that $p(-2) = 97 = R$.

Solving Applied Problems

EXAMPLE 6 **Finding the Width of a Rectangle**

Suppose that the polynomial $A(x) = 6x^2 + 7x - 5$ where $x > 1$ (in feet) represents the area of a rectangular plot of farmland and the width is represented by the polynomial $W(x) = 2x - 1$. Find a polynomial representing the length of the farmland.

Solution The area of a rectangle is the product of length times width. To find the length, we must divide the area by the width.

Notice that we may not use synthetic division here because the leading coefficient of the divisor is not 1.

$$
\require{enclose}
\begin{array}{r}
3x + 5 \\
2x - 1 \enclose{longdiv}{6x^2 + 7x - 5} \\
\underline{6x^2 - 3x} \\
10x - 5 \\
\underline{10x - 5} \\
0
\end{array}
$$

Thus, a polynomial representing the length is $3x + 5$.

EXAMPLE 7 **Using Synthetic Division**

The revenue of a manufacturing company is represented by the polynomial $R(x) = x^3 + 5x^2 + 12x + 12$ and the number of items made is represented by the polynomial $N(x) = x + 2$ where x is a cost variable and $x > 0$. What is the polynomial representing the price of the items?

Solution Price times quantity equals revenue. We will divide revenue by quantity to find the price. Since the divisor here, $x + 2$, is of degree 1 with a leading coefficient of 1, we may use synthetic division.

$$
\begin{array}{r|rrrr}
-2 & 1 & 5 & 12 & 12 \\
   &   & -2 & -6 & -12 \\
\hline
   & 1 & 3 & 6 & 0
\end{array}
$$

The price is the polynomial $P(x) = x^2 + 3x + 6$.

PROBLEM SET 4.3

Mastering the Concepts

In Problems 1–4, perform each division and simplify.

1. (a) $(-54x^2y^3) \div (3xy^2)$
(b) $(60x^2y^4) \div (-20xy^2)$
2. (a) $(36x^4y^2) \div (-9x^2y)$
(b) $(-18x^5y) \div (-3x^4y)$
3. (a) $(24x^5 + 15x^3 + 12x^2) \div (-3x^2)$
(b) $(-81x^6 + 36x^4 + 27x^2) \div (9x)$
4. (a) $(25x^4y^2 - 15x^2y + 35x^3y^3) \div (-5x^2y)$
(b) $(-100x^3y^4 + 70x^2y^5 - 35x^6y^3) \div (-5x^2y^3)$

In Problems 5–20, use the long division process to find the quotient and remainder and check the results.

5. $(x^2 - 7x + 10) \div (x - 5)$
6. $(2x^2 + x - 6) \div (x + 2)$
7. $(2x^2 - 5x - 6) \div (2x - 1)$
8. $(6x^2 + 11x + 7) \div (3x + 1)$
9. $(8x^2 - 14x + 5) \div (2x - 3)$
10. $(12x^2 + 6x - 9) \div (4x - 2)$
11. $(x^3 + 3x^2 + 4x + 7) \div (x - 2)$

12. $(x^3 + 7x^2 + 3x + 5) \div (x - 1)$

13. $(x^4 + 2x^3 - 3x^2 + 5x + 1) \div (x + 1)$

14. $(x^4 - 3x^3 + 2x^2 - 6x + 3) \div (x - 3)$

15. $(x^4 + 5x^3 + 9x^2 + 5x - 4) \div (x^2 + 2x - 1)$

16. $(12x^2 - 9x^3 - 3x + 1) \div (1 - 3x^2)$

17. $(x^3 + 16x + 52 - 3x^2) \div (x^2 + 26 - 5x)$

18. $(3x^5 - 2x^4 - 10x^3 + 26x^2 - 25x - 2)$
    $\div (-5x + 3x^2 + 1)$

19. $(x^2y^2 + 10xy - 9y^4 + 30y^2) \div (x + 3y)$

20. $(6x^3y^3 - x^2y^4 - 5xy^5 + 2y^6) \div (3x - 2y)$

In Problems 21–32, use synthetic division to perform each division and find the quotient and remainder. Also, find each functional value.

21. $P(x) = (3x^2 + 7x - 10) \div (x + 2); P(-2)$

22. $P(x) = (3x^2 - 5x + 12) \div (x - 3); P(3)$

23. $P(x) = (2x^2 + 8x + 13) \div (x + 2); P(-2)$

24. $P(x) = (4x^2 + 5x - 20) \div (x - 4); P(4)$

25. $P(x) = (4x^3 + 4x^2 - 2x - 3) \div (x - 1); P(1)$

26. $P(y) = (5y^3 + 6y^2 - 12y - 14) \div (y + 1); P(-1)$

27. $P(x) = (4x^3 + 3x^2 - 3x - 1) \div (x - 1); P(1)$

28. $P(y) = (5y^3 + 7y^2 - 12y - 8) \div (y + 3); P(-3)$

29. $P(b) = (5b^3 - 6b^2 + 3b + 11) \div (b - 2); P(2)$

30. $P(x) = (-2x^3 + 3x^2 + 2x + 5) \div (x + 1); P(-1)$

31. $P(x) = (x^5 + x^3 + 2) \div (x - 2); P(2)$

32. $P(x) = (x^4 - 16) \div (x - 2); P(2)$

Applying the Concepts

33. **Area:** The area of a rectangle (in square feet) is represented by the expression
$$6x^2 + 11x - 35, x \geq 3.$$
Its width (in feet) is represented by the expression
$$3x - 5.$$
(a) Use long division to find an expression for the length of the rectangle.

(b) Find the width when $x = 3$ feet.

34. **Area:** A patio has the shape of a rectangle whose area (in square feet) is given by the polynomial expression
$$81x^2 + 18x - 35.$$
If its width is represented by the binomial $9x - 5$ feet,

(a) find its length.

(b) Find the length when $x = 4$ feet.

35. **Volume:** Suppose that the volume of a right circular cylinder (in cubic inches) such as a soup can is represented by the expression:

$8x^3 + 28x^2 + 14x - 15$,
$x > 0.5$.

Its base area (in inches) is represented by the expression
$$4x^2 + 4x - 3.$$
(a) Find an expression for the height of the cylinder.

(b) Find the height when $x = 2$ inches.

36. **Volume:** The volume of a rectangular chest (in cubic feet) is given by the polynomial expression

$4x^3 + 16x^2 + 17x + 5$.

If the height of the chest is represented by $x + 1$ (in feet) and its width is represented by $2x + 5$ (in feet),

(a) find its length.

(b) Find the length when $x = 2$ feet.

Developing and Extending the Concepts

37. Perform the indicated operations and simplify.

(a) $\dfrac{1}{x} \div \dfrac{1}{x}$ (b) $\dfrac{1}{x} \div \dfrac{1}{x} \div \dfrac{1}{x}$ (c) $\dfrac{1}{x} \div \dfrac{1}{x} \div \dfrac{1}{x} \div \dfrac{1}{x}$

(d) For what value(s) of x will the expressions in parts (b) and (c) be equal?

(e) For what value(s) of x will the expression in part (c) be positive?

(f) For what value(s) of x will the expression in part (c) be negative?

38. Use the pattern established in Problem 37 to find the value of the expression
$$\frac{1}{x} \div \frac{1}{x} \div \frac{1}{x} \div \frac{1}{x} \div \frac{1}{x} \div \frac{1}{x}.$$

39. Use long division to divide $(x^3 + y^3)$ by $(x + y)$, then verify the special factoring formula
$$x^3 + y^3 = (x + y)(x^2 - xy + y^2).$$

40. Use long division to divide $(x^5 - y^5)$ by $(x - y)$, then verify the factoring formula
$$x^5 - y^5 = (x - y)(x^4 + x^3y + x^2y^2 + xy^3 + y^4).$$

Objectives

1. Factor a Monomial From a Polynomial
2. Factor by Grouping
3. Solve Applied Problems

4.4 Basic Factoring—Common Factors and Grouping

To **factor** an expression means to write it as a product of other expressions. An expression may have many factors, for instance, the factors of 12 are:

$$1 \cdot 12 = 12 \qquad (-1)(-12) = 12$$
$$2 \cdot 6 = 12 \qquad (-2)(-6) = 12$$
$$3 \cdot 4 = 12 \qquad (-3)(-4) = 12$$

We can view the distributive property from *left to right* as multiplication. For instance,

$$2x(3x + 7) = 6x^2 + 14x.$$

If we reverse this statement as

$$6x^2 + 14x = 2x(3x + 7),$$

we are rewriting a sum as a product, that is, *factoring the sum*. Here, we say that $2x$ and $3x + 7$ are *factors* of the polynomial expression $6x^2 + 14x$.

Polynomials that cannot be factored are called **prime polynomials.** In this section, we learn how to factor a monomial from a polynomial and how to factor by grouping.

Factoring a Monomial From a Polynomial

To factor a monomial from a polynomial, we determine the greatest common factor (GCF) of each term in the polynomial. The **greatest common factor** (GCF) of two or more expressions is the greatest factor that divides (without remainder) each expression. For example, the GCF of 15, 20, and 25 is 5 since 5 is the largest number that divides 15, 20, and 25. We express each term of an expression as the product of the GCF and its other factors. Then we use the distributive property to factor out the GCF. For example, to factor $15x + 10$, the GCF is 5 and we use the distributive property

$$15x + 10 = 5(3x) + 5(2) = 5(3x + 2)$$

so that 5 and $3x + 2$ are factors of $15x + 10$. We check this factorization by multiplying 5 by $3x + 2$ and obtaining the original polynomial. In general, we factor a monomial from a polynomial by factoring the GCF out of each of the terms of the polynomial.

EXAMPLE 1   **Finding the GCF of Each Expression**

(a) x^{10}, x^4, x^7, x^3 (b) $14x^4y^2, 21x^4y^3, 28x^5$

Solution   (a) Notice that x^3 is the lowest power of x that appears in any of the four terms.

$$x^{10} = x^3(x^7), \quad x^4 = x^3(x), \quad x^7 = x^3(x^4), \quad \text{and} \quad x^3 = x^3(1)$$

Therefore the GCF is x^3.

(b) The lowest power of x that appears in any of the three terms is x^4. Since y does not appear in the last term, it is not part of the GCF. The GCF of the numbers 14, 21, and 28 is 7. We write:

$$14x^4y^2 = 7x^4(2y^2), \quad 21x^4y^3 = 7x^4(3y^3), \quad \text{and} \quad 28x^5 = 7x^4(4x).$$

Therefore, the GCF of these terms is $7x^4$.

EXAMPLE 2   **Factoring Out Common Factors**

Factor each expression.

(a) $8x^2 + 12x$

(b) $3x^2y - 6xy$

(c) $12xy^4 + 4x^3y^2 - 6x^2y^5$

Solution   (a) $8x^2 + 12x = 4x(2x) + 4x(3)$ Express each term as a product of the GCF and the other factors

$\qquad\qquad\qquad\quad = 4x(2x + 3)$ Factor out the GCF

(b) $3x^2y - 6xy = 3xy(x) + 3xy(-2)$ The GCF is $3xy$

$\qquad\qquad\qquad = 3xy(x - 2)$ Factor out the GCF

(c) $12xy^4 + 4x^3y^2 - 6x^2y^5$

$\quad = 2xy^2(6y^2) + 2xy^2(2x^2) + 2xy^2(-3xy^3)$ $2xy^2$ is the GCF

$\quad = 2xy^2(6y^2 + 2x^2 - 3xy^3)$ Factor out the GCF

Each factorization should be checked by multiplication.

EXAMPLE 3   **Factoring Out a Binomial Factor**

Factor each expression.

(a) $5x(x + 2) + 2(x + 2)$

(b) $3b(5b + 7) - c(5b + 7)^2$

Solution   (a) $5x(x + 2) + 2(x + 2) = (5x + 2)(x + 2)$ $x + 2$ is the GCF

(b) $3b(5b + 7) - c(5b + 7)^2 = [3b - c(5b + 7)](5b + 7)$ $5b + 7$ is the GCF

$\qquad\qquad\qquad\qquad\qquad = (3b - 5bc - 7c)(5b + 7)$

Factoring by Grouping

In some situations, we may be able to factor a polynomial with no apparent common factor by considering the polynomial when the terms are properly *grouped*. The process is called *factoring by grouping*. The procedure is illustrated in the next example.

EXAMPLE 4   **Factoring by Grouping**

Factor each expression.

(a) $3xm - 2x + 3ym - 2y$ (b) $5y^2 - 10y - 3y + 6$ (c) $x^3 - 7x^2 + 2x - 14$

Solution
(a) $3xm - 2x + 3ym - 2y$
$= (3xm - 2x) + (3ym - 2y)$ Group terms with a common factor
$= x(3m - 2) + y(3m - 2)$ Factor x and y from each group
$= (x + y)(3m - 2)$ $3m - 2$ is the GCF

(b) $5y^2 - 10y - 3y + 6$
$= [5y \cdot y + 5y(-2)] + [-3y + (-3)(-2)]$ Group terms with common factors
$= 5y(y - 2) - 3(y - 2)$ Factor $5y$ and -3 from each group
$= (5y - 3)(y - 2)$ $y - 2$ is the GCF

(c) $x^3 - 7x^2 + 2x - 14$
$= x(x^2) - 7(x^2) + (2)x - 7(2)$ Group terms with common factors
$= x^2(x - 7) + 2(x - 7)$ $x - 7$ is a common factor
$= (x - 7)(x^2 + 2)$

Solving Applied Problems

Some applied problems can be written in factored form as the next example shows.

EXAMPLE 5   **Finding the Sale Price in Factored Form**

A clothing store has discounted the price of a dress by 30%. Because of excess inventory, the sale price is reduced by 20%. If x represents the original price of the dress in dollars, then the final sale price is represented by the model

$$S(x) = x - 0.30x - 0.20(x - 0.30x), x > 0$$

(a) Write the expression for the sale price in factored form.

(b) For what percentage of the original price is the dress selling? Do you think the dress is selling at 50% of its original price?

Solution
(a) $S(x) = x - 0.30x - 0.20(x - 0.30x)$
$= 1(x - 0.30x) - 0.20(x - 0.30x)$
$= (x - 0.30x)(1 - 0.20)$ $x - 0.30x$ is the common factor
$= 0.7x(0.80) = 0.56x$ Simplify

(b) $S(x) = 0.56x$ means that the selling price is 56% of the original price. That is, there is a 44% discount on the dress.

PROBLEM SET 4.4

Mastering the Concepts

In Problems 1–4, list all the positive factors of each expression. Assume all variables represent positive numbers.

1. (a) $18x^2$
   (b) $-26x^2$

2. (a) $42x^2$
   (b) $-52x^2$

3. (a) $62y$
   (b) $-80y^2$

4. (a) $65x^2$
   (b) $-63y$

In Problems 5–16, find the greatest common factor for the given pair of numbers or expressions.

5. 14 and 21

6. 44 and 60

7. 10, 15, and 40

8. 9, 18, and 24

9. x^2 and x^3

10. y^4 and y^7

11. $6x^2$ and $18x^3$

12. $35x^3$ and $15x^2$

13. $14x^3$ and $49x^5$

14. $12x^2$ and $27x^4$

15. $2x^2$, $4x$, and $8x^3$

16. $4a^2$, $8a^3$, and $12a$

In Problems 17–34, use common factoring to factor each polynomial.

17. (a) $5x + 10$
(b) $7t + 21$

18. (a) $7x - 70$
(b) $16x + 24$

19. (a) $3x - 63$
(b) $16y - 24$

20. (a) $3c - 21$
(b) $12y - 18$

21. (a) $8x - 12$
(b) $30 - 12y$

22. (a) $15x - 35$
(b) $18y + 12$

23. (a) $3x^2 + 24x^3$
(b) $9x - 4x^2$

24. (a) $5a^2 - 4a^5$
(b) $6r + 30r^2$

25. (a) $14x + 11x^2$
(b) $9y^2 - 13y^3$

26. (a) $16a^4 + 12a^3$
(b) $27x^3 - 18x^2$

27. (a) $12u^2 + 24$
(b) $10y^3 - 12y^2$

28. (a) $64a^2 + 96$
(b) $100r^2 + 125r$

29. (a) $a^3 - 4a^2 - a$
(b) $y^3 - 7y^2 - 3y$

30. (a) $x^3 + 3x^2 + 7x$
(b) $2u^2 - 8u + 10$

31. (a) $16y^3 + 12y^2 - 8y^4$
(b) $y^6 + 7y^5 - 11y^3$

32. (a) $-10c^4 + 5c^3 + 35c$
(b) $4x^3 + 12x^2 - 8x$

33. (a) $12a^3b^2 + 36a^2b^3$
(b) $6ab^2 + 30a^2b$

34. (a) $12x^3y - 48x^2y^2$
(b) $4x^3 - 2x^2 + x$

In Problems 35–40, factor each expression by grouping.

35. (a) $ax + ay + bx + by$
(b) $x^2a + x^2b + a + b$

36. (a) $ax^2 + b + bx^2 + a$
(b) $yz + 2y - z - 2$

37. (a) $ab^2 - b^2c - ad + cd$
(b) $2x^2 - yz^2 - x^2y + 2z^2$

38. (a) $x^3 + x^2 - 5x - 5$
(b) $x^2 - ax + bx - ab$

39. (a) $x^2 - 5x + 2x - 10$
(b) $x^2 + 3x + 4x + 12$

40. (a) $2y^2 + 5y + 6y + 15$
(b) $x^2 - 4x + 3x - 12$

Applying the Concepts

41. Geometry: The total surface area A of a right circular cylinder is given by

$$A = 2\pi r^2 + 2\pi rh$$

where r is the radius and h is the height. Write this formula in completely factored form.

42. Business: If P dollars are invested at a simple interest rate r for t years, then the accumulated value A is given by the formula

$$A = P + Prt.$$

Write this formula in a completely factored form.

In Problems 43 and 44, find an expression for the area of the shaded region. Write each expression in factored form.

43.

44.

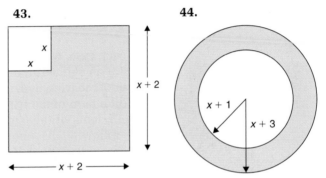

Developing and Extending the Concepts

In Problems 45–50, use common factoring to factor each polynomial.

45. (a) $4xy^2z + x^2y^2z - x^3y^3$
(b) $9m^2n + 18mn^2 - 27mn$

46. (a) $x^3y^2 + x^2y^3 + 2xy^4$
(b) $8xy^2 + 24x^2y^3 + 4xy^3$

47. (a) $3x^2y^2 + 6x^2z^2 - 9x^2$
(b) $3x(2a + b) + 5y(2a + b)$

48. (a) $2a^3b - 8a^2b^2 - 8ab^3$
(b) $(2m + 3)x - (2m + 3)$

49. (a) $x(y - z) - (z - y)$
(b) $4x^{-3} + 6x^{-5}$

50. (a) $10x^{-4} + 15x^{-6}$
(b) $m(x - y) + (y - x)$

In Problems 51–56, use factoring by grouping to factor each expression.

51. $8x^2 - 6xy + 20xz - 15yz$

52. $8xy + 21zw - 6zy - 28xw$

53. $ax^2y^2 - 5x^2y^2 + 4at^3 - 20t^3$

54. $xz^2 + 5z^2 + x + 5 + xz + 5z$

55. $3x^2 + 3 + 2x^2y + 2y + x^2z + z$

56. $5xyz - 5x + y^2z - y + yz^2 - z$

Objectives

1. Factor Trinomials of the Form $x^2 + bx + c$
2. Factor Trinomials of the Form $ax^2 + bx + c$
3. Factor Trinomials by Grouping
4. Solve Applied Problems

4.5 Factoring Trinomials

In section 4.2, we used the FOIL method to find the product of two binomials. Now we consider a trinomial of the form $ax^2 + bx + c$ and develop a procedure to find its binomial factors.

Factoring Trinomials of the Form $x^2 + bx + c$

The product of two binomials is often a trinomial. For example, consider the products in the following table.

Factored Form	FOIL	Trinomial Form
$(x + 4)(x + 7)$	$x^2 + 7x + 4x + 28$	$x^2 + 11x + 28$
$(x - 4)(x - 7)$	$x^2 - 7x - 4x + 28$	$x^2 - 11x + 28$
$(x - 4)(x + 7)$	$x^2 + 7x - 4x - 28$	$x^2 + 3x - 28$

In each case, we notice that the product of the first terms of the binomials is the first term of the trinomial. The product of the last terms of the binomials is the last term of the trinomial. The sum of the outer and inner terms of the binomial is the middle term of the trinomial. For the first example, the relationship between the coefficients of the trinomial and the coefficients of the factors is illustrated in the following diagrams.

$$x^2 + 11x + 28 = (x + 4)(x + 7)$$

The coefficients of the middle terms in the trinomial, namely 11, is obtained as follows.

$$x^2 + (7 + 4)x + 28 = (x + 4)(x + 7)$$

These diagrams illustrate how to factor the trinomial $x^2 + 11x + 28$. That is, we first find two integers whose product is 28 and whose sum is 11, then write the correct factorization as

$$x^2 + 11x + 28 = (x + 4)(x + 7).$$

EXAMPLE 1   **Factoring Trinomials**

Factor each trinomial.

(a) $x^2 + 9x + 20$ (b) $x^2 + x - 20$

Solution   (a) To factor this trinomial, observe that all the coefficients are positive, so, we need two positive integers whose product is 20, and whose sum is 9. Since the sum is positive, the integers are 4 and 5. Thus,

┌─ The product of 4 and 5 is 20.

$$x^2 + 9x + 20 = (x + 4)(x + 5)$$

└─ The sum of 4 and 5 is 9.

(b) Since the constant term of this trinomial is -20, we need two numbers with different signs so that their product is -20 and their sum is $+1$. The numbers are -4 and 5. Thus,

┌─ The product of -4 and 5 is -20.

$$x^2 + x - 20 = (x + 5)(x - 4)$$

└─ The sum of -4 and 5 is 1.

EXAMPLE 2   Factoring a Trinomial in Two Variables

Factor each trinomial.

(a) $x^2 - 3xy - 4y^2$ (b) $x^2 + 5xy + 7y^2$

Solution   (a) Notice that the last term of the trinomial contains y^2, so the middle term must contain a factor of y and the first term is x^2 so that the middle term must also have a factor of x. For the coefficient of the middle term, we must find two numbers whose product is -4 and whose sum is -3. The two numbers are -4 and 1. Thus we have:

┌─ The product of $-4y$ and $1y$ is $-4y^2$.

$$x^2 - 3xy - 4y^2 = (x + y)(x - 4y)$$

└─ The sum of $-4xy$ and $1xy$ is $-3xy$.

Thus, $x^2 - 3xy - 4y^2 = (x + y)(x - 4y)$ or $(x - 4y)(x + y)$ since multiplication is a commutative operation.

(b) We need two integers whose product is 7, and whose sum is 5. Since there are no such numbers, the polynomial

$$x^2 + 5xy + 7y^2$$

is not factorable using integer coefficients. Such a polynomial is called a *prime polynomial*.

Strategy for Factoring a Trinomial of the Form $x^2 + bx + c$

1. Write $x^2 + bx + c$ as a product of two binomials of the form
   $x^2 + bx + c = (x \_\_\_\_\_)(x \_\_\_\_\_)$
2. List pairs of factors of c so that the product of the last terms is c.
3. Try various combinations of these factors.

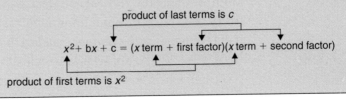

The following table will show how to shorten our work when factoring a trinomial.

TABLE 1

Trinomial	Signs of b and c	Signs of the Two Factors
$x^2 + bx + c$	signs of b and c are positive	positive
$x^2 - bx + c$	sign of b is negative and c is positive	negative
$x^2 - bx - c$	signs of b and c are negative	one sign is positive and one negative
$x^2 + bx - c$	sign of b is positive and of c is negative	one sign is positive and one negative

The above strategy suggests that to factor a trinomial of the form

$$x^2 + bx + c,$$

we take a shortcut by listing the factors of c, and then finding their sum which results in b. For example, in factoring

$$x^2 + 6x + 8$$

we are interested in factors of 8 whose sum is 6. Because both 6 and 8 are positive, we are looking for positive factors.

Factors of 8	$8 \cdot 1$	$4 \cdot 2$
Sum of factors	$8 + 1 = 9$	$4 + 2 = 6$

Because the desired sum is 6, we chose the factors 4 and 2. The factored form is

$$x^2 + 6x + 8 = (x + 4)(x + 2).$$

Schematically, we can confirm the factoring of this expression

$$x^2 + 6x + 8$$

by using the following cross method diagram.

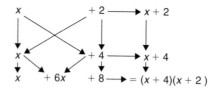

The above diagram is created as follows:

1. In the first column, enter any factors of the first term.
2. In the second column, enter any factors of the last term.
3. Multiply along the vertical and diagonal lines.
4. Add across the horizontal lines.
5. If the trinomial at the bottom matches the original trinomial, the factorization is correct. Otherwise, rearrange the factors or choose different factors.

EXAMPLE 3

Factoring a Trinomial with -1 as the Leading Coefficient

Factor the trinomial $-x^2 + 7x + 8$.

Solution

Because it is easier to factor a trinomial with a positive leading coefficient, we begin by factoring -1 as a common factor. We have

$$-x^2 + 7x + 8 = -(x^2 - 7x - 8).$$

We will use a table to find two factors of -8 whose sum is -7.

Factors of -8	$-1(8)$	$1(-8)$	$-2(4)$	$2(-4)$
Sum of factors	$-1 + 8 = 7$	$1 - 8 = -7$	$-2 + 4 = 2$	$2 - 4 = -2$

Notice that only one set of factors satisfies both conditions and the factorization is:

$$-x^2 + 7x + 8 = -(x + 1)(x - 8).$$

Factoring Trinomials of the Form $ax^2 + bx + c$

To factor a trinomial with a leading coefficient that is not 1, the strategy is somewhat the same as the one we used for a trinomial with a leading coefficient of 1. However, there are more possible factors and each must be checked. For example, to factor a trinomial of the form

$$3x^2 - 10x + 8$$

we can rely on our experience with the FOIL method by looking for two binomials whose product is $3x^2 - 10x + 8$. Since the product of the first term is $3x^2$, we try the form

$$(3x - \underline{\quad})(x - \underline{\quad}).$$

Next, we will use a table to fill the blanks with two integers that have a product of 8 and give a middle term of $-10x$. Then try different possibilities to find the correct factorization as the next example shows.

EXAMPLE 4

Factoring Trinomials with $a \neq 1$

Factor each trinomial.

(a) $2x^2 + 11x + 12$ (b) $3x^2 - 10x + 8$ (c) $6r^2 + 7r - 3$

Solution

(a) We begin by factoring the first term, $2x^2$, as

$$2x^2 = 2x \cdot x.$$

Because the signs of the middle and last terms are positive, the signs in both binomials are also positive. So the trinomial can be expressed in the form

$$2x^2 + 11x + 12 = (2x + \underline{\quad})(x + \underline{\quad}).$$

Knowing that the factors of the last term, 12, must be positive, we list all the possibilities in Table 2.

TABLE 2

Trinomial	Possible Combinations of Binomial Factors	Product of First Term	Product of Last Term	Sum of Products of Outer and Inner Terms
	$(2x + 12)(x + 1)$	$2x^2$	$+ 12$	$2x(1) + 12x = 14x$
	$(2x + 6)(x + 2)$	$2x^2$	$+ 12$	$2x(2) + 6x = 10x$
	$(2x + 4)(x + 3)$	$2x^2$	$+ 12$	$2x(3) + 4x = 10x$
$2x^2 + 11x + 12$	$(2x + 3)(x + 4)$	$2x^2$	$+ 12$	$2x(4) + 3x = 11x$

We see that $(2x + 3)(x + 4)$ produces the correct middle term.

(b) For $3x^2 - 10x + 8$, the possible factors of $3x^2$ are $1x$ and $3x$. Because the sign of the last term is positive and the sign of the middle term is negative, the signs in the binomials must be the same and negative.

$$(3x - \underline{})(x - \underline{}).$$

TABLE 3

Trinomial	Possible Combinations of Binomial Factors	Product of First Term	Product of Last Term	Sum of Products of Outer and Inner Terms
	$(3x - 8)(x - 1)$	$3x^2$	$+ 8$	$3x(-1) + (-8)x = -11x$
	$(3x - 1)(x - 8)$	$3x^2$	$+ 8$	$3x(-8) + (-1)x = -25x$
	$(3x - 2)(x - 4)$	$3x^2$	$+ 8$	$3x(-4) + (-2)x = -14x$
$3x^2 - 10x + 8$	$(3x - 4)(x - 2)$	$3x^2$	$+ 8$	$3x(-2) + (-4)x = -10x$

Table 3 shows that the correct factorization is

$$(3x - 4)(x - 2).$$

(c) In $6r^2 + 7r - 3$, the first term, $6r^2$, has factors $2r(3r)$ or $r(6r)$. Because the sign of the last term is negative, the signs of the second terms of the binomial are different. The factors of -3 are -1 and 3 or 1 and -3. Because there are several possible combinations of factors shown in Table 4.

TABLE 4

Trinomial	Possible Combinations of Binomial Factors	Product of First Term	Product of Last Term	Sum of Products of Outer and Inner Terms
	$(6r - 3)(r + 1)$	$6r^2$	-3	$6r(1) + (-3)r = 3r$
	$(6r - 1)(r + 3)$	$6r^2$	-3	$6r(3) + (-1)r = 17r$
	$(3r - 3)(2r + 1)$	$6r^2$	-3	$3r(1) + (-3)2r = -3r$
$6r^2 + 7r - 3$	$(3r - 1)(2r + 3)$	$6r^2$	-3	$3r(3) + (-1)2r = 7r$
	$(2r - 3)(3r + 1)$	$6r^2$	-3	$2r(1) + (-3)3r = -7r$

The correct factorization is $(2r + 3)(3r - 1)$.

The strategy for factoring

$$ax^2 + bx + c$$

is illustrated in the following diagram which shows the relationship between the coefficients of the trinomial and the coefficients of the binomial factors.

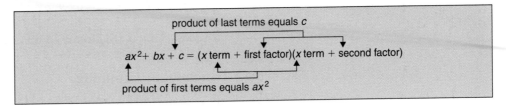

The diagram illustrates how a trinomial of the form $ax^2 + bx + c$ is factored. We begin by writing

$$ax^2 + bx + c = (\underline{\quad}x + \underline{\quad})(\underline{\quad}x + \underline{\quad}).$$

Then we fill in the blanks with numbers so that:

1. The product of the first terms is ax^2.
2. The product of the last terms is c.
3. The sum of the outer and inner terms is bx.

We try all possible choices of factors of a and c until we find a combination that gives us the desired middle term.

EXAMPLE 5   **Factoring by Substitution**

Factor the trinomial $3x^6 - 7x^3 - 6$.

Solution   We begin by making the substitution

$$u = x^3 \text{ and so } u^2 = (x^3)^2 = x^6.$$

With this substitution the original trinomial becomes

$$3u^2 - 7u - 6.$$

The cross method diagram can also be used to show the factoring of this trinomial.

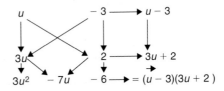

Returning to the original variable, x, the factorization is $(x^2 - 3)(3x^2 + 2)$.

Factoring Trinomials by Grouping

The grouping techniques can be used to factor a trinomial of the form

$$ax^2 + bx + c.$$

We must still find two numbers, p and q, so that

$$p + q = b \text{ and } pq = ac.$$

Then we rewrite the trinomial as

$$ax^2 + px + qx + c$$

and factor by grouping. This is illustrated in the following example.

EXAMPLE 6   **Factoring a Trinomial by Grouping**

Factor the trinomial $6x^2 + 19x + 10$.

Solution   Table 5 helps us find two possible numbers which are factors of 6(10) and whose sum is + 19 and rewrite the trinomial.

TABLE 5

Factors of 60	Sum of Factors Equals 19
1(60)	1 + 60 = 61
2(30)	2 + 30 = 32
3(20)	3 + 20 = 23
4(15)	4 + 15 = 19
5(12)	5 + 12 = 17
6(10)	6 + 10 = 16

$$6x^2 + 19x + 10 = (6x^2 + 15x) + (4x + 10)$$

Be sure to check every factorization.

$$= 3x(2x + 5) + 2(2x + 5) \quad \text{Factor } 3x \text{ and } 2 \text{ from the grouped terms}$$

$$= (3x + 2)(2x + 5)$$

The graph of a polynomial can be used as a tool to factor each factorable polynomial. For instance, note that the graphs in Figure 1 show that each polynomial in example on page 220 intercepts the x axis in two places.

Figure 1

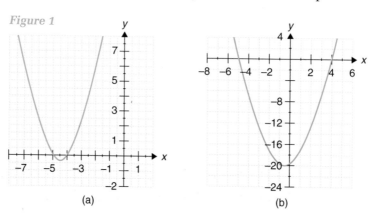

(a)   (b)

In Figure 1a, the graph of the polynomial $x^2 + 9x + 20$ crosses the x axis at $x = -4$ and $x = -5$. The graph in Figure 1b shows that the graph of the polynomial $x^2 + x - 20$ crosses the x axis at $x = -5$ and $x = 4$. Following this pattern, we see a strong correlation between the factors of a polynomial and the x intercepts of its graph. Table 6 demonstrates this relationship.

TABLE 6

Polynomial	x Intercepts of the Graph	Factorization
$x^2 + 9x + 20$	$x = -4, x = -5$	$(x + 4)(x + 5)$
$x^2 + x - 20$	$x = -5, x = 4$	$(x - 4)(x + 5)$

To determine whether or not a trinomial of the form

$$ax^2 + bx + c, a \neq 0$$

is factorable, first evaluate the expression

$$D = b^2 - 4ac.$$

If D is the square of an integer, such as 1, 4, 9, 16, 25, etc., then the trinomial is factorable, otherwise it is not. For example, to determine if the trinomial $3x^2 + 2x - 5$ is factorable, we evaluate the expression $b^2 - 4ac$ for $a = 3$, $b = 2$, and $c = -5$ so that

$$b^2 - 4ac = (2)^2 - 4(3)(-5)$$
$$= 4 - (-60) = 64 = 8^2.$$

Therefore, the trinomial is factorable as follows

$$3x^2 + 2x - 5 = (3x + 5)(x - 1).$$

Figure 2 shows the graph and $-5/3$ and 1 are the zeros of the function $f(x) = 3x^2 + 2x - 5$.

Figure 2

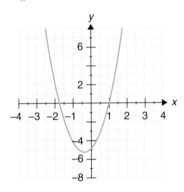

Solving Applied Problems

Formulas from geometry and other sciences can be written in a factored form as the next example shows.

EXAMPLE 7   **Area of a Field**

A flower seed company uses a rectangular piece of land as a test plot. If x is the number of rows plowed in this plot, then the area (in square meters) plowed is given by the model:

$$A(x) = 15x^2 + 13x - 6, \quad x > 0.$$

(a) Completely factor the right side of this formula.

(b) Find the area, length, and width when 5 rows are plowed.

Solution   (a) To factor the trinomial, we find two numbers with different signs whose product is $-90 = 15(-6)$ and sum is 13. These numbers are 18 and -5 (Table 7).

TABLE 7

Factors of -90	Sum of Factors Equal 13
$-1(90)$	$-1 + 90 = 89$
$-2(45)$	$-2 + 45 = 43$
$-3(30)$	$-3 + 30 = 27$
$-5(18)$	$-5 + 18 = 13$
$-6(15)$	$-6 + 15 = 9$
$-9(10)$	$-9 + 10 = 1$

$$A(x) = 15x^2 + 18x - 5x - 6$$
$$= 3x(5x + 6) - 1(5x + 6)$$
$$= (3x - 1)(5x + 6)$$

(b) Since the area of a rectangle is length times width, $A = lw$, one of the factors represents the length and the other factor represents the width.

Replacing x by 5, we have:

$$A(5) = [3(5) - 1][5(5) + 6]$$
$$= [15 - 1][25 + 6] = (14)(31) = 434 \text{ m}^2.$$

So if 5 rows are plowed, the area of the field is 434 square meters. The length of the field is 31 meters and the width is 14 meters.

PROBLEM SET 4.5

Mastering the Concepts

In Problems 1–26, factor each trinomial.

1. (a) $x^2 + 7x + 12$
   (b) $x^2 - 4x + 3$
2. (a) $x^2 - 8x + 7$
   (b) $y^2 + 7y + 10$
3. (a) $x^2 + 9x + 14$
   (b) $x^2 - 9x + 20$
4. (a) $y^2 - 8y + 15$
   (b) $y^2 + 7y + 6$
5. (a) $d^2 - 15d + 56$
   (b) $t^2 - 22t + 121$
6. (a) $y^2 - 16y + 63$
   (b) $y^2 - 11y + 24$
7. (a) $x^2 - 5x - 6$
   (b) $t^2 + t - 12$
8. (a) $x^2 + 5x - 14$
   (b) $y^2 - 2y - 35$
9. (a) $y^2 - 3y - 28$
   (b) $d^2 + 8d - 33$
10. (a) $b^2 - 11b - 26$
    (b) $t^2 - 4t - 21$
11. (a) $y^2 + y - 30$
    (b) $u^2 - 12u - 13$
12. (a) $t^2 + 2t - 48$
    (b) $y^2 - 6y - 16$
13. (a) $u^2 + 6u - 27$
    (b) $t^2 - 4t - 96$
14. (a) $x^2 - 20x - 44$
    (b) $u^2 - u - 132$
15. (a) $x^2 - 7xy - 30y^2$
    (b) $b^2 - 9bd - 36d^2$
16. (a) $u^2 + 2uv - 63v^2$
    (b) $c^2 - cd - 6d^2$
17. (a) $x^2 - 20xy - 44y^2$
    (b) $y^2 + yz - 110z^2$
18. (a) $m^2 + 13mn + 42n^2$
    (b) $u^2 + 15uv + 36v^2$
19. (a) $3m^2 - 10m + 8$
    (b) $6x^2 - x - 12$
20. (a) $6b^2 + 7b - 3$
    (b) $9y^2 - 30y + 24$
21. (a) $2y^2 + 11y + 12$
    (b) $4x^2 + x - 3$
22. (a) $32x^2 - 4xy - 21y^2$
    (b) $36y^2 - 40yz - 21z^2$
23. (a) $14x^2 + 29x - 15$
    (b) $6x^2 - 13x + 6$
24. (a) $10x^2 + 29xy - 21y^2$
    (b) $12x^2 + 29xy + 15y^2$

25. (a) $-y^2 + 6y + 27$

    (b) $-t^2 + 2t + 15$

26. (a) $-7x^2 + 6x + 13$

    (b) $-12x^2 + 35x + 3$

In Problems 27–30, match each function with the given graph, then use the graph to factor the function.

27. $f(x) = x^2 - 3x - 4$

28. $g(x) = x^2 - 3x - 40$

29. $h(x)^2 = 2x^2 - x - 10$

30. $k(x) = 4x^2 - 7x - 2$

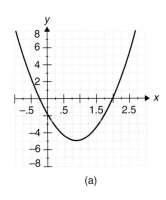

(a)

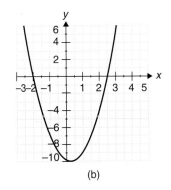

(b)

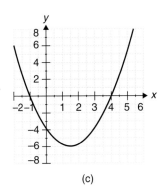

(c)

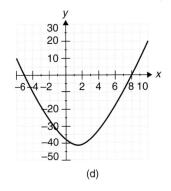

(d)

In Problems 31–34, find $D = b^2 - 4ac$ to determine which trinomial is factorable. Factor when possible.

31. (a) $x^2 - 4x - 15$

    (b) $2x^2 - 3x - 2$

32. (a) $3x^2 - 5x + 2$

    (b) $15x^2 + 17x - 4$

33. (a) $x^2 - xy - 6y^2$

    (b) $6x^2 - 5x + 2$

34. (a) $6x^2 + 19xz + 10z^2$

    (b) $x^2 + 10x + 20$

Applying the Concepts

35. Electric Circuits: Two electric circuits have a total resistance R (in Ohms) given by

$$R(x) = 2x^2 - 3x + 1.$$

Write the function in factored form.

36. Water Flow: The rate of flow, f, of water (in gallons per minute) in a 3-inch pipe with a friction loss of 10 pounds per square inch is given by the model

$$f(x) = 2x^2 + x - 10.$$

Write this function in a factored form.

37. Distance Traveled: The distance s (in feet) traveled in t seconds by a projectile fired vertically upward at a speed of 256 feet per seconds and 240 feet above ground is given by the model

$$s(t) = 16t^2 - 256t + 240.$$

Write this expression in a factored form.

38. Manufacturing: A fabricator produces a temple of plastic with area modeled by the function

$$A(x) = 2x^2 + 35x - 100.$$

Factor this expression.

Developing and Extending the Concepts

In Problems 39–54, factor each expression completely.

39. $3x^4 + 13x^3y - 10x^2y^2$

40. $7x^4 - 10x^3y + 3x^2y^2$

41. $-3x^2 - 16xy + 12y^2$

42. $12x^2 + 25xy + 12y^2$

43. $2(x + 1)^2 + 7(x + 1) + 5$

44. $12(x + 3)^2 - 5(x + 3) - 2$

45. $5(x + 2)^2 - 12(x + 2) + 4$

46. $6(x + 5)^2 - 17(x + 5) + 5$

47. $2(x - 3)^2 + 5(x - 3) + 2$

48. $-9(x - 7)^2 - 37(x - 7) - 4$

49. $x^4 + 5x^2 + 6$

50. $5y^4 - y^2 - 6$

51. $2y^6 - 7y^3 + 3$

52. $2x^{10} + 7x^5 + 6$

53. $2x^4 + x^2y - 6y^2$

54. $x^8 + 3x^4y^3 - 10y^6$

4.6 Factoring by Special Forms

In this section, we factor certain forms of polynomials by using special products in a *reverse order*. Success in factoring depends on our ability to recognize multiplication patterns in the polynomials to be factored. These forms occur frequently in algebra and should be memorized.

Factoring Special Binomials

We begin with three multiplications whose product leads to special forms of factoring. They are known as the *difference of two squares*, the *sum of two cubes*, and the *difference of two cubes*; they are listed in Table 1 below.

TABLE 1

Name	Pattern
1. difference of two squares	$a^2 - b^2 = (a - b)(a + b)$
2. sum of two cubes	$a^3 + b^3 = (a + b)(a^2 - ab + b^2)$
3. difference of two cubes	$a^3 - b^3 = (a - b)(a^2 + ab + b^2)$

EXAMPLE 1   **Factoring by Using the Difference of Two Squares**

Determine which binomial fits the form of the difference of two squares, then factor the binomial.

(a) $9x^2 - 25$ (b) $25x^2 - 64$ (c) $16x^2 + 49$

Solution   Only $9x^2 - 25$ and $25x^2 - 64$ in (a) and (b) are differences of squares.

$$a^2 - b^2 = (a - b)(a + b)$$

(a) Using $a = 3x$ and $b = 5$, we have:

$$9x^2 - 25 = (3x)^2 - (5)^2 = (3x - 5)(3x + 5).$$

(b) Using $a = 5x$ and $b = 8$, we have:

$$25x^2 - 64 = (5x)^2 - (8)^2 = (5x - 8)(5x + 8).$$

(c) $16x^2 + 49$ does not fit the pattern since it is the sum of two squares, not the difference. To justify this, try the following products:

$$(4x - 7)(4x + 7) = 16x^2 - 49$$
$$(4x - 7)(4x - 7) = (4x - 7)^2 = 16x^2 - 56x + 49$$
$$(4x + 7)(4x + 7) = (4x + 7)^2 = 16x^2 + 56x + 49.$$

Notice that none of them is equal to $16x^2 + 49$; so $16x^2 + 49$ is not factorable.

EXAMPLE 2   **Factoring Completely**

Factor each binomial expression completely.

(a) $5x - 5xy^4$ (b) $x^4 - 10x^2 + 9$

Solution (a) $5x - 5xy^4 = 5x(1 - y^4)$ Common factor $5x$

$= 5x(1 - (y^2)^2)$ Difference of two squares

$= 5x(1 - y^2)(1 + y^2)$ Factored as the difference of two squares

$= 5x(1 - y)(1 + y)(1 + y^2)$ Factor the difference of two squares

(b) The trinomial $x^4 - 10x^2 + 9$ is quadratic in form. If we let $t = x^2$, the expression can be written as

$(x^2)^2 - 10(x^2) + 9$ x^4 rewritten as $(x^2)^2$

$= t^2 - 10t + 9$ Substitute $t = x^2$

$= (t - 1)(t - 9)$ Factor trinomial

$= (x^2 - 1)(x^2 - 9)$ Substitute $t = x^2$

$= (x - 1)(x + 1)(x - 3)(x + 3)$ Factor difference of squares

EXAMPLE 3   **Factoring Forms Containing the Difference of Squares**

Factor each expression completely.

(a) $(x + 2)^2 - 4y^2$ (b) $16x^2 - (3x + 2)^2$ (c) $(x - 2)^2 - (2x - 1)^2$

Solution (a) $(x + 2)^2 - 4y^2 = [(x + 2) - 2y][(x + 2) + 2y]$ Difference of squares

$= (x + 2 - 2y)(x + 2 + 2y)$ Remove parentheses

(b) $16x^2 - (3x + 2)^2 = [4x - (3x + 2)][4x + (3x + 2)]$ Difference of squares

$= [4x - 3x - 2][4x + 3x + 2]$ Remove parentheses

$= (x - 2)(7x + 2)$ Simplify

(c) $(x - 2)^2 - (2x - 1)^2 = [(x - 2) - (2x - 1)][(x - 2) + (2x - 1)]$

$= [x - 2 - 2x + 1][x - 2 + 2x - 1]$ Remove parentheses

$= (-x - 1)(3x - 3)$ Simplify

$= -1(x + 1)3(x - 1)$ Common factors

$= -3(x + 1)(x - 1)$

Each of these factorizations may be checked by multiplying the original expression and then factoring. In part (c):

$(x - 2)^2 - (2x - 1)^2 = x^2 - 4x + 4 - (4x^2 - 4x + 1)$ Square each binomial

$= x^2 - 4x + 4 - 4x^2 + 4x - 1$ Subtract

$= -3x^2 + 3$ Simplify

$= -3(x^2 - 1)$ Factor out -3

$= -3(x - 1)(x + 1)$ Difference of squares

In the previous 3 examples, we factored polynomials as the difference between two squares. Now we factor polynomials as the sum or the difference of two cubes. The results in Table 1 of these forms can be verified as follows.

Consider product 2: $(a + b)(a^2 - ab + b^2)$

$$= a^3 - a^2b + ab^2 + a^2b - ab^2 + b^3$$
$$= a^3 + b^3$$

For product 3: $(a - b)(a^2 + ab + b^2)$

$$= a^3 + a^2b + ab^2 - a^2b - ab^2 - b^3$$
$$= a^3 - b^3$$

EXAMPLE 4   **Factoring the Sum and Difference of Two Cubes**

Factor completely.

(a) $y^3 + 8$ (b) $27x^3 + y^3$ (c) $64x^3 - 125y^3$

Solution   (a) Since $y^3 + 8 = y^3 + 2^3$, this is in the form of the sum of cubes.

$$a^3 + b^3 = (a + b)(a^2 - ab + b^2)$$
$$y^3 + 2^3 = (y + 2)(y^2 - 2y + 4)$$

(b) $27x^3 + y^3 = (3x)^3 + y^3$

$$= (3x + y)[(3x)^2 - 3xy + y^2]$$
$$= (3x + y)(9x^2 - 3xy + y^2)$$

(c) $64x^3 - 125y^3 = (4x)^3 - (5y)^3$

$$= (4x - 5y)[(4x)^2 + (4x)(5y) + (5y)^2]$$
$$= (4x - 5y)(16x^2 + 20xy + 25y^2)$$

Notice that the second factor in each of these factorizations is a second degree polynomial which cannot be factored.

EXAMPLE 5   **Factoring the Sum and Difference of Two Cubes**

Factor each expression.

(a) $\dfrac{1}{8x^3} + 27$ (b) $(x - 2)^3 - (x + 2)^3$

Solution   Each of these expressions fits the pattern of the sum and difference of cubes.

(a) $\dfrac{1}{8x^3} + 27 = \left(\dfrac{1}{2x}\right)^3 + 3^3$

$$= \left(\dfrac{1}{2x} + 3\right)\left[\left(\dfrac{1}{2x}\right)^2 - \dfrac{1}{2x}(3) + 3^2\right]$$

$$= \left(\dfrac{1}{2x} + 3\right)\left[\dfrac{1}{4x^2} - \dfrac{3}{2x} + 9\right]$$

(b) $(x - 2)^3 - (x + 2)^3 = [(x - 2) - (x + 2)][(x - 2)^2 + (x - 2)(x + 2) + (x + 2)^2]$

$$= (x - 2 - x - 2)(x^2 - 4x + 4 + x^2 - 4 + x^2 + 4x + 4)$$
$$= -4(3x^2 + 4)$$

Factoring Perfect Square Trinomials

Recall from section 4.2, the special products

$$a^2 + 2ab + b^2 = (a + b)^2 \text{ and } a^2 - 2ab + b^2 = (a - b)^2.$$

Notice what happens in these forms:

1. Both the first term and the last term of the trinomial (a^2 and b^2) are perfect squares.
2. The middle term is twice the product of a and b with the same sign. That is, either $2ab$ or $-2ab$.

By recognizing a perfect square trinomial, we are able to factor by inspection.

EXAMPLE 6

Factoring Perfect Square Trinomials

Factor each expression, if possible.

(a) $x^2 + 8x + 16$ (b) $9x^2 - 30xy + 25y^2$ (c) $(t + 2)^2 - 12(t + 2) + 36$

Solution

(a) Notice that x^2 and 16 are perfect squares and $2(x)(4)$ is the middle term. Thus,

$$x^2 + 8x + 16 = (x + 4)^2.$$

(b) We can use the same methods to factor a trinomial in two variables.

$$9x^2 - 30xy + 25y^2 = (3x)^2 - 2(3x)(5y) + (5y)^2$$
$$= (3x - 5y)^2$$

(c) To factor this polynomial, think of $t + 2$ as a and 6 as b. Thus the pattern is:

$$a^2 - 2ab + b^2 = (a - b)^2$$
$$(t + 2)^2 - 12(t + 2) + 36 = (t + 2)^2 - 2(6)(t + 2) + 6^2$$
$$= [(t + 2) - 6]^2 = (t - 4)^2.$$

To check this last factorization, we simplify the expression and then factor.

$$(t + 2)^2 - 12(t + 2) + 36 = t^2 + 4t + 4 - 12t - 24 + 36$$
$$= t^2 - 8t + 16 = (t - 4)^2$$

Miscellaneous Factoring

There are situations where the terms of a polynomial can be grouped in a way that will lead to factorization, as the following examples show.

EXAMPLE 7

Using Grouping to Factor Special Polynomials

Factor each expression.

(a) $x^2 - 6x + 9 - 4y^2$ (b) $x^4 + 6x^2 - 4x^2 + 9$

Solution

(a) $x^2 - 6x + 9 - 4y^2 = (x - 3)^2 - (2y)^2$ First three terms are a perfect square

$$= [(x - 3) - 2y][(x - 3) + 2y]$$ Factor the difference of squares

$$= (x - 3 - 2y)(x - 3 + 2y)$$ Simplify

(b) We begin by using a substitution, $y = x^2$ and $y^2 = x^4$.

$$x^4 + 6x^2 - 4x^2 + 9$$

$$= y^2 + 6y - 4y + 9$$

$$= y^2 + 6y + 9 - 4y \qquad \text{Rearrange so that we have a perfect square}$$

$$= (y + 3)^2 - 4y \qquad \text{Factor the perfect square}$$

$$= (x^2 + 3)^2 - 4x^2 \qquad \text{Substitute } y \text{ for } x^2$$

$$= [(x^2 + 3) - 2x][(x^2 + 3) + 2x] \qquad \text{Factor the difference of squares}$$

$$= (x^2 - 2x + 3)(x^2 + 2x + 3) \qquad \text{Simplify}$$

Solving Applied Problems

Many applied problems from geometry and other sciences can be written in a factored form.

EXAMPLE 8

Writing Surface Area and Volume in Factored Form

The surface area A and the volume V of a hollow sphere like a soccer ball are expressed by the formulas:

$$A = 4\pi r^2 - 4\pi s^2 \text{ and } V = \tfrac{4}{3}\pi r^3 - \tfrac{4}{3}\pi s^3$$

where r is the outside radius and s is the inside radius of the sphere. Factor these expressions for A and V.

Solution

(a) $A = 4\pi r^2 - 4\pi s^2 = 4\pi(r^2 - s^2)$ Common factor 4π

$\qquad = 4\pi(r - s)(r + s)$ Difference of squares

(b) $V = \dfrac{4}{3}\pi r^3 - \dfrac{4}{3}\pi s^3 = \dfrac{4}{3}\pi(r^3 - s^3)$ Common factor $\dfrac{4}{3}\pi$

$\qquad = \dfrac{4}{3}\pi(r - s)(r^2 + rs + s^2)$ Difference of cubes

 **PROBLEM SET 4.6**

Mastering the Concepts

In Problems 1–10, factor each expression completely.

1. (a) $x^2 - 100$
   (b) $100 - x^2$
2. (a) $x^2 - 36$
   (b) $36 - x^2$
3. (a) $16x^2 - 49$
   (b) $16x^2 - 81y^4$
4. (a) $25x^2 - 64$
   (b) $25x^2 - 9y^2$
5. (a) $16u^2 - 25v^2$
   (b) $25m^2 - 49n^2$
6. (a) $(a - b)^2 - 100c^2$
   (b) $144p^2 - (q - 3)^2$
7. (a) $49x^2 - 16y^4$
   (b) $16x^4y^4 - 1$
8. (a) $81x^4 - 1$
   (b) $256x^4 - y^4$
9. $x^4 - 81$
10. $y^4 - 256$

In Problems 11–20, factor each expression completely.

11. (a) $x^3 - 1$
    (b) $y^3 + 125$
12. (a) $27w^3 + z^3$
    (b) $x^3y^3 - 64$
13. (a) $w^3 - 8y^3z^3$
    (b) $64z^3 + 27b^3$
14. (a) $8x^3 + 125y^3$
    (b) $125x^3 - y^6$
15. (a) $x^9 - 1$
    (b) $y^9 + 512$
16. (a) $u^6 - 27$
    (b) $(x + 2)^3 - y^3$
17. (a) $(x - 3)^3 + y^3$
    (b) $(2x + 1)^3 + 8$
18. (a) $t^6 - 64$
    (b) $y^6 - x^6$
19. $y^9 - x^9$
20. $8x^9 + y^6$

In Problems 21–30, factor each expression completely.

21. (a) $x^2 + 6x + 9$ (b) $y^2 - 8y + 16$

22. (a) $y^2 + 12xy + 36x^2$ (b) $4x^2 - 12x + 9$

23. (a) $4x^2 - 8x + 4$ (b) $12x^2 - 12x + 3$

24. (a) $2x^2 - 20xy + 50y^2$ (b) $9x^2 - 24x + 16$

25. (a) $4x^4 - 4x^2 + 1$ (b) $9x^4 - 12x^2 + 4$

26. (a) $4x^4 - 16x^2 + 16$ (b) $12x^4 + 12x^2 + 3$

27. (a) $x^8 + 2x^4 + 1$ (b) $x^6 - 4x^3 + 4$

28. (a) $x^6 - 6x^3y + 9y^2$ (b) $x^6 + 8x^3y + 16y^2$

29. $4x^2 - 4xy^2 + y^4$

30. $4x^2 + 12xy^2 + 9y^4$

In Problems 31–40, factor each expression completely.

31. (a) $x^4 + 2x^2y^2 + y^4$
   (b) $16x^4 - y^4$

32. (a) $9x^4 - y^4$
   (b) $25x^4 + 25x^2y^2 + 4y^4$

33. (a) $5x^3 - 55x^2 + 140x$
   (b) $x^2yz^2 - xyz^2 - 12yz^2$

34. (a) $16x^4 - x^2 + 6xy - 9y^2$
   (b) $x^4 - x^2 + 4x - 4$

35. (a) $x^2 + 30x + 225$
   (b) $x^2 - 30xy + 225y^2$

36. (a) $x^2 - 22x + 121$
   (b) $x^2 + 22xy + 121y^2$

37. $81x^4 + 3x$

38. $128x^4 - 2x$

39. $2bx - 4cx + 6by - 12cy$

40. $8cx - 8dx - 12cy + 12dy$

Applying the Concepts

41. A flower seed company uses a rectangular piece of land as a test plot. If x is the number of rows plowed in this plot, then the area (in square meters) plowed is given by the polynomial:

$$A(x) = 15x^2 + 13x - 6.$$

(a) Factor the right side of this formula completely.

(b) Find the area $A(x)$, the length, and width when 5 rows are plowed.

42. Total Revenue: The total revenue (in dollars) generated by selling $4x - 5$ units of a commodity is given by the function

$$r(x) = 64x^3 - 125, \ x \geq 2.$$

Use factoring to find an expression for price per unit in terms of x. [Hint: total revenue = (price of units sold) $\times$ (number of units sold).]

Developing and Extending the Concepts

In Problems 43–52, factor each expression completely.

43. (a) $3x^5 - 48xy^4$ (b) $3t^4 - 24t$

44. (a) $54x^4 - 2xy^3$ (b) $36t - 4t(y + 3)^2$

45. (a) $64u^6 - 729$ (b) $3t^8 + 81t^2$

46. (a) $2u^7 - 128u$ (b) $7x^7y + 7xy^7$

47. $2x^3 + x^2y - x^2 + 2xy + y^2 - y$

48. $2ax - b + 2bx - c + 2cx - a$

49. $x^2 - y^2 + 2x - 2zy + 1 - z^2$

50. $x^2 + y^2 - z^2 - 9 + 2xy - 6z$

51. $625x^4 - 81y^4$

52. $t^4 - 81y^4$

Objectives

1. Use the Zero Factor Property
2. Use Factoring to Solve Equations
3. Solve Equations Quadratic in Form
4. Interpret the x Intercepts of a Graph
5. Solve Applied Problems

4.7 Solving Polynomial Equations by Factoring

Suppose that P and Q represent two polynomial expressions such that

$$P = Q.$$

Such an equation is called a *polynomial equation.* Examples are:

$$2x^2 + 7x - 4 = 0$$

$$y^3 = 12y + 1$$

$$3t^4 + 5t^2 = 2t - 6$$

The degree of the highest term in an equation represents the *degree of the polynomial equation.* A second degree polynomial equation is called a *quadratic equation.* Examples of quadratic equations are:

$$25x^2 - 7x + 3 = 0$$

$$21x^2 = 4x - 1$$

$$(2x + 1)(x - 3) = 0$$

A polynomial equation of the form

$$ax^2 + bx + c = 0$$

where a, b, and c are real numbers and $a \neq 0$ is called a **quadratic equation** in **standard form.** In this section, we will solve polynomial equations using factoring.

Using the Zero Factor Property

To solve polynomial equations using factoring, we use the following zero factor property.

Zero Factor Property

If a and b are real numbers, or algebraic expressions, and if $ab = 0$ then

$$a = 0 \text{ or } b = 0.$$

In other words, if the product of two real numbers (or algebraic expressions) is zero, then at least one number (or expression) must be zero.

EXAMPLE 1   **Using the Zero Factor Property**

Use the zero factor property to solve the equation $(x - 2)(x + 5) = 0$.

Solution   Since the product of the factors equals zero, by the zero factor property, one or both factors must be zero. Setting each factor equal to zero, we have:

$$x - 2 = 0 \quad \text{or} \quad x + 5 = 0$$

$$x = 2 \qquad\qquad x = -5.$$

Check:

x	Left Side	Right Side	True/False
2	$(2 - 2)(2 + 5) = 0(7) = 0$	0	True
−5	$(-5 - 2)(-5 + 5) = -7(0) = 0$	0	True

Thus, the equation $(x - 2)(x + 5) = 0$ has exactly two solutions, 2 and -5.

Using Factoring to Solve Equations

The general strategy for solving a polynomial equation by factoring is outlined as follows.

Solving Polynomial Equations by Factoring

> 1. Write the equation in standard form, i.e., set the equation equal to zero.
> 2. Factor the non zero side of the equation completely.
> 3. Apply the zero factor property and set each factor equal to zero.
> 4. Solve the resulting equations.
> 5. Check each solution in the original equation.

EXAMPLE 2   **Solving Equations by Factoring**

Solve each equation.

(a) $2x^2 - x = 1$ (b) $5x^2 = 15x$

Solution   (a)

$2x^2 - x = 1$	Original equation
$2x^2 - x - 1 = 0$	Write in standard form
$(2x + 1)(x - 1) = 0$	Factor the quadratic expression
$2x + 1 = 0$ or $x - 1 = 0$	Apply the zero factor property
$2x = -1$ $\qquad$ $x = 1$	Solve
$x = -\dfrac{1}{2}$	

To apply the zero factor property, the equation must be written in standard form. That is, one side of the equation must be zero.

The solutions are $-\dfrac{1}{2}$ and 1.

Check:

x	Left Side	Right Side	True/False
$-\dfrac{1}{2}$	$2\left(-\dfrac{1}{2}\right)^2 - \left(-\dfrac{1}{2}\right) = 2\left(\dfrac{1}{4}\right) + \left(\dfrac{1}{2}\right) = 1$	1	True
1	$2(1)^2 - 1 = 2 - 1 = 1$	1	True

(b)

$5x^2 = 15x$	Original equation
$5x^2 - 15x = 0$	Write in standard form
$5x(x - 3) = 0$	Factor the quadratic expression
$5x = 0$ or $x + 3 = 0$	Apply the zero factor property
$x = 0$ $\qquad$ $x = -3$	Solve

The solutions are 0 and -3 and should be checked by substitution.

EXAMPLE 3   **Solving a Nonstandard Equation**

Solve the equation $(t + 6)(t - 2) = -7$.

Solution   Although this equation appears to be in factored form, one side is not equal to zero. We begin by multiplying the factors on the left side.

$(t + 6)(t - 2) = -7$	Original equation
$t^2 + 4t - 12 = -7$	Multiply factors

$$t^2 + 4t - 5 = 0 \qquad \text{Write in standard form}$$

$$(t + 5)(t - 1) = 0 \qquad \text{Factor quadratic expression}$$

$$t + 5 = 0 \quad \text{or} \quad t - 1 = 0 \qquad \text{Zero factor property}$$

$$t = -5 \qquad\qquad t = 1$$

Therefore, the solutions are -5 and 1. Be sure to check these solutions.

The zero factor property can be extended to three or more factors as the next example shows.

EXAMPLE 4   **Solving a Polynomial Equation**

Solve the equation $p^3 + 4p^2 - 9p - 36 = 0$.

Solution   We factor the left side by grouping.

$$p^3 + 4p^2 - 9p - 36 = 0 \qquad \text{Original equation}$$

$$p^2(p + 4) - 9(p + 4) = 0 \qquad \text{Group to remove common factors}$$

$$(p + 4)(p^2 - 9) = 0 \qquad \text{Common factor } p + 4$$

$$(p + 4)(p - 3)(p + 3) = 0 \qquad \text{Difference of squares}$$

$$p + 4 = 0 \quad \text{or} \quad p - 3 = 0 \quad \text{or} \quad p + 3 = 0 \qquad \text{Apply zero factor property}$$

$$p = -4 \qquad\qquad p = 3 \qquad\qquad p = -3$$

The solutions are -4, 3, and -3. Check these solutions in the original equation.

Notice that the third degree equation in example 4 has three factors and three solutions. This is not an accident. A polynomial equation can have at most as many real solutions as its degree.

Solving Equations Quadratic in Form

Some equations such as

$$x^4 - 5x^2 + 4 = 0 \text{ and } (x - 2)^2 - 3(x - 2) - 4 = 0$$

are not quadratic but can be treated as quadratic equations provided that a suitable substitution is made. After making the appropriate substitution, we can use the factor method to solve the resulting equations. The following table shows how the above examples can be written in quadratic form.

Original Equation	Suggested Substitution	Equivalent Quadratic Equation
$x^4 - 5x^2 + 4 = 0$	$u = x^2, u^2 = x^4$	$u^2 - 5u + 4 = 0$
$(x - 2)^2 - 3(x - 2) - 4 = 0$	$u = x - 2, u^2 = (x - 2)^2$	$u^2 - 3u - 4 = 0$

EXAMPLE 5   **Solving Equations Quadratic in Form**

Solve each equation.

(a) $x^4 - 5x^2 + 4 = 0$ (b) $(x - 2)^2 - 3(x - 2) - 4 = 0$

Solution   (a) Let $u = x^2$, $u^2 = x^4$, then substitute in the original equation.

$$x^4 - 5x^2 + 4 = 0 \qquad \text{Original equation}$$
$$u^2 - 5u + 4 = 0 \qquad \text{Substitute } u \text{ for } x^2$$
$$(u - 1)(u - 4) = 0 \qquad \text{Factor left side}$$
$$u - 1 = 0 \quad \text{or} \quad u - 4 = 0 \qquad \text{Apply zero property}$$
$$u = 1 \qquad\qquad u = 4$$
$$x^2 = 1 \qquad\qquad x^2 = 4 \qquad \text{Replace } u \text{ by } x^2$$
$$x = \pm 1 \qquad\qquad x = \pm 2$$

A common error is to solve for u and forget to solve for x.

The solutions are -2, -1, 1, and 2.

(b) Let $u = x - 2$ and $u^2 = (x - 2)^2$, then substitute.

$$(x - 2)^2 - 3(x - 2) - 4 = 0 \qquad \text{Original equation}$$
$$u^2 - 3u - 4 = 0 \qquad \text{Substitute } u \text{ for } x - 2$$
$$(u + 1)(u - 4) = 0 \qquad \text{Factor the left side}$$
$$u + 1 = 0 \qquad u - 4 = 0 \qquad \text{Apply the zero property}$$
$$u = -1 \qquad u = 4$$
$$x - 2 = -1 \qquad x - 2 = 4 \qquad \text{Replace } u \text{ by } x - 2$$
$$x = 1 \qquad x = 6$$

The solutions are 1 and 6.

Interpreting the x Intercepts of a Graph

A function f of the form

$$f(x) = ax^2 + bx + c, \, a \neq 0$$

is referred to as a **quadratic function.** In the next example, we will see how solutions to a quadratic equation determine the x intercepts of the graph of the corresponding function f. Equivalently, we use the graph of the function f to look for values of x for which $f(x) = 0$. The values of x for which

$$f(x) = 0$$

are the x intercepts and are called the **zeros** of f.

EXAMPLE 6   **Analyzing a Graph of a Quadratic Function**

Figure 1 shows the graph of the function

$$f(x) = x^2 - 7x + 10.$$

(a) Find $f(0)$, $f(2)$, and $f(5)$. Determine if these points are on the graph of f.

(b) Find the x intercepts of the graph of f.

(c) Use factoring to solve the equation $x^2 - 7x + 10 = 0$.

(d) What is the relationship between the solutions of the equation in part (c) and the x intercepts of the graph of f?

Figure 1

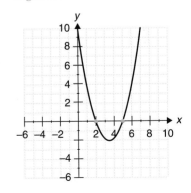

Solution

(a) $f(x) = x^2 - 7x + 10$

$f(0) = 0^2 - 7(0) + 10 = 10$

$f(2) = 2^2 - 7(2) + 10 = 4 - 14 + 10 = 0$

$f(5) = 5^2 - 7(5) + 10 = 25 - 35 + 10 = 0$

The points $(0,10)$, $(2,0)$, and $(5,0)$ are on the graph of f.

(b) The graph of f suggests that the x intercepts are 2 and 5. From part (a) $f(2) = f(5) = 0$.

(c) $x^2 - 7x + 10 = 0$

$(x - 5)(x - 2) = 0$ Factor the polynomial

$x - 5 = 0$ or $x - 2 = 0$ Zero factor property

$x = 5$ $x = 2$

(d) The x intercepts of the graph are the solution of the equation.

Solving Applied Problems

Now we solve some applied problems that are modeled by factorable quadratic equations. Recall the strategies developed by G. Polya for using mathematical models to solve applied problems.

> 1. Make a plan
> 2. Write the solution
> 3. Look back

In working these applied problems, the emphasis should be on writing the mathematical models.

EXAMPLE 7

Analyzing a Path of a Projectile

A projectile is shot vertically upward from the top of a building 80 feet high with an initial speed of 64 feet per second. Its height h (in feet) above the ground after t seconds is given by the model

$$h(t) = -16t^2 + 64t + 80, \quad t \geq 0.$$

After how many seconds will the projectile hit the ground?

Solution

We must find a value for t that results in a value of zero for h. That is, we set $h = 0$ and solve the resulting equation.

$-16t^2 + 64t + 80 = 0$ Height, $h = 0$

$t^2 - 4t - 5 = 0$ Divide by -16

$(t - 5)(t + 1) = 0$

$t - 5 = 0$ or $t + 1 = 0$

$t = 5$ or $t = -1$

Figure 2

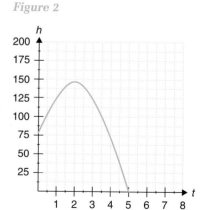

Since the time cannot be negative, 5 is the only meaningful solution for this model. This means the projectile will hit the ground about 5 seconds after it has been shot. Figure 2 shows the height of the projectile at each point in time. Notice the t intercept of the graph is 5 which is the solution of the equation.

EXAMPLE 8

Solving a Number Problem

The middle two digits in a Social Security Number are always even. Suppose two such numbers are consecutive and the product is 528. Find the numbers.

Solution

Make a Plan

The goal is to find two positive even integers which differ by 2 and have a product of 528. Any negative number is not part of the domain of the application.

Let x = the first even integer, then

$x + 2$ = the next even integer.

The following diagram is a visual description of the given.

First even positive integer x	$\bullet$ $\bullet$	Second even positive integer $x + 2$	$=$ $=$	Product of the two even integers 528

Translating the problem into an equation, we have

$$x(x + 2) = 528.$$

Write the Solution

Now we solve the equation.

$x(x + 2) = 528$	Original equation
$x^2 + 2x = 528$	Use the distributive property
$x^2 + 2x - 528 = 0$	Write in standard form
$(x + 24)(x - 22) = 0$	
$x + 24 = 0$ or $x - 22 = 0$	Apply the zero property
$x = -24$ $\quad$ $x = 22$	

We reject the solution -24 because it is not a positive integer. Since $x = 22$ and $x + 2 = 24$, the two consecutive even integers are 22 and 24.

Look Back

The numbers, 22 and 24, are two consecutive positive even integers. Their product, $(22)(24)$ is 528.

EXAMPLE 9   **Solving a Geometric Model**

A homeowner has a new pool which measures 20 feet by 40 feet. Now a walkway of uniform width must be added around the pool. The area of the border must be 544 square feet. Find the width of the border.

Solution   **Make a Plan**

Figure 3

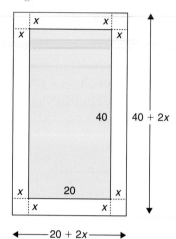

Let x (in feet) represent the width of the border. Figure 3 shows the pool and the border. Recall that the area, A, of a rectangle is

$$A = lw$$

here $20 + 2x = w$, the width including the border

and $40 + 2x = l$, the length including the border.

The area of the pool surface is $20(40) = 800$ square feet. The total area including the border less the area of the pool surface is 544 square feet.

The following diagram is a visual description of the given.

Length		Total Area	Width	less	Pool area		Border area
$2x + 40$	•		$2x + 20$	−	800	=	544
						=	

Translating the problem into an equation, we have

$$(2x + 40)(2x + 20) - 800 = 544.$$

Write the Solution

$(2x + 40)(2x + 20) - 800 = 544$	Original equation
$4x^2 + 120x + 800 - 800 = 544$	Simplify
$4x^2 + 120x = 544$	Simplify
$4x^2 + 120x - 544 = 0$	Write in standard form
$x^2 + 30x - 136 = 0$	Divide each side by 4
$(x - 4)(x + 34) = 0$	Factor
$x - 4 = 0$ or $x + 34 = 0$	Apply the zero property
$x = 4$ or $x = -34$	

Since the width of the border cannot be negative, the width of the border must be 4 feet.

Look Back

The total area of the pool and the border is given by

$[2(4) + 40][2(4) + 20] = (48)(28) = 1344$ less the area of the pool, 800, is 544.

Another way to check this problem is to add the areas of the rectangles that make the border. Four corners plus two lengths plus two widths.

$$4(4)(4) + 2(4)(40) + 2(4)(20) = 64 + 320 + 160 = 544$$

Figure 4

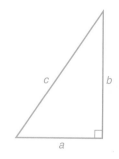

Recall the Pythagorean theorem: In any right triangle, the sum of the squares of the legs is equal to the square of the hypotenuse. Using the right triangle in Figure 4, we have

$$a^2 + b^2 = c^2.$$

EXAMPLE 10   **Finding the Sides of a Right Triangular Deck**

One level of a deck is to be built in the shape of a right triangle. One side is 14 meters longer than the other and the hypotenuse is 26 meters long. How long are the sides?

Solution   **Make a Plan**

The length of the sides differ by 14 meters, so we have:

$$\text{Let } x = \text{the length of the longer side}$$

$$x - 14 = \text{the length of the shorter side.}$$

Using the Pythagorean theorem with $a = x$, $b = x - 14$, and $c = 26$ (Figure 4), we form the following equation.

$$a^2 + b^2 = c^2$$

$$x^2 + (x - 14)^2 = 26^2$$

Write the Solution

$x^2 + (x - 14)^2 = 26^2$	Original equation
$x^2 + x^2 - 28x + 196 = 676$	Simplify
$2x^2 - 28x - 480 = 0$	Write in standard form
$x^2 - 14x - 240 = 0$	Divide each side by 2
$(x - 24)(x + 10) = 0$	Factor

$$x - 24 = 0 \quad \text{or} \quad x + 10 = 0$$

$$x = 24 \quad \text{or} \quad x = -10$$

Because the length must be positive, the length of the longer side is 24 meters, and the length of the shorter side is $x - 14 = 24 - 14 = 10$ meters.

Look Back

Notice that $10^2 + 24^2 = 100 + 576 = 676$ and $676 = 26^2$, 26 meters is the length of the hypotenuse.

◆ **PROBLEM SET 4.7**

Mastering the Concepts

In Problems 1–26, solve each equation by factoring.

1. (a) $x^2 + 7x + 6 = 0$
   (b) $x^2 - 12x + 11 = 0$
2. (a) $x^2 + 25x + 24 = 0$
   (b) $x^2 - 13x + 12 = 0$
3. (a) $c^2 - 6c + 8 = 0$
   (b) $y^2 - 37y + 36 = 0$
4. (a) $y^2 + y = 12$
   (b) $t^2 - 6t + 5 = 0$

5. (a) $y^2 - 6y + 9 = 0$
   (b) $9t^2 - 12t + 4 = 0$
6. (a) $x^2 - 24x + 144 = 0$
   (b) $49z^2 + 14z + 1 = 0$
7. (a) $b^2 - 7b - 8 = 0$
   (b) $u^2 + 5u - 66 = 0$
8. (a) $t^2 - t - 20 = 0$
   (b) $y^2 + 9y - 10 = 0$

9. (a) $x^2 - 2x = 35$

    (b) $x^2 + 2x = 35$

10. (a) $15x^2 - 19x + 6 = 0$

    (b) $y^2 - y = 12$

11. (a) $6x^2 + 11x = 10$

    (b) $6x^2 - 11x = 10$

12. (a) $6z^2 + 17z = 3$

    (b) $10t^2 - 11t = -3$

13. (a) $6u^2 + 7u = 20$

    (b) $10y^2 - 31y = 14$

14. (a) $4t^2 + 20 = 21t$

    (b) $4x^2 - 7 = 27x$

15. (a) $x^2 - 7x = 0$

    (b) $2m^2 - 9m = 0$

16. (a) $x^2 + 8x = 0$

    (b) $y^2 + 7y = 0$

17. $y(2y - 19) = 33$

18. $x(5x + 2) = 3$

19. $x(x - 2) = 9 - 2x$

20. $(x + 7)(x - 4) = 26$

21. $(2x + 1)(3x - 2) = 10$

22. $(2x - 1)(x + 2) = -3$

23. $4x^2 - (x - 1)^2 = 0$

24. $9y^2 - (y + 1)^2 = 0$

25. $6y^2 + 7y - 3 = -12y^2 - 54y + 4$

26. $16x^2 - 3x + 1 = -9x^2 + 17x - 3$

In Problems 27–34, solve each higher degree equation by factoring.

27. $(p - 1)(p + 2)(2p - 3) = 0$

28. $(3u + 1)(u + 3)(2u + 5) = 0$

29. $x(x - 2)(x + 3)(2x + 7) = 0$

30. $3x(2x + 1)(3x - 4)(4x - 5) = 0$

31. $x^3 - 11x^2 - 4x + 44 = 0$

32. $x^3 + 2x^2 - x - 2 = 0$

33. $2x^4 + 25x^3 + 12x^2 = 0$

34. $6x^4 + 5x^3 - 25x^2 = 0$

35. Figure 5 shows the graph of the function

$$f(x) = x^2 - 4x.$$

(a) Find $f(-4)$, $f(-2)$, $f(0)$, $f(2)$ and $f(4)$.

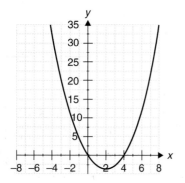

Figure 5

(b) Plot the ordered pairs obtained in part (a). Determine if these points lie on the graph of the function.

(c) Solve the equation $x^2 - 4x = 0$ algebraically.

(d) What is the relationship between the solutions of the equation in part (c) and the x intercepts of the graph of f?

36. Figure 6 shows the graph of $f(x) = 4x - x^2$. Repeat the instructions (a), (b), (c), and (d) from problem 35 for the function f.

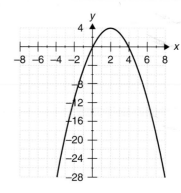

Figure 6

37. Figure 7 shows the graph of $f(x) = x^2 - 2x - 15$.

(a) Find $f(-5)$, $f(-3)$, $f(-2)$, $f(0)$, and $f(5)$.

(b) Plot the ordered pairs obtained from part (a). Determine if these points lie on the graph of the function.

(c) Solve the equation $x^2 - 2x - 15 = 0$ algebraically.

(d) What is the relationship between the solutions of the equation in part (c) and the x intercepts of the graph of f?

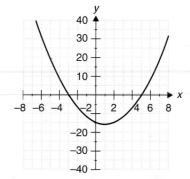

Figure 7

38. Figure 8 shows the graph of $f(x) = 15 - 2x - x^2$. Repeat the instructions (a), (b), (c), and (d) from problem 37.

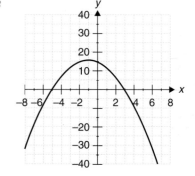

Figure 8

In Problems 39–42, match the function with its graph. Then use the x intercepts of the graph to solve the associated equation.

39. $f(x) = x^2 - 3x - 4$
$x^2 - 3x - 4 = 0$

40. $f(x) = x^2 - 9x + 18$
$x^2 - 9x + 18 = 0$

41. $f(x) = 10x - x^2 - 25$
$10x - x^2 - 25 = 0$

42. $f(x) = 6 + x - x^2$
$6 + x - x^2 = 0$

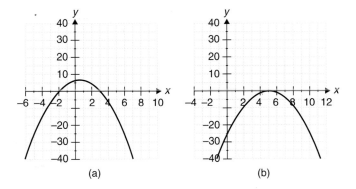

(a)   (b)

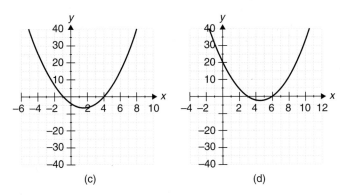

(c)   (d)

Applying the Concepts

43. Number Problem: (a) Find two consecutive positive integers whose product is 30. (b) Find two consecutive negative integers whose product is 56.

44. Number Problem: (a) Find two consecutive positive odd integers whose product is 63. (b) Find two negative consecutive even integers such that the sum of their squares is 100.

Figure 9

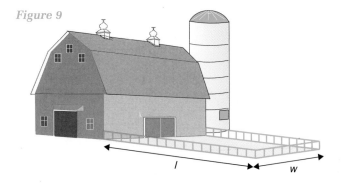

45. Gardening: A farmer has 60 feet of fencing to enclose a rectangular vegetable garden next to her barn, with the barn forming one side of the garden. What length and width are needed to enclose an area of 418 square feet and use all the fencing? (Figure 9)

46. Window Design: An artist is designing a rectangular stained glass window in such a way that the length of the window is 3 inches less than twice its width. What are the dimensions of the window if its area is 740 square inches?

47. Border of a Pool: A rectangular swimming pool 20 feet wide and 60 feet long has a concrete walkway of uniform width as a border. If the total area of the walkway is 516 square feet, how wide is the walkway? (Figure 10)

Figure 10

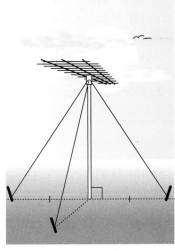

48. Advertising: A newspaper advertisement has the shape of a rectangle whose length is 1 centimeter more than its width. The newspaper charges $9 for each square centimeter. What are the dimensions of the ad if the total cost is $270?

49. Television Antenna: A rural homeowner has his television antenna held in place by three guy wires (Figure 11). Suppose that the distances to each of the stakes from the base of the antenna are the same. Also the distance from the base of the antenna to one stake is 7 feet shorter than the height of the antenna and the length of a guy wire is 13 feet. How long is the distance from the base of the antenna to a stake?

Figure 11

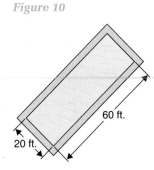

50. Camping Tent: The entrance of a tent with vertical walls has the shape of an isosceles triangle. If the base of the entrance is 2 feet less than the height, and the area of the entrance is 7.5 square feet, how tall is the entrance of the tent?

51. Geometry: The sum of the lengths of one side of a right triangle and the hypotenuse is

32 centimeters. The other leg is 1 centimeter shorter than the hypotenuse. Find the length of each side of the triangle.

52. **Television Screen:** When we speak of a 20 inch T.V. set, we refer to the diagonal of the screen measuring 20 inches. Suppose the width and height of the screen (in inches) are w and h respectively (Figure 12).

(a) What is the width and height of a 20 inch T.V. set if its width is 4 inches more than its height?

(b) What are the dimensions of a 25 inch T.V. set if its height is 5 inches less than its width?

Figure 12

Developing and Extending the Concepts

In Problems 53–56, solve each equation.

53. $(2x - 3)(x + 1) - 4(2x - 3) = 0$
54. $(1 - 2x)(1 + 2x) = (1 - 2x)^2$
55. $y(y + 1) - (y + 1) = 0$
56. $(t + 1)(t + 2) - 6t = (2t - 1)(t - 1)$

In Problems 57–60, solve each equation for the indicated unknown.

57. $x^2 - 2ax - 15a^2 = 0$, for x
58. $12y^2 - 10my = 12m^2$, for y
59. $(at - bt)^2 = t(b - a)$, for t
60. $6m^2 + mb = 2b^2$, for m

◈ CHAPTER 4 REVIEW PROBLEM SET

1. Consider the polynomial function

    $$g(x) = 5x^4 + x^3 + 2x^2 - x - 13.$$

    (a) What is the degree of this polynomial?
    (b) What are the coefficients of the polynomial?
    (c) What is the value of the polynomial when $x = 2$?

2. Consider the polynomial function

    $$f(x) = -2x^4 + 5x^3 - 2x^2 + 7x + 11.$$

    (a) What is the degree of this polynomial?
    (b) What are the coefficients of the polynomial?
    (c) What is the value of the polynomial when $x = -3$?

In Problems 3–16, perform each operation and simplify.

3. $(4x^3 + 3x^2) + (-5x^2 + 2x)$
4. $(3x^2 + 7x + 8) - (2x^2 + 3x + 2)$
5. $(-y^3 + 2y^2 + 3y - 7) - (-3y^3 + y^2 - 5y + 2)$
6. $(2x^3 + 4x^2 + 7x - 11) + (5x^2 - 3x + 5)$
7. $(a - b)(a^2 - 2ab + b^2)$
8. $(u^2 - u + 1)(2u^2 - 3u + 2)$
9. $(4y + 9x)(3y - 9x)$
10. $(4u - 5v)(u - 3v)$
11. $(3x + 7)^2$
12. $(m^2 - 3)(m^4 + 3m^2 + 9)$

13. $(5x^2 - 16x + 3) \div (x - 3)$
14. $(2y^2 + 17y + 21) \div (3 + 2y)$
15. $(x^3 - 6x^2 + 12x - 8) \div (x^2 - 4x + 4)$
16. $(x^4 - 3xy^3 - 2y^4) \div (x^2 - xy - y^2)$

In Problems 17–40, factor each expression completely.

17. $7x^2y - 21xy^3$
18. $7x(y - x) + 14x^2(y - x)$
19. $5am^2 - bn + 5an - bm^2$
20. $a^4 - b^4 - c^4 - 2b^2c^2$
21. $9u^3 - 81uv^2$
22. $m^6 - q^6$
23. $m^2 - 5m - 36$
24. $6x^2 - 29x + 35$
25. $2a^3b - 10a^2b^2 - 28ab^3$
26. $x^2yz - 6xy^2z - 16y^3z$
27. $m^2 - 3m - 10$ 28. $x^2 + 2x - 120$
29. $2y^2 + 7y + 3$ 30. $3x^2 - 13x + 12$
31. $5x^2 + 17x + 6$ 32. $2y^2 - 15y + 28$
33. $9t^2 - 12t + 4$ 34. $24y^2 - 14y - 5$
35. $36c^2 - d^2$ 36. $49x^2 - 100y^2$
37. $100c^2 - 16d^2$ 38. $36x^2 - 400$
39. $81w^4 - 16$ 40. $16r^6 - 121s^6$

In Problems 41–48, solve each equation.

41. $x^2 - 11x + 24 = 0$ **42.** $5x^2 + 3x - 14 = 0$

43. $4x^2 + 19x - 5 = 0$ **44.** $3x^2 + 4x + 1 = 0$

45. $(x + 6)(x + 5) = 2$ **46.** $(x + 1)(x - 2) = 28$

47. $81x^2 - 25 = 0$ **48.** $(2x + 5)^2 = 36$

In Problems 49–52, match the graph to the function. Find the x and y intercepts.

49. $f(x) = x^2 - 8x - 9$ **50.** $f(x) = -x^2 + 2x + 8$

51. $f(x) = -x^2 + 5x + 14$ **52.** $f(x) = 5x^2 - 10x$

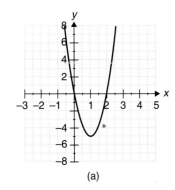

(a)

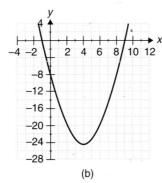

(b)

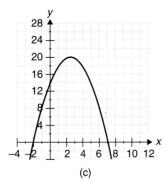

(c)

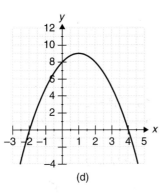

(d)

53. Management: Suppose that a company manufactures and sells x units of a product. Its profit P (in dollars) is given by the formula

$$P(x) = -x^2 + 170x - 6600.$$

Find the profit if the company makes and sells

(a) 50 units.

(b) 100 units.

(c) How many units will return 0 profit, i.e., find x so that $P(x) = 0$?

54. Height of an Arrow: An arrow is shot straight upward with an initial speed of 80 feet per second from a 96 foot high cliff. Its height, h, (in feet) after t seconds is given by the formula

$$h(t) = -16t^2 + 80t + 96.$$

What is the height of the arrow after

(a) 2 seconds?

(b) 4 seconds?

(c) When does the arrow hit the ground?

55. Geometry: The lengths of the height and base of a triangle are consecutive even numbers and the height is the shorter than the base.

(a) If x units is the height, find a formula for the area of the triangle.

(b) Find the area of the triangle if $x = 6$ inches.

(c) Find the height if the area is 84 square inches.

56. Compound Interest: Suppose that $1000 is invested in an account at an annual interest rate r for 2 years. The accumulated amount A in the account after 2 years is given by the formula

$$A(r) = 1000(1 + r)^2.$$

(a) Write the right side of the formula in expanded form.

(b) What is the amount if $r = 3\%$?

57. Geometry: Figure 1 shows a shaded area of a rectangular plot with two semicircles.

(a) Find an expression for the total area in Figure 1.

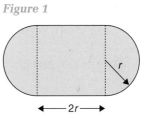

Figure 1

(b) Write the expression for the area in a factored form.

(c) What is the total area if $r = 2$ meters? Round to one decimal place.

CHAPTER 4 PRACTICE TEST

1. Find the degree of each polynomial function and state the coefficient of each term. Also find the value of the function for the indicated values.
   (a) $G(x) = 7x^2 - 3x + 5$; $G(-2)$ and $G(3)$
   (b) $h(x) = 4x^3 + x - 1$; $h(0)$ and $h(-1)$

2. Perform each operation and simplify.
   (a) $(3x^2 + 7x + 5) + (2x^2 - 5x - 2)$
   (b) $(6x^2 - 3x + 5) - (x^2 + 5x - 2)$
   (c) $(2x + 1)(2x^2 + x - 3)$
   (d) $(6x^3 + 19x^2 + 5x - 4) \div (3x - 1)$
   (e) $(7x^3 - 13x^2 + 6x + 3) \div (x - 2)$
   Use synthetic division for 7(e).

3. Find each product.
   (a) $(7x + 1)^2$
   (b) $(1 - 5x)^2$
   (c) $(3x - 2)(3x + 2)$
   (d) $(x + 2)(x^2 - 2x + 4)$

4. Factor each expression.
   (a) $9m^2 - 100n^2$
   (b) $a^2 - (b - c)^2$
   (c) $xy^2 - 4y^2 + 2x - 8$
   (d) $x^2 + 16x - 17$
   (e) $6y^2 - y - 7$
   (f) $2w^2 + 4w - 6$

5. Solve each equation.
   (a) $x^2 - 17x + 30 = 0$
   (b) $25x^2 - 49 = 0$
   (c) $x^4 - 13x^2 + 36 = 0$
   (d) $(x + 7)(x - 4) = 26$

6. A rectangular piece of cardboard measures 12 inches by 16 inches. An open box is formed by cutting congruent squares that measure x by x inches from each of the four corners of the cardboard and folding the sides.
   (a) Find the area $A(x)$ of the base of the box in terms of x.
   (b) Simplify the expression, then find the area of the base of the box if $x = 3$ inches.
   (c) Find the value of x when $A(x) = 32$ square inches.

Rational Expressions and Functions

CHAPTER 5

Chapter Contents

A tall evergreen casts a shadow of 29 meters and at the same time a 3 meter bush casts a shadow of 2.5 meters. Find the height of the evergreen. This will be solved in example 7 of section 5.5.

In this chapter, we will consider functions in the form of fractions in which the numerator and denominator are polynomials. We will consider the addition, subtraction, multiplication, and division of these new types of functions. We also solve equations involving fractions and explore their applications.

5.1 Rational Expressions and Functions

In this section, we focus on defining rational expressions and functions. Rational functions and expression model phenomena as diverse as the cost of manufacturing products as well as distance and time in uniform motion situations.

Defining Rational Expressions and Functions

Recall from section R4 that a rational number can be expressed in the form:

$$\frac{a}{b}, \text{ where } a \text{ and } b \text{ are integers and } b \neq 0.$$

In this section we extend our work to include expressions that can be written as a ratio of two polynomials.

> An algebraic expression of the form
>
> $$\frac{P}{Q} \text{ where } P \text{ and } Q \text{ are polynomials, with } Q \neq 0$$
>
> is called a **rational expression.**

Examples of rational expressions are:

$$\frac{2}{x}, \quad \frac{-3}{y-2}, \quad \frac{t+1}{t^2+t-3}, \quad \frac{c^2+4c-21}{c^2+c-6}, \quad \text{and} \quad \frac{16w^2-9}{7}.$$

A function expressed by a rational expression formula is referred to as a *rational function,* more formally, we have the following definition.

Definition of a Rational Function

> A **rational function** is one that can be written in the form
>
> $$f(x) = \frac{p(x)}{q(x)}$$
>
> where $p(x)$ and $q(x)$ are polynomials, and $q(x) \neq 0$.

The definition of a rational function excludes $q(x) = 0$ to avoid dividing by zero. For example, if

$$f(x) = \frac{3}{x-2}$$

then

$$f(2) = \frac{3}{2-2} = \frac{2}{0}$$

which is undefined because the denominator is zero. In general, a rational function is **undefined** for those values of the variable which make the denominator zero.

Unless told otherwise, we shall automatically assume that the variables in any rational expression are restricted to numerical values that will give a non zero denominator. Throughout this book, we shall not always list this restriction.

Finding Domains of Rational Functions

Figure 1

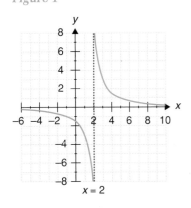

x = 2

For rational functions, the domain should not include values for which the denominator is zero. In this case, we say the domain of a rational function is all values of the variable for which the function is defined. Note that the denominator of

$$f(x) = \frac{3}{x - 2}$$

is zero precisely when $x - 2 = 0$ or $x = 2$. Therefore, the domain of f is the set of real numbers x such that $x \neq 2$. Figure 1 shows the graph of f and it also confirms that only 2 is excluded from the function's domain. The graph of the function moves toward the vertical line without touching or crossing it. This vertical line whose equation is $x = 2$ is not part of the graph, but is helpful in obtaining the graph of the rational function. It is called a *vertical asymptote* of the graph of f.

In general,

> A vertical line $(x = a)$ which a graph of a function approaches, but never touches, is called a **vertical asymptote** of the graph of f.

EXAMPLE 1   **Finding the Domain of Rational Functions**

Find the domain of each function and indicate the equation of the vertical asymptote.

(a) $f(x) = \dfrac{1}{x^2}$ (b) $g(x) = \dfrac{x}{x^2 - 2x - 3}$ (c) $h(x) = \dfrac{4}{x^2 + 1}$

Solution

(a) The denominator of f is zero when $x = 0$. Therefore, the domain of f consists of all real numbers x except 0. The graph in Figure 2a confirms that 0 is excluded and that $x = 0$ is the vertical asymptote.

(b) To avoid division by zero, we must determine the values of x that cause the denominator, $x^2 - 2x - 3$, of g to be zero.
That is,

$$x^2 - 2x - 3 = 0$$

$$(x - 3)(x + 1) = 0$$

$$x - 3 = 0 \qquad x + 1 = 0$$

$$x = 3 \qquad\qquad x = -1.$$

Thus the domain of g consists of all real numbers x where $x \neq -1$ and $x \neq 3$. The graph in Figure 2b confirms that -1 and 3 are excluded from the domain of g, and that the lines $x = -1$ and $x = 3$ are the vertical asymptotes of g.

(c) The denominator $x^2 + 1$ of the function h is never zero for any real value of x. The graph in Figure 2c confirms that there are no breaks in the curve and no value of x is excluded from the domain of h. Also the graph of h has no vertical asymptote.

Figure 2

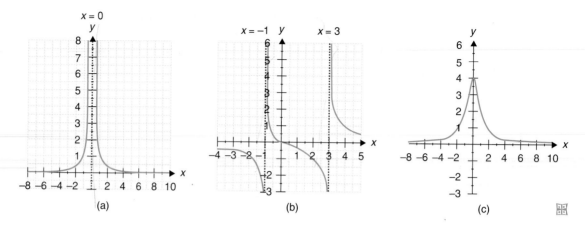

(a)   (b)   (c)

Reducing Rational Expressions

Rational expressions, like rational numbers, are said to be in **lowest terms, reduced to lowest terms,** or **simplified** if the numerator and denominator have no common factors (other than 1 and −1).

Thus, to *simplify* or *reduce* a rational expression to lowest terms,

1. we factor both the numerator and the denominator into prime factors, and
2. divide both the numerator and denominator by any common factors.

The procedure is done by using the following fundamental principle of fractions.

Fundamental Principle of Fractions

If $P(x)$, $Q(x)$, and $K(x)$ are polynomial expressions and $Q(x) \neq 0$, and $K(x) \neq 0$, then

$$\frac{P(x)K(x)}{Q(x)K(x)} = \frac{P(x)}{Q(x)}$$

Cancellation of common factors is usually indicated by slanted lines drawn through the canceled factors. For instance:

$$\frac{x^2 + 3x}{x^2 + 2x - 3} = \frac{x(x+3)}{(x-1)(x+3)} = \frac{x}{x-1}.$$

If one rational expression can be obtained from another either by canceling common factors or by multiplying the numerator and denominator of the expression by the same non zero factor, then the two rational expressions are said to be **equivalent** provided that they are both defined. Thus, the canceling of common factors above shows that

$$\frac{x^2 + 3x}{x^2 + 2x - 3} \quad \text{and} \quad \frac{x}{x-1}$$

are equivalent rational expressions where they are defined.

EXAMPLE 2   **Reducing Rational Expressions**

Reduce each rational expression to lowest terms.

(a) $\dfrac{12xy}{18y}$ (b) $\dfrac{28x^3y}{21xy^2}$ (c) $\dfrac{y^2 - 3y}{y^2 - 9}$ (d) $\dfrac{6p^2 - 7p - 3}{4p^2 - 8p + 3}$

Solution

(a) $\dfrac{12xy}{18y} = \dfrac{2 \cdot 6x\cancel{y}}{3 \cdot 6\cancel{y}} = \dfrac{2x}{3}$ For $y \neq 0$

(b) $\dfrac{28x^3y}{21xy^2} = \dfrac{4x^2(\cancel{7xy})}{3y(\cancel{7xy})} = \dfrac{4x^2}{3y}$ For $x \neq 0$ and $y \neq 0$

(c) $\dfrac{y^2 - 3y}{y^2 - 9} = \dfrac{y(\cancel{y - 3})}{(y + 3)(\cancel{y - 3})} = \dfrac{y}{y + 3}$ For $y \neq 3$ and $y \neq -3$

(d) $\dfrac{6p^2 - 7p - 3}{4p^2 - 8p + 3} = \dfrac{(3p + 1)(\cancel{2p - 3})}{(2p - 1)(\cancel{2p - 3})} = \dfrac{3p + 1}{2p - 1}$ For $p \neq \dfrac{1}{2}$ and $p \neq \dfrac{3}{2}$

It should be noticed that dividing both the numerator and denominator of a rational expression by common factors can alter the domain of the rational function defined by the expression. For example, the domain of the function

$$f(x) = \frac{x^2 - 4}{x - 2}$$

consists of all real numbers x such that $x \neq 2$. However, to make the domain of the given function f agree with the simplified form

$$f(x) = \frac{x^2 - 4}{x - 2} = \frac{(x + 2)(\cancel{x - 2})}{\cancel{x - 2}} = x + 2$$

we write

$$f(x) = x + 2, \ x \neq 2.$$

Figure 3

Figure 3 shows the graph of

$$f(x) = \frac{x^2 - 4}{x - 2}$$

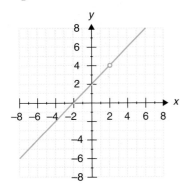

with 2 excluded from its domain. The open dot or hole at $x = 2$ shows that 2 is excluded from the domain of the function.

We have seen in example 2 above that simplifying a rational expression results from canceling identical factors in the numerator and denominator of the expression. However, in some cases, factors in the numerator and denominator have exactly opposite signs such as $x - 2$ and $2 - x$. One approach to simplifying an expression such as

$$\frac{2 - x}{x - 2}$$

involves factoring -1 from the numerator, so that

$$\frac{2 - x}{x - 2} = \frac{-1(-2 + x)}{x - 2} = \frac{-1(\cancel{x - 2})}{\cancel{x - 2}} = -1.$$

To generalize this situation, we suggest the following property.

Every rational expression has three signs associated with it—the *sign of the rational expression*, the *sign of the numerator*, and the *sign of the denominator*.

That is, if $\dfrac{P}{Q}$ is a rational expression where $Q \neq 0$, then

1. $-\dfrac{P}{Q} = \dfrac{-P}{Q} = \dfrac{P}{-Q}$

2. $\dfrac{P}{Q} = \dfrac{-P}{-Q} = -\dfrac{P}{-Q} = -\dfrac{-P}{Q}$

EXAMPLE 3 **Changing the Signs of Expressions**

Rewrite each rational expression with the denominator $x - 3$.

(a) $\dfrac{1}{3-x}$ (b) $\dfrac{-x}{3-x}$ (c) $-\dfrac{-1}{3-x}$

Solution (a) $\dfrac{1}{3-x} = \dfrac{(-1)(1)}{(-1)(3-x)}$ Change signs in numerator and denominator.

$= \dfrac{-1}{-3+x} = \dfrac{-1}{x-3}$ Rearrange terms in the denominator.

(b) $\dfrac{-x}{3-x} = \dfrac{(-1)(-x)}{(-1)(3-x)}$ Change signs in numerator and denominator.

$= \dfrac{x}{-3+x} = \dfrac{x}{x-3}$

(c) $-\dfrac{-1}{3-x} = \dfrac{-1}{(-1)(3-x)}$ Change signs in denominator and fraction.

$= \dfrac{-1}{-3+x} = \dfrac{-1}{x-3}$

Building Rational Expressions to Higher Terms

The fundamental principle of fractions can also be used to build rational expressions to higher terms. This technique will be useful when we add and subtract rational expressions.

EXAMPLE 4 **Building an Expression to Higher Terms**

Replace each question mark with an appropriate expression so that the resulting expressions are equivalent.

(a) $\dfrac{3}{4} = \dfrac{?}{20t}$ (b) $\dfrac{2a}{3b} = \dfrac{?}{3ab - 3b^2}$ (c) $\dfrac{x+3}{x-2} = \dfrac{x^2 + 5x + 6}{?}$

Solution (a) $\dfrac{3}{4} = \dfrac{(5t) \cdot 3}{(5t) \cdot 4} = \dfrac{15t}{20t}; ? = 15t$ Factor $20t = (5t) \cdot 4$

(b) $\dfrac{2a}{3b} = \dfrac{(a-b)(2a)}{(a-b)(3b)} = \dfrac{2a^2 - 2ab}{3ab - 3b^2}$ Factor $3ab - 3b^2 = (3b)(a-b)$

so that $? = 2a^2 - 2ab$.

(c) $\dfrac{x + 3}{x - 2} = \dfrac{(x + 3)(x + 2)}{(x - 2)(x + 2)}$

$= \dfrac{x^2 + 5x + 6}{x^2 - 4}$ Factor $x^2 + 5x + 6 = (x + 3)(x + 2)$ ▦

so that ? $= x^2 - 4$.

Recall from section R4 that two rational numbers are equal (or equivalent) if they have the same value. We can also check to see whether they were equal by cross multiplication. Two rational expressions are equal if their cross products are equal. That is,

$$\frac{P}{Q} = \frac{R}{S} \quad \text{if and only if} \quad PS = QR$$

For example, for all nonzero values of x and y,

$$\frac{3y}{xy} = \frac{3}{x}$$

since their cross products $(3y)x$ and $3(xy)$ are equal.

EXAMPLE 5   **Identifying Equal Rational Expressions**

Use cross multiplication to determine whether the rational expressions are equal.

(a) $\dfrac{14y}{7y + 21}$ and $\dfrac{2y}{y + 3}$ (b) $\dfrac{15x}{10x + 4}$ and $\dfrac{3x}{2x + 1}$

Solution   We compare the cross products

(a) $(14y)(y + 3) = 14y^2 + 42y$ and $(7y + 21)(2y) = 14y^2 + 42y$.

Since the cross products are equal, the rational expressions are equal.

(b) $(15x)(2x + 1) = 30x^2 + 15x$ and $(10x + 4)(3x) = 30x^2 + 12x$.

Since the cross products are not equal, the rational expressions are not equal.   ▦

Solving Applied Problems

Rational expressions can be used to describe real-world situations, as the next example shows.

EXAMPLE 6   **Solving a Revenue Problem**

The total revenue R (in thousands of dollars) from the sale of a certain product is approximated by the function

$$R(t) = \frac{1000(t^2 - t - 2)}{t^2 + 2t - 8}, t > 2$$

where t is the number of years of production.

(a) Reduce the right side of this function to lowest terms.

(b) Find the total revenue at the end of the fourth year.

Solution

(a) $R(t) = \dfrac{1000(t^2 - t - 2)}{t^2 + 2t - 8}$ Given

$= \dfrac{1000(t + 1)(t - 2)}{(t - 2)(t + 4)}$ Factor

$= \dfrac{1000(t + 1)}{t + 4}$ Reduce

(b) At the end of the fourth year, $t = 4$, so

$$R(4) = \frac{1000(4 + 1)}{4 + 4} = \frac{5000}{8} = 625.$$

Therefore, the total revenue is 625 thousand dollars ($625,000).

Check: $R(4) = \dfrac{1000(4^2 - 4 - 2)}{4^2 + 2(4) - 8} = 625$

PROBLEM SET 5.1

Mastering the Concepts

In Problems 1 and 2, find the value of each rational function at the specified numbers.

1. $f(x) = \dfrac{4x}{x + 3}$

(a) $f(1)$ (b) $f(-2)$ (c) $f(-3)$ (d) $f(0)$

2. $g(x) = \dfrac{x - 2}{2x - 5}$

(a) $g(2)$ (b) $g(-2)$ (c) $g\left(\dfrac{1}{2}\right)$ (d) $g\left(\dfrac{5}{2}\right)$

In Problems 3–6, find the domain of each function. Also indicate the equations of the vertical asymptotes.

3. (a) $f(x) = \dfrac{7x}{x - 3}$

(b) $g(x) = \dfrac{x + 2}{x + 6}$

4. (a) $f(x) = \dfrac{-11}{12 - 6x}$

(b) $g(x) = \dfrac{x + 2}{(x + 2)(x - 8)}$

5. (a) $f(x) = \dfrac{2x - 4}{(x - 6)(x + 7)}$

(b) $g(x) = \dfrac{x - 1}{(x + 3)(x - 2)}$

6. (a) $f(x) = \dfrac{x + 2}{x^2 + 2x - 3}$

(b) $g(x) = \dfrac{1 - x}{12 - 4x - x^2}$

In Problems 7–20, reduce each rational expression to lowest terms.

7. (a) $\dfrac{12xy^2}{18x^2y}$

(b) $\dfrac{8x^2 - 4x}{4x}$

8. (a) $\dfrac{14x^2y^3}{21x^2y^4}$

(b) $\dfrac{6u^2 - 3u}{3u}$

9. (a) $\dfrac{x^2y - xy^2}{x^2 - xy}$

(b) $\dfrac{x + 4x^2y}{1 + 4xy}$

10. (a) $\dfrac{x^2 - 5x}{2x - 10}$

(b) $\dfrac{x^2 - 3x}{x^2y - 3xy}$

11. (a) $\dfrac{x^2 + x}{x^2 - x}$

(b) $\dfrac{x^2 - 9}{x^2 + 3x}$

12. (a) $\dfrac{x^2 - xy}{x^2y - xy^2}$

(b) $\dfrac{xy^2 + x^2y}{xy + y^2}$

13. (a) $\dfrac{y^2 + 6y + 8}{4y^2 + 16y}$

(b) $\dfrac{x^2 + 5x + 6}{3x^2 + 9x}$

14. (a) $\dfrac{y^2 + 4y}{y^2 + 6y + 8}$

(b) $\dfrac{4x^2 - 9}{6x^2 - 9x}$

15. (a) $\dfrac{x^2 + x - 12}{x^2 + 4x - 21}$

(b) $\dfrac{x^2 + 7x + 12}{x^2 + 5x + 6}$

16. (a) $\dfrac{3x^2 + 7x + 4}{3x^2 - 5x - 12}$

(b) $\dfrac{10x^2 + 29x - 21}{4x^2 + 12x - 7}$

17. (a) $\dfrac{x^2 + xy - 2y^2}{x^2 + 3xy + 2y^2}$

(b) $\dfrac{3x^2 - 12y^2}{x^2 + 4xy + 4y^2}$

18. (a) $\dfrac{2x^2 + 13x + 20}{2x^2 + 17x + 30}$

(b) $\dfrac{3 + 13m - 10m^2}{2m^2 + 5m - 12}$

19. (a) $\dfrac{x^3 - 4xy^2}{3x^3 - 2x^2y - 8xy^2}$

(b) $\dfrac{3x^2 - 6xy + 3y^2}{3x^2 - 3y^2}$

20. (a) $\dfrac{x^2 - 9}{c(x + 3) - b(x + 3)}$

(b) $\dfrac{4a^3 - 14a^2 + 12a}{24a + 4a^2 - 8a^3}$

In Problems 21–30, replace each question mark with an appropriate expression so that the resulting expressions are equivalent.

21. (a) $\dfrac{3}{4x} = \dfrac{?}{8x^2y}$

(b) $\dfrac{13x}{25} = \dfrac{?}{75y}$

22. (a) $\dfrac{7}{3x} = \dfrac{?}{6x^2y^2}$

(b) $\dfrac{5x}{4y} = \dfrac{20x^2y}{?}$

23. (a) $\dfrac{5m^3y}{7my^3} = \dfrac{?}{28m^5y^6}$

(b) $\dfrac{6}{11xy} = \dfrac{?}{33x^2y}$

24. (a) $\dfrac{3x}{4y} = \dfrac{?}{4xy + 4y^2}$

(b) $\dfrac{3}{x + y} = \dfrac{?}{5(x + y)^2}$

25. (a) $\dfrac{9(a - b)}{7a} = \dfrac{?}{7a^4(a - b)}$

(b) $\dfrac{mn}{m - n} = \dfrac{?}{m^3 - mn^2}$

26. (a) $\dfrac{a^3 - a}{a^2 - a} = \dfrac{a + 1}{?}$

(b) $\dfrac{1}{x - y} = \dfrac{?}{x^3y - xy^3}$

27. $\dfrac{2x - 3}{x^2 + 3x + 9} = \dfrac{?}{x^3 - 27}$

28. $\dfrac{y - 3}{y^2 - y + 1} = \dfrac{?}{y^3 + 1}$

29. $\dfrac{x + 2}{x - 2} = \dfrac{?}{2x^2 - 7x + 6}$

30. $\dfrac{3x + 2y}{5x + 3y} = \dfrac{?}{3y^2 - 4xy - 15x^2}$

In Problems 31–36, use cross multiplication to determine whether or not each pair of rational expressions is equal.

31. (a) $\dfrac{5}{4x}$ and $\dfrac{10}{8x}$

(b) $\dfrac{-7y}{6}$ and $\dfrac{21y}{-18}$

32. (a) $\dfrac{7a}{9b}$ and $\dfrac{8a}{10b}$

(b) $\dfrac{-14}{-9x}$ and $\dfrac{-12}{7x}$

33. $\dfrac{x + y}{a}$ and $\dfrac{7x + y}{7a}$

34. $\dfrac{x + 2}{x^2 - 4}$ and $\dfrac{3}{3x - 12}$

35. $\dfrac{x + 2y}{5}$ and $\dfrac{2x^2 - 8y^2}{10x - 20y}$

36. $\dfrac{x^2 - 9}{x - 3}$ and $\dfrac{2x + 6}{2}$

In Problems 37 and 38, rewrite each rational expression with the denominator $x - 4$.

37. (a) $\dfrac{-7}{4 - x}$

(b) $\dfrac{4}{-(x - 4)}$

38. (a) $\dfrac{-4x}{4 - x}$

(b) $\dfrac{3 - x}{4 - x}$

Applying the Concepts

39. Geometry: The height h of a closed vertical cylindrical tank of radius r is given by the formula:

$$h = \frac{s - 2\pi r^2}{2\pi r}$$

where s is the surface area of the tank. Determine the height of the tank if its surface area is 256π square feet and its radius is 6 feet.

40. Investment: In business the principal amount of an investment can be determined by using the formula:

$$P = \frac{A}{1 + rt}$$

where P is the principal amount of dollars invested at rate r, and A is the amount of money accumulated after t years. Find P when $A = \$580$, $r = 8\%$, and $t = 2$ years.

Developing and Extending the Concepts

In Problems 41–46, reduce each rational expression.

41. $\dfrac{3x^3 - 3xy^2}{3xy^2 + 3x^2y - 6x^3}$

42. $\dfrac{x^3 - 2x^2 + 5x - 10}{3x^5 + 15x^3 - x^2 - 5}$

43. $\dfrac{xz + xw - yz - yw}{xy + xz - y^2 - yz}$

44. $\dfrac{(2 + h)^3 - 8}{h}$

45. $\dfrac{x^{2n} - 9}{x^n + 3}$, where n is a positive integer.

46. $\dfrac{(1 + h)^2 + 2(1 + h) - 3}{h}$

In Problems 47–50, match each function with its appropriate graph.

47. $f(x) = \dfrac{3}{(x - 2)^2}$

48. $f(x) = \dfrac{3x}{x^2 - 9}$

49. $f(x) = \dfrac{5x}{x + 2}$

50. $f(x) = \dfrac{-2}{x^2 + 1}$

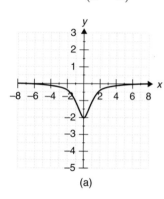

(a)

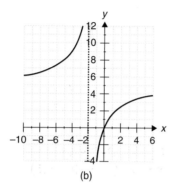

(b)

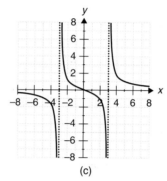

(c)

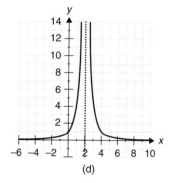

(d)

In Problems 51 and 52, simplify each function and graph it. Indicate the domain of each function.

51. $f(x) = \dfrac{x^2 - x - 6}{x - 3}$

52. $f(x) = \dfrac{x^2 - 4x + 3}{x - 1}$

Objectives

1. Multiply Rational Expressions
2. Divide Rational Expressions
3. Solve Applied Problems

5.2 Multiplication and Division of Rational Expressions

Now that you know how to simplify rational expressions, we will discuss multiplication and division of these expressions. Both multiplication and division are based upon the corresponding properties of rational numbers covered in section R5.

Multiplying Rational Expressions

To multiply two rational expressions, we multiply their numerators and divide by the product of the denominators. This is, if $\dfrac{P}{Q}$ and $\dfrac{R}{S}$ are rational expressions, then

$$\frac{P}{Q} \cdot \frac{R}{S} = \frac{P \cdot R}{Q \cdot S}$$

For instance,

$$\frac{5x}{4y} \cdot \frac{8y^2}{25x} = \frac{40xy^2}{100xy} = \frac{5x \cdot 2 \cdot 4y \cdot y}{4y \cdot 5 \cdot 5x} = \frac{2y}{5}.$$

The process of multiplying rational expressions can be shortened by using the following step-by-step procedure.

Multiplying Rational Expressions

> Step 1: Factor all numerators and denominators completely.
> Step 2: Cancel the common factors in the numerators and denominators.
> Step 3: Multiply the remaining factors in the numerators and in the denominators.

Using this procedure, the illustration on page 258 is multiplied as follows:

$$\frac{5x}{4y} \cdot \frac{8y^2}{25x} = \frac{\cancel{5x}}{\cancel{4y}} \cdot \frac{2 \cdot \cancel{4y} \cdot y}{5 \cdot \cancel{5x}} = \frac{2y}{5}.$$

EXAMPLE 1   **Multiplying Rational Expressions**

Multiply and simplify.

(a) $\dfrac{x+5}{x-3} \cdot \dfrac{x^2-x-6}{x^2-25}$ (b) $\dfrac{c}{c-1} \cdot \dfrac{c^2-1}{c^2}$ (c) $\dfrac{6t-6}{t^2+2t} \cdot \dfrac{t^2+4t+4}{2t^2+2t-4}$

Solution   (a) $\dfrac{x+5}{x-3} \cdot \dfrac{x^2-x-6}{x^2-25} = \dfrac{x+5}{x-3} \cdot \dfrac{(x-3)(x+2)}{(x-5)(x+5)}$ Factor numerators and denominators

$= \dfrac{\cancel{x+5}}{\cancel{x-3}} \cdot \dfrac{\cancel{(x-3)}(x+2)}{(x-5)\cancel{(x+5)}}$ Cancel common factors

$= \dfrac{x+2}{x-5}$

(b) $\dfrac{c}{c-1} \cdot \dfrac{c^2-1}{c^2} = \dfrac{c}{(c-1)} \cdot \dfrac{(c-1)(c+1)}{c \cdot c}$ Factor numerators and denominators

$= \dfrac{\cancel{c}}{\cancel{(c-1)}} \cdot \dfrac{\cancel{(c-1)}(c+1)}{\cancel{c} \cdot c}$ Cancel common factors

$= \dfrac{c+1}{c}$

c is not a factor of the numerator so the expression cannot be reduced.

6 − 6t = −6(t − 1).

(c) $\dfrac{6-6t}{t^2+2t} \cdot \dfrac{t^2+4t+4}{2t^2+2t-4} = \dfrac{-2 \cdot 3(t-1)(t+2)^2}{t(t+2)2(t+2)(t-1)}$ Factor

$= \dfrac{-1 \cdot 2 \cdot 3\cancel{(t-1)}\cancel{(t+2)^2}}{t\cancel{(t+2)}2\cancel{(t+2)}\cancel{(t-1)}} = \dfrac{-3}{t}$ Cancel

Dividing Rational Expressions

We divide two rational expressions in the same way as we divided rational numbers in section R5. To divide rational expressions, we follow these steps:

Division of Rational Expressions

> Step 1: Take the reciprocal of the divisor.
> Step 2: Proceed as in multiplication.

That is,

$$\text{if } \frac{P}{Q} \text{ and } \frac{R}{S} \text{ are rational expressions and } R \neq 0,$$

then

$$\frac{P}{Q} \div \frac{R}{S} = \frac{P}{Q} \cdot \frac{S}{R} = \frac{P \cdot S}{Q \cdot R}.$$

EXAMPLE 2 **Dividing Rational Expressions**

Divide and simplify the results.

(a) $\dfrac{3a^2}{5b} \div \dfrac{2a^2}{6b^3}$ (b) $\dfrac{a+b}{2} \div \dfrac{(a+b)^2}{6}$ (c) $\dfrac{x^2+5x+6}{x^2-4} \div \dfrac{x^2+4x+4}{x^2-4x+4}$

Solution (a) $\dfrac{3a^2}{5b} \div \dfrac{2a^2}{6b^3} = \dfrac{3a^2}{5b} \cdot \dfrac{6b^3}{2a^2}$ Reciprocal of the divisor

$$= \frac{3 \cdot a^2 \cdot 2 \cdot 3 \cdot b^2 \cdot b}{5 \cdot b \cdot 2 \cdot a^2} = \frac{9b^2}{5}$$ Factor and reduce

(b) $\dfrac{a+b}{2} \div \dfrac{(a+b)^2}{6} = \dfrac{a+b}{2} \cdot \dfrac{6}{(a+b)^2}$ Reciprocal of divisor

$$= \frac{a+b}{2} \cdot \frac{2 \cdot 3}{(a+b)(a+b)} = \frac{3}{a+b}$$ Factor and reduce

(c) $\dfrac{x^2+5x+6}{x^2-4} \div \dfrac{x^2+4x+4}{x^2-4x+4} = \dfrac{x^2+5x+6}{x^2-4} \cdot \dfrac{x^2-4x+4}{x^2+4x+4}$

$$= \frac{(x+2)(x+3)}{(x-2)(x+2)} \cdot \frac{(x-2)(x-2)}{(x+2)(x+2)}$$

$$= \frac{(x-2)(x+3)}{(x+2)(x+2)} = \frac{(x-2)(x+3)}{(x+2)^2}$$

The next example demonstrates how we follow the order of operations to multiply and divide rational expressions.

EXAMPLE 3 **Multiplying and Dividing Rational Expressions**

Perform the indicated operations and simplify the result.

(a) $\dfrac{w^2-w}{z^2-z} \div \dfrac{w^2}{w-1} \cdot \dfrac{z^2w-zw}{w-1}$ (b) $\dfrac{w^2-w}{z^2-z} \div \left[\dfrac{w^2}{w-1} \cdot \dfrac{z^2w-zw}{w-1} \right]$

Solution (a) We follow the order of operations. Here, first division and then multiplication.

$$\frac{w^2-w}{z^2-z} \div \frac{w^2}{w-1} \cdot \frac{z^2w-zw}{w-1} = \frac{w^2-w}{z^2-z} \cdot \frac{w-1}{w^2} \cdot \frac{z^2w-zw}{w-1}$$

$$= \frac{w(w-1)}{z(z-1)} \cdot \frac{w-1}{w \cdot w} \cdot \frac{zw(z-1)}{w-1} = w-1$$

(b) Because of the brackets, the divisor is the product of the two expressions inside the brackets.

$$\frac{w^2-w}{z^2-z} \div \left[\frac{w^2}{w-1} \cdot \frac{z^2w-zw}{w-1}\right] = \frac{w^2-w}{z^2-z} \div \left[\frac{w^2(z^2w-zw)}{(w-1)^2}\right] \qquad \text{Multiply}$$

$$= \frac{\cancel{w}(w-1)}{z(z-1)} \cdot \frac{(w-1)^2}{w^2 \cdot \cancel{w} \cdot z(z-1)} \qquad \text{Reciprocal}$$

$$= \frac{(w-1)^3}{w^2z^2(z-1)^2} \qquad \text{Reduce}$$

Now we state two basic operations defined for rational functions.

Product and Quotient of Rational Functions

For two rational functions f and g, the **product** and **quotient** functions $f \cdot g$ and $\dfrac{f}{g}$ are defined as follows:

1. **Product:** $(f \cdot g)(x) = f(x) \cdot g(x)$

2. **Quotient:** $\left(\dfrac{f}{g}\right)(x) = \dfrac{f(x)}{g(x)}$

When determining the domain of the product and quotient of rational functions, we do so *before* simplifying each expression.

EXAMPLE 4   **Multiplying and Dividing Rational Functions**

(a) Let $f(x) = \dfrac{2x-5}{x-4}$ and $g(x) = \dfrac{x^2-5x+4}{2x-5}$. Find $(f \cdot g)(x)$ and its domain.

(b) Let $f(x) = 3x-6$ and $g(x) = x^2+x-6$. Find $\dfrac{f}{g}(x)$ and its domain.

Solution   (a) $(f \cdot g)(x) = f(x) \cdot g(x) = \dfrac{2x-5}{x-4} \cdot \dfrac{x^2-5x+4}{2x-5}$

$$= \frac{2x-5}{x-4} \cdot \frac{(x-1)(x-4)}{2x-5} \qquad \text{Factor}$$

$$= \frac{\cancel{2x-5}}{\cancel{x-4}} \cdot \frac{(x-1)\cancel{(x-4)}}{\cancel{2x-5}} \qquad \text{Cancel common factors}$$

$$= x-1$$

Since the denominator of a rational expression cannot be zero, x cannot have values 4 or 5/2. Therefore, the domain of $f \cdot g$ is all real values of x except 4 and 5/2.

(b) $\left(\dfrac{f}{g}\right)(x) = \dfrac{f(x)}{g(x)}$

$= \dfrac{3x - 6}{x^2 + x - 6}$ Substitute expressions for $f(x)$ and $g(x)$

$= \dfrac{3(x - 2)}{(x - 2)(x + 3)} = \dfrac{3}{x + 3}$ Cancel common factors

In any rational expression the denominator may not be zero. Hence, the domain of $\left(\dfrac{f}{g}\right)(x)$ is all real values of x except 2 or -3.

Solving Applied Problems

Multiplication and division of rational expressions can be used to model examples in the real world.

EXAMPLE 5   **Modeling a Geometric Problem**

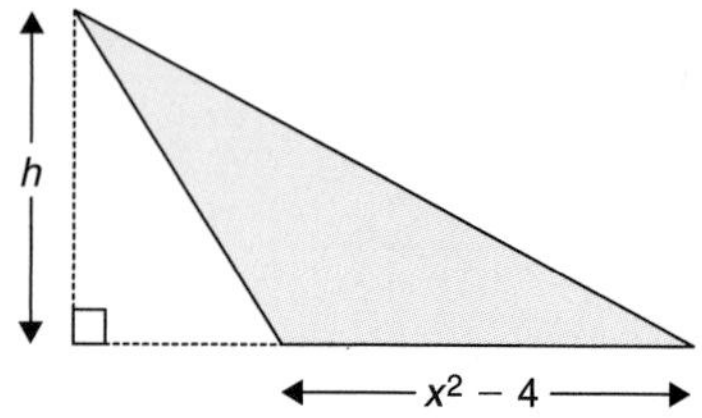

Figure 1 shows a garden in the form of a triangle. The area A of the garden is modeled by the function

$$A(x) = x^2 + 7x - 18, \quad x > 0.$$

The base b of the triangular garden is given by

$$b(x) = x^2 - 4, \quad x > 0.$$

Figure 1

h

$\leftarrow\!\!\!\!-\ x^2 - 4\ -\!\!\!\!\rightarrow$

(a) Find an expression that represents the height h of the triangular garden.

(b) If $x = 5$ feet, find the height h.

Solution   (a) The area A of a triangle is given by the formula

$$A = \frac{1}{2}bh$$

where b is the length of the base and h is the height. Solving for h, we have

$h = 2A \div b$

$= 2(x^2 + 7x - 18) \div (x^2 - 4)$ Substitute polynomials

$= 2(x^2 + 7x - 18) \cdot \dfrac{1}{x^2 - 4}$ Reciprocal

$= 2(x + 9)(x - 2) \cdot \dfrac{1}{(x - 2)(x + 2)}$ Factor

$= \dfrac{2(x + 9)}{x + 2}$ Reduce

(b) If $x = 5$, we have

$$h = \frac{2(5 + 9)}{5 + 2}$$

$$= \frac{2(14)}{7} = 4 \text{ feet.}$$

PROBLEM SET 5.2

Mastering the Concepts

In Problems 1–20, perform each multiplication or division and reduce the result to lowest terms.

1. (a) $\dfrac{10}{9x} \cdot \dfrac{12x}{15y}$

   (b) $\dfrac{3x}{7} \cdot \dfrac{14y}{27x^2}$

2. (a) $\dfrac{2x}{3y} \cdot \dfrac{12y}{5x}$

   (b) $\dfrac{3x}{5yz} \cdot \dfrac{4y}{9x}$

3. $\dfrac{7x^3}{8y^4} \cdot \dfrac{16y}{21x^2}$

4. $\dfrac{5x^2y}{3a^2b} \cdot \dfrac{6ab}{10x^2}$

5. $\dfrac{-5xy^2}{3z} \cdot 6xz^2$

6. $-4a \cdot \dfrac{5b}{6bc}$

7. $\dfrac{4xy}{x-y} \cdot \dfrac{x-y}{8xy}$

8. $\dfrac{x+2}{3x} \cdot \dfrac{9x^2}{x+2}$

9. $\dfrac{x+4}{x^2} \cdot \dfrac{5x}{3x+12}$

10. $\dfrac{3x+3y}{xy} \cdot \dfrac{5x}{7x+7y}$

11. (a) $\dfrac{11x}{5} \div \dfrac{22x}{25}$

    (b) $\dfrac{9x}{4} \div \dfrac{36x}{15}$

12. (a) $\dfrac{8x}{9y} \div \dfrac{4x}{3y}$

    (b) $\dfrac{9x}{4} \div \dfrac{15}{36x}$

13. $15xy \div \dfrac{x^2}{5xz}$

14. $\dfrac{5z^2}{3xy} \div 2z$

15. $\dfrac{a}{4bc} \div 16ab$

16. $\dfrac{6ab^2}{5} \div \dfrac{15a}{3b^2}$

17. $7xy \div \dfrac{-12xy^2}{3z}$

18. $-11x \div \dfrac{22x^2}{5z}$

19. $\dfrac{5x^2y}{x-2} \div \dfrac{15xy}{x-2}$

20. $\dfrac{3a+3}{a-1} \div \dfrac{7a+7}{9a-9}$

In Problems 21–46, perform each operation and simplify the result.

21. $\dfrac{m^2+mn}{mn} \cdot \dfrac{7n}{m^2-n^2}$

22. $\dfrac{m^2-4}{m+3} \cdot \dfrac{m^2-9}{m+2}$

23. $\dfrac{x^2-x}{x-1} \cdot \dfrac{x}{5x-5}$

24. $\dfrac{6x-12}{7x-21} \div \dfrac{2x-4}{x^2-9}$

25. $\dfrac{x^2-144}{x+4} \cdot \dfrac{x^2-16}{x+12}$

26. $\dfrac{3y^2-12}{3y^2-3} \cdot \dfrac{y-1}{2y+4}$

27. $\dfrac{2x-y}{x^2-y^2} \div \dfrac{4x^2-y^2}{x+y}$

28. $\dfrac{5x+10}{x^3} \div \dfrac{x+2}{x^4}$

29. $\dfrac{1-9a^2}{a+1} \div \dfrac{6a+2}{2a+2}$

30. $\dfrac{3x+6}{5x+5} \cdot \dfrac{10x+10}{x^2-6x-16}$

31. $\dfrac{3y^2-y-2}{3y^2+y-2} \cdot \dfrac{3y^2-5y+2}{3y^2+5y+2}$

32. $\dfrac{2x-y}{x^2-y^2} \div \dfrac{4x^2-y^2}{x+y}$

33. $\dfrac{2x^2-5x-3}{3x^2-5x-2} \div \dfrac{2x^2+11x+5}{x^2+3x-10}$

34. $\dfrac{2a-6}{6a^2-15a} \cdot \dfrac{18a^3-45a^2}{4a^2-12a}$

35. $\dfrac{t^2-8t+15}{t^2+2t-35} \cdot \dfrac{t^2+9t+14}{15-2t-t^2}$

36. $\dfrac{6y^2+7y-3}{9y^2-25} \div \dfrac{12y^2-y-1}{12y^2+20y}$

37. $\dfrac{x^3-4x^2+3x}{x+2} \div (x^2-3x)$

38. $\dfrac{y^2-3}{y^3-4y} \cdot \dfrac{y^2-4}{y^4-9}$

39. $\dfrac{a^3+b^3}{2a^3+2a^2b} \cdot \dfrac{6a^2-6b^2}{3a^3-3a^2b+3ab^2}$

40. $\dfrac{8x^3-27}{9x^2-3x+1} \div \dfrac{4x^2+6x+9}{27x^3+1}$

41. $\dfrac{8x^2+2xy}{15x+6y} \cdot \dfrac{10x^2-xy-2y^2}{8x^2-2xy-y^2}$

42. $\dfrac{x^3-5x^2+x-5}{3x^4-3x^3+3x^2-3x} \cdot \dfrac{4x^3-4x^2+2x-2}{2x^3-10x^2+2x-10}$

43. $\dfrac{6x^2+xy-2y^2}{6x^2-5xy+y^2} \div \dfrac{3x^2+17xy+10y^2}{6x^2+13xy-5y^2}$

44. $\dfrac{10x^2+10xy-xy^2-y^3}{x^3+x^2+xy^2+y^2} \div \dfrac{x^3+x+x^2y+y}{x^3+x+x^2+1}$

45. $\dfrac{27x^3-1}{9x^2+3x+1} \div \dfrac{x^2-6x+9}{(x-3)^2}$

46. $(x-7) \div \dfrac{x^2-2x-35}{2x+3}$

In Problems 47 and 48, find $(f \cdot g)(x)$ and the domain of the product.

47. $f(x) = \dfrac{x-3}{7x+7}$ and $g(x) = \dfrac{x^2+2x+1}{x^2-9}$

48. $f(x) = \dfrac{3x^4+6x^2}{6x^2+14x+4}$ and $g(x) = \dfrac{3x^2+x}{x^2+2}$

In Problems 49 and 50, find

$\left(\dfrac{f}{g}\right)(x)$ and the domain of $\dfrac{f}{g}$.

49. $f(x) = 2x^3 - 7x^2 + 3x$

and

$g(x) = 4x^2 - 1$

50. $f(x) = 2x^3 - 5x^2 - 12x$

and

$g(x) = 2x^2 + x - 3$

Applying the Concepts

51. Fuel Economy: On a recent trip a car was driven continuously for 17.5 hours. It traveled x miles and used y gallons of gasoline. The average rate of speed was

$$r = \frac{x}{17.5}$$

miles per hour and the

average rate of fuel consumption was

$$f = \frac{y}{17.5}$$

gallons per hour.

(a) Find an expression for the average mileage in miles per gallon.

(b) Evaluate the expression from part (a) for $x = 900$ miles and $y = 45$ gallons.

52. Pumping Rate: A newly dug well pumps water into a reservoir at a rate of 55 gallons per minute. At this rate, find:

(a) the time (in hours) to pump x gallons into the reservoir

(b) the time (in hours) to pump 412,500 gallons into the reservoir.

Developing and Extending the Concepts

In Problems 53–64, perform the indicated operations and simplify.

53. $\left[\dfrac{27x^2}{16y^2} \cdot \dfrac{20y^3}{9x}\right] \div \dfrac{18x^3}{5y}$

54. $\left[\dfrac{10x^2}{17y^3} \cdot \dfrac{34y}{25x^3}\right] \div \dfrac{2y^3}{15x^4}$

55. $\left[\dfrac{x-1}{x^2-4} \cdot \dfrac{2x+4}{x^2-1}\right] \div \dfrac{2x+2}{x-2}$

56. $\left[\dfrac{7x^2-28}{3x^2-12x} \cdot \dfrac{9x^3+6x^2}{2x^2-4x}\right] \div \dfrac{x^3+2x^2}{2x^2-8x}$

57. $\left[\dfrac{x^2-1}{x+2} \cdot \dfrac{x^2-4}{x-3}\right] \div \dfrac{x^2-3x+2}{x^2-9}$

58. $\left[\dfrac{y^2-3y+2}{y^2-4} \cdot \dfrac{y+2}{y-1}\right] \div \left[\dfrac{y^2-4}{y-2}\right]$

59. $\dfrac{x^2-1}{x+1} \div \left[\dfrac{x^2-1}{x^2+2x+1} \cdot \dfrac{x^2+1}{x^2-2x+1}\right]$

60. $\dfrac{a^2-ab}{ab+b^2} \div \left[\dfrac{a^2-b^2}{a^2+2ab+b^2} \cdot \dfrac{a^2b+ab^2}{a^2-2ab+b^2}\right]$

61. $\dfrac{b-1}{21-4a-a^2} \cdot \dfrac{b-2}{b-b^3} \div \dfrac{2-b}{a^2+6a-7}$

62. $\dfrac{7}{5v^2(v+3)} \div \left(\dfrac{v^2-5v+6}{8v^2} \cdot \dfrac{21}{5v^2-45}\right)$

63. $\left(\dfrac{x^2-1}{x^2}\right) \cdot \dfrac{2}{x-1} \div \dfrac{x^2+2x+1}{x^3}$

64. $\dfrac{2y^2+5y-3}{6y^2-5y-6} \div \left(\dfrac{2y^2+9y-5}{12y^2-y-6} \div \dfrac{2y^2+y-6}{4y^2+9y-9}\right)$

65. Perform the indicated operations and simplify.

(a) $\dfrac{1}{x} \div \dfrac{1}{x}$

(b) $\dfrac{1}{x} \div \dfrac{1}{x} \div \dfrac{1}{x}$

(c) $\dfrac{1}{x} \div \dfrac{1}{x} \div \dfrac{1}{x} \div \dfrac{1}{x}$

(d) For what value(s) of x will the expressions in parts (b) and (c) be equal?

(e) For what value(s) of x will the expression in part (c) be positive?

(f) For what value(s) of x will the expression in part (c) be negative?

66. Use the pattern established in problem 65 to find the value of the expression

$$\dfrac{1}{x} \div \dfrac{1}{x} \div \dfrac{1}{x} \div \dfrac{1}{x} \div \dfrac{1}{x} \div \dfrac{1}{x}$$

5.3 Addition and Subtraction of Rational Expressions

When we add or subtract rational expressions, we follow the same procedure as we did when adding or subtracting rational numbers.

Adding and Subtracting Rational Expressions with Common Denominators

To *add* or *subtract* rational expressions with common denominators, we add or subtract the numerators and repeat the common denominator.

For example, we add $\dfrac{3}{7x}$ and $\dfrac{2}{7x}$ as follows:

$$\frac{3}{7x} + \frac{2}{7x} = \frac{3 + 2}{7x}$$

$$= \frac{5}{7x}.$$

This addition can also be justified by using the distributive property.

$$\frac{3}{7x} + \frac{2}{7x} = 3 \cdot \frac{1}{7x} + 2 \cdot \frac{1}{7x}$$

$$= (3 + 2)\frac{1}{7x}$$

$$= 5 \cdot \frac{1}{7x}$$

$$= \frac{5}{7x}.$$

Similarly, we subtract $\dfrac{2}{7x}$ from $\dfrac{3}{7x}$ as follows:

$$\frac{3}{7x} - \frac{2}{7x} = \frac{3 - 2}{7x}$$

$$= \frac{1}{7x}.$$

In general, to add or subtract rational expressions, we use the following properties:

Addition and Subtraction

If $\dfrac{P}{Q}$ and $\dfrac{R}{Q}$ are rational expressions, then

Addition: $\dfrac{P}{Q} + \dfrac{R}{Q} = \dfrac{P + R}{Q}$ Subtraction: $\dfrac{P}{Q} - \dfrac{R}{Q} = \dfrac{P - R}{Q}$

EXAMPLE 1

Adding and Subtracting with Common Denominators

Perform each operation and simplify the results.

(a) $\dfrac{2t}{t+7} + \dfrac{14}{t+7}$ (b) $\dfrac{25}{5-y} - \dfrac{y^2}{5-y}$ (c) $\dfrac{3x-1}{x^2+x-12} - \dfrac{14-2x}{x^2+x-12}$

Solution

Since we have a common denominator, we add (subtract) the numerators and repeat the denominator. Thus,

(a) $\dfrac{2t}{t+7} + \dfrac{14}{t+7} = \dfrac{2t+14}{t+7} = \dfrac{2(t+7)}{t+7} = 2$ Factor and reduce

(b) $\dfrac{25}{5-y} - \dfrac{y^2}{5-y} = \dfrac{25-y^2}{5-y}$ Subtract the numerators

$\qquad\qquad = \dfrac{(5-y)(5+y)}{5-y} = 5+y$ Factor and simplify

The fraction bar is a grouping symbol so we subtract the entire numerator.

(c) $\dfrac{3x-1}{x^2+x-12} - \dfrac{14-2x}{x^2+x-12} = \dfrac{3x-1-(14-2x)}{x^2+x-12} = \dfrac{3x-1+2x-14}{x^2+x-12}$

$\qquad\qquad = \dfrac{5x-15}{x^2+x-12} = \dfrac{5(x-3)}{(x+4)(x-3)} = \dfrac{5}{x+4}$ Factor and simplify

Finding the Least Common Denominator

We can gain insight into adding or subtracting rational expressions with different denominators by looking at the *least common denominator* (LCD). Recall from section R4 that the **least common denominator** is the simplest denominator that is divisible by all the original denominators. Each expression in the sum or difference is then *built up* to an equivalent expression having the LCD as a denominator. The following strategy outlines how to identify the LCD.

Finding the LCD of Two Rational Expressions

1. **Factor** each denominator completely into a product of prime factors.
2. **List** each different prime factor to the highest power which appears in any factored denominator.
3. The LCD is the **product** of these prime factors.

EXAMPLE 2

Finding the Least Common Denominator

Find the LCD of each pair of rational expressions.

(a) $\dfrac{5}{3x}$ and $\dfrac{2}{x^2}$ (b) $\dfrac{11}{12x^2y}$ and $\dfrac{7}{18x^3y^2}$ (c) $\dfrac{4}{y^2-y}$ and $\dfrac{3}{y^2-1}$

Solution

(a) We factor each denominator into powers of primes.

$$3x = 3 \cdot x \quad \text{and} \quad x^2 = x \cdot x$$

We list each prime factor with its highest power. The LCD is the product of these factors.

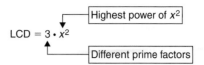

$$\text{LCD} = 3 \cdot x^2$$

Highest power of x^2

Different prime factors

(b) Factor each denominator into powers of prime factors for $\dfrac{11}{12x^2y}$ and $\dfrac{7}{18x^3y^2}$.

$$12x^2y = 2^2 \cdot 3x^2y \qquad \text{and} \qquad 18x^3y^2 = 2 \cdot 3^2x^3y^2$$

The different factors are 2, 3, x, and y with highest powers of 2^2, 3^2, x^3, and y^2.

$$\text{LCD} = 2^2 \cdot 3^2x^3y^2 = 36x^3y^2 = 36x^3y^2$$

(c) First factor the denominators of $\dfrac{4}{y^2 - y}$ and $\dfrac{3}{y^2 - 1}$. The following table helps find the LCD.

Denominators	Prime Factors		
$y^2 - y$	y	$y - 1$	
$y^2 - 1$		$y - 1$	$y + 1$
LCD	$y(y - 1)(y + 1)$		

The different factors are y, $y - 1$, and $y + 1$. Each is to the first power.

$$\text{LCD} = y\,(y - 1)(y + 1)$$

Adding and Subtracting Rational Expressions with Different Denominators

To add or subtract two fractions with different denominators, we first obtain a common denominator by using the fundamental principle of fractions so that they have a common denominator. The procedure used for obtaining a common denominator involves writing each expression as the product of prime factors. Thus, we combine rational expression with different denominators using the following steps:

Adding and Subtracting Rational Expression with Different Denominators

Recall: Building a rational expression means to multiply the numerator and denominator by the same factors.

1. Find the LCD of the rational expressions.
2. Rewrite each rational expression as an equivalent rational expression with the LCD. This is done by multiplying both the numerator and denominator of each rational expression by a factor needed to obtain the LCD.
3. Add or subtract the numerators while maintaining the LCD.
4. Factor, if possible, to simplify.

EXAMPLE 3   **Adding Rational Expressions**

Add the expressions.

(a) $\dfrac{7}{10x^2} + \dfrac{4}{15x}$ 　　(b) $\dfrac{1}{b-3} + \dfrac{b-2}{3-b}$ 　　(c) $\dfrac{5}{x^2(x+3)} + \dfrac{7}{x(x+3)^2}$

Solution   (a) Factoring the denominators, we have

$$10x^2 = 2 \cdot 5x^2 \text{ and } 15x = 3 \cdot 5x.$$

The LCD is $2 \cdot 3 \cdot 5x^2 = 30x^2$.

Now we write equivalent expressions with $30x^2$ as the denominator.

$$\frac{7}{10x^2} + \frac{4}{15x} = \frac{7}{2 \cdot 5x^2} \cdot \frac{3}{3} + \frac{4}{3 \cdot 5x} \cdot \frac{2x}{2x} \quad \text{Build each rational expression to the LCD}$$

$$= \frac{21}{30x^2} + \frac{8x}{30x^2}$$

$$= \frac{21 + 8x}{30x^2} \quad \text{Add, cannot be factored nor simplified}$$

(b) Recall that the signs in a rational expression can be changed without changing the value of the expression as follows:

$$\frac{b-2}{3-b} = \frac{-(b-2)}{-(3-b)} = \frac{2-b}{b-3}.$$

Thus, in changing the signs of the rational expression, we have two expressions with the same denominator.

$$\frac{1}{b-3} + \frac{b-2}{3-b} = \frac{1}{b-3} + \frac{2-b}{b-3}$$

$$= \frac{1+2-b}{b-3} = \frac{3-b}{b-3} = \frac{(-1)(b-3)}{b-3} = -1 \quad \text{Factor } 3-b = -1(b-3)$$

(c) The factors of these denominators are x and $x+3$. The highest power of x is 2 and of $x+3$ is also 2, so that the LCD is $x^2(x+3)^2$. Thus, we have:

$$\frac{5}{x^2(x+3)} + \frac{7}{x(x+3)^2} = \frac{5(x+3)}{x^2(x+3)(x+3)} + \frac{7x}{xx(x+3)^2}$$

$$= \frac{5(x+3) + 7x}{x^2(x+3)^2} \quad \text{Add numerators, repeat LCD}$$

$$= \frac{5x + 15 + 7x}{x^2(x+3)^2} \quad \text{Simplify numerator}$$

$$= \frac{12x + 15}{x^2(x+3)^2}$$

EXAMPLE 4 **Subtracting Rational Expressions**

Subtract the rational expressions.

(a) $\dfrac{x + 3}{x^2 - x - 2} - \dfrac{2x - 1}{x^2 + 2x - 8}$ (b) $\dfrac{x - 2}{2x^2 + 7x - 15} - \dfrac{-x - 1}{2x^2 - 5x + 3}$

Solution (a) First we factor each denominator and find the LCD using the table.

Denominators	Prime factors		
$x^2 - x - 2$	$x - 2$	$x + 1$	
$x^2 + 2x - 8$	$x - 2$		$x + 4$
LCD	$(x - 2)(x + 1)(x + 4)$		

The LCD of the fractions is $(x - 2)(x + 1)(x + 4)$, and we have:

$$\dfrac{x + 3}{x^2 - x - 2} - \dfrac{2x - 1}{x^2 + 2x - 8} = \dfrac{x + 3}{(x - 2)(x + 1)} - \dfrac{2x - 1}{(x - 2)(x + 4)} \quad \text{Factor denominators}$$

$$= \dfrac{x + 3}{(x - 2)(x + 1)} \cdot \dfrac{x + 4}{x + 4} - \dfrac{2x - 1}{(x - 2)(x + 4)} \cdot \dfrac{x + 1}{x + 1} \quad \text{Build fractions}$$

$$= \dfrac{(x + 3)(x + 4) - (2x - 1)(x + 1)}{(x - 2)(x + 1)(x + 4)} \quad \text{Subtract}$$

$$= \dfrac{(x^2 + 7x + 12) - (2x^2 + x - 1)}{(x - 2)(x + 1)(x + 4)}$$

$$= \dfrac{x^2 + 7x + 12 - 2x^2 - x + 1}{(x - 2)(x + 1)(x + 4)} \quad \text{Change signs}$$

$$= \dfrac{-x^2 + 6x + 13}{(x - 2)(x + 1)(x + 4)}.$$

(b) Factor the denominators and find the LCD.

Denominators	Prime Factors		
$2x^2 + 7x - 15$	$2x - 3$	$x + 5$	
$2x^2 - 5x + 3$	$2x - 3$		$x - 1$
LCD	$(2x - 3)(x + 5)(x - 1)$		

The LCD of the fractions is $(2x - 3)(x + 5)(x - 1)$, so we have:

$$\dfrac{x - 2}{2x^2 + 7x - 15} - \dfrac{-x - 1}{2x^2 - 5x + 3} = \dfrac{x - 2}{(2x - 3)(x + 5)} - \dfrac{-x - 1}{(2x - 3)(x - 1)}$$

$$= \dfrac{(x - 2)(x - 1) - (-x - 1)(x + 5)}{(2x - 3)(x + 5)(x - 1)} \quad \text{Rewrite with LCD}$$

$$= \dfrac{x^2 - 3x + 2 + x^2 + 6x + 5}{(2x - 3)(x + 5)(x - 1)} \quad \text{Simplify}$$

$$= \dfrac{2x^2 + 3x + 7}{(2x - 3)(x + 5)(x - 1)}.$$

EXAMPLE 5   **Adding and Subtracting More than Two Expressions**

Combine into a single rational expression and simplify.

$$\frac{x}{x+1} - \frac{x}{x-1} + \frac{2}{x^2-1}$$

Solution   The table shows how the LCD is obtained. Therefore, the LCD is $(x+1)(x-1)$ and we have

Denominators	Prime Factors	
$x+1$	$x+1$	
$x-1$		$x-1$
x^2-1	$x+1$	$x-1$
LCD	$(x+1)(x-1)$	

$$\frac{x}{x+1} - \frac{x}{x-1} + \frac{2}{x^2-1} = \frac{x}{x+1} - \frac{x}{x-1} + \frac{2}{(x-1)(x+1)}$$

Factor denominator

$$= \frac{x}{(x+1)} \cdot \frac{x-1}{x-1} - \frac{x}{(x+1)} \cdot \frac{x+1}{x+1} + \frac{2}{(x+1)(x-1)}$$

Build fractions

$$= \frac{x(x-1) - x(x+1) + 2}{(x+1)(x-1)}$$

Combine

$$= \frac{x^2 - x - x^2 - x + 2}{(x+1)(x-1)}$$

Simplify

$$= \frac{-2x+2}{(x-1)(x+1)}$$

$$= \frac{-2(x-1)}{(x-1)(x+1)} = \frac{-2}{x+1}.$$

Factor and reduce

If we let

$$f(x) = \frac{x}{x+1} - \frac{x}{x-1} + \frac{2}{x^2-1}$$

and

$$g(x) = -\frac{2}{x+1}$$

we can verify algebraically that $f(x) = g(x)$ for all values of x where each is defined. We observe that the domain of f consists of all real numbers except $x = -1$ and $x = 1$ since those values of x would make the denominators equal to zero. Whereas, the domain of g consists of all real numbers except $x = -1$ because -1 would make the denominator zero. Figures 1a and 1b show the graphs of f and g respectively.

Figure 1

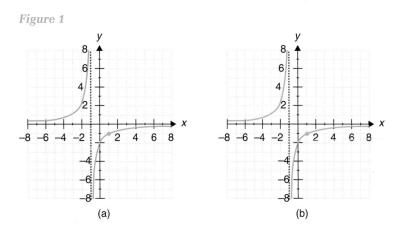

(a)   (b)

Notice that the shape of the graphs is exactly the same except for the hole in the graph of f at $x = 1$. Therefore, algebraically and graphically f and g are equivalent functions where they are defined.

Solving Applied Problems

In applied problems, denominators can often include units of measure. For example, in the formula

$$t = \frac{d}{r}$$

where t is time, d is distance, and r is rate. Rate can be measured in many different ways, miles per hour, feet per second, meters per minute, and many other combinations. In order to add or subtract fractions with such denominators, the units of measure must be the same (i.e., a common denominator).

EXAMPLE 6   **Finding Common Denominators Involving Dimensions**

Consider traveling x miles at 50 miles per hour and then $2x$ miles at 50 feet per second.

(a) Write a rational expression for total time, then simplify it.

(b) Evaluate the expression for $x = 12$ miles.

Solution   (a) First convert 50 feet per second to miles per hour so as to be consistent with the other units.

$$\frac{50 \text{ ft}}{1 \text{ sec}} = \left(\frac{50 \text{ ft}}{1 \text{ sec}}\right)\left(\frac{1 \text{ mi}}{5280 \text{ ft}}\right)\left(\frac{3600 \text{ sec}}{1 \text{ hr}}\right) \approx \frac{34 \text{ mi}}{1 \text{ hr}}$$

Using the formula, $t = \dfrac{d}{r}$, total time is

$$t = \frac{x}{50} + \frac{2x}{34}.$$

Denominators	Prime Factors		
50	2	5^2	
34	2		17
LCD	$2 \cdot 5^2 \cdot 17$		

Since the units are the same, the table helps us find the LCD $= 2 \cdot 5^2 \cdot 17 = 850$.

$$t = \frac{x \cdot 17}{50 \cdot 17} + \frac{2x \cdot 25}{34 \cdot 25} = \frac{17x + 50x}{850} = \frac{67x}{850}$$

(b) for $x = 12$ miles, we have $\dfrac{67(12)}{850} \approx 0.95$ hours

The total time is about 0.95 hours which is about 57 minutes.

PROBLEM SET 5.3

Mastering the Concepts

In Problems 1–44, perform each operation and simplify the results.

1. $\dfrac{5}{8x} + \dfrac{1}{8x}$

2. $\dfrac{7}{12t} + \dfrac{5}{12t}$

3. $\dfrac{6}{5x} - \dfrac{14}{5x}$

4. $\dfrac{3}{2y^2} - \dfrac{7}{2y^2}$

5. $\dfrac{5}{7x} + \dfrac{3}{7x}$

6. $\dfrac{2}{3t} + \dfrac{5}{3t}$

7. $\dfrac{7}{12t} - \dfrac{1}{12t}$

8. $\dfrac{15}{16y} - \dfrac{3}{16y}$

9. $\dfrac{t-1}{4k} + \dfrac{t+1}{4k}$

10. $\dfrac{3x-1}{5k} + \dfrac{2x+1}{5k}$

11. $\dfrac{2a}{a+3} + \dfrac{3}{a+3}$

12. $\dfrac{2m}{m-3} - \dfrac{6}{m-3}$

13. $\dfrac{9}{y-5} + \dfrac{6}{y-5}$

14. $\dfrac{12u}{4u+3} - \dfrac{-9}{4u+3}$

15. $\dfrac{2x}{x^2-4} - \dfrac{4}{x^2-4}$

16. $\dfrac{2t}{4t^2-9} - \dfrac{-3}{4t^2-9}$

17. $\dfrac{4}{16-x^2} + \dfrac{x}{16-x^2}$

18. $\dfrac{36}{6-v} + \dfrac{-v^2}{6-v}$

19. $\dfrac{14y-3}{2y^2+y-1} - \dfrac{2y+3}{2y^2+y-1}$

20. $\dfrac{2m+5}{m^2+m-2} + \dfrac{m+1}{m^2+m-2}$

21. $\dfrac{3x}{5y} + \dfrac{1}{7}$

22. $\dfrac{4}{x} - \dfrac{5}{3}$

23. $\dfrac{y}{4} + 3$

24. $y - \dfrac{1}{y}$

25. $\dfrac{a}{b^2} + \dfrac{a^3}{b^3} - \dfrac{1}{b}$

26. $\dfrac{1}{u} - \dfrac{t}{u} + \dfrac{t^2}{u^2}$

27. $\dfrac{5}{t^2} + \dfrac{3}{2t} + \dfrac{2}{3}$

28. $\dfrac{7}{4x^2} + \dfrac{5}{12x} + \dfrac{1}{6}$

29. $\dfrac{3}{x+2} + \dfrac{5}{x-1}$

30. $\dfrac{5}{x+3} - \dfrac{5}{3x}$

31. $\dfrac{9}{y-5} + \dfrac{6}{y-3}$

32. $\dfrac{2a}{a-3} + \dfrac{3}{a+7}$

33. $\dfrac{x}{3x+2} + \dfrac{1}{x-4}$

34. $\dfrac{2u}{4u+3} + \dfrac{u}{2u+5}$

35. $\dfrac{3}{x^2+x} + \dfrac{2}{x^2-1}$

36. $\dfrac{3v}{4v^2-1} + \dfrac{1}{2v+1}$

37. $\dfrac{m-5}{m^2-5m-6} + \dfrac{m+4}{m^2-6m}$

38. $\dfrac{3y}{y^2+7y+10} - \dfrac{y}{y^2+y-20}$

39. $\dfrac{3x+1}{2x^2-5x-12} - \dfrac{x+4}{6x^2+7x-3}$

40. $\dfrac{2}{a^2-4} + \dfrac{7}{a^2-4a-12}$

41. $\dfrac{c}{c^2-9} - \dfrac{c-1}{c^2-5c+6}$

42. $\dfrac{2t+3}{3t^2+2t-8} + \dfrac{3t+4}{2t^2+t-6}$

43. $\dfrac{2x}{2x^2-5x-12} - \dfrac{3x}{5x^2-18x-8}$

44. $\dfrac{x}{6x^2+13x+6} + \dfrac{2x}{4x^2+8x+3}$

In Problems 45–52, combine into a single rational expression and simplify.

45. $\dfrac{7}{4x^3y^2} + \dfrac{5}{12xy^3} - \dfrac{6}{6x^2y}$

46. $\dfrac{5}{x^2y} - \dfrac{3}{2xy^2} + \dfrac{8}{4x^3y^3}$

47. $\dfrac{2a}{a^2-1} + \dfrac{1}{a+1} - \dfrac{1}{a-1}$

48. $\dfrac{2}{u+3} - \dfrac{1}{u-3} + \dfrac{2u}{u^2-9}$

49. $\dfrac{x-3}{x+3} - \dfrac{x-3}{3-x} + \dfrac{x^2}{9-x^2}$

50. $\dfrac{v}{v+2} - \dfrac{v}{v-2} + \dfrac{v^2}{v^2-4}$

51. $\dfrac{2t^2-t}{3t^2-27} - \dfrac{t-3}{3t-9} + \dfrac{6t^2}{9-t^2}$

52. $\dfrac{4}{a^2-4} + \dfrac{2}{a^2-4a+4} + \dfrac{1}{a^2+4a+4}$

Applying the Concepts

53. Traveling Time: Consider traveling x miles at 60 miles per hour and then y miles at 88 feet per second.

(a) Write a rational expression for the total time, then simplify it.

(b) Evaluate the expression for $x = 700$ miles and $y = 1000$ miles.

54. Work Rate: You can do a job in x hours, while it takes your friend $x + 2$ hours to complete the same job. Write and simplify an expression for the part of the job you and your friend, working together, can complete in 1 hour.

55. Filling a Pool: An inlet pipe can fill a pool in x hours, while an outlet pipe can empty it in $x + 3$ hours. Write and simplify an expression for the part of the pool that is unfilled after 1 hour if both pipes are open.

56. Work Rate: A printing press can print a morning newspaper in x hours, a second press takes $x + 1$ hours, while a third press takes $x + 2$ hours to print it. In one hour the three presses working together can print

$$\frac{1}{x} + \frac{1}{x+1} + \frac{1}{x+2} \text{ of the job.}$$

(a) Combine and simplify this expression.

(b) If $x = 6$ hours, find the value of this expression.

57. Manufacturing: A manufacturer estimates that the revenue (in dollars) from the sale of x units of a certain product is given by the expression:

$$x^2 + 4x\left[4 - \frac{x}{16(x+4)}\right].$$

(a) Rewrite this expression as a single rational expression.

(b) What is the revenue if 100 units were produced and sold?

58. Electrical Circuit: If three resistors of resistances R_1, R_2, and R_3 are connected in parallel, then the combined resistance R is given by the formula:

$$\frac{1}{R} = \frac{1}{R_1} + \frac{1}{R_2} + \frac{1}{R_3}.$$

(a) Combine and simplify the right side of the formula if R_1 is x Ohms, R_2 is $x + 4$ Ohms and R_3 is $2x + 3$ Ohms.

(b) Find the value of R (in Ohms) if $x = 4$ Ohms.

In Problems 59–62, find (a) $(f + g)(x)$ and (b) $(f - g)(x)$.

59. $f(x) = \dfrac{x}{x^2-9}$ and $g(x) = \dfrac{3}{x^2+x-6}$

60. $f(x) = \dfrac{x}{x-1}$ and $g(x) = \dfrac{2x+2}{x^2-1}$

61. $f(x) = \dfrac{x}{x^2 - 4}$ and $g(x) = \dfrac{x - 3}{x^2 + 5x + 6}$

62. $f(x) = \dfrac{x - 1}{x^2 + x - 6}$ and $g(x) = \dfrac{x - 2}{x^2 + 4x + 3}$

Developing and Extending the Concepts

In Problems 63 and 64, use a calculator to evaluate the expressions

(i) $\dfrac{1}{x} + \dfrac{1}{y}$ and   (ii) $\dfrac{y + x}{xy}$ for the given values.

63. (a) $x = 1.5, y = 2.5$
    (b) $x = -0.12558, y = 1.0285$

64. (a) $x = 0.25, y = -1.25$
    (b) $x = -1.31, y = 1.25$

In Problems 65–68, combine into a single rational expression and simplify.

65. $\left(\dfrac{x}{y} + \dfrac{y}{x}\right)\left(\dfrac{x}{y} - \dfrac{y}{x}\right)$

66. $\left(\dfrac{1}{x + 2} - \dfrac{1}{x - 2}\right)\left(\dfrac{1}{x + 2} + \dfrac{1}{x - 2}\right)$

67. $\left(\dfrac{x^2 - 9}{x}\right)\left(\dfrac{1}{x + 3} + \dfrac{1}{x - 3}\right)$

68. $\left(\dfrac{x - 2}{x + 1} - \dfrac{x - 2}{x + 2}\right)\left(\dfrac{x^2 + 3x + 2}{x - 2}\right)$

69. In Calculus e^x can be approximated by the expression

$$1 + x + \dfrac{x^2}{2} + \dfrac{x^3}{6} + \dfrac{x^4}{24}.$$

(a) Find the value of this expression at $x = 1$.

(b) Simplify this expression and evaluate it at $x = 1$.

(c) Use a calculator to approximate e^1. How does this approximation compare to the values in parts (a) and (b)?

70. Identify the missing operation sign to make the expressions equal.

$$\dfrac{3x + 1}{x + 4} \ ? \ \dfrac{2x + 5}{x + 4} = \dfrac{x - 4}{x + 4}$$

71. (a) Write $\dfrac{1}{x} + \dfrac{1}{y}$ as a single rational expression.

(b) Evaluate the expression in part (a) for $x = 2$ and $y = 3$.

(c) Evaluate $\dfrac{2}{x + y}$ for $x = 2$ and $y = 3$.

(d) Are the two values in part (b) and (c) equal? Explain.

72. (a) Write $\dfrac{1}{x + 3} \cdot \dfrac{1}{y + 5}$ as a single rational expression.

(b) Evaluate the expression in part (a) for $x = 2$ and $y = 3$.

(c) Evaluate $\dfrac{1}{xy + 3y + 5x + 15}$ for $x = 2$ and $y = 3$.

(d) Are the two values in part (b) and (c) equal? Explain.

Objectives

1. Simplify Complex Fractions
2. Simplify Expressions with Negative Exponents
3. Solve Applied Problems

5.4 Complex Fractions

So far, we have worked with rational expressions of the form

$$\dfrac{P}{Q} \text{ where } Q \neq 0.$$

In this section, we look at a form of a fraction with rational expressions in its numerator and/or denominator. These fractions are referred to as *complex rational expressions*.

Definition–Complex Fraction

A **complex fraction** is one that has a rational expression in its numerator or its denominator, or both its numerator and denominator.

Examples of complex fractions are

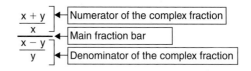

$$\frac{\frac{5}{7}}{3}, \quad \frac{\frac{2}{x}}{11}, \quad \frac{3 + \frac{3}{y}}{9}, \quad \frac{9}{\frac{w}{w-5}}, \quad \frac{\frac{5}{t}}{7 + \frac{2}{t}}, \quad \text{and} \quad \frac{p + \frac{3}{p}}{\frac{2}{2-p} - \frac{p}{2-p}}.$$

The parts of a complex fraction are:

$$\left. \begin{array}{c} \dfrac{x+y}{x} \\ \hline \dfrac{x-y}{y} \end{array} \right.$$

← Numerator of the complex fraction

← Main fraction bar

← Denominator of the complex fraction

Simplifying Complex Fractions

To simplify a complex fraction means to write the expression without a fraction in its numerator and its denominator. There are two methods to accomplish this.

Method 1 Simplifying Complex Fractions

1. Find the LCD of all denominators appearing within the complex fraction.
2. Multiply the numerator and the denominator of the complex fraction by the LCD found in step 1.
3. Simplify the resulting expression when possible.

EXAMPLE 1 **Simplifying a Complex Fraction**

Use method 1 to simplify the complex fraction $\dfrac{\frac{9}{x}}{\frac{12}{x^2}}$.

Solution The LCD of the numerator and the denominator is x^2.

$$\frac{\frac{9}{x}}{\frac{12}{x^2}} = \frac{\frac{9}{x}}{\frac{12}{x^2}} \cdot \frac{x^2}{x^2} \qquad \text{Multiply numerator and denominator by } x^2$$

$$= \frac{9x}{12}$$

$$= \frac{3 \cdot 3x}{3 \cdot 4}$$

$$= \frac{3x}{4} \qquad \text{Factor and reduce}$$

EXAMPLE 2   **Simplify a Complex Fraction**

Use method 1 to simplify the fraction $\dfrac{c - \dfrac{1}{d}}{1 - \dfrac{c}{d}}$.

Solution

$$\frac{c - \dfrac{1}{d}}{1 - \dfrac{c}{d}} = \frac{d}{d} \cdot \frac{\left(c - \dfrac{1}{d}\right)}{\left(1 - \dfrac{c}{d}\right)}$$

Multiply numerator and denominator by the LCD, d

$$= \frac{d(c) - d\left(\dfrac{1}{d}\right)}{d(1) - d\left(\dfrac{c}{d}\right)}$$

$$= \frac{cd - 1}{d - c}$$

Distribute and simplify

EXAMPLE 3   **Simplify a Complex Fraction**

Use method 1 to simplify the fraction $\dfrac{\dfrac{a}{b} - \dfrac{b}{a}}{\dfrac{1}{a} + \dfrac{1}{b}}$.

Solution   Multiply the numerator and denominator of the complex fraction by the LCD, ab.

$$\frac{\dfrac{a}{b} - \dfrac{b}{a}}{\dfrac{1}{a} + \dfrac{1}{b}} = \frac{\left(\dfrac{a}{b} - \dfrac{b}{a}\right)ab}{\left(\dfrac{1}{a} + \dfrac{1}{b}\right)ab}$$

Multiply by the LCD

$$= \frac{\left(\dfrac{a}{b}\right)ab - \left(\dfrac{b}{a}\right)ab}{\left(\dfrac{1}{a}\right)ab + \left(\dfrac{1}{b}\right)ab}$$

Distributive property

$$= \frac{a^2 - b^2}{b + a} = \frac{(a - b)(a + b)}{b + a} = a - b$$

Factor and reduce

Complex fractions can also be simplified by a second method as follows:

Method 2 Simplifying Complex Fractions

1. Express the complex fraction as a quotient of two rational expressions. Add or subtract expressions to get one rational expression in its numerator and/or denominator.
2. Find the quotient of the numerator and denominator by multiplying the numerator of the fraction by the reciprocal of its denominator.
3. Simplify when possible.

EXAMPLE 4   **Simplifying a Complex Fraction**

Use method 2 to simplify the fraction

$$\frac{\dfrac{1}{x-2}-\dfrac{1}{x-1}}{3+\dfrac{1}{x^2-3x+2}}.$$

Solution   Subtract the fractions in the numerator to get one rational expression in the numerator. Then add the fractions in the denominator to get one rational expression in the denominator. The common denominator of this complex fraction is $(x-2)(x-1)$.

$$\frac{\dfrac{1}{x-2}-\dfrac{1}{x-1}}{3+\dfrac{1}{x^2-3x+2}}=\frac{\dfrac{x-1}{(x-2)(x-1)}-\dfrac{x-2}{(x-2)(x-1)}}{\dfrac{3(x^2-3x+2)}{x^2-3x+2}+\dfrac{1}{x^2-3x+2}}$$

$$=\frac{\dfrac{x-1-(x-2)}{(x-2)(x-1)}}{\dfrac{3x^2-9x+6+1}{x^2-3x+2}}$$

$$=\frac{\dfrac{1}{(x-2)(x-1)}}{\dfrac{3x^2-9x+7}{x^2-3x+2}}$$

Multiply by the reciprocal of the denominator and then reduce.

$$=\frac{1}{x^2-3x+2}\div\frac{3x^2-9x+7}{x^2-3x+2}=\frac{1}{\cancel{x^2-3x+2}}\cdot\frac{\cancel{x^2-3x+2}}{3x^2-9x+7}=\frac{1}{3x^2-9x+7}$$

EXAMPLE 5   **Evaluating the Difference Quotient**

Let $f(x)=\dfrac{3}{x}$. Find and simplify $\dfrac{f(a+h)-f(a)}{h}$.

Solution $\dfrac{f(a+h)-f(a)}{h}=\dfrac{\dfrac{3}{a+h}-\dfrac{3}{a}}{h}$ ⠀⠀⠀Substituted $f(a+h)=\dfrac{3}{a+h}$ and $f(a)=\dfrac{3}{a}$

$$=\frac{\dfrac{3a}{a(a+h)}-\dfrac{3(a+h)}{a(a+h)}}{h}$$ ⠀⠀⠀$a(a+h)$ is the LCD

$$=\frac{\dfrac{3a-3(a+h)}{a(a+h)}}{h}=\frac{\dfrac{-3h}{a(a+h)}}{h}$$ ⠀⠀⠀Simplify numerator

$$=\frac{-3\cancel{h}}{a(a+h)}\cdot\frac{1}{\cancel{h}}$$

$$=\frac{-3}{a(a+h)}$$ ⠀⠀⠀Reduce

Consider the function.

$$f(x) = \frac{x + \dfrac{3x}{x - 3}}{1 + \dfrac{9}{x^2 - 9}} = \frac{x + \dfrac{3x}{x - 3}}{1 + \dfrac{9}{(x - 3)(x + 3)}}$$

$$= \frac{(x - 3)(x + 3)\left(x + \dfrac{3x}{x - 3}\right)}{(x - 3)(x + 3)\left(1 + \dfrac{9}{(x - 3)(x + 3)}\right)}$$

$$= \frac{x^3 - 9x + 3x^2 + 9x}{x^2 - 9 + 9} = \frac{x^3 + 3x^2}{x^2} = x + 3$$

Figure 1a displays the graph of the function

$$f(x) = \frac{x + \dfrac{3x}{x - 3}}{1 + \dfrac{9}{x^2 - 9}}$$

whereas Figure 1b displays the graph of the resulting function $y = x + 3$.

Figure 1

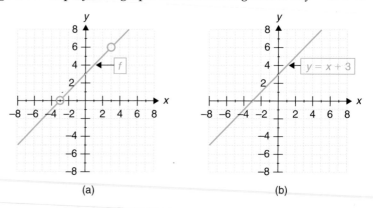

(a)   (b)

Figure 1 gives a visual support for the argument that $f(x)$ is equivalent to $x + 3$ everywhere $f(x)$ is defined. Note that there are two holes in the graph in Figure 1a at $x = -3$ and at $x = 3$. The function f is not defined for these values.

Simplifying Expressions with Negative Exponents

Recall from the properties of exponents that

$$\frac{a^m}{a^n} = a^{m-n}.$$

If we allow m to equal n, then we have

$$\frac{a^m}{a^m} = a^{m-m} = a^0.$$

But we know that $\dfrac{a^m}{a^m} = 1$.

Comparing the two expressions, we have the definition of the **zero exponent.**

Definition of Zero Exponent

$$a^0 = 1 \text{ for any real number } a \neq 0$$

For example,

$$7^0 = 1, \quad (12)^0 = 1, \quad (x + 3)^0 = 1, \quad \text{and} \quad (x^2y^3)^0 = 1.$$

Recall also from the properties of exponents that

$$a^m \cdot a^n = a^{m+n}.$$

What happens if we allow one of the exponents to be negative and apply the above property? For example, if $m = 4$ and $n = -4$. Then

$$a^m \cdot a^n = a^4 \cdot a^{-4} = a^{4 + (-4)} = a^0 = 1$$

so $\qquad a^4 \cdot a^{-4} = 1.$

That is, a^{-4} must be the multiplicative inverse or reciprocal of a^4. In symbols we write,

$$a^{-4} = \frac{1}{a^4}.$$

This reasoning leads to the following definition.

Negative Exponent

$$a^{-n} = a^{-n} = \frac{1}{a^n} \text{ for } a \text{ non zero and } n \text{ a positive integer.}$$

If n is a positive integer and $a \neq 0$, then

$$\frac{1}{a^{-n}} = \frac{1}{\dfrac{1}{a^n}} = 1 \div \frac{1}{a^n} = a^n.$$

EXAMPLE 7

Using the Properties of Exponents

Rewrite each expression without negative exponents.

(a) y^{-5} (b) $(-4)^{-3}$ (c) $\dfrac{1}{x^{-7}}$ (d) $\left(\dfrac{3}{4}\right)^{-2}$

Solution

Here we restrict all variables so that they represent non zero real numbers.

(a) $y^{-5} = \dfrac{1}{y^5}$ (b) $(-4)^{-3} = \dfrac{1}{(-4)^3} = \dfrac{1}{-64} = -\dfrac{1}{64}$

(c) $\dfrac{1}{x^{-7}} = x^7$ (d) $\left(\dfrac{3}{4}\right)^{-2} = \dfrac{1}{\left(\dfrac{3}{4}\right)^2} = \left(\dfrac{4}{3}\right)^2 = \dfrac{16}{9}$

EXAMPLE 8   **Simplifying an Expression with Negative Exponents**

Rewrite the expression with only positive exponents and simplify the result.

$$\frac{(x + h)^{-1} - x^{-1}}{x^{-1}}$$

Solution   First we rewrite the expression with positive exponents

$$\frac{(x + h)^{-1} - x^{-1}}{x^{-1}} = \frac{\dfrac{1}{x + h} - \dfrac{1}{x}}{\dfrac{1}{x}}.$$

Multiply both the numerator and denominator of this expression by the LCD: $x(x + h)$.

$$\frac{\dfrac{1}{x + h} - \dfrac{1}{x}}{\dfrac{1}{x}} = \frac{\left[\dfrac{1}{x + h} - \dfrac{1}{x}\right] \cdot x(x + h)}{\left[\dfrac{1}{x}\right] \cdot x(x + h)}$$

$$= \frac{x(x + h)\left[\dfrac{1}{x + h}\right] - x(x + h)\left[\dfrac{1}{x}\right]}{x(x + h)\left[\dfrac{1}{x}\right]} \qquad \text{Distributive property}$$

$$= \frac{x - (x + h)}{x + h} = \frac{-h}{x + h} \qquad \text{Simplified}$$

Solving Applied Problems

Sometimes, formulas contain rational expressions as the next example shows.

EXAMPLE 9   **Solving an Electronic Problem**

In electronics, a resistor is any device that offers resistance to the flow of electrical current. When a light bulb is on, it acts as a resistor. Suppose that two light bulbs are placed in a parallel circuit for a household. The total resistance R (in Ohms) is given by the formula

$$R = \frac{1}{R_1^{-1} + R_2^{-1}}$$

where R_1 and R_2 represent the resistance of each light bulb.

(a) Simplify the complex rational expression.

(b) Find the total resistance R when $R_1 = 3$ Ohms and $R_2 = 4$ Ohms.

Solution   (a) The total resistance R can be written as follows:

$$R = \frac{1}{\dfrac{1}{R_1} + \dfrac{1}{R_2}} = \frac{1(R_1 R_2)}{\left[\dfrac{1}{R_1} + \dfrac{1}{R_2}\right](R_1 R_2)} \qquad \text{Multiply by the LCD, } R_1 R_2$$

$$= \frac{R_1 R_2}{(R_1 R_2)\dfrac{1}{R_1} + (R_1 R_2)\dfrac{1}{R_2}} \qquad \text{Distributive property}$$

$$= \frac{R_1 R_2}{R_2 + R_1} \qquad \text{Simplify}$$

(b) Substituting 3 for R_1 and 4 for R_2 we have

$$R = \frac{3(4)}{4 + 3}$$

$$= \frac{12}{7} \text{ Ohms.}$$

Negative exponents and complex rational expressions can be used to solve applied problems including *installment loans for cars* or *mortgages* as well as electronics problems. The formula

$$m = \frac{Pr}{12\left[1-\left(1 + \frac{r}{12}\right)^{-12t}\right]}$$

is used to calculate the equal *monthly car* or *home mortgage payments, m,* required to pay off a borrowed *principal* amount of *P* dollars, at an annual *interest rate* of *r,* paid over a *term* of *t years.* The *total amount of interest, I,* charged (in dollars) is the difference between the total amount paid and the principal owed, *P.* This is given by the formula

$$I = 12mt - P.$$

EXAMPLE 10   **Determine a Car Payment**

Suppose you want to buy a car and need to borrow $8,000 from the bank. The bank charges an annual interest rate of 6.5% on automobile loans with monthly payments over a term of 36 months (3 years). Calculate

(a) the monthly payment and

(b) the total interest charged.

Solution   (a) Substituting $P = 8,000$, $r = 0.065$, and $t = 3$ into the monthly car payment formula, we have

$$m = \frac{Pr}{12\left[1-\left(1 + \frac{r}{12}\right)^{-12t}\right]}$$

$$= \frac{8000(0.065)}{12\left[1-\left(1 + \frac{0.065}{12}\right)^{-12(3)}\right]}$$

$$= 245.19.$$

Therefore, the monthly car payment is $245.19.

(b) Substitute $m = 245.19$ and $t = 3$ in the formula

$$I = 12mt - P \text{ we get}$$

$$I = 12(245.19)(3) - 8000 = 826.84$$

Therefore, the total interest charged is $826.84.

◆ PROBLEM SET 5.4

Mastering the Concepts

In Problems 1–10, simplify each rational expression by rewriting the expression as a simple rational expression reduced to lowest terms. Use method 1.

1. (a) $\dfrac{\dfrac{2}{3x}}{\dfrac{x}{2y}}$ (b) $\dfrac{\dfrac{x}{y}}{\dfrac{z}{y}}$

2. (a) $\dfrac{\dfrac{3}{8x}}{\dfrac{5}{12x}}$ (b) $\dfrac{\dfrac{4}{15y}}{\dfrac{12}{5y}}$

3. (a) $\dfrac{\dfrac{-1}{5b}}{\dfrac{2a}{3b}}$ (b) $\dfrac{\dfrac{-x}{7}}{\dfrac{y}{14}}$

4. (a) $\dfrac{\dfrac{-xy}{t}}{\dfrac{yz}{x}}$ (b) $\dfrac{\dfrac{4}{x^2}}{\dfrac{y}{x}}$

5. (a) $\dfrac{\dfrac{-y}{5x}}{\dfrac{1}{5x^2}}$ (b) $\dfrac{\dfrac{3x}{11y}}{\dfrac{-1}{22y^2}}$

6. (a) $\dfrac{\dfrac{x^2y^2}{x}}{\dfrac{5xy^3}{16}}$ (b) $\dfrac{\dfrac{5ab^3}{16}}{\dfrac{a^2b^2}{8}}$

7. $\dfrac{\dfrac{1}{1+a}-1}{3a}$

8. $\dfrac{2-\dfrac{1}{y}}{y-1}$

9. $\dfrac{3+\dfrac{1}{x}}{2-\dfrac{3}{x}}$

10. $\dfrac{5+\dfrac{1}{x}}{1-\dfrac{3}{x}}$

In Problems 11–18, simplify each rational expression by rewriting the expression as a simple expression reduced to lowest terms. Use method 2.

11. (a) $\dfrac{\dfrac{2}{3x}}{\dfrac{4}{5x}}$ (b) $\dfrac{\dfrac{17y}{6}}{\dfrac{34y}{9}}$

12. (a) $\dfrac{\dfrac{x}{y}}{x}$ (b) $\dfrac{x}{\dfrac{x}{y}}$

13. (a) $\dfrac{\dfrac{-a}{5b^2}}{\dfrac{6a^2}{2y}}$ (b) $\dfrac{\dfrac{1}{b^2}}{\dfrac{3}{b}}$

14. (a) $\dfrac{\dfrac{x^2}{y^2}}{\dfrac{y}{z}}$ (b) $\dfrac{\dfrac{-x^2}{y}}{\dfrac{-y^2}{z}}$

15. $\dfrac{x-\dfrac{y}{x}}{x-\dfrac{x}{y}}$

16. $\dfrac{\dfrac{x}{y}-3}{\dfrac{x}{z}-5}$

17. $\dfrac{\dfrac{1}{5}-\dfrac{1}{x}}{x-5}$

18. $\dfrac{4-\dfrac{x}{y}}{1-\dfrac{4y}{x}}$

In Problems 19–36, simplify each rational expression by any appropriate method.

19. $\dfrac{\dfrac{2}{x}+\dfrac{7}{y}}{\dfrac{8}{x}-\dfrac{6}{y}}$ *$\dfrac{2y+7y}{8y-6x}$*

20. $\dfrac{\dfrac{x}{3}-\dfrac{y}{4}}{\dfrac{y}{5}-\dfrac{x}{6}}$

21. $\dfrac{1+\dfrac{2}{m}}{1-\dfrac{4}{m^2}}$

22. $\dfrac{\dfrac{9}{y^2}-4}{\dfrac{3}{y}-2}$

23. $\dfrac{\dfrac{c}{2d}-\dfrac{3d}{6c}}{\dfrac{6}{3c}-\dfrac{2}{d}}$

24. $\dfrac{\dfrac{2}{1+v}}{3+\dfrac{1}{1+v}}$

25. $\dfrac{\dfrac{1}{1+t}-2}{3-\dfrac{1}{1+t}}$

26. $\dfrac{\dfrac{x^2-y^2}{x+y}}{1+\dfrac{x}{y}}$

27. $\dfrac{1+x-\dfrac{1}{1-x}}{1-x-\dfrac{1}{1+x}}$

28. $\dfrac{y+x+\dfrac{y^2}{x-y}}{x-y+\dfrac{y^2}{y+x}}$

29. $\dfrac{\dfrac{u}{v}+1+\dfrac{v}{u}}{\dfrac{u^3-v^3}{uv}}$

30. $\dfrac{\dfrac{a}{b}-1+\dfrac{b}{a}}{\dfrac{a^3+b^3}{a^2b+ab^2}}$

31. $\dfrac{\dfrac{3}{1-x}+\dfrac{x}{x-1}}{\dfrac{1}{x-1}}$

32. $\dfrac{\dfrac{a}{b}-2+\dfrac{b}{a}}{\dfrac{a}{b}+2+\dfrac{b}{a}}$

33. $\dfrac{\dfrac{2}{t}+\dfrac{1}{1+t}}{\dfrac{3}{t+1}-\dfrac{1}{t}}$

34. $\dfrac{\dfrac{x^2-y^2}{x+y}}{1-\dfrac{x}{y}}$

35. $\dfrac{m+\dfrac{n}{n-m}}{n+\dfrac{m}{m-n}}$

36. $\dfrac{\dfrac{1}{a+b}-\dfrac{1}{a-b}}{\dfrac{4}{a^2-b^2}}$

In Problems 37–46, rewrite each expression as a complex rational expression, then simplify. Answers should have only positive exponents.

37. (a) $\dfrac{y^{-4}}{y^{-2}}$ (b) $\dfrac{x^4}{x^{-2}}$ **38.** (a) $\dfrac{1}{y^{-3}}$ (b) $\dfrac{\dfrac{1}{x^{-2}}}{\dfrac{3}{x^{-3}}}$

39. $\dfrac{8}{2^{-2} - 4^{-2}}$ **40.** $\dfrac{x^{-1} + y^{-1}}{xy}$

41. $\dfrac{a^{-1}b^{-2}}{a + b}$ **42.** $(a^{-1} + b^{-1})^{-1}$

43. $\dfrac{x^{-2} - y^{-2}}{x^{-1} - y^{-1}}$ **44.** $\dfrac{x^{-1} + y^{-1}}{x^{-2} - y^{-2}}$

45. $\dfrac{(x + 1)^{-1} + 1}{3(x + 1)^{-1}}$ **46.** $\dfrac{1 - 4(1 - x)^{-1}}{2 + 8(x - 1)^{-1}}$

In Problems 47–50, find and simplify

$$\frac{f(a + h) - f(a)}{h}.$$

47. $f(x) = \dfrac{3}{x}$ **48.** $f(x) = \dfrac{2}{x + 1}$

49. $f(x) = \dfrac{-3}{x - 1}$ **50.** $f(x) = \dfrac{5}{x^2}$

Applying the Concepts

51. Dimensional Analysis: Simplify each expression.

(a) $\dfrac{\dfrac{70 \text{ miles}}{35 \text{ miles}}}{1 \text{ hour}}$ (b) $\dfrac{\dfrac{100 \text{ miles}}{20 \text{ miles}}}{2 \text{ hours}}$

(c) $\dfrac{\dfrac{140 \text{ days}}{7 \text{ days}}}{1 \text{ week}}$ (d) $\dfrac{\dfrac{\dfrac{45 \text{ miles}}{1 \text{ hour}}}{150 \text{ gallons}}}{1 \text{ hour}}$

52. Photography: A camera lens has a characteristic measurement, f, called its *focal length*. When an object is in focus, its distance from the lens is p and the distance from the lens to the film (or image) is q. The symbols p, q, and f are related by the equation:

$$f = \frac{1}{\dfrac{1}{p} + \dfrac{1}{q}}.$$

(a) If $p = x + 10$ units and $q = x - 2$ units, simplify the complex rational expression.

(b) If $p = 36$ centimeters and $q = 12$ centimeters, find f.

53. Car Payments: You are buying a car and need to borrow $16,000 from the credit union at an annual interest of 10.8%.

(a) What is the monthly payment over a term of 4 years?

(b) What is the total interest charged?

54. Mortgage Payments: A home is sold for $180,000 in which 5/9 of its price is mortgaged at an annual interest rate of 8.5%.

(a) What is the monthly payment on the mortgage over a period of 30 years?

(b) What is the total interest charged?

Developing and Extending the Concepts

In Problems 55 and 56, simplify each rational expression.

55. $\dfrac{\dfrac{x - 1}{x + 1} - \dfrac{x + 1}{x - 1}}{\dfrac{x - 1}{x + 1} + \dfrac{x + 1}{x - 1}}$ **56.** $\dfrac{\dfrac{2u - v}{2u + v} + \dfrac{2u + v}{2u - v}}{\dfrac{2u - v}{2u + v} - \dfrac{2u + v}{2u - v}}$

In Problems 57 and 58, rewrite each expression as a complex rational expression and simplify.

57. $\dfrac{2y^{-2} - y^{-1} - 1}{2y^{-2} - 3y^{-1} - 2}$

58. $\dfrac{x^{-3} + y^{-3}}{x^{-1} + y^{-1}} \cdot \dfrac{8}{x^{-2} - x^{-1}y^{-1} + y^{-2}}$

In Problems 59 and 60, simplify each expression.

59. (a) $1 + \dfrac{1}{2 + \dfrac{1}{2}}$

(b) $1 + \dfrac{1}{2 + \dfrac{1}{2 + \dfrac{1}{2}}}$

(c) $1 + \dfrac{1}{2 + \dfrac{1}{2 + \dfrac{1}{2 + \dfrac{1}{2}}}}$

(d) $1 + \dfrac{1}{x + \dfrac{1}{x + \dfrac{1}{x + \dfrac{1}{x}}}}$

60. (a) $1 + \dfrac{1}{1 + \dfrac{1}{2 + 2}}$

(b) $1 + \dfrac{1}{1 + \dfrac{1}{2 + \dfrac{2}{1 + 1}}}$

(c) $1 + \dfrac{1}{1 + \dfrac{1}{2 + \dfrac{2}{1 + \dfrac{1}{2}}}}$

(d) $1 + \dfrac{1}{1 + \dfrac{1}{x + \dfrac{x}{1 + \dfrac{1}{x}}}}$

In Problems 61 and 62, find and simplify $\dfrac{f(a + h) - f(a)}{h}$.

61. $f(x) = \dfrac{-7}{(x - 1)^2}$

62. $f(x) = \dfrac{2}{(x + 3)^2}$

Objectives

1. Solve Equations Involving Rational Expressions
2. Solve Applied Problems

5.5 Equations Involving Rational Expressions

In section 1.1 we presented techniques for solving equations containing fractions whose denominators did not contain variables. The technique was to multiply each side of the equation by the LCD of the fractions. In this section, we will proceed in a similar way to solve equations containing rational expressions with variables in the denominator.

Solving Equations Involving Rational Expressions

The strategy used in section 1.1 allows us to solve equations involving variables in the denominators such as

$$\frac{1}{2x} + \frac{2}{x} = \frac{5}{4} \quad \text{and} \quad \frac{1}{x - 1} = \frac{2}{x + 1}.$$

The following guidelines are helpful in solving equations involving rational expressions.

Strategy for Solving Rational Expressions

1. Eliminate values of the variable which make any of the denominators zero.
2. Find the LCD of all rational expressions in the equation.
3. Multiply each side of the equation by the LCD to eliminate fractions.
4. Remove parentheses (if any) and combine all like terms on each side of the equation.
5. Solve the transformed equation.
6. Check the solution in the *original* equation.

Whenever a variable appears in any denominator, we must *check* the solution to avoid values of the variable that may cause the denominators of the original equation to be zero. Division by zero is not defined and such values are called **extraneous** solutions.

EXAMPLE 1   **Solving an Equation Containing Rational Expressions**

Solve the equation $\dfrac{4}{5t} + 2 = \dfrac{4}{t} + \dfrac{8}{5}$.

Solution

$$\frac{4}{5t} + 2 = \frac{4}{t} + \frac{8}{5}$$

Note that $t \neq 0$ for the expressions to be defined.

$$5t\left(\frac{4}{5t} + 2\right) = 5t\left(\frac{4}{t} + \frac{8}{5}\right)$$

Multiply each side by $5t$, the LCD

$$5t\left(\frac{4}{5t}\right) + 5t(2) = 5t\left(\frac{4}{t}\right) + 5t\left(\frac{8}{5}\right)$$

Distributive property

$$4 + 10t = 20 + 8t$$

Simplify each side

$$4 + 2t = 20$$

$$2t = 16$$

$$t = 8$$

Check:

t	Left Side	Right Side
8	$\dfrac{4}{5(8)} + 2 = \dfrac{4}{40} + 2 = \dfrac{1}{10} + 2 = 2.1$	$\dfrac{4}{8} + \dfrac{8}{5} = \dfrac{1}{2} + \dfrac{8}{5} = \dfrac{5}{10} + \dfrac{16}{10} = \dfrac{21}{10}$ or 2.1

Therefore, 8 is the solution.

EXAMPLE 2   **Solving an Equation Containing Rational Expressions**

Solve the equation $\dfrac{x}{2x - 1} = \dfrac{2}{3}$.

Solution

$$\frac{x}{2x - 1} = \frac{2}{3}$$

Note $2x - 1 \neq 0$ so $x \neq \dfrac{1}{2}$

$$3(2x - 1)\frac{x}{2x - 1} = 3(2x - 1)\frac{2}{3}$$

Multiply each side by $3(2x-1)$, the LCD

$$3x = 2(2x - 1)$$

Cancel common factors.

$$3x = 4x - 2$$

$$-x = -2$$

$$x = 2$$

Multiply both sides by -1

Check:

x	Left Side	Right Side
2	$\dfrac{2}{2(2) - 1} = \dfrac{2}{4 - 1} = \dfrac{2}{3}$	$\dfrac{2}{3}$

Therefore, 2 is the solution.

When two fractions are equal as in Example 2, *cross multiplication* can often be used to solve the equation. Thus,

$$\frac{x}{2x - 1} = \frac{2}{3}$$

has the same solution as the cross product

$$3x = 2(2x - 1)$$

provided $x \neq \frac{1}{2}$.

EXAMPLE 3 **Solving Equations Involving Rational Expressions**

Solve the equation $\dfrac{8}{x^2 - 9} - \dfrac{1}{x + 3} = \dfrac{1}{x - 3}$.

Solution Factor $x^2 - 9$ and find the LCD to be $(x - 3)(x + 3)$. The restricted values are $x \neq 3, -3$.

$$\frac{8}{(x - 3)(x + 3)} - \frac{1}{x + 3} = \frac{1}{x - 3}$$

$$(x - 3)(x + 3)\left[\frac{8}{(x - 3)(x + 3)} - \frac{1}{x + 3}\right] = \left[\frac{1}{x - 3}\right](x - 3)(x + 3) \qquad \text{Multiply by the LCD}$$

$$(x - 3)(x + 3)\left[\frac{8}{(x - 3)(x + 3)}\right] - (x - 3)(x + 3)\left[\frac{1}{x + 3}\right] = x + 3 \qquad \text{Distributive property}$$

$$8 - (x - 3) = x + 3 \qquad \text{Simplify}$$

$$8 - x + 3 = x + 3$$

$$-2x = -8$$

$$x = 4$$

A check in the original equation will show that 4 is the solution.

EXAMPLE 4 **Solving an Equation Involving Rational Expressions**

Solve the equation

$$\frac{4}{2p - 1} - \frac{3}{p + 3} = \frac{2p + 13}{2p^2 + 5p - 3}.$$

Solution The LCD of the fractions is $(2p - 1)(p + 3)$. The restricted values are $p \neq \frac{1}{2}, -3$.

$$\frac{4}{2p - 1} - \frac{3}{p + 3} = \frac{2p + 13}{(2p - 1)(p + 3)}$$

$$(2p - 1)(p + 3)\left[\frac{4}{2p - 1} - \frac{3}{p + 3}\right] = (2p - 1)(p + 3)\left[\frac{2p + 13}{(2p - 1)(p + 3)}\right] \qquad \text{Multiply by the LCD}$$

$$(2p - 1)(p + 3)\left[\frac{4}{2p - 1}\right] - (2p - 1)(p + 3)\left[\frac{3}{p + 3}\right] = 2p + 13 \qquad \text{Distributive property}$$

$$4(p + 3) - 3(2p - 1) = 2p + 13 \qquad \text{Reduce}$$

$$4p + 12 - 6p + 3 = 2p + 13$$

$$-2p + 15 = 2p + 13$$

$$-4p = -2$$

$$p = \frac{1}{2}$$

Check: Our apparent solution is 1/2, but 1/2 is a restricted value as we noted at the beginning. If p is replaced by 1/2 in the original equation the denominator on each side has a value of zero which is meaningless.

$$\frac{4}{2p-1} - \frac{3}{p+3} = \frac{2p+13}{2p^2+5p-3}$$

$$\frac{4}{2\left(\frac{1}{2}\right)-1} - \frac{3}{\frac{1}{2}+3} = \frac{2\left(\frac{1}{2}\right)+13}{2\left(\frac{1}{2}\right)^2+5\left(\frac{1}{2}\right)-3}$$

$$\frac{4}{0} - \frac{3}{\frac{1}{2}+3} = \frac{14}{0}$$

Therefore, 1/2 is not a value which makes the equation true. In fact, it makes the rational expressions undefined. Hence, the original equation has *no solution* and 1/2 is called an **extraneous solution.**

EXAMPLE 5   **Solving an Equation Involving Rational Functions**

Consider the function f defined by

$$f(x) = x + \frac{5}{x}.$$

Find all values of x so that $f(x) = -6$.

Solution   Since $f(x) = x + \frac{5}{x}$, we need to find all values of x so that

$$x + \frac{5}{x} = -6.$$

$$x\left[x + \frac{5}{x}\right] = -6(x) \qquad \text{Multiply each side by the LCD, } x$$

$$x^2 + 5 = -6x$$

$$x^2 + 6x + 5 = 0$$

$$(x+1)(x+5) = 0$$

$$x+1 = 0 \quad \text{or} \quad x+5 = 0$$

$$x = -1 \text{ or} \qquad x = -5$$

Figure 1

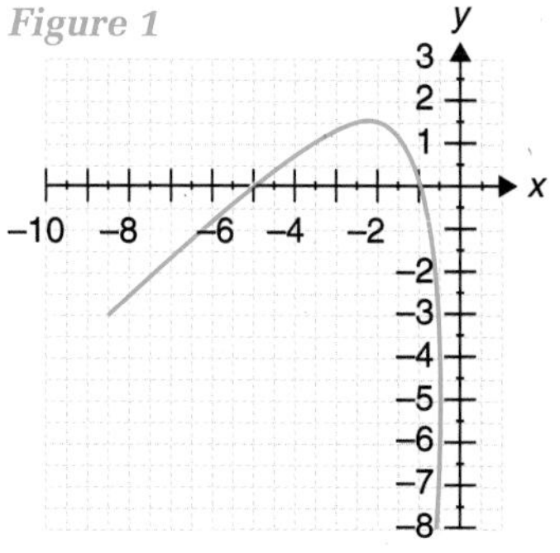

A check will show that these values satisfy the equation. Figure 1 shows that -5 and -1 are the x intercepts of the graph of

$$g(x) = x + \frac{5}{x} + 6$$

which confirms the solution of the above equation.

Solving Applied Problems

Recall that a **ratio** is the quotient of two numbers. For example,

The "fraction 3/5" is read as "the *ratio* of 3 to 5".

An equation in which two ratios are equal is called a **proportion.** Examples are:

$$\frac{5}{7} = \frac{15}{21} \quad \text{and} \quad \frac{4}{13} = \frac{20}{65}.$$

Proportions are a type of rational equation and can be solved by *cross multiplication.* That is,

Cross Multiplication

> If $\frac{a}{b} = \frac{c}{d}$, then $ad = bc$, $b \neq 0$ and $d \neq 0$.

EXAMPLE 6 **Solving a Proportion Problem**

Solve for x $\quad \dfrac{x + 1}{x + 10} = \dfrac{x - 2}{x + 4}.$

Solution The equation is not defined for $x = -10$ or -4. By cross multiplication, the proportion

$$\frac{x + 1}{x + 10} = \frac{x - 2}{x + 4}$$

is equivalent to

$$(x + 1)(x + 4) = (x + 10)(x - 2)$$
$$x^2 + 5x + 4 = x^2 + 8x - 20$$
$$5x - 8x = -4 - 20$$
$$-3x = -24$$
$$\boxed{x = 8.}$$

Check:

x	Left Side	Right Side
8	$\dfrac{x + 1}{x + 10} = \dfrac{8 + 1}{8 + 10} = \dfrac{9}{18} = \dfrac{1}{2}$	$\dfrac{x - 2}{x + 4} = \dfrac{8 - 2}{8 + 4} = \dfrac{6}{12} = \dfrac{1}{2}$

In geometry, proportions point out important facts about *similar triangles.* Two triangles are said to be *similar* if all pairs of corresponding sides are in proportion and corresponding angles are equal. For example, in Figure 2 the two triangles *ABC* and *DEF* are similar, so that

Figure 2

$$\frac{AB}{DE} = \frac{AC}{DF} = \frac{BC}{EF}$$

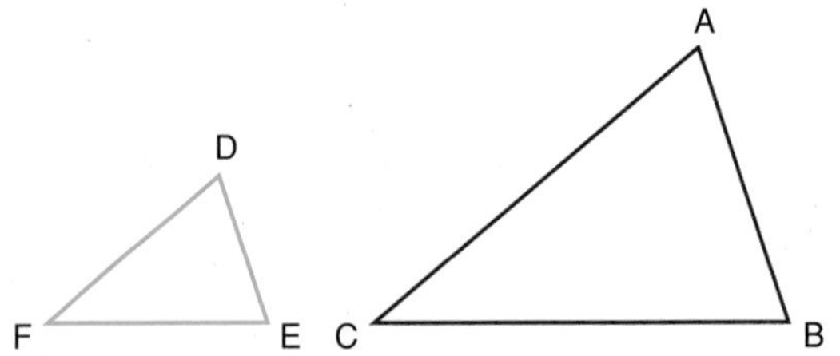

EXAMPLE 7 **Using Similar Triangles**

An evergreen casts a shadow of 29 meters and at the same time a 3 meter bush casts a shadow of 2.5 meters. Find the height of the evergreen.

Solution   Figure 3 shows two similar right triangles determined by the tree with its shadow and the bush with its shadow. Let h (in meters) be the height of the tree. The triangles are similar and thus the sides are proportional so we write:

Figure 3

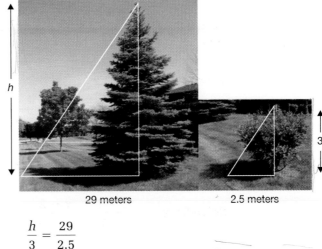

29 meters   2.5 meters

$$\frac{h}{3} = \frac{29}{2.5}$$

$$2.5h = 3(29) \qquad \text{Cross multiplication}$$

$$h = \frac{3(29)}{2.5} = 34.8$$

The tree is 34.8 meters tall.

Many applied problems in electronics and optics use equations and formulas involving rational expressions, as the next example shows.

EXAMPLE 8

Figure 4

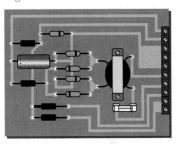

Finding the Resistance of a Circuit

In an electric circuit, the total resistance, R, of two resistors connected in parallel is determined by the formula

$$\frac{1}{R} = \frac{1}{R_1} + \frac{1}{R_2}$$

where R_1 and R_2 are the resistances (measured in Ohms) of the resistors (Figure 4).

(a) Solve this formula for R.

(b) If the total resistance R in a circuit is 60 Ohms and R_1 has twice the resistance as R_2, what is the resistance of each resistor?

Solution   (a)

$$\frac{1}{R} = \frac{1}{R_1} + \frac{1}{R_2} \qquad \text{LCD is } R \cdot R_1 \cdot R_2$$

$$R \cdot R_1 \cdot R_2 \left[\frac{1}{R} \right] = R \cdot R_1 \cdot R_2 \left[\frac{1}{R_1} + \frac{1}{R_2} \right] \qquad \text{Multiply by the LCD}$$

$$R_1 \cdot R_2 = R \cdot R_1 \cdot R_2 \left[\frac{1}{R_1} \right] + R \cdot R_1 \cdot R_2 \left[\frac{1}{R_2} \right] \qquad \text{Distributive property}$$

$$R_1 R_2 = R R_2 + R R_1$$

$$R_1 R_2 = R(R_2 + R_1)$$

$$\frac{R_1 R_2}{R_1 + R_2} = R$$

(b) $R = 60$ Ohms and $R_1 = 2R_2$, then

$$\frac{2R_2R_2}{2R_2 + R_2} = 60 \qquad \text{Substitute } 2R_2 \text{ for } R_1$$

$$\frac{2R_2^2}{3R_2} = 60$$

$$\frac{2R_2}{3} = 60$$

$$R_2 = (60)\frac{3}{2}$$

$$R_2 = 90$$

Thus R_2 is 90 Ohms and R_1 is 180 Ohms.

PROBLEM SET 5.5

Mastering the Concepts

In Problems 1–30, note what values of the variable make the equation undefined. Then solve the equation and check the solutions.

1. $\dfrac{4}{3x} - \dfrac{3}{x} = \dfrac{10}{3}$

2. $\dfrac{1}{2y} + \dfrac{2}{y} = \dfrac{1}{8}$

3. $\dfrac{3}{t} - \dfrac{4}{5} = -\dfrac{1}{5}$

4. $\dfrac{5}{u} + 3 = \dfrac{4}{3}$

5. $\dfrac{5}{x} + \dfrac{3}{8} = \dfrac{7}{16}$

6. $\dfrac{3}{8u} - \dfrac{1}{5u} = \dfrac{7}{10}$

7. $\dfrac{2}{3x} + 1 = \dfrac{4}{3x} + \dfrac{1}{3x}$

8. $\dfrac{1}{y} + \dfrac{2}{y} = 3 - \dfrac{3}{y}$

9. $\dfrac{2}{3y} = \dfrac{5}{9}$

10. $\dfrac{3}{5x} - \dfrac{1}{2x} = \dfrac{3}{10}$

11. $\dfrac{3}{x + 4} = \dfrac{3}{5}$

12. $\dfrac{6}{u + 2} = \dfrac{3}{4}$

13. $\dfrac{4x}{4x + 5} = \dfrac{3}{5}$

14. $\dfrac{4x + 5}{5x + 2} = \dfrac{3}{8}$

15. $\dfrac{3}{x + 7} = \dfrac{5}{x}$

16. $\dfrac{1}{2x} = \dfrac{1}{3x + 1}$

17. $\dfrac{8}{x - 3} = \dfrac{12}{x + 3}$

18. $\dfrac{3}{y - 2} = \dfrac{4}{y + 1}$

19. $\dfrac{y}{y - 2} - 1 = \dfrac{1}{y - 3}$

20. $\dfrac{3}{y - 3} + \dfrac{5}{2} = \dfrac{1}{y - 3}$

21. $\dfrac{2t}{t - 3} + \dfrac{3}{t - 3} = -2$

22. $\dfrac{2}{x + 1} + \dfrac{5x}{x + 1} = 4$

23. $\dfrac{y}{y + 1} + 3 = \dfrac{4}{y + 1}$

24. $\dfrac{x}{x - 1} - 1 = \dfrac{3}{x - 1}$

25. $\dfrac{1}{x + 1} + \dfrac{2}{3(x + 1)} = \dfrac{1}{3}$

26. $\dfrac{2}{3x - 2} - \dfrac{16x}{3(3x - 2)} = -2$

27. $\dfrac{1}{x - 1} - \dfrac{2}{1 - x} = 0$

28. $\dfrac{4}{4 - x} + \dfrac{1}{x - 4} = \dfrac{2}{x - 4}$

29. $\dfrac{3x}{x + 4} + \dfrac{4}{x + 2} = 3$

30. $\dfrac{2x}{x + 3} + \dfrac{4}{x + 4} = 2$

In Problems 31–36, solve each equation. Indicate whether the equation has an extraneous solution.

31. $\dfrac{x}{x - 2} + \dfrac{2}{3} = \dfrac{2}{x - 2}$

32. $\dfrac{y}{y - 6} - 3 = \dfrac{6}{y - 6}$

33. $\dfrac{3x}{x - 4} = 5 + \dfrac{12}{x - 4}$

34. $\dfrac{x}{x - 3} = 2 + \dfrac{3}{x - 3}$

35. $\dfrac{y + 1}{y} = \dfrac{y - 1}{y}$

36. $\dfrac{4}{t - 5} + \dfrac{6}{t + 5} = \dfrac{-40}{(t - 5)(t + 5)}$

In Problems 37–40, solve each equation for the indicated variable.

37. $\dfrac{ay}{b} = c + \dfrac{d}{b}$ for y

38. $\dfrac{3}{y} - \dfrac{4}{b} = -\dfrac{1}{b}$ for y

39. $\dfrac{b - x}{3} = \dfrac{2a - b}{4} - \dfrac{2x}{3}$ for x

40. $\dfrac{u + 2c}{3} + \dfrac{u - 3c}{2} = \dfrac{5}{12}$ for u

In Problems 41–44, find all values of t so that $f(t)$ has the indicated values.

41. $f(x) = x + \dfrac{7}{x}$, $f(t) = -8$

42. $f(x) = x - \dfrac{3}{x}$, $f(t) = 2$

43. $f(x) = \dfrac{6x}{5} - \dfrac{x}{4}$, $f(t) = 19$

44. $f(x) = \dfrac{2x}{x + 1} + \dfrac{5}{2x}$, $f(t) = 2$

Applying the Concepts

45. Jet Stream Speed: On a recent trip between two cities, it took the same time to fly 700 miles with the jet stream as it took to fly 550 miles against the jet stream. How fast is the jet stream if the plane flies 500 miles per hour in still air?

46. Jet Stream Speed: A commercial airliner has a speed of 450 miles per hour in still air. If the airliner flew a distance of 1020 miles with the jet stream in the same time it took to fly 780 miles against the jet stream, how fast is the jet stream?

47. Average Speed: A car traveled 126 miles in the same amount of time that a van traveled 99 miles. If the car went 15 miles per hour faster than the van, how fast did each vehicle travel?

48. Current Speed: A canoeist can paddle a canoe 6 kilometers per hour in still water. If the canoeist can paddle 1 kilometer upstream in the same amount of time that they can paddle 1.5 kilometers downstream, how fast is the current?

In Problems 49–52, (a) find what values of the variable that make the equation undefined, (b) solve the equation and check the solutions.

49. $\dfrac{2}{y^2 + y - 2} + \dfrac{3y}{y^2 + 5y + 6} = \dfrac{5y}{y^2 + 2y - 3}$

50. $\dfrac{4}{2t^2 - 7t - 15} + \dfrac{2}{t^2 - 25} = \dfrac{2}{2t^2 + 13t + 15}$

51. $\dfrac{1}{y} = \dfrac{8}{2y - y^2} + \dfrac{y + 2}{y - 2}$

52. $\dfrac{u + 4}{6u^2 + 5u - 6} - \dfrac{u}{3u - 2} = -\dfrac{u}{2u + 3}$

Developing and Extending the Concepts

In Problems 53 and 54, and solve each formula for the indicated variable.

53. (a) $\dfrac{1}{p} + \dfrac{1}{q} = \dfrac{1}{f}$ for p

   (b) $R = \dfrac{2V - 2r}{l}$ for V

54. (a) $\dfrac{1}{r_1} + \dfrac{1}{r_2} + \dfrac{1}{r_3} = \dfrac{1}{R}$ for R

   (b) $\dfrac{E}{E_1} = \dfrac{R + R_1}{R_1}$ for E_1

55. Suppose that the triangles ABC and DEF are similar (Figure 5), that is, corresponding sides are proportional.

   (a) Write an equation that represents equal ratios.

   (b) Solve the resulting equation for x.

   (c) Find the lengths of the sides of the triangles.

Figure 5

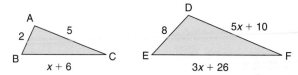

56. Hovercraft Speed: A hovercraft travels between Dover, England, and Calais, France, a distance of 38 kilometers. One day the wind was blowing so hard that the trip from Dover to Calais against the wind took twice as long as the return trip with the wind. If the wind was clocked at 10.7 kilometers per hour, how fast can the hovercraft travel with no wind?

57. Earthquake: The secondary seismic waves from the epicenter of an earthquake reached a recording station 150 kilometers away at the same time the primary seismic wave p reached another station 250 kilometers away from the epicenter.

   (a) How fast does each wave travel if the p wave travels 2 kilometers per second faster than the s wave?

   (b) Use the velocities of the two seismic waves from part (a) to find the distance to a recording station that recorded the s wave 2.5 seconds after the p wave.

58. Ecology: In a wildlife management study of a park, ecologists are to estimate the population of alligators. One year a researcher captured and tagged 300 alligators, then released them. Later, another researcher captured 1000 alligators and observed that 27 of them were tagged. The capture-recapture method of estimating animal populations assumes that the ratio of tagged alligators to all alligators in the park is the same as the ratio of tagged alligators in subsequent samplings to the number in the sampling. Based on this method, approximately how many alligators are there in the park? (Round to the nearest hundred.)

Objectives

1. Solve Number Problems
2. Solve Work Problems
3. Solve Motion Problems

5.6 Applications and Models Involving Rational Expressions

In this section, we consider applications and models whose translations from words into symbols will result in equations involving rational expressions.

Solving Number Problems

Many applications, such as encoding, involve numbers and lead to equations containing rational expressions as the next example shows.

EXAMPLE 1

Solving a Number Problem

One day the security code on a locked door has two numbers. One number is five times as large as a second number. The sum of their reciprocals is 2/5. Find the two numbers.

Solution

Make a Plan

The key here is the word reciprocal.

$$\text{Let } x = \text{the first number, then}$$

$$5x = \text{the second number.}$$

Their reciprocals are $\dfrac{1}{x}$ and $\dfrac{1}{5x}$ respectively. Since zero has no reciprocal, $x \neq 0$.

The following diagram is a visual description of the *given*.

Reciprocal of the first number	+	Reciprocal of the second number	=	$\dfrac{2}{5}$
$\dfrac{1}{x}$	+	$\dfrac{1}{5x}$	=	$\dfrac{2}{5}$

The equation which describes this situation is

$$\frac{1}{x} + \frac{1}{5x} = \frac{2}{5}.$$

Write the Solution

$$\frac{1}{x} + \frac{1}{5x} = \frac{2}{5} \qquad x \neq 0.$$

$$5x\left(\frac{1}{x} + \frac{1}{5x}\right) = 5x\left(\frac{2}{5}\right) \qquad \text{Multiply each side by } 5x, \text{ the LCD}$$

$$5 + 1 = 2x$$

$$6 = 2x$$

$$x = 3$$

$$\text{and } 5x = 15$$

Look Back

We check the result by adding the reciprocals of 3 and 15.

$$\frac{1}{3} + \frac{1}{15} = \frac{5}{15} + \frac{1}{15} = \frac{6}{15} = \frac{2}{5}$$

The solution checks in the original problem.

Solving Work Problems

Problems concerning work done by two or more people or machines working together are referred to as *work problems.* The work done at a constant rate yields equations involving rational expressions. If a job can be completed in t hours, then $1/t$ of the job can be completed in 1 hour.

$$\frac{1 \text{ job}}{t \text{ hours}} = \frac{1}{t} \text{ job/hour}$$

The key to writing the equation is knowing that all the parts of the job add up to 1 complete job; that is, the whole is equal to the sum of its parts. In these problems, we will assume that work is done at a constant rate regardless of how many are working.

EXAMPLE 2   **Solving a Work Problem**

Two cranes, operating together, can unload cargo ships at a rate of 1 ship every 4 hours. If one crane takes twice as long as the other to unload a ship, how long would it take each crane to unload a ship by itself?

Solution   **Make a Plan**

Let t = time (in hours) the faster crane unloads a ship, then

$2t$ = time (in hours) the slower crane unloads a ship.

One method of solving this type of problem is to think in terms of how much of the job is done by one crane in 1 hour. Therefore, the amount of the job completed in 1 hour is $1/t$ for the faster crane and $1/2t$ for the slower one. We can organize the given information in the table below.

Cranes	Part of Job Completed in 1 Hour	Number of Hours	Part of Job Completed in 4 Hours
Faster Crane	$\frac{1}{t}$	4	$\frac{4}{t}$
Slower Crane	$\frac{1}{2t}$	4	$\frac{4}{2t} = \frac{2}{t}$

The following diagram is a visual description of the *given.*

Part of job completed by faster crane	+	Part of job completed by slower crane	=	One complete job
$\frac{4}{t}$	+	$\frac{2}{t}$	=	1

Write the Solution

The equation is

$$\frac{4}{t} + \frac{2}{t} = 1 \qquad \text{Given}$$

$$\frac{6}{t} = 1 \qquad \text{Combine like terms}$$

$$t = 6$$

and $2t = 12$.

The faster crane would take 6 hours to unload the ship by itself, and the slower crane would take 12 hours to unload the ship alone.

Look Back

If the faster crane takes 6 hours to unload the ship, in 4 hours it has done

$$\left(\frac{1 \text{ job}}{6 \text{ hours}} \right) (4 \text{ hours}) = \frac{2}{3} \text{ job}$$

and the slower crane has done $\left(\dfrac{1 \text{ job}}{12 \text{ hours}} \right) (4 \text{ hours}) = \dfrac{1}{3}$ job.

Together they have completed the job,

$$\frac{2}{3} + \frac{1}{3} = 1$$

and the faster crane operates twice as fast as the slower one.

Solving Motion Problems

In section 1.3, we presented *motion problems* involving linear equations. Here, we solve motion problems involving rational expressions. Recall that

$$\text{time} = \frac{\text{distance}}{\text{rate}}.$$

EXAMPLE 3   **Solving a Jet Stream Speed Problem**

The *jet stream* is a narrow current of air encountered between altitudes of 25,000 and 35,000 feet. One day a flight from Los Angeles to New York (2460 miles) flying with the jet stream took the same amount of time that a flight from New York to Denver (1626 miles) took flying against the jet stream. How fast was the jet stream? Assume that both planes fly 550 miles per hour in still air.

Solution   **Make a Plan**

We can use the formula $t = \dfrac{d}{r}$ to find the time for both flights. If we

let r = speed (in miles per hour) of the jet stream, then

$550 + r$ = speed (in miles per hour) of the plane with the jet stream, and

$550 - r$ = speed (in miles per hour) of the plane against the jet stream.

The table below summarizes this information.

	Rate (miles per hour)	Distance (miles)	Time = $\dfrac{\text{Distance}}{\text{Rate}}$ (hours)
With the Jet Stream **Los Angeles to New York**	$550 + r$	2460	$\dfrac{2460}{550 + r}$
Against the Jet Stream **New York to Denver**	$550 - r$	1626	$\dfrac{1626}{550 - r}$

The problem states that both flights take the same time. The following diagram is a visual description of this *given*.

Time traveling with the jet stream	=	Time traveling against the jet stream

Thus, we can form an equation as follows:

$$\frac{2460}{550 + r} = \frac{1626}{550 - r}.$$

Write the Solution

$$\frac{2460}{550 + r} = \frac{1626}{550 - r} \qquad r \neq 550$$

$$2460(550 - r) = 1626(550 + r) \qquad \text{Cross multiply}$$

$$1{,}353{,}000 - 2460r = 894{,}300 + 1626r \qquad \text{Distributive property}$$

$$458{,}700 = 4086r$$

$$r = 112.26$$

The jet stream was traveling about 112.26 miles per hour.

Look Back

Time traveling with the jet stream is $\dfrac{2460}{550 + 112.26} = \dfrac{2460}{662.26} = 3.71$ hours.

Time traveling against the jet stream is $\dfrac{1626}{550 - 112.26} = \dfrac{1626}{437.74} = 3.71$ hours.

Since the times are the same, the solution is acceptable.

EXAMPLE 4

Solving a Uniform Motion Problem

A train travels from Spokane, Washington, to Glacier National Park, Montana, which is a 200 mile trip. If the speed of the train is 10 miles per hour less on the return trip, the train will take 1 hour longer than on the trip from Spokane to Glacier. How fast did the train travel in each direction?

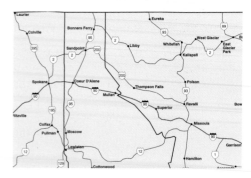

Solution

Make a Plan

Let r = speed of the train from Spokane to Glacier

$r - 10$ = speed of train from Glacier to Spokane.

We can organize the given information in the following table.

Direction	Rate	Time	Distance
Spokane to Glacier	r	$\dfrac{200}{r}$	200
Glacier to Spokane	$r - 10$	$\dfrac{200}{r - 10}$	200

The following diagram is a visual description of the given information.

Time traveled from Spokane to Glacier	+	1 hour	=	Time traveled from Glacier to Spokane

Write the Solution

The equation resulting from the diagram is as follows.

$$\frac{200}{r} + 1 = \frac{200}{r - 10} \qquad \text{The LCD is } r(r - 10)$$

$$r(r - 10)\left[\frac{200}{r} + 1\right] = \left[\frac{200}{r - 10}\right]r(r - 10) \qquad \text{Multiply by the LCD}$$

$$200(r - 10) + r(r - 10) = 200r \qquad \text{Distributive property}$$

$$200r - 2000 + r^2 - 10r = 200r$$

$$r^2 - 10r - 2000 = 0$$

$$(r - 50)(r + 40) = 0$$

$$r - 50 = 0 \quad \text{or} \quad r + 40 = 0$$

$$r = 50 \quad \text{or} \quad r = -40$$

Look Back

We must exclude the solution -40 since the speed cannot be negative. Thus, the speed from Spokane to Glacier was 50 miles per hour and from Glacier to Spokane was 40 miles per hour. Traveling at 50 miles per hour, the train took 4 hours to go from Spokane to Glacier. The return trip at 40 miles per hour took 5 hours. All the conditions are satisfied.

◈ **PROBLEM SET 5.6**

Mastering the Concepts

1. **Number Problem:** One number is four times as large as another. Find the two numbers if the sum of their reciprocals is 5/28.

2. **Number Problem:** If 7/10 is added to four times the reciprocal of a number, the result is 3/4. Find the number.

3. **Number Problem:** Find a number such that if this number is multiplied by the numerator and added to the denominator of the fraction 2/3, the result is the new fraction 7/5.

4. **Number Problem:** Find two consecutive integers so that the sum of their reciprocals is 15/56.

5. **Number Problem:** Find two integers such that one integer is three times another, and the sum of their reciprocals is 2/3.

6. **Number Problem:** Find a number such that when 5 is added to four times the reciprocal of the number, the sum is 27/5.

7. **Sales Tax:** In the state of Michigan, the sales tax on a car that sold for $12,000 is $720. At this rate, how much is the sales tax on a car that sells for $18,500?

8. **Billboard Lighting:** The lighting for a billboard is generally provided by solar energy. Suppose that 5 energy panels generate a total of 12 watts of power, how many panels are needed to generate 240 watts of power?

9. **Politics:** In a recent poll, 3000 people were asked their opinion about health care reform. If 37.5% of the people polled were men, how many women were polled?

10. **Journalism:** A national magazine pays freelance journalists by the word for published articles. Suppose a magazine pays $168 for an article that is 1050 words long. At this rate, how many additional words are needed in order for the journalist to earn $240 for an article?

Geometry: Two triangles are said to be *similar* if the corresponding angles are equal and the corresponding sides are proportional.

Figure 1

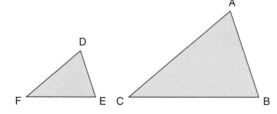

That is, in Figure 1, angle A = angle D, angle C = angle F, and angle E = angle B. Also, $\dfrac{\overline{AC}}{\overline{DF}} = \dfrac{\overline{AB}}{\overline{DE}} = \dfrac{\overline{BC}}{\overline{EF}}$.

In Problems 11 and 12, solve for x if the triangles ABC and DEF are similar.

11. $|\overline{AC}| = 4$, $|\overline{DF}| = 8$, $|\overline{BC}| = 3x + 1$,
    $|\overline{EF}| = 3x + 5$, $|\overline{AB}| = 6$, $|\overline{DE}| = 12$.

12. $|\overline{AC}| = 3x$, $|\overline{DF}| = 2x + 1$, $|\overline{BC}| = 18$,
    $|\overline{EF}| = 30$, $|\overline{AB}| = 15$, $|\overline{DE}| = 25$.

13. **Roofing a House:** A roofing crew can roof a house in 10 hours. If a second crew was added, the job would take 6 hours. How long would it take the second crew to roof the house alone?

14. **Wallpapering an Office:** Working together, two workers can hang wallpaper for an office in 6 hours. If one worker can hang the wallpaper in 9 hours, how long would it take the other worker to do the office alone?

15. **Filling a Reservoir:** An inlet pump can fill a reservoir in 10 hours. An outlet pump can drain the same reservoir in 30 hours. How long would it take to fill the reservoir if both pumps were in use at the same time?

16. **Landscaping:** A landscaper can prepare a new rock path and the surroundings in 10 hours. If the landscaper hired a worker to assist, together they can do the job in 6 hours. How long would it take the assistant to complete this job alone?

17. **Snow Removal:** Suppose the snow plows in a city can clear a 12 inch snowfall in 10 hours whereas the snow blowers take 6 hours. How long would it take them working together to clear the snow?

18. **Pollution Level:** At a garbage burning plant, one furnace reaches an air-quality alert level of pollution in 36 hours, whereas a second furnace reaches the same level in 45 hours. How long can both furnaces operate together without reaching that pollution alert level?

19. **Car Rental:** At a special weekend sale, a car rental company advertised that a fleet of midsize cars can be rented for $225 and the same number of minivans may be rented for $325. If a minivan cost $10 more to rent than a midsize car, what is the weekly rental rate for each type of vehicle?

20. **Engineering:** Engineers define the gear ratio of two gears, A and B, to be the ratio of the number of teeth in gear A to the number of teeth in gear B (Figure 2). Suppose that the gear ratio of gear A to B is 3/2 and the sum of the teeth in both gears is 85. Find the number of teeth in each gear.

Figure 2

21. **Investment:** An investment of $1500 earns $90 at the end of each year. At this rate, how much additional money must be invested to earn $120 at the end of each year?

22. **Monthly Car Payments:** A credit union advertised that a monthly car payment is $29.50 per $1000 loaned. At this rate, what is the monthly payment for a $9000 car loan?

23. **Catering:** A chef can prepare a meal for 25 people in 5 hours. One assistant takes 20 hours to prepare the same meal. How many assistants must be hired to prepare the same meal for 45 people in 6 hours?

24. **Basketball:** In basketball, the formula used to determine the success rate of free throws by a player is given by:

$$\frac{\text{Free Throws Made}}{\text{Free Throws Attempted}}$$

free throw line

Suppose that a basketball player made 32 free throws in 45 attempts for a current rate of 71%. How many consecutive free throws must the player make to reach an average of 75%?

25. **Baseball:** In baseball, the batting average of a player is given by the formula:

$$\frac{\text{Total hits}}{\text{Total times at bat}}.$$

In a season, a player has 112 hits out of 444 times at bat which is a batting average of 0.252. How many consecutive times must the player hit safely to raise this average to 0.267?

26. **Club Membership:** In 1995, a tennis club had 360 members and a number of nonmembers. By 2005, the number of members increased by 300 and the number of nonmembers increased by 700. If the percentage of members was the same in 2005 as it was in 1995, how many nonmembers were playing in the club in 1995?

27. **Travel Speed:** A commuter plane travels 90 miles per hour faster than a passenger bus. The commuter plane travels 525 miles in the same time it takes the bus to travel 210 miles. Find the speed of each.

28. **Travel Speed:** A canoeist travels at a speed of 4 miles per hour in still water. If the canoeist can  travel 2 miles upstream in the same time it takes to travel 6 miles downstream, what is the speed of the current?

29. **Jogger's Speed:** Two joggers travel a distance of 10 miles at a constant speed. If one runs twice as fast as the other, the faster runner would take 1 hour and 12 minutes less time than the slower jogger to run the 10 miles. Find the joggers' speeds.

30. **Filling a Tub:** A cold water faucet can fill a tub in 18 minutes. A hot water faucet can fill the same tub in 22 minutes. How long will it take both faucets together to fill the tub?

31. **Moving Sidewalk:** The moving sidewalk at an airport moves at a speed of 6 feet per second. A

traveler walking on the moving sidewalk covers 110 feet in the same time that it takes to walk 50 feet on the ground. How fast does the traveler walk on the ground in feet per second and in miles per hour?

32. **Skiing:** Two skiers competed on a 5 mile ski trail. The faster skier gave the slower skier a head start of 0.6 miles and still won the race by 2 minutes. If the faster skier's average speed is 3 miles per hour faster than the slower skier's average speed, what was the slower skier's average speed?

33. **Designing a Web Site:** Two people working together can design a web site for a small business in 4 hours. The slower worker requires 6 hours more than the faster worker to design the site for a client. How many hours does it take the faster worker to design the Web site alone?

34. **Disaster Relief:** A hurricane strikes a rural area leaving the residents without water, food, or shelter. Three crews arrive with the same supplies. One crew can dispense all their supplies alone in x hours. It takes the second crew 1

hour more than the first, and the third crew takes twice as long as the second crew to dispense all their supplies. If it takes all three crews working together one hour to dispense all their supplies, how long does it take each crew to dispense their supplies alone?

35. **Travel Time:** Two cars left Rapid City, South Dakota, at the same time. One traveled to the Badlands National Park

which is about 105 miles east of Rapid City. The second car traveled to Sioux Falls, South Dakota, approximately 350 miles east of Rapid City. It takes 3.5 hours longer to travel from Rapid City to Sioux Falls than it takes to travel from Rapid City to the Badlands. If both cars travel at the same speed, how long is each on the road?

36. **Filling Time:** In an orange grove, a pond is dug. The pond is to be filled by a pipe from the district water and another pipe from a spring on high ground. The first pipe can fill the pond in 3 hours, and the second can fill the pond in 5 hours. A third pipe used for irrigation can empty the pond in 10 hours. If all three pipes are used together, how long will it take to fill the pond?

❖ CHAPTER 5 REVIEW PROBLEM SET

In Problems 1–6, reduce each rational expression to lowest terms.

1. (a) $\dfrac{3m + 3n}{9m^2 - 9n^2}$

   (b) $\dfrac{4y^2 - 9}{2y^2 - 3y}$

2. (a) $\dfrac{x^2 - 5x}{x^2 - 25}$

   (b) $\dfrac{a^2 - 4b^2}{a^3 + 2a^2b}$

3. $\dfrac{x^2 - 7x + 12}{x^2 + 3x - 18}$

4. $\dfrac{3y^2 - 17y - 28}{3y^2 + 10y + 8}$

5. $\dfrac{x^4 - 8x}{x^4 + 2x^3 + 4x^2}$

6. $\dfrac{ax + ay - bx - by}{bx - by - ax + ay}$

In Problems 7 and 8, replace each question mark with an appropriate expression so that the resulting expressions are equivalent.

7. $\dfrac{3a}{2a - 2b} = \dfrac{?}{2a^2 - 2b^2}$

8. $\dfrac{2t + 1}{t - 2} = \dfrac{?}{3t^2 - 4t - 4}$

In Problems 9–12, find the domain of each function and reduce each expression.

9. $f(x) = \dfrac{x^2 + 5x + 6}{x^2 - x - 6}$

10. $f(x) = \dfrac{x^3 - 64}{x^2 - 16}$

11. $f(x) = \dfrac{x^2 - 4}{x^2 + 3x - 10}$

12. $f(x) = \dfrac{x^3 - 2x^2 - 3x}{x^4 - 4x^3 - 5x^2}$

In Problems 13–30, perform each operation and simplify.

13. $\dfrac{t + 1}{2t + 1} - \dfrac{t - 1}{1 - 2t}$

14. $\dfrac{3}{9 - t^2} - \dfrac{t}{9 - t^2}$

15. $\dfrac{3x}{4x^2 - 9} - \dfrac{x}{2x^2 + x - 6}$

16. $\dfrac{7}{x^2 - y^2} + \dfrac{2}{x^2 - 2xy + y^2}$

17. $\dfrac{2}{2m^2 + m - 3} + \dfrac{1}{2m^2 + 5m + 3}$

18. $\dfrac{4x}{3x^2 + 5x - 2} - \dfrac{x}{2x^2 + 5x + 2}$

19. $\dfrac{4x^2 - 64}{2x^3 - 16x} \cdot \dfrac{x - 4}{x + 4}$

20. $\dfrac{4x^2 - 11x - 3}{6x^2 - 5x - 6} \cdot \dfrac{6x^2 - 13x + 6}{4x^2 + 13x + 3}$

21. $\dfrac{x^2 + 5x + 6}{x^2 + x - 2} \cdot \dfrac{x^2 + 3x - 4}{x^2 + 7x + 12}$

22. $\dfrac{x^2 - y^2}{x^2 + 2xy + y^2} \cdot \dfrac{3x + 3y}{6x}$

23. $\dfrac{m^2 - m - 2}{m^2 - m - 6} \div \dfrac{m^2 - 2m}{2m + m^2}$

24. $\dfrac{y^2 - 5y - 6}{y^2 + y - 2} \div \dfrac{y^2 + y - 12}{y^2 + 3y - 4}$

25. $\dfrac{2x^2 - x - 6}{3x^2 - 11x - 4} \div \dfrac{2x^2 + 5x + 3}{3x^2 + 7x + 2}$

26. $\dfrac{8x^2 + 18x + 9}{6x^2 - 7x + 2} \div \dfrac{4x^2 + 7x + 3}{2x^2 + 9x - 5}$

27. $\dfrac{2 + 2(y - 1)^{-1}}{2(y - 1)^{-1}}$

28. $\dfrac{x - x^{-1}}{x - 2 + x^{-1}}$

29. $\dfrac{\dfrac{1}{x} - \dfrac{1}{y}}{\dfrac{x^2 - y^2}{xy}}$

30. $\dfrac{\dfrac{1}{2x - 3} - \dfrac{1}{2x + 3}}{\dfrac{x}{4x^2 - 9}}$

In Problems 31–40, solve each equation.

31. $\dfrac{3x}{5} - \dfrac{2x}{3} = 1$

32. $\dfrac{t}{3} + \dfrac{8}{5} = \dfrac{8 - t}{5}$

33. $\dfrac{2x + 1}{4} - \dfrac{3x - 4}{5} = \dfrac{1}{2}$

34. $\dfrac{x - 3}{10} + \dfrac{x - 1}{6} = \dfrac{3}{5}$

35. $\dfrac{t - 4}{3} + \dfrac{2(t + 3)}{5} = 5$

36. $\dfrac{y - 3}{4} + \dfrac{y - 5}{6} = \dfrac{2y - 1}{3}$

37. $\dfrac{1}{a} + \dfrac{1}{b} = \dfrac{1}{c}$, for a **38.** $y = \dfrac{3 - x}{x - 1}$, for x

39. $\dfrac{x + 1}{4} + \dfrac{3 + x}{5} = \dfrac{7}{4}$

40. $\dfrac{5x - 1}{3} - \dfrac{4x + 3}{5} = \dfrac{-1}{15}$

In Problems 41 and 42, for the given functions find:

(a) $(f + g)(x)$ (b) $(f - g)(x)$

(c) $(f \cdot g)(x)$ (d) $\left(\dfrac{f}{g}\right)(x)$.

41. $f(x) = \dfrac{x + 2}{x - 2}$ and $g(x) = \dfrac{x - 2}{x + 2}$

42. $f(x) = \dfrac{16 - x^2}{8x}$ and $g(x) = \dfrac{4 - x}{2x}$

In Problems 43 and 44, for the given function find

$$\dfrac{f(a + h) - f(a)}{h}.$$

43. $f(x) = 2x + 7$ **44.** $f(x) = \dfrac{1}{x}$

In Problems 45–48, match the graph to the function.

45. $f(x) = \dfrac{x + 1}{x - 1}$ **46.** $g(x) = \dfrac{1}{x}$

47. $h(x) = \dfrac{x^2 - 4}{x - 2}$ **48.** $k(x) = \dfrac{x}{x - 3}$

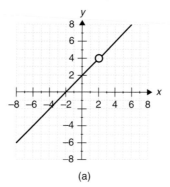
(a)

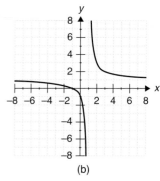
(b)

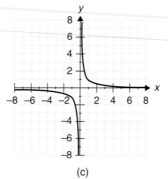
(c)

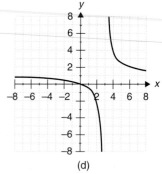
(d)

49. Management: A company's production P in thousands of units per year after t years from its starting date is given by the formula

$$P(t) = \dfrac{21t^2 - 84t + 168}{t^2 - 4}.$$

What is the company's production
(a) After 3 years from its starting date?
(b) After 5 years? Round to two decimal places.

50. Physics: The average speed v in miles per hour for a round trip made by an airplane is given by the formula

$$v = \frac{2}{\dfrac{1}{v_1} + \dfrac{1}{v_2}}$$

where v_1 (in miles per hour) is the average speed going to a destination and v_2 (in miles per hour) is the average speed returning.

(a) What is the average speed for a round trip of an airplane with $v_1 = 450$ miles per hour and $v_2 = 500$ miles per hour? Round to a whole number.

(b) Simplify the complex fraction.

51. Home Mortgage: A home is purchased for $215,000, with a 15% down payment, and an annual interest rate of 7.25%.

(a) What is the monthly payment on 30 year mortgage? [See Example 1 on page 281 in Section 5.4]

(b) What is the monthly payment on a 15 year mortgage?

52. Painting: A painter can paint a racquetball court in 8 hours. His assistant needs an additional 2 hours to paint the same court working by himself. How long will it take both of them working together to paint the court?

53. Boating: The speed of a motorboat in still water is 24 miles per hour. If the boat travels 27 miles upstream in the same time it takes to travel 45 miles downstream, what is the speed of the current?

CHAPTER 5 PRACTICE TEST

1. Simplify each expression.

(a) $\dfrac{3x^2 - 5x - 2}{ax + 2b - bx - 2a}$

(b) $\dfrac{8x^2 - 14xy + 3y^2}{12x^2 + 5xy - 2y^2}$

2. Perform each operation and simplify.

(a) $\dfrac{x^2 - 4}{x} \cdot \dfrac{2x}{3x - 6}$

(b) $\dfrac{x^2 + x - 2}{x + 2} \div \dfrac{x^2 + 2x - 3}{x + 3}$

(c) $\dfrac{x}{x^2 - 1} + \dfrac{3}{4x - 4}$

(d) $\dfrac{b}{ab - b^2} - \dfrac{a}{b^2 + ab}$

3. Simplify each expression.

(a) $\dfrac{1 + \dfrac{1}{x - 2}}{1 - \dfrac{1}{3x - 6}}$

(b) $\dfrac{x^{-1}}{x^{-1} + y^{-1}}$

4. Given the functions $f(x) = \dfrac{x - 4}{x + 4}$ and $g(x) = \dfrac{x + 4}{x - 4}$, find:

(a) $(f + g)(x)$

(b) $(f - g)(x)$

(c) $(f \cdot g)(x)$

(d) $\left(\dfrac{f}{g}\right)(x)$.

5. Find the domain of each function and match it to its graph.

(a) $f(x) = \dfrac{x}{x - 1}$

(b) $g(x) = \dfrac{x - 1}{x}$

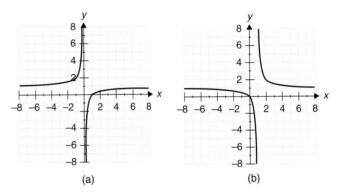

(a)   (b)

6. Working together a couple can wash the walls in a room in 2 hours. If the faster person can do the job alone in 3 hours, how long would it take the other person to do the job alone?

7. A trucker entered a toll road at 1:45 in the afternoon. The trucker exited the toll road at 3:00 pm 85 miles later and was given a ticket for speeding. What was the average speed of the truck?

Exponents, Radicals, and Complex Numbers

6

Chapter Contents

This is a picture of a Blackfeet native-designed teepee which is in the shape of a cone. When erecting this teepee, the poles will allow a height of 17 feet and the volume must be 1442 cubic feet. To know how much space is needed on the ground, one needs to know the radius of the base of the teepee. The formula needed to find the radius is

$$r = \sqrt{\frac{3V}{\pi h}}$$

where r is the radius of the base, V is the volume, and h is the height. This new symbol will be discussed in this chapter and this problem will be solved in example 6 of section 6.2.

So far, we have encountered some examples of mathematical applications involving integer exponents. In this chapter, we reexamine some of these applications and models and raise some new questions concerning rational numbers as exponents. We shall see that rational exponents require a consideration of roots and radicals, and these in turn will lead us to investigate a new number system called the *complex numbers*. Functions and their graphs will continue to play a role in the chapter.

Objectives

1. Extend the Properties of Positive Exponents
2. Solve Exponential Equations
3. Solve Applied Problems
4. Graph an Exponential Function

6.1 Exponential Expressions, Functions, and Equations

Extending the Properties of Positive Exponents

In section R5, we presented the properties of positive integral exponents. In this section, we will extend these properties to include all integer exponents: positive, negative, and zero exponents. First, let us consider how to define a zero exponent and a negative integer exponent. Recall the quotient property from section R5 which states:

$$\frac{a^n}{a^m} = a^{n-m}, a \neq 0 \text{ and } n > m.$$

We know that $\frac{3^4}{3^4} = 1$ and according to the quotient rule $\frac{3^4}{3^4} = 3^{4-4} = 3^0$.

So the value of 3^0 must be equal to 1. This is true for any base or exponent.

Similarly, consider $\frac{5^4}{5^6} = \frac{5 \cdot 5 \cdot 5 \cdot 5}{5 \cdot 5 \cdot 5 \cdot 5 \cdot 5 \cdot 5} = \frac{1}{5^2}$ and $\frac{5^4}{5^6} = 5^{4-6} = 5^{-2}$.

Therefore, 5^{-2} must be equal to $\frac{1}{5^2}$.

In general, we have the following definition.

Zero and Negative Exponents

If a is a real number such that $a \neq 0$ and n is an integer, then

1. $a^0 = 1$ 2. $a^{-n} = \frac{1}{a^n}$ 3. $\frac{1}{a^{-n}} = a^n$.

The following results are direct consequences of the preceding definitions.

If $a \neq 0$ and $b \neq 0$, then

$$\left(\frac{a}{b}\right)^{-n} = \left(\frac{b}{a}\right)^n \quad \text{and} \quad \frac{a^{-n}}{b^{-m}} = \frac{b^m}{a^n}.$$

EXAMPLE 1   **Rewriting Expressions with Nonnegative Exponents**

Rewrite each expression without negative exponents. Assume all variables are restricted to values for which all expressions represent real numbers.

(a) 7^{-1} (b) $(0.25)^{-1}$ (c) $\left(\frac{p}{q}\right)^{-3}$ (d) $(a + 2)^0$ (e) $\frac{1}{a^{-3}}$ (f) $\frac{3^{-2}}{2^{-2}}$ (g) $3^{-1} + 4^{-1}$

Solution

(a) $7^{-1} = \dfrac{1}{7}$ Definition of negative exponent

(b) $(0.25)^{-1} = \left(\dfrac{1}{4}\right)^{-1}$ Convert decimal to fraction

$\qquad\qquad = \dfrac{4}{1} = 4$ Definition of negative exponent

(c) $\left(\dfrac{p}{q}\right)^{-3} = \left(\dfrac{q}{p}\right)^{3} = \dfrac{q^3}{p^3}$ Negative exponent, power of quotient

(d) $(a + 2)^0 = 1$ Definition of zero exponent

(e) $\dfrac{1}{a^{-3}} = a^3$ Negative exponent

(f) $\dfrac{3^{-2}}{2^{-2}} = \dfrac{2^2}{3^2} = \dfrac{4}{9}$ Negative exponent

(g) $3^{-1} + 4^{-1} = \dfrac{1}{3} + \dfrac{1}{4} = \dfrac{4 + 3}{12} = \dfrac{7}{12}$ Negative exponent

Because of the nature of the definitions for zero and negative exponents, the properties of exponents still hold when the exponents are integers.

Properties of Integer Exponents

Let a and b be non zero real numbers, and let m and n be integers. Then each of the following is true for all values of a and b for which the expressions are defined:

(i) $a^m \cdot a^n = a^{m+n}$ Product property

(ii) $(a^m)^n = a^{mn}$ Power property

(iii) $(ab)^m = a^m b^m$ Power of product property

(iv) $\left(\dfrac{a}{b}\right)^m = \dfrac{a^m}{b^m}$ Power of quotient property

(v) $\dfrac{a^m}{a^n} = a^{m-n}$ Quotient property

EXAMPLE 2

Using Properties of Exponents

Rewrite each expression using only positive exponents and simplify the results.

(a) $x^6 \cdot x^{-2}$ (b) $(p^4)^{-3}$ (c) $(6y^3)^{-2}$ (d) $\left(\dfrac{3ab^0}{a^2(8b)^0}\right)^{-2}$

Solution

(a) $x^6 \cdot x^{-2} = x^{6-2} = x^4$ Product property

(b) $(p^4)^{-3} = p^{-12} = \dfrac{1}{p^{12}}$ Power property

(c) $(6y^3)^{-2} = 6^{-2} \cdot (y^3)^{-2}$ 　　　Power of product property

$$= \frac{1}{6^2} \cdot \frac{1}{(y^3)^2}$$ 　　　Negative exponents

$$= \frac{1}{36} \cdot \frac{1}{y^6} = \frac{1}{36y^6}$$ 　　　Power property

(d) $\left(\dfrac{3ab^0}{a^2(8b)^0} \right)^{-2} = \left(\dfrac{3a(1)}{a^2(1)} \right)^{-2}$ 　　　Definition of zero exponent

$$= \left(\frac{3}{a} \right)^{-2}$$ 　　　Quotient property

$$= \left(\frac{a}{3} \right)^2 = \frac{a^2}{9}$$ 　　　Negative exponent and power of quotient

Solving Exponential Equations

Up until now, we have focused on the properties of exponential expressions. Now we consider a strategy for solving exponential equations such as

$$2^x = 8, \quad 2^{2x} = 32, \quad \text{and} \quad 5^{2x} = 25.$$

For instance, to solve the exponential equation

$$5^{2x} = 25$$

first we rewrite the equation in the form $5^{2x} = 5^2$. Then observe that because the base of the exponentials is the same, the exponents must be equal. That is, $2x = 2$ or $x = 1$. To solve this equation, we used the following property:

Exponential Property

Let a be a positive real number such that $a \neq 1$ and suppose that m and n are real numbers. Then

$$a^m = a^n \text{ if and only if } m = n.$$

EXAMPLE 3　　**Solving Exponential Equations**

Solve each equation.

(a) $2^x = 16$ 　　　　(b) $4^{-x} = 256$

Solution　　(a) First we express 16 as a power of 2 so that both sides of the equation will have the same base. We rewrite the equation as

$$2^x = 16$$ 　　　Given

$$2^x = 2^4$$ 　　　$16 = 2^4$

$$x = 4$$ 　　　If the bases are equal, the powers are equal.

The solution is 4. Check this in the original equation.

(b) $4^{-x} = 256$ Given

    $4^{-x} = 4^4$ $256 = 4^4$

    $-x = 4$ or $x = -4$

The solution is -4.

Solving Applied Problems

In section R5, we saw that work in science and engineering often involves the use of very large as well as very small numbers. It is often convenient to represent such numbers in *scientific notation*. For example,

$$0.003 = 3 \times 10^{-3}$$
$$0.00007 = 7 \times 10^{-5}$$
$$0.00000059 = 5.9 \times 10^{-7}.$$

EXAMPLE 4 **Using Scientific Notation**

A biologist determined that the radius r of a cell (in the form of a sphere) is 1.5×10^{-4} millimeters. Use the formula $V = \frac{4}{3}\pi r^3$ to find the volume of this cell. Round off to two decimal places.

Solution We replace r by 1.5×10^{-4} in the formula $V = \frac{4}{3}\pi r^3$ to obtain

$$V = \frac{4}{3}\pi(1.5 \times 10^{-4})^3$$
$$= 1.41 \times 10^{-11}.$$

The volume of the cell is approximately 1.41×10^{-11} cubic millimeters.

Investments earning compound interest grow faster than simple interest investments of the same percentage rates because the interest earned each period is added to the principal, that is, the interest is reinvested. Consider the calculation of interest on a January deposit of $1000 in a bank account which earns 4% interest compounded quarterly:

Date	Principal	Rate of Interest	Time	Interest Earned
January	$1000	4% per year	0 yrs	$0
March	$1000	4% per year	$\frac{1}{4}$ yrs	$1000(4%)$\frac{1}{4}$ = $10
June	$1000 + 10 = $1010	4% per year	$\frac{1}{4}$ yrs	$1010(4%) \cdot \frac{1}{4}$ = $10.10
September	$1010 + 10.10 = $1020.10	4% per year	$\frac{1}{4}$ yrs	$1020.10(4%) \cdot \frac{1}{4}$ = $10.20

Thus at the end of three quarters, the account has a balance of

$$\$1020.10 + 10.20 = \$1030.30$$

and has earned $30.30 interest. Bankers use the formula

$$S = P\left(1 + \frac{r}{n}\right)^{nt}$$

called the **compound interest formula** to model the **future value** S (in dollars) of the **present value** or **principal** P (in dollars) invested for a **term** of t years at an **annual interest rate** r **compounded** n times per year. Using this formula in the above example, the balance is:

$$S = 1000\left(1 + \frac{0.04}{4}\right)^{\left(4 \cdot \frac{3}{4}\right)} = \$1030.30.$$

EXAMPLE 5 **Hourly Wage Increase**

Suppose that the starting hourly wage of a part time employee in a fast food restaurant is $5.50. After one year, the employee receives a 6% annual raise and may choose

(a) a 6% raise in 4 installments over the course of the year or

(b) a 6% raise in one installment each year.

Find the hourly wage for each method after 3 years.

Solution (a) For a 6% raise in 4 installments over the course of the year, we replace P by 5.50, n by 4, t by 3 and r by 0.06 in the formula.

$$S = P\left(1 + \frac{r}{n}\right)^{nt}$$

$$S = 5.50\left(1 + \frac{0.06}{4}\right)^{(4)(3)} \approx 6.58$$

Therefore, the hourly wage of the employee after 3 years is approximately $6.58.

(b) If a 6% raise is given in one installment per year, all the variables are the same except for $n = 1$. We have

$$S = P\left(1 + \frac{r}{n}\right)^{nt}$$

$$S = 5.50\left(1 + \frac{0.06}{4}\right)^{(1)(3)} \approx 6.55$$

Therefore, receiving the same increase over 4 quarterly installments returns more money than over single yearly installments.

Graphing an Exponential Function

A function of the form $\qquad\qquad f(x) = b^x$

where $b > 0$, $b \neq 1$ and x is a real number is called an **exponential function with base** b. The algebraic form of linear functions which we have studied is very different from exponential functions. In the next example, we examine graphs of exponential functions and show their distinctive features.

EXAMPLE 6

Graphing Exponential Functions

(a) For the function $f(x) = 2^x$, fill in the table for the given input values.

x	$f(x) = 2^x$	(x,y)
-10	$f(-10) =$	
-4	$f(-4) =$	
-2	$f(-2) =$	
-1	$f(-1) =$	
0	$f(0) =$	
1	$f(1) =$	
2	$f(2) =$	
10	$f(10) =$	

Plot the ordered pairs and show they fit on the graph of $f(x) = 2^x$.

(b) Use the graph to indicate the domain and range of the function.

Solution

(a)

x	$f(x) = 2^x$	(x,y)
-10	$f(-10) = 2^{-10} \approx 0.00098$	$(-10, 0.00098)$
-4	$f(-4) = 2^{-4} = 0.0625$	$(-4, 0.0625)$
-2	$f(-2) = 2^{-2} = 0.25$	$(-2, 0.25)$
-1	$f(-1) = 2^{-1} = 0.5$	$(-1, 0.5)$
0	$f(0) = 2^0 = 1$	$(0, 1)$
1	$f(1) = 2^1 = 2$	$(1, 2)$
2	$f(2) = 2^2 = 4$	$(2, 4)$
10	$f(10) = 2^{10} = 1024$	$(10, 1024)$

Figure 1 shows the plotted points from the above table and connecting them with a smooth curve.

Figure 1

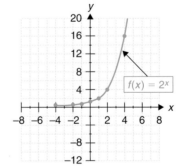

(b) The domain of the function consists of all real numbers. The range of the function consists of all positive real numbers, or $(0, 8)$.

EXAMPLE 7   **Solving Exponential Equations Graphically**

Solve the exponential equation $2^x - 16 = 0$ graphically.

Solution   Figure 2 shows the graph of the function $f(x) = 2^x - 16$. The point $(4,0)$ is the x intercept and tells us that $f(4) = 0$. The solution to the equation is 4.

Figure 2

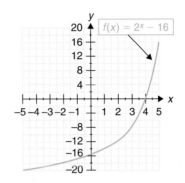

PROBLEM SET 6.1

Mastering the Concepts

In Problems 1–12, rewrite each expression in a simplified form using only nonnegative exponents. (Assume all variables are restricted to values for which all expressions represent real numbers.)

1. (a) 4^0 (b) 8^0 (c) $(-3)^0$
2. (a) $(-2)^0$ (b) 7^0 (c) $(0.3)^0$
3. (a) $(2x)^0$ (b) $(7^{-1} + 3y)^0$ (c) $(0.17z)^0 + 6$
4. (a) $(-3y^2)^0$ (b) $(3 + x^2)^0$ (c) $2 + (3x - x)^0$
5. (a) 5^{-2} (b) 2^{-4} (c) $(-7)^{-2}$
6. (a) $(-3)^{-4}$ (b) $(-10)^{-1}$ (c) $(-5)^{-1}$
7. (a) $\left(\dfrac{4}{3}\right)^{-2}$ (b) $\left(\dfrac{1}{3}\right)^{-2}$ (c) $\left(\dfrac{1}{x}\right)^{-1}$
8. (a) $\left(\dfrac{1}{p}\right)^{-2}$ (b) $\left(\dfrac{2}{y}\right)^{-3}$ (c) $\dfrac{10^0}{10^{-3}}$
9. (a) $\dfrac{1}{5^{-2}}$ (b) $\dfrac{1}{x^{-4}}$ (c) $\dfrac{1}{y^{-3}}$
10. (a) $\dfrac{1}{4^{-3}}$ (b) $\dfrac{3}{t^{-2}}$ (c) $\dfrac{5}{y^{-4}}$
11. (a) $\dfrac{2^{-3}}{3^{-2}}$ (b) $\dfrac{x^{-4}}{y^{-2}}$ (c) $\dfrac{a^{-2}}{b^{-5}}$
12. (a) $\dfrac{5^{-2}}{3^{-2}}$ (b) $\dfrac{x^{-2}}{y^{-4}}$ (c) $\dfrac{a^{-3}}{b^{-4}}$

In Problems 13–26, use the properties of exponents to rewrite each expression so that it contains only positive exponents, and simplify the results. (Assume all variables are restricted to values for which all expressions represent real numbers.)

13. (a) $7^{-1} \cdot 7^3$ (b) $5^{-3} \cdot 5^{-2}$
14. (a) $9^{-3} \cdot 9$ (b) $3^{-4} \cdot 3^7$
15. (a) $x^{-3} \cdot x^{-2}$ (b) $y^{-7} \cdot y^{-2}$
16. (a) $(2^3)^{-2}$ (b) $(5^2)^{-1}$
17. (a) $[(-4)^2]^{-2}$ (b) $(p^{-2})^{-4}$
18. (a) $(3x)^{-4}$ (b) $(3y^{-4})^2$
19. (a) $(5^{-1}p^4)^{-2}$ (b) $(p^{-1}q^{-2})^5$
20. (a) $\left(\dfrac{5}{x}\right)^{-2}$ (b) $\left(\dfrac{3}{p^{-1}}\right)^{-1}$
21. (a) $\left(\dfrac{p^3}{q^2}\right)^{-2}$ (b) $\left(\dfrac{x^{-2}}{y^{-2}}\right)^{-1}$
22. (a) $\dfrac{8^{-3}}{8^{-5}}$ (b) $\dfrac{7^{-3}}{7^{-8}}$
23. $\left(\dfrac{b^{-1}}{3a^{-1}}\right)\left(\dfrac{3a}{b}\right)^{-1}$ **24.** $\left(\dfrac{a^{-1}c^2}{a^{-2}c^{-1}}\right)^{-2}$
25. $\left(\dfrac{x^{-1}y^{-2}}{x^{-2}y^3}\right)^2$ **26.** $\left(\dfrac{p^{-2}q^{-1}r^{-3}}{q^3r^{-1}}\right)^{-4}$

In Problems 27 and 28, solve each exponential equation graphically.

27. (a) $3^x = 9$ **28.** (a) $4^x = 64$
   (b) $2^x = 8$ (b) $5^x = 25$

In Problems 29–32, solve each equation algebraically.

29. (a) $2^{-3x} = 8$
    (b) $2^{4x} = 16$
30. (a) $5^{-y} = 125$
    (b) $6^{-x} = 36$
31. (a) $3^{-x} = 27$
    (b) $4^{-x} = 16$
32. (a) $3^{2x} = 81$
    (b) $5^{-2x} = 25$

In Problems 33–36, use a calculator to simplify each expression. Round off to two decimal places and write each answer in scientific notation.

33. $\dfrac{(9.75 \times 10^8) \cdot (1.5 \times 10^{-3})}{(7.5 \times 10^{-2}) \cdot (1.1 \times 10^{-4})}$

34. $\dfrac{(5.7 \times 10^4)^2 \cdot (6.65 \times 10^{-5})^3}{(3.3 \times 10^{-7}) \cdot (3.8 \times 10^{-6})^4}$

35. $\dfrac{(1.12 \times 10^{-3}) \cdot (8.25 \times 10^{-5})}{(2.35 \times 10^{-6})^2}$

36. $\dfrac{(1.86 \times 10^5) \cdot (2.4 \times 10^{-9})}{(3.6 \times 10^{-7}) \cdot (4.8 \times 10^{-11})^3}$

Applying the Concepts

37. **Biology:** A certain virus is shaped like a sphere. If its radius is 0.000013 centimeters, find the volume of the virus. Express the answer in scientific notation and round off to 2 decimal places. $\left(V = \dfrac{4}{3}\pi r^3 \right)$

38. **Computer Calculations:** A computer can do an arithmetic operation in 0.00000036 seconds. If the electricity in its circuits travels at the speed of light (186,000 miles per second), how far will the electricity travel in the time it takes the computer to complete 7 arithmetic operations? Express the answer in scientific notation and round off to two decimal places.

39. **Travel Time:** A satellite leaves the earth traveling at a uniform speed of 310,000 miles per hour. How long does it take for the satellite to reach the sun, if the average distance from the earth to the sun is 93,000,000 miles? Express the answer in scientific notation.

40. **Chemistry:** One gram of hydrogen contains 6.023×10^{23} atoms. What is the weight of one atom of hydrogen? Express the answer in scientific notation and round off to two decimal places.

41. **Hourly Wage Increase:** An employee's starting wage is $7.50 per hour. Find the hourly wage after 5 years if the total annual raise is 6%, and the raise is given in 2 installments over the course of the year.

42. **Compound Interest:** A deposit of $5000 is placed in a savings account. Find the balance in

the account after 3.5 years if it is compounded quarterly at a rate of 7.2%.

43. **Compound Interest:** Find the amount of money in an account if $2000 is deposited at 4% interest compounded monthly for 3 years.

44. **Compound Interest:** Find the amount of money in a forgotten savings account if $5.00 is invested at 5% interest compounded quarterly for 50 years.

Developing and Extending the Concepts

In Problems 45 and 46, each of the following expressions is obtained in calculating compound interest. Write a story which would give rise to this expression and then use a calculator to evaluate it. Round off each answer to two decimal places.

45. (a) $1000\left(1 + \dfrac{0.08}{4} \right)^{4(3)}$

    (b) $5000\left(1 + \dfrac{0.04}{12} \right)^{12(3)}$

46. (a) $10{,}000\left(1 + \dfrac{0.06}{12} \right)^{12(4)}$

    (b) $4000\left(1 + \dfrac{0.05}{52} \right)^{52(3)}$

In Problems 47–50,
   (a) evaluate each function for $x = -2, -1, 0, 1, 2, 3$.
   (b) Plot these points to sketch the graph of each function.
   (c) Use the graph to indicate the domain and range of the function.

47. $f(x) = 3^x$
48. $g(x) = 3^{-x}$
49. $h(x) = 4^x$
50. $F(x) = 4^{-x}$

In Problems 51–54, match each function with its graph. Then find the domain and range of the function.

51. $f(x) = 3^x$
52. $G(x) = -3^x$
53. $h(x) = 3^{-x}$
54. $k(x) = 3^x + 10$

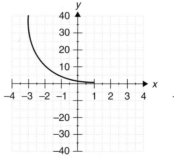

(a)

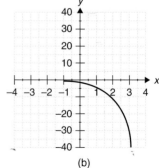

(b)

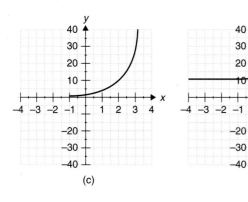

(c)

(d)

In Problems 55–58, solve each equation graphically.

55. (a) $3^x - 81 = 0$

(b) $2^x - 32 = 0$

56. (a) $4^x - 64 = 0$

(b) $5^x - 25 = 0$

57. (a) $2^{-3x} - 8 = 0$

(b) $2^{4x} - 16 = 0$

58. (a) $5^{-x} - 25 = 0$

(b) $6^{-x} - 36 = 0$

In Problems 59–62, rewrite each expression without negative exponents, and simplify. (Assume all variables are restricted to values for which all expressions represent real numbers.)

59. (a) $\dfrac{(a + b)^{-2}}{(a + b)^{-3}}$

(b) $x^{-2n} \cdot x^{2n + 6}$

60. (a) $x^{3 + n} \cdot x^{2 - n}$

(b) $\dfrac{(x + 3y)^{-12}}{(x + 3y)^{10}}$

61. (a) $\left(\dfrac{3^{-1} x^{-2} y^{-5}}{9^{-1} x^{-1} y^{-3}} \right)^{-3}$

(b) $\left(\dfrac{u^{-4} v^3 z^{-3}}{u^3 (vz)^{-2}} \right)^{-4}$

62. (a) $\left[\left(\dfrac{a^{-1}}{b} \right) \cdot \left(\dfrac{b^{-1}}{a} \right)^{-1} \right]^{-2}$

(b) $\left(\dfrac{a^{-4} b^2 c^{-6}}{(ab)^{-2}(bc)^{-3}} \right)^{-1}$

At times, we rewrite expressions with negative exponents. In Problems 63 and 64, rewrite the expression using negative exponents.

63. (a) $\dfrac{1}{x^2 y^4}$

(b) $\dfrac{5}{(2 + a)^5}$

64. (a) $\dfrac{6}{x^7 y}$

(b) $\dfrac{2}{(x + 1)^7}$

6.2 Radical Expressions and Functions

Objectives

1. Find the *n*th Root
2. Evaluate Radicals Using Absolute Value
3. Evaluate Radical Functions
4. Graph Radical Functions
5. Solve Applied Problems

Finding the *n*th Root

We know how to raise a number to a power such as $5^2 = 25$. Now let us consider the reverse, such as what number squared is 25. This is called finding the *root of a number*. The following table shows some types of roots.

Number	Type of Root	Root
$25 = 5^2$	square root of 25	5
$25 = (-5)^2$	square root of 25	-5
$-8 = (-2)^3$	cube root of -8	-2
$16 = 2^4$	fourth root of 16	2

In general, we have the following definition

Definition of an *n*th Root

Let a and b be real numbers, and n be a positive integer, then

a is an *n*th root of b if $a^n = b$.

Some numbers have more than one nth root.

1. Any positive number has two square roots. For example, both 5 and -5 are square roots of 25 because

$$25 = 5^2 \text{ and } 25 = (-5)^2$$

2. Negative numbers have no real square root, since no real number squared can be negative.
3. Zero has only one square root, namely 0.

For cube roots, the situation is different, every number (positive or negative) has only one real cube root. Unlike square roots, it is possible to have a cube root of a negative number. To avoid ambiguity about which root of a number is considered, we generalize the above result.

Number of Real nth Roots of a Number a.

If a Is	n Is Even	n Is Odd
1. positive	two real nth roots	one real nth root
2. negative	no real roots	one real nth root

Radical notation is commonly used to indicate a root. We will use $\sqrt[n]{a}$ to represent the **nth root of a.** The symbol $\sqrt{}$ is called a **radical** sign, n is called the **index** of the radical and a is called the **radicand.** The entire expression $\sqrt[n]{a}$ is called the **radical expression.**

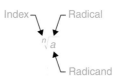

When the index is 2, it is usually omitted. That is, we use $\sqrt{}$ rather than $\sqrt[2]{}$ to represent square roots. If there are two nth roots, then $\sqrt[n]{a}$ denotes the positive root, called the **principal nth root.** We describe the nth root as follows:

Do not confuse $-\sqrt{16}$ with $\sqrt{-16}$. Because $-\sqrt{16} = -4$, whereas $\sqrt{-16}$ is not a real number.

nth Root $\sqrt[n]{a}$.

Radicand a	Index n Is Even	Index n Is Odd
positive	$\sqrt[n]{a}$ is the positive nth root of a	$\sqrt[n]{a}$ is the nth root of a
negative	$\sqrt[n]{a}$ is not a real number	$\sqrt[n]{a}$ is the nth root of a

Many radicals such as $\sqrt{3}$ and $\sqrt{14}$ are not *rational numbers.* To evaluate $\sqrt{3}$ and $\sqrt{14}$ on a calculator, we get nonterminating, and not repeating decimals. Using a calculator we have:

$$\sqrt{3} = 1.7320508$$

$$\sqrt{14} = 3.7416574$$

Each result is nonterminating and nonrepeating decimal.

Table 1 below shows perfect square roots.

TABLE 1

Perfect Square	Radical Expression	Principal Square Root
1	$\sqrt{1}$	1
4	$\sqrt{4}$	2
9	$\sqrt{9}$	3
16	$\sqrt{16}$	4
25	$\sqrt{25}$	5
36	$\sqrt{36}$	6
49	$\sqrt{49}$	7
64	$\sqrt{64}$	8
81	$\sqrt{81}$	9
100	$\sqrt{100}$	10
121	$\sqrt{121}$	11
144	$\sqrt{144}$	12

Radical expression that have indices of 2, 4, 6, 8, . . . or any *even* integer are called **even roots.** For example,

$$\sqrt[4]{16} = 2 \text{ because } 2^4 = 16$$

$$\sqrt[4]{81} = 3 \text{ because } 3^4 = 81$$

and

$$\sqrt[6]{64} = 2 \text{ because } 2^6 = 64$$

Radical expressions that have indices of 3, 5, 7, . . . or any odd integer are called **odd roots.** Table 2 below shows some odd roots.

TABLE 2

Cube Roots	Fifth Roots
$\sqrt[3]{1} = 1$	$\sqrt[5]{1} = 1$
$\sqrt[3]{8} = 2$	$\sqrt[5]{32} = 2$
$\sqrt[3]{27} = 3$	$\sqrt[5]{243} = 3$
$\sqrt[3]{64} = 4$	$\sqrt[5]{-1} = -1$
$\sqrt[3]{125} = 5$	$\sqrt[5]{-32} = -2$
$\sqrt[3]{216} = 6$	$\sqrt[5]{-243} = -3$

Once again there is an important difference between

$$-\sqrt[4]{81} \quad \text{and} \quad \sqrt[4]{-81}, \quad \text{because}$$

$$-\sqrt[4]{81} = -\sqrt[4]{81} = -3$$

whereas,

$$-\sqrt[4]{-81} \text{ is not a real number.}$$

EXAMPLE 1   **Finding the Roots of a Number**

Find the principal roots of the given numbers.

(a) $\sqrt{49}$ (b) $-\sqrt{49}$ (c) $\sqrt{-9}$ (d) $\sqrt[3]{27}$

(e) $\sqrt[3]{-27}$ (f) $\sqrt[5]{32}$ (g) $\sqrt[7]{-128}$

Solution

	$\sqrt[n]{b} = a$	Reason
(a)	$\sqrt{49} = 7$	$7^2 = 49$
(b)	$-\sqrt{49} = -7$	$-\sqrt{49}$ is the opposite of $\sqrt{49}$
(c)	$\sqrt{-9}$ not real	Any real number squared is nonnegative
(d)	$\sqrt[3]{27} = 3$	$3^3 = 27$
(e)	$\sqrt[3]{-27} = -3$	$(-3)^3 = -27$
(f)	$\sqrt[5]{32} = 2$	$2^5 = 32$
(g)	$\sqrt[7]{-128} = -2$	$(-2)^7 = -128$

It is important to note that a radical with an even index must have a nonnegative radicand; whereas a radical with an odd index will always be a real number as a radical. Remember that $\sqrt[n]{0} = 0$.

Evaluating Radicals Using Absolute Value

Recall that the notation $\sqrt{x^2}$ indicates the positive square root of x^2 only. For example,

$$\sqrt{(-3)^2} = \sqrt{9} = 3.$$

When variables are present in the radicand and it is unclear whether the variable represents a positive number or a negative number, absolute value bars are sometimes needed to ensure that the result is a positive number. With this in mind, we write

$$\sqrt{x^2} = |x|.$$

Similar situations may occur when the index is any even positive integer. When the index is any *odd* positive integer, absolute value bars are not necessary. In general,

If a is a real number and n is a positive integer, $n \geq 2$, then:

1. $\sqrt[n]{a^n} = |a|$ if n is even   2. $\sqrt[n]{a^n} = a$ if n is odd.

For example,

$$\sqrt{5^2} = |5| = 5 \quad \text{and} \quad \sqrt{0^2} = |0| = 0.$$

EXAMPLE 2   **Evaluating Radical Expressions**

Evaluate each expression (if possible).

(a) $\sqrt{(-7)^2}$ (b) $\sqrt[3]{(-5)^3}$ (c) $\sqrt[4]{-2^4}$ (d) $\sqrt[4]{x^8}$

(e) $\sqrt{9x^2}$ (f) $\sqrt{(x-1)^2}$ (g) $\sqrt[3]{27\,y^3}$ (h) $\sqrt[5]{(x+3)^5}$

Solution

	$\sqrt[n]{b} = a$	Reason
(a)	$\sqrt{(-7)^2} = \lvert -7 \rvert = 7$	Index of the radical is even
(b)	$\sqrt[3]{(-5)^3} = -5$	Index of the radical is odd
(c)	$\sqrt[4]{-2^4}$ not real	Even index with negative radicand
(d)	$\sqrt[4]{x^8} = x^2$	$(x^2)^4 = x^8,\ x^2 \geq 0$
(e)	$\sqrt{9x^2} = \lvert 3x \rvert = 3\lvert x \rvert$	Even root
(f)	$\sqrt{(x-1)^2} = \lvert x - 1 \rvert$	Even root
(g)	$\sqrt[3]{27y^3} = 3y$	Odd root
(h)	$\sqrt[5]{(x+3)^5} = x + 3$	Odd root

EXAMPLE 3   **Evaluating Radical Expressions**

Simplify each expression.

(a) $\sqrt{x^2 - 4x + 4}$ (b) $\sqrt{4x^2 + 12xy + 9y^2}$

Solution   (a) Note that $x^2 - 4x + 4$ is a perfect square trinomial, so that

$$\sqrt{x^2 - 4x + 4} = \sqrt{(x-2)^2} = \lvert x - 2 \rvert$$

(b) Again we write $4x^2 + 12xy + 9y^2$ as $(2x + 3y)^2$, so that

$$\sqrt{4x^2 + 12xy + 9y^2} = \sqrt{(2x + 3y)^2} = \lvert 2x + 3y \rvert$$

We can raise a number to the nth power and take the principal nth root of a number. For instance,

$$\left(\sqrt{9} \right)^2 = 3^2 = 9 \text{ and } \sqrt{9^2} = \sqrt{81} = 9$$

$$\left(\sqrt[3]{27} \right)^3 = 3^3 = 27 \text{ and } \sqrt[3]{27^3} = \sqrt[3]{19{,}683} = 27.$$

In general,

> If a has a principal nth root, then
>
> $$\left(\sqrt[n]{a} \right)^n = a \text{ and } \sqrt[n]{a^n} = a.$$

Evaluating Radical Functions

A function f of the form

$$f(x) = \sqrt{x}$$

is called a **square root function.**
A function g of the form

$$g(x) = \sqrt[3]{x}$$

is called a **cube root function.**
To evaluate these functions, sometimes, it is necessary to use a calculator.

EXAMPLE 4 **Evaluating Radical Functions**

Let $f(x) = \sqrt{-3x + 1}$ and $g(x) = \sqrt[3]{4x + 7}$

Find each value.

(a) $f(-5)$ (b) $f(5)$ (c) $g(5)$ (d) $g(-5)$

Solution

(a) $f(-5) = \sqrt{-3(-5) + 1} = \sqrt{15 + 1} = \sqrt{16} = 4$

(b) $f(5) = \sqrt{-3(5) + 1} = \sqrt{-15 + 1} = \sqrt{-14}$ not a real number

(c) $g(5) = \sqrt[3]{4(5) + 7} = \sqrt[3]{20 + 7} = \sqrt[3]{27} = 3$

(d) $g(-5) = \sqrt[3]{4(-5) + 7} = \sqrt[3]{-20 + 7} = \sqrt[3]{-13} = -2.35$ approximately

EXAMPLE 5 **Analyzing Graphs of Radical Functions**

(a) In each case, substitute values of x to obtain the corresponding values for y.

x	0	1	4	5	9
$y = \sqrt{x}$					

x	−8	−1	0	2	8
$y = \sqrt[3]{x}$					

(b) Plot these points from the tables of part (a), and indicate whether or not it fits on the graphs of the functions $y = \sqrt{x}$ and $y = \sqrt[3]{x}$ in Figure 1.

(c) What do you think of the graph of $F(x) = -\sqrt{x}$ looks like?

Solution

(a) We substituted values of x and found the corresponding values for $y = \sqrt{x}$ and $y = \sqrt[3]{x}$.

x	0	1	4	5	9
$y = \sqrt{x}$	0	1	2	2.24	3

x	−8	−1	0	2	8
$y = \sqrt[3]{x}$	−2	−1	0	1.26	2

(b) By plotting the points in the above table and connecting them with a smooth curve, we see that these points fit on the graphs of $y = \sqrt{x}$ and $y = \sqrt[3]{x}$ in Figures 1a and 1b.

(c) By examining the graphs in Figure 1a, we see that the graph of $F(x) = -\sqrt{x}$ is similar to the graph in Figure 1a, but it is reflected across the x axis (Figure 1c). The domain of F is $[0, \infty)$.

Figure 1

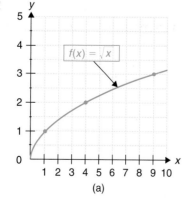
(a)

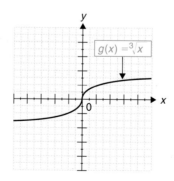
(b)

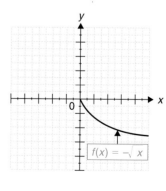
(c)

Graphing Radical Functions

Figure 1(a) shows the graph of the *square root function.*

$$f(x) = \sqrt{x}.$$

Figure 1(b) shows the graph of the *cube root function*

$$g(x) = \sqrt[3]{x}.$$

By studying the graph of $f(x) = \sqrt{x}$ in Figure 1(a), we observe that the domain of f is all nonnegative real numbers or $[0, \infty)$ and the range is also $[0, \infty)$.

Looking at the graph of $g(x) = \sqrt[3]{x}$, we see that both of the domain and range are all real numbers.

Solving Applied Problems

EXAMPLE 6

Finding the Radius of a Cone

This is a picture of a Blackfeet native designed teepee which is in the shape of a cone. When erecting this teepee, the poles will allow a height of 17 feet and the volume must be 1442 cubic feet. To know how much space is needed on the ground, one needs to know the radius of the base of the teepee. Find the radius using the formula

$$r = \sqrt{\frac{3V}{\pi h}}$$

where r is the radius of the base, V is the volume and h is the height.

Solution

Substituting $V = 1442$ cubic feet, $h = 17$ feet, and $\pi \approx 3.14$, we have

$$r = \sqrt{\frac{3V}{\pi h}} = \sqrt{\frac{3(1442)}{3.14(17)}} \approx 9.00 \text{ feet.}$$

The periodic motion of a swinging pendulum was first noticed by the Arab mathematician, Ibn Yunus (c. 950–1009). In 1656, the first pendulum clock was built by Christian Huygens. The length of time T (in seconds) of one swing back and forth of a pendulum of length L (in feet) is described by the model

$$T = \frac{\pi}{2}\sqrt{\frac{L}{2}}.$$

EXAMPLE 7

Finding the Length of a Pendulum

In designing a grandfather clock that keeps correct time, it was determined that the pendulum took exactly 2 seconds to swing back and forth. To determine the length of the pendulum:

(a) Use the formula $T = \dfrac{\pi}{2}\sqrt{\dfrac{L}{2}}$ to complete the following table. (Round off to two decimal places.)

L(feet)	0	1	2.50	2.75	3.00	3.25	3.50	3.75	4.00
T(seconds)									

(b) Use the table to solve the equation

$$\frac{\pi}{2}\sqrt{\frac{L}{2}} = 2$$

by approximating the value of L when $T = 2$.

(c) Plot the points found in the table.

(d) Compare the results of the table and the graph of the function to find the length L that will keep "correct time".

Solution

(a) The output values are given in the following table:

Figure 2

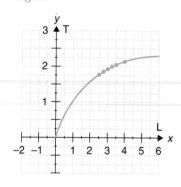

L(feet)	0	1	2.50	2.75	3.00	3.25	3.50	3.75	4.00
T(sec)	0	1.11	1.76	1.84	1.92	2.00	2.08	2.15	2.22

(b) The table shows that a time of 2 seconds corresponds to a length of 3.25 feet.

(c) The graph is in Figure 2.

(d) The table value and the graphical value seem to agree on the length of the pendulum to obtain the correct time.

PROBLEM SET 6.2

Mastering the Concepts

In Problems 1 and 2, complete each statement.

1. (a) Since $5^2 = 25$, then ____ is a square root of 25.

(b) If $4^3 = 64$, then ____ is a cube root of 64.

2. (a) Since $3^4 = 81$, then ____ is a fourth root of 81.

(b) If $(2x)^5 = 32x^5$, then ____ is a fifth root of $32x^5$.

In Problems 3–12, find the principal roots of the given number. State the reason when it is not a real number.

3. (a) $\sqrt{16}$

(b) $\sqrt[3]{-8}$

4. (a) $\sqrt{144}$

(b) $-\sqrt{100}$

5. (a) $\sqrt{0.04}$

(b) $-\sqrt{0.36}$

6. (a) $\sqrt[3]{-0.008}$

(b) $-\sqrt{0.09}$

7. (a) $\sqrt[4]{81}$
(b) $\sqrt[4]{-81}$

8. (a) $\sqrt[3]{-64}$
(b) $\sqrt{-64}$

9. (a) $-\sqrt[4]{\dfrac{16}{81}}$
(b) $\sqrt{\dfrac{-9}{25}}$

10. (a) $\sqrt[5]{\dfrac{1}{32}}$
(b) $-\sqrt[5]{\dfrac{-1}{32}}$

11. (a) $-\sqrt[3]{\dfrac{1}{8}}$
(b) $-\sqrt[3]{\dfrac{-27}{64}}$

12. (a) $-\sqrt{\dfrac{-9}{25}}$
(b) $-\sqrt{\dfrac{81}{144}}$

In Problems 13–26, simplify each expression (if possible). Assume all variables represent positive real numbers.

13. (a) $-\sqrt{11^2}$
(b) $\sqrt{(-11)^2}$

14. (a) $-\sqrt{5^4}$
(b) $\sqrt{(-5)^4}$

15. (a) $\sqrt[3]{(-3)^3}$
(b) $-\sqrt[3]{(-3)^3}$

16. (a) $\sqrt[4]{-9}$
(b) $-\sqrt[4]{(-3)^4}$

17. (a) $\sqrt[3]{-8x^3}$
(b) $\sqrt{(-6y)^2}$

18. (a) $-\sqrt[3]{-8y^3}$
(b) $\sqrt{(-16)^2}$

19. (a) $\sqrt[3]{8x^9\,y^6}$
(b) $\sqrt{4x^2\,y^4}$

20. (a) $-\sqrt[3]{-125\,y^6}$
(b) $\sqrt[4]{16y^{12}}$

21. (a) $(\sqrt{5x})^2$
(b) $\left(\sqrt[3]{(-3x)}\right)^3$

22. (a) $-\sqrt[4]{-16y^4}$
(b) $\sqrt[3]{-27x^3\,y^9}$

23. (a) $\left(\sqrt{y^7}\right)^2$
(b) $\left(\sqrt[3]{-3x}\right)^3$

24. (a) $\left(\sqrt[4]{4x}\right)^4$
(b) $\left(\sqrt[5]{3y}\right)^5$

25. (a) $\left(\sqrt[3]{-2x}\right)^3$
(b) $\left(\sqrt[4]{-5t^2}\right)^4$

26. (a) $\left(\sqrt[5]{2y^2}\right)^5$
(b) $\left(\sqrt[6]{-4y}\right)^6$

In Problems 27–32, simplify each expression. Assume that variables represent any real number.

27. (a) $\sqrt{4x^2}$
(b) $-\sqrt{25x^2}$

28. (a) $\sqrt{(-5x)^2}$
(b) $-\sqrt{36x^2}$

29. (a) $\sqrt{(-3x)^2}$
(b) $\sqrt[3]{y^6}$

30. (a) $\sqrt{100x^4}$
(b) $\sqrt{x^{12}}$

31. (a) $\sqrt[5]{x^5}$
(b) $\sqrt{(x-3)^2}$

32. (a) $-\sqrt{121x^{20}}$
(b) $-\sqrt{144y^{12}}$

33. For what values of x will
$$\sqrt{(3x-7)^2} = 3x - 7?$$

34. For what values of x will
$$\sqrt{x^2 + 6x + 9} = x + 3?$$

In Problems 35 and 36, use absolute value to simplify each expression.

35. $\sqrt{x^2 - 10x + 25}$

36. $\sqrt{4x^2 + 20x + 25}$

37. Can you find the domain of each function without looking at the graphs? If so, find the domains.
(a) $f(x) = \sqrt{x - 3}$
(b) $g(x) = \sqrt{x + 3}$

38. Find a square root function whose domain is given
(a) Dom: $x \ge 2$
(b) Dom: $x \le 3$

In Problems 39–44, evaluate each function at the indicated input values. Round off to two decimal places.

39. $f(x) = \sqrt{2x + 3}$
(a) $x = -1$
(b) $x = 0$
(c) $x = 2$

40. $f(x) = -\sqrt{3x + 1}$
(a) $x = 0$
(b) $x = 1$
(c) $x = 3$

41. $g(x) = \sqrt[3]{x - 8}$
(a) $x = -1$
(b) $x = 0$
(c) $x = 3$

42. $g(x) = \sqrt[3]{2x + 5}$
(a) $x = -1$
(b) $x = 3$
(c) $x = 7$

43. $f(x) = -\sqrt{x^2 + 4}$
(a) $x = -5$
(b) $x = 0$
(c) $x = 5$

44. $f(x) = \sqrt{x^3 - 5}$
(a) $x = 2$
(b) $x = 3$
(c) $x = 5$

In Problems 45 and 46,
(a) evaluate each function for $x = 0, 1, 4, 9, 16$.
(b) Plot these points to sketch the graph of each function.
(c) Use the graph to indicate the domain and range of the function.

45. $f(x) = \sqrt{x} + 2$

46. $g(x) = \sqrt{x} - 2$

Applying the Concepts

In Problems 47 and 48 the annual rate of return r on an investment of p dollars that is worth s dollars after t years is modeled by the formula
$$r = \left(\frac{s}{p}\right)^{1/t} - 1.$$

47. What is the rate of return for an investment of $1200 that is worth $1323 in 2 years?

48. What is the rate of return for a bond of $2800 that is worth $3150 after 3 years?

49. **Biology:** A biologist estimates that the population P of a certain culture of bacteria is given by the model
$$P(t) = 1000\sqrt[4]{t^5 + 10t^2 + 9}$$
where t is the time in hours since the culture was started. Find the population of the bacteria

after (a) 2 hours, (b) 3 hours. Round off to the nearest integer.

50. **Electronics:** In electronics, the resonant frequency f (in hertz) for a tuned circuit is given by the formula

$$f = \frac{1}{2\pi\sqrt{LC}}$$

where L is the inductance and C is the capacitance. Find the frequency f for a tuned circuit in which $L = 3.57 \times 10^{-8}$ and $C = 1.21 \times 10^{-10}$ (round off to two decimal places, $\pi \approx 3.14$).

51. **Physics:** The length of time T (in seconds) for a pendulum to complete one swing back and forth is given by the model

$$T = 2\pi\sqrt{\frac{L}{32}}$$

where L (in feet) is the length of the pendulum. Find the time of a complete swing back and forth for:

(a) a 2.4 foot pendulum to the nearest tenth of a second.

(b) The Hong Kong Bank of Canada pendulum in Vancouver, with a length of 90 feet.

52. **Physiology:** A manufacturer of scientifically designed walking shoes uses the equation

Figure 6

$$T = 2\pi\sqrt{0.068L}$$

where T is the time (in seconds) for one swing of a leg of length L (in meters) (Figure 6). Find the time of the swing of a leg of length 0.81 meters.

53. **Satellite Orbit:** In order for a satellite to maintain an orbit of h kilometers, its velocity v (in kilometers/second) must be related to its height by the equation

$$v = \frac{626.4}{\sqrt{h + R}}$$

where $R = 6372$ kilometers is the radius of the earth. What is the speed of the satellite when it is in orbit 105 kilometers from the earth's surface?

54. **Physiology:** Physiologists define the threshold weight of a human being as the weight beyond which the risk of death increases significantly. For middle-aged males, their height h in feet is related to the threshold weight w in pounds by

$$h = 4.56\sqrt[3]{\frac{w}{100}} \text{ where } 160 < w < 220.$$

(a) Estimate h when $w = 175$.

(b) If the threshold weight of a man is 200 pounds, what is his height?

Developing and Extending the Concepts

55. Let x be a real number; what restrictions are needed on n for each of the following equations to be true?

(a) $\sqrt[n]{x^n} = x$ (b) $\sqrt[n]{x^n} = |x|$

56. Indicate the first step needed to simplify the expression

$$\sqrt{x^2 + 6x + 9}.$$

Then explain how to determine the domain of $f(x)\sqrt{x^2 + 6x + 9}$.

In Problems 57–60, use the given function to find the output values.

57. $f(x) = \sqrt{1 - x}$
    (a) $f(-1)$ (b) $f(0)$ (c) $f(1)$

58. $k(x) = \sqrt{x} - \sqrt{x - 1}$
    (a) $k(-1)$ (b) $k(0)$ (c) $k(2)$

59. $h(x) = \sqrt{-x}$
    (a) $h(-1)$ (b) $h(0)$ (c) $h(1)$

60. $g(x) = \sqrt[3]{x}$
    (a) $g(-8)$ (b) $g(0)$ (c) $g(8)$

In Problems 61 and 62, determine the domain of each function algebraically and verify graphically.

61. (a) $f(x) = \sqrt{x - 4}$ 62. (a) $f(x) = \sqrt[3]{2x - 6}$
    (b) $g(x) = \sqrt{4 - x}$ (b) $g(x) = \sqrt[4]{x + 1}$

63. $f(x) = \sqrt{x - 3}$
    (a) Complete the table

x	3	4	7	19
$f(x)$				

(b) Plot the points obtained in the table in part (a) and connect them with a smooth curve.

(c) Use the graph to indicate the domain and range of *f*.

(d) Find the solution of the equation $\sqrt{x-3} = 2$ by identifying the *x* coordinate of the point of intersection of $y = \sqrt{x-3}$ and $y = 2$.

64. $g(x) = \sqrt{x+1}$

(a) Complete the table

x	−1	3	8	15
g(x)				

(b) Plot the points obtained in the table in part (a) and connect them with a smooth curve.

(c) Use the graph, *g*, to indicate the domain and range of *g*.

(d) Find the solution of the equation $\sqrt{x+1} = 3$ by identifying the *x* coordinate of the point of intersection of $y = \sqrt{x+1}$ and $y = 3$.

65. What can you say about a non zero radicand if its cube root is:

(a) doubled?   (b) tripled?

(c) quadrupled?   (d) What is the pattern?

66. What can you say about a non zero radicand if its square root is:

(a) doubled?   (b) tripled?

(c) quadrupled?   (d) What is the pattern?

Objectives

1. Define Expressions of the Form $a^{\frac{1}{n}}$
2. Define Expressions of the Form $a^{\frac{m}{n}}$
3. Convert Forms
4. Use Properties of Rational Exponents
5. Solve Applied Problems
6. Graph an Exponential Function

The restriction on $a^{\frac{1}{n}}$ is the same as the restriction on $\sqrt[n]{a}$.

6.3 Rational Exponents and Functions

Defining Expressions of the Form $a^{\frac{1}{n}}$

If all exponent properties hold even if some of the exponents are not integers, then we would have

$$(7^{\frac{1}{2}})^2 = 7^{\frac{2}{2}} = 7^1 = 7 \quad \text{and} \quad (5^{\frac{1}{3}})^3 = 5^{\frac{3}{3}} = 5^1 = 5.$$

That is, $7^{\frac{1}{2}}$ is a number whose square is 7, so $7^{\frac{1}{2}}$ is another name for a square root of 7. Similarly, $5^{\frac{1}{3}}$ is a number whose cube is 5, and so $5^{\frac{1}{3}}$ is another name for a cube root of 5. In general, we have the following definition.

Definition of Rational Exponents

Let *a* be a real number and *n* be an integer such that $n \geq 2$.

If the principal *n*th root of *a* exists, we define

$$a^{\frac{1}{n}} = \sqrt[n]{a}.$$

We need to be careful using the above definition when *n* is an even number and *a* is negative because in such a case $a^{\frac{1}{n}}$ is not a real number.

EXAMPLE 1   **Evaluating Expressions with Rational Exponents**

Evaluate each expression (if possible).

(a) $4^{\frac{1}{2}}$ (b) $-4^{\frac{1}{2}}$ (c) $(-4)^{\frac{1}{2}}$ (d) $8^{\frac{1}{3}}$

(e) $-8^{\frac{1}{3}}$ (f) $(-8)^{\frac{1}{3}}$ (g) $0^{\frac{1}{7}}$

Solution   We will evaluate each expressions using both radicals and rational exponents.

Radical Form	Rational Form
(a) $4^{\frac{1}{2}} = \sqrt{4} = 2$	$4^{\frac{1}{2}} = (2^2)^{\frac{1}{2}} = 2^{\frac{2}{2}} = 2$
(b) $-4^{\frac{1}{2}} = -\sqrt{4} = -2$	$-4^{\frac{1}{2}} = -(2^2)^{\frac{1}{2}} = -2^{\frac{2}{2}} = -2$
(c) $(-4)^{\frac{1}{2}} = \sqrt{-4}$ not real	Radicand must be nonnegative
(d) $8^{\frac{1}{3}} = \sqrt[3]{8} = 2$	$8^{\frac{1}{3}} = (2^3)^{\frac{1}{3}} = 2^{\frac{3}{3}} = 2$
(e) $-8^{\frac{1}{3}} = -\sqrt[3]{8} = -2$	$-8^{\frac{1}{3}} = -(2^3)^{\frac{1}{3}} = -2^{\frac{3}{3}} = -2$
(f) $(-8)^{\frac{1}{3}} = \sqrt[3]{-8} = -2$	$(-8)^{\frac{1}{3}} = ([-2]^3)^{\frac{1}{3}} = (-2)^{\frac{3}{3}} = -2$
(g) $0^{\frac{1}{7}} = \sqrt[7]{0} = 0$	$0^{\frac{1}{7}} = (0^7)^{\frac{1}{7}} = 0^{\frac{7}{7}} = 0$

Defining Expressions of the Form $a^{\frac{m}{n}}$

Having defined expressions of the form $a^{\frac{1}{n}}$, we now are able to extend our definition to include expressions of the form $a^{\frac{m}{n}}$, where $\frac{m}{n}$ is any rational number. Observe, for example, the two ways that $27^{\frac{2}{3}}$ can be evaluated on the assumption that the properties of exponents are to hold for rational numbers:

$$27^{\frac{2}{3}} = (27^{\frac{1}{3}})^2 = (\sqrt[3]{27})^2 = 3^2 = 9$$

or

$$27^{\frac{2}{3}} = (27^2)^{\frac{1}{3}} = \sqrt[3]{27^2} = \sqrt[3]{729} = 9.$$

This suggests that we define $a^{\frac{m}{n}}$ to be $(\sqrt[n]{a})^m$ whenever the roots make sense. More formally, we have the following definition.

Definition of $a^{\frac{m}{n}}$ and $a^{-\frac{m}{n}}$

Suppose that m and n are integers with $n \geq 2$, and that the fraction $\frac{m}{n}$ is reduced to lowest terms. Let a be a non zero real number for which $\sqrt[n]{a}$ is defined. Then

$$a^{\frac{m}{n}} = (a^{\frac{1}{n}})^m = (\sqrt[n]{a})^m = \sqrt[n]{a^m} \quad \text{and} \quad a^{-\frac{m}{n}} = \frac{1}{a^{\frac{m}{n}}} = \frac{1}{\sqrt[n]{a^m}}.$$

EXAMPLE 2   **Evaluating Expressions with Rational Exponents**

Evaluate each expression using both radical and rational exponents.

(a) $4^{\frac{3}{2}}$ (b) $4^{\frac{-3}{2}}$ (c) $27^{\frac{4}{3}}$

(d) $(-8)^{\frac{-2}{3}}$ (e) $-8^{\frac{-2}{3}}$ (f) $(-4)^{\frac{3}{2}}$

Solution

$-2^2 = -4$ and
$(-2)^2 = +4$

Radical Form	Rational Form
(a) $4^{\frac{3}{2}} = \left(4^{\frac{1}{2}}\right)^3 = \left(\sqrt{4}\right)^3 = 2^3 = 8$	$4^{\frac{3}{2}} = \left(2^2\right)^{\frac{3}{2}} = 2^{2\cdot\frac{3}{2}} = 2^3 = 8$
(b) $4^{\frac{-3}{2}} = \left(4^{\frac{1}{2}}\right)^{-3} = \sqrt{4^{-3}} = \sqrt{\frac{1}{64}} = \frac{1}{8}$	$4^{\frac{-3}{2}} = \left(2^2\right)^{\frac{-3}{2}} = 2^{-3} = \frac{1}{8}$
(c) $27^{\frac{4}{3}} = (27^{\frac{1}{3}})^4 = \left(\sqrt[3]{27}\right)^4 = 3^4 = 81$	$27^{\frac{4}{3}} = \left(3^3\right)^{\frac{4}{3}} = 3^{3\cdot\frac{4}{3}} = 3^4 = 81$
(d) $(-8)^{\frac{-2}{3}} = \left[(-8)^{\frac{1}{3}}\right]^{-2} = \left(\sqrt[3]{-8}\right)^{-2} = (-2)^{-2} = \frac{1}{4}$	$(-8)^{\frac{-2}{3}} = \left[(-2)^3\right]^{\frac{-2}{3}} = (-2)^{3\cdot\frac{-2}{3}} = (-2)^{-2} = \frac{1}{4}$
(e) $-8^{\frac{-2}{3}} = -\left(8^{\frac{1}{3}}\right)^{-2} = -\left(\sqrt[3]{8}\right)^{-2} = -2^{-2} = -\frac{1}{4}$	$-8^{\frac{-2}{3}} = -\left(2^3\right)^{\frac{-2}{3}} = -2^{-2} = -\frac{1}{4}$
(f) not real, nth root is even	$(-4)^{\frac{3}{2}} = \left[(-4)^{\frac{1}{2}}\right]^3$ base is negative

Converting Forms

In the preceding definitions, we introduced the exponential notation $a^{\frac{1}{n}}$ and the radical notation $\sqrt[n]{a}$ to represent the nth root of a. We also used the notations $\left(\sqrt[n]{a}\right)^m$ and $a^{\frac{m}{n}}$ to represent the powers of an nth root. In many situations, it is important to be able to convert expressions from radical forms to exponential forms and vice versa.

EXAMPLE 3   **Converting Forms**

Convert each radical expression to exponential form and each exponential expression to radical form.

(a) $7^{\frac{1}{2}}$ (b) $4x^{\frac{3}{5}}$ (c) $y^{\frac{-2}{3}}$

(d) $\sqrt{5}$ (e) $\sqrt[3]{x^2}$ (f) $\sqrt[5]{(3xy^2)^3}$

Solution   (a) $7^{\frac{1}{2}} = \sqrt{7}$ (b) $4x^{\frac{3}{5}} = 4\left(\sqrt[5]{x}\right)^3 = 4\sqrt[5]{x^3}$ (c) $y^{\frac{-2}{3}} = \dfrac{1}{y^{\frac{2}{3}}} = \dfrac{1}{\sqrt[3]{y^2}}$

(d) $\sqrt{5} = 5^{\frac{1}{2}}$ (e) $\sqrt[3]{x^2} = x^{\frac{2}{3}}$ (f) $\sqrt[5]{(3xy^2)^3} = (3xy^2)^{\frac{3}{5}}$

Using the Properties of Rational Exponents

The properties of exponents listed in section 6.1 also apply to rational exponents, provided that all the quantities are defined. Therefore, we list these properties to indicate that they are true for all rational exponents.

Properties of Rational Exponents

Let a and b be non zero real numbers, and let m and n be rational numbers. Then each of the following is true for all values of a and b for which both sides of the equation are defined:

(i) $a^m \cdot a^n = a^{m+n}$ Product property

(ii) $(a^m)^n = a^{mn}$ Power to a power property

(iii) $(ab)^m = a^m b^m$ Power of product property

(iv) $\left(\dfrac{a}{b}\right)^m = \dfrac{a^m}{b^m}$ Power of quotient property

(v) $\dfrac{a^m}{a^n} = a^{m-n}$ Quotient property

In addition to the above properties, we have

$$1. \quad a^0 = 1, a \neq 0 \quad \text{and} \quad a^{-m} = \frac{1}{a^m}$$

The following useful relationship is an immediate consequence of the exponent properties.

$$a^{\frac{m}{n}} = \left(a^{\frac{1}{n}}\right)^m = \left(a^m\right)^{\frac{1}{n}}$$

EXAMPLE 4

Using the Properties of Exponents

Simplify each expression. Write the result using only positive exponents. Assume all variables represent real numbers.

(a) $7^{-\frac{1}{2}} \cdot 7^{\frac{5}{2}}$ (b) $\left(x^{\frac{3}{5}}\right)^{-10}$ (c) $\dfrac{x^{\frac{2}{3}}}{x^{-\frac{1}{5}}}$ (d) $\left(125x^{-9}\right)^{\frac{-2}{3}}$ (e) $\left(\dfrac{32}{x^{-5}}\right)^{\frac{-2}{5}}$

Solution

(a) $7^{-\frac{1}{2}} \cdot 7^{\frac{5}{2}} = 7^{-\frac{1}{2}+\frac{5}{2}} = 7^{\frac{4}{2}} = 7^2 = 49$ Product property

(b) $\left(x^{\frac{3}{5}}\right)^{-10} = x^{\frac{3}{5}(-10)} = x^{-6} = \dfrac{1}{x^6}$ Power to a power property

(c) $\dfrac{x^{\frac{2}{3}}}{x^{-\frac{1}{5}}} = x^{\frac{2}{3}-\frac{-1}{5}} = x^{\frac{10}{15}+\frac{3}{15}} = x^{\frac{13}{15}}$ Quotient property

(d) $\left(125x^{-9}\right)^{\frac{-2}{3}} = \left(125^{\frac{-2}{3}}\right)\left(x^{-9}\right)^{\frac{-2}{3}}$ Power of product property

$= \left(5^3\right)^{\frac{-2}{3}} x^6$ Power to a power property

$= 5^{-2} x^6 = \dfrac{x^6}{25}$ Power property

(e) $\left(\dfrac{32}{x^{-5}}\right)^{\frac{-2}{5}} = \dfrac{\left(32\right)^{\frac{-2}{5}}}{\left(x^{-5}\right)^{\frac{-2}{5}}}$ Power of quotient property

$= \dfrac{\left(2^5\right)^{\frac{-2}{5}}}{x^2} = \dfrac{2^{-2}}{x^2} = \dfrac{1}{4x^2}$ Power to a power property

EXAMPLE 5 **Using the Properties of Exponents**

Suppose that x and y represent positive numbers. Simplify each expression and write the result using only positive exponents.

(a) $\dfrac{x^{-\frac{1}{5}} y^{\frac{5}{6}}}{x^{\frac{4}{5}} y^{\frac{1}{3}}}$

(b) $x^{\frac{-2}{3}}\left(x^{\frac{4}{3}} + x^{-\frac{1}{3}} y^{\frac{1}{5}}\right)$

Solution

(a) $\dfrac{x^{-\frac{1}{5}} y^{\frac{5}{6}}}{x^{\frac{4}{5}} y^{\frac{1}{3}}} = x^{-\frac{1}{5}-\frac{4}{5}} \cdot y^{\frac{5}{6}-\frac{1}{3}}$ Quotient property

$= x^{\frac{-5}{5}} \cdot y^{\frac{3}{6}} = \dfrac{y^{\frac{1}{2}}}{x}$ Product property

(b) $x^{\frac{-2}{3}}\left(x^{\frac{4}{3}} + x^{-\frac{1}{3}} y^{\frac{1}{5}}\right) = x^{\frac{-2}{3}} \cdot x^{\frac{4}{3}} + x^{\frac{-2}{3}} \cdot x^{-\frac{1}{3}} \cdot y^{\frac{1}{5}}$ Distributive property

$= x^{\frac{4}{3}-\frac{2}{3}} + x^{-1} y^{\frac{1}{5}}$ Product property

$= x^{\frac{2}{3}} + \dfrac{y^{\frac{1}{5}}}{x}$

Solving Applied Problems

The compound interest model is also applicable if the money is invested for a fraction of a year such as 2 1/2 years, 3 2/5 years and so on. The next example illustrates the use of the compound interest model for this case.

EXAMPLE 6 **Earning Compound Interest**

Suppose that $1000 is deposited in a bank that pays 3.5% annual interest compounded monthly. What is the value S (money accumulated) of the account after a term of 8 months or $\dfrac{2}{3}$ years?

Solution

Monthly compounding means that $n = 12$. Substituting 1000 for P, 0.035 for r, 12 for n, and $\dfrac{2}{3}$ for t in the compound interest formula, we have

$$S = P\left(1 + \frac{r}{n}\right)^{nt}$$

$$S = 1000\left(1 + \frac{0.035}{12}\right)^{12\left(\frac{2}{3}\right)} = \$1023.57.$$

Therefore, the accumulated amount of money after 2/3 year will be $1023.57.

Graphing an Exponential Function

We now explore graphical results of the definition of $a^{\frac{1}{n}}$. Remember that we can only input nonnegative values if n is even and any real number if n is odd. The next example explores a graph when $n = 3$, an odd number.

EXAMPLE 7   **Graphing an Exponential Function**

Graph the function $f(x) = x^{\frac{1}{3}} = \sqrt[3]{x}$.

(a) Find $f(-27)$, $f(-8)$, $f(-1)$, $f(0)$, $f(1)$, $f(8)$, $f(27)$, and $f(64)$.

(b) Plot the points in part (a) and connect them with a smooth curve.

(c) Indicate the domain and range of this function.

(d) Use part (c) to solve the equation $x^{\frac{1}{3}} = 4$.

Solution   (a) The following table shows the output values for the given input values.

Figure 1

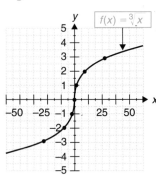

x	$y = f(x) = x^{\frac{1}{3}}$	(x,y)
-27	-3	$(-27,-3)$
-8	-2	$(-8,-2)$
-1	-1	$(-1,-1)$
0	0	$(0,0)$
1	1	$(1,1)$
8	2	$(8,2)$
27	3	$(27,3)$
64	4	$(64,4)$

(b) Figure 1 shows the plotted points and the smooth curve connecting them.

(c) The graph suggests that the domain of f is the set of all real numbers and the range is the set of real numbers.

(d) The point with 4 as the second coordinate is (64,4). The solution to the equation $x^{\frac{1}{3}} = 4$ is 64.

 PROBLEM SET 6.3

Mastering the Concepts

In Problems 1–14, find the exact value of each expression.

1. (a) $36^{\frac{1}{2}}$
   (b) $49^{-\frac{1}{2}}$
   (c) $27^{\frac{1}{3}}$

2. (a) $-16^{-\frac{1}{2}}$
   (b) $-8^{-\frac{1}{3}}$
   (c) $64^{-\frac{1}{3}}$

3. (a) $\left(-27\right)^{\frac{1}{3}}$
   (b) $81^{\frac{1}{4}}$
   (c) $-16^{\frac{1}{4}}$

4. (a) $\left(-64\right)^{\frac{1}{3}}$
   (b) $49^{\frac{1}{2}}$
   (c) $81^{-\frac{1}{4}}$

5. (a) $\left(-\dfrac{8}{27}\right)^{-\frac{1}{3}}$
   (b) $\left(\dfrac{81}{25}\right)^{-\frac{1}{2}}$
   (c) $\left(\dfrac{9}{16}\right)^{\frac{1}{2}}$

6. (a) $\left(\dfrac{1}{16}\right)^{-\frac{1}{4}}$
   (b) $\left(\dfrac{1}{64}\right)^{-\frac{1}{3}}$
   (c) $\left(\dfrac{16}{9}\right)^{-\frac{1}{2}}$

7. (a) $8^{\frac{2}{3}}$
   (b) $-8^{\frac{4}{3}}$

9. (a) $\left(-32\right)^{\frac{3}{5}}$
   (b) $128^{\frac{2}{7}}$

11. (a) $\left(-16\right)^{\frac{3}{2}}$
    (b) $\left(-36\right)^{\frac{3}{2}}$

13. (a) $(-8)^{-\frac{4}{3}}$
    (b) $(-27)^{-\frac{2}{3}}$

8. (a) $\left(-8\right)^{\frac{4}{3}}$
   (b) $\left(-27\right)^{\frac{2}{3}}$

10. (a) $81^{\frac{3}{4}}$
    (b) $\left(-64\right)^{\frac{4}{3}}$

12. (a) $27^{\frac{8}{12}}$
    (b) $25^{\frac{7}{14}}$

14. (a) $9^{-\frac{3}{2}}$
    (b) $(-32)^{-\frac{4}{5}}$

In Problems 15–20, convert to radical form. Do not simplify. Assume all variables represent positive real numbers. Use only positive exponents.

15. (a) $5^{\frac{1}{3}}$

(b) $6^{-\frac{1}{4}}$

17. (a) $x^{\frac{2}{7}}$

(b) $y^{\frac{3}{4}}$

19. (a) $(8m)^{\frac{2}{3}}$

(b) $(4xy^2)^{\frac{2}{5}}$

16. (a) $11^{\frac{1}{2}}$

(b) $7^{-\frac{1}{5}}$

18. (a) $5y^{\frac{3}{7}}$

(b) $4t^{\frac{2}{3}}$

20. (a) $(x + y)^{\frac{2}{3}}$

(b) $(x^2 + y^2)^{-\frac{1}{2}}$

In Problems 21–26, convert to rational exponents. Do not simplify. Assume all variables represent positive real numbers.

21. (a) $\sqrt{7}$

(b) $\sqrt{3}$

23. (a) $\sqrt[3]{m}$

(b) $\sqrt[5]{x^3}$

25. (a) $\sqrt[5]{(7x^2)^2}$

(b) $\sqrt[4]{(3xy)^2}$

22. (a) $\sqrt{15}$

(b) $\sqrt{17}$

24. (a) $\sqrt[3]{x^2}$

(b) $\sqrt[4]{x^3}$

26. (a) $\sqrt{(x + y)^3}$

(b) $\sqrt{(x^2 + y^2)}$

In Problems 27–40, simplify each expression and write each answer using only positive exponents.

27. (a) $2^{\frac{1}{3}} \cdot 2^{\frac{2}{3}}$

(b) $x^{-\frac{2}{3}} \cdot x^{\frac{5}{3}}$

29. (a) $(5^{\frac{1}{7}})^{14}$

(b) $(x^{\frac{-7}{4}})^{\frac{20}{7}}$

31. (a) $(8p^9)^{\frac{4}{3}}$

(b) $\left(\dfrac{125}{y^3}\right)^{-\frac{1}{3}}$

33. (a) $\dfrac{x^{\frac{2}{3}}}{x^{\frac{-1}{3}}}$

(b) $\left(\dfrac{x^{-3}}{y^2}\right)^{\frac{-1}{6}}$

35. $\left(\dfrac{x^{-10}y^8}{x^{-12}y^{-4}}\right)^{\frac{-1}{2}}$

37. $\dfrac{(125x^{-5}y^7)^{\frac{-1}{3}}}{(64x^8y^2)^{\frac{-1}{6}}}$

39. (a) $x^{\frac{1}{2}}(x^{\frac{3}{2}} + x^{\frac{-1}{2}})$

(b) $x^{\frac{-1}{2}}(x + x^{\frac{1}{2}})$

28. (a) $5^{\frac{1}{2}} \cdot 5^{-\frac{3}{2}}$

(b) $y^{\frac{2}{15}} \cdot y^{\frac{7}{60}}$

30. (a) $(8^{-\frac{2}{3}})^{-6}$

(b) $(y^{-2})^{\frac{-15}{2}}$

32. (a) $(32u^{-5})^{\frac{-3}{5}}$

(b) $\left(\dfrac{x^{-5}}{32}\right)^{\frac{4}{5}}$

34. (a) $\dfrac{x^{\frac{1}{3}}}{x^{-\frac{1}{6}}}$

(b) $\dfrac{y^{\frac{3}{2}}}{y^{\frac{-7}{2}}}$

36. $\left(\dfrac{5^2 4^6}{5^{-4}4^7}\right)^{\frac{-1}{2}}$

38. $\left(\dfrac{a^3 b^{\frac{3}{2}}}{a^{-3}b^{\frac{1}{2}}}\right)^{\frac{1}{6}}$

40. (a) $x^{\frac{5}{3}}(x^{\frac{-2}{3}} + x^{\frac{1}{3}})$

(b) $x^{\frac{-2}{3}}(x^{\frac{2}{3}} + x^{\frac{-1}{3}})$

In Problems 41–44, (a) find the values of each function at the indicated values.

(b) Plot the points in part (a) and connect them with a smooth curve.

(c) Indicate the domain and range of each function.

41. $f(x) = x^{\frac{2}{3}}$ $f(-27), f(-8), f(0), f(1),$ and $f(27)$.

42. $g(x) = x^{\frac{3}{2}}$ $g(0), g(1), g(4), g(9),$ and $g(16)$.

43. $h(x) = x^{\frac{-2}{3}}$ $h(-27), h(-1/8), h(0), h(1),$ and $h(27)$.

44. $F(x) = x^{\frac{-3}{2}}$ $F(1/4), F(1), F(4), F(9),$ and $F(16)$.

Applying the Concepts

45. Shoe Design: A man's shoe size s is approximately related to his height h in feet by the model

$$s \approx 0.769\, h^{\frac{3}{2}} \text{ where } 5 < h \le 9.$$

(a) If a man is 6 feet tall, what is his shoe size? Round to the nearest half size.

(b) According to the Bible, Goliath was 6 cubits tall (approximately 9 feet). What shoe size would he have worn? [Source: *Guinness Book of Records*]

46. Weight of a Whale: The International Whaling Commission determined that the relationship between the weight W in tons and length L in feet of a sperm whale is given by the model

$$W = 0.000137\, L^{3.18}.$$

Determine the weight of a sperm whale which is 67.5 feet long. Round off to the nearest ton.

47. Compound Interest Earned: Suppose that $3000 is invested in a risky bond that earned 8% annual interest. Find the amount of money in the account

(a) after 2½ years if the interest is compounded weekly.

(b) after 3⅔ years if the interest is compounded monthly.

48. Rate of Inflation: Because of inflation, the value of a home often increases with time. State and local agencies use the formula

$$r = \left(\dfrac{S}{C}\right)^{\frac{1}{t}} - 1$$

to assess the annual rate of inflation r for tax purposes, where C (in dollars) is the cost of the home when new, and S (in dollars) is the value of the home after t years. Suppose that a house cost $29,500 in 1969. In 2005, the same house is worth $125,000. What is the annual rate of inflation?

Developing and Extending the Concepts

In Problems 49–54, use the properties of rational exponents to simplify each expression. Express the answer using positive exponents. Assume all variables represent positive real numbers.

49. $\left(\dfrac{x^{\frac{2}{3}}}{y^{\frac{-3}{4}}}\right)^{12} \cdot \left(x^{\frac{-3}{8}}y^{\frac{1}{4}}\right)^{-2}$

50. $\left(\dfrac{81p^{-12}}{q^{16}}\right)\left(\dfrac{p^{\frac{-2}{3}}}{q^{\frac{1}{3}}}\right)^{3}$

51. $\dfrac{(3x^{3}y)^{\frac{1}{3}}(3xy^{5})^{\frac{2}{3}}}{(9x^{2}y)^{\frac{1}{2}}}$

52. $\dfrac{(5x^{2}y^{3})^{\frac{3}{4}}(5x^{2}y^{3})^{\frac{1}{4}}}{(5x^{2}y^{3})^{-2}}$

53. (a) $\left(x^{\frac{1}{2}} - y^{\frac{-1}{2}}\right)\left(x^{\frac{1}{2}} + y^{\frac{-1}{2}}\right)$

(b) $\left(x^{\frac{3}{2}} - y^{\frac{3}{2}}\right)\left(x^{\frac{3}{2}} + y^{\frac{3}{2}}\right)$

54. (a) $\left(a^{\frac{-1}{2}} - 3b^{\frac{-1}{2}}\right)\left(a^{\frac{-1}{2}} + 3b^{\frac{-1}{2}}\right)$

(b) $\left(x^{\frac{3}{2}} - 2y^{\frac{3}{2}}\right)\left(x^{\frac{3}{2}} + 2y^{\frac{3}{2}}\right)$

In Problems 55 and 56, (a) for each function find the indicated values, if they are real. Round off to two decimal places if necessary.

(b) Plot the points in part (a) and connect them with a smooth curve.

(c) Indicate the domain and range of each function.

55. $f(x) = x^{\frac{2}{5}}$
(a) $f(-32)$
(b) $f(-20)$
(c) $f(0)$
(d) $f(1)$
(e) $f(32)$

56. $g(x) = x^{\frac{3}{5}}$
(a) $g(-4)$
(b) $g(-1)$
(c) $g(0)$
(d) $g(1)$
(e) $g(4)$

57. Figure 2 shows the graph of $g(x) = x^{\frac{3}{2}}$ (solid curve) and $h(x) = x^{\frac{3}{2}} - 8$ (dashed curve) on the same plane.

(a) Indicate the domain and range of g and h.

(b) Solve the equation $x^{\frac{3}{2}} = 8$ graphically by finding the x intercept of the graph of h.

Figure 2

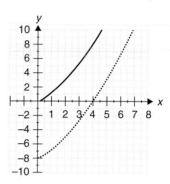

58. Figure 3 shows the graph of $f(x) = x^{\frac{-2}{3}}$ (solid curve) and $g(x) = x^{\frac{-2}{3}} - 1$ (dashed curve) on the same plane.

(a) Indicate the domain and range of f and g.

(b) Solve the equation $x^{\frac{-2}{3}} = 1$ graphically by finding the x intercept of the graph of g.

Figure 3

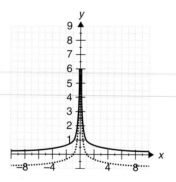

Objectives

1. Simplify Radicals
2. Solve Applied Problems

6.4 Simplifying Radical Expressions

Simplifying Radicals

In this section, we simplify radical expressions by using certain properties of radicals. For instance, we see that

$$\sqrt{4} \cdot \sqrt{9} = 2 \cdot 3 = 6 \text{ and } \sqrt{4 \cdot 9} = \sqrt{36} = 6.$$

Thus, we have

$$\sqrt{4} \cdot \sqrt{9} = \sqrt{4 \cdot 9}.$$

Similarly,

$$\sqrt[3]{\frac{27}{8}} = \frac{3}{2} \text{ and } \frac{\sqrt[3]{27}}{\sqrt[3]{8}} = \frac{3}{2}$$

so that

$$\sqrt[3]{\frac{27}{8}} = \frac{\sqrt[3]{27}}{\sqrt[3]{8}}.$$

These examples are based on the following properties for multiplying and dividing radicals.

Properties of Radicals

Let a and b be real numbers and n be a positive integer with $n \geq 2$. Then, provided that $\sqrt[n]{a}$ and $\sqrt[n]{b}$ are real numbers,

(i) $\qquad \sqrt[n]{ab} = \sqrt[n]{a} \cdot \sqrt[n]{b} \qquad$ Product property

(ii) $\qquad \sqrt[n]{\frac{a}{b}} = \frac{\sqrt[n]{a}}{\sqrt[n]{b}} \qquad$ Quotient property

To see that these properties are true, notice that

(i) $\sqrt[n]{ab} = \sqrt[n]{a} \cdot \sqrt[n]{b}$ is the relationship $(ab)^{\frac{1}{n}} = a^{\frac{1}{n}} \cdot b^{\frac{1}{n}}$; and

(ii) $\sqrt[n]{\frac{a}{b}} = \frac{\sqrt[n]{a}}{\sqrt[n]{b}}$ is the relationship $\left(\frac{a}{b}\right)^{\frac{1}{n}} = \frac{a^{\frac{1}{n}}}{b^{\frac{1}{n}}}.$

Additional examples of the product and quotient properties:

$$\sqrt{18} = \begin{cases} \sqrt{1} \cdot \sqrt{18} \\ \sqrt{2} \cdot \sqrt{9} \\ \sqrt{3} \cdot \sqrt{6} \end{cases} \quad \text{and} \quad \sqrt[3]{x^4} = \begin{cases} \sqrt[3]{x} \cdot \sqrt[3]{x^3} \\ \sqrt[3]{x^2} \cdot \sqrt[3]{x^2} \end{cases}$$

$$\text{Also,} \begin{cases} \frac{\sqrt{12}}{\sqrt{3}} = \sqrt{\frac{12}{3}} \\ \frac{\sqrt{x^3}}{\sqrt{x}} = \sqrt{\frac{x^3}{x}} \end{cases} \quad \text{and} \quad \begin{cases} \sqrt{\frac{4}{49}} = \frac{\sqrt{4}}{\sqrt{49}} \\ \sqrt{\frac{x^4}{9}} = \frac{\sqrt{x^4}}{\sqrt{9}} \end{cases}$$

We can use these properties of radicals to simplify radical expressions by reading these properties from left to right or from right to left. A *simplest radical* is useful in comparing radicals without changing their value. A radical expression (or rational exponent expression) is said to be in **simplest form** or **simplified** if all of the following conditions are true.

Simplified Radical Form	Simplified Rational Exponent Form
1. There are no radicands containing polynomial or prime integer factors to a power greater than or equal to the index of the radical.	1. All rational exponents are between 0 and 1.
2. No fraction appears within a radicand.	2. No fraction appears as the base of a rational exponent.
3. No radical appears in a denominator of a fraction.	3. No rational exponents appear in the denominator of a fraction.
4. The exponents of factors under the radical and the index of the radical have no common factors.	4. Rational exponents are reduced to lowest terms.

For the remainder of this chapter, we shall impose restrictions on variables so that each radical expression represents a real number.

EXAMPLE 1

Applying the Radical Properties

Simplify each radical expression by using the radical and rational exponent properties.

(a) $\sqrt{72}$

(b) $\sqrt[3]{-24}$

(c) $\sqrt{\dfrac{7}{16}}$

(d) $\sqrt[3]{\dfrac{5}{27}}$

Solution

$4 \cdot 9 \cdot 2 = 72$, using perfect square factors we write

$$\sqrt{72} = \sqrt{4 \cdot 9 \cdot 2}$$
$$= \sqrt{4} \cdot \sqrt{9} \cdot \sqrt{2}$$
$$= 2 \cdot 3\sqrt{2} = 6\sqrt{2}.$$

Radicals	Rational Expressions	Properties
(a) $\sqrt{72} = \sqrt{36 \cdot 2}$ $= \sqrt{36} \cdot \sqrt{2}$ $= 6\sqrt{2}$	$(72)^{\frac{1}{2}} = (36 \cdot 2)^{\frac{1}{2}}$ $= 36^{\frac{1}{2}} \cdot 2^{\frac{1}{2}}$ $= 6 \cdot 2^{\frac{1}{2}}$	Product
(b) $\sqrt[3]{-24} = \sqrt[3]{(-8) \cdot 3}$ $= \sqrt[3]{-8} \cdot \sqrt[3]{3}$ $= -2\sqrt[3]{3}$	$(-24)^{\frac{1}{3}} = (-8 \cdot 3)^{\frac{1}{3}}$ $= (-8)^{\frac{1}{3}} \cdot 3^{\frac{1}{3}}$ $= -2 \cdot 3^{\frac{1}{3}}$	Product
(c) $\sqrt{\dfrac{7}{16}} = \dfrac{\sqrt{7}}{\sqrt{16}} = \dfrac{\sqrt{7}}{4}$	$\left(\dfrac{7}{16}\right)^{\frac{1}{2}} = \dfrac{7^{\frac{1}{2}}}{16^{\frac{1}{2}}} = \dfrac{7^{\frac{1}{2}}}{4}$	Quotient
(d) $\sqrt[3]{\dfrac{5}{27}} = \dfrac{\sqrt[3]{5}}{\sqrt[3]{27}} = \dfrac{\sqrt[3]{5}}{3}$	$\left(\dfrac{5}{27}\right)^{\frac{1}{3}} = \dfrac{5^{\frac{1}{3}}}{27^{\frac{1}{3}}} = \dfrac{5^{\frac{1}{3}}}{3}$	Quotient

EXAMPLE 2

Applying the Radical Properties

Use the properties of radicals and rational exponents to simplify each expression. Assume that x and y are positive real numbers.

(a) $\sqrt{4x^3}$

(b) $\sqrt[3]{24x^4y^7}$

(c) $\dfrac{\sqrt{12x^2}}{\sqrt{3}}$

(d) $\dfrac{\sqrt[3]{40xy^4}}{2\sqrt[3]{5x}}$

Solution

Radicals	Rational Exponents	Properties
(a) $\sqrt{4x^3} = \sqrt{4x^2 \cdot x}$ $= \sqrt{4x^2} \cdot \sqrt{x} = 2x \cdot \sqrt{x}$	$(4x^3)^{\frac{1}{2}} = (4x^2 \cdot x)^{\frac{1}{2}}$ $= (4x^2)^{\frac{1}{2}} \cdot x^{\frac{1}{2}} = 2x \cdot x^{\frac{1}{2}}$	Product
(b) $\sqrt[3]{24x^4y^7} = \sqrt[3]{(8x^3y^6)(3xy)}$ $= \sqrt[3]{8x^3y^6} \cdot \sqrt[3]{3xy}$ $= 2xy^2 \sqrt[3]{3xy}$	$(24x^4y^7)^{\frac{1}{3}} = (2^3x^3y^6 \cdot 3xy)^{\frac{1}{3}}$ $= (2^3x^3y^6)^{\frac{1}{3}} (3xy)^{\frac{1}{3}}$ $= 2xy^2(3xy)^{\frac{1}{3}}$	Product
(c) $\dfrac{\sqrt{12x^2}}{\sqrt{3}} = \sqrt{\dfrac{12x^2}{3}}$ $= \sqrt{4x^2} = 2x$	$\dfrac{(12x^2)^{\frac{1}{2}}}{3^{\frac{1}{2}}} = \left(\dfrac{12x^2}{3}\right)^{\frac{1}{2}}$ $= (4x^2)^{\frac{1}{2}} = 2x$	Quotient and product
(d) $\dfrac{\sqrt[3]{40xy^4}}{2\sqrt[3]{5x}} = \dfrac{1}{2}\sqrt[3]{\dfrac{40xy^4}{5x}}$ $= \dfrac{1}{2}\sqrt[3]{8y^4} = \dfrac{1}{2}\sqrt[3]{(8y^3) \cdot y}$ $= \dfrac{1}{2}2y\sqrt[3]{y} = y\sqrt[3]{y}$	$\dfrac{(40xy^4)^{\frac{1}{3}}}{2(5x)^{\frac{1}{3}}} = \dfrac{1}{2}\left(\dfrac{40xy^4}{5x}\right)^{\frac{1}{3}}$ $= \dfrac{1}{2}(8y^4)^{\frac{1}{3}} = \dfrac{1}{2}(2^3y^3)^{\frac{1}{3}} \cdot y^{\frac{1}{3}}$ $= \dfrac{1}{2}2y \cdot y^{\frac{1}{3}} = y \cdot y^{\frac{1}{3}}$	Quotient and product

In order to multiply or divide radical expressions with different indices, we begin by building their indices to a common index of individual radicals. This is accomplished by converting each expression to exponential form, then using the LCD of the rational exponents. Traditionally, if an expression is in radical form, the simplified expression should also be in radical form.

EXAMPLE 3   **Simplifying Radical Expressions**

Use rational exponents to rewrite each expression as a single expression in a simple form. Assume $x \geq 0$.

(a) $\sqrt[3]{\sqrt{64}}$ (b) $\sqrt{3} \cdot \sqrt[3]{5}$ (c) $\sqrt[4]{x^3} \cdot \sqrt[3]{x^2}$

Solution

(a) $\sqrt[3]{\sqrt{64}} = \sqrt[3]{8} = 2$

(b) $\sqrt{3} \cdot \sqrt[3]{5} = 3^{\frac{1}{2}} \cdot 5^{\frac{1}{3}}$ Write radicals as rational exponents

$= 3^{\frac{3}{6}} \cdot 5^{\frac{2}{6}}$ LCD of $\dfrac{1}{2}$ and $\dfrac{1}{3}$ is 6

$= (3^3 5^2)^{\frac{1}{6}}$ Power of product property

$= (675)^{\frac{1}{6}} = \sqrt[6]{675}$ Return to radical expression

(c) $\sqrt[4]{x^3} \cdot \sqrt[3]{x^2} = x^{\frac{3}{4}} \cdot x^{\frac{2}{3}}$ Rewrite as rational exponents

$= x^{\frac{3}{4}+\frac{2}{3}}$ Product property

$= x^{\frac{9}{12}+\frac{8}{12}} = x^{\frac{17}{12}}$ LCD of 4 and 3 is 12

$= x^{\frac{12}{12}} \cdot x^{\frac{5}{12}}$ Factor perfect 12th root

$= x\sqrt[12]{x^5}$ Return to radical form

Solving Applied Problems

Applications and models from different fields involve radical expressions as the next example shows.

EXAMPLE 4   **Finding the Lengths of Rope**

A city is planting trees of various heights in a park. Each tree needs to be staked with a rope so that the prevailing winds will not topple the tree.

(a) If the rope to the tree is tied x feet above the ground and staked half that distance from the trunk, write a simplified formula for the length of rope needed.

(b) If x is 9 feet, how much rope is needed?

Solution   (a) Recall that the Pythagorean Theorem is:

$$a^2 + b^2 = c^2$$

where a and b are the legs and c is the hypotenuse. From Figure 1, the legs have lengths x and $\dfrac{x}{2}$, so that the hypotenuse is

Figure 1

$$c^2 = x^2 + \left(\frac{x}{2}\right)^2$$

$$c^2 = x^2 + \frac{x^2}{4} \qquad \text{Power of quotient property}$$

$$c^2 = \frac{4x^2}{4} + \frac{x^2}{4} = \frac{5x^2}{4}$$

$$c = \sqrt{\frac{5x^2}{4}} = \frac{x}{2}\sqrt{5} \qquad \text{Perfect square factors}$$

(b) Substitute 9 for x in the above formula.

$$c = \frac{9}{2}\sqrt{5} \approx 10 \text{ feet}$$

EXAMPLE 5   **Finding the Radius of a Can**

A canning company wants to reduce the volume of all their soup cans by 20% by reducing the radius and not the height of the cans.

(a) Find a formula for the radius of the new cans.

(b) If the original cans had a height of 4.25 inches and a radius of 3 inches, what should be the radius of the new cans? (Round to 3 decimal places.)

(c) Check your answer. (Use $3.14 \approx \pi$.)

Solution   (a) The formula for the volume and radius of a soup can (right cylinder) is

$$V = \pi r^2 h \text{ and } r = \sqrt{\frac{V}{\pi h}}$$

where r is the radius and h is the height. Let the old radius be r, the new radius is x, the height, h, does not change, and the new volume is $0.8V$. Thus, we have:

$$x = \sqrt{\frac{0.8V}{\pi h}}$$

$$x = \sqrt{\frac{0.8\pi r^2 h}{\pi h}} \qquad \text{Substitute } V = \pi r^2 h$$

$$x = \sqrt{0.8 r^2} \qquad\qquad \text{Reduce common factors}$$

$$x = r\sqrt{0.8} \qquad\qquad \sqrt{r^2} = r \text{ since } r > 0$$

(b) According to this formula, the new radius should be:

$$x = 3\sqrt{0.8} \approx 2.683 \text{ inches.}$$

(c) The original volume is $V = \pi r^2 h = \pi(3)^2 4.25 \approx 120.1$ cubic inches.

The reduced volume is $0.8V = 0.8(120.1) = 96.1$ cubic inches.

The volume of the new can is $\pi x^2 h = 3.14(2.683)^2 \, (4.25) \approx 96.1$ cubic inches. ▦

 PROBLEM SET 6.4

Mastering the Concepts

In Problems 1–20, use the properties of radicals to simplify each expression. Write the answer in the simplest radical form. Assume that all variables represent positive real numbers.

1. (a) $\sqrt{27}$ (b) $\sqrt[3]{-54}$

2. (a) $\sqrt{162}$ (b) $\sqrt[3]{-250}$

3. (a) $\sqrt{\dfrac{5}{4}}$ (b) $\sqrt[3]{-\dfrac{5}{8}}$

4. (a) $\sqrt{\dfrac{5}{16}}$ (b) $\sqrt[3]{-\dfrac{11}{125}}$

5. (a) $\sqrt{48x}$ (b) $\sqrt[3]{24y^3}$

6. (a) $\sqrt{98p^3}$ (b) $\sqrt[3]{54t^4}$

7. (a) $\sqrt{32c^5}$ (b) $\sqrt[3]{-32p^6}$

8. (a) $\sqrt[3]{-16y^4}$ (b) $\sqrt[3]{250x^5}$

9. (a) $\sqrt{\dfrac{7}{4x^2}}$ (b) $\sqrt[5]{-32p^6}$

10. (a) $\sqrt{\dfrac{3}{25x^4}}$ (b) $\sqrt[5]{\dfrac{x}{y^{10}}}$

11. (a) $\sqrt{\dfrac{3}{25x^2}}$ (b) $\sqrt{\dfrac{5}{9x^4}}$

12. (a) $\sqrt{\dfrac{17}{y^4}}$ (b) $\sqrt[7]{\dfrac{3}{-a^{14}}}$

13. (a) $\dfrac{\sqrt{8a^3}}{\sqrt{2a}}$ (b) $\dfrac{\sqrt{14x^3}}{\sqrt{7x}}$

14. (a) $\dfrac{\sqrt{45x^3}}{\sqrt{3x}}$ (b) $\dfrac{\sqrt[3]{48x^4}}{\sqrt[3]{6x}}$

15. (a) $\dfrac{\sqrt[3]{10x^6}}{\sqrt[3]{2}}$ (b) $\dfrac{\sqrt{x^5}}{\sqrt{x}}$

16. (a) $\dfrac{\sqrt[3]{54x^7}}{\sqrt[3]{2x}}$ (b) $\dfrac{\sqrt[3]{128x^4}}{\sqrt[3]{2x}}$

17. $\sqrt[3]{125a^3b^6c^9}$ 18. $\sqrt[3]{-8a^6b^9c^3}$

19. $-\sqrt[3]{-27x^3y^9z^{12}}$ 20. $\sqrt[4]{81x^{12}y^4z^{16}}$

In Problems 21–28, use rational exponents to rewrite each expression as a single radical expression in simple form. Assume all variables represent positive real numbers.

21. (a) $\sqrt[3]{\sqrt{1024}}$ (b) $\sqrt{\sqrt[3]{x^6}}$

22. (a) $\sqrt{\sqrt[4]{p^8}}$ (b) $\sqrt[5]{\sqrt[3]{p^{15}}}$

23. (a) $\sqrt{3} \cdot \sqrt[3]{4}$ (b) $\sqrt[3]{4} \cdot \sqrt{7}$

24. (a) $\sqrt{5} \cdot \sqrt[3]{6}$ (b) $\sqrt[4]{7} \cdot \sqrt{3}$

25. (a) $\sqrt{x} \cdot \sqrt[3]{x}$ (b) $\sqrt[4]{x} \cdot \sqrt{x}$

26. (a) $\sqrt[3]{x} \cdot \sqrt[4]{x}$ (b) $\sqrt[3]{x} \cdot \sqrt[5]{x}$

27. (a) $\sqrt{5x} \cdot \sqrt[3]{5x}$ (b) $\sqrt[3]{4x} \cdot \sqrt{2y}$

28. (a) $\sqrt{3x} \cdot \sqrt[3]{3x}$ (b) $\sqrt[3]{5x} \cdot \sqrt{3y}$

Applying the Concepts

29. **Advertising:** A small company determines that if it spends x dollars on radio advertising, y dollars on TV commercials, and z dollars on newspaper advertising then its annual sale

of p units of its product is modeled by the formula

$$p = \sqrt[5]{64x^3y^6z^5}.$$

(a) Express p in a simplified radical form.

(b) Use the above model to determine the number of units sold if the company spends annually $300 on radio advertising, $400 on TV commercials, and $200 on newspaper advertising. Round off the answer to the nearest thousand.

30. **Geometry:** The volume V of a spherical hot air balloon is related to its surface area A by the formula

$$V = 0.094 \sqrt{A^3}.$$

(a) Use the formula $A = 4\pi r^2$, where r is the radius of the balloon, to express V in terms of r in a simplified radical form.

(b) What is the volume of a balloon whose radius is 5.3 feet? Round off the answer to two decimal places.

Developing and Extending the Concepts

31. (a) Use a calculator to evaluate $\sqrt[4]{7}$. Round off to six decimal places.

(b) Use a calculator to evaluate $\sqrt{\sqrt{7}}$. Round off to six decimal places.

(c) Compare the results of parts (a) and (b).

32. What is the restriction on x in order for each equation to be true?

(a) $\sqrt{x} \cdot \sqrt{x} = x$

(b) $\sqrt{-x} \cdot \sqrt{x} = -x$

(c) $\sqrt[3]{x} \cdot \sqrt[3]{x} \cdot \sqrt[3]{x} = x$

33. (a) Use a calculator to evaluate each expression. Round off each result to three decimal places.

$$\sqrt{3}, \ \sqrt{\sqrt{3}}, \ \sqrt{\sqrt{\sqrt{3}}}, \ \sqrt{\sqrt{\sqrt{\sqrt{3}}}}$$

(b) If we continue to take the square root, what number does the value of the expression seem to be approaching?

34. What restrictions on a, b, and c should be made in order for each of these expressions

$$\sqrt{a} \cdot \sqrt{b} \cdot \sqrt{c} \text{ and } \sqrt{abc}$$

to be real numbers and $\sqrt{a} \cdot \sqrt{b} \cdot \sqrt{c} = \sqrt{abc}$?

In Problems 35–40, use the properties of radicals to simplify each expression. Assume that all variables represent positive real numbers.

35. $\sqrt[8]{x^{12}} \cdot \sqrt[8]{x^5 y^{-8}} \cdot \sqrt[3]{x^2 y^9}$

36. $\sqrt[3]{25y^2} \cdot \sqrt[3]{5y^4} \cdot \sqrt[3]{y^9}$

37. $\dfrac{\sqrt{324x^5 y} \cdot \sqrt{9x^2}}{\sqrt{25x^2 y}}$

38. $\dfrac{\sqrt{m^2 n} \cdot \sqrt{mn^4}}{\sqrt{mn^3}}$

39. $\dfrac{\sqrt[3]{p^2 q^3} \cdot \sqrt[3]{125\,p^3 q^2}}{\sqrt[3]{8p^3 q^4}}$

40. $\dfrac{\sqrt[3]{a^2 b^4} \cdot \sqrt[3]{a^4 b} \cdot \sqrt[3]{a^3 b^2}}{\sqrt[3]{ab^3} \cdot \sqrt[3]{a^2 b^7}}$

Objectives

1. Add and Subtract Radical Expressions
2. Multiply Radical Expressions
3. Divide Radical Expressions— Rationalize the Denominator
4. Solve Applied Problems

6.5 Operations with Radicals

In this section we develop skills for adding, subtracting, multiplying, and dividing radical expressions.

Adding and Subtracting Radical Expressions

Recall that the sum or difference of like terms of a polynomial can be simplified. To simplify these sums or differences, we use the distributive property. For example, $7x + 3x = (7 + 3)x = 10x$. Radicals with the same index and the same radicand are **like** radicals. For instance, $3\sqrt{6}$ and $-5\sqrt{6}$ are like radicals. We can add and subtract radicals in much the same way that we add or subtract like terms of a polynomial. For example,

$$3\sqrt{6} - 5\sqrt{6} = (3 - 5)\sqrt{6} = -2\sqrt{6}$$

and

$$7\sqrt{3} + 2\sqrt{3} = (7 + 2)\sqrt{3} = 9\sqrt{3}.$$

In each case we applied the distributive property. The expression

$$\sqrt{7} + \sqrt{5}$$

is the sum of two unlike radicals which cannot be simplified and cannot be added. Although these radicals cannot be added, they can be approximated on a calculator, $\sqrt{7} + \sqrt{5} \approx 4.8818$ rounded to four decimals.

Notice the difference
$\sqrt{9} + \sqrt{16} = 7$ *but*
$\sqrt{9 + 16} = \sqrt{25} = 5.$

Sometimes unlike radicals can be simplified and then added as in this example,

$$\sqrt{9} + \sqrt{16} = 3 + 4 = 7.$$

EXAMPLE 1   **Combining Like Radicals**

Combine like radicals. Assume $x \geq 0$.

(a) $12\sqrt{3} + 5\sqrt{3}$ (b) $7\sqrt[3]{2} - 3\sqrt[3]{2}$

(c) $3\sqrt{3x} - 6\sqrt{3x} + 5\sqrt{3x}$ (d) $3\sqrt{x} - 2\sqrt[3]{x} + 4\sqrt{x} + 7\sqrt[3]{x}$

Solution   (a) $12\sqrt{3} + 5\sqrt{3} = (12 + 5)\sqrt{3} = 17\sqrt{3}$

(b) $7\sqrt[3]{2} - 3\sqrt[3]{2} = (7 - 3)\sqrt[3]{2} = 4\sqrt[3]{2}$

(c) $3\sqrt{3x} - 6\sqrt{3x} + 5\sqrt{3x} = (3 - 6 + 5)\sqrt{3x} = 2\sqrt{3x}$

(d) $3\sqrt{x} - 2\sqrt[3]{x} + 4\sqrt{x} + 7\sqrt[3]{x} = 3\sqrt{x} + 4\sqrt{x} - 2\sqrt[3]{x} + 7\sqrt[3]{x}$

$$= 7\sqrt{x} + 5\sqrt[3]{x}$$

Sometimes, two or more terms containing different radicals such as

$$5\sqrt{8} + 11\sqrt{18}$$

can be combined when they are simplified as the next example shows.

EXAMPLE 2   **Combining Terms Containing Different Radicals**

Write radical expressions in simplest form, then combine like radicals.

(a) $5\sqrt{8} + 11\sqrt{18}$ (b) $2\sqrt[3]{54} - 2\sqrt[3]{16}$ (c) $4\sqrt{12} + 5\sqrt{8} - \sqrt{50}$

Solution   (a) $5\sqrt{8} + 11\sqrt{18} = 5\sqrt{4 \cdot 2} + 11\sqrt{9 \cdot 2}$ Perfect square factors

$$= 5\sqrt{4} \cdot \sqrt{2} + 11\sqrt{9} \cdot \sqrt{2}$$ Product property

$$= 5 \cdot 2\sqrt{2} + 11 \cdot 3\sqrt{2}$$

$$= 10\sqrt{2} + 33\sqrt{2} = 43\sqrt{2}$$

(b) $2\sqrt[3]{54} - 2\sqrt[3]{16} = 2\sqrt[3]{27 \cdot 2} - 2\sqrt[3]{8 \cdot 2}$ Perfect cube factors

$$= 2\sqrt[3]{27} \cdot \sqrt[3]{2} - 2\sqrt[3]{8} \cdot \sqrt[3]{2}$$

$$= 2 \cdot 3 \cdot \sqrt[3]{2} - 2 \cdot 2 \cdot \sqrt[3]{2}$$

$$= 6\sqrt[3]{2} - 4\sqrt[3]{2} = 2\sqrt[3]{2}$$

(c) $4\sqrt{12} + 5\sqrt{8} - \sqrt{50} = 4\sqrt{4 \cdot 3} + 5\sqrt{4 \cdot 2} - \sqrt{25 \cdot 2}$

$$= 4\sqrt{4} \cdot \sqrt{3} + 5\sqrt{4} \cdot \sqrt{2} - \sqrt{25} \cdot \sqrt{2}$$

$$= 4 \cdot 2\sqrt{3} + 5 \cdot 2\sqrt{2} - 5\sqrt{2}$$

$$= 8\sqrt{3} + 10\sqrt{2} - 5\sqrt{2}$$

$$= 8\sqrt{3} + 5\sqrt{2}$$ These are unlike radicals and cannot be added.

Multiplying Radical Expressions

Multiplication of radical expressions is very similar to the multiplication procedures used to multiply polynomials. We will also make use of the Product Property,

$$\sqrt[n]{a} \cdot \sqrt[n]{b} = \sqrt[n]{ab} \quad \text{or} \quad a^{\frac{1}{n}} \cdot b^{\frac{1}{n}} = (ab)^{\frac{1}{n}}.$$

EXAMPLE 3   **Multiplying Radical Expressions**

Perform each multiplication and simplify where possible. Assume $x \geq 0$.

(a) $\sqrt{5}(3\sqrt{7} + 2\sqrt{5})$ (b) $(\sqrt{3} - \sqrt{2})(2\sqrt{3} + \sqrt{2})$

(c) $(\sqrt{2} - 5\sqrt{3})^2$ (d) $(\sqrt{10x} - \sqrt{2})(\sqrt{10x} + \sqrt{2})$

Solution   (a) $\sqrt{5}(3\sqrt{7} + 2\sqrt{5}) = \sqrt{5} \cdot 3\sqrt{7} + \sqrt{5} \cdot 2\sqrt{5}$ Distributive property

$$= 3\sqrt{5 \cdot 7} + 2(\sqrt{5})^2$$

$$= 3\sqrt{35} + 2 \cdot 5$$

$$= 3\sqrt{35} + 10$$

(b) $(\sqrt{3} - \sqrt{2})(2\sqrt{3} + \sqrt{2}) = 2\sqrt{3} \cdot \sqrt{3} + \sqrt{3} \cdot \sqrt{2} - \sqrt{2} \cdot 2\sqrt{3} - \sqrt{2} \cdot \sqrt{2}$

$$= 2 \cdot 3 + \sqrt{6} - 2\sqrt{6} - 2 = 6 - \sqrt{6} - 2 = 4 - \sqrt{6}$$

(c) $(\sqrt{2} - 5\sqrt{3})^2 = (\sqrt{2})^2 - 2\sqrt{2} \cdot 5\sqrt{3} + (5\sqrt{3})^2$ Special product $(a - b)^2$

$$= 2 - 10\sqrt{6} + 75 = 77 - 10\sqrt{6}$$

(d) $(\sqrt{10x} - \sqrt{2})(\sqrt{10x} + \sqrt{2}) = (\sqrt{10x})^2 - (\sqrt{2})^2$ Special product

$$= 10x - 2$$

Dividing Radical Expressions—Rationalizing the Denominator

We now consider several types of quotients involving radicals. The property

$$\frac{\sqrt[n]{a}}{\sqrt[n]{b}} = \sqrt[n]{\frac{a}{b}}$$

is used when we divide two radical expressions. For example,

$$\frac{\sqrt{18}}{\sqrt{2}} = \sqrt{\frac{18}{2}} = \sqrt{9} = 3,$$

$$\frac{\sqrt[3]{54y^4}}{\sqrt[3]{2y}} = \sqrt[3]{\frac{54y^4}{2y}} = \sqrt[3]{27y^3} = 3y,$$

$$\text{and} \quad \frac{\sqrt{3}}{\sqrt{21}} = \sqrt{\frac{3}{21}} = \sqrt{\frac{1}{7}} = \frac{1}{\sqrt{7}}.$$

Beware that
$\sqrt[n]{a} + \sqrt[n]{b} \neq \sqrt[n]{a + b}.$
For example,
$\sqrt{16} + \sqrt{9} \neq \sqrt{16 + 9}$
$4 + 3 \neq \sqrt{25} = 5$

Recall that to write the last expression in simplest radical form, no radical should appear in the denominator. To accomplish this, we multiply the numerator and denominator by $\sqrt{7}$ so that the denominator becomes a perfect power for the given index.

$$\frac{1}{\sqrt{7}} = \frac{1}{\sqrt{7}} \cdot \frac{\sqrt{7}}{\sqrt{7}} = \frac{\sqrt{7}}{7}$$

This equality can be confirmed by calculating

$$\frac{1}{\sqrt{7}} \approx 0.37796 \text{ and } \frac{\sqrt{7}}{7} \approx 0.37796.$$

The process of removing radicals from the denominator so that the denominator contains only rational numbers is called **rationalizing the denominator.**

EXAMPLE 4   **Rationalizing the Denominator**

Rationalize the denominator and simplify. Assume that $x > 0$.

(a) $\dfrac{6\sqrt{5}}{\sqrt{6}}$ (b) $\sqrt{\dfrac{27}{8}}$ (c) $\dfrac{2}{\sqrt[3]{5}}$ (d) $\dfrac{5}{\sqrt{3x}}$

Solution

(a) $\dfrac{6\sqrt{5}}{\sqrt{6}} = \dfrac{6\sqrt{5}}{\sqrt{6}} \cdot \dfrac{\sqrt{6}}{\sqrt{6}}$ Multiply by $\dfrac{\sqrt{6}}{\sqrt{6}}$ to simplify

$$= \frac{6\sqrt{30}}{6} = \sqrt{30}$$

(b) $\sqrt{\dfrac{27}{8}} = \dfrac{\sqrt{27}}{\sqrt{8}} = \dfrac{\sqrt{9 \cdot 3}}{\sqrt{4 \cdot 2}} = \dfrac{3\sqrt{3}}{2\sqrt{2}}$ Factor

$$= \frac{3\sqrt{3}}{2\sqrt{2}} \cdot \frac{\sqrt{2}}{\sqrt{2}} = \frac{3\sqrt{6}}{4}$$ Multiply by $\dfrac{\sqrt{2}}{\sqrt{2}}$ to simplify

(c) $\dfrac{2}{\sqrt[3]{5}} = \dfrac{2}{\sqrt[3]{5}} \cdot \dfrac{\sqrt[3]{25}}{\sqrt[3]{25}} = \dfrac{2\sqrt[3]{25}}{\sqrt[3]{125}} = \dfrac{2\sqrt[3]{25}}{5}$ Multiply by $\dfrac{\sqrt[3]{25}}{\sqrt[3]{25}}$ to simplify

(d) $\dfrac{5}{\sqrt{3x}} = \dfrac{5}{\sqrt{3x}} \cdot \dfrac{\sqrt{3x}}{\sqrt{3x}}$

$$= \frac{5\sqrt{3x}}{\sqrt{9x^2}} = \frac{5\sqrt{3x}}{3x}$$

To rationalize a denominator that contains binomial expressions involving radicals such as

$$\frac{7}{\sqrt{5} - \sqrt{3}}$$

we recall that

$$(a - b)(a + b) = a^2 - b^2.$$

This suggests that we multiply both the numerator and denominator of the expression by the **conjugate** or **rationalizing factor** of the denominator. The expression $\sqrt{5} + \sqrt{3}$ is the conjugate or rationalizing factor of $\sqrt{5} - \sqrt{3}$. Thus, we rationalize the denominator of the above expression as follows:

$$\frac{7}{\sqrt{5} - \sqrt{3}} = \frac{7}{(\sqrt{5} - \sqrt{3})} \cdot \frac{\sqrt{5} + \sqrt{3}}{\sqrt{5} + \sqrt{3}} = \frac{7(\sqrt{5} + \sqrt{3})}{(\sqrt{5})^2 - (\sqrt{3})^2}$$

$$= \frac{7(\sqrt{5} + \sqrt{3})}{5 - 3} = \frac{7\sqrt{5} + 7\sqrt{3}}{2}.$$

By using a calculator, we see that the decimal approximation of

$$\frac{7}{\sqrt{5} - \sqrt{3}} \approx 13.888 \text{ and } \frac{7\sqrt{5} + 7\sqrt{3}}{2} \approx 13.888$$

which confirms the equality of the expressions.

We rationalize the denominator of $\dfrac{5\sqrt{x}}{x + \sqrt{y}}$ as follows:

$$\frac{5\sqrt{x}}{x + \sqrt{y}} = \frac{5\sqrt{x}}{x + \sqrt{y}} \cdot \frac{x - \sqrt{y}}{x - \sqrt{y}}$$

$$= \frac{5\sqrt{x}\left(x - \sqrt{y}\right)}{x^2 - \left(\sqrt{y}\right)^2}$$

$$= \frac{5x\sqrt{x} - 5\sqrt{x} \cdot \sqrt{y}}{x^2 - 4}$$

$$= \frac{5x\sqrt{x} - 5\sqrt{xy}}{x^2 - y}$$

EXAMPLE 5   **Rationalizing the Denominator**

Rationalize the denominator of each expression.

(a) $\dfrac{5}{2 - \sqrt{3}}$ (b) $\dfrac{3\sqrt{2}}{\sqrt{5} + \sqrt{2}}$

Solution   (a) $\dfrac{5}{2 - \sqrt{3}} = \dfrac{5}{2 - \sqrt{3}} \cdot \dfrac{2 + \sqrt{3}}{2 + \sqrt{3}}$ Multiply by the conjugate

$$= \frac{5\left(2 + \sqrt{3}\right)}{(2)^2 - \left(\sqrt{3}\right)^2}$$ Special product

$$= \frac{10 + 5\sqrt{3}}{4 - 3} = 10 + 5\sqrt{3}$$

(b) $\dfrac{3\sqrt{2}}{\sqrt{5} + \sqrt{2}} = \dfrac{3\sqrt{2}}{\sqrt{5} + \sqrt{2}} \cdot \dfrac{\sqrt{5} - \sqrt{2}}{\sqrt{5} - \sqrt{2}}$ Multiply by the conjugate

$$= \frac{3\sqrt{2}\left(\sqrt{5} - \sqrt{2}\right)}{\left(\sqrt{5}\right)^2 - \left(\sqrt{2}\right)^2}$$ Special product

$$= \frac{3\sqrt{2}\left(\sqrt{5} - \sqrt{2}\right)}{5 - 2} = \frac{3\sqrt{2}\left(\sqrt{5} - \sqrt{2}\right)}{3}$$

$$= \sqrt{2}\left(\sqrt{5} - \sqrt{2}\right) = \sqrt{10} - 2$$

Check each by using a calculator.

EXAMPLE 6   **Simplifying Radical Expression**

Simplify $5\sqrt{3} - \dfrac{4}{\sqrt{27}} + \sqrt{48}$.

Solution

$$5\sqrt{3} - \dfrac{4}{\sqrt{27}} + \sqrt{48} = 5\sqrt{3} - \dfrac{4}{\sqrt{27}} \cdot \dfrac{\sqrt{3}}{\sqrt{3}} + \sqrt{16 \cdot 3}$$

$$= 5\sqrt{3} - \dfrac{4\sqrt{3}}{\sqrt{81}} + 4\sqrt{3}$$

$$= 5\sqrt{3} - \dfrac{4\sqrt{3}}{9} + 4\sqrt{3}$$

$$= 9\sqrt{3} - \dfrac{4\sqrt{3}}{9} = \dfrac{81\sqrt{3} - 4\sqrt{3}}{9} = \dfrac{77\sqrt{3}}{9}.$$

A decimal approximation confirms that the calculator value of each expression is 14.82.

Solving Applied Problems

Radical expressions can be used to model real world problems.

EXAMPLE 7   **Modeling a Geometry Problem**

Figure 1 shows a piece of carpet in the form of a trapezoid whose area A is given by the formula

$$A = \frac{1}{2}(a + b)h$$

where a and b are the lengths of the parallel sides and h is the height.

(a) If $a = \sqrt{x}$ and $b = \sqrt{x} + 2$, find an expression for h in terms of x and A. Write the expression in a simplified radical form.

(b) Find h if $x = 9$ inches and $A = 24$ square inches.

Figure 1

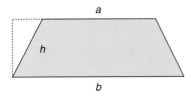

a

h

b

Solution   (a)

$$A = \frac{1}{2}(a + b)h$$

$$2A = (a + b)h \qquad \text{Multiply each side by 2}$$

$$(a + b)h = 2A \qquad \text{Reflect the equation}$$

$$h = \frac{2A}{a + b} \qquad \text{Divide by } a + b$$

Replacing a by $\sqrt{x}$ and b by $\sqrt{x} + 2$, we have

$$h = \frac{2A}{\sqrt{x} + \sqrt{x} + 2} = \frac{2A}{2\sqrt{x} + 2} = \frac{2A}{2(\sqrt{x} + 1)} = \frac{A}{\sqrt{x} + 1}$$

$$= \frac{A}{\sqrt{x} + 1} \cdot \frac{\sqrt{x} - 1}{\sqrt{x} - 1} = \frac{A(\sqrt{x} - 1)}{x - 1}. \qquad \text{Rationalize the denominator}$$

(b) $h = \dfrac{A(\sqrt{x} - 1)}{x - 1} = \dfrac{24(\sqrt{9} - 1)}{9 - 1} = \dfrac{24(3 - 1)}{8} = \dfrac{24(2)}{8} = 6$

Therefore, the height is 6 inches.

PROBLEM SET 6.5

Mastering the Concepts

In Problems 1–16, write each expression in simplest radical form and combine like terms. Assume all variables represent positive real numbers.

1. (a) $5\sqrt{7} + 3\sqrt{7}$
   (b) $9\sqrt{11} + 8\sqrt{11} - 3\sqrt{11}$
2. (a) $8\sqrt{3} + 2\sqrt{3}$
   (b) $8\sqrt[3]{4} - 3\sqrt[3]{4} + 2\sqrt[3]{4}$
3. (a) $7\sqrt{5} + 3\sqrt{5} - 2\sqrt{5}$
   (b) $8\sqrt[5]{2} - 4\sqrt[5]{2} + 3\sqrt[5]{2}$
4. (a) $\sqrt{x} - 4\sqrt{x} + 2\sqrt{x}$
   (b) $3\sqrt[3]{y} - \sqrt[3]{y} + 2\sqrt[3]{y}$
5. $\sqrt{x} + 3\sqrt{x} + 5\sqrt{y}$
6. $3\sqrt{x} - 2\sqrt{y} + 7\sqrt{y}$
7. $\sqrt{18} + \sqrt{8}$
8. $3\sqrt{18} + 4\sqrt{2}$
9. $2\sqrt{50} - 3\sqrt{128} + 4\sqrt{2}$
10. $5\sqrt{3} + 2\sqrt{12} - 2\sqrt{27}$
11. $4\sqrt{20} + 2\sqrt{45} + \sqrt{80}$
12. $\sqrt{72} - 2\sqrt{8} + \sqrt{2}$
13. $3\sqrt[3]{192} + 4\sqrt[3]{24} - 2\sqrt[3]{3}$
14. $\sqrt{75x} - \sqrt{3x} - \sqrt{12x}$
15. $\sqrt{27x} - 3\sqrt{12y} + 5\sqrt{3y}$
16. $4\sqrt{12} + 2\sqrt{27} - \sqrt{48}$

In Problems 17–28, find each product and simplify. Assume all variables represent positive real numbers.

17. (a) $\sqrt{2}(\sqrt{3} - \sqrt{2})$ (b) $\sqrt{3}(2 - 5\sqrt{3})$
18. (a) $\sqrt{3}(4 - \sqrt{2})$ (b) $\sqrt{6}(\sqrt{10} - \sqrt{15})$
19. (a) $\sqrt{3}(\sqrt{x} - \sqrt{6})$
    (b) $\sqrt{n}(4 - 3\sqrt{n})$
20. (a) $\sqrt{a}(\sqrt{a} - 3)$
    (b) $\sqrt{y}(5 - 2\sqrt{y})$
21. (a) $\sqrt{11}(2\sqrt{3} - 4\sqrt{11})$
    (b) $(\sqrt{5} - 1)(2\sqrt{5} + 3)$
22. (a) $\sqrt{5}(4\sqrt{5} - 3\sqrt{2})$
    (b) $(3 - \sqrt{2})(2 + \sqrt{3})$
23. (a) $(4\sqrt{2} - \sqrt{3})(5\sqrt{2} + 2\sqrt{3})$
    (b) $(\sqrt{3} - 2)^2$

24. (a) $(3\sqrt{6} + 4\sqrt{2})(\sqrt{6} - \sqrt{2})$
    (b) $(\sqrt{5} - \sqrt{3})^2$
25. (a) $(2\sqrt{x} - 3)^2$
    (b) $(\sqrt{5} - 3)(\sqrt{5} + 3)$
26. (a) $(1 + 3\sqrt{y})^2$
    (b) $(\sqrt{7} - 2)(\sqrt{7} + 2)$
27. (a) $(3\sqrt{5} - \sqrt{3})(3\sqrt{5} + \sqrt{3})$
    (b) $(3\sqrt{x} - 11)(3\sqrt{x} + 11)$
28. (a) $(4\sqrt{x} - 5)(4\sqrt{x} + 5)$
    (b) $(2\sqrt{x} + \sqrt{7})(2\sqrt{x} - \sqrt{7})$

In Problems 29–44, rationalize each denominator and simplify. Assume all variables represent positive real numbers.

29. (a) $\dfrac{2}{\sqrt{3}}$ (b) $\dfrac{9}{\sqrt{21}}$
30. (a) $\dfrac{1}{\sqrt{7}}$ (b) $\dfrac{5}{\sqrt{6}}$
31. (a) $\dfrac{2\sqrt{3}}{\sqrt{6}}$ (b) $\sqrt{\dfrac{5}{8}}$
32. (a) $\dfrac{4\sqrt{3}}{\sqrt{7}}$ (b) $\sqrt{\dfrac{18}{5}}$
33. (a) $\dfrac{8}{7\sqrt{11}}$ (b) $\dfrac{10}{3\sqrt{5x}}$
34. (a) $\dfrac{4}{3\sqrt{5x}}$ (b) $\dfrac{7}{2\sqrt{3x}}$
35. $\dfrac{5}{\sqrt[3]{9}}$ 36. $\dfrac{5}{\sqrt[3]{7}}$
37. $\dfrac{3}{\sqrt{5} - 2}$ 38. $\dfrac{4}{\sqrt{n} + 3}$
39. $\dfrac{7}{\sqrt{10} + 3}$ 40. $\dfrac{\sqrt{6}}{\sqrt{6} - 2}$
41. $\dfrac{3}{\sqrt{5} + \sqrt{2}}$ 42. $\dfrac{\sqrt{6}}{\sqrt{6} - \sqrt{3}}$
43. $\dfrac{8}{2\sqrt{7} - \sqrt{5}}$ 44. $\dfrac{8}{3\sqrt{5} - \sqrt{3}}$

In Problems 45–50, simplify each expression, then use a decimal approximation to confirm the results.

45. $\dfrac{\sqrt{5} - 2}{\sqrt{7}}$

46. $\dfrac{\sqrt{11} + 3}{\sqrt{8}}$

47. $\dfrac{\sqrt{7} + \sqrt{3}}{2\sqrt{3}}$

48. $\dfrac{\sqrt{7} - \sqrt{5}}{3\sqrt{5}}$

49. (a) $\dfrac{12 + 3\sqrt{27}}{18}$

(b) $\dfrac{5}{\sqrt{x}} - \sqrt{\dfrac{4}{x}} + \sqrt{x}$

50. (a) $\dfrac{-15 + 3\sqrt{63}}{21}$

(b) $\dfrac{3}{\sqrt{y}} - \sqrt{\dfrac{25}{y}} + \dfrac{2}{\sqrt{y^3}}$

Applying the Concepts

51. Geometry: The length of a rectangle is $\dfrac{12}{\sqrt{6}}$ inches and its width is $9\sqrt{\dfrac{3}{2}}$ inches. Find, in simplified radical form, the:
(a) area, $A = lw$
(b) perimeter, $P = 2l + 2w$.
(c) Find a two decimal calculator approximation for parts (a) and (b).

52. Geometry: Figure 2 shows a triangle with sides of lengths $\sqrt{8}$ centimeters, $\sqrt{32}$ centimeters, and $\sqrt{50}$ centimeters.

Figure 2

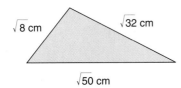

$\sqrt{8}$ cm $\sqrt{32}$ cm
$\sqrt{50}$ cm

(a) Find a simplified radical expression for the perimeter.
(b) Find a two decimal calculator approximation for the perimeter.

Developing and Extending the Concepts

53. Find a two decimal calculator approximation for each of the following expressions.
(a) $\left(\sqrt{63} + \sqrt{2}\right)\left(\sqrt{63} - 3\sqrt{2}\right)$ (b) $57 - 6\sqrt{14}$
(c) Show algebraically that
$$\left(\sqrt{63} + \sqrt{2}\right)\left(\sqrt{63} - 3\sqrt{2}\right) = 57 - 6\sqrt{14}.$$

54. Find a two decimal calculator approximation for each of the following expressions.
(a) $\dfrac{\sqrt{3} + 1}{1 - \sqrt{3}}$ (b) $-2 - \sqrt{3}$

(c) Show algebraically that
$$\dfrac{\sqrt{3} + 1}{1 - \sqrt{3}} = -2 - \sqrt{3}.$$

55. Perform each operation and simplify.
(a) $5\sqrt{\dfrac{1}{12}} - 2\sqrt{\dfrac{1}{3}} + \dfrac{8}{\sqrt{3}}$

(b) $8\sqrt{\dfrac{25}{7}} - 3\sqrt{\dfrac{16}{7}} + 5\sqrt{\dfrac{4}{7}}$

(c) $\sqrt{\dfrac{5}{3}} - \dfrac{15}{\sqrt{15}} + \dfrac{7\sqrt{15}}{3}$

(d) Use a two decimal calculator approximation to support the results of (a), (b), and (c).

56. Rationalize the denominator of
$$\dfrac{1}{\sqrt{7} - \sqrt{5} + \sqrt{2}} \quad \text{(Hint: rationalize twice)}$$

In Problems 57–60, for each given function find:
(a) the indicated values, if they are real numbers. Round off to two decimal places.
(b) Match each function with the given graphs.
(c) Indicate the domain and range of each function.

57. $f(x) = \sqrt{x - 1} + \sqrt{x + 1}$, find $f(1)$, $f(2)$, $f(3)$, and $f(5)$.

58. $g(x) = \sqrt{x - 1} - \sqrt{2x}$, find $g(0)$, $g(1)$, $g(2)$, and $g(4)$.

59. $h(x) = \sqrt{1 - x} + \sqrt{-x}$, find $h(-16)$, $h(-3)$, $h(-1)$, $h(0)$.

60. $k(x) = \sqrt[3]{-x} + \sqrt[3]{1 - x}$, find $k(-1)$, $k(1)$, $k(2)$, and $k(8)$.

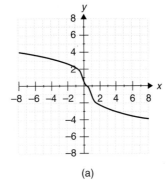

(a)

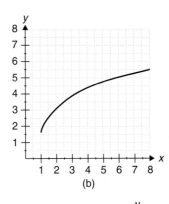

(b)

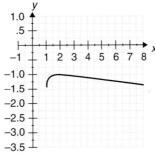

(c)

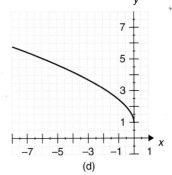

(d)

In Problems 61–64, perform each operation and simplify. Assume all variables represent positive real numbers.

61. (a) $\sqrt{m^3} - 2m\sqrt{m^5} + 3m\sqrt{m^7}$

    (b) $\sqrt[5]{32xy^{13}} - 8\sqrt[5]{x^2y^6} + \sqrt[5]{x^{11}y^3}$

62. (a) $a\sqrt[3]{a^3b} + a^2\sqrt{32ab} - \sqrt{162a^5b}$

    (b) $\dfrac{3\sqrt[4]{m^9}}{m} - 5\sqrt[4]{m^5} + \sqrt[8]{m^{10}}$

63. (a) $\dfrac{\sqrt{18x^3}}{3y} - \dfrac{x\sqrt{32x}}{2y} + \dfrac{6x\sqrt{2x}}{y}$

    (b) $x\sqrt[3]{27x} - 17\sqrt[3]{-x^4} + \dfrac{8\sqrt[3]{x^7}}{x}$

64. (a) $\left(\sqrt{x+3} - 2\right)\left(\sqrt{x+3} + 2\right)$

    (b) $\left(\sqrt{3y-1} - 5\right)\left(\sqrt{3y-1} + 5\right)$

65. $f(x) = \dfrac{2}{\sqrt{x+2} - \sqrt{x}}$

    (a) By rationalizing the denominator, show that $f(x)$ can also be written in the form $g(x) = \sqrt{x+2} + \sqrt{x}$.

    (b) Find the values of f and g for $x = 0, 2, 4, 8$, and 10, rounded to two decimal places.

66. (a) Rationalize the denominator of $\dfrac{3}{\sqrt[3]{5} - \sqrt[3]{2}}$.

    (Hint: recall the special product $(a - b)(a^2 + ab + b^2) = a^3 - b^3$.)

    (b) Use a three decimal calculator approximation to support the results of part (a).

In calculus it is sometimes desirable to rationalize the numerator of an expression. In this case, we multiply both the numerator and denominator of the expression by the conjugate of the numerator. For example, to rationalize the numerator of $\dfrac{\sqrt{6} - 2}{2}$, multiply by the conjugate of the numerator.

$$\frac{\sqrt{6} - 2}{2} = \frac{\sqrt{6} - 2}{2} \cdot \frac{\sqrt{6} + 2}{\sqrt{6} + 2}$$

$$= \frac{\left(\sqrt{6}\right)^2 - (2)^2}{2\left(\sqrt{6} + 2\right)} \qquad \text{Special product}$$

$$= \frac{6 - 4}{2\left(\sqrt{6} + 2\right)} = \frac{2}{2\left(\sqrt{6} + 2\right)} = \frac{1}{\sqrt{6} + 2}$$

A decimal approximation confirms that the calculator value of each expression is 0.225.

In Problems 67 and 68, rationalize the numerator of the following expressions.

67. $\dfrac{\sqrt{10} - 2}{2}$

68. $\dfrac{\sqrt{x} + \sqrt{3}}{x - 3}$

69. **Oceanography:** The cross section of a *nautilus shell* has a series of compartments similar to those in Figure 3a. As the sea creature grows, it vacates a chamber, builds a wall, and occupies the outer chamber. To build a wire model of this shell, calculate the total length of wire needed as an exact irrational number and as a two decimal place approximation. The frame must show the fourteen compartments as in Figure 3b with the given sides measuring 1 inch.

Figure 3

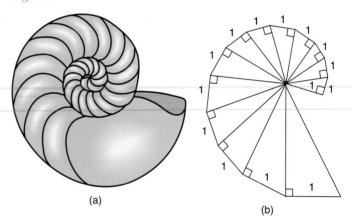

(a)   (b)

Objectives

1. Solve Equations Involving Radicals
2. Solve Equations Involving Rational Exponents
3. Solve Applied Problems

6.6 Equations with Radicals and Rational Exponents

In this section we present techniques for solving equations involving radicals and rational exponents.

Solving Equations Involving Radicals

Equations of the form

$$\sqrt{x} = 3, \sqrt{5t + 1} = 4 \text{ and } \sqrt[3]{5x + 2} = 3$$

are examples of **radical equations.** To solve an equation involving radicals, we use the following power rule.

Power Rule

> If both sides of an equation are raised to the same power, then the solutions of the original equation are included in the solutions of the new equation. That is, all solutions of
>
> $$P = Q \text{ are included in the solutions of } P^n = Q^n,$$
>
> where n is a positive integer.

This rule does not guarantee that by raising both sides of an equation to a power, the new equation is equivalent to the original equation. That is, the two equations may not have the same solutions. For example, the solution of the equation $x = 2$ is 2. However, if we square each side of this equation, we obtain the new equation $x^2 = 4$, whose solutions are -2 and 2. Notice that the solution of the equation $x = 2$ is also a solution of $x^2 = 4$. However, the converse is not necessarily true. The solutions of the original equation are contained in the solutions to the new equation. Because of this, *each solution of the new equation must be checked in the original equation for a possible extra or false solution.* Such an extra solution is called an **extraneous solution.**

EXAMPLE 1   **Solving a Radical Equation**

Solve the equation $\sqrt{5x + 1} = 4$ and check for any extraneous solution.

Solution   Use the power rule and square each side of the equation to get

$$\left(\sqrt{5x + 1}\right)^2 = 4^2$$

$$5x + 1 = 16$$

$$5x = 15$$

$$x = 3.$$

Check: If we replace x by 3, we have $\sqrt{5(3) + 1} = \sqrt{15 + 1} = \sqrt{16} = 4$.

Hence, 3 is the solution.

The solution of the radical equation in Example 1 is generalized in the following strategy for solving equations involving radicals.

Strategy for Solving Equations Involving Radicals

> **Step 1.** Write the equation so that one radical that contains variables is by itself on one side of the equation.
> **Step 2.** Raise each side of the equation to a power equal to the index of the radical.
> **Step 3.** Simplify each side of the new equation.
> **Step 4.** Repeat the above steps if the equation still contains a radical term and solve the new equation.
> **Step 5.** Whenever a radical has an *even* index, all proposed solutions must be checked in the original equation. If the index of the radical is odd, checking the proposed solutions is optional.

It is always prudent to check all solutions.

EXAMPLE 2   **Solving Equations Involving Radicals**

Solve each equation.

(a) $\sqrt{2x + 5} - 3 = 0$ (b) $\sqrt[3]{3t - 1} - 2 = 0$ (c) $\sqrt{4y^2 - 3} = 2y + 1$

Solution

(a)
$$
\begin{aligned}
\sqrt{2x + 5} - 3 &= 0 && \text{Original equation} \\
\sqrt{2x + 5} &= 3 && \text{Isolate radical} \\
\left(\sqrt{2x + 5}\right)^2 &= 3^2 && \text{Square each side} \\
2x + 5 &= 9 && \text{Simplify} \\
2x &= 4 \\
x &= 2
\end{aligned}
$$

Check: Substitute 2 for x in the original equation.

$$\sqrt{2(2) + 5} - 3 = \sqrt{4 + 5} - 3 = \sqrt{9} - 3 = 0 \qquad \text{Therefore, 2 is the solution.}$$

(b)
$$
\begin{aligned}
\sqrt[3]{3t - 1} - 2 &= 0 && \text{Original equation} \\
\sqrt[3]{3t - 1} &= 2 && \text{Isolate radical} \\
\left(\sqrt[3]{3t - 1}\right)^3 &= 2^3 && \text{Cube each side} \\
3t - 1 &= 8 && \text{Simplify} \\
3t &= 9 \\
t &= 3
\end{aligned}
$$

Since the index is odd, checking for extraneous solutions is optional. Therefore, 3 is the solution. It is always good practice to check solutions.

(c)
$$
\begin{aligned}
\sqrt{4y^2 - 3} &= 2y + 1 && \text{Original equation} \\
\left(\sqrt{4y^2 - 3}\right)^2 &= (2y + 1)^2 && \text{Square each side} \\
4y^2 - 3 &= 4y^2 + 4y + 1 && \text{Simplify} \\
-4y &= 3 + 1 \\
-4y &= 4 \\
y &= -1
\end{aligned}
$$

Check: Substitute -1 for y in the original equation.

Left Side	Right Side	True/False
$\sqrt{4(-1)^2 - 3} = \sqrt{4 - 3} = \sqrt{1} = 1$	$2(-1) + 1 = -2 + 1 = -1$	False

The solution $y = -1$ is an extraneous solution because it does not make the original equation true. Therefore, this equation has no solution.

Solving Equations Involving Rational Exponents

Equations involving rational exponents such as

$$x^{\frac{3}{2}} = 8, \quad y^{\frac{2}{3}} = 4, \quad \text{and} \quad (x - 3)^{\frac{2}{3}} = 4$$

can be solved by a procedure similar to that for solving equations involving radicals. To solve an equation of the form

$$x^{\frac{m}{n}} = a, \text{ for } x$$

we raise each side of the equation to the $\frac{n}{m}$th power. Remember that an equation containing rational exponents (in lowest terms) with n even must be checked for extraneous solutions.

100

EXAMPLE 3   **Solving Equations Involving Rational Exponents**

Solve each equation.

(a) $x^{\frac{3}{2}} = 8$

(b) $(x - 3)^{\frac{2}{3}} = 4$

Solution   (a) $x^{\frac{3}{2}} = 8$ Original equation with n even.

$\left(x^{\frac{3}{2}}\right)^{\frac{2}{3}} = 8^{\frac{2}{3}}$ Raise each side to the $\frac{2}{3}$ power.

$x = 8^{\frac{2}{3}} = (2^3)^{\frac{2}{3}}$ Simplify

$x = 2^2 = 4$

This solution *must* be checked in the original equation because n, in the original equation, is even.

Check: Substituting 4 for x results in

$4^{\frac{3}{2}} = (2^2)^{\frac{3}{2}} = 2^3 = 8$ Therefore, 4 is the solution.

(b) $(x - 3)^{\frac{2}{3}} = 4$ Original equation

$\left[(x - 3)^{\frac{2}{3}}\right]^{\frac{3}{2}} = 4^{\frac{3}{2}}$ Raise each side to the $\frac{3}{2}$ power.

$x - 3 = \left((\pm 2)^2\right)^{\frac{3}{2}}$ The $\frac{1}{2}$ power of 4 is ± 2 because of the roots of the equation.

$x - 3 = (\pm 2)^3$

$x - 3 = 8$ or $x - 3 = -8$

$x = 11$ or $x = -5$

Check:

x	Left Side	True/False
11	$(11 - 3)^{2/3} = 8^{2/3} = 2^2 = 4$	True
-5	$(-5 - 3)^{2/3} = (-8)^{2/3} = (-2)^2 = 4$	True

Therefore, 11 and -5 are both solutions.

Solving Applied Problems

Many applied problems from different fields can be solved using equations involving radicals or rational exponents.

EXAMPLE 4

Finding the Rate of Growth Over Time

In 1950 a hand woven Native American rug sold for approximately $50. In 2004, the same type of woven sells for approximately $1000. What is the rate of growth compounded annually?

Solution

The formula for compound interest is:

$$S = P\left(1 + \frac{r}{n}\right)^{nt}$$

where P is the initial amount, r is the rate of interest, n is the number of times compounded, t is time in years, and S is the final amount. We substitute $P = 50$, $n = 1$, $t = 54$, $S = 1000$, and solve the equation for r.

$$1000 = 50\left(1 + \frac{r}{1}\right)^{1(54)}$$

$20 = (1 + r)^{54}$ Isolate the exponential expression

$20^{\frac{1}{54}} = \left((1 + r)^{54}\right)^{\frac{1}{54}}$ Raise each side to the reciprocal power

$20^{\frac{1}{54}} = 1 + r$

$r = 20^{\frac{1}{54}} - 1 \approx 5.7\%$

Check: $50(1 + 0.057)^{54} \approx 997.74$ The more decimals used in the estimate for r, the closer the value for S will be to $1000.

EXAMPLE 5

Air Resistance on a Car

The amount of power (in horse power) necessary to overcome air resistance on a particular car is modeled by the formula

$$P = 1.11 \times 10^{-5} v^3$$

where v is velocity in kilometers per hour.

(a) Solve this formula for the velocity.

(b) Find the velocities for $P = 2$ horsepower and 4 horsepower.

(c) If the power is doubled, do you think the velocity will be doubled?

Solution

(a) $P = 1.11 \times 10^{-5} v^3$ Given

$v^3 = \dfrac{P}{1.11 \times 10^{-5}}$ Isolate the exponential

$v = \sqrt[3]{\dfrac{P}{1.11 \times 10^{-5}}}$ Take the cube root of both sides

$v = \sqrt[3]{9.009 \times 10^4 P}$ Simplify the radicand

(b) $v = \sqrt[3]{9.009 \times 10^4 P} = \sqrt[3]{9.009 \times 10^4 (2)} \approx 56$ kilometers per hour.

$v = \sqrt[3]{9.009 \times 10^4 P} = \sqrt[3]{9.009 \times 10^4 (4)} \approx 71$ kilometers per hour.

(c) The relationship between velocity and power is a cubic not a linear function. That is, as power increases the velocity does not increase in the same proportion. Only in a linear function is the change between input and output always in the same proportion. We called this the slope of the linear function.

◈ PROBLEM SET 6.6

Mastering the Concepts

In Problems 1–28, solve each equation. Check each solution.

1. (a) $\sqrt{x} - 3 = 0$

(b) $\sqrt{x - 1} - 3 = 0$

2. (a) $\sqrt{2x} - 4 = 0$

(b) $\sqrt{2x - 2} - 4 = 0$

3. (a) $\sqrt{x} = 5$

(b) $\sqrt{-x} = 2$

4. (a) $\sqrt{x} = 7$

(b) $\sqrt{-x} = 3$

5. $\sqrt{x - 2} = 3$

6. $\sqrt{x + 1} = 4$

7. $\sqrt{t + 1} - 2 = 0$

8. $\sqrt{y - 3} - 5 = 0$

9. $\sqrt{2y + 5} - 4 = 0$

10. $\sqrt{6t - 3} - 3 = 0$

11. $8 - \sqrt{y - 1} = 6$

12. $2 + \sqrt{7m - 5} = 6$

13. $\sqrt{x + 1} + 4 = 0$

14. $\sqrt{5x - 4} - 4 = 0$

15. $\sqrt{y + 5} - \sqrt{2y - 3} = 0$

16. $\sqrt{5x - 2} + 7 = 10$

17. $\sqrt{t + 14} - \sqrt{5t} = 0$

18. $\sqrt{4x + 5} - 3\sqrt{x} = 0$

19. $\sqrt{3y + 1} = \sqrt{y + 1}$

20. $\sqrt{2x + 10} - 2\sqrt{x} = 0$

21. $\sqrt[3]{t + 2} = -2$

22. $\sqrt[3]{3y - 4} = 2$

23. $\sqrt[3]{3x - 4} = \sqrt[3]{x + 10}$

24. $3\sqrt[3]{5 - x} = 2\sqrt[3]{18 - 3x}$

25. $\sqrt{x^2 + 3x} = x + 1$

26. $\sqrt{4x^2 + 13} = 2x + 1$

27. $\sqrt{6y + 2} - \sqrt{5y + 3} = 0$

28. $\sqrt{30x + 24} - 6\sqrt{x} = 0$

In Problems 29–42, solve each equation.

29. $x^{2/3} = 4$

30. $y^{2/5} = 1$

31. $y^{1/5} = -3$

32. $y^{1/7} = -2$

33. $x^{2/3} = 16$

34. $x^{3/5} = 8$

35. $x^{4/7} = 16$

36. $x^{3/7} = -8$

37. $x^{-3/4} = 27$

38. $x^{1/6} = 2$

39. $(x - 3)^{2/5} = 1$

40. $(x - 1)^{5/2} = 32$

41. $(t - 1)^{2/3} = 4$

42. $(5x - 7)^{4/3} = 16$

Applying the Concepts

43. Geometry: The radius r of a sphere whose surface area is A is given by the formula

$$r = \sqrt{\frac{A}{4\pi}}.$$

(a) Find the surface area of a spherical ball whose radius is 4.9 inches. Round off the answer to two decimal places.

(b) It is estimated that the radius of the moon is 1080 miles. Use the above formula to determine the surface area of the moon. Round off to the nearest square mile.

44. Law Enforcement: Law enforcement use the model

$$V = \sqrt{10.5x}$$

to estimate the speed v of a car (in miles per hour) from the distance x (in feet) it skidded before it came to a stop on a wet pavement. How fast was a car moving if it skidded 150 feet before it came to a stop? Round off to the nearest whole number.

45. Utility Pole: A wire is to be attached to support a 30 foot telephone pole. The wire must be anchored exactly 10 feet from the base of the pole and 15 feet above the base of the pole (Figure 1). The length of the attached wire is represented by the expression

$$\sqrt[4]{x - 2}.$$

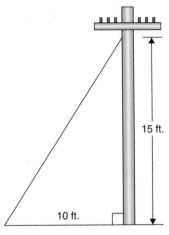

Figure 1

15 ft.

10 ft.

(a) Find the length of the attached wire. Round off to one decimal place.

(b) Formulate a radical equation and solve for x.

46. Cable Problem: A cable television company has a connector box located at point A on the shore of a lake. Nearby are two resorts B and C; B is located on the opposite shore from point A, and C is on land

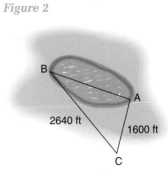

Figure 2

B

A

2640 ft

1600 ft

C

(Figure 2). The owners of the resorts would like to be serviced by the cable company. The distance from A to C is 1600 feet, the distance from B to C is 2640 feet and ABC forms a right triangle with angle A being a right angle.

(a) Find the length of the underwater cable between A and B such that BC is the length of the hypotenuse.

(b) It costs $1 per foot to run cable overland and $2 per foot to run underwater cable. Which is the cheapest route to wire resort B: run the cable underwater directly from A to B or to run the cable overland from A to C to B?

47. **Physics:** On earth a pendulum must be 39 inches long to be a "seconds" pendulum. That is, it takes two seconds for one complete swing back and forth of the pendulum. On the moon the pendulum must be 6.5 inches long to be a "seconds" pendulum. Find the value of g on earth and on the moon using the pendulum formula,

$$T = 2\pi\sqrt{\frac{L}{g}}$$

where T is the time (in seconds) for one complete swing back and forth, L is the pendulum length (in feet), and g is the gravitational constant.

48. **Speed of Sound:** The speed of sound, s, in air at 0° Celsius is 1087 feet per second. As the temperature T in degrees Celsius rises, the speed of sound will be modeled by the equation:

$$s = 1087\sqrt{\frac{T + 273}{273}}.$$

(a) Suppose you see a puff of smoke from a car that backfired 1500 feet away from you. If $T = 20°$ Celsius, how long does it take the sound to reach you?

(b) Suppose you hear thunder 2 seconds after you see lightening strike a tree which is 2250 feet from you. What is the temperature?

49. **Zoology:** Zoologists use the model

$$L = \sqrt[3]{\frac{w}{400}}$$

to relate the length L (in meters) of a lizard's body to its weight (in pounds). Determine the weight of a 1.75 meter monitor lizard. Round to the nearest pound.

50. **Population:** The population P (in thousands of people) of a small town is increasing according to the model

$$P = 15 + \sqrt{3t + 4}, \; t \geq 0$$

where t is time in years. When will the population reach twenty thousand people?

Developing and Extending the Concepts

In Problems 51–56, solve the following equations involving radicals. Check solutions for equations with even root indexes.

51. $\sqrt[4]{2x - 1} - 3 = 0$

52. $\sqrt[4]{9t^2 + 7} = \sqrt{3t + 1}$

53. $\sqrt[4]{y^2 - 7y + 1} = \sqrt{y - 5}$

54. $\sqrt[4]{t^2 + 1} = \sqrt{t + 1}$

55. $\sqrt[3]{x^3 + 6x^2} = x + 2$

56. $\sqrt[3]{x^2 + 2x - 6} = \sqrt[3]{x^2}$

In Problems 57–60, solve and check each equation. Use your results to match each pair of equations with one of the displays.

57. (a) $\sqrt{x} - 4 = 0$
    (b) $\sqrt{-x} - 4 = 0$

59. (a) $1 - \sqrt{2x + 1} = 0$
    (b) $\sqrt{2x + 1} - 1 = 0$

58. (a) $\sqrt{x - 1} - 5 = 0$
    (b) $\sqrt{1 - x} - 5 = 0$

60. (a) $-5 - \sqrt{3x - 1} = 0$
    (b) $\sqrt{3x - 1} - 5 = 0$

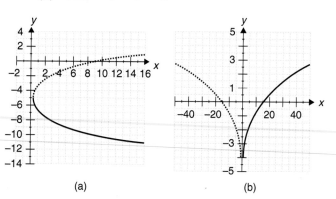

(a)

(b)

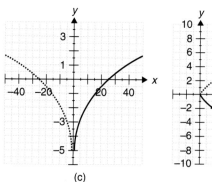

(c)

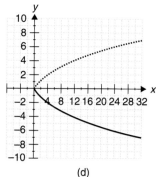

(d)

Objectives

1. Simplify Numbers of the Form $\sqrt{-b}$, $b > 0$
2. Define Complex Numbers
3. Perform Operations with Complex Numbers
4. Find Powers of i
5. Solve Applied Problems

6.7 Complex Numbers

Until this section, we emphasized that square roots of negative numbers, such as $\sqrt{-9}$, are not real numbers. In this section, we introduce a "new" number system that includes square roots of negative numbers. In 1545 the Italian physician and mathematician Girolamo Cardano (1501–1576) discovered that solutions to cubic equations involved square roots of negative numbers. Gradually, other mathematicians began to accept the idea of calculating with square roots of negative numbers. In 1637, the French philosopher and mathematician René Descartes (1596–1650) introduced the terms "real" and "imaginary". Later, in 1748, the Swiss mathematician Leonhard Euler (1707–1783) used the symbol "i" for $\sqrt{-1}$. Finally, in 1832 the great German mathematician Carl Friedrich Gauss (1777–1855) introduced the term "complex number". These numbers have been used in a wide variety of useful applications in electronics, engineering, and physics.

Simplifying Numbers of the Form $\sqrt{-b}$, $b > 0$

To launch our study of square roots of negative numbers we introduce the **imaginary unit**

$$i = \sqrt{-1}.$$

For the time being, regard i or $\sqrt{-1}$ as an "invented number" with the property that

$$i^2 = -1.$$

Using i, we can define the **principal square root** of a negative number as follows:

$$\text{If } b > 0, \text{ then } \sqrt{-b} = \sqrt{(-1)b} = \sqrt{-1} \cdot \sqrt{b} = i\sqrt{b}.$$

EXAMPLE 1   **Simplifying Square Roots of Negative Numbers**

Find the principal square root.

(a) $\sqrt{-4}$ (b) $\sqrt{-7}$ (c) $-\sqrt{-25}$ (d) $\sqrt{-16} + \sqrt{-9}$ (e) $\dfrac{\sqrt{-48}}{\sqrt{-3}}$

Solution

(a) $\sqrt{-4} = \sqrt{4(-1)} = \sqrt{4} \cdot \sqrt{-1} = 2i$

(b) $\sqrt{-7} = \sqrt{7(-1)} = \sqrt{7} \cdot \sqrt{-1} = \sqrt{7}i$

(c) $-\sqrt{-25} = -\sqrt{25(-1)} = -\sqrt{25} \cdot \sqrt{-1} = -5i$

(d) $\sqrt{-16} + \sqrt{-9} = \sqrt{16(-1)} + \sqrt{9(-1)}$
$$= \sqrt{16} \cdot \sqrt{-1} + \sqrt{9} \cdot \sqrt{-1}$$
$$= 4i + 3i = 7i$$

(e) $\dfrac{\sqrt{-48}}{\sqrt{-3}} = \dfrac{\sqrt{48} \cdot \sqrt{-1}}{\sqrt{3} \cdot \sqrt{-1}} = \sqrt{\dfrac{48}{3}} = \sqrt{16} = 4$

To perform multiplication with square roots of negative numbers, we must first write the number in the i form. That is, if a and b are positive real numbers, then

$$\sqrt{-a} \cdot \sqrt{-b} = \sqrt{(-1)a} \cdot \sqrt{(-1)b} = \sqrt{-1} \cdot \sqrt{a} \cdot \sqrt{-1} \cdot \sqrt{b}$$
$$= i\sqrt{a} \cdot i\sqrt{b} = i^2\sqrt{ab} = -\sqrt{ab}.$$

EXAMPLE 2   **Multiplying Square Roots of Negative Numbers**

Find each product.

(a) $\sqrt{-4} \cdot \sqrt{-9}$ (b) $\sqrt{-3}\left(\sqrt{-27} - \sqrt{-4}\right)$

Solution

When multiplying square roots of negative numbers, be sure to write them in the i form before multiplying.

(a) $\sqrt{-4} \cdot \sqrt{-9} = \sqrt{4} \cdot i \cdot \sqrt{9}i = 2i \cdot 3i = 6i^2 = -6$

(b) $\sqrt{-3}\left(\sqrt{-27} - \sqrt{-4}\right) = \sqrt{-3} \cdot \sqrt{-27} - \sqrt{-3} \cdot \sqrt{-4}$ Distributive Property

$= \sqrt{3}\sqrt{-1} \cdot \sqrt{27}\sqrt{-1} - \sqrt{3}\sqrt{-1} \cdot \sqrt{4}\sqrt{-1}$ Factor

$= \sqrt{3}i \cdot \sqrt{27}i - \sqrt{3}i \cdot \sqrt{4}i$

$= \sqrt{81}i^2 - 2\sqrt{3}i^2$ $\sqrt{12} = 2\sqrt{3}$

$= 9(-1) - 2(-1)\sqrt{3} = -9 + 2\sqrt{3}$

Defining Complex Numbers

Numbers such as $2i$, $\sqrt{5}i$, and $-3i$ are called *pure imaginary numbers*. We form a new set of numbers by adding these imaginary numbers to real numbers.

Definition of a Complex Number

> Any number, real or imaginary, of the form
>
> $$a + bi$$
>
> where a and b are real numbers and $i = \sqrt{-1}$, is called a **complex number.**

For example,

$$4 + 3i,\ 2 - 7i,\ 3 + \sqrt{5}\,i, \text{ and } \frac{3}{4} - 2\sqrt{7}i$$

are complex numbers.

The form $a + bi$ is called the **standard form** for complex numbers. The number a is called the **real part** of the complex number. The number b is called the **imaginary part** of the complex number.

Every real number a is a complex number, because the real number a can be written as

$$a = a + 0i.$$

Similarly, if b is a real number, then

$$bi = 0 + bi.$$

Two complex numbers $a + bi$ and $c + di$ are **equal** if and only if their real parts are equal and their imaginary parts are equal. That is,

$$a + bi = c + di \qquad \text{if and only if} \qquad a = c \text{ and } b = d.$$

EXAMPLE 3   **Equality of Two Complex Numbers**

Find x and y so that $5x - \sqrt{-36} = -15 + 3yi$.

Solution   First, we write the left side in standard form

$$5x - \sqrt{-36} = 5x - 6i$$

so,

$$5x - 6i = -15 + 3yi.$$

Thus, the real parts must be equal: $5x = -15$

and

$$x = -3.$$

The imaginary parts must be equal: $-6i = 3yi$

and

$$y = -2.$$

Performing Operations with Complex Numbers

To add (or subtract) complex numbers, we add (or subtract) their real parts and then add (or subtract) their imaginary parts. The following rule is similar to the rule for combining like terms of a polynomial.

Rules for Adding and Subtracting Complex Numbers

For any real numbers a, b, c, and d.

1. $(a + bi) + (c + di) = (a + c) + (b + d)i$
2. $(a + bi) - (c + di) = (a - c) + (b - d)i$

Notice that the sum and difference of complex numbers is again a complex number. Also, the commutative, associative, and distributive properties for real numbers are also true for complex numbers.

EXAMPLE 4   **Adding and Subtracting Complex Numbers**

Perform each operation and simplify.

(a) $(5 + 6i) + (9 + 3i)$ (b) $(4 - 2i) - (-3 + i)$

Solution   (a) $(5 + 6i) + (9 + 3i) = (5 + 9) + (6i + 3i)$ group real and imaginary parts

$$= (5 + 9) + (6 + 3)i$$

$$= 14 + 9i$$

(b) $(4 - 2i) - (-3 + i) = 4 - 2i + 3 - i$

$$= (4 + 3) + (-2 - 1)i$$

$$= 7 - 3i$$

Complex numbers of the form $a + bi$ have the same form as a binomial, so we can multiply two complex numbers in the same way as we multiply binomials.

EXAMPLE 5   **Multiplying Complex Numbers**

Perform each multiplication.

(a) $(4i)(3 + 7i)$ (b) $(4 + 3i)(2 - 4i)$ (c) $(1 + 2i)(1 - 2i)$

Solution   (a) $(4i)(3 + 7i) = 4i(3) + 4i(7i)$ Distributive property

$$= 12i + 28i^2$$

$$= 12i + 28(-1)$$

$$= -28 + 12i$$

(b) We multiply by using the FOIL method.

$$(4 + 3i)(2 - 4i) = 4 \cdot 2 - 4 \cdot 4i + 3i \cdot 2 - 3i \cdot 4i$$

$$= 8 - 16i + 6i - 12i^2$$

$$= 8 - 12(-1) - 16i + 6i \qquad i^2 = -1$$

$$= 20 - 10i$$

(c) This product has the form of the special product $(a - bi)(a + bi) = a^2 - b^2$.

$$(1 + 2i)(1 - 2i) = 1^2 - (2i)^2$$

$$= 1 - 4i^2$$

$$= 1 - 4(-1)$$

$$= 1 + 4 = 5$$

In example 5c, note that the product of two complex numbers can be a real number. This will happen with pairs of complex numbers are of the form $a + bi$, and $a - bi$ which are called **complex conjugates.** In general, the product of a complex number and its conjugate is always a real number. That is,

$$(a - bi)(a + bi) = a^2 - b^2i^2$$

$$= a^2 - b^2(-1)$$

$$= a^2 + b^2.$$

To perform the division

$$\frac{3i}{4 - 3i}$$

we look for a complex number in standard form that is equivalent to $\dfrac{3i}{4 - 3i}$. To do this, we need to eliminate i from the denominator, so we multiply the numerator and the denominator by the complex conjugate of the denominator, which is $4 + 3i$.

$$\frac{3i}{4 - 3i} = \frac{3i}{(4 - 3i)} \cdot \frac{4 + 3i}{4 + 3i}$$

$$= \frac{12i + 9i^2}{16 - 9i^2} = \frac{-9 + 12i}{16 + 9}$$

$$= \frac{-9 + 12i}{25} = \frac{-9}{25} + \frac{12}{25}i$$

Dividing Complex Numbers

*There are no remainders
when dividing complex
numbers.*

> To find the real and imaginary parts of a quotient of complex numbers, multiply
> the numerator and denominator by the complex conjugate of the denominator.

EXAMPLE 6   **Dividing Complex Numbers**

Perform each division and express the result in the form $a + bi$.

(a) $\dfrac{1}{2 + 3i}$

(b) $\dfrac{2 - 3i}{1 - 4i}$

Solution   (a) $\dfrac{1}{2 + 3i} = \dfrac{1}{2 + 3i} \cdot \dfrac{2 - 3i}{2 - 3i}$ Multiply by the conjugate

$$= \frac{2 - 3i}{4 - 9i^2} = \frac{2 - 3i}{4 - 9(-1)}$$

$$= \frac{2 - 3i}{4 + 9} = \frac{2 - 3i}{13} = \frac{2}{13} - \frac{3}{13}i$$

(b) $\dfrac{2 - 3i}{1 - 4i} = \dfrac{2 - 3i}{1 - 4i} \cdot \dfrac{1 + 4i}{1 + 4i}$ Multiply by the conjugate

$$= \frac{2 + 8i - 3i - 12i^2}{1 - 16i^2} = \frac{2 + 5i - 12(-1)}{1 - 16(-1)}$$

$$= \frac{14 + 5i}{17} = \frac{14}{17} + \frac{5}{17}i$$

Finding Powers of *i*

Positive integer exponents have the same meaning, in terms of repeated multiplication, for both complex numbers and real numbers. Therefore, we can extend the definition of positive integer exponents to include complex numbers. In particular by
using $i = \sqrt{-1}$ and $i^2 = 1$ we can find other powers of *i*. The following table lists
some of these powers of *i*. Notice the recurring pattern for every fourth power of *i*.

Power of *i*	Factor Powers of *i*	Value
i^1	i	i
i^2	i^2	-1
i^3	$i^2 \cdot 1 = (-1)i$	$-i$
i^4	$i^2 \cdot i^2 = (-1)(-1)$	1
i^5	$i^4 \cdot i = 1i$	i
i^6	$i^4 \cdot i^2 = 1(-1)$	-1
i^7	$i^4 \cdot i^3 = 1(-i)$	$-i$
i^8	$i^4 \cdot i^4 = 1 \cdot 1$	1
i^9	$i^8 \cdot i = 1 \cdot i$	i

EXAMPLE 7   **Simplifying Powers of i**

Find each power of i.

(a) i^{18} (b) i^{27} (c) i^{105} (d) i^{-3}

Solution   Because $i^4 = 1$, we will factor each expression as fourth powers of i.

(a) $i^{18} = i^{16} \cdot i^2 = (i^4)^4 \cdot i^2 = 1^4(-1) = -1$

(b) $i^{27} = i^{24} \cdot i^3 = (i^4)^6 \cdot i^3 = 1^6(-i) = -i$

(c) $i^{105} = i^{104} \cdot i = (i^4)^{26} \cdot i = 1^{26} \cdot i = i$

(d) We note that $i^{-3} = \dfrac{1}{i^3}$. By multiplying the numerator and denominator by i, we obtain a real number in the denominator, so that

$$i^{-3} = \frac{1}{i^3} = \frac{1}{i^3} \cdot \frac{i}{i} = \frac{i}{i^4} = \frac{i}{1} = i.$$

Solving Applied Problems

A famous puzzle was solved by Cardano using complex numbers.

EXAMPLE 8   **Solving a Puzzle Problem**

Find two numbers whose sum is 10 and whose product is 40.

Solution   In 1545, the Italian physician and mathematician Girolemo Cardano (1501–1576) suggested that the answer to this puzzle were the numbers $5 + \text{RM}15$ and $5 - \text{RM}15$ which today can be written as $5 + \sqrt{-15}$ and $5 - \sqrt{-15}$. Now we will add and multiply these complex numbers to demonstrate that they are the solutions to the puzzle.

$$(5 + \sqrt{-15}) + (5 - \sqrt{-15}) = (5 + \sqrt{15}i) + (5 - \sqrt{15}i) = 10, \text{ and}$$

$$(5 + \sqrt{-15})(5 - \sqrt{-15}) = (5 + \sqrt{15}i)(5 - \sqrt{15}i) = 5^2 - (\sqrt{15})^2 i^2$$

$$= 25 + 15 = 40$$

In 1797 the Norwegian surveyor Caspar Wessel used complex numbers to represent direction and distance on a two dimensional grid. The idea started from attempting to order complex numbers such as the following:

$$5, -3i, -3 + 3i, \text{ and } 4 + 3i$$

Figure 1

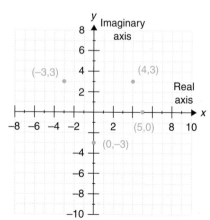

from smallest to largest. Every attempt had failed because there is no way of locating complex numbers on a number line. However, Gauss discovered that complex numbers can be graphed on a special coordinate system called the *complex plane*. The horizontal line (x axis) represents the real part of the complex number, and is called the **real axis.** The vertical line (y axis) represents the imaginary part and is called the **imaginary axis.** The graph of the complex number

$$z = a + bi$$

is associated with the ordered pair (a,b) in the Cartesian coordinate system. Figure 1 shows the graph of the complex numbers $5, -3i, -3 + 3i$, and $4 + 3i$.

EXAMPLE 9   **Adding Complex Numbers**

Calculate $(2 + 4i) + (5 - 2i)$ and graph.

Figure 2

Solution $(2 + 4i) + (5 - 2i) =$

$(2 + 5) + (4 - 2)i = 7 + 2i$

Figure 2 shows the points associated with these complex numbers. Notice that if we connect these points with the origin, the figure formed is a parallelogram. The point diagonally opposite the origin in the parallelogram is the sum of these complex numbers. This is the basis for vector addition used in navigation.

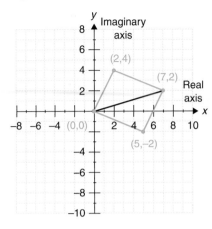

 PROBLEM SET 6.7

Mastering the Concepts

In Problems 1–10, find the principal square roots.

1. (a) $\sqrt{-16}$
   (b) $\sqrt{-81}$
2. (a) $\sqrt{-25}$
   (b) $\sqrt{-36}$
3. (a) $-\sqrt{-8}$
   (b) $\sqrt{-72}$
4. (a) $-\sqrt{-12}$
   (b) $-\sqrt{-18}$
5. (a) $-\sqrt{-\dfrac{9}{16}}$
   (b) $-\sqrt{-\dfrac{25}{4}}$
6. (a) $-\sqrt{-\dfrac{3}{4}}$
   (b) $-\sqrt{-\dfrac{5}{16}}$
7. (a) $\dfrac{\sqrt{-8}}{\sqrt{-2}}$
   (b) $\dfrac{\sqrt{-32}}{\sqrt{-2}}$
8. (a) $\dfrac{\sqrt{-50}}{\sqrt{-2}}$
   (b) $\dfrac{\sqrt{-72}}{\sqrt{-2}}$
9. $\sqrt{-8} + \sqrt{-72}$
10. $\sqrt{-18} + \sqrt{-50}$

In Problems 11–14, find each product.

11. $\sqrt{-4} \cdot \sqrt{-25}$
12. $\sqrt{-8} \cdot \sqrt{-2}$
13. $\sqrt{-4}\left(\sqrt{-5} + \sqrt{-6}\right)$
14. $\sqrt{-49}\left(\sqrt{-3} - 2\sqrt{-2}\right)$

In Problems 15 and 16, find x and y so that each statement is true.

15. $4x + 8i = 20 + 2yi$
16. $(4x - 3) + 21i = 5 + (2y - 3)i$

In Problems 17–38, perform each operation and express the answer in the form $a + bi$.

17. (a) $(-3 + 6i) + (2 + 3i)$
    (b) $(10 - 24i) + (3 + 7i)$
18. (a) $(-2 + 5i) + (-5 + i)$
    (b) $(6 + 8i) + (6 - 8i)$
19. (a) $(-2 - 3i) - (-3 - 2i)$
    (b) $(3 + 2i) - (3 - 2i)$
20. (a) $(7 + 24i) - (-3 - 4i)$
    (b) $(10 - 8i) - (10 + 8i)$
21. $(5 - 7i) - (5 - 13i)$
22. $(6 - 8i) - (5 + 3i)$
23. $[(4 + 3i) - (3 + 6i)] + (2 - 3i)$
24. $(8 - 5i) + [(3 + 7i) - (2 - 5i)]$
25. $-7i(-2 + 4i)$ 26. $3i(2 + 3i)$
27. $(2 - 4i)(3 + 2i)$ 28. $(1 + 2i)(4 - i)$
29. $(2 + 3i)(3 - 5i)$ 30. $(-3 - 2i)(2 + 5i)$
31. $(3 - 7i)(2 + 3i)$ 32. $(1 - 6i)(2 + 5i)$
33. $(-7 - 2i)^2$ 34. $(1 + i)^2$
35. $(5 - 4i)(5 + 4i)$ 36. $(3 - 2i)(3 + 2i)$
37. $(5 - 12i)(5 + 12i)$ 38. $(2 - 7i)(2 + 7i)$

In Problems 39–44, perform each division and express the answer in the form $a + bi$.

39. (a) $\dfrac{1}{4 + 3i}$ (b) $\dfrac{1 - i}{-1 + i}$

40. (a) $\dfrac{1}{3 - 5i}$ (b) $\dfrac{3 + 4i}{-2 + 5i}$

41. (a) $\dfrac{7}{5-4i}$ (b) $\dfrac{7+2i}{1-5i}$

42. (a) $\dfrac{-3}{3+4i}$ (b) $\dfrac{4-i}{2-7i}$

43. (a) $\dfrac{2+i}{2+3i}$ (b) $\dfrac{5-i}{1+2i}$

44. (a) $\dfrac{i}{-1-i}$ (b) $\dfrac{1+i}{2-3i}$

In Problems 45–48, find each power of i.

45. (a) i^{29} (b) $(-i)^{37}$ (c) i^{54}
46. (a) i^{49} (b) i^{65} (c) $(-i)^{76}$
47. (a) i^{108} (b) i^{-15} (c) i^{-7}
48. (a) i^{-18} (b) i^{-5} (c) i^{-14}

Applying the Concepts

49. Alternating Current: In 1893, Charles P. Steinmetz, an American electrical engineer, developed a theory of alternating currents based on complex numbers. He showed that *Ohm's Law* for alternating currents takes the form

$$E = IZ$$

where E is the *voltage*, I is the *current*, and Z is the *impedance* given by the equation

$$Z = R + (X_L - X_C)i$$

where X_L is the *inductive reactance* and X_C is the *capacitative reactance*. Find E if $I = 1 - i$, $R = 3$, $X_L = 2$, and $X_C = 1$.

50. Alternating Current: Use the equation

$$E = I[R + (X_L - X_C)i$$

to find I if $E = 3 + 2i$, $R = 4$, $X_L = 3$, and $X_C = 2$.

Developing and Extending the Concepts

In Problems 51–54, write each expression in the form $a + bi$.

51. $1 + i + (-i)^2 + i^3$ **52.** $i - i^2 + (-i)^3 - i^4$
53. $5i^7 + 3i^2 - 2i$ **54.** $2i^{84} - 3i^{100} + 2i^2$

55. Let $f(z) = z^2 - 2z + 2$, find and simplify $f(1 + i)$.
56. Let $f(z) = z^2 - 2z + 5$, find and simplify $f(1 + 2i)$.
57. Write the expression $(7 - 3i)^{-2}$ in the form $a + bi$.
58. Simplify the expression
$$i + i^2 + i^3 + i^4 + \ldots + i^{98} + i^{99} + i^{100}.$$

59. (a) Draw on a coordinate plane a parallelogram whose vertices are $A = (0,0)$, $B = (1,5)$, $C = (8,7)$, and $D = (7,2)$.
(b) Find the length of the diagonal $\overline{AC}$.
(c) Find the sum of the complex numbers represented by the ordered pairs B and D.

60. Physics: A can of iced tea is h centimeters high. A hole is pierced in the side of the can y centimeters from the bottom (Figure 3). The stream of tea coming from that hole first hits the table d centimeters from the can, where d is given by the equation

$$d = \sqrt{y(h-y)},$$
$$0 < y < h.$$

Figure 3

How far from the can will the tea hit the table if the can is 12 centimeters tall and the hole is pierced in the can 8 centimeters above the table? Round off to two decimal places.

 CHAPTER 6 REVIEW PROBLEMS

In Problems 1–8, rewrite each expression in a simplified form using only non negative exponents. Assume all variables are restricted to values for which all expressions represent real numbers.

1. (a) $(-7)^0 + \dfrac{1}{2^{-1}}$
(b) $\dfrac{1 + 3^{-2}}{2^{-2}}$
(c) $\dfrac{5^{-3}}{2^3} + \dfrac{2^{-3}}{5^3}$
(d) $\left(\dfrac{x^2}{y^3}\right)^{-2}$

2. (a) $(3)^0 + \dfrac{1}{7^{-2}}$
(b) $\dfrac{1}{2^{-3} + 3^{-2}}$
(c) $\dfrac{2^{-1}}{3^2} + \dfrac{3^{-1}}{2^4}$
(d) $\left(\dfrac{x^{-3}}{y^2}\right)^2$

3. (a) $3^{-11} \cdot 3^{15}$
(b) $(2^{-\frac{1}{3}})^9$
(c) $\dfrac{b^{-6}}{b^{-11}}$

4. (a) $9^{-4/7} \cdot 9^{11/7}$
(b) $(3x^{-2})^{-2}$
(c) $(t^{-4})^{-3}$

5. (a) $x^{-3/5} \cdot x^{5/4}$
(b) $\dfrac{x^{-5}}{x^{-2}}$
(c) $(2x^{-3}y^{-4})^{-3}$

6. (a) $(3^{-\frac{1}{2}})^{-4}$
(b) $\dfrac{7^{-4}}{7^{-3}}$
(c) $(4^{-1}x^{-3})^2$

7. (a) $\dfrac{x^{-3}y^2}{x^{-2}y^{-3}}$
(b) $\dfrac{a^3b^{-4}}{a^{-4}b^2}$
(c) $\left(\dfrac{x^4}{2y^{-2}}\right)^{-2}$

8. (a) $\dfrac{m^{-4}n^{-3}}{m^{-7}n^{-2}}$
(b) $\dfrac{a^4b}{a^{-3}b^4}$
(c) $\left(\dfrac{x^3y^{-2}}{(xy)^{-1}}\right)^{-1}$

In Problems 9 and 10, solve each exponential equation.

9. (a) $2^{-x} = 4$ **10.** (a) $3^{-2x} = 27$

     (b) $5^{2x} = 25$ (b) $7^{-x} = 49$

In Problems 11 and 12, find the principal root of the given number. State the reason when the number is not real.

11. (a) $\sqrt{81}$ (b) $\sqrt[3]{-64}$ (c) $\sqrt[4]{-16}$

12. (a) $\sqrt{25}$ (b) $\sqrt[3]{-216}$ (c) $\sqrt[4]{-64}$

In Problems 13 and 14, find the exact value of each expression.

13. (a) $16^{-3/4}$ (b) $-(-8)^{1/3}$ (c) $36^{-5/2}$

14. (a) $(-8)^{-2/3}$ (b) $-(-27)^{1/3}$ (c) $(-32)^{-3/5}$

In Problems 15 and 16, convert each expression to rational exponents.

15. (a) $\sqrt[3]{-7}$ **16.** (a) $\sqrt[4]{3x^2 + 1}$

     (b) $\sqrt[5]{x^4}$ (b) $\sqrt[3]{(a + b)^2}$

In Problems 17 and 18, convert each expression to radical form.

17. (a) $7^{-2/3}$ **18.** (a) $5^{3/4}$

     (b) $(x + 1)^{2/5}$ (b) $(x^2 + 7)^{-3/4}$

In Problems 19–30, use the properties of radicals to simplify each expression. Assume all variables represent positive real numbers.

19. (a) $\sqrt{125}$ (b) $\sqrt[3]{-125}$ (c) $\sqrt[4]{32t^5}$

20. (a) $\sqrt{8}$ (b) $\sqrt[3]{-40}$ (c) $\sqrt[3]{54x^7}$

21. (a) $\sqrt{\dfrac{4}{25}}$ (b) $\sqrt[3]{-\dfrac{7}{27}}$ (c) $\sqrt[5]{\dfrac{3}{32}}$

22. (a) $\sqrt[4]{\dfrac{5}{81}}$ (b) $\sqrt[3]{-\dfrac{11}{8}}$ (c) $\sqrt[5]{-\dfrac{3}{1024}}$

23. (a) $\sqrt[5]{\dfrac{t^5}{32}}$ (b) $\sqrt[4]{\dfrac{16x^8}{81}}$ (c) $\sqrt[3]{-\dfrac{5}{x^9}}$

24. (a) $\sqrt{\dfrac{11}{x^4}}$ (b) $\sqrt[3]{-\dfrac{7}{x^6}}$ (c) $\sqrt[5]{\dfrac{3}{x^{10}}}$

25. (a) $\dfrac{\sqrt{125x^7}}{\sqrt{5x}}$ (b) $\sqrt{\sqrt{16x^8}}$

26. (a) $\dfrac{\sqrt[3]{-24m^{11}}}{\sqrt[3]{3m^2}}$ (b) $\sqrt[3]{\sqrt[3]{x^{18}}}$

27. (a) $\sqrt{8x^3y^4} \cdot \sqrt{2xy^2}$ (b) $\sqrt[3]{4x^2} \cdot \sqrt[3]{2x}$

28. (a) $\sqrt[3]{\dfrac{x^3 x^2}{x^8}}$ (b) $\sqrt[7]{\dfrac{u^2 u^7}{u^{16}}}$

29. (a) $\left(\dfrac{16x^4}{y^8}\right)^{-\frac{3}{4}}$ (b) $\left(\dfrac{x^5 y^{10}}{32}\right)^{-\frac{3}{5}}$

30. (a) $\left(\dfrac{8m^{-6}}{27m^{-12}}\right)^{-\frac{1}{3}}$ (b) $\left(\dfrac{x^{-1}y^{-\frac{1}{3}}}{x^{-2}y^{-\frac{2}{3}}}\right)^{-3}$

In Problems 31–40, perform each operation and simplify. Assume all variables represent positive real numbers.

31. (a) $7\sqrt{2} - 3\sqrt{2} + 4\sqrt{2}$

     (b) $\sqrt{128} + \sqrt{8}$

32. (a) $\sqrt{5} - 6\sqrt{5} + 2\sqrt{5}$

     (b) $\sqrt{48} - \sqrt{12}$

33. (a) $\sqrt{63x} + 2\sqrt{112x} - \sqrt{252x}$

     (b) $\sqrt[3]{16y} - \sqrt[3]{54y} + \sqrt[3]{250y}$

34. (a) $\sqrt{3}\left(\sqrt{2} + \sqrt{5}\right)$

     (b) $\sqrt{5}\left(\sqrt{7} - \sqrt{2}\right)$

35. (a) $\left(\sqrt{10} + \sqrt{2}\right)^2$

     (b) $\left(2\sqrt{x} - 3\sqrt{y}\right)^2$

36. (a) $\left(2\sqrt{6} - 3\sqrt{2}\right)^2$

     (b) $\left(1 - \sqrt{2}\right)^3$

37. (a) $\left(\sqrt{8} - \sqrt{2}\right)\left(\sqrt{8} + \sqrt{2}\right)$

     (b) $\left(\sqrt{a} - b\right)\left(\sqrt{a} + b\right)$

38. (a) $\left(\sqrt{3t} + \sqrt{5}\right)\left(\sqrt{3t} - \sqrt{5}\right)$

     (b) $\left(\sqrt{t} + \sqrt{5y}\right)\left(\sqrt{t} - \sqrt{5y}\right)$

39. (a) $\left(\sqrt{x} - \sqrt{y}\right)\left(2\sqrt{x} + \sqrt{y}\right)$

     (b) $\left(3\sqrt{x} - y\right)\left(\sqrt{x} + 2y\right)$

40. (a) $(\sqrt{8} + \sqrt{18})(\sqrt{6} + 1)$

     (b) $\left(\sqrt{x} + \sqrt{y}\right)\left(2\sqrt{x} - 3\sqrt{y}\right)$

In Problems 41–46, rationalize each denominator and simplify. Assume all variables represent positive real numbers. Use decimal approximations to confirm the results in Problems 41–44.

41. (a) $\dfrac{2}{\sqrt{7}}$ (b) $\dfrac{3}{5\sqrt{2}}$

42. (a) $\dfrac{-3}{\sqrt{5}}$ (b) $\dfrac{2}{7\sqrt{3}}$

43. (a) $\dfrac{1}{2 + \sqrt{3}}$ (b) $\dfrac{\sqrt{7}}{\sqrt{2} - \sqrt{3}}$

44. (a) $\dfrac{\sqrt{2}}{\sqrt{3} + \sqrt{7}}$ (b) $\dfrac{\sqrt{8} + 3}{3\sqrt{2} + 2}$

45. $\dfrac{\sqrt{x} + 1}{\sqrt{x} - 1}$ **46.** $\dfrac{\sqrt{x} + \sqrt{3}}{\sqrt{x} - \sqrt{3}}$

In Problems 47 and 48, rationalize the numerator and simplify.

47. (a) $\dfrac{\sqrt{5}+3}{2}$ (b) $\dfrac{\sqrt{7}+\sqrt{3}}{\sqrt{3}}$

48. (a) $\dfrac{\sqrt{3}-1}{\sqrt{3}+1}$ (b) $\dfrac{5-\sqrt{3}}{\sqrt{5}}$

In Problems 49 and 50, simplify each radical.

49. (a) $2\sqrt{12}-3\sqrt{48}$
   (b) $\sqrt[3]{-27x^3y^6}$

50. (a) $(4\sqrt{5}-\sqrt{6})(2\sqrt{5}+3)$
   (b) $\sqrt[3]{64x^6y^7}$

In Problems 51–54, solve each equation.

51. (a) $x^{1/6}=2$ (b) $(x-2)^{3/2}=8$

52. (a) $y^{2/3}=4$ (b) $(t-3)^{1/5}=1$

53. (a) $\sqrt{x}-2=3$ (b) $(x+1)^2=4$

54. (a) $\sqrt[3]{x+2}=-3$ (b) $(2x-1)^2-9=0$

In Problems 55–60, write each expression in the form of $a+bi$, where a and b are real numbers.

55. (a) $4\sqrt{-9}-2\sqrt{-25}$ (b) $\sqrt{-16}\cdot\sqrt{-9}$

56. (a) $\sqrt{-16}\cdot\sqrt{-25}$ (b) $\dfrac{\sqrt{-25}}{\sqrt{-16}}$

57. (a) $5i(1-7i)$ (b) $(2-i)(2+i)$

58. (a) $(2-3i)^2$ (b) $\dfrac{2}{3+4i}$

59. (a) $\dfrac{3+2i}{4+i}$ (b) $\dfrac{1-i}{2-3i}$

60. (a) $\dfrac{2+i}{-5+i}$ (b) $\dfrac{3+\sqrt{-16}}{2-\sqrt{-9}}$

In Problems 61–66, evaluate each function at the indicated input values. Round off to two decimal places if necessary.

61. $f(x)=\left(\dfrac{1}{2}\right)^x$

   (a) $x=-2$ (b) $x=0$
   (c) $x=3.1$ (d) $x=-2.4$

62. $g(x)=\left(\dfrac{2}{3}\right)^x$

   (a) $x=-1$ (b) $x=0$
   (c) $x=2.3$ (d) $x=-3.1$

63. $f(x)=\sqrt{3x+1}$
   (a) $x=2$ (b) $x=0$
   (c) $x=2.4$ (d) $x=-\dfrac{1}{3}$

64. $g(x)=\sqrt[3]{4x+9}$
   (a) $x=-1$ (b) $x=0$
   (c) $x=2.5$ (d) $x=-\dfrac{9}{4}$

65. $f(x)=x^{-2/3}$
   (a) $x=-8$ (b) $x=0$
   (c) $x=3.1$ (d) $x=-4.3$

66. $g(x)=\sqrt{2-x}+\sqrt{-x}$
   (a) $x=-2$ (b) $x=0$
   (c) $x=-4$ (d) $x=-3.4$

67. Compound Interest: A deposit of $1000 is placed in a savings account. Find the balance in the account after 4.5 years if it is compounded monthly at a rate of 4.7%.

68. Navigation: The horizon distance d (in miles) that can be viewed on the surface of the ocean at a viewing height h (in feet) above the surface of the ocean can be approximated by the model

$$d(h)=1.4\sqrt{h}$$

   (a) Find the horizon distance from a balloon at an altitude of 1200 feet to view a ship on the ocean.

   (b) Is it possible to view a sailboat on the ocean at a distance of 20 miles from a blimp at an altitude of 205 feet?

69. Geometry: The radius r of a circle inscribed in a triangle of sides a, b, and c (Figure 4) is given by the formula

$$r=\sqrt{\frac{(s-a)(s-b)(s-c)}{s}},\ s=\frac{1}{2}(a+b+c).$$

Figure 4

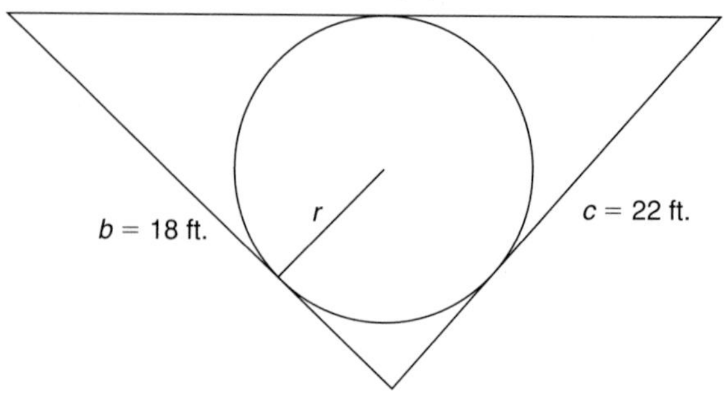

Find the radius of a circular flower bed that can be inscribed in a triangular plot with sides 18 feet, 22 feet, and 26 feet. Round off the answer to two decimal places.

70. Investment: The annual rate of return r on an investment of p dollars that is worth s dollars after t years is modeled by the formula

$$r = \left(\frac{s}{p}\right)^{\frac{1}{t}} - 1.$$

Find the annual rate of return of a stock that was purchased for \$22.55 and sold 6 years later for \$37.75.

In Problems 71–74,

(a) Graph each function.

(b) Use the graph to indicate the domain and range of each function.

(c) Find the x intercept of each graph to determine the solution of the associated equation.

71. $f(x) = \sqrt{2x + 2} - 4$; $\sqrt{2x + 2} = 4$

72. $g(x) = \sqrt{x + 3} - 1$; $\sqrt{x + 3} = 1$

73. $h(x) = \sqrt[3]{x - 1} - 1$; $\sqrt[3]{x - 1} = 1$

74. $k(x) = \sqrt[3]{x + 1} - 2$; $\sqrt[3]{x + 1} = 2$

CHAPTER 6 PRACTICE TEST

1. Rewrite each expression in a simplified form using only non negative exponents.

(a) $(-10)^0$ (b) $\left(\frac{-3}{7}\right)^{-2}$ (c) $\frac{1}{3^{-4}}$

(d) $x^{-11} x^7$ (e) $\left(\frac{x^{-2}y^2}{x^{-3}y^{-3}}\right)^{-1}$

2. Solve each exponential equation.

(a) $2^x = 16$ (b) $3^{-4x} = 81$ (c) $5^{2y} = 125$

3. Find the principal roots of the given numbers. When it is not a real number, state the reason.

(a) $\sqrt[4]{16}$ (b) $\sqrt{-25}$
(c) $-\sqrt{25}$ (d) $\sqrt[3]{-27}$

4. Simplify each expression (if possible). Assume all variables represent positive real numbers.

(a) $-\sqrt{3^2}$ (c) $\sqrt{9x^2}$
(b) $\left(\sqrt[3]{5x}\right)^3$ (d) $\left(\sqrt[4]{-3y}\right)^4$

5. Simplify $\sqrt{x^2 - 18x + 81}$

6. Let $f(x) = \sqrt[3]{x - 8}$. Find
(a) $f(0)$ (b) $f(-19)$.

7. Find the exact value of each expression.
(a) $4^{-1/2}$ (b) $16^{3/4}$
(c) $(-27)^{2/3}$ (d) $(-8)^{-4/3}$

8. (a) Convert $\sqrt[7]{x^2}$ to an expression with a rational exponent.
(b) Convert $(x)^{3/11}$ to a radical form.

9. Simplify each expression. Write each answer using only positive exponents.
(a) $5^{1/2} \cdot 5^{-3/2}$ (b) $(3^{1/7})^{14}$
(c) $\frac{4^{1/3}}{4^{-1/6}}$

10. Let $f(x) = x^{2/3}$. Find
(a) $f(-8)$ (b) $f(0)$ (c) $f(27)$.

11. Use the properties of radicals to simplify each expression. Assume all variables represent positive real numbers.
(a) $\sqrt[3]{54}$ (b) $\sqrt{\frac{3}{25x^2}}$
(c) $\sqrt[3]{-81x^4}$ (d) $\frac{\sqrt{45x^3}}{\sqrt{3x}}$

12. Rewrite $\sqrt{3} \cdot \sqrt[3]{4}$ as a single radical expression.

13. Perform each operation and simplify.
(a) $\sqrt{48} + \sqrt{12}$ (b) $\sqrt{3}\left(\sqrt{15} + \sqrt{6}\right)$
(c) $\left(\sqrt{7} + 2\sqrt{3}\right)^2$ (d) $\left(\sqrt{7} - \sqrt{2}\right)\left(\sqrt{7} + \sqrt{2}\right)$

14. Rationalize the denominator and simplify.
(a) $\frac{7}{\sqrt{21}}$ (b) $\frac{\sqrt{3} + 1}{\sqrt{3} + 2}$

15. Solve each equation.
(a) $x^{2/3} = 4$ (b) $x^{3/2} = 8$
(c) $\sqrt{x - 1} = 3$ (d) $4x^2 = 25$

16. Write in the form of $a + bi$ and simplify.
(a) $3 + \sqrt{-9}$ (b) $(2 + 3i) + (-7 - i)$
(c) $(2 - 5i)(1 + 2i)$ (d) $\frac{1 + i}{3 - 2i}$
(e) $\frac{1}{i^{-4}}$

17. An employee's starting wage is \$5.50 per hour. Find the hourly wage after 3 years if the total annual raise is 5% and the raise is given in 2 installments over the course of the year.

Quadratic Equations and Functions

CHAPTER **7**

Chapter Contents

An important part of the study of mathematics is learning to analyze real world situations and solve problems. Often, these situations obey basic algebraic models described by certain functions. For instance, the water jet fountain in the midfield terminal of the Detroit Metro Airport shoots water streams from different sources periodically. The height h of the water stream x feet from the source is approximated by the model

$$h = -0.423x^2 + 2.92x.$$

In example 6 of section 7.6, we use quadratic functions to describe this phenomenon.

In this chapter, we discuss various methods for solving quadratic equations and inequalities. Quadratic functions and properties of their graphs as well as graphs of circles conclude this chapter.

Objectives

1. Solve Equations by Extracting Roots
2. Solve Quadratic Equations by Completing the Square
3. Solve Applied Problems

7.1 Solving Quadratic Equations by Root Extraction and Completing the Square

Recall from section 4.5 that a second-degree equation in one variable that takes the form

$$ax^2 + bx + c = 0$$

where a, b, and c are real numbers, $a \neq 0$, is called a **standard form of a quadratic equation.** We solved some of these equations by factoring and applying the following zero property.

$$PQ = 0 \text{ if and only if } P = 0 \text{ or } Q = 0$$

We review a few examples.

EXAMPLE 1   **Solve Equations by Factoring**

Solve each equation by factoring.

(a) $6x^2 + 11x - 10 = 0$ (b) $2(x^2 + 3x) + 3 = -2(x^2 + 5x) + x - 11$

Solution   (a) The equation is already in standard form.

$$6x^2 + 11x - 10 = 0$$

$(3x - 2)(2x + 5) = 0$ Factor the left side

$3x - 2 = 0$ or $2x + 5 = 0$ Apply the zero property

$3x = 2$ $2x = -5$

$x = \dfrac{2}{3}$ $x = -\dfrac{5}{2}$

(b) First, we write the equation in standard form.

$$2(x^2 + 3x) + 3 = -2(x^2 + 5x) + x - 11$$

$2x^2 + 6x + 3 = -2x^2 - 10x + x - 11$ Simplify each side

$2x^2 + 6x + 3 = -2x^2 - 9x - 11$

$4x^2 + 15x + 14 = 0$ Standard form

$(4x + 7)(x + 2) = 0$ Factor

$4x + 7 = 0$ or $x + 2 = 0$ Zero property

$4x = -7$ $x = -2$

$x = -\dfrac{7}{4}$

The solutions for part (a) and (b) should be checked by substitution in the original equations.

Solving Equations by Extracting Roots

So far, we solved quadratic equations by factoring, but not every quadratic equation can be solved easily by factoring. In this section, we develop other methods for solving quadratic equations based on the following property.

Extracting Roots Property

For any real numbers a and p, where $a \neq 0$

$$au^2 = p \text{ is equivalent to } u^2 = \frac{p}{a}$$

so that

$$u^2 = \frac{p}{a} \text{ if and only if } u = \sqrt{\frac{p}{a}} \quad \text{or} \quad u = -\sqrt{\frac{p}{a}}.$$

The statement $u = \sqrt{\frac{p}{a}}$ or $u = -\sqrt{\frac{p}{a}}$ can also be written as $u = \pm\sqrt{\frac{p}{a}}$.

EXAMPLE 2   **Solving Equations by Extracting Roots**

Solve each equation.

(a) $x^2 = 9$ (b) $2x^2 = 24$ (c) $(2y + 3)^2 = 36$

Solution

(a) $x^2 = 9$
$x = \pm\sqrt{9}$
$x = \pm 3$

The solutions are 3 and -3.

(b) $2x^2 = 24$
$x^2 = 12$
$x = \pm\sqrt{12}$
$x = \pm\sqrt{4(3)} = \pm 2\sqrt{3}$

The solutions are $2\sqrt{3}$ and $-2\sqrt{3}$.

(c) $(2y + 3)^2 = 36$
$2y + 3 = \pm\sqrt{36}$
$2y + 3 = \pm 6$

$2y + 3 = 6$ or $2y + 3 = -6$
$2y = 3$ $\qquad 2y = -9$
$y = \frac{3}{2}$ $\qquad y = -\frac{9}{2}$

The solutions are $\frac{3}{2}$ and $-\frac{9}{2}$.

EXAMPLE 3   **Extracting Roots Where the Solutions are Complex**

Solve each equation.

(a) $x^2 = -25$ (b) $(y - 3)^2 + 8 = 0$

Solution

(a) $x^2 = -25$
$x = \pm\sqrt{-25}$
$x = \pm 5i$

The solutions are $5i$ and $-5i$.

(b) $(y - 3)^2 + 8 = 0$
$(y - 3)^2 = -8$
$y - 3 = \pm\sqrt{-8}$
$y = 3 \pm 2i\sqrt{2}$

The solutions are $3 + 2i\sqrt{2}$ and $3 - 2i\sqrt{2}$.

Check: $(\pm 5i)^2 = 25i^2 = -25$ $(3 \pm 2i\sqrt{2} - 3)^2 + 8 = (\pm 2i\sqrt{2})^2 + 8$

$$= 8i^2 + 8 = -8 + 8 = 0 \;\boxplus$$

Sometimes it may be necessary to change the form before we apply the above property as the next example shows.

EXAMPLE 4   **Obtaining Complex Solutions**

Solve the equation $2(5x - 1)^2 + 3 = -15$.

Solution $2(5x - 1)^2 + 3 = -15$ Given

$2(5x - 1)^2 = -18$

$(5x - 1)^2 = -9$ Isolate perfect square

$5x - 1 = \pm\sqrt{-9} = \pm 3i$ Extract the root

$5x = 1 \pm 3i$

$$x = \frac{1 \pm 3i}{5}$$

The solutions are $\dfrac{1 \pm 3i}{5}$ and the reader is encouraged to check these solutions. $\boxplus$

Solving Quadratic Equations by Completing the Square

Now we can solve any quadratic equation by writing the equation in the form $(x + k)^2 = p$. This can be accomplished by expressing the left side of the equation as a perfect square trinomial involving a square of a binomial of the form $(x + k)^2$. This process is called **completing the square.** Completing the square relies on recognizing perfect square trinomials. This involves finding the third term of a perfect square trinomial when given the first two terms. For example, we recognize

$$x^2 + 6x$$

as the first two terms of the perfect square trinomial

$$x^2 + 6x + 9 = (x + 3)^2.$$

Table 1 shows additional examples.

TABLE 1

First Two Terms	Trinomial	Perfect Square
$x^2 + 8x$	$x^2 + 8x + 16$	$(x + 4)^2$
$x^2 + 12x$	$x^2 + 12x + 36$	$(x + 6)^2$
$x^2 - 14x$	$x^2 - 14x + 49$	$(x - 7)^2$
$x^2 - 10x$	$x^2 - 10x + 25$	$(x - 5)^2$

Note that in each of the perfect square trinomials in Table 1, the constant term is produced by dividing the coefficient of the middle term by 2, then squaring the resulting quantity. For example, we can recognize that

$$x^2 + 2kx$$

is the first two terms of the perfect square trinomial

$$x^2 + 2kx + k^2 = (x + k)^2.$$

Observe that the term k^2 is produced by squaring one-half the coefficient of the middle term, $2k$. That is, we complete the square by adding

$$\left[\frac{1}{2}(2k)\right]^2 = k^2$$

so that

$$x^2 + 2kx + \left[\frac{1}{2}(2k)\right]^2 = x^2 + 2kx + k^2 = (x + k)^2.$$

coefficient of the middle term

For example, given $x^2 - 18x$, we can complete the square as follows:

$$x^2 - 18x + \left[\frac{1}{2}(-18)\right]^2 = x^2 - 8x + 81 = (x - 9)^2.$$

coefficient of the x term

EXAMPLE 5   **Solving Quadratic Equations with $a = 1$ by Completing the Square**

Solve the equations by completing the square. Give both the exact solution and a two decimal place approximation.

(a) $x^2 + 4x - 2 = 0$ (b) $x^2 + 2x + 5 = 0$

Solution

(a) $x^2 + 4x - 2 = 0$ Given

$x^2 + 4x = 2$ Add 2 to both sides

$x^2 + 4x + 4 = 2 + 4$ Complete the square, add $\left[\frac{1}{2}(4)\right]^2 = 4$ to both sides

$(x + 2)^2 = 6$ Write left side as a perfect square trinomial

$x + 2 = \pm\sqrt{6}$ Use the square root property

$x = -2 \pm\sqrt{6}$

The exact solutions are $-2 + \sqrt{6}$ and $-2 - \sqrt{6}$. Two decimal place approximations of the solutions are 0.45 and −4.45 respectively.

(b) $x^2 + 2x + 5 = 0$ Given

$x^2 + 2x = -5$ Add −5 to both sides

$x^2 + 2x + 1 = -5 + 1$ Complete the square, add $\left[\frac{1}{2}(2)\right]^2 = 1$ to both sides

$(x + 1)^2 = -4$ Write the left side as a perfect square trinomial

$x + 1 = \pm\sqrt{-4}$ Square root property

$x = -1 \pm 2i$

The solutions are $-1 + 2i$ and $-1 - 2i$. You are encouraged to check these solutions in the original equation.

Observe the visual support for our answers in Example 5. Figure 1a shows that the graph of $y = x^2 + 4x - 2$ has two x intercepts, one between -5 and -4 and the other between 0 and 1. Figure 1b shows that the graph of $y = x^2 + 2x + 5$ does not intersect the x axis. This supports our solution of complex numbers.

Figure 1

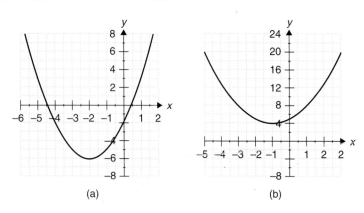

(a)   (b)

A trinomial may have a leading coefficient other than 1. In this case, we divide both sides of the equation by the leading coefficient before completing the square. This is illustrated in the next example.

EXAMPLE 6   **Solving an Equation by Completing the Square**

Solve the equation $2x^2 + 3x - 6 = 0$.

Solution

$$2x^2 + 3x - 6 = 0 \qquad \text{Original equation}$$

$$2x^2 + 3x = 6 \qquad \text{Add 6 to both sides}$$

$$x^2 + \frac{3}{2}x = \frac{6}{2} \qquad \text{Divide both sides by 2}$$

$$x^2 + \frac{3}{2}x + \frac{9}{16} = 3 + \frac{9}{16} \qquad \text{Add } \left[\frac{1}{2}\left(\frac{3}{2}\right)\right]^2 = \frac{9}{16} \text{ to both sides}$$

$$\left(x + \frac{3}{4}\right)^2 = \frac{57}{16} \qquad \text{Write the left side as a perfect square trinomial}$$

$$x + \frac{3}{4} = \frac{\pm\sqrt{57}}{4} \qquad \text{Use the square root property}$$

$$x = \frac{-3 \pm \sqrt{57}}{4}$$

Figure 2

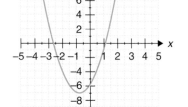

The exact solutions are $\dfrac{-3 + \sqrt{57}}{4}$ and $\dfrac{-3 - \sqrt{57}}{4}$. Two decimal place approximations of the solutions are 1.14 and -2.64 respectively. Figure 2 gives visual support for our answer. The x intercepts of the graph are between -3 and -2 and 1 and 2.

Solving Applied Problems

Quadratic equations arise from applications involving the Pythagorean Theorem and other models.

EXAMPLE 7   **Finding the Length of a Ladder**

A ladder leans against a building. The top of the ladder is 16 feet up the wall of the building and the foot of the ladder is 9 feet away from the base of the building. Find the length of the ladder rounded to one decimal place.

Figure 3

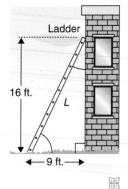

Solution   Figure 3 illustrates this situation. Using the Pythagorean Theorem:

$$L^2 = 9^2 + 16^2$$

$$L^2 = 81 + 256$$

$$L^2 = 337$$

$$L = \sqrt{337} \approx 18.4 \text{ feet.}$$

EXAMPLE 8   **Solving a Fuel Consumption Problem**

The model

$$m = -\frac{1}{40}(v^2 - 70v) - \frac{5}{8}, \quad 10 \le v \le 40$$

describes the fuel economy m (in miles per gallon) of a particular car being driven at a speed v (in miles per hour).

(a) Use completing the square to rewrite the above equation in the form

$$m = -\frac{1}{40}(v - h)^2 + k.$$

(b) Make a table of values of v and m to suggest at what speed, v, will the miles per gallon, m, be the greatest.

Figure 4

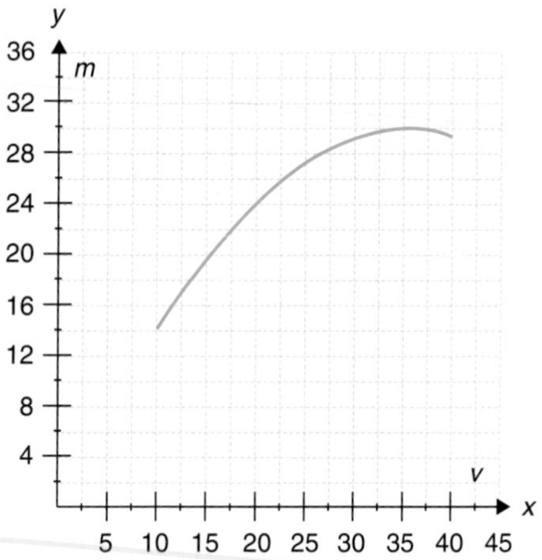

(c) Using the graph in Figure 4, indicate where the highest point on the graph occurs.

Solution   (a) By completing the square, we have

$$m = -\frac{1}{40}(v^2 - 70v + 35^2) + \frac{1}{40}(35)^2 - \frac{5}{8}$$

$$= -\frac{1}{40}(v - 35)^2 + \frac{245}{8} - \frac{5}{8}$$

$$= -\frac{1}{40}(v - 35)^2 + 30$$

(b)

Speed v	25	27	29	31	33	35	37	39
Mileage m	27.5	28.4	29.1	29.6	29.9	30.0	29.9	29.6

From the table, the maximum mileage is 30.0 miles per gallon at 35 miles per hour.

(c) The graph in Figure 4 visually suggests that (35,30) is the highest point on the graph. That is, at 35 miles per hour the fuel economy is 30 miles per gallon. Notice that the values of h and k in part (a) are 35 and 30 respectively. This is not a coincidence and will be explored later in this chapter.

PROBLEM SET 7.1

Mastering the Concepts

In Problems 1–4, use factoring to solve each equation.

1. (a) $3x^2 - x - 2 = 0$
   (b) $3x^2 = 15x$
   (c) $-x^2 + 5x + 6 = 0$
2. (a) $2x^2 + 5x - 3 = 0$
   (b) $2y^2 - 3y - 54 = 0$
   (c) $2t^2 - 13t - 24 = 0$
3. (a) $-x^2 + 2x + 24 = 0$
   (b) $x^2 + 14x + 49 = 0$
   (c) $15x^2 - 14x - 8 = 0$
4. (a) $(y - 7)(y + 5) = -20$
   (b) $2(x^2 - 3) - 3x = 2(x + 3)$
   (c) $-2x^2 + 24x - 22 = 0$

In Problems 5–10, solve each equation by root extraction. Check the solutions.

5. (a) $x^2 - 64 = 0$
   (b) $9x^2 - 4 = 0$
   (c) $4x^2 + 36 = 0$
   (d) $2y^2 + 18 = 0$
6. (a) $(2m + 1)^2 - 36 = 0$
   (b) $(x - 1)^2 - 9 = 0$
   (c) $(y + 2)^2 + 25 = 0$
   (d) $(y + 3)^2 + 16 = 0$
7. (a) $(3u - 6)^2 + 36 = 0$
   (b) $(3x - 2)^2 + 49 = 0$
8. (a) $(2x - 3)^2 - 6 = 10$
   (b) $(4u + 5)^2 - 21 = 4$
9. $x^2 + 6x + 9 = 81$
10. $x^2 - 10x + 25 = 100$

In Problems 11–14, what term should be added to the given expression so it becomes a perfect square trinomial?

11. (a) $x^2 + 6x$
    (b) $t^2 - 8t$
12. (a) $y^2 - 10y$
    (b) $z^2 + 12z$
13. (a) $x^2 + 7x$
    (b) $t^2 - 11t$
14. (a) $y^2 + 5y$
    (b) $x^2 + 9x$

In Problems 15–30, solve each equation by completing the square. Give both the exact solutions and the two decimal approximations.

15. (a) $x^2 + 2x - 3 = 0$
    (b) $x^2 + 4x - 7 = 0$
16. (a) $x^2 - 12x - 7 = 0$
    (b) $x^2 + 6x - 11 = 0$
17. (a) $x^2 - 5x - 6 = 0$
    (b) $x^2 + 3x - 4 = 0$
18. (a) $x^2 - 7x = 8$
    (b) $x^2 + 11x = 12$
19. $3m^2 - 12m - 3 = 0$
20. $3u^2 - 18u - 7 = 0$
21. $4y^2 - 8y - 3 = 0$
22. $25y^2 - 25y - 14 = 0$
23. $5m^2 - 8m - 17 = 0$
24. $9x^2 - 27x - 14 = 0$
25. $4y^2 + 7y - 5 = 0$
26. $3t^2 - 8t - 4 = 0$
27. $7m^2 + 5m - 1 = 0$
28. $25x^2 - 50x - 21 = 0$
29. $16y^2 - 24y = 5$
30. $9t^2 - 30t = -21$

In Problems 31 and 32, the function and graph are given.

(a) Find the output values for the given input values of the function whose graph is given.
(b) Rewrite each equation in the equivalent form $f(x) = a(x - h)^2 + k$ by completing the square.
(c) Use the algebraic results from part (b) to find the coordinates of the lowest or the highest point on the graph.
(d) Find the x intercept of each function graphically (to two decimal places) and algebraically (exact value).

31. (a) $f(x) = x^2 - 2x - 5, f(0), f(1)$

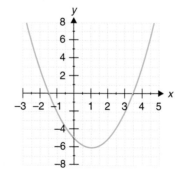

(b) $f(x) = x^2 - 6x + 7$, $f(0)$, $f(3)$

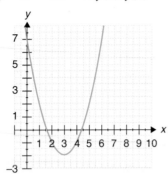

32. (a) $f(x) = x^2 + 5x - 3$, $f(0)$, $f\left(\dfrac{5}{4}\right)$

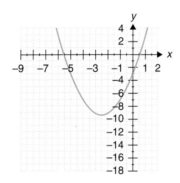

(b) $f(x) = x^2 + 3x - 2$, $f(0)$, $f\left(\dfrac{3}{4}\right)$

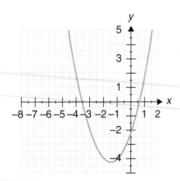

In Problems 33 and 34, find the intercepts of each graph of the given function.

33. (a) $f(x) = x^2 - 3x - 10$
(b) $g(x) = 6x^2 - 7x - 10$

34. (a) $f(x) = -x^2 - 3x + 18$
(b) $g(x) = 2x^2 + 22x + 60$

Applying the Concepts

35. Suspension Bridge Cable: The model

$$y = 0.0375x^2 - 1.5x + 30, \qquad 0 \le x \le 40$$

describes a parabolic cable for a suspension bridge, where y (in feet) is the height of the cable

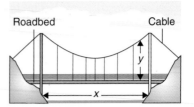

Figure 5

at a point above the roadbed x (in feet) from the left support tower (Figure 5).

(a) Use completing the square to rewrite this model in the form $y = a(x - h)^2 + k$.

(b) Indicate where the lowest point on the cable occurs, and visually check on the diagram the smallest height of the cable from the roadbed at the center of the bridge.

36. Flight of a Baseball: A ballplayer hits a high pop-up straight above home plate. If the bat meets the ball 4 feet above the ground and sends it up at a speed of 80 feet per second, then the height y (in feet) of the ball after t seconds is predicted by the model

$$y = -16t^2 + 80t + 4.$$

(a) Write the equation for this model in the form $y = a(t - h)^2 + k$ by completing the square.

(b) Use part (a) to indicate the highest point reached by the ball.

Developing and Extending the Concepts

37. Determine the value of k so that the trinomial is a perfect square.
(a) $kx^2 - 10x + 1$
(b) $9x^2 + kx + 25$

38. Solve the equation $x^2 - 2x - 15 = 0$ (a) by factoring and (b) by completing the square.

In Problems 39–44, solve each formula for the indicated variable. Assume all symbols represent positive real numbers.

39. (a) $A = s^2$, for s (area of a square)
(b) $S = 4\pi r^2$, for r (surface area of a sphere)

40. (a) $T = 2\pi\sqrt{\dfrac{L}{g}}$, for L (period of a pendulum)
(b) $V = s^3$, for s (volume of a cube)

41. (a) $A = (1 + r)^2$, for r (compound interest)
(b) $h = \dfrac{1}{2}gt^2$, for t (free falling object)

42. (a) $F = \dfrac{m_1 m_2}{r^2}$, for r (law of gravity)

 (b) $a^2 + b^2 = c^2$, for a (Pythagorean Theorem)

43. $-gt^2 + vt = s$, for t (falling object)

44. $S = 2\pi r^2 + 2\pi rh$, for r (surface area of a cylinder)

In Problems 45–50, use completing the square to obtain the solutions of each equation.

45. $3x^2 - 6x + 5 = 0$

46. $3x^2 + 12x + 31 = 0$

47. $9x^2 + 30x + 26 = 0$

48. $7x^2 - 2x + 5 = 0$

49. $4x^2 + 8x + 7 = 0$

50. $10x^2 + 5x + 2 = 0$

51. Let $f(x) = x^2 + 6x - 3$, find and simplify

 (a) $f(-3 + 2\sqrt{3})$

 (b) $f(-3 - 2\sqrt{3})$

52. Let $f(x) = 2x^2 + 3x + 2$, find and simplify

 (a) $f\left(\dfrac{-3 + \sqrt{7}\,i}{4}\right)$

 (b) $f\left(\dfrac{-3 - \sqrt{7}\,i}{4}\right)$

In Problems 53 and 54, one solution of each quadratic equation is given.

 (a) Find a value of a or b which gives this solution.

 (b) Use completing the square to determine the other solution for the values of a or b from part (a).

53. $ax^2 + 7x - 2 = 0$; $x = 2$

54. $9x^2 + bx - 11 = 0$; $x = 1$

In Problems 55–58, (a) Use completing the square to rewrite each function in the form

$$f(x) = a(x - h)^2 + k.$$

 (b) Indicate the coordinates of the lowest point or the highest point on the graph of f.

 (c) Find the x intercepts.

 (d) Match the function with its graph.

55. $f(x) = x^2 - 6x + 8$

56. $f(x) = x^2 - 12x + 12$

57. $f(x) = 4x^2 - 20x + 20$

58. $f(x) = -4x^2 + 12x - 3$

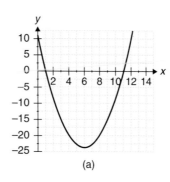

(a)

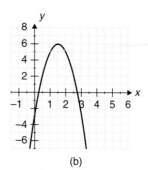

(b)

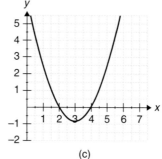

(c)

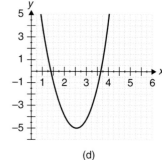

(d)

Objectives

1. Derive the Quadratic Formula

2. Use the Discriminant to Classify Solutions

3. Solve Applied Problems

7.2 The Quadratic Formula

We indicated in section 7.1 that every quadratic equation can be written in the standard form

$$ax^2 + bx + c = 0,\ a \neq 0.$$

We also pointed out that this form of the equation could be solved using the method of completing the square. We will use these ideas to produce a formula that can be used to solve any quadratic equation. It is called the *quadratic formula*.

Deriving the Quadratic Formula

We will begin with a quadratic equation in standard form and solve it using the method of completing the square.

$$ax^2 + bx + c = 0$$ Standard form of quadratic equation

$$ax^2 + bx = -c$$ Subtract c from each side

$$x^2 + \frac{b}{a}x = \frac{-c}{a}$$ Divide each side by a

$$x^2 + \frac{b}{a}x + \frac{b^2}{4a^2} = \frac{-c}{a} + \frac{b^2}{4a^2}$$ Add $\left[\frac{1}{2}\left(\frac{b}{a}\right)\right]^2 = \frac{b^2}{4a^2}$ to each side

$$\left(x + \frac{b}{2a}\right)^2 = \frac{b^2}{4a^2} - \frac{c}{a}$$ Express the left side as the square of a binomial

$$\left(x + \frac{b}{2a}\right)^2 = \frac{b^2}{4a^2} - \frac{4ac}{4a^2}$$ Rewrite with a common denominator

$$\left(x + \frac{b}{2a}\right)^2 = \frac{b^2 - 4ac}{4a^2}$$ Add fractions on right side

$$\sqrt{\left(x + \frac{b}{2a}\right)^2} = \sqrt{\frac{b^2 - 4ac}{4a^2}}$$ Take the square root of both sides

$$\left|x + \frac{b}{2a}\right| = \sqrt{\frac{b^2 - 4ac}{4a^2}}$$

$$x + \frac{b}{2a} = \pm\sqrt{\frac{b^2 - 4ac}{4a^2}}$$

$$x = \frac{-b}{2a} \pm \frac{\sqrt{b^2 - 4ac}}{2a}$$ Subtract $\frac{b}{2a}$ from each side

$$x = \frac{-b \pm \sqrt{b^2 - 4ac}}{2a}$$ Combine the fractions

It is important to note that the quadratic formula can always be used to solve a quadratic equation in the above form and will produce the same results as any other method. This is accomplished by substituting the values for a, b, and c in the formula.

The Quadratic Formula

A quadratic equation must be written in standard form before using the quadratic formula.

> If $ax^2 + bx + c = 0$ and $a \neq 0$, then the solutions of the equation are given by the **Quadratic Formula**
> $$x = \frac{-b \pm \sqrt{b^2 - 4ac}}{2a}$$

EXAMPLE 1   **Using the Quadratic Formula**

Solve the equation $2x^2 - 6x + 3 = 0$ by the quadratic formula. Find the exact value and a two decimal point approximation of the solution.

Solution   Since the equation is written in standard form, we begin by identifying a, b, and c. Substituting $a = 2$, $b = -6$, and $c = 3$ in the formula, we have

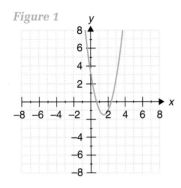

Figure 1

$$x = \frac{-b \pm \sqrt{b^2 - 4ac}}{2a}$$

$$x = \frac{-(-6) \pm \sqrt{(-6)^2 - 4(2)(3)}}{2(2)}$$

$$x = \frac{6 \pm \sqrt{36 - 24}}{4}$$

$$x = \frac{6 \pm \sqrt{12}}{4}$$

$$= \frac{6 \pm 2\sqrt{3}}{4}$$

$$= \frac{3 \pm \sqrt{3}}{2}.$$

The exact solutions of the equation are $\dfrac{3 - \sqrt{3}}{2}$ and $\dfrac{3 + \sqrt{3}}{2}$.

The two decimal approximations of these solutions are 0.63 and 2.37 respectively. Figure 1 gives visual support for our answer, because the x intercepts of the graph of $y = 2x^2 - 6x + 3$ appear to be between 0 and 1 and between 2 and 3.

EXAMPLE 2   **Solving Equations Using the Quadratic Formula**

Solve each equation using the quadratic formula.

(a) $3x^2 + 4x - 4 = 0$ (b) $x^2 - 2x = -1$

Solution   (a) Since the equation $3x^2 + 4x - 4 = 0$ is in standard form, $a = 3$, $b = 4$, and $c = -4$. Substituting these values in the formula, we have

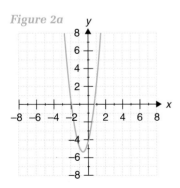

Figure 2a

$$x = \frac{-b \pm \sqrt{b^2 - 4ac}}{2a}$$

$$x = \frac{-4 \pm \sqrt{4^2 - 4(3)(-4)}}{2(3)}$$

$$x = \frac{-4 \pm \sqrt{16 + 48}}{6}$$

$$x = \frac{-4 \pm \sqrt{64}}{6}.$$

$$= \frac{-4 \pm 8}{6}.$$

Therefore the solutions are $\dfrac{-4 + 8}{6} = \dfrac{2}{3}$ and $\dfrac{-4 - 8}{6} = -2.$

Figure 2a confirms the algebraic solutions of these equations and shows the x intercepts of the graph of $y = 3x^2 + 4x - 4 = 0$ are between 0 and 1 and the other at -2.

(b) The standard form of the equation $x^2 - 2x = -1$ is

$$x^2 - 2x + 1 = 0.$$

Substituting $a = 1$, $b = -2$, and $c = 1$ in the quadratic formula, we have

$$x = \frac{-b \pm \sqrt{b^2 - 4ac}}{2a}$$

$$x = \frac{-(-2) \pm \sqrt{(-2)^2 - 4(1)(1)}}{2(1)}$$

$$x = \frac{2 \pm \sqrt{4 - 4}}{2}$$

$$x = \frac{2 \pm 0}{2} = 1.$$

Figure 2b

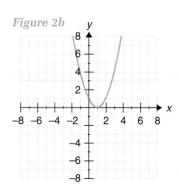

Therefore, there is only one solution, $x = 1$. Figure 2b shows one x intercept of the graph of $y = x^2 - 2x + 1$ at 1.

EXAMPLE 3   **Solving an Equation that Produces Complex Numbers**

Solve the equation

$$(x - 1)(x + 3) = -5.$$

Solution   The standard form of the equation $(x - 1)(x + 3) = -5$ is

$$x^2 + 2x + 2 = 0$$

where $a = 1$, $b = 2$ and $c = 2$. Substituting for a, b, and c in the quadratic formula, we have,

Figure 3

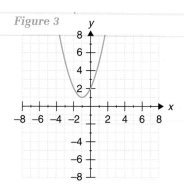

$$x = \frac{-b \pm \sqrt{b^2 - 4ac}}{2a}$$

$$x = \frac{-2 \pm \sqrt{2^2 - 4(1)(2)}}{2(1)}$$

$$x = \frac{-2 \pm \sqrt{4 - 8}}{2} = \frac{-2 \pm \sqrt{-4}}{2}$$

$$x = \frac{-2 \pm 2i}{2} = -1 \pm i.$$

Therefore, the two complex solutions are $-1 + i$ and $-1 - i$. Graphically, Figure 3 shows that there are no x intercepts for the graph of

$$y = x^2 + 2x + 2$$

because the solutions are complex numbers.

Using the Discriminant to Classify Solutions

The expression

$$b^2 - 4ac$$

that occurs under the radical symbol in the quadratic formula is called the **discriminant.** It provides interesting information about the number and types of solutions of a quadratic equation. In examples 2 and 3, we solved three quadratic equations. Our results are summarized in the following table.

Equation in Standard Form	Discriminant $b^2 - 4ac$	Sign of Discriminant	Type of Solutions
$3x^2 + 4x - 4 = 0$	64	positive	two real solutions
$x^2 - 2x + 1 = 0$	0	zero	one real solution
$x^2 + 2x + 2 = 0$	-4	negative	two complex solutions

The use of the discriminant can be generalized in the following test.

Discriminant Test for $ax^2 + bx + c = 0, a \neq 0$

If the discriminant is negative and a, b, and c are real numbers, the two solutions will be complex conjugates of each other.

Discriminant $b^2 - 4ac$	Type of Solutions
positive	two real solutions
zero	one real solution
negative	two complex solutions

EXAMPLE 4

Applying the Discriminant Test

Use the discriminant test to determine the number and type of solutions of each equation.

(a) $4x^2 - 5x + 1 = 0$　　　(b) $4x^2 - 12x + 9 = 0$　　　(c) $2x^2 + 3x + 2 = 0$

Solution

(a) The discriminant $b^2 - 4ac = (-5)^2 - 4(4)(1) = 9$ is positive.

Thus, the equation has two real solutions.

(b) The discriminant $b^2 - 4ac = (-12)^2 - 4(4)(9) = 0$ is zero.

Thus, the equation has one real solution.

(c) The discriminant $b^2 - 4ac = 3^2 - 4(2)(2) = -7$ is negative.

Thus, the equation has two complex solutions.

Use the quadratic formula to check these results.

Solving Applied Problems

Although the quadratic formula was known to the Babylonians around 2000 B.C., the idea of the discriminant was not discovered until the 17th century by Sir Isaac Newton. The discussion of the type of solutions of a quadratic equation came after the invention of the Cartesian coordinate system. In the next example, we will use the quadratic formula and the discriminant to help solve applied problems.

EXAMPLE 5

Solving a Diving Problem

An Olympic competitor stands on a 10 meter diving platform, and jumps from the platform at an initial upward speed of 2.2 meters per second. The competitor's height h (in meters) above the surface of the water after t seconds is given by the model

$$h(t) = -4.9t^2 + 2.2t + 10, \qquad t \geq 0.$$

Use the quadratic formula to determine how many seconds it takes for the competitor to reach the following heights:

(a) the surface of the water.

(b) ten meters above the surface of the water.

(c) twelve meters above the surface of the water. Is this possible?

Round off each answer to two decimal places.

Solution

(a) The diver reaches the surface of the water when $h(t) = 0$, so the equation is

$$0 = -4.9t^2 + 2.2t + 10, t \geq 0.$$

Using the quadratic formula with $a = -4.9$, $b = 2.2$, and $c = 10$, we have

$$t = \frac{-b \pm \sqrt{b^2 - 4ac}}{2a}$$

$$t = \frac{-2.2 \pm \sqrt{(2.2)^2 - 4(-4.9)(10)}}{2(-4.9)}$$

$$t = 1.67 \text{ and } -1.22.$$

The solutions are 1.67 and −1.22. We reject −1.22 since time is positive in this problem. This means the diver hits the surface of the water after 1.67 seconds.

(b) The diver reaches 10 meters above the surface of the water when $h(t) = 10$, so the equation is

$$10 = -4.9t^2 + 2.2t + 10 \text{ or } 4.9t^2 - 2.2t + 0 = 0.$$

Using the quadratic formula with $a = 4.9$, $b = -2.2$, and $c = 0$, we have

$$t = \frac{-b \pm \sqrt{b^2 - 4ac}}{2a}$$

$$t = \frac{-(-2.2) \pm \sqrt{(-2.2)^2 - 4(4.9)(0)}}{2(4.9)}$$

$$t = 0.45 \text{ and } 0.$$

So the solutions are 0.45 and 0. This means the diver was initially at a height of 10 meters at $t = 0$, the start, and again at $t = 0.45$ on the way down to the water.

(c) For $h = 12$, we have

$$12 = -4.9t^2 + 2.2t + 10 \text{ or } 4.9t^2 - 2.2t + 2 = 0.$$

Using the quadratic formula with $a = 4.9$, $b = -2.2$, and $c = 2$, we have

$$t = \frac{-b \pm \sqrt{b^2 - 4ac}}{2a}$$

$$t = \frac{-(-2.2) \pm \sqrt{(-2.2)^2 - 4(4.9)(2)}}{2(4.9)}$$

$$t = 0.22 \pm 0.60i.$$

Figure 4

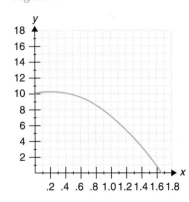

The solutions are $0.22 \pm 0.60i$. The solutions are complex because the discriminant is negative. This means the diver will never reach a height of 12 meters. Figure 4 shows the relationship between time and the heights indicated in parts (a) and (b).

PROBLEM SET 7.2

Mastering the Concepts

In Problems 1–20, solve each equation by the quadratic formula. Approximate irrational solutions to two decimal places. Write complex solutions in the form $a + bi$.

1. (a) $x^2 - 5x + 4 = 0$
   (b) $x^2 - 2x - 3 = 0$
2. (a) $x^2 - 2x - 24 = 0$
   (b) $x^2 + 10x - 24 = 0$
3. (a) $3x^2 + 8x + 4 = 0$
   (b) $3x^2 - 19x + 20 = 0$
4. (a) $6x^2 - 13x - 8 = 0$
   (b) $6x^2 + 11x + 4 = 0$
5. $4t^2 - 8t + 3 = 0$
6. $2u^2 - 5u - 3 = 0$
7. $6y^2 - 7y - 5 = 0$
8. $6p^2 + 3p - 5 = 0$
9. $6z^2 - 7z - 2 = 0$
10. $2x^2 - 5x + 1 = 0$
11. $5y^2 - 2y + 7 = 0$
12. $9m^2 - 2m + 11 = 0$
13. $12p^2 - 4p + 3 = 0$
14. $15y^2 + 2y + 8 = 0$
15. $3x^2 - x + 7 = 0$
16. $6x^2 + 17x - 14 = 0$
17. $t(t + 8) + 8(t - 8) = 0$
18. $(t - 2)(t - 3) = 5$
19. $(y - 1)(y - 5) = 9$
20. $(t + 1)(2t - 1) = 4$

In Problems 21–28, use the discriminant to determine the number and type of solutions to each equation.

21. $x^2 + 6x - 7 = 0$
22. $x^2 - 9x + 20 = 0$
23. $t^2 + 3t + 5 = 0$
24. $x^2 + 15x + 37 = 0$
25. $9z^2 + 30z + 25 = 0$
26. $2x^2 + x + 5 = 0$
27. $2x^2 + x - 5 = 0$
28. $20x^2 - 11x + 3 = 0$

Applying the Concepts

29. **Free Falling Object:** An object is tossed vertically upward from a platform 4 feet above the ground. The height h (in feet) above the ground of the object after t seconds is given by the model

    $$h(t) = -16t^2 + 192t + 4, \qquad 0 \le t \le 12.$$

    Use the quadratic formula to determine how many seconds it takes the object to reach the following heights:
    (a) 500 feet   (b) 580 feet   (c) 600 feet

30. **Insect Population:** A biologist determines that the population P of insects in an experiment after t weeks is given by the model

    $$P(t) = 100 + 24t + 2t^2, \qquad 0 \le t \le 40.$$

    When will the number of insects reach 3000?

31. **Height of a Diver:** During the 1992 Olympic Summer Games in Barcelona, Spain, Mark Lenzi of the United States won a gold medal in the springboard diving competition. If Mark springs into the air at an initial speed of 6.1 meters per second from a diving board 10 meters above the surface of the water, his height h (in meters) after t seconds is given by the model

    $$h(t) = -4.88t^2 + 6.1t + 10, \, t \ge 0.$$

    After how many seconds is Mark 11 meters above the surface of the water?

32. **Water in a Pond:** The volume of water (in cubic meters) in an irrigation pond after t hours of pumping from a well is approximated by the model

    $$V(t) = 3t^2 - 14t, \, t \ge 5.$$

    For how long must the water be pumped so that the pond will contain 185 cubic meters of water?

33. **Revenue Function:** Sometimes businesses can increase their revenues by lowering their prices. As they lower their prices, the number of units sold increases. Assume that a company's sales of its cellular telephones generate daily revenue R in dollars according to the function

    $$R(x) = (128 - 0.1x)x, \, x \ge 0$$

    where x is the quantity sold. Use the quadratic formula to find the quantity x the company must sell per day to have a daily revenue of $40,960.

34. **Profit Function:** Cellular phone sales (see Problem 33) generate a daily profit P in dollars according to the function

    $$P(x) = (128 - 0.1x)x - (36x + 1000), \, x \ge 0.$$

(a) Find the break–even point: that is, the level of sales that generates no profit and no loss. (Hint: Solve for x when $P(x) = 0$.)

(b) Use the quadratic formula to find the sales x that will yield a daily profit of $1,131.

Developing and Extending the Concepts

In Problems 35 and 36, the solution of a quadratic equation is given.

(a) Compare the values to the symbols a, b, and c used in the quadratic formula $x = \dfrac{-b \pm \sqrt{b^2 - 4ac}}{2a}$ and identify a, b, and c.

(b) Evaluate the expression without the use of a calculator.

(c) Substitute a, b, and c in the equation $ax^2 + bx + c = 0$ to find the original equation.

35. (a) $x = \dfrac{-9 \pm \sqrt{9^2 - 4(1)(8)}}{2(1)}$

(b) $x = \dfrac{-(-4) \pm \sqrt{(-4)^2 - 4(1)(3)}}{2(1)}$

36. (a) $x = \dfrac{-(-5) \pm \sqrt{(-5)^2 - 4(3)(2)}}{2(3)}$

(b) $x = \dfrac{-(-7) \pm \sqrt{(-7)^2 - 4(3)(-6)}}{2(3)}$

In Problems 37–40, use the quadratic formula to solve each equation. Round the answers to two decimal places.

37. $0.17u^2 - 0.55u - 3.87 = 0$

38. $1.47y^2 - 3.82y - 5.71 = 0$

39. $1.32x^2 + 2.78x - 9.321 = 0$

40. $8.84x^2 - 71.41x - 94.03 = 0$

In Problems 41–44, use the quadratic formula to solve each equation for the indicated unknown.

41. $nx^2 + mnx - m^2 = 0$, for x.

42. $2ay^2 - 7ay - 4ab^2 = 0$, for y.

43. $LI^2 + RI + \frac{1}{C} = 0$, for I.

44. $Mx^2 + 2Rx + K = 0$, for x.

In Problems 45–48, find the value of k so that the quadratic equation will have:

(a) one real solution,   (b) two real solutions, or

(c) no real solutions.

45. $-2x^2 - 3x + k = 0$ **46.** $-4x^2 + k = 0$

47. $kx^2 - 4x + 1 = 0$ **48.** $kx^2 - 15x + 5 = 0$

In Problems 49 and 50, use the fact that

$$x_1 = \frac{-b + \sqrt{b^2 - 4ac}}{2a} \text{ and } x_2 = \frac{-b - \sqrt{b^2 - 4ac}}{2a}$$

are solutions to the quadratic equation $ax^2 + bx + c = 0$, $a \neq 0$, to answer each question.

(a) Find the sum of the solutions, x_1 and x_2, and show that this sum is equal to $-b/a$.

(b) Find the product of the solutions, x_1 and x_2, and show that this product is equal to c/a.

49. $3x^2 - 5x - 2 = 0$; $x_1 = -1/3$ and $x_2 = 2$.

50. $2x^2 + 9x - 5 = 0$; $x_1 = 1/2$ and $x_2 = -5$.

In Problems 51–54, write a quadratic equation with integral values for a, b, and c, having the given solutions. Check your answers by using the results of Problems 49 and 50.

51. (a) -6 and 3

(b) -4 and -1

52. (a) 2 only solution

(b) -3 only solution

53. (a) $-\sqrt{5}$ and $\sqrt{5}$

(b) $2i$ and $-2i$

54. (a) $3 - 2i$ and $3 + 2i$

(b) $3 - \sqrt{11}$ and $3 + \sqrt{11}$

In Problems 55–58,

(a) Find the x intercepts of each quadratic function.

(b) Find the coordinates of the lowest or highest points on the graph by writing each function in the form

$$f(x) = a(x - h)^2 + k.$$

(c) Match the function with its graph.

55. $f(x) = x^2 - 6x - 8$

56. $f(x) = -x^2 + 6x + 8$

57. $f(x) = x^2 - 5x - 4$

58. $f(x) = -x^2 + 5x + 4$

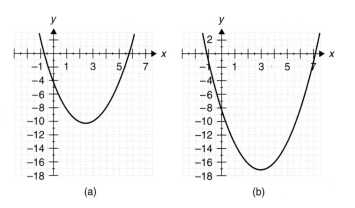

(a)   (b)

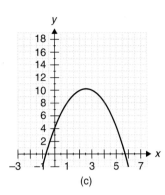

(c)

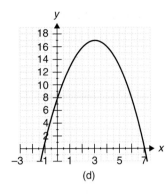

(d)

62. (a) $f(x) = 2x^2 - x - 4$

(b) $g(x) = -2x^2 + x + 4$

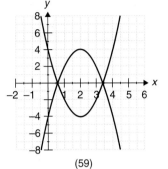

(59)

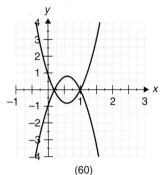

(60)

In Problems 59–62, the graphs of the functions f and g are displayed on the plane. Use the graph to estimate where the graphs of f and g intersect. Confirm your results by setting $f(x) = g(x)$ and solve the resulting equation by the quadratic formula.

59. (a) $f(x) = 2x^2 - 8x + 4$

(b) $g(x) = -2x^2 + 8x - 4$

60. (a) $f(x) = -5x^2 + 6x - 1$

(b) $g(x) = 5x^2 - 6x + 1$

61. (a) $f(x) = 3x^2 - x - 1$

(b) $g(x) = -3x^2 + x + 1$

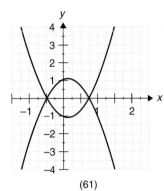

(61)

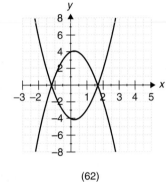

(62)

Objectives

1. Solve Equations Involving Radicals
2. Solve Equations Reducible to Quadratic Form
3. Solve Applied Problems

7.3 Equations Reducible to Quadratic Form

In this section, we discuss various types of equations that can be solved by using methods of solving quadratic equations. These equations are: *radical equations* and *equations quadratic in form*.

Solving Equations Involving Radicals

In section 6.6, we solved simple equations involving radicals. Here, we solve radical equations that lead to quadratic equations. Recall that the strategy for solving equations with radicals includes isolating the radicals and then raising both sides of the equation to a power that will clear the radicals. When we raise each side of an equation to an even power, *we must check for extraneous roots.*

The following example involves two radicals. In this case, we will square both sides of the equation twice in order to eliminate the radical expressions.

EXAMPLE 1   **Solving an Equation Involving Radicals**

Solve the equation $\sqrt{3x - 2} - \sqrt{x} = 2$.

Solution   We will isolate one radical on one side of the equation.

$$\sqrt{3x - 2} - \sqrt{x} = 2 \qquad \text{Original equation}$$

$$\sqrt{3x - 2} = \sqrt{x} + 2 \qquad \text{Isolate one radical}$$

$$\left(\sqrt{3x - 2}\right)^2 = \left(\sqrt{x} + 2\right)^2 \qquad \text{Square each side of the equation}$$

$$3x - 2 = x + 4\sqrt{x} + 4 \qquad \text{Simplify}$$

$$2x - 6 = 4\sqrt{x} \qquad \text{Isolate the radical}$$

$$x - 3 = 2\sqrt{x} \qquad \text{Divide each side by 2}$$

$$(x - 3)^2 = (2\sqrt{x})^2 \qquad \text{Square each side of the equation}$$

$$x^2 - 6x + 9 = 4x \qquad \text{Simplify}$$

$$x^2 - 10x + 9 = 0 \qquad \text{Write in standard form}$$

$$(x - 9)(x - 1) = 0 \qquad \text{Factor the left side}$$

$$x - 9 = 0 \quad \text{or} \quad x - 1 = 0 \qquad \text{Apply the zero factor property}$$

$$x = 9 \qquad\qquad x = 1$$

Figure 1

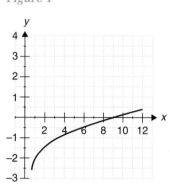

Check:

x	Left Side	Right Side	True/False
9	$\sqrt{3(9) - 2} - \sqrt{9} = \sqrt{25} - 3 = 5 - 3 = 2$	2	True
1	$\sqrt{3(1) - 2} - \sqrt{1} = \sqrt{1} - 1 = 1 - 1 = 0$	2	False

Figure 1 shows the visual support for our answer. The graph has one x intercept at $x = 9$, confirming our result: 9 is a solution. The number 1 is not an x intercept, and so it is an extraneous solution.

Solving Equations Reducible to Quadratic Form

Some equations are not quadratic but can be treated as quadratic equations provided that a suitable substitution is made. After making the appropriate substitution, we can use methods for solving quadratic equations to solve the resulting equations. The following table shows some examples of equations that can be written in quadratic form.

Original Equation	Suggested Substitution	Equivalent Quadratic Equation
$x^4 - 5x^2 + 4 = 0$	$u = x^2,\ u^2 = x^4$	$u^2 - 5u + 4 = 0$
$(x - 2)^2 - 3(x - 2) - 4 = 0$	$u = x - 2,\ u^2 = (x - 2)^2$	$u^2 - 3u - 4 = 0$
$\dfrac{1}{t^2} - \dfrac{5}{t} + 4 = 0$	$u = \dfrac{1}{t},\ u^2 = \dfrac{1}{t^2}$	$u^2 - 5u + 4 = 0$

EXAMPLE 2   **Solving Equations Quadratic in Form**

Solve each equation

(a) $x^4 - 5x^2 + 4 = 0$ (b) $(x - 2)^2 - 3(x - 2) - 4 = 0$

Solution

A common error is to solve for u and forget to solve for x.

(a) Let $u = x^2$, $u^2 = x^4$, then substitute in the original equation.

$x^4 - 5x^2 + 4 = 0$	Original equation
$u^2 - 5u + 4 = 0$	Substitution u for x^2
$(u - 1)(u - 4) = 0$	Factor left side
$u - 1 = 0$ or $u - 4 = 0$	Apply zero property
$u = 1$ $u = 4$	
$x^2 = 1$ $x^2 = 4$	Replace u by x^2
$x = \pm 1$ $x = \pm 2$	

The solutions are

$$-2, -1, \text{ and } 1 \text{ and } 2.$$

The graph in Figure 2 of the function

$$y = x^4 - 5x^2 + 4$$

indicates that the x intercepts are at $x = -2, -1, 1,$ and 2. The solutions are $-2, -1, 1,$ and 2.

Figure 2

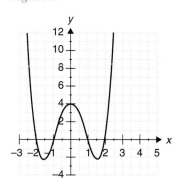

(b) Let $u = x - 2$ and $u^2 = (x - 2)^2$, then substitute

$(x - 2)^2 - 3(x - 2) - 4 = 0$	Original equation
$u^2 - 3u - 4 = 0$	Substitute u for $x - 2$
$(u + 1)(u - 4) = 0$	Factor the left side
$u + 1 = 0$ or $u - 4 = 0$	Apply the zero property
$u = -1$ $u = 4$	
$x - 2 = -1$ $x - 2 = 4$	Replace u by $x - 2$
$x = 1$ $x = 6$	

The solutions are

$$1 \text{ and } 6.$$

Quite often when we make a substitution, we place a restriction on the new variable. This way we avoid possible extraneous solutions. The following table lists some restrictions imposed on the substitution.

Original Equation	Substitution	Restriction	New Equation
$x - 15\sqrt{x} = 100$	$u = \sqrt{x}$	$u \geq 0$	$u^2 - 15u = 100$
$\dfrac{28}{x^2} + \dfrac{1}{x} - 2 = 0$	$u = \dfrac{1}{x}$	$u \neq 0$	$28u^2 + u - 2 = 0$
$\dfrac{1}{x} - \dfrac{1}{\sqrt{x}} - 12 = 0$	$u = \dfrac{1}{\sqrt{x}}$	$u > 0$	$u^2 - u - 12 = 0$

EXAMPLE 3 **Solving an Equation with an Extraneous Solution**

Solve the equation $x - 15\sqrt{x} = 100$.

Solution

Let $u = \sqrt{x}$ and $u^2 = x$.	Restriction: $\sqrt{x} \geq 0$ so $u \geq 0$
$x - 15\sqrt{x} = 100$	Original equation
$u^2 - 15u - 100 = 0$	Substitute u for $\sqrt{x}$
$(u - 20)(u + 5) = 0$	Factor the left side

$u - 20 = 0$	or	$u + 5 = 0$	Apply zero factor property
$u = 20$		$u = -5$	
$\sqrt{x} = 20$		$\sqrt{x} = -5$	Replace u by $\sqrt{x}$
$x = 400$		$x = 25$	Square both sides

A check shows that 25 is not a solution. The restriction on u, $u \geq 0$, prohibits using $u = -5$ but allows $u = 20$. Therefore, there is only one solution, namely $x = 400$.

The following examples are equations involving rational exponents that can be reduced to equations which are quadratic in form.

EXAMPLE 4 **Solving Equations with Rational Exponents**

Solve the equation $z^{2/3} - z^{1/3} - 12 = 0$.

Solution

Let $u = z^{1/3}$ and $u^2 = (z^{1/3})^2 = z^{2/3}$. Since this substitution involves cube roots there are no restrictions.

$z^{2/3} - z^{1/3} - 12 = 0$	Original equation
$u^2 - u - 12 = 0$	Substitute u for $z^{1/3}$
$(u + 3)(u - 4) = 0$	Factor the left side

$u + 3 = 0$	or	$u - 4 = 0$	Apply the zero factor property
$u = -3$		$u = 4$	
$z^{1/3} = -3$		$z^{1/3} = 4$	Replace u by $z^{1/3}$
$z = (-3)^3$		$z = 4^3$	Cube each side
$z = -27$		$z = 64$	

Therefore, the solutions are -27 and 64.

Solving Applied Problems

Now we examine applications and models that arise from equations that are quadratic in form.

EXAMPLE 5 **Solving an Earth Science Application**

In earth science, it is determined that at an elevation h (in meters) above sea level, the temperature T (in Celsius) at which water boils is given by the model

$$h(T) = 580(100 - T)^2 + 1000(100 - T), \qquad 95° \leq T \leq 100°.$$

(a) Using this model, determine at what temperature will water boil at sea level.

(b) At what temperature will water boil at the top of a mountain 4320 meters above sea level?

Solution   (a) At sea level, $h = 0$, so we solve the equation

$$0 = 580(100 - T)^2 + 1000(100 - T).$$

$580u^2 + 1000u = 0$		Let $u = 100 - T$
$58u^2 + 100u = 0$		Divide each side by 10
$u(58u + 100) = 0$		Factor the left side

$$u = 0 \qquad \text{or} \qquad 58u + 100 = 0$$
$$u = 0 \qquad\qquad\qquad u = -1.72$$
$$100 - T = 0 \qquad \text{or} \qquad 100 - T = -1.72$$
$$T = 100 \qquad\qquad\qquad T = 101.72$$

We reject $T = 101.72$ since it is not within the restriction of the model. The temperature at which water boils at sea level is 100°C.

(b) At the top of a mountain 4320 meters high, we have

$$4320 = 580(100 - T)^2 + 1000(100 - T).$$

$580u^2 + 1000u = 4320$	Let $u = 100 - T$
$29u^2 + 50u - 216 = 0$	Divide each side by 20

$$u = \frac{-b \pm \sqrt{b^2 - 4ac}}{2a}$$

$$u = \frac{-50 \pm \sqrt{50^2 - 4(29)(-216)}}{2(29)}$$

$$u = -3.72 \qquad \text{or} \qquad u = 2$$
$$100 - T = -3.72 \qquad\qquad 100 - T = 2$$
$$T = 103.72 \qquad \text{or} \qquad T = 98$$

We reject 103.72 since it is not in the restriction of the model. Therefore, the temperature at which water boils on top of the mountain is 98°C. ▦

PROBLEM SET 7.3

Mastering the Concepts

In Problems 1–6, solve each equation. Check your results.

1. $\dfrac{12}{x} - 7 = \dfrac{12}{1 - x}$

2. $\dfrac{1}{1 - x} + \dfrac{1}{2} = \dfrac{6}{x^2 - 1}$

3. $\dfrac{3 - 2x}{4x} - 4 = \dfrac{3}{4x - 3}$

4. $\dfrac{5}{4x + 16} - 1 = \dfrac{3}{4x - 8}$

5. $\dfrac{5}{x + 4} - \dfrac{3}{x - 2} = 4$

6. $\dfrac{2x - 5}{2x + 1} = \dfrac{7}{4} - \dfrac{6}{2x - 3}$

In Problems 7–14, solve each equation. Check your results.

7. $\sqrt{x} = \sqrt{x + 16} - 2$

8. $\sqrt{x + 2} = 5 - \sqrt{x - 3}$

9. $\sqrt{x + 12} = 2 + \sqrt{x}$

10. $\sqrt{6x + 7} = 1 + \sqrt{3x + 3}$

11. $\sqrt{y + 5} = 1 + \sqrt{y}$

12. $\sqrt{7 - 4t} = 1 + \sqrt{3 - 2t}$

13. $\sqrt{1 + 5x} = 1 + \sqrt{3x}$

14. $\sqrt{m + 4} + 1 = \sqrt{m + 11}$

In Problems 15–28, reduce each equation to a quadratic form by making a suitable substitution. Check for extraneous solutions.

15. (a) $x^4 - 7x^2 + 12 = 0$

 (b) $y^{-2} - 7y^{-1} + 12 = 0$

 (c) $t - 7\sqrt{t} + 12 = 0$

16. (a) $x^4 - 13x^2 + 36 = 0$

 (b) $y^{-2} - 13y^{-1} + 36 = 0$

 (c) $z - 13\sqrt{z} + 36 = 0$

17. (a) $x^4 - 17x^2 + 16 = 0$

 (b) $x^{\frac{2}{3}} - 17x^{\frac{1}{3}} + 16 = 0$

 (c) $z - 17\sqrt{z} + 16 = 0$

18. (a) $y^4 - 10y^2 + 9 = 0$

 (b) $x^{\frac{2}{3}} - 10x^{\frac{1}{3}} + 9 = 0$

 (c) $z - 10\sqrt{z} + 9 = 0$

19. (a) $x^4 - 2x^2 + 1 = 0$

 (b) $y^6 - 2y^3 + 1 = 0$

 (c) $(3 - \sqrt{x})^2 - 2(3 - \sqrt{x}) + 1 = 0$

20. (a) $y^4 - 16y^2 + 64 = 0$

 (b) $y^6 - 16y^3 + 64 = 0$

 (c) $(t^2 - 8)^2 - 16(t^2 - 8) + 64 = 0$

21. $(x + 2)^2 = 9(x + 2) - 20$

22. $(3x - 1)^2 - 9(3x - 1) = 10$

23. $(y^2 + 2y)^2 - 2(y^2 + 2y) = 3$

24. $(y^2 + 2y)^2 - 14(y^2 + 2y) = 15$

25. $(3 - \sqrt{x})^2 + 2(3 - \sqrt{x}) = 3$

26. $(4 + \sqrt{x})^2 - 2(4 + \sqrt{x}) = 24$

27. $\left(3x - \dfrac{2}{x}\right)^2 + 6\left(3x - \dfrac{2}{x}\right) + 5 = 0$

28. $\left(y - \dfrac{5}{y}\right)^2 - 2\left(y - \dfrac{5}{y}\right) = 8$

Applying the Concepts

29. **Population:** The population P (in thousands) of a small town is expected to increase according to the mathematical model

$$P = 3t + 4 + \sqrt{3t + 4}, \, t \geq 0$$

where t is time in years. When will the population be 56,000?

30. **Travel Time:** A motorboat leaves a dock traveling at a constant rate making a 48 kilometer trip to an island downstream and then returns to the dock. Suppose that the current rate is 4 kilometers per hour. The time T (in hours) that it took for the entire trip is given by the equation

$$T = \frac{48}{r - 4} + \frac{48}{r + 4}$$

where r (in kilometers per hour) is the speed of the boat in still water. Find r when T is 5 hours.

Developing and Extending the Concepts

In Problems 31 and 32, solve each equation involving negative rational exponents.

31. $x^{1/2} - 1 - 12x^{-1/2} = 0$

32. $x^{1/3} - 1 - 12x^{-1/3} = 0$

In Problems 33–40, find a substitution that will create a quadratic equation, then solve the equation.

33. $x^2 + 6x - 6(x^2 + 6x - 2)^{1/2} = -3$

34. $2x^2 + x - 4(2x^2 + x + 4)^{1/2} = 1$

35. $\dfrac{m}{m - 1} - 4 = \dfrac{5m - 5}{m}$

36. $\dfrac{p^2}{p + 2} + \dfrac{2p + 4}{p^2} = 3$

37. $\dfrac{m + 1}{m} + 2 = \dfrac{3m}{m + 1}$

38. $\dfrac{q^2 + 2q + 1}{q^2} - \dfrac{2q}{q + 1} = \dfrac{6q}{q + 1}$

39. $\dfrac{x^2 + 1}{x} + \dfrac{4x}{x^2 + 1} - 4 = 0$

40. $\left(\dfrac{2x^2 + 1}{x}\right)^2 + 5\left(\dfrac{2x^2 + 1}{x}\right) + 6 = 0$

In Problems 41–48, find x which makes $f(x) = 0$.

41. $f(x) = x - 2\sqrt{x} - 8$

42. $f(x) = x - 10\sqrt{x} + 9$

43. $f(x) = x^4 - 29x^2 + 100$

44. $f(x) = x^4 - 17x^2 + 16$

45. $f(x) = (2x + 3)^2 - 18(2x + 3) + 65$

46. $f(x) = (5x - 1)^2 - 13(5x - 1) + 30$

47. $f(x) = x^{2/3} - 8x^{1/3} + 15$

48. $f(x) = x^{2/3} - 2x^{1/3} - 8$

49. Let $f(x) = \sqrt{2x - 1} + \sqrt{x + 3}$, find x such that $f(x) = 3$.

50. Let $f(x) = \sqrt{5x - 1} + \sqrt{x + 3}$, find x such that $f(x) = 4$.

In Problems 51–56, given the graph of each function,

(a) Estimate the x intercepts of each graph.

(b) Algebraically, find the solution to the equation $f(x) = 0$.

51. $f(x) = x^4 - 18x^2 + 81$

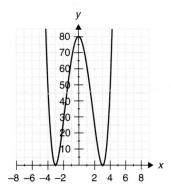

52. $f(x) = -x^4 + 5x^2 - 4$

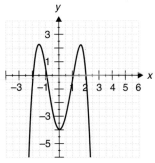

53. $f(x) = -x^4 + 8x^2 + 9$

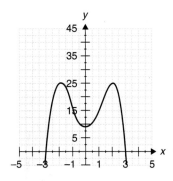

54. $f(x) = x^4 - 12x^2 + 27$

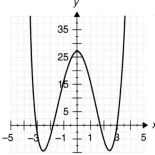

55. $f(x) = \sqrt{5x + 6} + \sqrt{3x - 2} - 6$

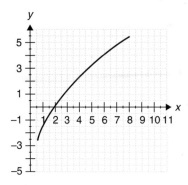

56. $f(x) = \sqrt{3x + 7} + \sqrt{x + 2} - 1$

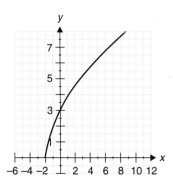

Objectives

1. Solve a Variety of Applied Problems
2. Solve Models by Using the Pythagorean Theorem
3. Solve Models Involving Rational Expressions
4. Solving Applied Problems

7.4 Quadratic Applications and Models

So far, we have solved many applied problems that were modeled by a given quadratic equation. In this section, we examine additional applied problems where the models have to be created and the techniques to solve them were developed in section 1.3. Recall the strategies developed by G. Polya for using mathematical models to solve applied problems.

1. Make a plan
2. Write the solution
3. Look back

Solving a Variety of Applied Problems

In solving these applied problems, the emphasis should be on writing the mathematical models.

EXAMPLE 1

Solving a Number Problem

Find two consecutive odd, positive integers whose product is 195.

Solution

Make a Plan

The goal is to find two positive odd integers that differ by 2 and have a product of 195. Any negative number is not part of the domain of the application.

$$\text{Let } x = \text{the first odd integer, then}$$

$$x + 2 = \text{the next odd integer.}$$

The following diagram is a visual description of the given.

First odd positive integer	•	Second odd positive integer	=	Product of the two integers
x	•	$x + 2$	=	195

Translating the problem into an equation, we have

$$x(x + 2) = 195$$

Write the Solution

Now we solve the equation

$x(x + 2) = 195$		Original equation
$x^2 + 2x = 195$		Use the distributive property
$x^2 + 2x - 195 = 0$		Write in standard form
$(x + 15)(x - 13) = 0$		
$x + 15 = 0 \quad$ or $\quad x - 13 = 0$		Apply the zero property
$x = -15 \qquad\qquad x = 13$		

We reject the solution -15 because it is not a positive integer.

Since $x = 13$ and $x + 2 = 15$, the two consecutive odd integers are 13 and 15.

Look Back

The numbers, 13 and 15, are two consecutive positive odd integers. Their product, $(13)(15)$ is 195.

EXAMPLE 2

Solving a Geometric Model

An artist is planning an acrylic painting in the form of a rectangle whose dimensions are 2 feet by 3 feet, surrounded by a stark border of uniform width. To achieve vitality, the artist wants the total area of the rectangle including the border to be 12 square feet. Find the width of the border that will accomplish this.

Solution

Make a Plan

Let x (in feet) represent the width of the border. Figure 1 shows the painting and the border. Recall that the area, A, of a rectangle is

$$A = lw$$

here $2 + 2x = w$, the width
and $3 + 2x = l$, the length.

The following diagram is a visual description of the given.

Length	•	Width	=	Total Area
$2x + 3$	•	$2x + 2$	=	12

Translating the problem into an equation, we have

$$(2x + 3)(2x + 2) = 12$$

Figure 1

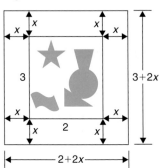

Write the Solution

$(2x + 3)(2x + 2) = 12$	Original equation
$4x^2 + 10x + 6 = 12$	Simplify
$4x^2 + 10x - 6 = 0$	Write in standard form
$2x^2 + 5x - 3 = 0$	Divide each side by 2
$(2x - 1)(x + 3) = 0$	Factor
$2x - 1 = 0$ or $x + 3 = 0$	Apply the zero property
$x = \dfrac{1}{2}$ or $x = -3$	

Since the width of the border cannot be negative, the width of the border must be 1/2 foot.

Look Back

The total area of the painting and the border is given by

$$[2(1/2) + 3][2(1/2) + 2] = (4)(3) = 12.$$

Figure 2

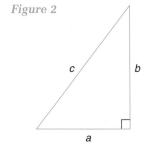

Solving Models by Using the Pythagorean Theorem

Recall the Pythagorean Theorem: in any right triangle, the sum of the squares of the legs is equal to the square of the hypotenuse. Using the right triangle in Figure 2, we have,

$$a^2 + b^2 = c^2$$

EXAMPLE 3

Finding the Lengths of the Rafters of a Solar Collector

In order to support a solar panel collector at an angle that will make it most efficient, the roof trusses of a house are designed as right triangles. The rafters of the truss are the legs of the right triangle, and the base of the truss is the hypotenuse (Figure 3a). A solar collector is mounted on the same side as the shorter rafter. If

Figure 3

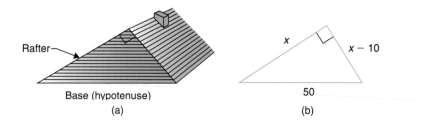

Rafter

Base (hypotenuse)

(a)

x

$x - 10$

50

(b)

that rafter is 10 feet shorter than the other rafter, and the base of each truss is 50 feet long, how long are the rafters?

Solution

Make a Plan

First, we must make sure that we understand the problem. The rafters are connected to each other at a right angle, making the base of the triangular truss the hypotenuse of a right triangle (Figure 3b). The Pythagorean Theorem describes the relationship between these sides. The rafters are not the same size, one rafter being 10 feet shorter than the other.

Let x = the length of the long rafter and

$x - 10$ = the length of the short rafter.

Using the Pythagorean Theorem with $a = x$, $b = x - 10$, and $c = 50$ (Figure 3b), we form the following equation.

$$a^2 + b^2 = c^2$$

$$x^2 + (x - 10)^2 = 50^2.$$

Write the Solution

$x^2 + (x - 10)^2 = 50^2$	Original equation
$x^2 + x^2 - 20x + 100 = 2500$	Simplify
$2x^2 - 20x - 2400 = 0$	Write in standard form
$x^2 - 10x - 1200 = 0$	Divide each side by 2
$(x + 30)(x - 40) = 0$	Factor
$x + 30 = 0$ or $x - 40 = 0$	
$x = -30$ or $x = 40$	

Because the length must be positive, the length of the long rafter is 40 feet and the length of the short rafter is $x - 10 = 40 - 10 = 30$ feet.

Look Back

$30^2 + 40^2 = 900 + 1600 = 2500 = 50^2$ and 50 feet is the length of the base.

Solving Models Involving Rational Expressions

The next two examples are concerned with interest and distance. They are appropriately placed in this section because the rational expressions involved yield quadratic equations. They are analogous since the formula for interest, $I = Prt$, and the formula for distance, $d = rt$ involve the same mathematical operations.

EXAMPLE 4   **Solving an Investment Problem**

In planning a $10,000 investment portfolio, an investor decides to invest in two one-year ventures. In one year, the first earned $300 in interest and the second earned $500 in interest. If the interest rate on the second venture is 1% higher than that of the first, what is each interest rate?

Solution   **Make a Plan**

The simple interest formula is given by

$$I = Prt, \text{ if } t = 1, \text{ then } I = Pr \text{ and } P = \frac{I}{r}.$$

The following table summarizes the given information for $t = 1$ year.

Let $r =$ the interest rate for the first investment and

$r + 0.01 =$ the interest rate for the second investment.

Investment	Interest (dollars)	Rate (percent)	Principal (dollars)
First	$300	r	$\dfrac{300}{r}$
Second	$500	$r + 0.01$	$\dfrac{500}{r + 0.01}$

The total investment of $10,000 is the sum of the two principals.

The following diagram is a visual description of the given.

Principal in first venture $\dfrac{300}{r}$	+	Principal in second venture $\dfrac{500}{r + 0.01}$	=	Total principal
	+		=	$10,000

Translating the problem into an equation, we have

$$\frac{300}{r} + \frac{500}{r + 0.01} = 10,000.$$

Write the Solution

$$\frac{300}{r} + \frac{500}{r + 0.01} = 10,000 \qquad \text{Original equation}$$

$$r(r + 0.01)\left[\frac{300}{r} + \frac{500}{r + 0.01}\right] = 10,000r(r + 0.01) \qquad \text{Multiply by the LCD}$$

$$300(r + 0.01) + 500r = 10,000r(r + 0.01) \qquad \text{Simplify}$$

$$300r + 3 + 500r = 10,000r^2 + 100r \qquad \text{Distributive property}$$

$$10,000r^2 - 700r - 3 = 0 \qquad \text{Write in standard form}$$

$$r = \frac{-b \pm \sqrt{b^2 - 4ac}}{2a}$$

Quadratic formula

$$r = \frac{-(-700) \pm \sqrt{(-700)^2 - 4(10{,}000)(-3)}}{2(10{,}000)}$$

$$r \approx -0.004 \text{ or } r \approx 0.074$$

Therefore, the interest rate for the first investment is $0.074 = 7.4\%$ and the rate for the second is $0.074 + 0.01 = 0.084 = 8.4\%$. The negative value for r is not appropriate for this situation.

Look Back

To check, $\dfrac{300}{0.074} + \dfrac{500}{0.074 + 0.01} = 10{,}006 \approx 10{,}000$. The more decimal points in the approximation used for r, the closer the sum of the principals will be to $\$10{,}000$.

Notice how we can use the mathematical techniques we learned earlier to help us solve equations applicable to new models. Here, we multiplied each side of the equation by the LCD to clear the fractions, then we solved the resulting quadratic equation.

EXAMPLE 5   **Solving a Rate Problem**

A bicyclist trained for a 100 kilometer race every weekend. Her average speed for the second weekend was 5 kilometers per hour faster than her average speed for the first. By doing this, she cut her riding time by 40 minutes; what was her average speed for the first weekend?

Solution   **Make a Plan**

Solving the distance formula, $d = rt$, for t yields $t = \dfrac{d}{r}$.

Let r = rate in kilometers per hour for the first weekend, then

$r + 5$ = rate in kilometers per hour for the second weekend.

The following table summarizes the given information.

	Distance (kilometers)	Rate (kilometers/hour)	Time (hours)
First Weekend	100	r	$\dfrac{100}{r}$
Second Weekend	100	$r + 5$	$\dfrac{100}{r + 5}$

We form an equation by noting that the time saved is 40 minutes or 2/3 hour.

$$\frac{100}{r} - \frac{100}{r + 5} = \frac{2}{3}$$

Write the Solution

$$\frac{100}{r} - \frac{100}{r + 5} = \frac{2}{3} \qquad \text{Original equation}$$

$$3r(r + 5)\left[\frac{100}{r} - \frac{100}{r + 5}\right] = \frac{2}{3}3r(r + 5) \qquad \text{Multiply by the LCD}$$

$$300(r + 5) - 300r = 2r(r + 5) \qquad \text{Use the distributive property}$$

$$300r + 1500 - 300r = 2r^2 + 10r \qquad \text{Simplify}$$

$$2r^2 + 10r - 1500 = 0 \qquad \text{Write in standard form}$$

$$r^2 + 5r - 750 = 0 \qquad \text{Divide each side by 2}$$

$$(r + 30)(r - 25) = 0$$

$$r + 30 = 0 \quad \text{or} \quad r - 25 = 0 \qquad \text{Apply the zero property}$$

$$r = -30 \qquad\qquad r = 25$$

Therefore, the bicyclists average speed was 25 kilometers per hour for the first weekend.

Look Back

To check, $\dfrac{100}{25} - \dfrac{100}{25 + 5} = 4 - \dfrac{100}{30} = \dfrac{12}{3} - \dfrac{10}{3} = \dfrac{2}{3}.$

PROBLEM SET 7.4

Mastering the Concepts

1. **Number Problem:** (a) Find two consecutive positive integers whose product is 552. (b) Find two consecutive negative integers whose product is 342.

2. **Number Problem:** (a) Find two consecutive odd integers whose product is 783. (b) Find two consecutive even integers such that the sum of their squares is 340.

3. **Window Design:** An artist is designing a rectangular stained glass window in such a way that the length of the window is 3 inches less than twice its width. What are the dimensions of the window if its area is 740 square inches?

4. **Gardening:** A homeowner has 60 feet of fencing to enclose a rectangular vegetable garden next to her house, with the house forming one side of the garden. What length and width are needed to enclose an area of 418 square feet and use all the fencing?

5. **Border of a Pool:** A rectangular swimming pool 20 feet wide and

60 feet long has a concrete walkway of uniform width as a border. If the total area of the walkway is 516 square feet, how wide is the walkway? (Figure 4)

6. **Advertising:** A newspaper advertisement has the shape of a rectangle whose length is 2 centimeters more than its width. The newspaper charges $9 for each square centimeter. What are the dimensions of the ad if the total cost is $216?

Figure 5

7. **Television Antenna:** A rural homeowner has his television antenna held in place by three guy wires (Figure 5). Suppose that the distances to each of the stakes from

Figure 4

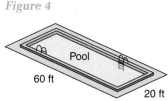

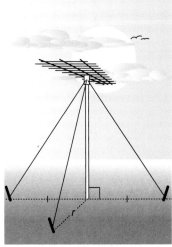

the base of the antenna are the same. Also the distance from the base of the antenna to one stake is 7 feet shorter than the height of the antenna and the length of a guy wire is 13 feet. How long is the distance from the base of the antenna to a stake?

8. **Television Screen:** When we speak of a 20 inch T.V. set, we refer to the diagonal of the screen measuring 20 inches. Suppose the width and height of the screen (in inches) are w and h respectively (Figure 6).

Figure 6

   (a) What is the width and height of a 20 inch T.V. set if its width is 4 inches more than its height?

   (b) What are the dimensions of a 25 inch T.V. set if its height is 5 inches less than its width?

9. **Camping Tent:** The entrance of a tent with vertical walls has the shape of an isosceles triangle (Figure 7). If the base of the entrance is 2 feet less than the height, and the area of the entrance is 24 square feet, how tall is the wall of the tent?

Figure 7

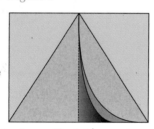

10. **Geometry:** The sum of the lengths of one side of a right triangle and the hypotenuse is 32 centimeters. The other leg is 1 centimeter shorter than the hypotenuse. Find the length of each side of the triangle.

11. **Investment Portfolio:** An inheritance of $15,000 was invested in two one-year investments. In one year, the first investment earned $540 in interest and the second earned $480 in interest. If the interest rate on the second investment is 2% higher than that of the first, what is each interest rate?

12. **Investment Portfolio:** An investor decided to invest $12,000 into two one-year investments, one a time deposit and the other a bond. Suppose that the interest earned in one year on the time deposit is $560, and the interest earned in one year on the bond is $450. If the interest rate on the bond is 1% higher than that of the time deposit, what is each interest rate?

13. **Commuting:** A commuter takes a train to college 10 miles away from home. The return train travels along the same route at a rate that is 10 miles per hour faster. If the commuter spends a total of 50 minutes per day on the train, what is the rate of each train?

14. **Car Trips:** On a 50 mile trip, a motorist traveled 10 miles in heavy traffic and then 40 miles in less congested traffic. The average speed of the motorist in heavy traffic was 20 miles per hour less than the average speed in light traffic. What was each rate of speed if the total time for the trip is 1.5 hours?

Developing and Extending the Concepts

15. **Age Difference:** Two brothers were born in consecutive years. The product of their present ages is 156. How old are they now?

16. **Ecology:** In a predator-prey model, such as foxes and rabbits, the number of the prey (rabbits) is 10 times the number of predators (foxes). If the product of their populations is 36,000, what is the predator population?

17. **Birth Dates:** Two friends discovered that they were both born in March but 16 days apart. The product of the day in March when they were born is 377. On what day in March was each born?

18. **Designing a Book:** A book designer determines that 88 square inches of print is needed on a page. In order to accomplish this, the height of the page should be two-thirds of the width. In addition, the top and bottom margins are to be 1 inch each, and the side margins are to be 2 inches each. What are the dimensions of the page?

19. **Drainage Pipe:** The cross-sectional area of a circular drainage pipe under a road is 260 square inches.

   (a) What is the diameter of the pipe in inches?

   (b) What is the diameter of the pipe in feet?

20. **Geometry:** The surface area A of a box with a square base whose width is w and height l is given by the formula

   Figure 8

   $$A = 2w^2 + 4lw.$$

   Find the width w of the base if $l = 5$ inches and $A = 78$ square inches (Figure 8).

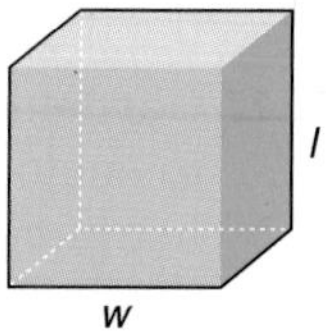

21. **Geometry:** The lengths of the shorter side, the longer side, and the hypotenuse of a right triangle are represented by consecutive even integers. Find the length of each side of the triangle.

22. **Geometry:** If increasing the lengths of the sides of a square by 8 feet results in a square whose

area is 25 times the area of the original square, what was the length of a side of the original square?

23. **Microwave Oven:** Suppose that the width and height of the interior of a microwave oven are each 1 foot less than its depth. What is the depth of the interior of the oven if its interior capacity is 2 cubic feet?

24. **Rink Area:** A rectangular skating rink was originally 100 feet long and 70 feet wide. Due to increased demand, the owners decided to enlarge the rink to 13,000 square feet. If they add rectangular strips of equal width to one side and one end, maintaining the rectangular shape of the rink, how wide should these strips be?

25. **Golden Rectangle and Golden Ratio:** A **golden rectangle** is a rectangle possessing proportions in a certain ratio w/l, called the **golden ratio,** which has long been of interest to artists and architects. The ancient Greeks used this ratio in their architecture even before the building of the Parthenon in Athens in the fifth century B.C. (Figure 9a).

Figure 9a

For instance, in a golden rectangle of width w and length l, the condition under which w/l is the golden ratio is that

$$\frac{w}{l} = \frac{l - w}{w}.$$

(a) The ratio w/l in the rectangle in Figure 9b is $1/x$. For the small rectangle, the ratio of $\frac{w}{l}$ is $\frac{x-1}{1}$. Use this relation to find the exact golden ratio.

Figure 9b

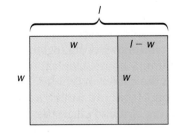

(b) Find the length l that yields a golden rectangle with a width of 5 inches.

26. **Geometry:** A diagonal of a polygon is a line segment between two nonadjacent vertices. Figure 10 shows all the diagonals of a polygon with eight vertices. A polygon with n vertices has a total of $\frac{n(n-3)}{2}$ diagonals for $n \geq 3$.

Figure 10

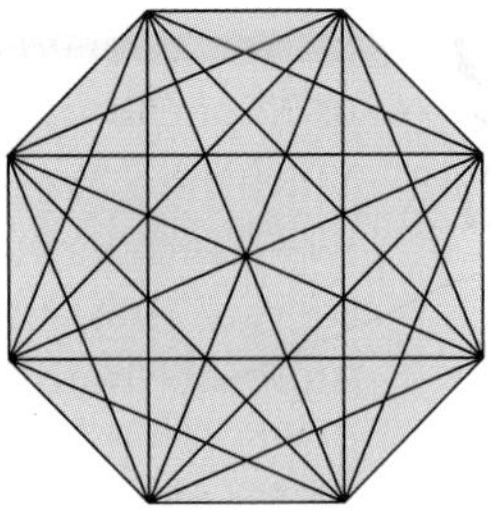

(a) How many vertices are on a polygon with 14 diagonals?

(b) Is there a polygon that has 17 diagonals?

27. **Compound Interest:** A firm invests $10,000 in a savings account for 2 years at a rate of interest r compounded annually. If a total of $12,100 is in the account at the end of the second year, what is the annual interest rate r?

28. **Surface Area of an Oil Drum:** A standard oil drum is a cylinder of height 34 inches with a closed top and bottom. Find the radius r of such a drum if its total surface area is 990π square inches.

$$S = 2\pi r^2 + 2\pi rh$$

29. **Transmission Distance:** Two campus security officers with two-way radios leave the same point and walk at right angles to each other. One walks at 4 miles per hour, the other at 3 miles per hour. How long can they communicate with one another if each radio has a maximum range of 2 miles?

30. **Bicyclist:** Two bicyclists, each moving at a constant speed, leave the same point and travel at right angles to each other (Figure 11). One travels 5 kilometers per hour faster than the other. In 3 1/2 hours the bicyclists are 98 kilometers apart. How fast does each travel? Round to the nearest kilometer per hour.

Figure 11

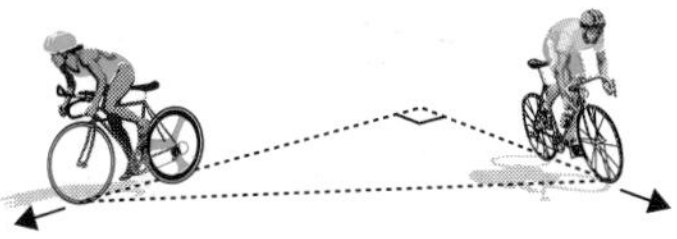

31. **Air Travel:** Two planes leave the same point and fly at the same altitude, one due north and the other due east. After 2 hours they are 1000 miles apart. If one plane averages 100 miles per hour more than the other, how fast are the planes flying?

32. **Car Travel:** A sales representative travels 280 miles at a certain speed. Traveling the same distance going home, the sales representative goes 10 miles per hour faster, and takes one hour less to travel the same distance. What is the sales person's speed each way rounded to the nearest mile per hour?

33. **Marathon Training:** In training for a marathon, two friends run 13.5 miles every Saturday. They can cut their training time by 18 minutes if they increase their average speed by 1.5 miles per hour. What is their current average speed for the 13.5 mile run?

34. **Truck Speed:** While moving, you drive a loaded rental truck 225 miles. After emptying the truck, you drive it another 150 miles to the nearest drop-off point at a speed of 5 miles per hour faster than before. If you drive a total of 8 hours, what was your speed with the empty truck?

35. **Rowing Rate:** A rowing crew takes 1 1/2 hours to complete a round trip, rowing 10 kilometers with the current and 10 kilometers against the current. If the rate of the current is 5 kilometers per hour, find the rate at which the crew can row in still water.

36. **Navigation:** The current in a shipping channel of the Detroit River flows at a rate of about 4 miles per hour. If it takes a tanker 5 hours to travel 24 miles up the river and back, how fast does the tanker travel in still water? Round to the nearest whole number.

37. **Work:** Two conveyor belts can unload a shipment of red peppers in 4 hours. Working alone, the slower belt takes 6 hours longer than the faster belt to unload the peppers. How many hours would each belt take to complete the job?

38. **Work:** A carpenter working alone can finish a job in 2 hours less time than an apprentice. Together they can complete the job in 7 hours. Find the time that each requires to complete the job alone. Round to one decimal place.

39. **Volume of a Sandbox:** *Figure 12*
    A child's sandbox will be made by cutting a 10-inch square from each corner of a square sheet of metal and turning up the sides (Figure 12). The sandbox will hold

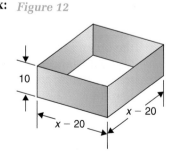

    20 cubic feet of sand. What size piece of metal should be used? (Hint: 1 cubic foot = 12^3 cubic inches.)

40. **Volume of Rain:** A 40-foot-long rectangular sheet of aluminum, 12 inches wide, is to be made into a rain gutter by turning up two sides so that they are

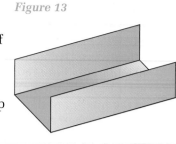

Figure 13

    perpendicular to the bottom (Figure 13). How much should be turned up in order to give the gutter a capacity of 37 gallons? (Hint: 1 gallon = 231 cubic inches.)

41. **Filling a Tank:** Two pipes that are connected to the same tank can fill the tank in 3 hours when used together. The larger pipe can fill the tank 2 hours faster than the small one. How long will it take the small pipe to fill the tank alone?

42. **Filling a Pool:** Two hoses are connected to a swimming pool. The pool can be filled in 4 hours if both hoses are used. How many hours are required for each hose to fill the pool alone, if the smaller hose requires 3 hours more than the larger hose?

43. **Marketing:** The marketing manager of a department store found that, on the average, 200 tennis rackets were sold monthly at the unit price of $80. However, for each $5 reduction in price, an extra 30 rackets were sold. How many rackets were sold at the reduced price if the total revenue last month was exactly $19,250? (Hint: Revenue = (price)(quantity).)

44. **Surface Area of a Silo:** A silo is a structure shaped like a right circular cylinder of radius r and height h topped by a half sphere with radius r (Figure 14). The surface area S of the silo is given by the formula.

Figure 14

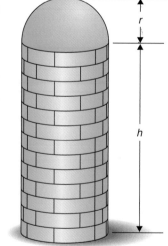

    $$S = 2\pi rh + 2\pi r^2$$

    Find the base radius of the silo if $h = 30$ feet and $S = 800\pi$ square feet.

45. **Janitorial Services:**
    Two janitors working together can clean a fitness center in 5 hours. One experienced janitor can clean the center in 5 hours less than the new person. How long would it take each janitor working alone to clean the fitness center?

46. Physics: During a grand remodeling project, you decide to rewire your kitchen. You want one outlet for the toaster and a second outlet for the lights. The maximum total resistance allowed by the circuit breaker is $R = 8$ Ohms, where the total resistance R is given by the formula

$$\frac{1}{R} = \frac{1}{R_1} + \frac{1}{R_2}.$$

If the light's total resistance R_2 is 5 Ohms greater than the toaster's resistance R_1, what is the toaster's resistance? Round to one decimal point.

47. Economics: A proprietor of a service center determines the demand for a popular brand of gasoline in a given week is given by the model

$q_1 = \dfrac{1200}{p}$, and the supply for the week is given

by the model $q_2 = p - 110$, where q_1 and q_2 denote the number in thousands of gallons demanded and supplied each week respectively,

at the price of p cents per gallon. Find the price per gallon at which the supply is equal to the demand.

48. Landscaping: A landscape designer is planning to plant a rectangular tulip bed on a circular plot of land so that the diagonals of the rectangle intersect at the center of the circle (Figure 15). If the length of the rectangular bed is 4 meters more than its width, and if the radius of the circular plot is 7 meters, what is the length of the rectangular bed? Round to one decimal place.

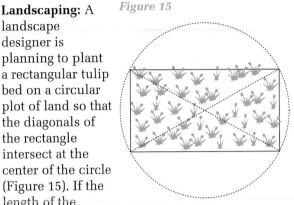

Figure 15

Objectives

1. Solve Quadratic Inequalities
2. Solve Rational Inequalities
3. Solve Applied Problems

7.5 Nonlinear Inequalities

In section 1.5, we discussed linear inequalities in one variable. In this section, we study quadratic as well as polynomial inequalities in one variable. In addition, we solve rational inequalities.

Solving Quadratic Inequalities

An inequality that can be written so that one side is a quadratic expression and the other side is zero is called a **quadratic inequality** in **standard form**. Examples are:

$$x^2 + 3x - 10 < 0 \qquad 6x^2 - 5x + 1 > 0$$
$$3x^2 + 2x - 6 \leq 0 \qquad x^2 - 8x + 7 \geq 0.$$

A **solution** of a quadratic inequality in one variable is the set of values of the variable that makes the inequality a true statement. Let us consider the values of x that make the quadratic inequality

$$x^2 + 3x - 10 < 0$$

a true statement. For instance, if we substitute -2 for x in the inequality

$$x^2 + 3x - 10 < 0$$

we obtain

$$(-2)^2 + 3(-2) - 10 = -12 < 0$$

which is a true statement. Thus -2 satisfies the inequality. Now, we ask: is -2 the only solution to the inequality? The answer is no. There are other values

Figure 1

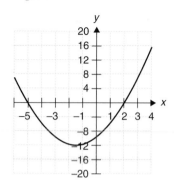

that satisfy the inequality. The values can be found by examining the signs in Figure 1 of the graph of the quadratic function

$$f(x) = x^2 + 3x - 10.$$

If you trace the points on the graph from left to right, you will find that the signs of the y coordinates are positive until you get to the x intercept at $x = -5$. Then the sign of the y coordinates become negative until $x = 2$. Then the y coordinates become positive again. Thus, the x intercepts are the points where the y values change signs. These intercepts are obtained by solving the equation

$$x^2 + 3x - 10 = 0$$

$$(x + 5)(x - 2) = 0$$

$$x = -5 \text{ or } x = 2.$$

The graph in Figure 1 shows that $f(x) < 0$ when $-5 < x < 2$.

It is not always necessary to graph the quadratic function and determine from the graph which values of the variable satisfy the inequality. In some cases, it may be inconvenient to graph the function, so we provide an alternate method to solve these inequalities. In the above illustration, the x intercepts -5 and 2 of the graph of

$$f(x) = x^2 + 3x - 10$$

(also called *boundary values*) divide the number line into three open intervals.

$$(-\infty, -5) \ (-5, 2), \text{ and } (2, \infty)$$

Figure 2a

which we designate by A, B, and C respectively (Figure 2a).

It is important to note that if $x^2 + 3x - 10$ is negative for one value of x in a given interval, then the expression is negative for *any* value of x in that interval. The same holds true if the value of

$$x^2 + 3x - 10$$

is positive for a particular value of x in an interval. To determine if $x^2 + 3x - 10 < 0$ in each interval, we choose a test number from each interval and substitute its value for x in the inequality $x^2 + 3x - 10 < 0$. If the resulting inequality is true, the interval containing the test point is a solution interval. The following table lists the results for test values -6, 0, and 3 in the intervals A, B, and C respectively.

Interval	Test Value	Test Value of $x^2 + 3x - 10$	Sign of $x^2 + 3x - 10$	True/False
A	-6	$(-6)^2 + 3(-6) - 10 = 8$	positive	False
B	0	$(0)^2 + 3(0) - 10 = -10$	negative	True
C	3	$(3)^2 + 3(3) - 10 = 8$	positive	False

Figure 2b

The values of x in the interval $(-5, 2)$ satisfy the inequality. Therefore, the solution set is $(-5, 2)$ (Figure 2b). The end points are not included because the inequality is strict. Notice that the solution $(-5, 2)$ is consistent with the graph in Figure 1. The method illustrated above can also be used to solve any inequality whose left side is a polynomial of degree higher than two. The values of x for which $f(x) = 0$ are the x intercepts of the graph of f and are called the **boundary values** or **cut points**.

Strategy for Solving a Polynomial Inequality—The Cut Point Method

Step 1. Write the inequality as an equation in standard form, then solve the resulting equation to determine the cut points (or the boundary values).

Step 2. Arrange the solutions obtained from Step 1 in increasing order on a number line. These solutions will divide the number line into intervals.

Step 3. Choose a test number in each interval and determine if its value satisfies the original inequality. Also test each boundary value.

Step 4. Draw a number line showing the information in Step 3. Then read the solution set for the original inequality by identifying all the intervals where the inequality is true.

EXAMPLE 1   **Solving a Quadratic Inequality**

Solve the inequality $x^2 + x \geq 12$. Write the solution in interval notation and show the solution on a number line.

Solution   Step 1: $x^2 + x - 12 \geq 0$ Write in standard form

$x^2 + x - 12 = 0$ Solve related equation

$(x + 4)(x - 3) = 0$

$x = -4 \text{ or } x = 3$

Figure 3a

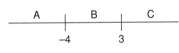

Step 2: The numbers -4 and 3 separate the number line into three intervals which we denote by A, B, and C (Figure 3a).

Step 3: Select a test number in each interval and the cut points to determine if they satisfy the original inequality. These values are listed in the following table.

Interval	Test Value	Test Value of $x^2 + x - 12$	Sign of $x^2 + x - 12$	True/False
A	-5	$(-5)^2 + (-5) - 12 = 8$	positive	True
	-4	$(-4)^2 + (-4) - 12 = 0$	zero	True
B	0	$0^2 + 0 - 12 = -12$	negative	False
	3	$3^2 + 3 - 12 = 0$	zero	True
C	4	$4^2 + 4 - 12 = 8$	positive	True

Figure 3b

Step 4. Figure 3b illustrates the information obtained in Step 3. The solution includes both intervals A and C, as well as the cut (boundary) points -4 and 3. The solution set is $x \leq -4$ or $x \geq 3$. In interval notation, the solution set is $(-\infty, -4]$ or $[3, \infty)$.

The graph of the function

Figure 4

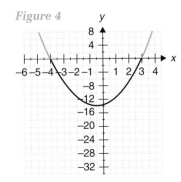

$$f(x) = x^2 + x - 12$$

in Figure 4 confirms that $f(x) \geq 0$, when $x \leq -4$ or $x \geq 3$.

EXAMPLE 2

Solving an Inequality

Solve the inequality $2x^2 + 3x - 2 \le x^2 + 2x$. Express the solution in interval notation.

Solution

Writing the inequality in standard form, we have

$$2x^2 + 3x - 2 \le x^2 + 2x$$

$$x^2 + x - 2 \le 0.$$

$$x^2 + x - 2 = 0 \qquad \text{Solve the equation for } x \text{ intercepts}$$

$$(x + 2)(x - 1) = 0$$

$$x = -2 \text{ or } x = 1$$

Figure 5a

The cut points or boundary values are -2 and 1. The numbers -2 and 1 separate the number line into three intervals (Figure 5a). We use -3, 0, and 2 and the cut points -2 and 1 as test values. The results are shown in the following table.

Figure 5b

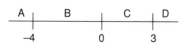

Interval	Test Value	Test Value of $x^2 + x - 2$	Sign of $x^2 + x - 2$	True/False
A	-3	$(-3)^2 + (-3) - 2 = 4$	positive	False
	-2	$(-2)^2 + (-2) - 2 = 0$	zero	True
B	0	$0^2 + 0 - 2 = -2$	negative	True
	1	$1^2 + 1 - 2 = 0$	zero	True
C	2	$2^2 + 2 - 2 = 4$	positive	False

Figure 6

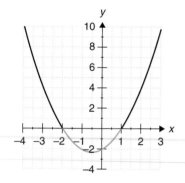

Figure 5b illustrates the solution of $-2 \le x \le 1$ which includes the points -2 and 1. In interval notation, the solution set is $[-2, 1]$. The graph of the function

$$f(x) = x^2 + x - 2$$

in Figure 6, shows that $f(x) \le 0$, when $-2 \le x \le 1$, which is consistent with the above solution.

EXAMPLE 3

Solving a Polynomial Inequality

Solve the inequality

$$x^3 + x^2 - 12x > 0.$$

Write the solution in interval notation, and illustrate the solution on a number line.

Solution

First, we solve the equation for the cut points.

$$x^3 + x^2 - 12x = 0$$

$$x(x + 4)(x - 3) = 0$$

$$x = 0, x = -4, \text{ and } x = 3.$$

Figure 7a

The solutions -4, 0, and 3 separate the number line into four intervals, which we denote by A, B, C, and D (Figure 7a). The following table shows the results of the test values -5, -2, 2, and 4 as well as the cut points -4, 0, and 3.

Interval	Test Value	Test Value of $x^3 + x^2 - 12x$	Sign of $x^3 + x^2 - 12x$	True/False
A	-5	$(-5)^3 + (-5)^2 - 12(-5) = -40$	negative	False
	-4	$(-4)^3 + (-4)^2 - 12(-4) = 0$	zero	False
B	-2	$(-2)^3 + (-2)^2 - 12(-2) = 20$	positive	True
	0	$0^3 + 0^2 - 12(0) = 0$	zero	False
C	2	$2^3 + 2^2 - 12(2) = -12$	negative	False
	3	$3^3 + 3^2 - 12(3) = 0$	zero	False
D	4	$4^3 + 4^2 - 12(4) = 32$	positive	True

Figure 7b

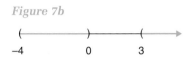

The solution is illustrated in intervals B and D on the number line in Figure 7b. This solution is described by the inequalities $-4 < x < 0$ or $x > 3$. Notice that the cut (boundary) points are not included in the solution set. In interval notation, we write

$$(-4,0) \text{ or } (3,\infty).$$

Figure 8 shows the graph of

$$f(x) = x^3 + x^2 - 12x$$

is positive when $-4 < x < 0$ or $x > 3$, which confirms the solution set is $(-4,0)$ or $(3,\infty)$.

Figure 8

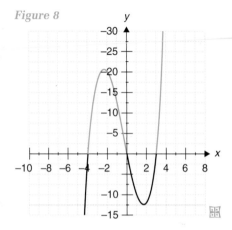

Solving Rational Inequalities

Inequalities such as

$$\frac{x - 4}{x + 2} \leq 0 \quad \text{or} \quad \frac{x + 3}{5x - 7} > 0$$

contain rational expressions are called **rational inequalities in standard form.** These inequalities are solved using the same procedure as quadratic inequalities with one modification. In solving a rational inequality, we must determine all values of x that make the denominator zero.

EXAMPLE 4

Solving a Rational Inequality

Solve the inequality $\dfrac{x - 4}{x + 2} \leq 0$.

Write the solution in set notation and illustrate the solution on the number line.

Solution

Notice that a rational expression can change its algebraic sign only if the numerator or denominator changes algebraic sign. Hence, intervals where a rational expression is positive are separated from intervals where that expression is negative by values of x for which the numerator or denominator is zero. The numerator of

$$\frac{x - 4}{x + 2}$$

Figure 9a

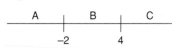

is zero when $x = 4$; its denominator is zero when $x = -2$. The numbers -2 and 4 separate the number line into three intervals (Figure 9a). Next we select a test value in each interval as well as the cut points -2 and 4. These values are listed in the following table.

Interval	Test Value	Test Value of $\dfrac{x-4}{x+2}$	Sign of $\dfrac{x-4}{x+2}$	True/False
A	-3	$\dfrac{-3-4}{-3+2} = 7$	positive	False
	-2	$\dfrac{-2-4}{-2+2} = $ Undefined	undefined	False
B	0	$\dfrac{0-4}{0+2} = -2$	negative	True
	4	$\dfrac{4-4}{4+2} = 0$	zero	True
C	5	$\dfrac{5-4}{5+2} = \dfrac{1}{7}$	positive	False

Figure 9b

Figure 9b illustrates the information obtained from the above table. The solution set is

$$-2 < x \le 4 \text{ or } [-2,4].$$

Figure 10

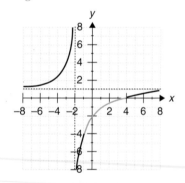

Observe that 4 belongs to the solution set because the quotient $\dfrac{x-4}{x+2}$ is equal to zero when $x = 4$. The value -2 is must be excluded from the solution set because the denominator is zero when $x = -2$ and this causes the quotient to be undefined. Figure 10 shows the graph of f is below the x axis when $-2x < 4$.

Solving Applied Problems

The next example shows how to use inequalities to solve applied problems and models.

EXAMPLE 5   **Solving a Fireworks Problem**

A firework shell is launched from a mortar on the ground with an initial speed of 64 feet per second. The height h (in feet) of the shell after t seconds from being launched is given by the model

$$h(t) = 64t - 16t^2.$$

During what time interval will the shell be at least 48 feet above the ground?

Solution   When the shell is at least 48 feet above the ground, the time t will satisfy the inequality

$$h(t) = 64t - 16t^2 \ge 48.$$

Figure 11

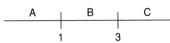

We solve the inequality by first solving the associated equation.

$$64t - 16t^2 = 48 \quad \text{Original equation}$$
$$16t^2 - 64t + 48 = 0 \quad \text{Standard form}$$
$$16(t - 3)(t - 1) = 0$$
$$t = 3, 1$$

Figure 12

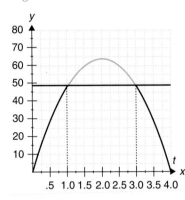

The numbers 1 and 3 separate the number line into three intervals, A, B, and C (Figure 11).

The solution set is [1,3], which means that the shell containing the firework is at least 48 feet high when the time is between and including 1 and 3 seconds. Figure 12 shows the graph of the original model, $h(t) = 64t - 16t^2$. Notice that the h coordinate of each point on the graph represents the height of the shell for each value of time. The points we are interested in are on or above the line $h = 48$. Those values of t belong to the interval [1,3].

PROBLEM SET 7.5

Mastering the Concepts

In Problems 1–14, solve each inequality using the cut point strategy. Write the solution set in interval notation and show the solution set on a number line.

1. $(x - 4)(x + 3) < 0$
2. $(x - 2)(x + 3) < 0$
3. $(x + 5)(x - 2) \le 0$
4. $(x - 4)(x + 1) \le 0$
5. $(x - 1)(x + 3) \ge 0$
6. $(x - 3)(x + 1) \ge 0$
7. $x^2 - x - 2 < 0$
8. $x^2 - x - 6 < 0$
9. $x^2 - 3x + 2 \ge 0$
10. $x^2 - 5x - 6 \ge 0$
11. $2x^2 - 5 \ge -x^2$
12. $2x^2 - 5 \le x^2 - 7x - 12$
13. $5x^2 + 6x + 1 \ge -4x^2$
14. $12x^2 + 7x + 1 > -4x^2 - x$
15. $16x^2 - 24x \le -9$
16. $49x^2 + 1 \le -14x$

In Problems 17–22, find the domain of each function.

17. $f(x) = \sqrt{x^2 + 2x - 3}$ (see Problem 5)
18. $g(x) = \sqrt{x^2 - 2x - 3}$ (see Problem 6)
19. $f(x) = \dfrac{1}{\sqrt{2 + x - x^2}}$ (see Problem 7)
20. $h(x) = \dfrac{1}{\sqrt{6 + x - x^2}}$ (see Problem 8)
21. $f(x) = \sqrt{x^2 - 3x + 2}$ (see Problem 9)
22. $g(x) = \sqrt{x^2 - 5x - 6}$ (see Problem 10)

In Problems 23–32, solve each inequality using the cut point strategy. Write the solution set in interval notation and show the solution set on a number line.

23. $\dfrac{1}{x - 3} > 0$
24. $\dfrac{1}{x + 2} > 0$
25. $\dfrac{x - 1}{x + 4} \le 0$
26. $\dfrac{x + 3}{x - 3} \le 0$
27. $\dfrac{x + 1}{2 - x} \ge 0$
28. $\dfrac{2 - x}{x + 3} \ge 0$
29. $\dfrac{2x - 1}{x + 2} > 0$
30. $\dfrac{x}{x - 5} > 0$
31. $\dfrac{2x}{x + 4} \ge 0$
32. $\dfrac{-x}{3 - x} \le 0$

In Problems 33–38, find the domain of each function

33. $f(x) = \dfrac{1}{\sqrt{x - 3}}$ (see Problem 23)
34. $g(x) = \dfrac{1}{\sqrt{x + 2}}$ (see Problem 24)
35. $f(x) = \sqrt{\dfrac{1 - x}{x + 4}}$ (see Problem 25)
36. $g(x) = \sqrt{\dfrac{x + 3}{3 - x}}$ (see Problem 26)
37. $f(x) = \sqrt{\dfrac{x + 1}{2 - x}}$ (see Problem 27)
38. $g(x) = \sqrt{\dfrac{2 - x}{x + 3}}$ (see Problem 28)

In Problems 39–42, solve each inequality by examining the graph of the corresponding function. Write the solution in interval notation and show the solution on a number line.

39. $f(x) = x(x - 3)(x + 3)$;
$f(x) < 0$

40. $f(x) = (x - 1)(x + 2)(x - 3)$;
$f(x) > 0$.

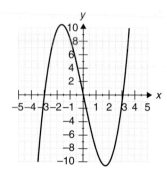 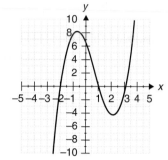

41. $f(x) = x^3 - 9x$;
$f(x) \geq 0$

42. $f(x) = 4x - x^3$;
$f(x) \leq 0$

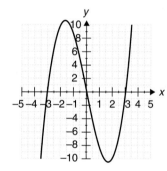 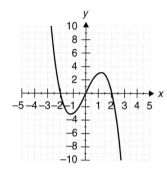

Applying the Concepts

43. Total Profit: A company determines that its total profit P (in dollars) is determined by the function

$$P(x) = 360x - 2x^2$$

where x is the number of items produced and sold.
(a) Find all values of x for which the company makes a profit.
(b) Find all values of x for which the company takes a loss.
(c) Find all values of x for which the company makes a profit of at least $16,000.

44. Height of a Golf Ball: The height h (in feet) of a golf ball above a fairway depends on the time t (in seconds) it has been in flight. A golfer hits a tee shot that has an approximate height after t seconds and is modeled by the function

$$h(t) = 80t - 16t^2.$$

(a) For what time interval will the height of the golf ball be more than 64 feet above the ground?
(b) For what time interval will the height of the golf ball be less than 20 feet above the ground?

45. Height of a Thrown Ball: A ball is thrown vertically upward from the roof of an 80-foot building with an initial speed of 32 feet per second. Its height h (in feet) above the ground after t seconds is modeled by the function

$$h(t) = -16t^2 + 32t + 80.$$

(a) For what time interval will the height exceed 87 feet?
(b) For what time interval will the height be less than 47 feet?

46. Storage: The formula for the surface area A of an open cylinder is given by

$$A = \pi r^2 + 2\pi rh$$

where r is the radius and h is the height. Suppose a tomato soup can has a height of 6 inches.

(a) For what measures of the radius will the surface area exceed 200 square inches?
(b) For what measures of the radius will the surface area be less than 400 square inches?

Developing and Extending the Concepts

In Problems 47–60, solve each inequality algebraically. Write the solution in interval notation and show the solution on a number line.

47. $2x^3 - 8x \leq 0$

48. $(x + 2)(x^2 - 5x + 4) < 0$

49. $(x + 3)(x^2 - x - 2) > 0$

50. $(x + 4)(x^2 - 2x - 8) \geq 0$

51. $\dfrac{(x - 1)(x - 3)}{x + 4} \leq 0$

52. $\dfrac{(x + 2)(x - 1)}{x - 5} > 0$

53. $\dfrac{x + 1}{(x - 2)(x + 3)} > 0$

54. $\dfrac{x - 2}{(x + 2)(x + 1)} \leq 0$

55. $\dfrac{x + 1}{x - 3} \leq 1$

56. $\dfrac{x + 2}{x - 1} \geq 2$

57. $\dfrac{1}{x} \leq 3$

58. $\dfrac{1}{x} \geq 2$

59. $\dfrac{(x + 1)(x + 3)}{x - 4} > 0$

60. $\dfrac{(x + 2)(x - 3)}{x - 2} > 0$

Objectives

1. Graph Functions of the Form $f(x) = ax^2$
2. Graph Functions by Shifting
3. Write Functions in the Form $f(x) = a(x - h)^2 + k$
4. Solve Applied Problems—Maximum and Minimum

7.6 Quadratic Functions

In Chapter 4, we graphed quadratic functions by point plotting. So far in this chapter, we discussed the x intercepts of the graphs of these functions. Here in this section, we study graphs of quadratic functions of the form

$$f(x) = ax^2 + bx + c, a \neq 0$$

in more detail. We consider the role played by the constant numbers a, b, and c in determining the shape and the position of the graphs. For instance, people often wonder why a golf player requires so many golf clubs. The reason is, the length of the club, its weight, and the angle of the striking face are designed to produce the right *parabolic* path of the golf ball at each point in the game.

Graphing Functions of the Form $f(x) = ax^2$

The simplest quadratic function is of the form

$$f(x) = ax^2$$

and the curve is called a **parabola.** In fact, the graph of any quadratic function is a *parabola*. The most basic quadratic functions are of the form

$$f(x) = x^2 \text{ and } g(x) = -x^2.$$

These functions can be graphed by choosing some input values of x and computing the corresponding output values. The resulting ordered pairs are plotted and connected with smooth curves.

EXAMPLE 1

Graphing Quadratic Functions by Point Plotting

Produce the graph of the function

$$f(x) = x^2$$

by selecting input values and obtaining the corresponding output values, then plot these points.

Solution

First we create a table showing the input and output values for the functions.

x	$f(x) = x^2$	$(x, f(x))$
-2	$f(-2) = (-2)^2 = 4$	$(-2, 4)$
-1	$f(-1) = (-1)^2 = 1$	$(-1, 1)$
0	$f(0) = (0)^2 = 0$	$(0, 0)$
1	$f(1) = (1)^2 = 1$	$(1, 1)$
2	$f(2) = (2)^2 = 4$	$(2, 4)$

Figure 1

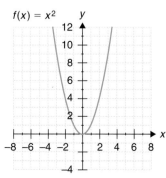

Figure 1 shows the graph of the function f by plotting the points from the table and connecting them with a smooth curve.

Most quadratic functions have graphs that look like the ones in Figure 1. To examine the effect of the coefficient a when graphing

$$f(x) = ax^2,$$

Figure 2

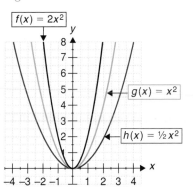

we consider Figure 2 which shows the graphs of

$$f(x) = 2x^2, \; g(x) = x^2, \text{ and } h(x) = \frac{1}{2}x^2.$$

Notice the effect that the value of a has on the shape of each graph. The graph of $h(x) = 1/2\,x^2$ is *flatter* than the graph of $g(x) = x^2$, while the graph of $f(x) = 2x^2$ is *thinner* than $g(x) = x^2$. We can verify that each of the graphs is correct by plotting the input and output values in the following table.

x	$f(x) = 2x^2$	$g(x) = x^2$	$h(x) = \frac{1}{2}x^2$
-2	8	4	2
-1	2	1	$\frac{1}{2}$
0	0	0	0
1	2	1	$\frac{1}{2}$
2	8	4	2

Figure 3

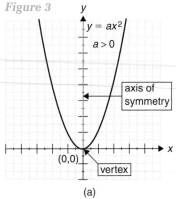

(a)

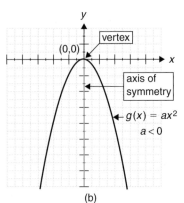

(b)

In general, all graphs of parabolas in Figure 3 are of the form

$$f(x) = ax^2,$$

and have the same basic shape. They have the following geometric characteristics shown in Table 1. The **vertex** is the highest or lowest point on the curve. The **axis of symmetry** is the line through the vertex which divides the parabola into two equal pieces. The sign of the coefficient a determines whether a parabola opens upward or downward. If $a > 0$, the parabola opens **upward** (Figure 3a) and if $a < 0$ the parabola opens **downward** (Figure 3b).

TABLE 1

Function	Graph Opens	Vertex	Axis of Symmetry	Domain	Range
$y = ax^2, a > 0$	Upward Figure 3a	(0,0) lowest point	y axis	R	$[0,\infty)$
$y = ax^2, a < 0$	Downward Figure 3b	(0,0) highest point	y axis	R	$(-\infty,0]$

Graphing Functions by Shifting

Figure 4a shows the graphs of

$$y = x^2$$

$$f(x) = x^2 + 2$$

and

$$g(x) = x^2 - 2$$

Notice in Figure 4a that the graphs of $f(x)$ and $g(x)$ are identical to the graph of $y = x^2$ except that $f(x)$ is shifted 2 units vertically up and $g(x)$ is shifted 2 units vertically down.

Figure 4b shows the graphs of

$$y = x^2$$

$$j(x) = (x + 2)^2$$

and

$$m(x) = (x - 2)^2$$

where $j(x)$ and $m(x)$ are identical to $y = x^2$ except that $j(x)$ is shifted 2 units horizontally to the left and $m(x)$ is shifted 2 units horizontally to the right.

Figure 4c shows the graphs of

$$y = x^2$$

and

$$p(x) = (x - 2)^2 + 2$$

where $p(x)$ is identical to $y = x^2$ except that $p(x)$ is shifted 2 units horizontally to the right and 2 units vertically upward.

Figure 4

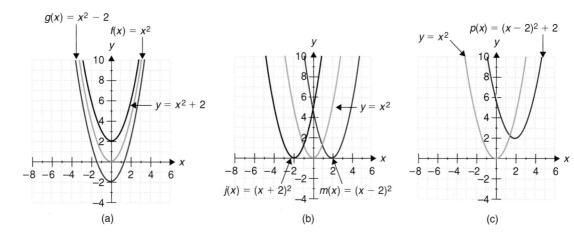

(a)   (b)   (c)

We generalize the above results by comparing the graph of $f(x) = a(x - h)^2 + k$ with the graph of $y = ax^2$ and examine the effects of the numbers h and k.

Graphing $f(x) = a(x - h)^2 + k$

Function	Vertex	Axis of Symmetry	Geometric Description of Graphs	Example
$g(x) = ax^2 + k$	$(0,k)$	$x = 0$	Looks like $y = ax^2$, shifted vertically k units upward if $k > 0$ and $\|k\|$ units downward if $k < 0$.	Figure 4a
$m(x) = a(x - h)^2$	$(h,0)$	$x = h$	Looks like $y = ax^2$, shifted horizontally h units to the right if $h > 0$ and $\|h\|$ units to the left if $h < 0$	Figure 4b
$p(x) = a(x - h)^2 + k$	(h,k)	$x = h$	Looks like $y = ax^2$, shifted h units horizontally to the right and k units vertically upward when $h > 0$ and $k > 0$.	Figure 4c

For example, Table 2 shows the description of the graph of each of the following functions.

TABLE 2

Function	Vertex	Axis of Symmetry	Geometric Description of Graph
$f(x) = 3(x - 2)^2 + 4$	$(2,4)$	$x = 2$	Opens upward and shifted right 2 units and up 4 units
$g(x) = -\frac{2}{3}(x-4)^2 + 5$	$(4,5)$	$x = 4$	Opens downward and shifted right 4 units and up 5 units
$h(x) = 2(x + 3)^2 - 2$	$(-3,-2)$	$x = -3$	Opens upward and shifted left 3 units and down 2 units
$k(x) = -2(x + \frac{2}{3})^2 - 7$	$(-\frac{2}{3},-7)$	$x = -\frac{2}{3}$	Opens downward and shifted left 2/3 units and down 7 units

EXAMPLE 2 **Graphing Functions of the Form $f(x) = a(x - h)^2 + k$**

Figure 5 shows the graphs of

$$y = 2x^2, f(x) = 2(x + 2)^2 + 1, \text{ and } g(x) = -2(x + 2)^2 + 1$$

on the same coordinate system.

(a) Discuss the transformations of each graph.

(b) Find the vertex of each parabola and its axis of symmetry.

(c) Find the range of each function.

Solution

(a) The graph of $f(x) = 2(x + 2)^2 + 1$ looks like $y = 2x^2$ but shifted 2 units to the left and 1 unit upward. The graph of $g(x) = -2(x + 2)^2 + 1$ looks like $y = 2x^2$ but shifted 2 units to the left, 1 unit upward, and reflected across the x axis.

(b) and (c) The vertex, axis of symmetry and the range of each function are found in the following table.

Figure 5

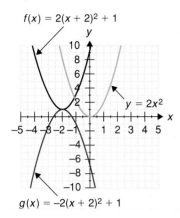

$f(x) = 2(x + 2)^2 + 1$

$g(x) = -2(x + 2)^2 + 1$

Function	Vertex	Axis of Symmetry	Range
$f(x) = 2(x + 2)^2 + 1$	$(-2,1)$	$x = -2$	$[1,\infty)$
$g(x) = -2(x + 2)^2 + 1$	$(-2,1)$	$x = -2$	$(-\infty,1]$

Writing Functions in the Form $f(x) = a(x - h)^2 + k$

In order to describe graphs of parabolas using the above procedure, we need to rewrite the function

$$f(x) = ax^2 + bx + c \text{ into the form } f(x) = a(x - h)^2 + k.$$

To accomplish this, we complete the square to obtain a perfect square trinomial, which can be factored as the square of a binomial. The technique is shown in the next example.

EXAMPLE 3

Completing the Square of a Function

Rewrite

$$f(x) = 2x^2 - 12x + 10 \text{ into the form } f(x) = a(x - h)^2 + k$$

by completing the square. Find the vertex and the axis of symmetry.

Solution

We complete the square to find the values of *a*, *h*, and *k*.

Figure 6

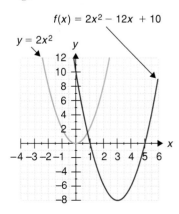

$f(x) = 2x^2 - 12x + 10$

$y = 2x^2$

$\begin{aligned}
f(x) &= 2x^2 - 12x + 10 &&\text{Given function}\\
&= 2(x^2 - 6x) + 10 &&\text{Factor 2 from the first two terms}\\
&= 2(x^2 - 6x + 9) - 2(9) + 10 &&\text{Add and subtract } 2\left[\tfrac{1}{2}(6)\right]^2 = 2(9)\\
&= 2(x - 3)^2 - 18 + 10 &&\text{Factor } x^2 - 6x + 9 \text{ as a perfect square}\\
&= 2(x - 3)^2 - 8
\end{aligned}$

Here $a = 2$, $h = 3$, and $k = -8$. Figure 6 shows the graphs of

$$y = 2x^2 \text{ and } f(x) = 2x^2 - 12x + 10$$

(for comparison) on the same coordinate system. Observe that the graph of $f(x)$ is obtained by shifting the graph of $y = 2x^2$, 3 units to the right and 8 units downward.

The vertex is $(3,-8)$ and the axis of symmetry is $x = 3$.

EXAMPLE 4   **Completing the Square of a Function**

Rewrite

$$f(x) = -2x^2 - 8x - 11 \text{ into the form } f(x) = a(x - h)^2 + k$$

by completing the square. Find the vertex and the axis of symmetry.

Solution   We complete the square to find the values of a, h, and k.

$f(x) = -2x^2 - 8x - 11$	Given function
$= -2(x^2 + 4x) - 11$	Factor -2 from the first two terms
$= -2(x^2 + 4x + 4) + 2(4) - 11$	Add and subtract $-2\left[\dfrac{1}{2}(4)\right]^2 = -8$
$= -2(x + 2)^2 + 8 - 11$	Factor $x^2 + 4x + 4$ as a perfect square
$= -2(x + 2)^2 - 3$	

Here $a = -2$, $h = -2$, and $k = -3$. Figure 7 shows the graphs of $y = -2x^2$ and $f(x) = -2x^2 - 8x - 11$ (for comparison) on the same coordinate system. Observe that the graph of $f(x)$ is obtained by shifting the graph of $y = -2x^2$, 2 units to the left and 3 units downward. The vertex is $(-2, -3)$ and the axis of symmetry is $x = -2$.

Figure 7

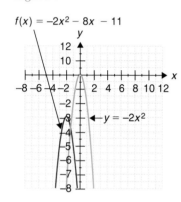

A short way to change the form of the function

$$f(x) = ax^2 + bx + c \text{ into the form } f(x) = a(x - h)^2 + k$$

is to let $h = -\dfrac{b}{2a}$ and $k = f\left(\dfrac{-b}{2a}\right)$.

then substitute the values obtained into $f(x) = a(x - h)^2 + k$.

For example, we could have found the vertex of the parabola in Example 4 directly by writing

$$h = \frac{-b}{2a} = \frac{-8}{2(-2)} = -2, \text{ and}$$

$$k = f\left(\frac{-b}{2a}\right) = f(-2) = -2(-2)^2 - 8(-2) - 11$$

$$= -3$$

so that the vertex is $(-2, -3)$.

Figure 8

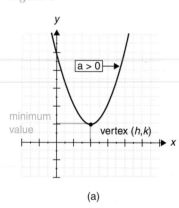

(a)

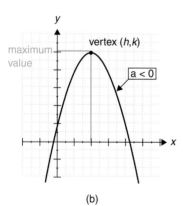

(b)

Solving Applied Problems—Maximum and Minimum

We indicated that if $a > 0$, the parabola $f(x) = ax^2 + bx + c$ opens upward. In this case, we say the y coordinate of the vertex is the **minimum value of f** (Figure 8a). Similarly, if $a < 0$, the parabola opens downward, and we say that the y coordinate of the vertex is the **maximum value of f** (Figure 8b).

Applied problems and models in which a quantity has to be *maximized* or *minimized* can be solved by finding the coordinates of the vertex, as the next example shows.

EXAMPLE 5

Finding the Maximum Height

A cliff diver's path from the top of a cliff 110 feet above the surface of the water after t seconds is approximated by the model

$$h(t) = -t^2 + 2t + 110, \qquad t \geq 0$$

where h is the height in feet above the water.

(a) Find the maximum height of the diver.

(b) Find the time it takes for the diver to reach the maximum height.

Solution

(a) To find the maximum height of the diver, we use

$$h(t) = -t^2 + 2t + 110$$

and complete the square to find the vertex.

$h(t) = -t^2 + 2t + 110$	Original equation
$= -(t^2 - 2t) + 110$	Factor -1 from the two terms
$= -(t^2 - 2t + 1) + 1 + 110$	Complete the square
$= -(t - 1)^2 + 111$	Perfect square trinomial

The vertex of the parabola is $(1,111)$. Since $a = -1$, a negative number, the y coordinate 111, of the vertex is the maximum height of the diver.

(b) The maximum height occurs when $t = 1$. That is, the maximum height 111 feet occurs after 1 second.

EXAMPLE 6

Modeling a Water Jet Fountain

The water jet fountain in the midfield terminal of the Detroit Metro Airport periodically shoots water streams from different sources. The height h (in feet) of the water x feet from a source is given by the model

$$h = -0.423x^2 + 2.92x.$$

(a) Find the horizontal span of one arc of water measured from the source to its landing point.

(b) Find the value of x for which the height h is a maximum.

(c) Find the maximum height of the water stream.

Figure 9

Solution

(a) The total length of one arc is the x intercept of the graph of the function h (Figure 9), so that

$$-0.423x^2 + 2.92x = 0$$
$$x(-0.423x + 2.92) = 0$$
$$x = 0 \qquad -0.423x + 2.92 = 0$$
$$x = 6.9$$

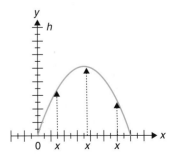

Therefore, the total length of an arc is approximately 6.9 feet.

(b) The value of x for which the height h is a maximum is given by the x coordinate of the vertex, $x = -\dfrac{b}{2a}$ where $a = -0.423$ and $b = 2.92$. We have

$$x = -\frac{b}{2a} = -\frac{2.92}{2(-0.423)} = 3.45.$$

The x value for the maximum height is approximately 3.45 feet.

(c) The h coordinate of the vertex is the maximum height of the water stream. That is,

$$h(3.45) = -0.423(3.45)^2 + 2.92(3.45) = 5.04.$$

The maximum height of the water stream is approximately 5.04 feet.

EXAMPLE 7   **Designing a Gazebo**

An architect is designing a rectangular gazebo edged with 144 feet of fancy railing.

(a) Create a model that expresses the area A in terms of the length x (in feet).

(b) Use the graph in Figure 10 to approximate the maximum value of A and the value of x at which A is a maximum.

(c) Find the exact dimensions of the gazebo that will yield the largest area.

(d) Complete the following table and plot the ordered pairs (length, area) = (x, A). Determine whether these points lie on the graph. Use the data to estimate the dimensions of the gazebo.

Length, x	42	40	38	36	32	28	24
Area, A							

Solution   We use the formula for area A of a rectangle

$$A = xy$$

where x is the length and y is the width, both in feet.

(a) We eliminate y, expressing the area A strictly as a function of the length x, by using the fact that there is 144 feet of fencing material.

$2x + 2y = 144$	Perimeter must be 144
$x + y = 72$	Divide by 2
$y = 72 - x$	Solve for y

Figure 10

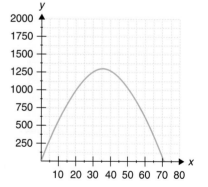

Substituting $72 - x$ for y in the area formula, we obtain,

$$A = xy = x(72 - x)$$
$$A(x) = 72x - x^2.$$

(b) Figure 10 shows the graph of this function. From the graph, we see that the coordinate of the maximum point is approximately (35,1290).

(c) By completing the square, we write $A(x)$ as

$$A(x) = -x^2 + 72x$$
$$= -(x^2 - 72x + 36^2) + 36^2$$
$$= -(x - 36)^2 + 1296.$$

so that the coordinate of the vertex is exactly (36, 1296) and the x coordinate of the vertex is 36. Solving for y we have,

$$y = 72 - 36 = 36.$$

Therefore, the dimensions of the gazebo with largest area are 36 feet by 36 feet.

(d) In order to better understand the problem, we look at some dimensions. Using the following table, several possible values for x and y are chosen so that a rectangular base can be enclosed with 144 feet of railing.

Length, x	42	40	38	36	32	28	24
Width, y	30	32	34	36	40	44	48
Area, A	1260	1280	1292	1296	1280	1232	1152

By plotting the data points (x,A) we observe that all these points fit on the graph in Figure 10. The table also shows that 1296 is greater than any other value for A. Therefore, the solution is confirmed.

EXAMPLE 8   **Analyzing a Cost Function**

Suppose that the manufacturing cost C (in dollars) per radio and the number x of radios manufactured per hour, are related by the data points in the following table.

Number of Radios Manufactured, x	Cost per Radio, C (in dollars)
3	23
4	18
6	14
7	15
9	23

(a) Construct a scatter gram that exhibits the data.

(b) Assuming that the relationship is approximately quadratic, find a quadratic function $C(x) = ax^2 + bx + c$ that fits these data.

(c) Connect the points in the scatter gram and find the vertex of the parabola algebraically.

(d) Assuming the function in part (b) provides an accurate relationship between C and x for this data, find the value of x that predicts the minimum cost C. What is the minimum cost for this value of x?

Solution

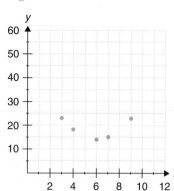

(a) We plot the data points (x,C) as $(3,23)$, $(4,18)$, $(6,14)$, $(7,15)$, and $(9,23)$ and observe that they do not fit a linear function (Figure 11a).

(b) Assuming that a quadratic function of the form

$$C(x) = ax^2 + bx + c$$

fits these data points, we substitute any three of the given values of x and C into the above equation. For instance, for the points $(3,23)$, $(6,14)$ and $(9,23)$, we have

$$23 = a(3)^2 + b(3) + c$$
$$14 = a(6)^2 + b(6) + c$$
$$23 = a(9)^2 + b(9) + c$$

By simplifying, we see that we need to solve the system

$$\begin{cases} 23 = 9a + 3b + c \\ 14 = 36a + 6b + c. \\ 23 = 81a + 9b + c \end{cases}$$

Solving this system for a, b, and c, we have

$$a = 1, b = -12, \text{ and } c = 50.$$

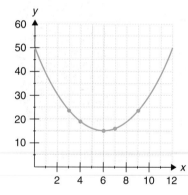

Thus the function $C(x) = x^2 - 12x + 50$ fits the given data.

(c) Figure 11b shows the graph of C. The vertex is $(6,14)$ because

$$x = \frac{b}{2a} = -\frac{-12}{2(1)} = 6 \text{ and } C(6) = (6)^2 - 12(6) + 50 = 14.$$

(d) By plotting all the given points, we confirm that these points fit on the graph. It appears from the graph that the minimum number of radios per hour that should be manufactured is 6 at a minimum cost of $14 each.

◆ PROBLEM SET 7.6

Mastering the Concepts

In Problems 1–4, sketch the graphs of each set of functions on the same coordinate system.

1. (a) $y = x^2$
   (b) $y = 2x^2$
   (c) $y = x^2/2$

2. (a) $y = x^2$
   (b) $y = 4x^2$
   (c) $y = x^2/4$

3. (a) $y = -x^2$
   (b) $y = -3x^2$
   (c) $y = -x^2/3$

4. (a) $y = -x^2$
   (b) $y = -4x^2$
   (c) $y = -x^2/4$

In Problems 5–8, sketch the graphs of each set of functions on the same coordinate system.

5. (a) $y = x^2$
   (b) $y = x^2 + 2$
   (c) $y = x^2 - 3$

6. (a) $y = -x^2$
   (b) $y = -x^2 + 1$
   (c) $y = -x^2 + 4$

7. (a) $y = -x^2$
   (b) $y = -(x + 2)^2$
   (c) $y = -(x - 1)^2$

8. (a) $y = x^2$
   (b) $y = (x + 3)^2$
   (c) $y = (x - 2)^2$

In Problems 9–20, sketch the graph of each quadratic function. Identify the vertex, the axis of symmetry, and the domain and range of f.

9. $f(x) = (x - 2)^2 + 1$
10. $f(x) = (x + 2)^2 - 3$
11. $f(x) = -(x - 4)^2 + 2$
12. $f(x) = (x + 4)^2 - 1$
13. $f(x) = 1 - (x + 2)^2$
14. $f(x) = -1 - (x - 2)^2$
15. $f(x) = 2(x + 3)^2 + 2$
16. $f(x) = -2(x + 3)^2 - 3$
17. $f(x) = -3(x - 1)^2 + 1$
18. $f(x) = 3(x - 1)^2 + 5$

19. $f(x) = -\dfrac{1}{2}(x + 3)^2 + 1$

20. $f(x) = -\dfrac{1}{2}(x - 3)^2 + 1$

In Problems 21–28, by completing the square, rewrite $f(x)$ in the form $f(x) = a(x - h)^2 + k$. Sketch the graph and find the vertex, the axis of symmetry, and the range of f.

21. (a) $f(x) = x^2 - 2x$
    (b) $f(x) = -x^2 + 2x$
22. (a) $f(x) = x^2 + 4x$
    (b) $f(x) = -x^2 - 4x$
23. $f(x) = x^2 - 6x + 10$
24. $f(x) = x^2 + 6x + 10$
25. $f(x) = -2x^2 - 4x - 2$
26. $f(x) = 2x^2 + 4x + 2$
27. $f(x) = 2x^2 + 6x + 9$
28. $f(x) = 2x^2 - 6x - 9$

In Problems 29–34, find the x and y intercepts, the vertex, and the axis of symmetry. Sketch the graph of f and find the range.

29. (a) $f(x) = 4 - x^2$
    (b) $f(x) = x^2 - 4$
30. (a) $f(x) = 1 - x^2$
    (b) $f(x) = x^2 - 1$
31. $f(x) = x^2 - 10x + 25$
32. $f(x) = 10x - x^2 - 25$
33. $f(x) = x^2 - 2x - 3$
34. $f(x) = -x^2 + 2x + 3$

In Problems 35–40, without graphing, find the vertex and indicate whether this function has a maximum value or a minimum value.

35. $f(x) = 3(x - 7)^2 + 5$
36. $f(x) = 5(x + 2)^2 + 3$
37. $f(x) = -2(x + 6)^2 + 1$
38. $f(x) = -4(x - 1/2)^2 + 2$

39. $f(x) = \dfrac{2}{3}(x - 1)^2 + 2$

40. $f(x) = -\dfrac{2}{5}(x + 5)^2 + 1$

Applying the Concepts

41. Height of a Rocket: A model rocket is launched and then it accelerates until the propellant burns out, after which it coasts upward to its highest point. The height d (in feet) of the rocket above the ground t seconds after the burn out, is modeled by the function

$$d(t) = -16t^2 + 192t, \ t \geq 0.$$

(a) Calculate and explain the meaning of each output value, $d(5)$, $d(10)$ and $d(15)$.

(b) Write $d(t)$ in the form

$$d(t) = a(t - h)^2 + k.$$

(c) Find the vertex of $d(t)$, then use the y coordinate of the vertex to determine the maximum height attained by the rocket.

(d) How many seconds after the burn out does it take the rocket to reach the maximum height?

(e) Find the time for the rocket to hit the ground.

(f) Sketch the graph of d that describes the situation.

42. Minimum Cost: The cost C (in dollars) to produce x calculators is modeled by the function

$$C(x) = 2x^2 - 800x + 92{,}000,$$
$$0 \leq x \leq 210.$$

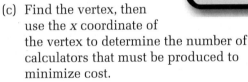

(a) Calculate and explain the meaning of each output value $C(0)$, $C(10)$, $C(100)$, $C(200)$.

(b) Write $C(x)$ in the form $C(x) = a(x - h)^2 + k$.

(c) Find the vertex, then use the x coordinate of the vertex to determine the number of calculators that must be produced to minimize cost.

(d) Determine the minimum cost.

(e) Sketch the graph of the function that describes the situation.

43. Maximum Revenue: A vending machine company that operates juice machines determines that the daily revenue R (in dollars) from selling juices at a price p (in dollars) per bottle is modeled by the following function.

$$R(p) = 270p - 150p^2, \ 0 \leq p \leq \$1.10$$

(a) Calculate the following values and explain the meaning of each. $R(0)$, $R(0.45)$, $R(0.75)$, $R(0.85)$.

(b) What is the price of a bottle of juice if the daily revenue is $120?

(c) Write $R(p)$ in the form $R(p) = a(x - h)^2 + k$.

(d) Find the vertex, then use the x coordinate of the vertex to determine the price per bottle to attain the maximum revenue.

44. **Volcanic Eruption:** In the 1968 volcanic eruption in Costa Rica, large blocks, thrown out by volcanic explosions, showered the surrounding area. The trajectories of their flight indicate that the blocks were propelled upward like mortar shells. The height d (in meters) of the blocks is approximated by the model

$$d(t) = -7.8t^2 + 358t$$

where t is in seconds (Source: William Melson, Smithsonian Institution).

(a) Calculate and explain the meaning of each output value: $d(10)$, $d(20)$, $d(30)$, and $d(40)$.

(b) Write $d(t)$ in the form $d(t) = a(t - h)^2 + k$.

(c) Find the vertex of $d(t)$, then use the second coordinate of the vertex to determine the maximum height attained by the blocks.

(d) How many seconds does it take a block to reach the maximum height?

(e) Find the time for a block to land on the ground.

(f) Sketch the graph of d that describes the situation.

45. **Maximum Area:** A homeowner wishes to build a rectangular observation deck whose perimeter is 40 meters.

(a) Let x (in meters) be the length of one side of the observation deck. Explain why the formula

$$A(x) = (20 - x)x$$

describes the area of the deck.

(b) What are the restrictions on the model of part (a)?

(c) Sketch the graph of this function and indicate the vertex.

(d) Use the information from part (c) to find the dimensions of the largest observation deck.

(e) What is the maximum area of the deck?

46. **Maximum Area:** A rancher has 40 meters of fencing to enclose a rectangular pen next to a barn, with the barn wall forming one side of the fence.

(a) Let x (in meters) be the length of the side of the pen that lies along the barn. Explain why the formula

$$A(x) = x(20 - x/2)$$

describes the area of the pen.

(b) What is the domain of A? Why?

(c) Sketch the graph of this function and indicate the vertex.

(d) Use the information from part (c) to find the dimensions of the pen that produces a maximum enclosed area.

(e) What is the maximum area of the pen?

47. **Maximum Area:** A farmer has 600 feet of fencing to enclose two rectangular holding pens (Figure 12).

Figure 12

(a) Let x (in feet) be the width of each pen. Explain why the formula

$$A(x) = x(300 - 1.5x)$$

describes the total area A (in square feet) for the pens.

(b) What is the domain of A? Why?

(c) Sketch the graph of A and indicate the vertex.

(d) What is the maximum total area for the pens?

(e) What are the dimensions of the largest total area for the pens?

48. **Maximum Area:** A total of 800 meters of fence are to be used to fence a rectangular plot of land divided into three equal portions (Figure 13)

Figure 13

(a) Let x (in meters) be the width of each of each plot of land. Explain why the formula

$$A(x) = x(400 - 2x)$$

describes the entire rectangular area.

(b) What is the domain of A?

(c) Sketch the graph of A and indicate the vertex.

(d) What are the dimensions of the largest area?

(e) What is the maximum area?

In Problems 49 and 50, write an equation of the given parabola that has the shape of $f(x) = a(x - h)^2 + k$.

49. (a)

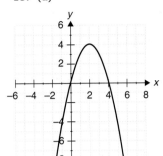

(b)

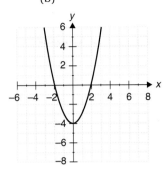

50. (a)

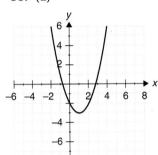

(b)

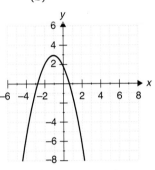

51. Find an equation of a quadratic function in the form

$$F(x) = a(x - h)^2 + k$$

if the graph of F has the same shape as the graph of

$$f(x) = 2(x + 1)^2 + 5$$

and $F(x)$ has a minimum value at the same point as

$$g(x) = -2(x - 3)^2 + 2$$

has a maximum value.

52. Graph the following functions on the same coordinate system.

$$y_1 = x^2 + x, \; y_2 = x^2 + 4x, \text{ and } y_3 = x^2 - 4x.$$

Find the vertex of each parabola.

53. Data Analysis: Find a quadratic function that best fits the data (0,20), (1,84), and (5,20). Sketch the graph.

54. Maximum Volume: Consider a sandbox being built by cutting corners from a 12 foot by 12 foot sheet of metal (Figure 14).

(a) Complete the following table.

Figure 14

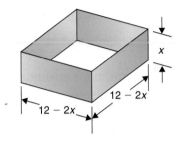

Height x	1	2	3	4	5	6
Length 12 − 2x						
Width 12 − 2x						
Volume x(12 − 2x)²						

(b) Use the table to find x that maximizes the volume.

(c) Graph V and find x that maximizes the volume.

55. (a) Find two numbers whose sum is 20 and whose product is a maximum.

(b) Find two number that differ by 6 and whose product is a minimum.

Objectives

1. Find the Distance Between Two Points
2. Graph Circles
3. Solve Applied Problems

7.7 Distance Formula and Circles

In this section, we find the distance between two points in the Cartesian coordinate system. Then we use this formula to derive an equation of a circle.

Finding the Distance Between Two Points

One useful feature of the Cartesian coordinate system is that there is a formula that gives the distance between two points in the plane. For instance, to find the distance between the points $P_1 = (3,2)$ and $P_2 = (11,8)$, we draw a right triangle as shown in Figure 1. Notice that the coordinates of P_3 are (11,2). The distance from

Figure 1

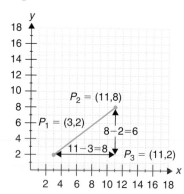

P_1 to P_3 is $11 - 3 = 8$ units, and the distance from P_3 to P_2 is $8 - 2 = 6$ units. Using the Pythagorean Theorem, we have

$$d^2 = 8^2 + 6^2 = 64 + 36 = 100.$$

So that $d = \sqrt{100} = 10$ units.

This result can be generalized in the following formula.

The Distance Formula

Let $P_1 = (x_1, y_1)$ and $P_2 = (x_2, y_2)$ be two points in the plane. The distance d between P_1 and P_2 is given by the formula

$$d = \sqrt{(x_2 - x_1)^2 + (y_2 - y_1)^2}.$$

The distance formula holds regardless of the positions of the points P_1 and P_2 in the plane. This is true, because

$$(x_2 - x_1)^2 = (x_1 - x_2)^2 \text{ and } (y_2 - y_1)^2 = (y_1 - y_2)^2$$

Figure 2

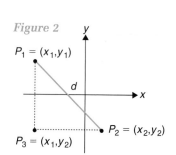

and the distance formula can also be written as

$$d = \sqrt{(x_1 - x_2)^2 + (y_1 - y_2)^2}.$$

To derive this formula, consider the case in which the points $P_1 = (x_1, y_1)$ and $P_2 = (x_2, y_2)$ do not lie on the same horizontal or vertical lines (Figure 2). To determine the length of $\overline{P_1 P_2}$, denoted by

$$d = |\overline{P_1 P_2}|$$

we draw a right triangle with $\overline{P_1 P_2}$ as the hypotenuse. The two sides intersect at P_3 whose coordinates are given by $P_3 = (x_1, y_2)$ (Figure 2). Then

$$|\overline{P_1 P_3}| = |y_2 - y_1| \text{ and } |\overline{P_3 P_2}| = |x_2 - x_1|.$$

Using the Pythagorean Theorem on the right triangle $P_1 P_2 P_3$, we have

$$d^2 = |\overline{P_1 P_2}|^2 = |\overline{P_3 P_2}|^2 + |\overline{P_1 P_3}|^2$$
$$= |x_2 - x_1|^2 + |y_2 - y_1|^2$$

but $|x_2 - x_1|^2 = |x_1 - x_2|^2$ and $|y_2 - y_1|^2 = |y_1 - y_2|^2$.

Thus,

$$d^2 = (x_2 - x_1)^2 + (y_2 - y_1)^2$$
$$d = \sqrt{(x_2 - x_1)^2 + (y_2 - y_1)^2}.$$

EXAMPLE 1   **Finding the Distance Between Two Points**

Find the distance between each pair of points.

(a) $P_1 = (2, -4)$ and $P_2 = (-2, -1)$

(b) $P_1 = (-1, -2)$ and $P_2 = (3, 4)$

Solution

(a) The distance d is given by

$$d = \sqrt{(x_2 - x_1)^2 + (y_2 - y_1)^2}$$
$$= \sqrt{(-2 - 2)^2 + (-1 - (-4))^2}$$
$$= \sqrt{(-4)^2 + 3^2} = \sqrt{25} = 5 \quad .$$

(b) $d = \sqrt{(x_2 - x_1)^2 + (y_2 - y_1)^2}$
$$= \sqrt{(3 - (-1))^2 + (4 - (-2))^2}$$
$$= \sqrt{4^2 + 6^2} = \sqrt{52} = 2\sqrt{13} \approx 7.21$$

EXAMPLE 2

Using the Distance Formula

Find the value of x, so that the distance between the points $P_1 = (1, 4)$ and $P_2 = (x, -1)$ is $\sqrt{29}$.

Solution

Let $\sqrt{29} = |\overline{P_1 P_2}|$, so that

$$\sqrt{29} = \sqrt{(x - 1)^2 + (-1 - 4)^2}$$

$$29 = (x - 1)^2 + (-5)^2 \qquad \text{Square both sides}$$

$$29 = x^2 - 2x + 1 + 25 \qquad \text{Square the binomial}$$

$$x^2 - 2x - 3 = 0 \qquad \text{Simplify}$$

$$(x - 3)(x + 1) = 0 \qquad \text{Factor}$$

$$x - 3 = 0 \quad \text{or} \quad x + 1 = 0$$

$$x = 3 \qquad\qquad x = -1.$$

There are two points $(3, -1)$ and $(-1, -1)$ which are $\sqrt{29}$ away from $(1, 4)$ (Figure 3).

Figure 3

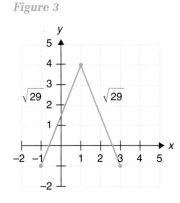

Graphing Circles

It is possible to recognize and graph equations of circles. We know from geometry that a circle consists of all points that are a fixed distance r, (its **radius**) from a fixed point C, (its **center**) (Figure 4a). Suppose that a circle is located in a Cartesian plane so that its center C is (h, k) and its radius is r. If a point $P = (x, y)$ is on the circle, its distance from $C = (h, k)$ has to be r units (Figure 4b). By the distance formula, x and y satisfy the equation

$$\sqrt{(x - h)^2 + (y - k)^2} = r \quad \text{or} \quad (x - h)^2 + (y - k)^2 = r^2.$$

Figure 4

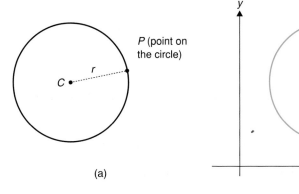

(a)

(b)

Conversely, any ordered pair (x,y) that satisfies this last equation defines a point $P = (x,y)$ that lies on the circle with center $C = (h,k)$ and radius r. Thus, we have the following general result:

Standard Equation of a Circle

Any point (x,y) on the circle with center (h,k) and radius r must satisfy the equation

$$(x - h)^2 + (y - k)^2 = r^2.$$

This equation is called the **standard equation of the circle.**

EXAMPLE 3   **Graphing a Circle**

Graph each circle. Indicate the center and radius.

(a) $x^2 + y^2 = 9$
(b) $(x - 1)^2 + (y - 2)^2 = 16$

Solution   (a) The equation $x^2 + y^2 = 9$ is written in standard form as

$$(x - 0)^2 + (y - 0)^2 = 9.$$

Its center is at $(0,0)$ and its radius is 3 (Figure 5a).

(b) The equation

$$(x - 1)^2 + (y - 2)^2 = 16$$

is in standard form with the center at $(1,2)$ and radius 4 (Figure 5b).

Figure 5

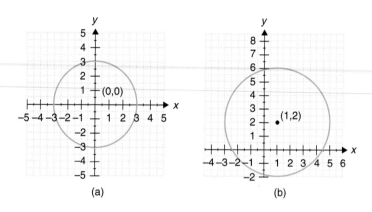

(a)   (b)

EXAMPLE 4   **Finding the Standard Equation of a Circle**

Find the standard equation for the circle of radius 4 with center at the point $(4,-1)$.

Solution   Substituting 4 for r and $(4,-1)$ for (h,k) in the equation

$$(x - h)^2 + (y - k)^2 = r^2$$

we get $(x - 4)^2 + (y - (-1))^2 = 4^2.$

The standard equation is:

$$(x - 4)^2 + (y + 1)^2 = 16.$$

EXAMPLE 5   **Finding the Equation of a Circle**

Find an equation of a circle with center at $(-1,3)$ and that contains the point $(4,11)$.

Solution   Substituting $(-1,3)$ for (h,k) and $(4,1)$ for (x,y) in the equation

$$(x - h)^2 + (y - k)^2 = r^2$$

$$(4 - (-1))^2 + (1 - 3)^2 = r^2$$

$$5^2 + (-2)^2 = r^2$$

$$25 + 4 = r^2$$

$$29 = r^2.$$

So that the standard equation for a circle with center at $(-1,3)$ and radius $\sqrt{29}$ is

$$(x + 1)^2 + (y - 3)^2 = (\sqrt{29})^2 \text{ or } (x + 1)^2 + (y - 3)^2 = 29.$$

At times, an alternative algebraic form of an equation of a circle is given as the next example shows.

EXAMPLE 6   **Finding the Center and Radius of a Circle**

Show that the graph of the equation

$$x^2 + y^2 + 6x - 2y - 15 = 0$$

is a circle by converting it to its standard equation. Determine the center and the radius, and sketch the circle.

Solution   We will create a perfect square trinomial in both x and y by isolating variable terms on one side, grouping the x terms and the y terms as follows.

$$(x^2 + 6x) + (y^2 - 2y) = 15$$

$$(x^2 + 6x + 9) + (y^2 - 2y + 1) = 15 + 9 + 1 \qquad \text{Complete the square for } x \text{ and } y$$

$$(x + 3)^2 + (y - 1)^2 = 25 \qquad \text{Factor and simplify}$$

We recognize the form of this last equation. Its graph is a circle of radius $r = 5$ and center $(h,k) = (-3,1)$ (Figure 6).

Figure 6

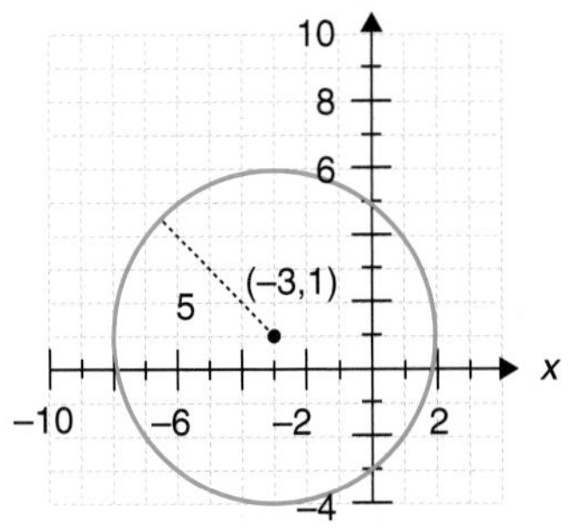

Solving Applied Problems

In the next example, we show the use of a circle in architecture.

EXAMPLE 7   **Solving an Architecture Problem**

The Norman style architecture of a window is characterized by massive construction, carved decorations, and rectangles surmounted by semicircular arches (Figure 7). Suppose that the rectangular part is 4 feet wide and 4.5 feet in length.

(a) Find an equation of the semicircle using the coordinate system with the origin at the midpoint of the width of the rectangle.

(b) What is the height of the arch 1.5 feet to the right of the origin of the coordinate system?

Solution   (a) Since the center of the circle is at the origin of the coordinate system, $(h,k) = (0,0)$ and the radius is $4/2 = 2$. The standard equation of this circle is:

$$x^2 + y^2 = 4.$$

Figure 7

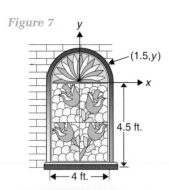

(b) The height of the arch 1.5 feet to the right of the origin is the value of the y coordinate of the point on the circle with an x coordinate of 1.5. This y coordinate is the solution of the equation

$$(1.5)^2 + y^2 = 4 \text{ so we have}$$
$$y^2 = 4 - 2.25$$
$$y = \sqrt{1.75} \approx 1.3.$$

The clearance is approximately 1.3 feet.

◆ PROBLEM SET 7.7

In Problems 1–6, find the distance between the two points with the given coordinates. Round off to two decimal places when necessary.

1. (a) $(7,10)$ and $(1,2)$
   (b) $(-3,-4)$ and $(-5,-7)$
2. (a) $(-2,5)$ and $(3,-1)$
   (b) $(4,-3)$ and $(6,2)$
3. (a) $(-5,0)$ and $(-2,-4)$
   (b) $(-4,7)$ and $(0,-8)$
4. (a) $(6,2)$ and $(6,-2)$
   (b) $(-4,-3)$ and $(0,0)$
5. (a) $(2.3,4.3)$ and $(4.2,-5.7)$
   (b) $(3.5,-2.8)$ and $(-7.3,-9.8)$
6. (a) $(3.5,1.7)$ and $(-2.5,9.1)$
   (b) $(-8.4,5.2)$ and $(3.7,4.8)$

In Problems 7–18, graph each circle. Indicate the center and radius.

7. (a) $x^2 + y^2 = 1$
   (b) $x^2 + y^2 = 9$
8. (a) $x^2 + y^2 = 25$
   (b) $x^2 + y^2 = 16$
9. $(x - 1)^2 + (y + 2)^2 = 4$
10. $(x - 4)^2 + (y + 1)^2 = 4$
11. $(x + 1)^2 + (y - 2)^2 = 4$
12. $(x + 4)^2 + (y - 2)^2 = 16$
13. $(x - 2)^2 + (y + 4)^2 = 16$
14. $(x + 2)^2 + (y - 4)^2 = 16$
15. $(x - 3)^2 + (y + 4)^2 = 25$
16. $(x + 3)^2 + (y - 1)^2 = 25$
17. $(x + 1/2)^2 + (y - 1/3)^2 = 1/4$
18. $(x - 1/4)^2 + (y + 1/2)^2 = 1/4$

In Problems 19–22, find an equation of a circle satisfying the given conditions.

19. (a) Center $(0,0)$, radius 4
    (b) Center $(0,0)$, radius 3

20. (a) Center $(3,4)$, radius 7
    (b) Center $(-5,2)$, radius 7
21. (a) Center $(-4,3)$, radius $4\sqrt{3}$
    (b) Center $(-5,-2)$, radius $2\sqrt{5}$
22. (a) Center $(-3,-1)$, radius $3\sqrt{2}$
    (b) Center $(1,-4)$, radius $2\sqrt{3}$

In Problems 23–28, write each equation in standard form by completing the square. Find the center and radius.

23. $x^2 + y^2 - 4x - 6y + 4 = 0$
24. $x^2 + y^2 + 8x - 6y - 15 = 0$
25. $x^2 + y^2 - 6x + 4y + 4 = 0$
26. $x^2 + y^2 - 8x - 10y + 5 = 0$
27. $x^2 + y^2 - 6x + 8y + 9 = 0$
28. $x^2 + y^2 + 4x + 4y - 8 = 0$

Applying the Concepts

29. **Geometry:** Apply the distance formula and the Converse of the Pythagorean Theorem to determine whether or not the triangle with vertices $A = (-3,1)$, $B = (3,10)$, and $C = (3,1)$ is a right triangle.

30. **Geometry:** Use the distance formula to determine whether or not the triangle with vertices $A = (5,-1)$, $B = (-6,5)$, and $C = (2,6)$ is an isosceles triangle. (Two of the three sides are equal)

31. **Gear Design:** Figure 8 shows two gears meshed together in such a way that the large gear is represented by the circle with equation $x^2 + y^2 = 64$. The smaller gear is tangent to the large one and is centered at the point $(-5,12)$. Find an equation of the small gear. Assume the point of tangency is on the line through the centers of the gears.

Figure 8

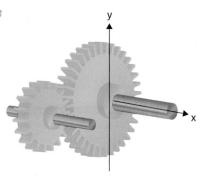

32. **Watering Trough Construction:** A cylindrical barrel is cut to make a watering trough in such a way that the trough is 8 inches high and 30 inches wide (Figure 9). What is the radius of the original barrel to one decimal place?

Figure 9

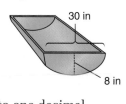

30 in

8 in

33. **Tunnel Design:** An arch in the shape of a semicircle with a diameter measuring 50 meters is to support a bridge over a river (Figure 10). The center of the arch is to be 25 meters above the surface of the river.

Figure 10

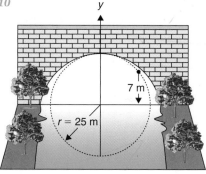

7 m

$r = 25$ m

(a) Find an equation of this semicircular arch.

(b) Find the point(s) on the arch if the water level rises 7 meters from the original surface of the river.

34. Find an equation of a circle so that the points $A = (-2, -3)$ and $B = (6, -3)$ are the ends of its diameter.

35. Find an equation of a circle whose center is at $(3, 6)$ and tangent to the x axis.

36. Without graphing, what can you say about the graph of

$$y = 3 - \sqrt{9 - (x - 2)^2}?$$

37. Find an equation of a circle whose center is at $(-2, 5)$ and whose circumference is 8π units.

38. Graph each circle:

(a) $(x + 2)^2 + y^2 = 9$ (b) $x^2 + (y - 3)^2 = 16$

39. Find the equation for the given graphs.

(a)

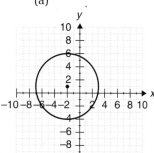

(b)

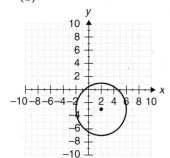

◈ CHAPTER 7 REVIEW PROBLEM SET

In Problems 1 and 2, solve each equation by factoring.

1. (a) $x^2 - 14x - 15 = 0$
   (b) $6x^2 - 15x = 0$ (c) $5x^2 = 2x + 24$

2. (a) $y^2 - 5y - 84 = 0$
   (b) $5t^2 - 10t = 0$ (c) $4y(y + 1) = 3$

In Problems 3 and 4, obtain complex solutions of each equation. Check the solutions.

3. (a) $(x - 1)^2 + 9 = 0$
   (b) $(3m - 1)^2 + 25 = 0$

4. (a) $\left(x - \dfrac{3}{5}\right)^2 + \dfrac{16}{25} = 0$
   (b) $(2t + 3)^2 + 49 = 0$

In Problems 5 and 6, what term should be added to the given expression so it becomes a perfect square trinomial?

5. (a) $u^2 - 8u$
   (b) $x^2 + 11x$

6. (a) $x^2 - 14x$
   (b) $y^2 + 17y$

In Problems 7 and 8, solve each equation by completing the square. Give both the exact solutions and the two decimal approximations.

7. (a) $x^2 - 6x + 7 = 0$
   (b) $4x^2 + 13 = 12x$
8. (a) $y^2 + 4y + 21 = 0$
   (b) $2t^2 + 3 = 8t$

In Problems 9 and 10, solve each equation by the quadratic formula. Approximate irrational solutions to two decimal places. Write complex solutions in the form $a + bi$.

9. (a) $x^2 + 10x - 13 = 0$
   (b) $(2x - 1)(x + 2) = 5$
10. (a) $3 - x - x^2 = 0$
    (b) $4x^2 - x - 1 = 0$

In Problems 11 and 12, write a quadratic equation with integral values for a, b, and c having the given solutions.

11. (a) $-4, -3$
    (b) $5 \pm \sqrt{7}$
    (c) $3 \pm 4i$
12. (a) $-5, -5$
    (b) $\pm\sqrt{11}$
    (c) $3 \pm \sqrt{5}i$

In Problems 13–18, solve each equation and check the results.

13. $\dfrac{3 - 2x}{4x} - 4 = \dfrac{3}{4x - 3}$

14. $\dfrac{2x - 5}{2x + 1} = \dfrac{7}{4} - \dfrac{6}{2x - 3}$

15. $\sqrt{5x + 5} + \sqrt{x - 4} = 5$
16. $\sqrt{5x + 1} = 1 + \sqrt{3x}$
17. $\sqrt{y - 3} + \sqrt{y} = 3$
18. $\sqrt{2x + 1} + \sqrt{8 - x} = 5$

In problems 19–30, solve each equation in quadratic form by making a suitable substitution. Check for extraneous solutions.

19. (a) $x^4 - 5x^2 + 4 = 0$
    (b) $y^{-2} - 5y^{-1} + 4 = 0$
20. (a) $x^4 - 8x^2 + 16 = 0$
    (b) $y^{-2} - 8y^{-1} + 16 = 0$
21. (a) $1 - 2(x + 3)^2 - 3(x + 3)^4 = 0$
    (b) $1 - 2t^{-2} - 3t^{-4} = 0$
22. (a) $16(x + 1)^4 - 17(x + 1)^2 + 1 = 0$
    (b) $16t^{-4} - 17t^{-2} + 1 = 0$
23. $(x^2 + x)^2 - 8(x^2 + x) + 12 = 0$
24. $(y^2 + 4y)^2 - 17(y^2 + 4y) = 60$
25. $t^2 + 2t + 3(t^2 + 2t)^{-1} = 4$

26. $2t^2 - t - 6(2t^2 - t)^{-1} + 1 = 0$
27. $t^{2/3} - 5t^{1/3} + 6 = 0$
28. $t^{2/3} - 2t^{1/3} = 8$
29. $3x + x^{1/2} - 4 = 0$
30. $x^{2/5} + 8x^{1/5} + 16 = 0$

In Problems 31–46, find all x intercepts of the graph of the function. Round off irrational solutions to 2 decimal places.

31. (a) $f(x) = 3x^2 - 2 + x$
    (b) $f(x) = x^2 - 8x - 9$
32. (a) $f(x) = 10x^2 + 19x - 15$
    (b) $f(x) = -8x^2 + 26x - 15$
33. (a) $f(x) = 16x^3 - 34x^2 - 15x$
    (b) $f(x) = 3x^3 + 2x^2 - 27x - 18$
34. (a) $f(x) = 6x^3 - 43x^2 + 20x$
    (b) $f(x) = 8x^3 - 34x^2 + 30x$
35. $f(x) = 2x^2 + 18x + 4$
36. $f(x) = -x^2 + 5x + 8$
37. $f(x) = 0.1x^2 + 0.6x - 1.2$
38. $f(x) = 1.74x^2 - 2.04x + 6.2$
39. $f(x) = 2x^{-2} + x^{-1} - 1$
40. $f(x) = 6x^{-2} - x^{-1} - 12$
41. $f(x) = x - 13\sqrt{x} + 40$
42. $f(x) = x^{2/3} + 11x^{1/3} + 28$
43. $f(x) = x^4 - 7x^2 + 12$
44. $f(x) = -3x^4 - 14x^2 + 5$

45. $f(x) = \dfrac{1}{x - 2} - 6\sqrt{\dfrac{1}{x - 2}} - 40$

46. $f(x) = \left(\dfrac{x - 1}{x}\right)^2 - 4\left(\dfrac{x - 1}{x}\right) - 12$

In Problems 47–50, solve each inequality. Write the solution set in interval notation and illustrate the solution on a number line.

47. (a) $x^2 + 2x - 15 < 0$
    (b) $(x - 1)(x + 1)(x - 2) > 0$
48. (a) $4x^2 - 3x \geq 10 + 3x^2$
    (b) $(x + 1)(x - 2)(x - 4) > 0$
49. (a) $\dfrac{3x - 1}{5x - 7} \leq 0$
    (b) $\dfrac{(x + 1)(x - 2)}{x - 5} \leq 0$
50. (a) $\dfrac{6x - 3}{8x + 4} \geq 0$
    (b) $\dfrac{x + 2}{(x - 2)(x + 7)} > 0$

In Problems 51 and 52,

(i)   write $f(x)$ in the form $f(x) = a(x - h)^2 + k$.
(ii) Find the vertex.
(iii) Find the intercepts.
(iv) Find the range.
(v)   Find the maximum or minimum value.
(vi) Sketch the graph of f.

51. (a) $f(x) = -3x^2 - 12x - 8$
   (b) $f(x) = x^2 + 2x + 5$
   (c) $f(x) = -x^2 + 4x - 8$
52. (a) $f(x) = -2x^2 - 20x + 47$
   (b) $f(x) = 2x^2 + 16x + 33$
   (c) $f(x) = -2x^2 - 4x + 1$

In Problems 53 and 54, write each equation in standard form. Find the center and radius.

53. (a) $x^2 + y^2 + 8x - 6y + 21 = 0$
   (b) $x^2 + y^2 - 6x - 4y + 4 = 0$
54. (a) $x^2 + y^2 + 2x - 8y + 16 = 0$
   (b) $x^2 + y^2 + 6y = 0$

In Problems 55 and 56, find the distance between each pair of points. Round off each answer to two decimal places.

55. (a) $(1,2)$ and $(-2,1)$
   (b) $(-4,7)$ and $(5,-3)$
56. (a) $(2,5)$ and $(8,3)$
   (b) $(-2,3)$ and $(-4,-9)$

In Problems 57 and 58, solve each equation for the indicated variable.

57. (a) $s = vt + \dfrac{gt^2}{2}$, for t
   (b) $s = \pi r^2 + 2\pi rh$, for r
58. (a) $s = \pi r^2 + 2\pi r$, for r
   (b) $2N = k^2 - 3k$, for k
59. **Number Problem:** Find a number x such that the reciprocal of the expression $x - \sqrt{3}$ is equal to the expression $x + \sqrt{3}$.
60. **Falling Objects:** An object is thrown vertically upward from the ground with an initial speed of 60 feet per second. Its height h (in feet) t seconds after it is released is given by the model

$$h = 60t - 16t^2, t \geq 0$$

   (a) When is the object 50 feet high?
   (b) When will the object hit the ground?
61. **Camping Tent:** An entrance of a tent is in the shape of an isosceles triangle whose height is

1 foot less than twice its base. Find the height and the base of the triangular entrance if its area is 28 square feet. Round off to two decimal places.

62. **Car Travel:** A trucker travels 280 miles at a certain speed. On the return trip, the trucker travels 5 miles per hour faster and as a result the trip takes 1 hour less. How fast is the trucker traveling in each part of the trip?
63. **Law Enforcement:** A law enforcement officer estimates that the braking distance d (in feet) for a particular make of car is given by the model

$$d(v) = 0.1v^2 + v, v \geq 0$$

where v is the speed of the car in miles per hour. What is the speed of the car at the time of an accident if it leaves skid marks measuring 120 feet?
64. **Jogging:** Two joggers leave the same point and travel at right angles to each other, each moving at a uniform speed. One jogs 1 mile per hour faster than the other. After 3 hours, they are 15 miles apart. Find the speed of each jogger.
65. **Investment Portfolio:** Suppose P_1 dollars is invested at an annual interest rate r (percent). After one year, an additional P_2 is added to the account at the same rate. At the end of the second year, the balance A in the account is given by the model

$$A = P_1(1 + r)^2 + P_2(1 + r).$$

An investor invests \$1000 in a saving account for 2 years. At the beginning of the second year, an additional \$2000 is invested. If a total of \$3368 is in the account at the end of the second year, what is the annual interest rate r?
66. **Boating:** A boat can travel 18 miles per hour in still water. It can travel 32 miles upstream and 32 miles downstream in a total of 4 hours. What is the speed of the current?
67. **Advertising:** A poster is to have 3 inch margins at the top and bottom and 2 inch margins on the sides. The area of the picture portion is 48 square inches and the total area of the poster is 160 square inches. What is the possible length and width of the poster?

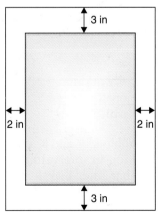

68. **Ticket Pricing:** Tickets to a concert cost \$30 for each customer. At this price, 800 people attend the concert. However, it is projected that for a $1 increase in ticket price, the average attendance will decrease by 10.

(a) At what price will the receipts be $25,840?

(b) What price will maximize receipts?

69. Launched Object: A projectile is shot straight upward from the ground with an initial speed of 64 feet per second. Its height (in feet) t seconds after it is shot is given by the model

$$h(t) = 64t - 16t^2, t \geq 0.$$

(a) Find the maximum height attained by the projectile.

(b) Find the time it takes for the projectile to return to earth.

70. Fencing a Pen: A farmer wishes to fence a rectangular pen with 200 feet of fencing. Find the dimensions of the pen with maximum area.

71. Ballistics: A ball is thrown vertically upward from the ground with an initial speed of 64 feet per second. Its height h (in feet) after t seconds is given by the model

$$h(t) = 64t - 16t^2, t \geq 0.$$

Find the time interval when the ball is at least 48 feet above the ground.

72. Number Problem: Find a number such that the sum of the number and its reciprocal is at most 5/2.

In Problems 73–76, graph each function. Use the x intercepts of the graph to solve the associated equation and inequality. Round off to two decimal places.

73. (a) $f(x) = x^2 - 0.75x - 0.25$
(b) $x^2 - 0.75x - 0.25 = 0$
(c) $x^2 - 0.75x - 0.25 > 0$

74. (a) $f(x) = 1.41x^2 + 5x + 1.41$
(b) $1.41x^2 + 5x + 1.41 = 0$
(c) $1.41x^2 + 5x + 1.41 \leq 0$

75. (a) $f(x) = 5x^2 - 13x - 8$
(b) $5x^2 - 13x - 8 = 0$
(c) $5x^2 - 13x - 8 \leq 0$

76. (a) $f(x) = 10x^2 - 13x - 3$
(b) $10x^2 - 13x - 3 = 0$
(c) $10x^2 - 13x - 3 > 0$

CHAPTER 7 PRACTICE TEST

Solve each equation. Round off answers to two decimal places.

1. (a) $x^2 - 2x - 15 = 0$
(b) $6x(x - 1) = 12$
(c) $2x^3 + 13x^2 + 21x = 0$

2. (a) $(x + 1)^2 + 4 = 0$
(b) $x^2 - 4x + 2 = 0$
(c) $3x^2 - 18x - 5 = 0$

3. (a) $3x^2 - 5x + 4 = 0$
(b) $9x^2 - 12x + 4 = 0$
(c) $3x^2 + 2x - 3 = 0$

4. $\sqrt{2x + 1} + \sqrt{x} = 1$

5. $x^{2/3} - 17x^{1/3} + 16 = 0$

6. Write a quadratic equation with integral coefficients a, b, and c having -3 and 6 as solutions.

In Problems 7 and 8, solve the inequality and write the solution in interval notation. Show the solution on a number line. Use this solution to determine the domain of function f.

7. $2x^2 + x - 1 > 0$ $f(x) = \dfrac{1}{\sqrt{2x^2 + x - 1}}$

8. $\dfrac{2x + 3}{x - 5} \geq 0$ $f(x) = \sqrt{\dfrac{2x + 3}{x - 5}}$

9. Sketch the graph of the function

$$f(x) = -2(x - 1)^2 + 3.$$

Identify the vertex, the axis of symmetry, the domain and range of f.

10. Rewrite $f(x) = 2x^2 - 8x - 3$ in the form $f(x) = a(x - h)^2 + k$. Sketch the graph and find the range of f.

11. (a) Find the radius and center of the circle

$$x^2 + y^2 + 6x - 2y - 15 = 0.$$

(b) Sketch the graph.

12. A homeowner is designing a rectangular garden in such a way that its length is 3 feet less than twice its width. What is the length and the width of the garden if its area is 104 square feet?

13. A cliff diver's path from the top of a cliff 120 feet above the surface of the water after t seconds is modeled by the function.

$$h(t) = -t^2 + 4t + 120, t \geq 0$$

where h is the height in feet above the water.

(a) Find the maximum height of the diver.

(b) Find the time it takes for the diver to reach the maximum height.

Exponential and Logarithmic Functions

Chapter Contents

You are probably familiar with bank accounts in which the interest is said to be *compounded.* In a compounded interest model, the interest earned is added to the principal each period. The essential variable in the model is the number of times a year the interest is compounded. The interest may be compounded quarterly, monthly, weekly, or daily. The mathematical model for this involves a new function called an *exponential function.* As you will see in this chapter, it is also possible for interest to be compounded continuously. A continuously compounding function can be used as a powerful modeling tool. As an illustration, in 1626, Peter Minuit of the Dutch West India company purchased Manhattan Island from the Native Americans (Algonquins) for $24. Comparing continuous and quarterly compounding at a rate of 4% per year, how much will Manhattan be worth after 374 years? The answer to this question is provided in example 6 in section 8.1.

In this chapter we introduce two closely related functions—*exponential and logarithmic functions*. These functions are widely used to develop mathematical models from real-life situations in fields where there is growth such as business, medicine, seismology, engineering, psychology, and economics.

Objectives

1. Evaluate Exponential Functions
2. Graph Exponential Functions
3. Graph Natural Exponential Functions
4. Solve Applied Problems

The restriction $b \neq 1$ is made to exclude the constant function $f(x) = 1^x = 1$.

8.1 Exponential Functions

When a *disease epidemic* like AIDS starts, the number of people with the disease is often expressed as a function of time from the start of the epidemic.

When a population grows over time, the *population increase* can be expressed as a function of time from the start of the population. When a person deposits money in an interest-bearing account, the *bank balance* increases as a function of time starting with the initial deposit. Each of these situations can be modeled by an *exponential function*.

Evaluating Exponential Functions

In section 6.1, we introduced the idea of exponential functions, and we used the graphs of the functions to solve simple exponential equations. First, we recall the definition of exponential functions:

A function of the form

$$f(x) = b^x$$

where b is a positive constant, $b \neq 1$, is called an **exponential function** with base b.

Additional examples of exponential functions are:

$$f(x) = 3^x, \, g(x) = \left(\frac{1}{2}\right)^x, \text{ and } h(x) = 500(1.07)^{-2x}.$$

Note that in exponential functions the variables are in the exponents.

In general,

an *exponential function* can also be written in the form

$$f(x) = ab^{cx}$$

where a, b, and c are constants and $b > 0$, $b \neq 1$, $c \neq 0$.

In Chapter 5, we gave the meaning of exponential expressions such as 3^x, where x is a rational number. That is,

$$3^4 = 3 \cdot 3 \cdot 3 \cdot 3 \quad \text{and} \quad 3^{5/2} = \left(3^{1/2}\right)^5 = \sqrt{3} \cdot \sqrt{3} \cdot \sqrt{3} \cdot \sqrt{3} \cdot \sqrt{3}.$$

If x is any irrational number, the expression 3^x can also be defined. For example, if we assign a value of $x = \sqrt{2}$ to the expression 3^x, a calculator gives an approximation for $\sqrt{2} \approx 1.41421356237$ and an approximation of $3^{\sqrt{2}} \approx 4.72880438784$. This approach allows us to say that the domain of this function is the entire set of real numbers. Also, the output of this function, $f(x) = 3^x$, is positive, since a positive number, 3, raised to any real number is positive. The properties of exponents still hold when exponents are real numbers.

EXAMPLE 1   **Evaluating Exponential Functions**

Consider the functions $f(x) = 5^{-x}$ and $g(x) = 1000(1.03)^x$. Find the output values for each function at these input values: $x = 1.2$, $\sqrt{3}$, and $-\sqrt{7}$. Round off each result to three decimal places.

Solution   We use a calculator and round off to 3 decimal places. By constructing a table of values for the two functions, we have:

x	$f(x) = 5^{-x}$	$g(x) = 1000(1.03)^x$
1.2	$5^{-1.2} = 0.145$	$1000(1.03)^{1.2} = 1{,}036.107$
$\sqrt{3}$	$5^{-\sqrt{3}} = 0.062$	$1000(1.03)^{\sqrt{3}} = 1{,}052.531$
$-\sqrt{7}$	$5^{+\sqrt{7}} = 70.681$	$1000(1.03)^{-\sqrt{7}} = 924.775$

Graphing Exponential Functions

We may graph exponential functions by selecting some values for x, then find the corresponding values for $y = f(x)$, and plotting these points. You may observe some characteristics of the graphs of these functions as the next example shows.

EXAMPLE 2   **Graphing Exponential Functions**

Sketch the graph of each exponential function by point plotting.

(a) $f(x) = 3^x$ (b) $g(x) = \left(\dfrac{1}{3}\right)^x$

Solution   Use the graphs to indicate the domain and range of each function.

For both (a) and (b), we compute output values for different input values for the two functions. Then we plot these points and connect them with a smooth curve since these functions are defined for all real input values.

Figure 1

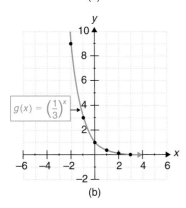

(a)

(b)

x	-3	-2	-1	0	1	2	3
$f(x) = 3^x = y$	$\dfrac{1}{27}$	$\dfrac{1}{9}$	$\dfrac{1}{3}$	1	3	9	27
$g(x) = \left(\dfrac{1}{3}\right)^x = y$	27	9	3	1	$\dfrac{1}{3}$	$\dfrac{1}{9}$	$\dfrac{1}{27}$

A number of features should be noted about the graphs of these functions (Figure 1a and 1b). First, the y intercept of each graph is 1, but neither has an x intercept. The graphs suggest that the domain of each function consists of all real numbers R, and the range of each consists of positive real numbers or $(0,\infty)$. We can also observe from Figure 1a that as the x values are increasing, the y values are also increasing; while Figure 1b shows that as the x values are increasing, the y values are decreasing. Table 1 summarizes these features.

TABLE 1

$f(x) = 3^x$	Domain R	Range $(0,\infty)$	Increasing
$g(x) = \left(\dfrac{1}{3}\right)^x$	Domain R	Range $(0,\infty)$	Decreasing

These graphs suggest some generalizations about exponential functions and their graphs. Note that the graph of

$$y = b^x, b > 1 \text{ (Figure 2a)}$$

resembles the graph of $f(x) = 3^x$ (Figure 1a), and the graph of

$$y = b^x, 0 < b < 1 \text{ (Figure 2b)}$$

resembles the graph of $g(x) = \left(\dfrac{1}{3}\right)^x$ (Figure 1b).

Figure 2

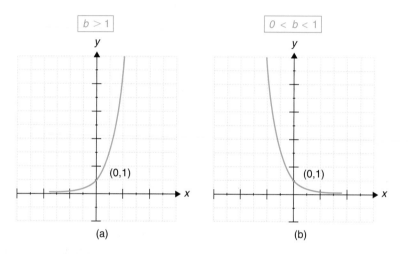

(a)   (b)

1. Domain: R	1. Domain: R
2. Range: $(0,\infty)$	2. Range: $(0,\infty)$
3. Increasing	3. Decreasing
4. y intercept: 1	4. y intercept: 1

In addition, we have the following property.

5. $b^u = b^v$ if and only if $u = v$

With this interpretation of b^x, the rules of exponents covered in earlier chapters can be extended to include *real numbers*.

That is, for $a > 0$, $b > 1$, and $b \neq$ and x and y are real numbers.

1. $b^x \cdot b^y = b^{x+y}$ 4. $b^0 = 1$

2. $(b^x)^y = b^{xy}$ 5. $b^{-x} = \dfrac{1}{b^x}$

3. $(ab)^x = a^x \cdot b^x$ 6. $\dfrac{b^x}{b^y} = b^{x-y}$

In section 6.1, we solved exponential equations such as

$$2^x = 16 \text{ and } 3^{-x} = 81.$$

Here we use property 5 above to solve some additional equations.

EXAMPLE 3

Solving Exponential Equations

Solve each equation.

(a) $2^x = 16$ (b) $4^x = 8$ (c) $4^{x+2} = 8^x$

Solution

All the numbers in parts (a), (b) and (c) shoud be expressed in powers of 2.

(a) $2^x = 16$	(b) $4^x = 8$	(c) $4^{x+2} = 8^x$	Given
$2^x = 2^4$	$(2^2)^x = 2^3$	$(2^2)^{x+2} = (2^3)^x$	Write as powers
$x = 4$	$2^{2x} = 2^3$	$2^{2x+4} = 2^{3x}$	
	$2x = 3$	$2x + 4 = 3x$	Property 5
	$x = 3/2$	$4 = x$	

Graphing Natural Exponential Functions

We have already seen how the value of the base b influences the graph of an exponential function. Now we choose an exponential function with an important base and analyze its graph.

Most types of natural growth or decay such as bacteria growth, drug concentration, or spread of diseases are modeled by exponential functions with a special base known as the *natural base.* The importance of this base was first recognized by the Swiss mathematician Leonhard Euler (pronounced "oiler"), 1707–1783, and in his honor it is denoted by the symbol e. Euler found that the value of e can be obtained by approximating the expression

$$\left(1 + \frac{1}{n}\right)^n$$

for different values of n. Table 2 shows output values of the expression rounded to six decimal places.

TABLE 2

n	1	2	4	12	365	8760
$\left(1 + \frac{1}{n}\right)^n$	2.000000	2.250000	2.441406	2.613035	2.714567	2.718127

The output values in the table suggest that as n increases, the expression

$$\left(1 + \frac{1}{n}\right)^n$$

keeps getting closer and closer to a very important fixed number, denoted by **e.** By using advanced methods and high speed computers, the numerical value of e has been calculated to thousands of decimal places. Rounded to six decimals places

$$e = 2.718282.$$

The constant e turns out to be an ideal base for exponential functions which model growth and decay of populations in nature as well as geometric patterns such as the spiral pattern of sunflower seeds. Because of these applications, such exponential functions are called *natural functions* and e is called the *natural number.*

Natural Exponential Function

For x a real number, the equation

$$f(x) = e^x$$

defines a function with base e, called a **natural exponential function.**

EXAMPLE 4   **Analyzing the Graph of the Natural Exponential Function**

Let $f(x) = e^x$.

(a) Find each output value rounded to 4 decimal places for the input values of

$$x = -3, -2, -1, 0, 1, 2, \text{ and } \sqrt{3}.$$

(b) Connect these points with a smooth curve.

(c) Indicate the domain and range of the function.

Solution   (a)

x	-3	-2	-1	0	1	2	$\sqrt{3}$
$f(x) = e^x$	0.0498	0.1353	0.3679	1.0000	2.7183	7.3201	5.6522

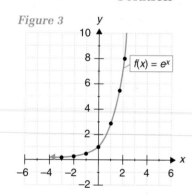

Figure 3

(b) Since $e > 1$, the graph of $f(x) = e^x$ has the same shape as the graph of $y = b^x$ for $b > 1$ (Figure 3). We plot the points found in the table and connect them with a smooth curve. Note that the plotted points fit on the curve.

(c) The graph suggests the domain of this function consists of real numbers $\mathbf{R}$ and the range consists of positive real numbers or $(0, \infty)$.

As with the graphs of other functions, the graph of an exponential function can be reflected, stretched vertically, or shifted horizontally and vertically to obtain graphs of other exponential functions.

EXAMPLE 5   **Analyzing Graphs**

Figure 4 shows the graphs of

$$f(x) = e^x, \, g(x) = e^{x-2}$$

and

$$h(x) = e^x - 2$$

on the same coordinate system.

(a) Make a table for the input values at

$$x = -2, -1, 0, 1, \text{ and } 2.$$

(b) Indicate how the graphs of g and h are obtained from the graph of f.

Solution   (a) First we create a table for the input values for each of these functions and round off the output values to 2 decimal places.

Figure 4

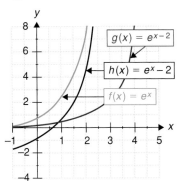

x	$f(x) = e^x$	$g(x) = e^{x-2}$	$h(x) = e^x - 2$
-2	0.14	0.02	-1.86
-1	0.37	0.05	-1.63
0	1.00	0.14	-1.00
1	2.72	0.37	0.72
2	7.39	1.00	5.39

(b) We can see how these graphs are generated if we plot the points from the above table and connect them with a smooth curve. Notice that the three graphs are identical in shape, but $g(x) = e^{x-2}$ is shifted 2 units to the right of $f(x) = e^x$. The graph of $h(x) = e^x - 2$ is obtained by shifting the graph of $f(x) = e^x$, 2 units downward.

Solving Applied Problems

Recall the **compound interest** formula that was covered in section 6.1.

$$S = P\left(1 + \frac{r}{n}\right)^{nt}$$

where S is the *future value*, P is the *present value* (also known as the *principal*), r is the *annual interest rate*, and t is the *number of years*. When interest is compounded n times per year for larger and larger values of n, we say that the interest is **compounded continuously.** In this situation, we use the formula

$$S = Pe^{rt}$$

for continuously compounded interest, where S, P, r, and t are defined as in the above formula.

EXAMPLE 6   **Purchasing Manhattan Island**

In 1626, Peter Minuit purchased Manhattan Island for $24 worth of trinkets and beads. Suppose that the $24 had been deposited in a forgotten savings account that paid 4% annual interest. Find the amount of this investment after 374 years if the interest is compounded (a) quarterly and (b) continuously.

Solution   (a) Using the formula

$$S = P\left(1 + \frac{r}{n}\right)^{nt}$$

with $P = 24$, $r = 0.04$, $n = 4$, and $t = 374$, we have

$$S = 24\left(1 + \frac{0.04}{4}\right)^{4(374)} \approx 69{,}982{,}011.38.$$

(b) For continuously compounded interest, we use the formula

$$S = Pe^{rt}$$

with $P = 24$, $r = 0.04$, $t = 374$. So that

$$S = 24 \, e^{0.04(374)} \approx 75{,}380{,}096.83$$

Therefore, in part (a) the $24 would have increased to $69,982,011.38.

In part (b), the $24 would have increased to $75,380,096.83.

Exponential functions represented by the functions

$$S(t) = Pb^t \text{ and } S(t) = Pb^{-t}$$

in which either the quantity S *increases* or *grows exponentially* as a function of t, or the quantity S *decreases* or *decays exponentially* as a function of t, serve as models for diverse phenomena called **exponential growth and decay.** The simplest growth model predicts a reasonably accurate description of growth of populations of people, bacteria, zebra mussels, and cell cultures.

EXAMPLE 7

Population Growth

Figure 5

The graph in Figure 5 is for the population P (in thousands) of a city in the United States which can be approximated by the exponential growth expression

$$P = 12.4(2.71)^{0.021t}, t \geq 0$$

where t is the number of years after 2005.

(a) Approximate the population in the year 2031 algebraically.

(b) In what year is the population 16 thousand?

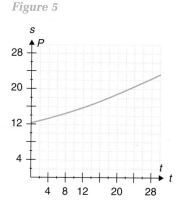

Solution

(a) In 2031, $t = 26$.

$$P = 12.4(2.71)^{0.021(26)} \approx 21.37 \text{ thousand.}$$

(b) Graphically, the population is 16 thousand when $t = 12$ in the year 2017.

EXAMPLE 8

Solving a Restaurant Management Problem

Figure 6

At a famous restaurant, the chef determines that the soup should be served to customers at a temperature of no less than 160°F. It has been determined that the cooling rate for this soup is 0.21°F per minute. According to a law from physics, known as *Newton's Law of Cooling,* the temperature T of the soup t minutes after it is removed from a boiling pot is given by the model

$$T = 68 + 144e^{-0.21t}, t \geq 0$$

whose graph is shown in Figure 6.

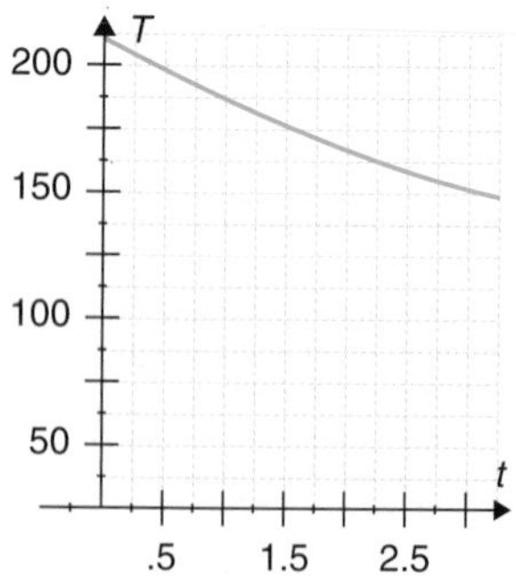

(a) What is the temperature T of the soup after 1 minute?

(b) Use the graph of this model to determine when the temperature of the soup is 175°F.

Solution

(a) When $t = 1$, we have

$$T = 68 + 144e^{-0.21(1)} = 185°F.$$

(b) From the graph, the temperature of 175°F corresponds to a time of approximately 1.5 minutes.

PROBLEM SET 8.1

In Problems 1–6, find the approximate value of each expression. Round off to 3 decimal places.

1. (a) $3^{-\sqrt{5}}$

   (b) $3^{2-\sqrt{2}}$

   (c) $\left(\dfrac{1}{3}\right)^{0.43}$

   (d) $\left(\dfrac{1}{3}\right)^{1+\pi}$

2. (a) $5^{-\pi}$

   (b) $5^{\sqrt{3}}$

   (c) $\left(\dfrac{1}{5}\right)^{0.37}$

   (d) $\left(\dfrac{1}{5}\right)^{-1.71}$

3. (a) $e^{1.3}$

   (b) $e^{-0.2}$

   (c) $\left(\dfrac{1}{e}\right)^{\sqrt{5}}$

   (d) $\left(\dfrac{1}{e}\right)^{-3.4}$

4. (a) $36e^{2.1}$

   (b) $36e^{0.72}$

   (c) $36\left(\dfrac{1}{e}\right)^{-0.4}$

   (d) $36\left(\dfrac{1}{e}\right)^{\pi}$

5. (a) $0.75(0.94)^{-1.2}$

   (b) $0.75(1.02)^{\pi}$

   (c) $200(1 + e^{1.7})$

6. (a) $0.81(0.74)^{-1.5}$

   (b) $300\left(1 + \dfrac{0.06}{4}\right)^{12}$

   (c) $40(3 + e^{-0.2})$

In Problems 7–10, each graph is a transformation of the graph of $y = 2^x$. Match the function with its graph.

7. $f(x) = 2^{x+1}$

8. $f(x) = 2^x + 1$

9. $f(x) = -2^x$

10. $f(x) = 2^x - 1$

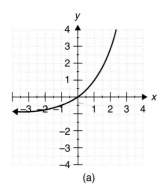

(a)

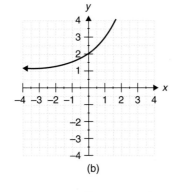

(b)

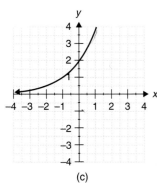

(c)

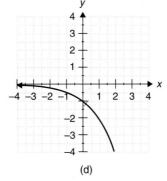

(d)

In Problems 11–14, each graph is a transformation of the graph of $y = e^{-x}$. Match each function with its graph.

11. $f(x) = e^{-x} - 1$

12. $f(x) = e^{-(x-1)}$

13. $f(x) = -e^{-x}$

14. $f(x) = e^{-x} + 1$

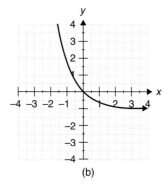

(a)

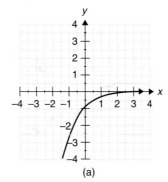

(b)

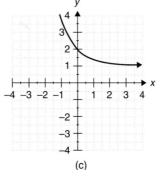

(c)

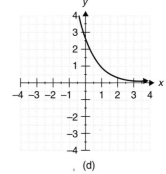

(d)

In Problems 15–20, use point plotting to graph each function. Indicate the domain and range.

15. $f(x) = 5^x$

16. $g(x) = \left(\dfrac{1}{5}\right)^x$

17. $f(x) = 4^{-x}$

18. $g(x) = 6^{-x}$

19. $f(x) = e^{-x}$

20. $g(x) = -\left(\dfrac{1}{2}\right)^x$

In Problems 21–26, solve each exponential equation.

21. (a) $3^x = 27$

    (b) $3^{x+1} = 27$

22. (a) $5^x = 125$

    (b) $5^{-x+2} = 125$

23. (a) $2^x = 4$

    (b) $2^{2x-3} = 4$

24. (a) $9^x = 81$

    (b) $9^{2x+3} = 81$

25. (a) $4^x = 8$

    (b) $4^{2x-1} = 8$

26. (a) $27^x = 9$

    (b) $27^{x-2} = 9$

In Problems 27–32, sketch the graph of each function by point plotting. Indicate the domain and range of each function.

27. (a) $f(x) = 3^x$
    (b) $g(x) = \left(\frac{1}{3}\right)^x$
    (c) $h(x) = 3^{x+2}$

28. (a) $f(x) = 4^x$
    (b) $g(x) = \left(\frac{1}{4}\right)^x$
    (c) $h(x) = 4^{x-2}$

29. (a) $f(x) = 3^x$
    (b) $g(x) = -3^x$
    (c) $h(x) = 3^x - 1$

30. (a) $f(x) = \left(\frac{1}{2}\right)^x$
    (b) $g(x) = -\left(\frac{1}{2}\right)^x + 2$
    (c) $h(x) = 2^{-x+1} + 3$

31. (a) $f(x) = e^x$
    (b) $g(x) = e^{-x} - 2$
    (c) $h(x) = 5e^x + 1$

32. (a) $f(x) = -e^x$
    (b) $g(x) = -\left(\frac{1}{e}\right)^x + 2$
    (c) $h(x) = -e^x + 2$

Applying the Concepts

33. **Compound Interest:** Find the value of a $12,000 investment at the end of 4 years with an annual interest rate of 8% compounded (a) quarterly (b) monthly (c) continuously.

34. **Compound Interest:** An investor has $2000 to deposit for 3 years. Shopping around, the investor discovers that interest is being calculated in four different ways: 4% quarterly, 3.95% monthly, 3.9% daily, and 3.9% continuously. Calculate the amount of money the investor would have with each method after 3 years. Which method would yield the most return?

35. **Population Growth:** In 1994, the population of Mexico City was 15.5 million and growing at an annual rate of approximately 0.7% or 0.007. The number of people, N, t years after 1994 is given by the model

    $$N = 15.5(1 + r)^t.$$

    Using this model, predict the population of the city in (a) the year 2005 (b) the year 2010. Round to one decimal place. (Source: *The Universal Almanac.*)

36. **Population Growth:** In 1798, the *English economist Thomas Malthus* invented a model for predicting the population growth P after t years

(assuming that the birth rate b and the death rate d are fixed). The model is

$$P = P_0 e^{kt}$$

where P_0 is the initial population and k is the annual growth (or decay), $k = b - d$ and t is the number of years since the initial population. This model is known as the *Malthusian model.* According to the U.S. Census Bureau, the population of the United States in 1990 was 244 million, the birth rate was 1.6% or 0.016, and the death rate was 0.86% or 0.0086. Suppose that the population grew according to the Malthusian model. Predict the population in the year 2005 and in the year 2010.

37. **Salvage Value:** A sport utility vehicle was purchased for $31,000. Its value is expected to depreciate by 18% of the previous year's value so that its value V (in dollars) after t years is given by the model

    $$V = 31,000(1 - r)^t$$

    where r is the annual rate of depreciation. Find the value of this vehicle after (a) 2 years, (b) 4 years, and (c) 5 years.

38. **Caffeine Consumption:** Caffeine is a chemical stimulant that is found in coffee, teas, and some soft drinks. An average person eliminates 10% of this compound each  hour after ingestion. Suppose an office worker drinks a cup of coffee that contains 50 milligrams of caffeine. The number of milligrams N of caffeine that remains in the worker's system after t hours is given by the model

    $$N = 50(1 - r)^t$$

    where r is the hourly rate for eliminating the compound after ingestion. How much caffeine remains in the worker's system after (a) 2 hours? (b) 2.5 hours?

39. **Housing Values:** State and local agencies frequently use the formula

    $$S = C(1 + r)^t$$

    to assess the value S (in dollars) of a property t years after it was purchased at a price of

C dollars, assuming an annual rate of inflation *r*. During the 1970s and 1980s, California real estate prices soared dramatically, with some areas seeing an average annual inflation rate as high as 14%. If a ranch house sold for $250,000 in March 1976, what was the price to the nearest dollar of the ranch in March of 1980, assuming 14% annual inflation?

40. **Cost of Living:** In 2000, the cost of tuition at a certain college was $10,000. The cost *C* (in dollars) after *t* years is modeled by the function

$$C(t) = 10,000e^{0.05t}$$

Use this model to predict the cost to the nearest dollar of tuition in the years (a) 2006 and (b) 2010.

41. **Environmental Science:** According to the *Bouguer-Lambert law,* the percentage *P* of light that penetrates ordinary sea water to a depth of *d* feet is given by the function

$$P = e^{-0.044d}.$$

What is the percentage of light penetration in sea water to a depth of (a) 4 feet? (b) 6 feet?

42. **Recycling:** In the state of Michigan, stores charge an additional $0.10 deposit for each bottle of beverage sold. Customers return these bottles for refunds, and then the stores return them to the beverage companies for recycling. Suppose that 65% of all bottles distributed will be recycled every year. If a company distributed 2.5 million bottles in 1 year, the number *N* (in millions) of recycled containers still in use after *t* years is given by the model

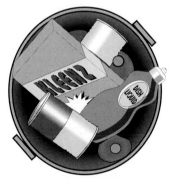

$$N = 2.5(0.65)^t.$$

How many recycled bottles (in millions) are still in use after (a) 1 year? (b) 3 years? (c) 5 years? Round off the answers to two decimal places.

43. **Radioactive Decay:** Certain radioactive elements decay exponentially. The decay model for a specific radioactive element is

$$A = A_o e^{-0.04463t}$$

where *A* is the amount present after *t* days and A_o is the amount present initially. Assume there is a block of 50 grams of the element at the start. How much of the element rounded to one decimal place remains after (a) 10 days? (b) 15 days? (c) 30 days?

44. **Baking a Cake:** Suppose a cake is baked in a 350°F oven and then removed to a room at 70°F. The temperature *T* of the cake after *t* minutes is given by the model

$$T = 70 + 280\, e^{-0.14t}.$$

What is the temperature of the cake after (a) 5 minutes? (b) 10 minutes? (c) 20 minutes?

45. **Cellular Phones:** A study shows that the number *N* (in millions) of cellular phone users after *t* years can be approximated by the model

$$N = 38(1.273)^t$$

where *t* = 0 represents the year 2000.

(a) Use this model to predict the number of users at the end of 2005 and 2009.

(b) Use the graph in Figure 7 of $N = 38(1.273)^t$ to estimate the date when there will be 500 million users. Round up to the nearest year.

Figure 7

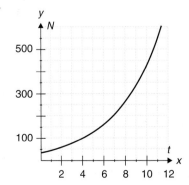

46. Internet Shopping: Retail revenue from shopping on the Internet is growing at a surprising rate. A study shows that the Internet sales S in billions of dollars after t years is approximated by the model

$$S = 41.3(1.52)^t$$

where $t = 0$ represents the year 2002.

(a) Use this model to predict the Internet sales in the years 2005 and 2008.

(b) Use the graph in Figure 8 to find when sales will reach $1000 billion. Round up to the nearest year.

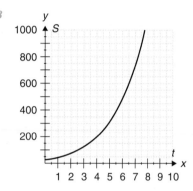

Figure 8

Developing and Extending the Concepts

In Problems 47 and 48, solve each equation.

47. (a) $2^{-4x+1} = 64$

(b) $5^{2x} = 25^{3x}$

48. (a) $25^{3x} = 5^{2x-3}$

(b) $2^{3x^2-2x} = 32$

49. Graph $f(x) = 2^x$ and $g(x) = 4^x$ on the same plane.

(a) What one point lies on both graphs?

(b) What is the domain and range of each graph?

(c) Compare the steepness of the graphs.

50. Graph $f(x) = 2^{-x}$ and $g(x) = 3^{-x}$ on the same plane.

(a) What one point lies on both graphs?

(b) What is the domain and range of each graph?

(c) Compare the steepness of the graphs.

51. Future Value: The formula

$$F = m\left[\frac{(1+i)^{kn} - 1}{i}\right]$$

represents the *future value* F (in dollars) of an annuity with deposits of m dollars made regularly k times each year for n years, with interest compounded k times per year at an annual rate r, where i is the periodic rate, $i = \dfrac{r}{k}$.

(a) How much is the account worth if $600 were deposited each month at an annual interest rate of 6% compounded monthly for a period of 20 years?

(b) How much money was deposited?

52. Retirement Installment: A common method for financing retirement is to invest a large amount of money at a given rate r compounded annually and then withdraw a fixed amount of money from this investment each year. The amount of possible withdrawal W (in dollars) if an investment P (in dollars) is expected to last for t years is given by the model

$$W = \frac{Pr(1+r)^t}{(1+r)^t - 1}$$

where r is the annual interest rate. If $400,000 is invested in an account at a rate of 7% with the intention of making withdrawals for 12 years, how much can be withdrawn each year?

53. Data Analysis: In 1991, the population P (in millions) of India was about 866 million and growing after t years according to the table

Year	1991	1992	1993	1994	1995	1996	1997
Population	866	883	894	917	934	952	970

(a) Construct a scatter gram that exhibits the data.

(b) Consider the function

$$P = 866e^{0.019t}.$$

Make a table of values using this function for $t = 0, 1, 2, 3, 4, 5,$ and 6. Round to whole numbers.

(c) Compare the values in (a) and (b).

54. Data Analysis: According to the Television Bureau of Advertising, Inc., the percent p of T.V. households with VCRs are patterned by the following table.

Year t	90	91	92	93	94	95	96	97
Percent	68.6	71.9	75.0	77.1	77.2	77.5	77.6	77.7

(a) Construct a scatter gram that exhibits the data.

(b) Consider the function

$$P = 77.8[1 + 63.4e^{-0.638t}]^{-1}$$

and find values of P for $t = 10, 11, 12, 13, 14, 15, 16,$ and 17.

(c) Compare the values in parts (a) and (b).

In Problems 55 and 56, (a) sketch the graphs of $f(x)$ and $g(x)$ on the same coordinate systems. (b) comment on the shapes of the graphs when $-1 \leq x \leq 1$.

55. $f(x) = e^x$ and $g(x) = 1 + x + \dfrac{x^2}{2}$

56. $f(x) = e^{-x}$ and $g(x) = 1 - x + \dfrac{x^2}{2}$

8.2 Logarithms

Objectives

1. Convert Logarithmic Forms
2. Solve Logarithmic Equations of the Form $\log_b u = k$
3. Evaluate Common and Natural Logarithms
4. Approximate Logarithms to Any Base
5. Approximate Antilogarithms
6. Solve Applied Problems

In example 3 of section 8.1, we solved the exponential equation

$$2^x = 16$$

by expressing 16 as a power of 2 so that each side of the equation has the same base. Because $16 = 2^4$, we rewrite

$$2^x = 16 = 2^4$$

or $x = 4$, and the solution of this equation is 4. In other words, the exponent of the base 2 that "produces" 16 is 4. In this section, we introduce the word *logarithm* to be used synonymously with the word *exponent,* and so we describe the phrase "the logarithm of 16 to the base 2 is 4" by writing

$$\log_2 16 = 4.$$

Thus, the equivalency between the exponential and logarithmic form in the above equations is expressed as follows:

$$\log_2 16 = 4 \text{ if and only if } 2^4 = 16.$$

In general, we have the following definition.

Definition of Logarithm

> If y is a positive real number, then the exponent x such that
> $$b^x = y$$
> is called the **logarithm of y with base b** and is denoted by
> $$\log_b y = x.$$

Thus, we can write

> $$\log_b y = x \text{ if and only if } b^x = y.$$

Converting Logarithmic Forms

We can switch back and forth between the exponential and logarithmic forms of equations, as the next examples illustrate.

EXAMPLE 1 **Converting Logarithmic to Exponential Form**

Change each logarithmic equation to exponential form.

(a) $\log_2 16 = 4$ (b) $\log_9 3 = \dfrac{1}{2}$ (c) $\log_2 \dfrac{1}{4} = -2$

Solution We compare each equation in Table 1 to the definition above so as to convert.

TABLE 1

Logarithmic Equation $\log_b y = x$	b	y	x	Exponential Equation $b^x = y$
(a) $\log_2 16 = 4$	2	16	4	$2^4 = 16$
(b) $\log_9 3 = \dfrac{1}{2}$	9	3	$\dfrac{1}{2}$	$9^{\frac{1}{2}} = 3$
(c) $\log_2 \dfrac{1}{4} = -2$	2	$\dfrac{1}{4}$	-2	$2^{-2} = \dfrac{1}{4}$

EXAMPLE 2 **Converting Exponential to Logarithmic Form**

Change each exponential equation to logarithmic form.

(a) $7^2 = 49$ (b) $\sqrt[3]{27} = 3$ (c) $5^{-2} = \dfrac{1}{25}$

Solution We identify each part of the definition in Table 2 and write the logarithmic equation.

TABLE 2

Exponential Equation $b^x = y$	b	y	x	Logarithmic Equation $\log_b y = x$
(a) $7^2 = 49$	7	49	2	$\log_7 49 = 2$
(b) $\sqrt[3]{27} = 27^{\frac{1}{3}} = 3$	27	3	$\dfrac{1}{3}$	$\log_{27} 3 = \dfrac{1}{3}$
(c) $5^{-2} = \dfrac{1}{25}$	5	$\dfrac{1}{25}$	-2	$\log_5 \dfrac{1}{25} = -2$

To gain a deeper understanding of the equivalency of the logarithmic and exponential equations, we consider evaluating logarithms to different bases.

EXAMPLE 3 **Evaluating Logarithms**

Find the exact value of each logarithm.

(a) $\log_3 81$ (b) $\log_{10} 0.1$ (c) $\log_2 4\sqrt{2}$

Solution

	Logarithmic Form $\log_b y = x$	Exponential Form $b^x = y$	Power of Base	Solution
(a)	$\log_3 81 = x$	$3^x = 81$	$3^x = 3^4$	$x = 4$
(b)	$\log_{10} 0.1 = x$	$10^x = 0.1$	$10^x = 10^{-1}$	$x = -1$
(c)	$\log_2 4\sqrt{2} = x$	$2^x = 4\sqrt{2}$	$2^x = 2^2 \cdot 2^{\frac{1}{2}} = 2^{\frac{5}{2}}$	$x = \dfrac{5}{2}$

Solving Logarithmic Equations of the Form $\log_b u = k$

We consider a few problems where we want to solve simple equations of the form

$$\log_b u = k$$

for any of the variables u, b, or k by converting them to exponential forms. All values were chosen so that they can be solved without a calculator.

EXAMPLE 4   **Solving Equations of the Form $\log_b u = k$**

Solve each equation for the indicated variable.

(a) $\log_2 8 = x$ (b) $\log_{27} y = \dfrac{2}{3}$ (c) $\log_b 36 = -2$

(d) $\log_2 (5x - 3) = 5$ (e) $\log_5 1 = x$

Solution   (a) $\log_2 8 = x$ — Given

$2^x = 8$ — Exponential form

$2^x = 2^3$ — Exponential with same base

$x = 3$ — $b^x = b^y$ if and only if $x = y$

(b) $\log_{27} y = \dfrac{2}{3}$ — Given

$27^{\frac{2}{3}} = y$ — Exponential form

$(3^3)^{\frac{2}{3}} = y$ — $3^3 = 27$

$3^2 = 9 = y$

(c) $\log_b 36 = -2$ — Given

$b^{-2} = 36$ — Exponential form

$(b^{-2})^{-\frac{1}{2}} = 36^{-\frac{1}{2}}$ — Raise to the reciprocal power

$b = (6^2)^{-\frac{1}{2}} = 6^{-1}$ — Power property

$b = \dfrac{1}{6}$ — Negative exponent

(d) $\log_2 (5x - 3) = 5$ — Given

$2^5 = 5x - 3$ — Exponential form

$32 = 5x - 3$

$35 = 5x$

$x = 7$

(e) $\log_5 1 = x$ — Given

$5^x = 1$ — Exponential form

$5^x = 5^0$ — $5^0 = 1$

$x = 0$ — Equate exponents

Example 4(e) above can be generalized in the following property:

Property

For any real number b, $b > 0$ and $b \neq 1$,

1. $\log_b 1 = 0$ and $\qquad\qquad$ 2. $\log_b b = 1$.

For example,

$$\log_{10} 1 = 0 \qquad\qquad \log_{10} 10 = 1$$
$$\log_7 1 \; = 0 \qquad\qquad \log_7 7 \;\; = 1$$
$$\log_3 1 \; = 0 \qquad\qquad \log_3 3 \;\; = 1$$
$$\ln 1 \;\; = 0 \qquad\qquad \ln e \;\;\; = 1$$

Evaluating Common and Natural Logarithms

Originally, logarithms were developed to assist in reducing computational strain in research in astronomy and other sciences. Today, with the availability of calculators and computers, logarithms have lost most of their importance as a computational device. However, they are still widely used in both theoretical and applied sciences. Of all possible logarithmic bases

the base 10 and the base e

are used almost exclusively. **Common logarithms** are logarithms with base 10. Their value for any positive number x is denoted by **log x**. It is customary to omit writing the numeral 10 to designate the base. Thus, we write

---

$$\log_{10} x = \log x.$$

In many practical applications involving logarithms, the number e is used as a base. Logarithms with a base e are called **natural logarithms,** and the symbol **ln x** is commonly used instead of $\log_e x$ to express their values. Thus, we write

---

$$\log_e x = \ln x.$$

Most calculators have a key labeled "log" and a key labeled "ln" to represent common and natural logarithms respectively. In fact, "log" and "ln" are both used in applications, so whenever you see either used in this book, without a base indicated, they should be interpreted as in the boxes above.

To find the *exact* value of some common logarithms, the use of calculator is not necessary. For example,

$$\log 100 = 2 \quad \text{because} \quad 10^2 = 100$$

$$\log 10 = 1 \quad \text{because} \quad 10^1 = 10$$

$$\log 0.1 = -1 \quad \text{because} \quad 10^{-1} = \frac{1}{10} = 0.1$$

$$\log 0.01 = -2 \quad \text{because} \quad 10^{-2} = \frac{1}{100} = 0.01.$$

In general,

$$\log 10^u = u.$$

For example,

$$\log_3 \sqrt{27} = \log_3 (3^3)^{\frac{1}{2}}$$
$$= \log_3 3^{\frac{3}{2}} = \frac{3}{2}$$

and

$$\log_3 \sqrt[3]{25} = \log_5 (5^2)^{\frac{1}{3}}$$
$$= \log_5 5^{\frac{2}{3}} = \frac{2}{3}$$

To find the common logarithm of a positive number that is not an integral power of 10, we can use a calculator with key labeled "log". The common logarithmic function is defined by the equation

$$f(x) = \log x.$$

EXAMPLE 5   **Approximating Common Logarithms**

Evaluate each logarithm. Round off each answer to 4 decimal places.

(a) log 364   (b) log 86.2   (c) log 0.568   (d) log 0.0841   (e) log(−7.142)

Solution   Using a calculator with a button labeled "log", round off to 4 decimal places, we have

(a) log 364 = 2.5611   (b) log 86.2 = 1.9355

(c) log 0.568 = −0.2457   (d) log 0.0841 = −1.0752

(e) log(−7.142) = "error"

Here the "error" is indicated because −7.142 is not a positive number as the definition of a logarithm requires. Scientific calculators display error messages when the logarithm is not defined. However, some graphing calculators display an ordered pair representing a complex number (section 6.7) as the value of log (−7.142) rather than an error message. We should interpret such a display as log (−7.142) is not defined.

To find the exact value of some natureal logarithms, we have:

$$\ln e^2 = 2 \qquad \text{because} \qquad e^2 = e^2$$
$$\ln e^5 = 5 \qquad \text{because} \qquad e^5 = e^5$$
$$\ln e^{-0.1} = -0.1 \qquad \text{because} \qquad e^{-0.1} = e^{-0.1}$$

In general,

$$\ln e^u = u.$$

The next example uses a calculator with a natural logarithm function key labeled "ln x" or just "ln" to evaluate each expression.

EXAMPLE 6   **Approximating Natural Logarithms**

Evaluate each logarithm. Round off each answer to 4 decimal places.

(a) $\ln 8132$ (b) $\ln 743$ (c) $\ln 52.31$ (d) $\ln 0.728$ (e) $\ln 0.0436$

Solution   Evaluate using a calculator with a button labeled "ln" and rounding off to 4 decimal places.

(a) $\ln 8132 = 9.0036$ (b) $\ln 743 = 6.6107$

(c) $\ln 52.31 = 3.9572$ (d) $\ln 0.728 = -0.3175$

(e) $\ln 0.0436 = -3.1327$

The natural logarithmic function is defined by the equation

$$g(x) = \ln x.$$

EXAMPLE 7   **Approximating Logarithms by a Calculator**

Approximate the value of each expression. Round off each answer to 4 decimal places.

(a) $\dfrac{\log 3}{\log 2.1}$ (b) $\log\left(\dfrac{3}{2.1}\right)$ (c) $\dfrac{\ln 4}{\ln 1.07}$ (d) $\ln\left(\dfrac{4}{1.07}\right)$ (e) $\log 3 - \log 2.1$

Solution   By using a calculator and rounding off to 4 decimal places, we have:

(a) $\dfrac{\log 3}{\log 2.1} = 1.4807$ First calculate log 3 and log 2.1 and then divide

(b) $\log\left(\dfrac{3}{2.1}\right) = 0.1549$ Divide 3 by 2.1 and then find the logarithm

(c) $\dfrac{\ln 4}{\ln 1.07} = 20.4895$ Find ln 4 and ln 1.07 and then divide

(d) $\ln\left(\dfrac{4}{1.07}\right) = 1.3186$ Divide the numbers first, then take the ln

(e) $\log 3 - \log 2.1 = 0.1549$ Take logs then subtract

Approximating Logarithms to Any Base

Suppose we wish to find $\log_3 7$. There is no logarithm base 3 on a calculator and 7 is not an obvious power of 3. So we must find another way. It is often useful to rewrite $\log_3 7$ in terms of the *common or natural* logarithm which can be directly evaluated by using a calculator. This is accomplished by using the following *change of base formula*.

Change of Base Formula

If a, b, and x are positive real numbers, and a and b are not equal to 1, then

$$\log_b x = \frac{\log_a x}{\log_a b}.$$

Using the change of base formula, we can write $\log_b x$ as

$$\log_b x = \frac{\log x}{\log b} \qquad \text{or} \qquad \log_b x = \frac{\ln x}{\ln b}.$$

EXAMPLE 8   **Approximating Logarithm to Any Base**

Approximate the value of each logarithm. Round off each answer to 4 decimal places.

(a) $\log_3 7$ (b) $\log_3 25$

Solution   We use the change of base formula.

$$\log_3 x = \frac{\log x}{\log 3} \qquad \text{or} \qquad \log_3 x = \frac{\ln x}{\ln 3}$$

We could choose either base and the ratio will be the same.

(a) $\log_3 7 = \dfrac{\log 7}{\log 3} = \dfrac{\ln 7}{\ln 3} = 1.7712$

(b) $\log_3 2.5 = \dfrac{\log 2.5}{\log 3} = \dfrac{\ln 2.5}{\ln 3} = 0.8340$

Approximating Antilogarithms

In order to use logarithms to solve applied problems, we sometimes need to determine x in equations such as $\log x = y$ or $\ln x = y$ where y is known. The x is known as the **antilogarithm.** Consider the following example.

EXAMPLE 9   **Approximating Solutions of Logarithmic Equations**

Solve each equation for x, the antilogarithm. Round off to 2 decimal places.

(a) $\log x = 1.7782$ (b) $\log x = -1.8742$

(c) $\ln x = 1.4142$ (d) $\ln x = -0.6713$

Solution   The following table shows how to convert each expression to exponential form.

Given	Exponential Form	Value
(a) $\log x = 1.7782$	$10^{1.7782} = x$	$x = 60.01$
(b) $\log x = -1.8742$	$10^{-1.8742} = x$	$x = 0.01$
(c) $\ln x = 1.4142$	$e^{1.4142} = x$	$x = 4.11$
(d) $\ln x = -0.6713$	$e^{-0.6713} = x$	$x = 0.51$

Solving Applied Problems

Applications such as measuring the strength of an earthquake and determining the acidity of a liquid are among several models that make use of logarithmic functions.

Seismologists measure the magnitude M of an earthquake using a logarithmic scale called the **Richter scale,** in honor of the American seismologist Charles F. Richter, 1900–1984. The energy E (in ergs) released by an earthquake of magnitude M can be approximated by the model

$$\log E = 11.4 + 1.5M.$$

EXAMPLE 10   **Finding Magnitudes of Earthquakes**

Find the magnitude of each earthquake.

(a) The 1990 earthquake in Iran with $E \approx 8.9 \times 10^{22}$ ergs.

(b) The 1992 earthquake in southern California with $E \approx 3.2 \times 10^{22}$ ergs.

Solution   We substitute each value of E into the model and solve for M.

$$\log E = 11.4 + 1.5M$$

(a) Replacing E by 8.9×10^{22} and using a calculator, we have,

$$\log (8.9 \times 10^{22}) = 11.4 + 1.5M$$
$$22.9494 = 11.4 + 1.5M$$
$$1.5M = 11.5494$$
$$M = 7.7.$$

Therefore, the magnitude of the earthquake in Iran was 7.7 on the Richter Scale.

(b) Replacing E by 3.2×10^{22}, we have,

$$\log(3.2 \times 10^{22}) = 11.4 + 1.5M$$
$$22.5051 = 11.4 + 1.5M$$
$$1.5M = 11.5151$$
$$M = 7.4.$$

Therefore, the magnitude of the earthquake in California was 7.4.

PROBLEM SET 8.2

Mastering the Concepts

In Problems 1–6, write each logarithmic equation as an exponential equation.

1. (a) $\log_9 81 = 2$
    (b) $\log_6 36 = 2$

2. (a) $\log_3 9 = 2$
    (b) $\log_6 216 = 3$

3. (a) $\log_{1/3} 9 = -2$
    (b) $\log_{10} \dfrac{1}{10} = -1$

4. (a) $\log_{4/9} \dfrac{27}{8} = \dfrac{-3}{2}$
    (b) $\log_4 2 = \dfrac{1}{2}$

5. (a) $\log_5 \sqrt{5} = \dfrac{1}{2}$
    (b) $\log_7 \sqrt[4]{7} = \dfrac{1}{4}$

6. (a) $\ln \dfrac{1}{e} = -1$
    (b) $\ln e^2 = 2$

In Problems 7–12, write each exponential equation as a logarithmic equation.

7. (a) $5^3 = 125$
    (b) $4^{-2} = 1/16$

8. (a) $10^3 = 1000$
    (b) $2^{-3} = 0.125$

9. (a) $\sqrt{9} = 3$
    (b) $\left(\dfrac{1}{8}\right)^{-2/3} = 4$

10. (a) $\sqrt[5]{32} = 2$
    (b) $100^{-3/2} = 0.001$

11. (a) $5^{\frac{1}{2}} = \sqrt{5}$
    (b) $4^{\frac{1}{3}} = \sqrt[3]{4}$

12. (a) $e^3 = c$
    (b) $e^{-4} = t$

In Problems 13–20, solve each equation for x.

13. (a) $\log_2 x = 2$

(b) $\log_3 x = 3$

15. (a) $\log_x 81 = 4$

(b) $\log_x 125 = 3$

17. (a) $\log_3 9 = x$

(b) $\log_3 27 = x$

19. (a) $\log_3 \dfrac{1}{9} = x$

(b) $\log_4 \dfrac{1}{16} = x$

14. (a) $\log_x 16 = 2$

(b) $\log_x 8 = 3$

16. (a) $\log_4 x = \dfrac{1}{2}$

(b) $\log_{25} x = \dfrac{1}{2}$

18. (a) $\log_3 1 = x$

(b) $\log_2 \dfrac{1}{16} = x$

20. (a) $\log_8 x = \dfrac{1}{3}$

(b) $\log_9 x = \dfrac{1}{2}$

In Problems 21–28, find the exact value of each logarithm.

21. (a) $\log_2 4$ (b) $\log_4 16$ (c) $\log_{\frac{1}{8}} 4$

22. (a) $\log_5 25$ (b) $\log_7 49$ (c) $\log_{\frac{1}{5}} 25$

23. (a) $\log_5 5$ (b) $\log_9 1$ (c) $\ln 1$

24. (a) $\log_9 81$ (b) $\log_9 \dfrac{1}{3}$ (c) $\log_b \sqrt{b}$

25. (a) $\log_9 1/9$ (b) $\log_3 9\sqrt{3}$ (c) $\ln e$

26. (a) $\log_5 \dfrac{1}{125}$ (b) $\log_3 243$ (c) $\log_5 5^b$

27. (a) $\log_4 \dfrac{1}{64}$ (b) $\log_3 \dfrac{1}{27}$ (c) $\log_{\frac{2}{3}}\left(\dfrac{3}{2}\right)$

28. (a) $\log_5 \dfrac{1}{125}$ (b) $\log_{10} 1000$ (c) $\log \sqrt[3]{10^2}$

In Problems 29–40, evaluate each logarithm. Round off each answer to 4 decimal places.

29. (a) $\log 317$

(b) $\log 3914$

31. (a) $\log 7.86$

(b) $\log 0.786$

33. (a) $\log 0.035$

(b) $\log 0.007$

35. (a) $\ln 54$

(b) $\ln 154$

37. (a) $\ln 0.071$

(b) $\ln 0.007$

39. (a) $\ln 0.0031$

(b) $\ln 0.0762$

30. (a) $\log 8.04$

(b) $\log 7.84$

32. (a) $\log 0.341$

(b) $\log 0.412$

34. (a) $\log 0.0014$

(b) $\log 0.0073$

36. (a) $\ln 71$

(b) $\ln 712$

38. (a) $\ln 5$

(b) $\ln \dfrac{1}{7}$

40. (a) $\ln 0.0137$

(b) $\ln 0.00413$

In Problems 41–48, find the exact value of each logarithm.

41. (a) $\log 1000$

(b) $\log \dfrac{1}{100}$

42. (a) $\log \sqrt[3]{10}$

(b) $\log \dfrac{1}{\sqrt{10}}$

43. (a) $\log \sqrt[4]{10}$

(b) $\log 0.001$

45. (a) $\ln e^3$

(b) $\ln e^{-1}$

47. (a) $\ln \sqrt[4]{e}$

(b) $\ln e^{-4}$

44. (a) $\log 10^4$

(b) $\log 10^{-5}$

46. (a) $\ln \sqrt[5]{e}$

(b) $\ln \dfrac{1}{\sqrt[3]{e}}$

48. (a) $(\ln e^2)^3$

(b) $\sqrt{\ln e^4}$

In Problems 49 and 50, approximate each value to 4 decimal places.

49. (a) $\dfrac{\log 7}{\log 34}$

(b) $\dfrac{\ln 5}{\ln 2.03}$

(c) $\log \dfrac{7}{13}$

(d) $\ln \dfrac{5}{17}$

50. (a) $\log 7 - \log 3.4$

(b) $\ln 13 - \ln 1.7$

(c) $\ln 1.3 + \ln 7.2$

(d) $\log 31 + 2\log 5$

In Problems 51–56, find each value of x, the antilogarithm. Round off each answer to 2 decimal places.

51. (a) $\log x = 0.4133$

(b) $\log x = 0.4871$

(c) $\log x = 1.2945$

53. (a) $\ln x = 2.9781$

(b) $\ln x = -1.6289$

(c) $\ln x = -1.8082$

55. (a) $\ln (2x + 4) = -0.4$

(b) $\log (3x - 2) = -0.7$

56. (a) $\ln (5x) = 0.17$

(b) $\log (5x) = 0.72$

52. (a) $\ln x = 1.725$

(b) $\ln x = 0.3517$

(c) $\ln x = 0.8991$

54. (a) $\log x = -1.8971$

(b) $\log x = 2.4881$

(c) $\log x = 0.2415$

In Problems 57–60, approximate each logarithmic value to 4 decimal places.

57. (a) $\log_5 7$

(b) $\log_6 11$

59. (a) $\log_{\frac{1}{2}} 6$

(b) $\log_{\frac{1}{5}} 3$

58. (a) $\log_4 5$

(b) $\log_4 15$

60. (a) $\log_7 \dfrac{1}{6}$

(b) $\log_{\frac{1}{7}} 6$

Applying the Concepts

61. Seismology: In 1964, a severe earthquake in Alaska that killed 131 people, released 10^{24} ergs. What was the magnitude of this earthquake? Round off the answer to one decimal place. ($\log E = 11.4 + 1.5M$)

62. **Seismology:** Nowadays, seismologists are using a different version of Richter's ideas in which the magnitude M of an earthquake is computed as a function of the energy intensity I. One such model adopted by scientists is

$$M = \log I.$$

Calculate the magnitude of the indicated earthquakes:

(a) Mexico City, 1985, where $I = 125{,}890{,}000$
(b) Armenia, 1988, where $I = 6{,}310{,}000$
(c) Guam, 1993, where $I = 100{,}000{,}000$.

63. **Audiology:** Audiologists generally agree that continuous exposure to a sound level in excess of 90 decibels for more than 5 hours daily may cause long-term hearing problems. Suppose that the **sound level** S (measured in decibels) is given by the *Fechner-Weber law* in the form of the model

$$S = 10 \log \frac{P}{P_0}$$

where P is the power (in watts) of the sound being measured, and P_0 is the power of sound of the lowest threshold of human hearing. Find the loudness in decibels for the indicated sounds:

(a) a supersonic jet where $\dfrac{P}{P_0} = 10^{11}$

(b) a machine in a factory where
$\dfrac{P}{P_0} = 11{,}000{,}000$

(c) a rock concert where $\dfrac{P}{P_0} = 263{,}000{,}000$.

64. **Chemistry:** In chemistry, the **pH of a substance** is defined by the model

$$pH = -\log[H+]$$

where $[H+]$ is the concentration form of hydrogen ions in the substance measured in moles per liter. The pH of distilled water is 7. A substance with a pH of less than 7 is called an acid, whereas a substance with a pH of more than 7 is called a base. Find the pH of each substance for the given concentration of hydrogen ions:

(a) acid rain whose $[H+]$ is 3.16×10^{-4}
(b) a vinegar whose $[H+]$ is 7.94×10^{-4}
(c) a lemon juice whose $[H+]$ is 5.01×10^{-3}
(d) a tomato whose $[H+]$ is 6.30×10^{-5}

65. **Chemistry:** Find the hydrogen ion concentration $[H+]$ in moles per liter of each substance. Write the answers in scientific notation as in Problem 64.

(a) vinegar; pH = 3.1
(b) beer; pH = 4.3
(c) lemon juice; pH = 2.3
(d) bile; pH = 7.9

66. **Advertising:** A bicycle store determines that the number N of bicycles sold in relation to the number x (in dollars) spent on advertising is given by the model

$$N = 51 + 100 \ln \left(\frac{x}{100} + 2 \right)$$

How many bikes will be sold if (a) nothing is spent on advertising? (b) \$1,000 is spent? (c) \$5,000 is spent?

67. **Spread of Flu:** A person with a flu virus visited a school campus. The number t (in days) it takes the virus to infect n people is given by the model

$$t = -16.3 \ln \left(\frac{5450}{n} - \frac{4}{11} \right), \quad 4000 \leq n \leq 14{,}000.$$

How many days will it take to infect (a) 6000 people? (b) 10,000 people? (c) 13,000 people?

68. **Doubling Investment:** The number of years, n, required to double your investment when it is invested at an annual interest rate r (in decimal form) compounded annually is given by the model

$$n = \frac{\ln 2}{\ln(1 + r)}.$$

Find the number of years it takes to double your money at each annual interest rate

(a) 4%   (b) 6%   (c) 8%.

In Problems 69–72, let $f(x) = \log x$ and $g(x) = \ln x$. Find the exact output value.

69. (a) $f\left(\sqrt[3]{10}\right)$ (b) $g\left(\dfrac{1}{\sqrt[3]{e}}\right)$ (c) $f\left(\dfrac{1}{\sqrt[4]{10}}\right)$

70. (a) $f(0.01)$ (b) $g(e^{-4})$ (c) $f\left(\dfrac{1}{0.01}\right)$

71. (a) $f\left(\dfrac{1}{10^3}\right)$ (b) $g\left(\dfrac{1}{e^3}\right)$ (c) $f(0.001)$

72. (a) $f\left(\sqrt[4]{\dfrac{1}{10}}\right)$ (b) $g\left(\sqrt[5]{\dfrac{1}{e^2}}\right)$ (c) $f\left(\sqrt[3]{\dfrac{1}{10^5}}\right)$

Developing and Extending the Concepts

In Problems 73–76, let $f(x) = \ln x$ and evaluate each logarithmic function by using a calculator. Round off each answer to 4 decimal places.

73. (a) $f(4) + f\left(\sqrt[3]{4}\right)$ (b) $f\left(4 + \sqrt[3]{4}\right)$

74. (a) $f\left(\dfrac{\sqrt{3}}{2}\right)$　　(b) $\dfrac{f(\sqrt{3})}{f(2)}$

75. (a) $f(\sqrt{7}-\sqrt{3})$　　(b) $f(\sqrt{7})-f(\sqrt{3})$

76. (a) $f(\sqrt{11})$　　(b) $\sqrt{f(11)}$

77. (a) Is log(ln 1) defined as a real number? Explain.

(b) Is ln(log 0.5) defined as a real number? Explain.

78. Let $f(x)=\log_4 x$, find and simplify $f(\log_2 16)$.

79. Suppose that $f(x)=2A\log x+3B$, where A and B are constant real numbers, and $f(1)=6$ and $f(10)=16$. Find A and B.

80. Let $f(x)=\log_2 x$, find and simplify $f(\log_5 625)$.

Objectives

1. Use the Properties of Logarithms
2. Solve Applied Problems

8.3 Properties of Logarithms

The properties that supported the historical computational use of logarithms are still helpful in simplifying logarithmic expressions. These properties are a direct consequence of the definition of logarithms. Recall from section 8.2 that

$$\log_b 1 = 0 \text{ and } \log_b b = 1.$$

Using the Properties of Logarithms

Since logarithms are exponents, by writing exponential equations in logarithmic form, we can develop the following additional properties.

Properties of Logarithms

Let b be a positive number, and $b \neq 1$, then

1. $\log_b b^x = x$
2. $b^{\log_b x} = x.$

These properties follow directly from the definition of logarithms, and some can be verified as follows:

1. $\log_b b^x = x$　　as an exponential form $b^x = b^x$
2. $b^{\log_b x} = x$　　because $\log_b x = \log_b x$ when written in logarithmic form

EXAMPLE 1　**Simplifying Logarithms**

Simplify each expression.

(a) $\log_5 5^3$　　　　(b) $6^{\log_6 7}$

Solution　Using the above properties, we have

(a) $\log_5 5^3 = 3$　　Property 1, $5^3 = 5^3$

(b) $6^{\log_6 7} = 7$　　Property 2, $\log_6 7 = \log_6 7$

In particular, if $b = 10$ or $b = e,$ we have the following properties:

1. $\log 1 = 0$　　　1. $\ln 1 = 0$
2. $\log 10 = 1$　　　2. $\ln e = 1$
3. $\log 10^x = x$　　3. $\ln e^x = x$
4. $10^{\log x} = x$　　4. $e^{\ln x} = x$

EXAMPLE 2   **Simplifying Logarithms**

Simplify each expression

(a) $e^{\ln 2}$ (b) $\ln e^4$ (c) $10^{\log 5}$ (d) $\log 10^4$ (e) $\ln e^{x^2 + 3x}$

Solution   Using the above properties for base 10 or base e, we have:

(a) $e^{\ln 2} = 2$ Property 4 where $x = 2$

(b) $\ln e^4 = 4$ Property 3 where $x = 4$

(c) $10^{\log 5} = 5$ Property 4 where $x = 5$

(d) $\log 10^4 = 4$ Property 3 where $x = 4$

(e) $\ln e^{x^2 + 3x} = x^2 + 3x$ Property 4 where the exponent is $x^2 + 3x$

So far we have introduced some basic properties of logarithms. We now introduce and explore additional basic properties. Because a logarithm is an exponent, logarithmic properties are restatements of exponential properties. These properties provide a basis for computational work with logarithms.

Basic Properties of Logarithms

For positive real numbers b, M, and N where $b \neq 1$, and r is any real number, then:

1. **The Product Property:** $\log_b MN = \log_b M + \log_b N$
2. **The Quotient Property:** $\log_b \dfrac{M}{N} = \log_b M - \log_b N$
3. **The Power Property:** $\log_b N^r = r \log_b N$

We verify these properties as follows:

1. Since $M = b^{\log_b M}$ and $N = b^{\log_b N}$,

$$MN = b^{\log_b M} \cdot b^{\log_b N} = b^{\log_b M + \log_b N}$$

By converting the equation to logarithmic form, we get
$\log_b MN = \log_b M + \log_b N$.

2. Since $\dfrac{M}{N} \cdot N = M$, we have

$$\log_b \left[\left(\frac{M}{N} N \right) \right] = \log_b M$$

$$\log_b \frac{M}{N} + \log_b N = \log_b M \qquad \text{Product property}$$

$$\log_b \frac{M}{N} = \log_b M - \log_b N. \qquad \text{Subtract } \log_b N \text{ from each side}$$

3. Since $N = b^{\log_b N}$, then

$$N^r = (b^{\log_b N})^r = b^{r \log_b N} \qquad \text{Power property for exponents}$$

$$\log_b N^r = r \log_b N. \qquad \text{Convert to logarithmic form}$$

The basic properties of logarithms for the special cases of the common and natural logarithms are stated in Table 1.

TABLE 1 Basic Properties of Common and Natural Logarithms

Common Logarithms Base 10	Natural Logarithms Base e
(i) $\log MN = \log M + \log N$	(i) $\ln MN = \ln M + \ln N$
(ii) $\log \dfrac{M}{N} = \log M - \log N$	(ii) $\ln \dfrac{M}{N} = \ln M - \ln N$
(iii) $\log N^r = r \log N$	(iii) $\ln N^r = r \ln N$

Warning: Make sure you understand the above properties, and beware that the properties for the product and quotient of logarithms assert that

1. The logarithm of a product is equal to the sum of the logarithms of the factors.
2. The logarithm of a quotient is equal to the difference of the logarithms of the dividend and divisor.

    (i) $\log_b (M + N) \neq \log_b M + \log_b N$
    (ii) $\log_b (M - N) \neq \log_b M - \log_b N$
    (iii) $\log_b MN \neq (\log_b M)(\log_b N)$
    (iv) $\log_b \dfrac{M}{N} \neq \dfrac{\log_b M}{\log_b N}$

EXAMPLE 3

Using Properties of Logarithms

Use the properties of logarithms to write each expression as a sum or a difference of multiples of logarithms.

(a) $\log_5 (7 \cdot 9)$ (b) $\ln (8 \cdot 15)$ (c) $\log_3 \dfrac{7}{19}$ (d) $\ln 5^{\frac{2}{3}}$

Solution

(a) $\log_5 (7 \cdot 9) = \log_5 7 + \log_5 9$ Product property

(b) $\ln (8 \cdot 15) = \ln 8 + \ln 15$ Product property

(c) $\log_3 \dfrac{7}{19} = \log_3 7 - \log_3 19$ Quotient property

(d) $\ln 5^{\frac{2}{3}} = \dfrac{2}{3}\ln 5$ Power property

The next example uses two or more of these properties in the same problem.

EXAMPLE 4

Using the Properties of Logarithms

Write each expression as a sum, difference, and/or multiple of logarithms.

(a) $\log_3 (5y^4)$ (b) $\log_2 \dfrac{x^3 y^2}{7}$

(c) $\ln[5(x + 3)^4]$ (d) $\ln\left[\dfrac{x(x + 3)^{\frac{2}{5}}}{6}\right]$

Solution

(a) $\log_3 (5y^4) = \log_3 5 + \log_3 y^4$ Product property

$= \log_3 5 + 4\log_3 y$ Power property

(b) $\log_2 \dfrac{x^3 y^2}{7} = \log_2(x^3 y^2) - \log_2 7$ Quotient property

$= \log_2 x^3 + \log_2 y^2 - \log_2 7$ Product property

$= 3\log_2 x + 2\log_2 y - \log_2 7$ Power property

(c) $\ln[5(x + 3)^4] = \ln 5 + \ln(x + 3)^4$ Product property

$= \ln 5 + 4\ln(x + 3)$ Power property

Remember that the logarithm of a sum cannot be simplified.

(d) $\ln\left[\dfrac{x(x + 3)^{\frac{2}{5}}}{6}\right] = \ln\left(x(x + 3)^{\frac{2}{5}}\right) - \ln 6$ Quotient property

$= \ln x + \ln (x + 3)^{\frac{2}{5}} - \ln 6$ Product property

$= \ln x + \dfrac{2}{5} \ln (x + 3) - \ln 6$ Power property

The following examples show how to change sums, differences, and/or multiples of logarithms to a single logarithm.

EXAMPLE 5   **Combining Logarithmic Expressions**

Write each expression as a single logarithm.

(a) $\log_3 15 + \log_3 13$ (b) $\ln (t - 2) + \ln (t + 2)$

(c) $3\log_2 x - \log_2 5x$ (d) $\log (x + 2) + 2\log (x + 4) - 4\log x$

Solution   We assume that all quantities whose logarithms are taken are positive.

(a) $\log_3 15 + \log_3 13 = \log_3 (15)(13)$ Product property

$= \log_3 195$

(b) $\ln(t - 2) + \ln(t + 2) = \ln (t - 2)(t + 2)$ Product property

$= \ln (t^2 - 4)$

(c) $3\log_2 x - \log_2 5x = \log_2 x^3 - \log_2 5x$ Power property

$= \log_2 \dfrac{x^3}{5x} = \log_2 \dfrac{x^2}{5}$ Quotient property

(d) $\log(x + 2) + 2\log(x + 4) - 4\log x$

$= \log (x + 2) + \log(x + 4)^2 - \log x^4$ Power property

$= \log (x + 2)(x + 4)^2 - \log x^4$ Product property

$= \log\left[\dfrac{(x + 2)(x + 4)^2}{x^4}\right]$ Quotient property

Solving Applied Problems

The properties of logarithms are used to rewrite formulas used in a variety of applications.

EXAMPLE 6   **Determining Atmospheric Pressure at Different Altitudes**

The amount of atmospheric pressure P (in pounds per square inch) at a height h (in miles) is given by the model

$$h = -\frac{1}{0.21}(\ln P - \ln 14.7).$$

(a) Rewrite the equation for this model for P in terms of h.

(b) Determine the atmospheric pressure 40 miles above sea level. Round off to four decimal places.

Solution

(a) $-\dfrac{1}{0.21}(\ln P - \ln 14.7) = h$ Original equation

$\ln P - \ln 14.7 = -0.21h$ Multiply each side by -0.21

$\ln \dfrac{P}{14.7} = -0.21h$ Quotient property

$\dfrac{P}{14.7} = e^{-0.21h}$ Convert to exponential form

$P = 14.7e^{-0.21h}$

(b) If $h = 40$ miles, then the above equation becomes

$$P = 14.7e^{-0.21(40)} = 0.0033.$$

Thus, at a height of 40 miles the atmospheric pressure is 0.0033 pounds per square inch.

 PROBLEM SET 8.3

Mastering the Concepts

In Problems 1–14, rewrite each expression as a sum or difference of multiples of logarithms. Assume all variables represent positive real numbers.

1. (a) $\log_3 5y$
    (b) $\log_2 7x$

2. (a) $\log_4 uv$
    (b) $\log_7 cd$

3. (a) $\log_5 \dfrac{x}{3}$
    (b) $\log_3 \dfrac{t}{11}$

4. (a) $\log_3 \dfrac{x}{15}$
    (b) $\log_4 \dfrac{p}{q}$

5. (a) $\log_4 x^{12}$
    (b) $\log_2 p^{2/3}$

6. (a) $\log_5 w^{13}$
    (b) $\log_2 u^{3/5}$

7. (a) $\ln \sqrt{Q}$
    (b) $\ln \sqrt[3]{P}$

8. (a) $\log \sqrt[5]{x}$
    (b) $\log \sqrt[4]{x}$

9. (a) $\log_3 \left(\dfrac{xy^2}{4}\right)$
    (b) $\log_3 \left(\dfrac{x^3 y}{4}\right)$

10. (a) $\ln (xy)^{-3/5}$
    (b) $\ln (x^2 y)^{4/3}$

11. (a) $\log (x^4 y)$
    (b) $\log \sqrt[5]{xy}$

12. (a) $\ln \sqrt{xy^3}$
    (b) $\ln \sqrt[7]{x^2 y}$

13. (a) $\log_5 \left(\dfrac{x^2}{y^4}\right)$
    (b) $\log_3 \left(\dfrac{x^7}{y^2}\right)$

14. (a) $\ln \dfrac{\sqrt{x}}{y^2}$
    (b) $\ln \dfrac{x^2}{\sqrt[3]{y}}$

In Problems 15–26, rewrite each expression as a single logarithm. Assume all variables represent positive real numbers.

15. (a) $\log_3 x + \log_3 4$
    (b) $\log_5 3 + \log_5 y$

16. (a) $\log_4 y + \log_4 7$
    (b) $\log u + \log v$

17. (a) $\ln 4 - \ln x$
    (b) $\log x - \log 3$

18. (a) $\log_4 4 - \log_4 v$
    (b) $\log w - \log u$

19. (a) $\log x^3 - \log x^2$
    (b) $\log_3 y^2 - \log_3 y$

20. (a) $\ln x^2 - \ln y$
    (b) $\log_3 u^2 + \log_3 v^3$

21. (a) $3\log_2 x + 2\log_2 x$
    (b) $2\log_3 y + \log_3 y^3$

22. (a) $2\ln x + 3\ln x^2$
    (b) $4\log y + \log y^2$

23. (a) $3\log z - 2\log y$
    (b) $3\log_2 4 - 2\log_2 v$

24. (a) $3\log u - 2\log v$
    (b) $5\log u - \log v$

25. $\log (x - 3) + \log (x + 3)$

26. $\ln (x + 5) + \ln (x - 5)$

In Problems 27–32, simplify each expression.

27. (a) $e^{2\ln 3}$
(b) $e^{-2\ln y}$

28. (a) $e^{-3\ln 2}$
(b) $e^{4\ln x}$

29. (a) $\ln e^{4x}$
(b) $\ln e^{-2y}$

30. (a) $\ln e^{\sqrt{x}}$
(b) $\ln e^{-4u}$

31. (a) $e^{1-\ln x}$
(b) $e^{2-3\ln x}$

32. (a) $e^{4+3\ln x}$
(b) $e^{x+2\ln x}$

Applying the Concepts

33. Chemistry: The pH of a solution is given by

$$\text{pH} = \log\left[\frac{1}{\text{H}^+}\right]$$

when H^+ is the concentration of the hydrogen ion. Use the properties of logarithms to express pH in terms of $\log \text{H}^+$.

34. Earthquakes: The seismologist Charles Richter established the measurement of the magnitude of an earthquake, in terms of its intensity, I, by means of the equation

$$M = \log I - \log I_o$$

where I_o is a certain minimum intensity. Use the properties of logarithms to write the right side of this equation as a single logarithm.

Developing and Extending the Concepts

In Problems 35–38, rewrite each expression in terms of $\log 2$ and $\log 5$.

35. (a) $\log 10$
(b) $\log 0.4$

36. (a) $\log 2.5$
(b) $\log \sqrt{20}$

37. (a) $\log 0.1$
(b) $\log 0.16$

38. (a) $\log 250$
(b) $\log 0.005$

39. Suppose that $\log_b 2 = 0.69$, $\log_b 3 = 1.10$, and $\log_b 5 = 1.61$. Find the value of each expression.
(a) $\log_b 7.5$ (b) $\log_b \sqrt{0.9}$ (c) $\log_b 1.5$

40. Simplify each expression
(a) $10^{\log 3x + \log 2x}$ (b) $e^{\ln 6x^4 - \ln 2x^2}$

In Problems 41–46, rewrite each expression as a sum or difference of multiples of logarithms. Assume all variables represent positive numbers.

41. (a) $\log_4 \sqrt[7]{x^4 y^5}$
(b) $\log_5 \sqrt[5]{x\, y^4}$

42. (a) $\log_7 \dfrac{x^3 \sqrt[4]{y}}{z^3}$
(b) $\log_4 \left(\dfrac{u^4 v^5}{\sqrt[4]{z^3}}\right)$

43. $\log\left(\dfrac{c^7 \sqrt[9]{d^2}}{5\sqrt{f}}\right)$

44. $\ln \sqrt{\dfrac{y}{y+3}}$

45. $\ln\left(4\sqrt[3]{x}\sqrt{y}\right)$

46. $\log\left(\dfrac{7}{\sqrt[3]{x^2 y^4}}\right)$

In Problems 47–52, rewrite each expression as a single logarithm.

47. (a) $\log_a \dfrac{x}{y} + \log_a \dfrac{y^2}{3x}$
(b) $\log_2 \dfrac{x^2}{y} - \log_2 \dfrac{x^4}{y^2}$

48. (a) $2\log_3 a - 3\log_3 b - 4\log_3 x$
(b) $5\log x - 5\log y - 7\log z$

49. $\ln \dfrac{a}{a-1} + \ln \dfrac{a^2-1}{a}$

50. $\log \dfrac{x+y}{z} - \log \dfrac{1}{x+y}$

51. $\log_3(x^2 - 25) - \log_3 (x^2 - 4x - 5)$

52. $\log (3x^2 - 5x - 2) - \log (x - 2)$

In Problems 53–56, use the properties of logarithms to combine the given expressions into a single logarithmic expression with a coefficient of 1. Assume all variables are positive.

53. $7 + 3\log_5 x$

54. $1 + 5 \log_3 \sqrt{y}$

55. $uv + \ln w$

56. $0.07x + \ln 100$

57. Show that $\log_3 (uv) = \dfrac{\ln u + \ln v}{\ln 3}$

58. Show that $\log_3 y + 2 = \log_3 9y$

Objectives

1. Graph Logarithmic Functions in Any Base
2. Graph Common and Natural Logarithms
3. Solve Applied Problems

8.4 Graphs of Logarithmic Functions

In this section, we explore logarithmic functions by graphing them. In doing so, we restate their relation to exponential functions.

Graphing Logarithmic Functions in Any Base

There are different ways to graph logarithmic functions. One way to graph $y = \log_b x$ is to convert its equation to the equivalent exponential form $x = b^y$. Then, we find some ordered pair solutions that satisfy the equation and use the **point-plotting** method to plot these point and connect them with a smooth curve.

EXAMPLE 1   **Graphing Logarithmic Functions**

Use the graph of each function to show the plotted points fit on the graph. From the graph, find the domain and range of each function.

(a) $y = \log_2 x$ (b) $y = \log_{1/2} x$

Solution   To graph these functions, we first convert

$$y = \log_2 x \text{ to its equivalent form } x = 2^y$$

and for

$$y = \log_{1/2} x \text{ we write } x = \left(\frac{1}{2}\right)^y.$$

Since each equation is solved for x, we choose y values and compute the corresponding x values. The following table shows such values for y and their corresponding x values.

Figure 1

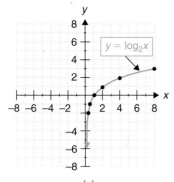

(a)

y	-3	-2	-1	0	1	2	3
$x = 2^y$	$\frac{1}{8}$	$\frac{1}{4}$	$\frac{1}{2}$	1	2	4	8
$x = \left(\frac{1}{2}\right)^y$	8	4	2	1	$\frac{1}{2}$	$\frac{1}{4}$	$\frac{1}{8}$

(a) Figure 1(a) shows the graph of

$$y = \log_2 x.$$

By plotting these points we see that fit on the smooth curve (Figure 1a). The domain of this function is all positive real numbers, $(0,\infty)$ and the range consists of all real numbers **R**.

(b) Figure 1b shows the graph of

$$y = \log_{1/2} x,$$

the plotted points fit on the smooth curve. The domain of the function is $(0,\infty)$ and the range is the set of real numbers **R**.

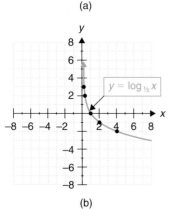

(b)

The previous example suggests that the graphs of all logarithmic functions have the following properties. Figure 2 shows the graph of $f(x) = \log_b x$ for $b > 1$ (Figure 2a) and for $0 < b < 1$ (Figure 2b).

Figure 2

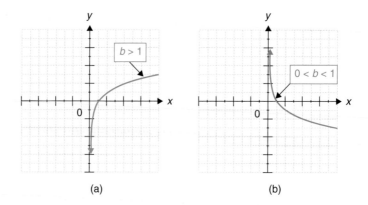

(a)   (b)

1. Domain: $(0,\infty)$	1. Domain: $(0,\infty)$
2. Range: **R**	2. Range: **R**
3. Increasing	3. Decreasing
4. *x* intercept: 1	4. *x* intercept: 1
5. *y* intercept: none	5. *y* intercept: none

The general behavior patterns for exponential and logarithmic functions of the same base are illustrated by the two graphs of the functions

$$y = 2^x \text{ and } y = \log_2 x$$

obtained by point-plotting on the same axes. Using these graphs, we make the following observations.

	Exponential Function	**Logarithmic Function**
Domain	**R**	$(0,\infty)$
Range	$(0,\infty)$	**R**
Points on Graph	(x,y)	(y,x)

switched

Figure 3

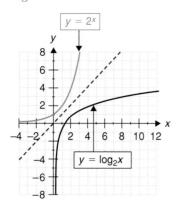

This important feature about the two graphs in Figure 3 is that they are reflections of each other across the line $y = x$. That is, if we fold the graph along the line $y = x$, the exponential and logarithmic functions will match point for point. Thus the graph of a logarithmic function can also be determined by reflecting the graph of its associated exponential function across the line $y = x$. We say that the two graphs are symmetric about the line $y = x$ and

$$y = 2^x \text{ and } y = \log_2 x$$

are *inverse functions*. We may therefore write:

if $f(x) = 2^x$, then the **inverse** of f, denoted by $f^{-1}(x)$, is $f^{-1}(x) = \log_2 x$.

In general,

If $f(x) = b^x$, $b > 0$, and $b \neq 1$, then $f^{-1}(x) = \log_b x$.

EXAMPLE 2 **Graphing Logarithmic Functions by Reflection**

Sketch the graph of $f(x) = 3^x$, then graph $g(x) = \log_3 x$ by reflecting the graph of f across the line $y = x$.

Figure 4

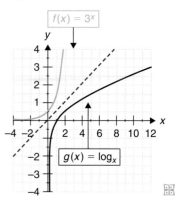

Solution Figure 4 shows the graph of $f(x) = 3^x$ and the line $y = x$. By reflecting the graph of $f(x) = 3^x$ across the line $y = x$, we obtain the graph of $g(x) = \log_3 x$. The domain of g is the set $(0, \infty)$, and the range of g consists of all real numbers **R**.

Graphing Common and Natural Logarithms

Two bases, 10 and e, are used in a variety of applications so that these functions require special attention.

EXAMPLE 3 **Analyzing a Graph of a Natural Logarithm Function**

Figure 5 displays the graph of

$$h(x) = \ln (x + 2).$$

Indicate the domain and range of h.

Figure 5

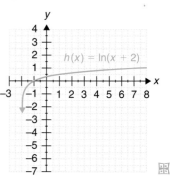

Solution The graph of $h(x) = \ln (x + 2)$ is obtained by shifting the graph of $y = \ln x$ horizontally 2 units to the left (Figure 5). The domain of h is $(-2, \infty)$ and its range consists of all real numbers **R**.

EXAMPLE 4 **Analyzing a Graph of a Common Logarithm Function**

Figure 6 displays the graphs of

$$f(x) = \log x \text{ and } g(x) = \log(-x)$$

Compare the graphs of f and g. Indicate the domain and range of g.

Figure 6

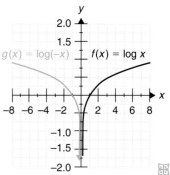

Solution The graph of $g(x) = \log (-x)$ is obtained by reflecting the graph of $f(x) = \log x$ across the y axis. The domain of g is $(-\infty, 0)$ and the range consists of all real numbers **R**.

Recall that since we can evaluate $\log_b x$ on a calculator by using the change of base formula, we can sketch the graph of such a logarithmic function on a grapher by using the change of base formula

$$h(x) = \log_b x = \frac{\log x}{\log b}.$$

Solving Applied Problems

Logarithmic functions are used to model phenomena such as advertising, manufacturing, growth and decay of populations, and other applications, as the next example shows.

EXAMPLE 5   **Finding the Time of a Skydiver**

The time t (in seconds) required for a skydiver in free fall (before the parachute opens) to reach a velocity v (in feet per second) is given by the function

$$t = -5.47 \ln(1 - 0.006v), \ 0 \le t \le 16.$$

Find the time required for the free fall velocity of the diver to be 100 feet per second, 130 feet per second, and 160 feet per second. Round off to two decimal places.

Solution   We make a table using a calculator to find t.

v	$t = -5.47 \ln(1 - 0.006v)$
100	$t = -5.47 \ln(1 - 0.006[100]) = 5.01$
130	$t = -5.47 \ln(1 - 0.006[130]) = 8.28$
160	$t = -5.47 \ln(1 - 0.006[160]) = 17.61$

Therefore, the times corresponding to the velocities in feet per second of 100, 130, and 160 are 5.01 seconds, 8.28 seconds, and 17.61 seconds respectively.

PROBLEM SET 8.4

Mastering the Concepts

1. Sketch the graph of $y = \log_3 x$:
    (a) by graphing $x = 3^y$.
    (b) by graphing $y = 3^x$ and reflecting the graph across the line $y = x$.
    (c) Find the domain and range of the function.

2. Sketch the graph of $y = \log_{\frac{1}{3}} x$:

    (a) by graphing $x = \left(\dfrac{1}{3}\right)^y = 3^{-y}$.

    (b) by graphing $y = \left(\dfrac{1}{3}\right)^x = 3^{-x}$ and reflecting
        the graph across the line $y = x$.

    (c) Find the domain and range of the function.

In Problems 3–6, each of the curves is the graph of a function of the form $y = \log_b x$. Find the base b if its graph contains the given point.

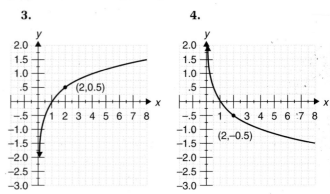

3.

4.

5.

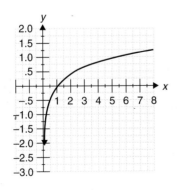

6.

In Problems 7–12, each of these points lies on the graph of $y = \log_b x$. Find the base b.

7. $(16,2)$

8. $(125,3)$

9. $\left(\dfrac{1}{8}, -3\right)$

10. $\left(\dfrac{1}{49}, -2\right)$

11. $\left(\dfrac{1}{27}, -3\right)$

12. $\left(\dfrac{1}{81}, -2\right)$

In Problems 13–20, sketch the graph of each function by converting each equation to exponential form. Then use the point-plotting method to sketch the graph. Find the domain and the x intercept.

13. $f(x) = \log_3 x$

14. $g(x) = \log_{\frac{1}{3}} x$

15. $f(x) = \log_4 x$

16. $g(x) = \log_{\frac{1}{4}} x$

17. $f(x) = \log_5 x$

18. $g(x) = \log_{\frac{1}{5}} x$

19. $f(x) = \log_6 x$

20. $g(x) = \log_{\frac{1}{6}} x$

In Problems 21–24, use the accompanying graph of the function $f(x) = \log_5 x$ in Figure 7 to assist you in matching each function with its graphs. Also use the graph to find the domain of each function.

Figure 7

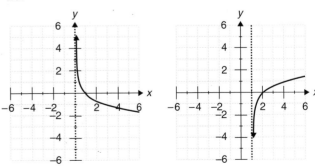

21. $f(x) = \log_5 (-x)$ **22.** $g(x) = \log_5 (x + 2)$

23. $h(x) = \log_5 (x - 1)$ **24.** $F(x) = -\log_5(x + 1)$

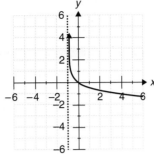
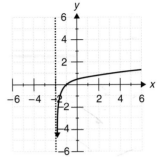

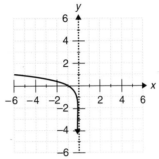
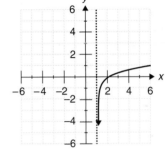

In Problems 25–28, use the accompanying graph of the function $f(x) = \log_3 x$ in Figure 8 to assist you in writing a rule for the logarithmic function whose graph is given.

Figure 8

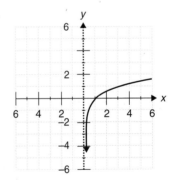

25.

26.

27. **28.**

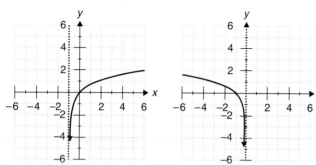

In Problems 29–36, use the graph of the first function and transformation of graphs to graph the second function. Also find the domain of each function.

29. $y = \log x$ and $f(x) = \log (x + 1)$
30. $y = \ln x$ and $f(x) = \ln (x + 1)$
31. $y = \ln x$ and $f(x) = \ln (x + 4)$
32. $y = \log x$ and $f(x) = \log (x - 2)$
33. $y = \log x$ and $f(x) = -\log (x + 1)$
34. $y = \log x$ and $f(x) = \log (1 - x)$
35. $y = \ln x$ and $f(x) = -\ln (-x)$
36. $y = \log x$ and $f(x) = -\log (x + 2)$

Applying the Concepts

37. **Advertising**: A company has determined that the number N of units of its product sold is related to the amount x (in thousand dollar increments) spent on advertising by the model

$$N = 3000 + 300 \log x, \quad x \geq 1.$$

(a) How many units are sold after spending 10 thousand dollars on advertising?

(b) Use the graph in Figure 9 to determine the number of units sold after spending $1000 on advertising.

(c) Use the graph to estimate how much should be spent on advertising if the number of units sold is 3200.

Figure 9

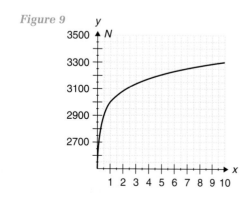

38. **Manufacturing:** A manufacturer of auto parts has determined that the weekly profit P (in dollars) is given by the model

$$P = 1830 + 900 \log x$$

where x is the number of parts produced each minute.

(a) What is the weekly profit (to the nearest dollar) if the number of parts produced each minute is 1000?

(b) Use the graph of P in Figure 10 to estimate the number of parts per minute (round off to the nearest whole number) that must be produced in order to have a weekly profit of approximately $4000.

Figure 10

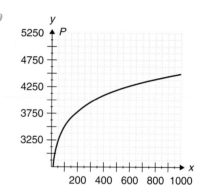

Developing and Extending the Concepts

In Problems 39–48, for each function use the change of base formula to write an equivalent function in terms of:

(a) $\log x$ (b) $\ln x$.

39. $f(x) = \log_3 x$ **40.** $f(x) = \log_5 x$
41. $g(x) = \log_{1/3} x$ **42.** $g(x) = \log_{3/7} x$
43. $h(x) = -2 \log_7 x$ **44.** $h(x) = -3 \log_4 x$
45. $F(x) = -3 \log_{2/3} x$ **46.** $F(x) = -\log_{2/3} x$
47. $g(x) = 2 + \log_3 x$ **48.** $(x) = 2 - \log_3 x$

In Problems 49–52, sketch the graphs of f and g on the same coordinate system. What conclusions can you draw?

49. $f(x) = 3^x$ and $g(x) = \log_3 x$

50. $f(x) = \left(\dfrac{1}{4}\right)^x$ and $g(x) = \log_{\frac{1}{4}} x$

51. $f(x) = e^x$ and $g(x) = \ln x$

52. $f(x) = e^{-x}$ and $g(x) = \ln\left(\dfrac{1}{x}\right)$

53. **Forgetting Curve:** A group of students had an average score of 70 on an examination. As part of an experiment, the students were tested on the same material at monthly intervals thereafter

Figure 11

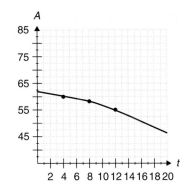

without further study. The researcher determined that the average score A (as a percent), after t months was approximated by the function

$$A(t) = 70 - 6 \ln(t + 1), \ t \geq 0.$$

(a) What was the average score after 4 months?

(b) Use the graph in Figure 11 to determine after how many months t was the average score 55.

54. Pollution: A lake polluted by bacteria is treated with a bactericidal agent. Environmentalists estimate that t days after the treatment, the

Figure 12

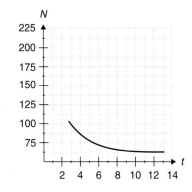

number N of viable bacteria per milliliter is given by the model

$$N(t) = 10t - 30 - 100 \ln(0.1t), \ 1 \leq t \leq 12.$$

(a) Find the number of bacteria per milliliter after 8 days.

(b) Use the graph in Figure 12 to estimate how many days after the treatment the number of bacteria per milliliter is 100.

In Problems 55 and 56, find $f^{-1}(x)$. Also sketch the graphs of f and f^{-1} on the same coordinate system.

55. $f(x) = 3^x$ **56.** $f(x) = \ln(4 - x)$

Objectives

1. Solve Special Exponential Equations

2. Solve Exponential Equations with Different Bases

3. Solve Logarithmic Equations

4. Solve Applied Problems

8.5 Exponential and Logarithmic Equations

Many applications and models make use of exponential and logarithmic functions. In this section, we develop methods to solve equations involving these functions.

Solving Special Exponential Equations

Recall from section 8.2 that we solved exponential equations such as $3^{2x + 4} = 27$ by using the exponential equality which requires that the bases be equal.

$$3^{2x + 4} = 27$$

So that

$$3^{2x + 4} = 3^3$$

$$2x + 4 = 3$$

$$2x = -1$$

$$x = -\frac{1}{2}$$

Thus, if the bases are equal or can be made equal, we have the tools to solve the equations. Now we must consider the situation where the bases are not equal nor can they be made equal easily.

Solving Exponential Equations with Different Bases

If we try to use the same approach as the one in the previous example to solve an exponential equation such as

$$3^x = 7,$$

we face the difficulty of expressing 7 as a power of 3. However we can solve this type of problem by using the properties of logarithms and the following property.

Logarithmic Equality

Let u, v, and b be positive numbers, with $b \neq 1$.

$$\log_b u = \log_b v \text{ if and only if } u = v.$$

To solve an equation such as $3^x = 7$, *we usually take the common or natural logarithm on both sides of the equation.* Let us consider some examples.

EXAMPLE 1 **Solving Exponential Equations with Different Bases**

Solve each equation.

(a) $5^x = 7$ (b) $e^{3x + 7} = 8$

Round off each answer to two decimal places.

Solution (a) We take the common logarithm of both sides of the equation:

$$5^x = 7 \qquad \text{Original equation}$$
$$\log 5^x = \log 7 \qquad \text{Take the common log of each side}$$
$$x \log 5 = \log 7 \qquad \text{Use the power property}$$
$$x = \frac{\log 7}{\log 5} \approx 1.21. \qquad \text{Simplify}$$

Therefore, the solution is approximately 1.21.

(b) Taking the natural logarithm of both sides of the equation, we have

$$e^{3x + 7} = 8 \qquad \text{Original equation}$$
$$\ln e^{3x + 7} = \ln 8 \qquad \text{Take ln of each side}$$
$$(3x + 7)\ln e = \ln 8 \qquad \text{Use the power property}$$
$$3x + 7 = \ln 8 \qquad \text{Note } \ln e = 1$$
$$x = \frac{\ln 8 - 7}{3} \approx -1.64. \qquad \text{Simplify}$$

Therefore, the solution is approximately -1.64.

Solving Logarithmic Equations

Recall from section 8.2 that to solve logarithmic equation involving logarithms such as

$$\log_2 (5x - 3) = 5,$$

we first convert the equation to the equivalent exponential form:

$$5x - 3 = 2^5$$

and then solve this equation.

$$5x - 3 = 32$$
$$5x = 35$$
$$x = 7$$

Also, solutions to logarithmic equations must be checked for extraneous roots. Logarithmic functions are defined only for positive real numbers.

Check: $\log_2 (5x - 3) = \log_2 [5(7) - 3] = \log_2 (32) = 5$ because $2^5 = 32$.

Next we will use the properties developed in section 8.3 to solve equations involving logarithms such as

$$\log x + \log (x + 1) = 2.$$

In this case, the properties of logarithms are needed to combine the terms of an equation before we may convert it to its equivalent exponential form.

EXAMPLE 2 **Solving a Logarithmic Equation**

Solve the equation $\log_2 (x^2 + 15x)^2 = 8$.

Solution First we use the power property to rewrite the equation as

$2 \log_2 (x^2 + 15x) = 8$	Power property
$\log_2 (x^2 + 15x) = 4$	Divide
$x^2 + 15x = 2^4$	Exponential equation
$x^2 + 15x = 16$	
$x^2 + 15x - 16 = 0$	Standard form
$(x + 16)(x - 1) = 0$	Factor
$x + 16 = 0 \qquad x - 1 = 0$	
$x = -16 \qquad x = 1$	

The solutions are -16 and 1. These solutions check in the original equation.

EXAMPLE 3 **Solving a Logarithmic Equation**

Remember that ln x has a base of e.

Solve the equation $\ln x + \ln(x - 2) = \ln(x + 4)$.

Solution

$\ln x + \ln(x - 2) = \ln(x + 4)$	Given
$\ln[x(x - 2)] = \ln(x + 4)$	Product property
$x(x - 2) = x + 4$	Logarithmic equality
$x^2 - 2x = x + 4$	
$x^2 - 3x - 4 = 0$	Standard form
$(x - 4)(x + 1) = 0$	Factored
$x - 4 = 0 \qquad x + 1 = 0$	
$x = 4 \qquad\quad x = -1$	

Check:

x	Left Side	Right Side	True/False
4	$\ln 4 + \ln(4 - 2) = \ln 4 + \ln 2 \approx 2.079$	$\ln (4 + 4) = \ln 8 \approx 2.079$	True
-1	$\ln (-1)$ not defined		False

The function $\ln x$ is not defined at $x = -1$, so -1 cannot be a solution for the equation. The solution of 4 makes the equation true and is the solution.

EXAMPLE 4   **Solving Logarithmic Equations**

Solve each equation.

(a) $\log x + \log (x + 3) = 1$

(b) $\log_4 (x + 3) - \log_4 x = 1$

Solution

Remember that log x has a base of 10.

(a) $\log x + \log (x + 3) = 1$ Original equation

$\log x(x + 3) = 1$ Product property

$x(x + 3) = 10^1$ Convert to exponential form

$x^2 + 3x - 10 = 0$ Standard form

$(x + 5)(x - 2) = 0$ Factor

$x + 5 = 0 \quad x - 2 = 0$

$x = -5 \quad x = 2$

Check:

x	Left Side	Right Side	True/False
-5	$\log(-5) + \log(-5 + 3)$ not defined	1	False
2	$\log (2) + \log (2 + 3) = \log (2) + \log (5) = 1$	1	True

Because the number -5 is not in the domain of the logarithmic function, it is not a solution and is called an extraneous root. The solution 2 makes the equation true.

(b) $\log_4 (x + 3) - \log_4 x = 1$ Given

$\log_4 \frac{x + 3}{x} = 1$ Quotient property

$\frac{x + 3}{x} = 4^1$ Convert to exponential form

$x + 3 = 4x$ Multiply each side by the LCD

$3 = 3x$

$1 = x$

Check:

x	Left Side	Right Side	True/False
1	$\log_4 (1 + 3) - \log_4 1 = \log_4 4 - \log_4 1 = 1$	1	True

Therefore, the solution is 1.

Solving Applied Problems

Recall that the functions

$$P(t) = P_0 e^{kt} \text{ and } P(t) = P_0 e^{-kt} \text{ for } k > 0$$

describe *exponential growth* and *exponential decay* models respectively. The **doubling time** for the model $P(t) = P_0 e^{kt}$ is the amount of time necessary for the population to double its size. The rate of decay k for the model $P(t) = P_0 e^{-kt}$ is measured in terms of its **half-life,** which is the period of time required for half of the quantity to decay.

EXAMPLE 5

Doubling Time of an Investment

An investment of $5000 offers an annual interest rate of 6% compounded quarterly. In t years, it will grow to the amount A given by the function

$$A(t) = 5000(1.015)^{4t}.$$

(a) How long will it take to accumulate $6,734 in the account?

(b) How long will it take for the amount of this investment to double in value?

Solution

(a) We replace A with 6,734 and solve the equation.

$$6{,}734 = 5000(1.015)^{4t} \qquad \text{Original equation}$$

$$(1.015)^{4t} = \frac{6734}{5000} \qquad \text{Divide each side by 5000}$$

$$\log(1.015)^{4t} = \log\left(\frac{6734}{5000}\right) \qquad \text{Take the logarithm of each side}$$

$$4t \log(1.015) = \log\left(\frac{6734}{5000}\right) \qquad \text{Power property}$$

$$t = \frac{\log\left(\dfrac{6734}{5000}\right)}{4 \log(1.015)} \approx 5$$

Therefore, it takes approximately 5 years to accumulate $6,734.

(b) Doubling an investment of $5,000 gives us a return of $10,000. Replacing A with $10,000, we have

$$10{,}000 = 5000(1.015)^{4t} \qquad \text{Original equation}$$

$$(1.015)^{4t} = 2 \qquad \text{Divide each side by 5000}$$

$$\log(1.015)^{4t} = \log 2 \qquad \text{Take the logarithm of each side}$$

$$4t \log (1.015) = \log 2 \qquad \text{Use the power property}$$

$$t = \frac{\log 2}{4 \log(1.015)} \approx 11.6.$$

Therefore, it takes approximately 11.6 years to double the amount.

EXAMPLE 6   **Finding the Half-Life of a Radioactive Material**

Potassium 42 is a radioactive isotope that is often used as a tracer by cardiologists. It decays at the rate of 5.4% per hour. The amount A remaining from an initial amount I, at t hours is given by the function

$$A(t) = Ie^{-0.054t}.$$

(a) Find the half-life of potassium 42. That is, find the time it takes for one-half of a sample of the isotope to decay. Round off to one decimal.

(b) How long will it take for 80% of the sample to decay? Round off to one decimal place.

Solution   (a) We must solve the equation $A(t) = Ie^{-0.054t}$ for t when $A(t) = 0.5I$.

$$0.5I = Ie^{-0.054t}$$

$$e^{-0.054t} = 0.5 \qquad\qquad \text{Divide both sides by } I$$

$$-0.054t = \ln 0.5 \qquad\qquad \text{Take the natural logarithm of both sides}$$

$$t = \frac{\ln 0.5}{-0.054} \approx 12.8$$

Thus, the half-life of the sample is approximately 12.8 hours.

(b) If 80% has decayed, then 20% is left. Solve the equation for t when $A(t) = 0.2I$.

$$0.2I = Ie^{-0.054t}$$

$$e^{-0.054t} = 0.2 \qquad\qquad \text{Divide both sides by } I$$

$$-0.054t = \ln 0.2 \qquad\qquad \text{Take the natural logarithm of both sides}$$

$$t = \frac{\ln 0.2}{-0.054} \approx 29.8$$

Therefore, it takes approximately 29.8 hours for 80% of the sample to decay. ▨

 PROBLEM SET 8.5

Mastering the Concepts

In Problems 1–20, solve each equation. Round off each answer to two decimal places.

1. $3^x = 5$
2. $3^{2x} = 27$
3. $9^x = 4$
4. $8^{x+2} = 11$
5. $7^x = 13$
6. $2^{5x} = 3^{x+1}$
7. $2^x = 13$
8. $3^{2x} = 5.70$
9. $4^{-x} = 3.4$
10. $5^{-x} = 8.9$
11. $7^{3x-2} = 48$
12. $e^{2x} = 3.7$
13. $e^{-4x} = 3.7$
14. $e^{2x+1} = 6.2$
15. $e^{2x+5} = 3.1$
16. $5e^{-0.14x} = 23$
17. $2e^{0.23x} = 43.8$
18. $4^{y+1} = 6.4$
19. $4^{8x-1} = 5^{1-3x}$
20. $e^{3-7x} = 4^{1-5x}$

In Problems 21–32, find the *exact* solution of each equation.

21. $\log_5 (2x - 7) = 2$
22. $\log_3 (2x + 3) = 2$
23. $\log_3 (x^2 + 2x) = 1$
24. $\log_2 (3x^2 - 2x) = 3$
25. $\log_2 (x^2 - 6x) = \log_2 40$
26. $\log_5 x = 5\log_5 2 - 2\log_5 4$
27. $\log_b x = 3\log_b 2 + \dfrac{1}{2} \log_b 25 - \log_b 20$
28. $\log_b x = \log_b 6 + \dfrac{1}{2} \log_b 16 - \log_b 216$
29. $\log (x^2 - 3x) = 1$
30. $\ln 7 + \ln x = 0$
31. $\ln 2 + \ln(x - 1) = 0$
32. $3\log x - \log(x^2) = 2$

In Problems 33–36, solve each equation. Round off each answer to two decimal places.

33. $\log_4 (3x + 1) = 0.6531$
34. $\log_2 (5x + 3) = 1.7435$
35. $\log (7x - 1) = 1.0374$
36. $\ln (8x - 3) = 0.7364$

In Problems 37–48, use the properties of logarithms to find the *exact* solution of each equation.

37. (a) $\log_4 x - \log_4 21 = \dfrac{1}{2}$
    (b) $\log_2 5x + \log_2 3 = 3$
38. (a) $\log_2 x + 2\log_2 3 = 3$
    (b) $\log_2 x - \dfrac{1}{2}\log_2 16 = 2$
39. $\log x + \log (x + 3) = 1$
40. $\log_{12} x + \log_{12} (x + 1) = 1$
41. $\log (3x - 2) - \log (2x + 1) = 0$
42. $\ln (3x + 4) - \ln (2x + 8) = 0$
43. $\log_5 x + \log_5 (x - 4) = 1$
44. $\log (x^2 - 9) - \log (x + 3) = 2$
45. $\log (x^2 - 4) - \log (x - 2) = 2$
46. $\log_4 x + \log_4 (6x + 10) = 1$
47. $\log_2 x + \log_2 (3x + 1) = 1$
48. $\log_3 x + \log_3 (x - 8) = 2$

Applying the Concepts

49. Investment Portfolio: Suppose that $7000 is invested in a savings account at an annual interest rate of 5% compounded quarterly. Use the model

$$A = 7000(1.0125)^{4t}$$

to determine the number of years for this investment to double. (Round off the answer to the nearest year.)

50. Investment Portfolio: Suppose that $5000 is invested in a mutual fund that earns an annual rate of return of 7% compounded continuously according to the model

$$A = 5000e^{0.07t}.$$

When will this portfolio triple itself?

51. Population Growth: In 1995, the population of Brazil was about 158 million people, and the population grows approximately according to the model

$$P = 158e^{0.025t}$$

where t is the number of years after 1995, with $t = 0$ representing 1995 and P is in millions of people.
(a) How many years after 1995 will the population of Brazil be 172 million? Round off to one decimal place.
(b) How many years after 1995 will the population double? Round off to one decimal place.

52. Population Growth: In 1990, the population of Albuquerque, New Mexico, was about 385,000, and the population grows approximately according to the model

$$P = 385(1.049)^t$$

where t is the number of years after 1990, with $t = 0$ representing 1990 and P is in thousands.
(a) In what year will the population be 1.3 million? Round off to the nearest year.
(b) In what year will the population double? Round off to the nearest year.

53. Salmon Fishing: In 1991, the total revenue for a West Coast state from salmon fishing was $18.8 million, and the yearly economic impact ever since is given by the model

$$R = 18.8(0.78)^t, \ 0 \le t \le 15$$

where t is the number of years after 1991 and revenue R is in millions of dollars. Use this model to predict when the revenue will be $20.1 million dollars. Round off to the nearest year.

54. Casino Gambling: Suppose that a "big time" gambler visited Las Vegas and decided to spend his $200 gambling allotment on a 25-cent slot machine that pays back 93%. His strategy is to set aside all his winnings each day for playing the next day. Suppose that his gambling strategy follows the model

$$A = 200(0.93)^t$$

where A is the number of dollars after t days of playing. How many days does he expect to play and end with a balance of $10? Round off to the nearest day.

55. Water Consumption: Suppose that the water consumption in a large city increases at the rate of 8% each year. After t years, the water consumption is given by the model

$$C = C_0(1.08)^t$$

where C (in billions of gallons) is the water consumption after t years and C_o is the water consumption when $t = 0$. In how many years will the water consumption triple? Round off the answer to the nearest year.

56. **Oceanography:** In oceanography, the *Bonguer-Lambert law* states that the intensity, P, of light (as a decimal) that penetrates the surface of the ocean to a depth of x feet is approximated by the model

$$P = e^{-0.015x}.$$

At what depth under the surface of the ocean will the intensity of light be 0.45% of the surface light intensity. Round off to the nearest foot.

57. **Ecology:** In 1986, the estimated number n of an endangered species of African antelope was 7500 and was decreasing after t years according to the model

$$n = 7500e^{-0.062t}.$$

In how many years will the population of antelope decline to 1500? Round off to the nearest tenth of a year.

58. **Archeology:** A wooden death mask was discovered in an ancient burial site in New Mexico. It was determined by archeologists that the amount A of carbon-14 left after t years is approximated by the model

$$A = Ie^{-0.000121t}$$

where I is the original amount of carbon-14. If 55% of the original carbon-14 of the wooden mask still remains, what is the approximate age of the mask? Round off to the nearest tenth of a year.

59. **Radioactive Decay:** The percentage P (in decimals) of radioactive carbon-14 found in a specimen t years after it begins to decay is modeled by the function

$$P = e^{-0.000121t}.$$

(a) After how many years of decay will 60% of it still be present?

(b) What is the half-life of this element?

60. **Radioactive Decay:** Certain radioactive elements decay exponentially. The decay model of cesium-137 after t years is given by the model

$$P = P_o e^{-0.023t}.$$

Find the half-life of this element.

61. **Recycling:** A beverage company estimates that about 60% of all aluminum cans distributed are recycled each year. If the company distributes 2.5 million cans, the number N of cans still in use after t years is given by the model

$$N = 2.5(0.60)^t.$$

How many years will it take for the number of cans in use to reach one-half of a million? Round off the answer to the nearest year.

62. **Present Value:** Suppose that $7225 is invested in a certificate of deposit (CD) and it will pay $10,000 at maturity. If the interest rate is 6.5% compounded continuously, how long will it take to reach maturity?

63. **Earth Science:** The atmospheric pressure P (in pounds per square inch) at an altitude of h feet above sea level is given by the model

$$\ln P = \ln 14.7 - 0.00005h.$$

At what altitude will the atmospheric pressure be 11.4 pounds per square inch? Round off to the nearest ten feet.

64. **Business:** A computer software company discovers that its annual revenue R (in thousands of dollars) is related to its research and development expenditures x (in tens of thousands of dollars) by the model

$$R = 3750 + 475\ln(2x + 2), \ x \geq 0.$$

How much money should be spent on research and development if the company's revenue is to be $5.5 million?

Developing and Extending the Concepts

In Problems 65–74, find the *exact* solution of each equation.

65. $3 + 4^{2x+1} = 35$

66. $\left(\dfrac{3}{5}\right)^{3-x} = \left(\dfrac{25}{9}\right)^{2x}$

67. $e^{x^2-1} = 1$

68. $\left(\sqrt{2}\right)^{3-2x} = \left(\sqrt[3]{2}\right)^{12x}$

69. $\dfrac{2^{x+1}}{8^x} = \dfrac{1}{2^{-x}}$

70. $\left(\dfrac{1}{5}\right)^{x^2} = (125)^{\frac{2}{3}x-1}$

71. $\log|3x + 1| = 0$

72. $\log(4x - 5) - \log(x - 2) = \log(2x + 1)$

73. $\ln x^2 = (\ln x)^2$

74. $(\ln x)^2 + \ln x^3 - 4 = 0$

75. Skydiving: After a skydiver jumps out of an airplane, the diver's speed S (in meters per second) is estimated by the model

$$S(t) = 20[1 - (0.6)^t], \ 0 \le t \le 15$$

where t represents the time of free fall in seconds.

(a) Estimate the speed of the skydiver after 6 seconds.

(b) Estimate the time when the speed of the skydiver is 19.66 meters per second.

76. Radioactive Decay: The number of years t it would take 200 milligrams of radium to decompose so that only x milligrams remain is given by the model

$$t = 2350[\ln 200 - \ln x].$$

Estimate the number of years it would take to have 18 milligrams remain.

77. Advertising: A company has determined that if its monthly advertising expenditure is x (in thousands of dollars), then the total monthly sales S (in thousands of dollars) is given by the model

$$S = 870\ln (x + 5).$$

What are the expected monthly sales total (to the nearest thousand dollars) when the monthly expenditure for advertising is (a) $10,000; (b) $50,000; and (c) $100,000?

78. Spread of AIDS: In 1985 the National Institute of Allergy and Infectious Diseases estimated that the number of cases of AIDS reported was 8250. The following table relates the time t (in years after 1985) to the number of cases reported.

Time t (in years)	Number N of Cases
0	8250
1	12,835
2	19,970
3	31,070
4	48,338
5	75,205
6	117,004

(a) Determine which of the following models fit these data:
$$N = 8250e^{0.84t} \quad N = 8250e^{0.442t} \quad N = 8250(2^{1.33t})$$

(b) Predict after how many years will 800,000 cases occur.

◆ CHAPTER 8 REVIEW PROBLEM SET

In Problems 1 and 2, find (a) $f(-3)$, (b) $f(-2)$, (c) $f(0)$, (d) $f(2.3)$, and (e) $f(\sqrt{2})$. Round off each answer to two decimal places.

1. $f(x) = 7^x$

2. $f(x) = e^{-2x}$

In Problems 3 and 4, sketch the graph of each function. Indicate the domain and range.

3. (a) $f(x) = 7^x$
   (b) $g(x) = e^{-x}$

4. (a) $f(x) = (0.3)^x$
   (b) $g(x) = e^{3x}$

In Problems 5–8, sketch the graph of each pair of functions on the same coordinate system. Compare the graphs of f and g.

5. (a) $f(x) = 7^x$
   (b) $g(x) = 7^x + 2$

6. (a) $f(x) = e^{-x}$
   (b) $g(x) = -e^{-x}$

7. (a) $f(x) = 2^x$
   (b) $g(x) = 2^{x-3}$

8. (a) $f(x) = 5^{-x}$
   (b) $g(x) = 5^{-x} - 3$

In Problems 9 and 10, simplify each expression. Round off each answer to two decimal places.

9. (a) $3^{\sqrt{2}} \cdot 3^{\sqrt{5}}$
   (b) $\left(7^{\sqrt{2}}\right)^{\sqrt{3}}$

10. (a) $7^{\sqrt{3}} \cdot 7^{-\sqrt{5}}$
   (b) $\left(0.3^{\sqrt{7}}\right)^{\sqrt{3}}$

In Problems 11 and 12, find the value of each expression. Round off each answer to two decimal places.

11. $\dfrac{(46,000)(0.105)}{12\left[1 - (1 + \frac{0.105}{12})^{-12(30)}\right]}$

12. $\dfrac{1,200,000}{1 + (1200 - 1)e^{-0.4(3)}}$

In Problems 13 and 14, write each equation in logarithmic form.

13. (a) $4^{-2} = \dfrac{1}{16}$
   (b) $5 = x^y$

14. (a) $\left(\dfrac{1}{3}\right)^{-2} = 9$
   (b) $3 = m^{-n}$

In Problems 15 and 16, write each equation in exponential form.

15. (a) $\log_\pi \pi = 1$

(b) $-2 = \log_7 \frac{1}{49}$

16. (a) $\log_{\frac{1}{3}} 1 = 0$

(b) $2 = \log_4 x$

In Problems 17 and 18, find the exact value of x.

17. (a) $\log_4 32 = x$

(b) $e^{x+4} = \frac{1}{e^{3x}}$

18. (a) $\log_x 128 = \frac{7}{3}$

(b) $\frac{3^{2x+1}}{27^x} = \frac{1}{9^x}$

In Problems 19 and 20, find the exact value of each logarithm.

19. (a) $\log_2 8$

(b) $\log_9 3$

(c) $\log_4 64$

(d) $\log_6 216$

20. (a) $\log_5 125$

(b) $\log_{125} 5$

(c) $\log_9 9$

(d) $\log_{100} 10{,}000$

In Problems 21–24, find each value. Round off each answer to 4 decimal places.

21. (a) $\log 7806$

(b) $\ln 7.601$

(c) $\log 0.0012$

(d) $\ln (0.0462)$

22. (a) $\log 312.5$

(b) $\ln 276$

(c) $\log 0.0072$

(d) $\ln(0.0043)$

23. (a) $\log(\ln 71)$

(b) $\ln(\log 13)$

(c) $\log(\ln 0.065)$

(d) $\ln(\log 0.065)$

24. (a) $\log(\ln 13)$

(b) $\ln(\log 27)$

(c) $\log(\ln 0.5)$

(d) $\ln(\log 0.05)$

In Problems 25 and 26, find x. Round off each answer to two decimal places.

25. (a) $\log x = 0.4133$

(b) $\ln x = 0.4871$

(c) $\log x = -1.7382$

26. (a) $\log x = 3.5412$

(b) $\ln x = -2.8913$

(c) $\ln x = 0.7813$

In Problems 27 and 28, find the exact value of x.

27. (a) $\log_4 16 = x$

(b) $\log_3 9\sqrt{3} = x$

(c) $\ln(17x - 33) = 0$

28. (a) $\log_{36} x = -1/2$

(b) $\log_{1/2} 16 = x$

(c) $\ln(3x - 11) = 0$

In Problems 29 and 30, rewrite each expression as a sum or difference of multiples of logarithms. Assume all variables are positive real numbers.

29. (a) $\log_6 7x$

(b) $\ln(4x^3)$

(c) $\log 4(xy^5)$

(d) $\ln \sqrt[4]{x^2 y^3}$

30. (a) $\ln 5x$

(b) $\log_3(5xy^2)$

(c) $\log_3 \sqrt[3]{xy}$

(d) $\ln \dfrac{\sqrt[3]{x}}{\sqrt[3]{54}}$

In Problems 31 and 32, rewrite each expression as a logarithm of one quantity. Assume all variables are positive real numbers.

31. (a) $\log \frac{3}{7} + \log \frac{14}{27}$

(b) $\ln(x + 1) - \ln x$

(c) $\ln (xy + y^2) - \ln(xz + yz)$

32. (a) $\log_3 \frac{5}{12} + \log_3 \frac{4}{15}$

(b) $5\ln x - \ln y$

(c) $\ln\left(\dfrac{x}{u} + x\right) - \ln\left(\dfrac{y}{u} + y\right)$

In Problems 33 and 34, simplify each expression. Assume all variables are positive real numbers.

33. (a) $e^{\ln x}$

(b) $e^{-2 - \ln 3}$

(c) $\ln e^{x^2 - 4}$

34. (a) $e^{-3\ln 2}$

(b) $e^{3 + 2\ln 2}$

(c) $\ln e^{x - x^2}$

In Problems 35–38, sketch the graphs of the following functions on the same coordinate system. Find the domain of each function.

35. (a) $f(x) = \log_3 x$

(b) $g(x) = -\log_3 x$

(c) $h(x) = \log_3 (-x)$

36. (a) $f(x) = \log_4 x$

(b) $g(x) = -\log_4 x$

(c) $h(x) = \log_4 (x + 1)$

37. (a) $f(x) = \ln x$

(b) $g(x) = \ln(x - 1)$

(c) $h(x) = \ln |x + 1|$

38. (a) $f(x) = -\ln x$

(b) $g(x) = \ln(-x)$

(c) $h(x) = \ln |x|$

In Problems 39–44, solve each equation for x. Round off when necessary to two decimal places.

39. (a) $2^{3x} = 8$

(b) $3^{x^2 - 2x} = 9$

(c) $8^{x^2} = 16^x$

40. (a) $3^{7x} = 81$

(b) $5^{x^2 - 3x} = 625$

(c) $3^{x^2 + 4x} = 9^{-2}$

41. (a) $2^x = 5$

(b) $7^{x + 1} = 11$

(c) $e^x = 5^x$

42. (a) $4^x = 3$

(b) $5^{x + 1} = 4$

(c) $8^{x^2} = 5^x$

43. (a) $\log x + \log (x - 48) = 2$

(b) $\ln(3x + 5) - \ln(2x + 6) = 0$

44. (a) $\log (4x + 1) - \log (2x + 9) = 0$

(b) $2\log (x + 2) + \log 12 = \log (x + 2)$

45. Compound Interest: An initial deposit of $700 earns 4.3% interest, compounded quarterly. How much money will be in the account after 4 years?

46. Compound Interest: A company invests $210,000 for its pension plan at 9.5% compounded monthly.
(a) What is the value of the plan after 5 years?
(b) If they decide to change plans after 3 years, what is the value of the plan after 3 years of investment? Round to the nearest dollar.

47. Population Growth: Suppose that at the start of an experiment, 3000 bacteria are present in a culture. After t hours, the number N in the culture is growing according to the model

$$N(t) = 3000e^{0.06t}.$$

(a) How many bacteria are present 8 hours later?
(b) After how many hours will the number of bacteria be 50,000? Round to the nearest whole number.

48. Cellular Phones: A telephone company estimates that the number N (in thousands) of cellular phones in use by its customers is given by the model

$$N(t) = 0.4(1.74)^t$$

where t is the number of years after 1997. Estimate the number of cellular phones which will be in use in 2011 and 2015. Round to the nearest whole number.

49. Real Estate Appreciation: Suppose that the value of real estate increases at a rate of 5% per year. The value V (in dollars) of a home purchased for P dollars after t years is modeled by the function

$$V(t) = P(1.05)^t.$$

If this trend continues, find the value of a home purchased for $75,000 and sold 10 years later. Round to the nearest hundred dollars.

50. Oceanography: The intensity I of light (in lumens) for a certain location at a distance x meters below the surface of an ocean is modeled by the function

$$I = I_0(0.4)^x$$

where I_0 is the intensity at the surface of the ocean. Find the intensity of light at a depth of 3 meters where I_0 is 8.

51. Tornadoes: The speed S of the wind (in miles per hour) near the center of a tornado is modeled by the equation

$$S(x) = 50 + 55\log x, \ 0 < x \le 20$$

where x is the distance (in miles) the tornado travels. On May 31, 1998, a tornado struck and destroyed the entire city of Spencer, South Dakota, covering a distance of approximately 11 miles. Approximate the speed of the tornado. Round to the nearest mile per hour.

52. Manufacturing: A manufacturer of computer monitors has determined that the weekly profit P (in dollars) is given by the model

$$P(x) = 2840 + 950\ln x$$

where x is the number of monitors produced each hour. What is the weekly profit (to the nearest dollar) if the number of monitors produced each hour is 500?

In Problems 53–58, use the given graph as a guideline to state the domain and range of each function.

53. $f(x) = e^x + e^{-x}$

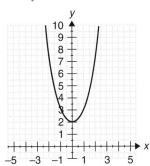

54. $f(x) = e^{-x^2}$

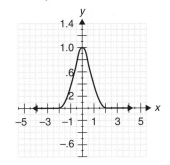

55. $f(x) = -2\log_{2/3} x$

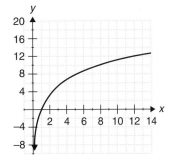

56. $f(x) = 4 + \log_3 x$

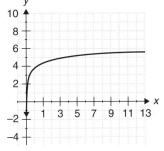

57. $f(x) = \ln x^2$

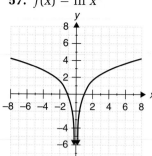

58. $f(x) = \ln \sqrt{x}$

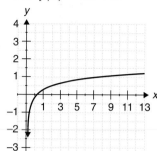

CHAPTER 8 PRACTICE TEST

1. For each function below, find $f(1.73)$, $f(-\sqrt{3})$, and $f(2 - \sqrt{2})$. Round off each answer to two decimal places.
 (a) $f(x) = 2^x$
 (b) $f(x) = (0.7)^x$
 (c) $f(x) = 4(1 + e^{-x})$

2. Sketch the graphs of the following functions on the same coordinate system. Describe the relationships between these functions.
 (a) $f(x) = 2^x$
 (b) $g(x) = -2^x$
 (c) $h(x) = 2^x + 3$

3. Write each equation in logarithmic form.
 (a) $8^{\frac{2}{3}} = 4$
 (b) $49^{-\frac{1}{2}} = \dfrac{1}{7}$

4. Write each equation in exponential form.
 (a) $\log_{25} \dfrac{1}{5} = -\dfrac{1}{2}$
 (b) $\log_{\frac{1}{2}} 16 = -4$

5. Find the *exact* value of each logarithm.
 (a) $\log_3 27$
 (b) $\log_{10} 0.001$
 (c) $\log_2 8\sqrt{2}$
 (d) $\log_5 0.04$

6. Solve each equation for x.
 (a) $\log_3 x = 2$
 (b) $\log_3 9\sqrt{3} = x$
 (c) $\left(\dfrac{1}{3}\right)^x = 9^{3x-1}$
 (d) $5^x = 11$
 (e) $\log_3(x + 2) - \log_3 x = 2$
 (f) $e^{3x} = 4$

7. (a) Find each value rounded to 4 decimal places.
  (i) $\log 17.6$
  (ii) $\ln(0.0892)$
  (iii) $\log_7 7^{32}$
 (b) Find x, round off each answer to two decimal places.
  (i) $\log x = 1.4023$
  (ii) $\ln x = 2.4163$
  (iii) $\log x = -1.3271$
  (iv) $\ln x = -2.3$

8. Sketch the graphs of the following functions on the same coordinate system. Find the domain of each function.
 (a) $f(x) = \log x$
 (b) $g(x) = \log(-x)$
 (c) $h(x) = \log(x + 1)$

9. Rewrite each expression as a sum or difference of multiples of logarithms. Assume all variables are positive real numbers.
 (a) $\log_3(8x)$
 (b) $\ln \dfrac{x}{7}$
 (c) $\ln(x^3 y^4)$
 (d) $\log \sqrt[3]{x^2 y}$

10. Rewrite each expression as a single logarithm. Assume all variables are positive real numbers.
 (a) $\log \dfrac{3}{8} + \log \dfrac{16}{15}$
 (b) $3\ln x - 2\ln y$
 (c) $\log_2(x + 2) + 3\log_2 x$

11. Find the value of each expression. Round off the answers to two decimal places.
 (a) $10^{-2.8192}$
 (b) $\log_7 23$
 (c) $e^{-0.72}$

12. Let $\log_2 3 = A$ and $\log_2 5 = B$. Write $\log_2 \dfrac{5\sqrt[3]{60}}{24}$ in terms of A and B.

13. Suppose that $5000 is invested in a high risk mutual fund that earns an annual rate of 17% compounded quarterly.
 (a) How much money is accumulated after 3 years?
 (b) When will this investment double itself?

14. The speed S of the wind (in miles per hour) near the center of a tornado is given by the model
$$S(x) = 45 + 55\log x, \quad 0 < x \le 20$$

where x is the distance (in miles) the tornado travels. On July 2, 1997 a tornado struck portions of the Detroit metropolitan area covering a distance of approximately 17 miles.

(a) Approximate the speed of the wind near the center of the tornado.

(b) Use the given graph to estimate the distance covered by a tornado whose speed is 100 miles per hour.

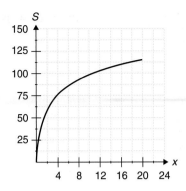

Topics in Algebra

Chapter Contents

Unlike the average classroom where the seats are arranged in rectangular fashion, the seats in many auditoriums are arranged in rows of increasing size. Below is the seating chart for the Macomb Center for the Performing Arts. In the first 12 rows of the center section, the first row has 11 seats, the second row has 12 seats, the third row has 13 seats, and so on in an arithmetic sequence. We will use the topics of this chapter to find the total number of seats in these rows. Example 6 of section 9.6 will explore the answer.

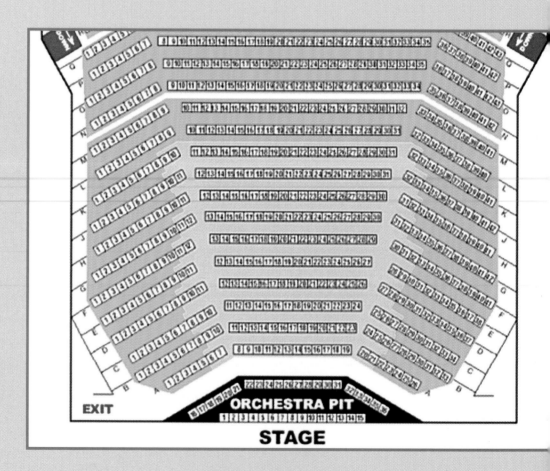

In this chapter, we develop an alternate method for solving systems of linear equations using determinants. We discuss the graphing of conic sections which are used in the graphical solution of nonlinear systems of equations. Finally, we will introduce three topics that are very important for advanced courses. The two main topics are *sequences* and *series* and they are the center of our attention in this chapter. With the growing availability of computers, these topics are playing increasingly important roles in real-world situations. The third topic is the *binomial theorem,* which provides a valuable technique for expanding binomials to any positive integer power.

Objectives

1. Find the Value of a Determinant
2. Apply Cramer's Rule to Linear Systems of Two Equations
3. Examine the Cases when D = 0 in Cramer's Rule
4. Apply Cramer's Rule to Linear Systems of Three Equations
5. Solve Applied Problems

9.1 Solution of Linear Systems Using Determinants and Cramer's Rule

So far, we have solved linear systems of equations by substitution, elimination, and graphing. In this section, we consider an alternative method for solving these systems called **Cramer's rule.**

Finding the Value of a Determinant

Cramer's rule is based on the idea of a *determinant.* If *a, b, c,* and *d* are any real numbers, the symbol

$$\begin{vmatrix} a & b \\ c & d \end{vmatrix} \text{ or } \det \begin{bmatrix} a & b \\ c & d \end{bmatrix}$$

is called a **2 by 2** (*also written as* **2 × 2**) **determinant** with *entries* or *elements a, b, c,* and *d.* Its value is defined by the number $ad - cb$; that is

$$\begin{vmatrix} a & b \\ c & d \end{vmatrix} = ad - cb.$$

EXAMPLE 1   **Finding the Value of a Determinant**

Find the value of each determinant.

(a) $\begin{vmatrix} 3 & 4 \\ 2 & 7 \end{vmatrix}$ (b) $\begin{vmatrix} 5 & 4 \\ 6 & -7 \end{vmatrix}$

(c) $\begin{vmatrix} 3 & 5 \\ 2 & 6 \end{vmatrix}$

Solution   (a) $\begin{vmatrix} 3 & 4 \\ 2 & 7 \end{vmatrix} = 3(7) - 2(4) = 21 - 8 = 13$

(b) $\begin{vmatrix} 5 & 4 \\ 6 & -7 \end{vmatrix} = 5(-7) - 6(4) = -35 - 24 = -59$

(c) $\begin{vmatrix} 3 & 5 \\ 2 & 6 \end{vmatrix} = 3(6) - 2(5) = 18 - 10 = 8$

Applying Cramer's Rule for Linear Systems of Two Equations

In 1750, Gabriel Cramer (1704–1752), a professor of mathematics in Geneva, Switzerland, proved the general rule for solving linear systems of equations which now bears his name. Today we write Cramer's rule as follows.

Cramer's Rule for Systems of Two Linear Equations

Consider the system of two linear equations:

$$\begin{cases} ax + by = h \\ cx + dy = k \end{cases}$$

The solution of this system is given by

$$x = \frac{D_x}{D} \quad \text{and} \quad y = \frac{D_y}{D}, D \neq 0$$

where

$$D = \begin{vmatrix} a & b \\ c & d \end{vmatrix}, D_x = \begin{vmatrix} h & b \\ k & d \end{vmatrix}, \text{and } D_y = \begin{vmatrix} a & h \\ c & k \end{vmatrix}.$$

In Cramer's rule, D is called the **coefficient determinant** because its entries are the coefficients of the unknowns in the system. Notice that D_x is obtained by replacing the *first* column of D (the coefficients of x) by the constant column on the right side of the system, and that D_y is obtained by replacing the *second* column of D (the coefficients of y) by the constant column on the right.

EXAMPLE 2   **Applying Cramer's Rule**

Solve the system

$$\begin{cases} 2x - y = 7 \\ x + 3y = 14 \end{cases}.$$

Solution   In this system

$$D = \begin{vmatrix} 2 & -1 \\ 1 & 3 \end{vmatrix} = 2(3) - 1(-1) = 7,$$

$$D_x = \begin{vmatrix} 7 & -1 \\ 14 & 3 \end{vmatrix} = 7(3) - 14(-1) = 35,$$

$$D_y = \begin{vmatrix} 2 & 7 \\ 1 & 14 \end{vmatrix} = 2(14) - 1(7) = 21.$$

Hence, by Cramer's rule,

$$x = \frac{D_x}{D} = \frac{35}{7} = 5 \text{ and } y = \frac{D_y}{D} = \frac{21}{7} = 3.$$

The solution is (5,3).

Examining the Cases when $D = 0$ in Cramer's Rule

Next we examine the two cases for which $D = 0$. These cases indicate geometrically that the system either involves two parallel lines or two coinciding lines.

EXAMPLE 3 **Examining Systems with $D = 0$**

Use Cramer's rule to solve the following systems.

(a) $\begin{cases} x - 2y = -6 \\ x - 2y = 2 \end{cases}$ (b) $\begin{cases} 3x - 4y = 20 \\ 6x = 8y + 40 \end{cases}$

Solution (a) In this system

$$D = \begin{vmatrix} 1 & -2 \\ 1 & -2 \end{vmatrix} = 1(-2) - 1(-2) = 0,$$

Figure 1

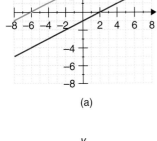

$$D_x = \begin{vmatrix} -6 & -2 \\ 2 & -2 \end{vmatrix} = -6(-2) - 2(-2) = 16, \text{ and}$$

$$D_y = \begin{vmatrix} 1 & -6 \\ 1 & 2 \end{vmatrix} = 1(2) - 1(-6) = 8.$$

Hence, $x = \dfrac{D_x}{D} = \dfrac{16}{0}$ and $y = \dfrac{D_y}{D} = \dfrac{8}{0}$, and so the system has no solution.

Actually, the system is inconsistent as can be seen from the graphs which show two parallel lines in Figure 1a.

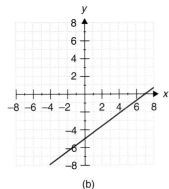

(a)

(b) First, we rewrite the system in the form

$$\begin{cases} 3x - 4y = 20 \\ 6x - 8y = 40 \end{cases}$$

and so,

$$D = \begin{vmatrix} 3 & -4 \\ 6 & -8 \end{vmatrix} = -24 - (-24) = 0,$$

$$D_x = \begin{vmatrix} 20 & -4 \\ 40 & -8 \end{vmatrix} = -160 - (-160) = 0, \text{ and}$$

$$D_y = \begin{vmatrix} 3 & 20 \\ 6 & 40 \end{vmatrix} = 120 - 120 = 0.$$

Hence, $x = \dfrac{D_x}{D} = \dfrac{0}{0}$ and $y = \dfrac{D_y}{D} = \dfrac{0}{0}$.

In this case, the system has an infinite number of solutions and the system is consistent and dependent. Figure 1b shows two lines which coincide. Keep in mind that Cramer's rule makes sense only when $D \neq 0$. If $D = 0$, the system is either inconsistent, or consistent and dependent.

Applying Cramer's Rule to Linear Systems of Three Equations

To extend Cramer's rule to linear systems of three equations in three variables, we begin by extending the definition of a determinant.

Definition of 3×3 Determinant

The **value** of a 3×3 determinant is defined as follows.

$$\begin{vmatrix} a_1 & a_2 & a_3 \\ b_1 & b_2 & b_3 \\ c_1 & c_2 & c_3 \end{vmatrix} = a_1 \begin{vmatrix} b_2 & b_3 \\ c_2 & c_3 \end{vmatrix} - a_2 \begin{vmatrix} b_1 & b_3 \\ c_1 & c_3 \end{vmatrix} + a_3 \begin{vmatrix} b_1 & b_2 \\ c_1 & c_2 \end{vmatrix}$$

We refer to this definition as the **expansion formula** for a 3×3 determinant.

In the expansion formula, notice that each entry in the first row is multiplied by the 2×2 determinant that remains when the row and column containing the multiplier are (mentally) crossed out. Thus,

$$a_1 \text{ is multiplied by } \begin{vmatrix} a_1 & a_2 & a_3 \\ b_1 & b_2 & b_3 \\ c_1 & c_2 & c_3 \end{vmatrix} = \begin{vmatrix} b_2 & b_3 \\ c_2 & c_3 \end{vmatrix}$$

$$a_2 \text{ is multiplied by } \begin{vmatrix} a_1 & a_2 & a_3 \\ b_1 & b_2 & b_3 \\ c_1 & c_2 & c_3 \end{vmatrix} = \begin{vmatrix} b_1 & b_3 \\ c_1 & c_3 \end{vmatrix}$$

$$\text{and } a_3 \text{ is multiplied by } \begin{vmatrix} a_1 & a_2 & a_3 \\ b_1 & b_2 & b_3 \\ c_1 & c_2 & c_3 \end{vmatrix} = \begin{vmatrix} b_1 & b_2 \\ c_1 & c_2 \end{vmatrix}$$

To **expand** a 3×3 determinant means to find its value by using the expansion formula. We emphasize: when expanding a 3×3 determinant, *do not forget the negative sign in front of the middle term.*

EXAMPLE 4   **Evaluating a 3×3 Determinant**

Use the expansion formula to evaluate the determinant.

$$\begin{vmatrix} 3 & 1 & 2 \\ -4 & 2 & 4 \\ 1 & 0 & 5 \end{vmatrix}$$

Solution

$$\begin{vmatrix} 3 & 1 & 2 \\ -4 & 2 & 4 \\ 1 & 0 & 5 \end{vmatrix} = 3\begin{vmatrix} 2 & 4 \\ 0 & 5 \end{vmatrix} - 1\begin{vmatrix} -4 & 4 \\ 1 & 5 \end{vmatrix} + 2\begin{vmatrix} -4 & 2 \\ 1 & 0 \end{vmatrix}$$

$$= 3[2(5) - 0(4)] - 1[-4(5) - 1(4)] + 2[-4(0) - 2(1)]$$

$$= 3(10) - (-24) + 2(-2) = 50$$

Now we can state Cramer's rule for solving a linear system of three equations in three variables.

Cramer's Rule for Solving 3×3 Linear Systems

The solution of the system

$$\begin{cases} a_1x + a_2y + a_3z = k_1 \\ b_1x + b_2y + b_3z = k_2 \\ c_1x + c_2y + c_3z = k_3 \end{cases}$$

is given by

$$x = \frac{D_x}{D}, \; y = \frac{D_y}{D}, \text{ and } z = \frac{D_z}{D}, \; D \neq 0$$

where

$$D = \begin{vmatrix} a_1 & a_2 & a_3 \\ b_1 & b_2 & b_3 \\ c_1 & c_2 & c_3 \end{vmatrix}, D_x = \begin{vmatrix} k_1 & a_2 & a_3 \\ k_2 & b_2 & b_3 \\ k_3 & c_2 & c_3 \end{vmatrix},$$

$$D_y = \begin{vmatrix} a_1 & k_1 & a_3 \\ b_1 & k_2 & b_3 \\ c_1 & k_3 & c_3 \end{vmatrix}, D_z = \begin{vmatrix} a_1 & a_2 & k_1 \\ b_1 & b_2 & k_2 \\ c_1 & c_2 & k_3 \end{vmatrix}.$$

EXAMPLE 5

Using Cramer's Rule to Solve a Linear System of Three Equations

Solve the system by Cramer's rule.

$$\begin{cases} x + 2y + 3x = 7 \\ 4x - y - 3z = 7 \\ 5x + 2y - z = 11 \end{cases}$$

Solution

Using Cramer's rule, we have

$$D = \begin{vmatrix} 1 & 2 & 3 \\ 4 & -1 & -3 \\ 5 & 2 & -1 \end{vmatrix} = 1 \begin{vmatrix} -1 & -3 \\ 2 & -1 \end{vmatrix} - 2 \begin{vmatrix} 4 & -3 \\ 5 & -1 \end{vmatrix} + 3 \begin{vmatrix} 4 & -1 \\ 5 & 2 \end{vmatrix}$$

$$= 1(7) - 2(11) + 3(13) = 24$$

$$D_x = \begin{vmatrix} 7 & 2 & 3 \\ 7 & -1 & -3 \\ 11 & 2 & -1 \end{vmatrix} = 7 \begin{vmatrix} -1 & -3 \\ 2 & -1 \end{vmatrix} - 2 \begin{vmatrix} 7 & -3 \\ 11 & -1 \end{vmatrix} + 3 \begin{vmatrix} 7 & -1 \\ 11 & 2 \end{vmatrix}$$

$$= 7(7) - 2(26) + 3(25) = 72$$

$$D_y = \begin{vmatrix} 1 & 7 & 3 \\ 4 & 7 & -3 \\ 5 & 11 & -1 \end{vmatrix} = 1 \begin{vmatrix} 7 & -3 \\ 11 & -1 \end{vmatrix} - 7 \begin{vmatrix} 4 & -3 \\ 5 & -1 \end{vmatrix} + 3 \begin{vmatrix} 4 & 7 \\ 5 & 11 \end{vmatrix}$$

$$= 1(26) - 7(11) + 3(9) = -24$$

$$D_z = \begin{vmatrix} 1 & 2 & 7 \\ 4 & -1 & 7 \\ 5 & 2 & 11 \end{vmatrix} = 1 \begin{vmatrix} -1 & 7 \\ 2 & 11 \end{vmatrix} - 2 \begin{vmatrix} 4 & 7 \\ 5 & 11 \end{vmatrix} + 7 \begin{vmatrix} 4 & -1 \\ 5 & 2 \end{vmatrix}$$

$$= 1(-25) - 2(9) + 7(13) = 48$$

so

$$x = \frac{D_x}{D} = \frac{72}{24} = 3, \; y = \frac{D_y}{D} = \frac{-24}{24} = -1, \text{ and } z = \frac{D_z}{D} = \frac{48}{24} = 2.$$

Thus, the solution is $(3, -1, 2)$.

Solving Applied Problems

Determinants can be used to solve applied problems from different fields.

EXAMPLE 6   **Solving a Puzzle Problem**

Suppose that you hire 4 people to do a small job that pays a sum of $125. What is the share of each individual if when the first person's share is increased by 4, the second's is decreased by 4, the third's is multiplied by 4, and the fourth is divided by 4, all these results are equal?

Solution   **Make a Plan**

Let x = the first person's share in dollars,

y = the second person's share in dollars,

z = the third person's share in dollars, and

$125 - x - y - z$ = the fourth person's share in dollars.

"All the results are equal" means that when the first person's share is increased by 4 that value is the same as the second person's share decreased by 4. The equation is:

$$x + 4 = y - 4.$$

The second person's share decreased by 4 is equal to the third person's share multiplied by 4. The equation is:

$$y - 4 = 4z.$$

The fourth person's share divided by 4 is the same as any one of the previous equations. We chose the last equation and the equation is:

$$4z = \frac{125 - x - y - z}{4}.$$

Then, we form the system as follows.

$$\begin{cases} x + 4 = y - 4 \\ y - 4 = 4z \\ 4z = \dfrac{125 - x - y - z}{4} \end{cases}$$

Write the Solution   First, we rewrite the system as follows.

$$\begin{cases} x - y & = -8 \\ \quad\ y - 4z = 4 \\ x + y + 17z = 125 \end{cases}$$

Then, we use Cramer's rule to solve the system.

$$D = \begin{vmatrix} 1 & -1 & 0 \\ 0 & 1 & -4 \\ 1 & 1 & 17 \end{vmatrix} = 25, \qquad D_x = \begin{vmatrix} -8 & -1 & 0 \\ 4 & 1 & -4 \\ 125 & 1 & 17 \end{vmatrix} = 400$$

$$D_y = \begin{vmatrix} 1 & -8 & 0 \\ 0 & 4 & -4 \\ 1 & 125 & 17 \end{vmatrix} = 600 \qquad D_z = \begin{vmatrix} 1 & -1 & -8 \\ 0 & 1 & 4 \\ 1 & 1 & 125 \end{vmatrix} = 125$$

so that,

$$x = \frac{D_x}{D} = \frac{400}{25} = 16, \ y = \frac{D_y}{D} = \frac{600}{25} = 24, \text{ and } z = \frac{D_z}{D} = \frac{125}{25} = 5.$$

Therefore, the shares are as follows:

The first = \$16, second = \$24, third = \$5, and the fourth = $125 - 16 - 24 - 5 = \$80$.

Look Back   The four conditions are satisfied:

the first share increased by 4 equals the second share decreased by 4:

$$16 + 4 = 24 - 4,$$

the second share decreased by 4 equals the third share multiplied by 4:

$$24 - 4 = 4(5),$$

the third share multiplied by 4 equals the fourth share divided by 4:

$$5(4) = \frac{80}{4}.$$

Finally, the sum of the shares is

$$125 = 16 + 24 + 5 + 80.$$

 PROBLEM SET 9.1

Mastering The Concepts

In Problems 1–8, evaluate each determinant.

1. (a) $\begin{vmatrix} 1 & -1 \\ 2 & 1 \end{vmatrix}$

   (b) $\begin{vmatrix} 2 & 3 \\ 5 & 1 \end{vmatrix}$

2. (a) $\begin{vmatrix} 3 & -2 \\ 8 & -1 \end{vmatrix}$

   (b) $\begin{vmatrix} 1 & 3 \\ 6 & -1 \end{vmatrix}$

3. (a) $\begin{vmatrix} 5 & -2 \\ 1 & 4 \end{vmatrix}$

   (b) $\begin{vmatrix} 5 & 1 \\ 2 & -3 \end{vmatrix}$

4. (a) $\begin{vmatrix} -10 & 1 \\ 3 & 2 \end{vmatrix}$

   (b) $\begin{vmatrix} 7 & 1 \\ -2 & 1 \end{vmatrix}$

5. $\begin{vmatrix} 1 & 2 & 4 \\ 2 & -3 & 1 \\ 3 & -1 & -2 \end{vmatrix}$

6. $\begin{vmatrix} 1 & 1 & 1 \\ 2 & -1 & -1 \\ 1 & -1 & 2 \end{vmatrix}$

7. $\begin{vmatrix} 1 & 1 & 1 \\ 2 & 3 & -1 \\ 3 & 5 & 1 \end{vmatrix}$

8. $\begin{vmatrix} 1 & 2 & 5 \\ 4 & 1 & 3 \\ 6 & 9 & -1 \end{vmatrix}$

In Problems 9–28, use Cramer's rule to solve each system.

9. $\begin{cases} 2x - y = 0 \\ x + y = 1 \end{cases}$

10. $\begin{cases} -3x + y = 3 \\ -2x - y = -5 \end{cases}$

11. $\begin{cases} 3x - 2y = -12 \\ 9x + 3y = -9 \end{cases}$

12. $\begin{cases} x + y = 30 \\ 2x - 2y = 25 \end{cases}$

13. $\begin{cases} 7x - 9y = -28 \\ 5x + 2y = 39 \end{cases}$

14. $\begin{cases} 8x - 2y = 52 \\ 3x - 5y = 45 \end{cases}$

15. $\begin{cases} 7x + 4y = 1 \\ 9x + 4y = 3 \end{cases}$

16. $\begin{cases} 3x + 7y = 16 \\ 2x + 5y = 13 \end{cases}$

17. $\begin{cases} -3x + y = 3 \\ -2x - y = -5 \end{cases}$

18. $\begin{cases} 3x + 2y = 7 \\ -2x + 7y = 12 \end{cases}$

19. $\begin{cases} 2x - y + z = 3 \\ -x + 2y - z = 1 \\ 3x + y + 2z = -1 \end{cases}$

20. $\begin{cases} x + y + z = 6 \\ 3x - y + 2z = 7 \\ 2x + 3y - z = 5 \end{cases}$

21. $\begin{cases} 2x - 3y = 4 \\ x + y - 2z = 1 \\ x - y - z = 5 \end{cases}$

22. $\begin{cases} 3x + 2y + 2z = 6 \\ x - 5y + 6z = 2 \\ 6x - 8y = 12 \end{cases}$

23. $\begin{cases} x - y + z = 3 \\ 2x + 3y - 2z = 5 \\ 3x + y - 4z = 12 \end{cases}$

24. $\begin{cases} x + y + z = 4 \\ x - y + 2z = 8 \\ 2x + y - z = 3 \end{cases}$

25. $\begin{cases} x + y + z = 2 \\ 2x + 3y - z = 3 \\ 3x + 5y + z = 8 \end{cases}$

26. $\begin{cases} x + 2y + 5z = 4 \\ 4x + y + 3z = 9 \\ 6x + y + z = 21 \end{cases}$

27. $\begin{cases} x + y + z = 6 \\ x - y + 2z = 12 \\ 2x + y + z = 1 \end{cases}$

28. $\begin{cases} x + 3y - z = 4 \\ 3x - 2y + 4z = 11 \\ 2x + y + 3z = 13 \end{cases}$

Applying the Concepts

In Problems 29–38, write a linear system of equations. Then use Cramer's rule to solve the system.

29. **Geometry:** Find the measure of each of two complementary angles if the measure of one angle is twice the measure of the other.

30. **Marketing:** A packinghouse dealer paid a total of $27,700 for two kinds of avocados – Hass and Bacon. The Hass were sold at a profit of 30% on the dealer's cost, but the Bacon started to spoil, resulting in a selling price of 3% loss on the dealer's cost. If a profit of $5963.70 was made on the total transaction, how much did the dealer pay for each kind of avocado?

31. **Student Loans:** A student's loans totaled $10,500. Part was a bank loan at 8.5% annual simple interest, the rest was a federal education loan at 8% annual simple interest. After one year, the interest accumulated from both loans was $860. What was the original amount of each loan?

32. **Boating:** It took a motorboat 3 hours to make a trip downstream when the current is 6 miles per hour. It took 5 hours for the return trip against the current. What is the speed of the boat in still water?

33. **Nutrition:** A nutritionist wants to create a special diet out of three substances: A, B, and C. The requirements in the diet are 480 units of calcium, 190 units of iron, and 360 units of vitamins. The calcium, iron, and vitamin contents in substances A, B, and C are listed in Table 1. How many ounces of each substance are needed?

TABLE 1

Substances	Calcium Units per Ounce	Iron Units per Ounce	Vitamin Units per Ounce
A	30	10	15
B	20	10	10
C	10	5	20

34. **Production:** Three assembly lines, A, B, and C, can produce 8400 TV dinners per day. Together, lines A and B can produce 4900 TV dinners, while lines B and C together can produce 5600 TV dinners. Find the number of TV dinners each assembly line can produce alone.

35. **Horticulture:** A horticulturist wishes to mix three types of fertilizer which contain 25%, 35%, and 40% nitrogen, respectively, in order to get a mixture of 4000 pounds of $35\frac{5}{8}\%$ nitrogen. The final mixture should contain three times as much of the 40% type as the 25% type. How much of each type is in the final mixture?

36. **Fast Food Chain:** A fast food chain sells three types of franchises, A, B, and C. Franchise A sells for $20,000, franchise B for $25,000, and franchise C for $30,000. In one year, the company sold 18 franchises for a total of $430,000. If the number of franchise A sold was twice the number sold of franchise C, how many of each type did the company sell that year?

37. **Resource Management:** A natural resource department supplies three types of food to support species A, B, and C of fish in a lake environment. Each week 25,000 units of Food I, 20,000 units of Food II, and 47,000 units of Food III are supplied to the lake. Suppose that a, b, and c denote the number of fish in the lake of species A, B, and C respectively. Assume that all food is consumed and the units of Food I, Food II, and Food III eaten by the fish are listed in Table 2. How many fish of each species can coexist in the lake?

TABLE 2

Species	Units of Food I	Units of Food II	Units of Food III
A	1	1	3
B	3	4	5
C	2	1	2

38. **Drinking Water:** A town uses three pumps to provide its drinking water. When pumps A, B, and C are all working, they can pump 74,000 gallons of water into a tank in 2 hours. If pump A runs for 4 hours and pump B runs for 2 hours, together they can pump 64,000 gallons of water. If pump B runs for 5 hours and pump C for 4 hours, together they can pump 120,000 gallons of water. How many gallons per hour does each pump provide?

Developing and Extending the Concepts

39. Find the value of $\begin{vmatrix} \sqrt{6} & -3\sqrt{5} \\ 2\sqrt{5} & 7\sqrt{6} \end{vmatrix}$.

40. Find the value of $\begin{vmatrix} \sqrt{7}-\sqrt{3} & 5+\sqrt{2} \\ 5-\sqrt{2} & \sqrt{7}+\sqrt{3} \end{vmatrix}$.

In Problems 41–46, solve each equation for x.

41. $\begin{vmatrix} x & x \\ 5 & 3 \end{vmatrix} = 2$

42. $\begin{vmatrix} x & -x \\ 5 & 3 \end{vmatrix} = 2$

43. $\begin{vmatrix} x+1 & x \\ x & x-2 \end{vmatrix} = -6$

44. $\begin{vmatrix} x-2 & 1 \\ x+1 & -2 \end{vmatrix} = 35$

45. $\begin{vmatrix} x & 4 & 5 \\ 0 & 1 & x \\ 5 & 2 & 0 \end{vmatrix} = 7$

46. $\begin{vmatrix} 5x & 0 & 1 \\ 2x & 1 & 2 \\ 3x & 2 & 3 \end{vmatrix} = 0$

47. (a) Find the value of $\begin{vmatrix} 1 & 0 & 2 \\ 4 & 6 & -1 \\ -1 & 0 & -1 \end{vmatrix}$.

(b) Interchange the first and the second rows, then evaluate the "new" determinant.

(c) Compare the answers of (a) and (b).

48. (a) Find the value of $\begin{vmatrix} 5 & 8 & 3 \\ 0 & 2 & 10 \\ 0 & 0 & 7 \end{vmatrix}$.

(b) Find the product of the three numbers on the diagonal from the upper left to the bottom right.

(c) Compare the answers of (a) and (b).

49. (a) Find the value of the determinant $\begin{vmatrix} 1 & 0 & 2 \\ 1 & 1 & 1 \\ 2 & 1 & -1 \end{vmatrix}$.

(b) Multiply the second row of part (a) by 10 and find the value of the determinant.

(c) Compare the answers of (a) and (b).

50. (a) Find the value of the determinant $\begin{vmatrix} 2 & 1 & -1 \\ 1 & 0 & 0 \\ 3 & 1 & 0 \end{vmatrix}$.

(b) Interchange the first and third rows of part (a) and find the value of the determinant.

(c) Compare the answers of (a) and (b).

Objectives

1. Graph an Ellipse with Center at (0,0)

2. Graph a Hyperbola with Center at (0,0)

3. Solve Applied Problems

9.2 Conic Sections: The Ellipse and Hyperbola

The term *Conic Section* (or *Conics*) refers to various figures or sections formed by intersecting a right circular cone with a plane. If a plane is perpendicular to the axis of the cone, a *circle* is formed (Figure 1a). When the plane is not perpendicular to the axis of the cone, its intersection forms a *parabola* (Figure 1b), or an *ellipse* (Figure 1c), or a *hyperbola* (Figure 1d).

In sections 7.6 and 7.7, we discussed the circle and the parabola. In this section, we introduce the *ellipse* and the *hyperbola*.

Figure 1

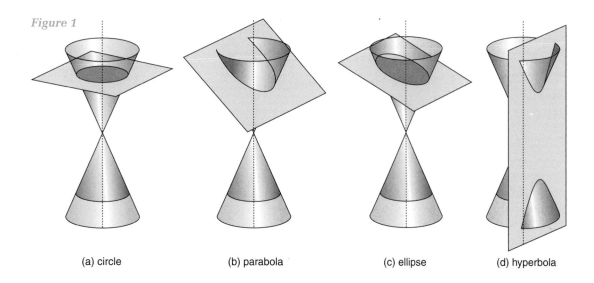

(a) circle   (b) parabola   (c) ellipse   (d) hyperbola

Graphing an Ellipse with Center at (0, 0)

Ellipses are useful in providing mathematical models of a variety of physical phenomena ranging from art to astronomy. The geometric definition of an ellipse follows:

Ellipse

> An **ellipse** is the set of all points P in a plane such that the sum of the distances from P to two fixed points is a constant. The fixed points are called the **foci** (plural of **focus**) of the ellipse.

We can use this definition to draw an ellipse by taking a string of length k and fastening its ends at two fixed points, F_1 and F_2. Figure 2a shows two fixed points, F_1 and F_2, and a string of length k stretched tightly about them to the point P. Hence, as P moves about, $|\overline{PF_1}| + |\overline{PF_2}|$ always has a constant value k. That is, as Figure 2b shows,

$$d_1 + m_1 = d_2 + m_2 = k \text{ , where } k \text{ is a constant.}$$

Thus, if a pencil point P is inserted into the string and moved so as to keep the string tight, it traces out an ellipse.

The ellipse in Figure 2b has foci $F_1 = (-c,0)$ and $F_2 = (c,0)$ and y intercepts at the points $V_1 = (0,b)$ and $V_2 = (0,-b)$, where $c > 0$ and $b > 0$. The line segment $\overline{V_3V_4}$ where $V_3 = (-a,0)$ and $V_4 = (a,0)$ is called the **major axis,** and the line segment $\overline{V_1V_2}$ is called the **minor axis.** The foci, F_1 and F_2, lie on the major axis. The **center** of an ellipse is the midpoint of the major (or minor) axis. The **vertices** of an ellipse are the endpoints of the major axis.

Figure 2

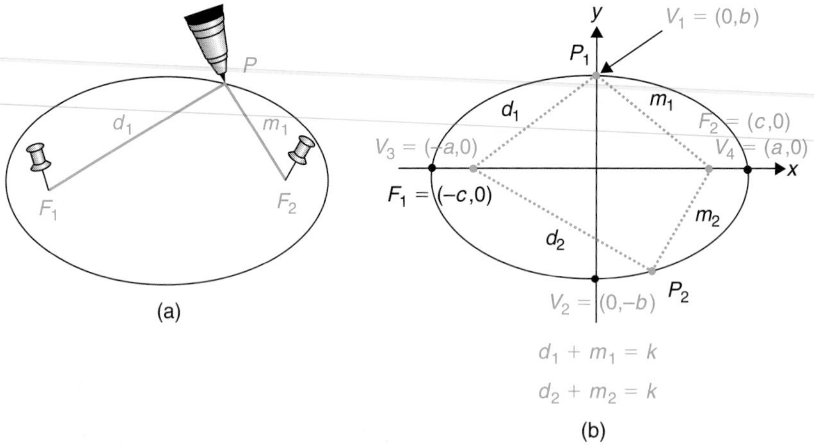

(a)

$d_1 + m_1 = k$

$d_2 + m_2 = k$

(b)

Figure 3

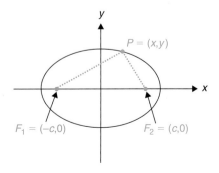

As with the parabola, we can use the definition of the ellipse to derive its standard equation. We choose the x axis as the line containing the foci $F_1 = (-c,0)$ and $F_2 = (c,0)$, where $c > 0$. The origin is the midpoint of the line segment $\overline{F_1F_2}$. Also, we assume that the constant sum of the distances between any point P on the ellipse and the foci is $2a$. (This constant is written in the form $2a$ so that the equation of the ellipse will have a simple form.) In triangle F_1PF_2 (Figure 3), we know from geometry that the sum of the lengths of the two sides of the triangle, $(|\overline{PF_1}| + |\overline{PF_2}| = 2a)$, is greater than the length of the third side, $(|\overline{F_1F_2}| = 2c)$.

Thus, $2a > 2c$ or $a > c$. If we let $P = (x,y)$, we get

$$d(P,F_1) + d(P,F_2) = 2a.$$

Using the distance formula and simplifying we get the following equations of the ellipse.

Equations of an Ellipse with Center at (0,0)

1. The **equation of an ellipse** centered at the origin with foci at $F_1 = (-c,0)$ and $F_2 = (c,0)$ with vertices at $V_1 = (-a,0)$ and $V_2 = (a,0)$ and y intercepts $-b$ and b is given by

$$\frac{x^2}{a^2} + \frac{y^2}{b^2} = 1, \text{ where } a^2 = b^2 + c^2 \text{ (Figure 4a).}$$

2. The **equation of an ellipse** centered at the origin with foci at $F_1 = (0,-c)$ and $F_2 = (0,c)$ with vertices at $V_1 = (0,-a)$ and $V_2 = (0,a)$ and x intercepts $-b$ and b is given by

$$\frac{x^2}{b^2} + \frac{y^2}{a^2} = 1, \text{ where } a^2 = b^2 + c^2 \text{ (Figure 4b).}$$

Figure 4

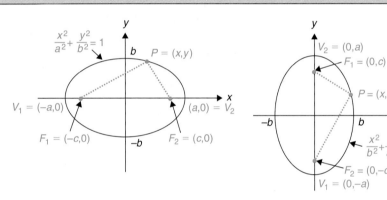

(a) Horizontal major axis   (b) Vertical major axis

EXAMPLE 1

Graphing an Ellipse

Identify the intercepts, the vertices, the foci, and sketch the graph of each ellipse.

(a) $4x^2 + 9y^2 = 36$ (b) $25x^2 + 9y^2 = 225$

Solution

(a) To find the x intercepts, let $y = 0$ and solve for x. So

$$x^2 = 9 \text{ or } x = \pm\, 3.$$

To find the y intercepts, let $x = 0$ and solve for y, so

$$y^2 = 4 \text{ or } y = \pm\, 2.$$

We write the equation in standard form by dividing each side by 36.

$$\frac{x^2}{9} + \frac{y^2}{4} = 1$$

Therefore, the vertices are

$$(-3,0) \quad \text{and} \quad (3,0).$$

Here $a = 3$ and $b = 2$, so that $c^2 = a^2 - b^2 = 9 - 4 = 5$ so, $c = \pm\sqrt{5}$. Hence, the foci are $\left(-\sqrt{5}, 0\right)$ and $\left(\sqrt{5}, 0\right)$. Figure 5a shows the graph of the ellipse.

(b) To find the x intercepts, let $y = 0$.

$$x^2 = 9 \text{ or } x = \pm 3$$

To find the y intercept, let $x = 0$.

$$y^2 = 25 \text{ or } y = \pm 5$$

We write the equation in standard form by dividing each side by 225.

$$\frac{x^2}{9} + \frac{y^2}{25} = 1$$

Therefore, the vertices are $(0,-5)$ and $(0,5)$.

Here, $a = 5$ and $b = 3$, so $c^2 = a^2 - b^2 = 25 - 9 = 16$ or $c = \pm 4$. Hence, the foci are $(0,-4)$ and $(0,4)$. Figure 5b shows the graph of the ellipse.

Figure 5

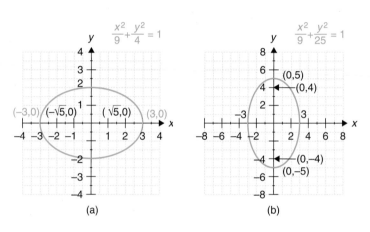

(a)   (b)

EXAMPLE 2   **Finding an Equation of an Ellipse**

Find an equation of the ellipse with foci

$$F_1 = (0,-4) \text{ and } F_2 = (0,4),$$

and the vertices

$$V_1 = (0,-6) \text{ and } V_2 = (0,6).$$

Sketch the graph.

Solution   The equation of the ellipse we are to determine is of the form

$$\frac{x^2}{b^2} + \frac{y^2}{a^2} = 1.$$

So we must find a^2 and b^2. Because the foci are on the y axis, the vertical axis is the major axis of the ellipse. Since the vertices are $(0,-6)$ and $(0,6)$, the value of a is 6. Also with the foci at $(0,-4)$ and $(0,4)$, this means that $c = 4$. We compute b^2 as follows.

$$b^2 = a^2 - c^2 = 36 - 16 = 20$$

Figure 6

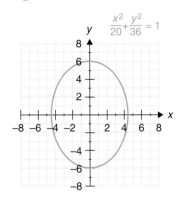

$$\frac{x^2}{20} + \frac{y^2}{36} = 1$$

Hence, an equation of the ellipse is

$$\frac{x^2}{20} + \frac{y^2}{36} = 1.$$

The graph is in Figure 6.

Graphing a Hyperbola with Center of (0,0)

As with the ellipse, we give a geometric definition of a hyperbola.

Hyperbola

> A **hyperbola** is the set of all points P in the plane such that the absolute value of the difference of the distances from P to two fixed points is a constant positive number. The two fixed points are called the **focal points,** or **foci,** of the hyperbola. The point halfway between the foci is the **center** of the hyperbola.

Figure 7 shows a geometric description of a hyperbola with foci F_1 and F_2 and with center C. The points P_1 and P_2 are on the hyperbola, and

$$|d_1 - m_1| = |d_2 - m_2| = k.$$

Suppose that a hyperbola is positioned in a Cartesian coordinate system in such a way that the center is at the origin and two foci $F_1 = (-c,0)$ and $F_2 = (c,0)$ lie on the x axis, with $c > 0$. To derive the standard equation of the hyperbola analytically, we let $P = (x,y)$ be a point on the hyperbola, and assume that the constant difference k equals $2a$ (Figure 8). Since $P = (x, y)$, we have

$$|d(P,F_1) - d(P,F_2)| = 2a.$$

Equations in standard form for hyperbolas may be derived from the geometric definition. As we shall see the hyperbola may intercept the x or the y axis depending on its standard equation.

Figure 7

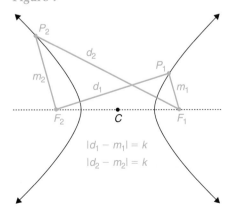

Figure 8

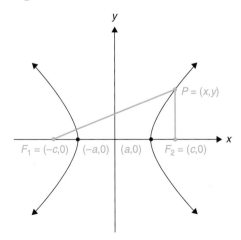

Standard Equations of a Hyperbola

1. The **standard equation for a hyperbola** centered at the origin with vertices at $V_1 = (-a,0)$ and $V_2 = (a,0)$ and foci at $F_1 = (-c,0)$ and $F_2 = (c,0)$ is given by

$$\frac{x^2}{a^2} - \frac{y^2}{b^2} = 1$$

where $a^2 + b^2 = c^2$ (Figure 9a).

2. The **standard equation for a hyperbola** centered at the origin with vertices at $V_1 = (0,-b)$ and $V_2 = (0,b)$ and foci at $F_1 = (0,-c)$ and $F_2 = (0,c)$ is given by

$$\frac{y^2}{b^2} - \frac{x^2}{a^2} = 1$$

where $a^2 + b^2 = c^2$ (Figure 9b).

Figure 9

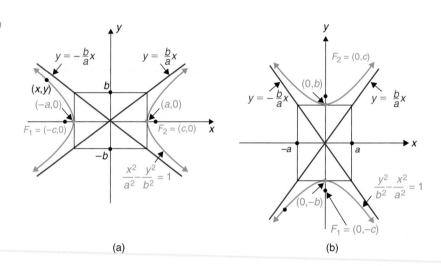

(a)   (b)

Some distinguishing features of a hyperbola with foci on the x axis are as follows:

1. The line segment that has the vertices as its end points is called the **transverse axis.**

2. The line segment that contains the center of the hyperbola at its end points and is perpendicular to the transverse axis is called the **conjugate axis.** Although the points $(0,-b)$ and $(0,b)$ are not on the hyperbola whose equation is

$$\frac{x^2}{a^2} - \frac{y^2}{b^2} = 1,$$

they are useful for describing its graph.

3. The line determined by the transverse axis is called the **axis of symmetry** of the hyperbola. For the hyperbola

$$\frac{x^2}{a^2} - \frac{y^2}{b^2} = 1,$$

the foci lie on the axis of symmetry, c units from the center, at points $(-c,0)$ and $(c,0)$.

4. The lines

$$y = -\frac{b}{a}x \text{ and } y = \frac{b}{a}x$$

are called the **asymptotes** of a hyperbola of either of the two forms

$$\frac{x^2}{a^2} - \frac{y^2}{b^2} = 1 \quad \text{or} \quad \frac{y^2}{b^2} - \frac{x^2}{a^2} = 1.$$

EXAMPLE 3   **Graphing a Hyperbola**

Identify the intercepts, the vertices, and the foci. Find the equations of the asymptotes and sketch the graph.

(a) $4x^2 - 9y^2 = 36$ (b) $4y^2 - 25x^2 = 100$

Solution   (a) The x intercepts are found by setting $y = 0$ and solving for x.

$$\frac{x^2}{9} = 1, \, x^2 = 9, \, x = \pm 3$$

The x intercepts are -3 and 3 so the vertices of the hyperbola are $(-3,0)$ and $(3,0)$. The transverse axis of the hyperbola is horizontal, and there are no y intercepts. We write the equation in standard form.

$$4x^2 - 9y^2 = 36 \qquad \text{Original equation}$$

$$\frac{x^2}{9} - \frac{y^2}{4} = 1 \qquad \text{Divide each side by 36.}$$

Figure 10

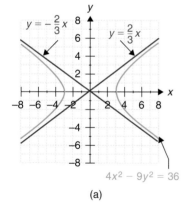

(a)

Here, $a = 3$, $b = 2$, so $c^2 = a^2 + b^2 = 9 + 4 = 13$ and $c = \sqrt{13}$.

So the foci are $(-\sqrt{13}, 0)$ and $(\sqrt{13}, 0)$. Figure 10a shows the graph of the hyperbola. The equations of the asymptotes are

$$y = \frac{b}{a}x = \frac{2}{3}x \text{ and } y = -\frac{b}{a}x = -\frac{2}{3}x.$$

(b) The y intercepts are found by setting $x = 0$ and solving for y.

$$4y^2 = 100, \, y^2 = 25, \, y = \pm 5$$

The y intercepts are -5 and 5 and the vertices of the hyperbola are $(0,-5)$ and $(0,5)$. The transverse axis of the hyperbola is vertical, and there are no x intercepts. We write the equation in standard form

$$4y^2 - 25x^2 = 100 \qquad \text{Original equation}$$

$$\frac{y^2}{25} - \frac{x^2}{4} = 1 \qquad \text{Divide each side by 100}$$

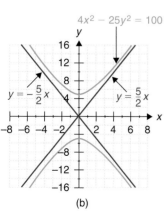

(b)

Here, $a = 2$, $b = 5$, so $c^2 = a^2 + b^2 = 4 + 25 = 29$ and $c = \sqrt{29}$.

So the foci are $(0,-\sqrt{29})$ and $(0,\sqrt{29})$. Figure 10b shows the graph of the hyperbola. The equations of the asymptotes are

$$y = \frac{b}{a}x = \frac{5}{2}x \text{ and } y = -\frac{b}{a}x = -\frac{5}{2}x.$$

EXAMPLE 4   **Finding an Equation of a Hyperbola**

Write an equation of a hyperbola centered at the origin, where vertices are $(-4,0)$ and $(4,0)$, with no y intercept, and whose foci are $(-5,0)$ and $(5,0)$. Sketch the graph and find the equations of the asymptotes.

Solution   Here $a = 4$ and $c = 5$, so that $c^2 = a^2 + b^2$ or $b^2 = c^2 - a^2$ and

$$b = \sqrt{25 - 16} = 3.$$

Therefore, an equation of the hyperbola is of the form

Figure 11

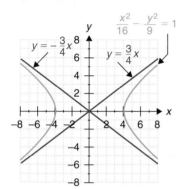

$$\frac{x^2}{a^2} - \frac{y^2}{b^2} = 1$$

so that

$$\frac{x^2}{16} - \frac{y^2}{9} = 1. \text{ (Figure 11)}$$

The equations of the asymptotes are given by

$$y = \frac{b}{a}x = \frac{3}{4}x \text{ and } y = -\frac{b}{a}x = -\frac{3}{4}x.$$

Solving Applied Problems

Ellipses and hyperbolas are of practical importance in many applications of mathematics as the next example shows.

EXAMPLE 5   **Solving an Ellipse Problem**

The park located between the White House and the Washington Monument in Washington, D.C. is elliptical in shape. The major axis is 458 meters long and its minor axis is 390 meters. What is the distance between the foci of this ellipse?

Solution   Let the center of this ellipse be at the origin. We know that $2a = 458$ and $2b = 390$ so that

$$a = 229 \text{ and } b = 195.$$

An equation of this ellipse in standard form is

$$\frac{x^2}{(229)^2} + \frac{y^2}{(195)^2} = 1$$

$c^2 = a^2 - b^2 = (229)^2 - (195)^2 = 14,416$, so that $c \approx 120$. Therefore, the foci of this ellipse are approximately at $(-120,0)$ and $(120,0)$. Hence, the distance between the two foci is $120 - (-120) = 240$ meters.

PROBLEM SET 9.2

Mastering the Concepts

In Problems 1–4, match each equation with its graph.

1. $4x^2 + 9y^2 = 36$
2. $x^2 - y^2 = 4$
3. $9x^2 + 4y^2 = 36$
4. $y^2 - x^2 = 4$

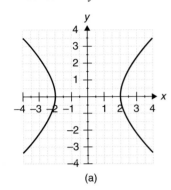

(a)

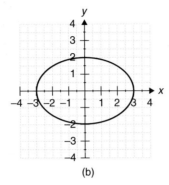

(b)

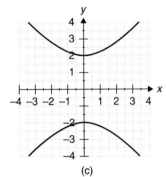

(c)

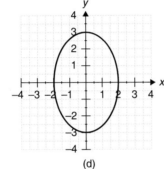

(d)

In Problems 5–12, identify the intercepts, the vertices, and the foci. Sketch the graph of each ellipse.

5. $9x^2 + 16y^2 = 144$
6. $16y^2 + 25x^2 = 400$
7. $4y^2 + 16x^2 = 64$
8. $16x^2 + 4y^2 = 64$
9. $4x^2 + 25y^2 = 100$
10. $25x^2 + 10y^2 = 250$
11. $15x^2 + 3y^2 = 45$
12. $25x^2 + 15y^2 = 150$

In Problems 13–22, identify the intercept, the vertices, and the foci of each hyperbola. Also sketch the graph and find the equations of the asymptotes.

13. $25x^2 - 9y^2 = 225$
14. $25y^2 - 9x^2 = 225$
15. $9x^2 - 4y^2 = 36$
16. $9y^2 - 4x^2 = 36$
17. $x^2 - 9y^2 = 9$
18. $9x^2 - y^2 = 1$
19. $16y^2 - 4x^2 = 48$
20. $36x^2 - 9y^2 = 1$
21. $4y^2 - 5x^2 = 20$
22. $3y^2 - x^2 = 1$

In Problems 23–26, find the standard equation of the conic that satisfies the given conditions. Sketch the graph.

23. An ellipse with center at the origin, foci at $F_1 = (-\sqrt{2}, 0)$, $F_2 = (\sqrt{2}, 0)$ and vertices at $V_1 = (-2,0)$ and $V_2 = (2,0)$.

24. An ellipse with center at the origin, foci at $F_1 = (0,-4)$, $F_2 = (0,4)$, and vertices at $V_1 = (0,-5)$ and $V_2 = (0,5)$.

25. A hyperbola with center at the origin, foci at $F_1 = (0,-\sqrt{2})$, $F_2 = (0,\sqrt{2})$, and asymptotes at $y = -x$ and $y = x$.

26. A hyperbola with center at the origin, vertices at $V_1 = (-3,0)$, $V_2 = (3,0)$, and asymptotes at $y = -\dfrac{4}{3}x$ and $y = \dfrac{4}{3}x$.

Applying the Concepts

27. Dimension of an Arch: An arch in the shape of the upper half of an ellipse is to support a bridge over a roadway 100 feet wide. The center of the arch is 30 feet above the center of the roadway, and the paved portion of the roadway is within a strip that is 80 feet wide (Figure 12).

Figure 12

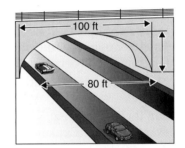

(a) Sketch the ellipse in an xy coordinate system, with horizontal major axis along the roadway and center at the origin.

(b) Find the standard equation of the ellipse in part (a).

(c) Confirm that the foci of the ellipse are located at the edges of the paved portion of the roadway.

28. Communication Satellite: A communication satellite travels in an elliptical orbit around the Earth. The radius of the Earth is about 6400 kilometers, and its center is located at one focus of the orbit. Suppose that the satellite moves in such a way that when it is closest to the center of the Earth, it is 2000 kilometers from the surface, and when it is farthest from the center, it is 10,000 kilometers from the surface.

(a) Sketch the orbit in an xy coordinate system with the major axis lying on the x axis and with center at the origin.

(b) Determine the standard equation of the ellipse in part (a).

(c) What is the height of the satellite above the surface of the Earth at a point corresponding to the focus (at the center of the earth) of the elliptical orbit?

29. **Locating Explosions:** Two microphones are located at two detection stations, A and B, 1150 meters apart on an east-west line (Figure 13). An explosion occurs at an unknown location P. The sound of the explosion is detected by the microphone at station A exactly 2 seconds before it is detected by the microphone at station B. Assume that sound travels at a uniform speed of 130 meters per second.

Figure 13

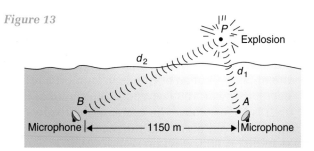

(a) Sketch the curve that describes the possible location of P in an xy coordinate system so that the origin is at the middle of the horizontal axis between A and B.

(b) Find the standard equation of the curve in part (a).

(c) Can the explosion at P be pinpointed at an exact location? Explain.

30. **Airplane Flight:** An airplane starts at a point directly north of the origin and flies northeasterly in a hyperbolic path. Transmitters located at the origin and 150 miles directly north of the origin send out synchronized radio signals. Instruments in the airplane measure the difference between the arrival times of the signals. By knowing the speed of the radio signals, it is determined that the distance from the plane to the transmitter at the origin is 50 miles farther than the distance to the other transmitter.

(a) Sketch the path of the airplane in a coordinate system that satisfies the given conditions.

(b) Find the standard equation of the curve in part (a).

(c) Find the location of the airplane when it is 40 miles directly east of the line between the two transmitters.

Developing and Extending the Concepts

31. (a) Sketch the graph of the equation of the ellipse
$$36x^2 + 100y^2 = 3600.$$

(b) Find the foci.

(c) Find the length of the major axis.

(d) Show that the points $(0,-6)$ and $(10,0)$ are on the graph.

(e) Verify that the sum of the distances from any of the above two points to each of the foci is equal to the length of the major axis.

32. (a) Sketch the graph of the equation of the hyperbola
$$4x^2 - 16y^2 = 144.$$

(b) Find the foci.

(c) Find the length of the transverse axis.

(d) Show that the points $(10,-4)$ and $(6\sqrt{2}, 3)$ lie on the graph.

(e) Verify that the difference of the distances from each of the above two points to each of the foci is equal to the length of the transverse axis.

In Problems 33–40, equations for ellipses and hyperbolas are given. Find an equation for the new ellipses and hyperbolas by shifting each center as specified.

33. $\dfrac{x^2}{9} + \dfrac{y^2}{4} = 1$, 2 units to the left and 1 unit down

34. $\dfrac{x^2}{4} + \dfrac{y^2}{9} = 1$, 3 units to the right and 2 units up

35. $\dfrac{x^2}{9} + \dfrac{y^2}{16} = 1$, 3 units to the right and 4 units up

36. $\dfrac{x^2}{9} + \dfrac{y^2}{1} = 1$, 2 units to the left and 3 units down

37. $\dfrac{x^2}{4} - \dfrac{y^2}{1} = 1$, 2 units to the left and 1 unit up

38. $\dfrac{x^2}{25} - \dfrac{y^2}{16} = 1$, 4 units to the right and 1 unit up

39. $\dfrac{y^2}{16} - \dfrac{x^2}{4} = 1$, 3 units to the right and 3 units down

40. $\dfrac{y^2}{4} - \dfrac{x^2}{1} = 1$, 3 units to the left and 2 units down

Objectives

1. Solve Nonlinear Systems of Equations
2. Solve Nonlinear Systems of Inequalities
3. Solve Applied Problems

9.3 Systems of Nonlinear Equations and Inequalities

So far, the systems of equations and inequalities we have solved have been linear. In this section, we consider the solution of systems in which at least one of these equations or inequalities is nonlinear. Some examples are shown below:

$$\begin{cases} x^2 + y = 6 \\ 2x - 5y = -6 \end{cases} \text{ and } \begin{cases} x + y \le 16 \\ x^2 + y \ge 4 \end{cases}.$$

We use substitutions and elimination to solve these systems algebraically.

Solving Nonlinear Systems of Equations

At times, we are able to solve nonlinear systems of equations by using various algebraic techniques such as substitution and elimination, as we did when solving linear systems.

EXAMPLE 1   **Solving a Nonlinear System by Substitution**

Solve the system by using substitution.

$$\begin{cases} x^2 - 2y = 0 \\ x + 2y = 6 \end{cases}$$

Solution   We solve the first equation for y to get

$$y = \frac{x^2}{2},$$

then we substitute the expression $\frac{x^2}{2}$ for y in the second equation to get

$$x + 2\left(\frac{x^2}{2}\right) = 6$$

$$x^2 + x - 6 = 0 \qquad \text{Write the equation in standard form}$$

$$(x + 3)(x - 2) = 0 \qquad \text{Factor}$$

$$x + 3 = 0 \qquad x - 2 = 0$$

$$x = -3 \qquad x = 2.$$

Next, we find y by substituting these numbers for x in the first equation.

$$x = -3 \qquad y = \frac{x^2}{2} = \frac{(-3)^2}{2} = \frac{9}{2}$$

$$x = 2 \qquad y = \frac{x^2}{2} = \frac{(2)^2}{2} = 2$$

Thus, the two solutions are $\left(-3, \dfrac{9}{2}\right)$ and $(2,2)$. The graph in Figure 1 shows the two points of intersection whose coordinates are

$$\left(-3, \frac{9}{2}\right) \text{ and } (2, 2).$$

Figure 1

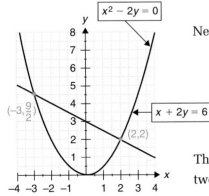

EXAMPLE 2   **Solving a Nonlinear System by Elimination**

Solve the system by elimination.

$$\begin{cases} x^2 + y^2 = 25 \\ x^2 + y = 13 \end{cases}$$

Solution   Using the elimination method and subtracting the second equation from the first, we get

$$y^2 - y = 12.$$

Solving this equation for y by factoring, we have

$$y^2 - y - 12 = 0 \qquad \text{Standard form}$$
$$(y + 3)(y - 4) = 0 \qquad \text{Factor}$$
$$y + 3 = 0 \qquad y - 4 = 0$$
$$y = -3 \qquad\qquad y = 4.$$

Substituting for y in the second equation, we have

$$y = -3, \quad x^2 + y = 13 \qquad\qquad y = 4, \quad x^2 + y = 13$$
$$x^2 + (-3) = 13 \qquad\qquad\qquad x^2 + 4 = 13$$
$$x^2 = 16 \qquad\qquad\qquad\qquad x^2 = 9$$
$$x = \pm 4 \qquad\qquad\qquad\qquad x = \pm 3.$$

Therefore, the solutions are

$$(-4, -3), (4, -3), (-3, 4), \text{ and } (3, 4).$$

Figure 2

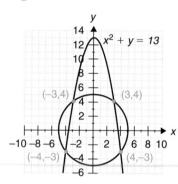

The graph in Figure 2 shows the four points of intersection whose coordinates are

$$(-4, -3), (4, -3), (-3, 4), \text{ and } (3, 4).$$

When the elimination method can be used, as in example 2, it is usually the most efficient way to solve the system. Otherwise, try the substitution method.

Solving Nonlinear Systems Of Inequalities

In section 3.5, we solved linear systems of inequalities with two variables by first graphing each of the inequalities on the same coordinate system. The points common to all graphs describe the solution set of the system. The same approach is used to solve nonlinear systems of inequalities as the next example shows.

EXAMPLE 3   **Graphing a Nonlinear System of Inequalities**

Graph the solution of the system of inequalities.

$$\begin{cases} y > \dfrac{x^2}{4} \\ x + y \le 4 \end{cases}$$

Figure 3

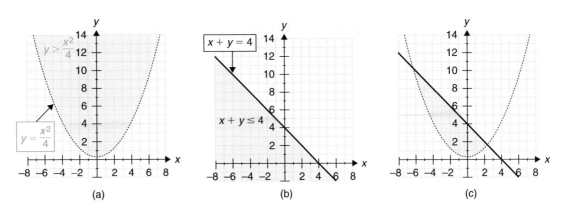

(a)   (b)   (c)

Solution   We begin by graphing each inequality separately and then determining where the regions overlap. Figure 3a shows the graph of $y > \dfrac{x^2}{4}$, as the shaded region above the curve $y = \dfrac{x^2}{4}$. Figure 3b shows the graph of the inequality $x + y \le 4$ as the shaded region on and below the line $x + y = 4$. Thus, the graph of the solution set of the system of inequalities consists of points common to both regions; that is, the shaded region in Figure 3c.

Solving Applied Problems

At times, we use nonlinear systems to solve applied problems as the next example shows.

EXAMPLE 4   **Finding the Dimensions of a Television Screen**

A 25 inch diagonal of a television screen is advertised as having a viewing area of 300 square inches. Find the width and height of the screen.

Solution   **Make A Plan**

Figure 4 shows the television in question. Since the viewing screen is almost rectangular, its area A is given by the formula

Figure 4

$$A = \text{length} \cdot \text{width}.$$

Let $x =$ the width (in inches) of the screen, and

$y =$ the height (in inches) of the screen.

Using the Pythagorean Theorem, we write

$$x^2 + y^2 = 25^2.$$

The system that describes this situation is given by

$$\begin{cases} x^2 + y^2 = 25^2 \\ xy = 300 \end{cases}.$$

Write the Solution   Write $xy = 300$ as $y = \dfrac{300}{x}$ and replace y in the first equation by $\dfrac{300}{x}$.

$$x^2 + \left(\frac{300}{x}\right)^2 = 625 \qquad\qquad \text{Original equation}$$

$$x^2 + \frac{90{,}000}{x^2} = 625$$

$$x^4 + 90{,}000 = 625x^2 \qquad\qquad \text{Multiply each side by } x^2$$

$$x^4 - 625x^2 + 90{,}000 = 0 \qquad\qquad \text{Write in standard form}$$

$$(x^2 - 400)(x^2 - 225) = 0 \qquad\qquad \text{Factor}$$

$$x^2 - 225 = 0 \qquad\qquad x^2 - 400 = 0$$

$$x^2 = 225 \qquad\qquad x^2 = 400$$

$$x = \pm 15 \qquad\qquad x = \pm 20$$

x cannot be negative in this situation, so x is either 15 or 20. If $x = 15$, then $y = 20$, and if $x = 20$, then $y = 15$. Because the width of a television screen is usually larger than its height, we conclude that the screen is 20 inches wide and 15 inches high.

Look Back   Although we have four solutions for each x and y, we use only the positive values. Here

$$x^2 + y^2 = 20^2 + 15^2 = 400 + 225 = 625, \text{ and}$$

$$xy = (20)(15) = 300.$$

 PROBLEM 9.3

Mastering The Concepts

In Problems 1–6, solve each system by substitution.

1. $\begin{cases} x - y = 1 \\ x^2 + y = 5 \end{cases}$

2. $\begin{cases} x - 2y = -5 \\ x^2 + y^2 = 25 \end{cases}$

3. $\begin{cases} 3x - y = 2 \\ x^2 + y^2 = 20 \end{cases}$

4. $\begin{cases} x + y = 6 \\ x^2 + y^2 = 20 \end{cases}$

5. $\begin{cases} 3x + 2y = 1 \\ 3x^2 - y^2 = 11 \end{cases}$

6. $\begin{cases} x - y^2 = 0 \\ x^2 + 2y^2 = 24 \end{cases}$

In Problems 7–12, solve each system by elimination.

7. $\begin{cases} x^2 - 2y^2 = 1 \\ x^2 + 4y^2 = 25 \end{cases}$

8. $\begin{cases} x^2 - 4y^2 = -15 \\ -x^2 + 3y^2 = 11 \end{cases}$

9. $\begin{cases} x^2 + 9y^2 = 33 \\ x^2 + y^2 = 25 \end{cases}$

10. $\begin{cases} x^2 + 5y^2 = 70 \\ 3x^2 - 5y^2 = 30 \end{cases}$

11. $\begin{cases} x^2 + 2y^2 = 20 \\ 2x^2 - y^2 = 20 \end{cases}$

12. $\begin{cases} 3x^2 - y^2 = 3 \\ 4x^2 + 3y^2 = 43 \end{cases}$

In Problems 13–20, match each system with its graph, then use an appropriate method to solve the system.

13. $\begin{cases} x^2 - 25y^2 = 20 \\ 2x^2 + 25y^2 = 88 \end{cases}$

14. $\begin{cases} 3x^2 - 8y^2 = 40 \\ 5x^2 + y^2 = 81 \end{cases}$

15. $\begin{cases} 2x^2 - 3y^2 = 6 \\ 3x^2 + 2y^2 = 35 \end{cases}$

16. $\begin{cases} x^2 - y^2 = 7 \\ x^2 + y^2 = 25 \end{cases}$

17. $\begin{cases} 6x + 2y = 2 \\ 3x^2 - y^2 = -73 \end{cases}$

18. $\begin{cases} 5x - 3y = 10 \\ x^2 - y^2 = 4 \end{cases}$

19. $\begin{cases} x + y = 6 \\ x^2 + y^2 - 4y = 12 \end{cases}$

20. $\begin{cases} x - y = -1 \\ x^2 + 3y^2 = 13 \end{cases}$

(a)

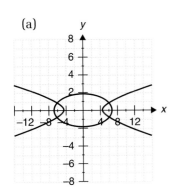

(b)

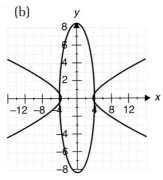

(c)

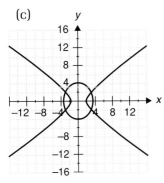

(d)

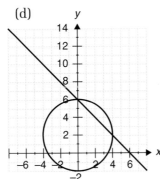

(e)

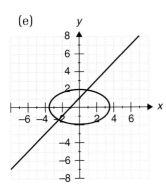

(f)

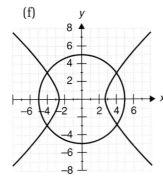

(g)

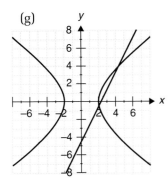

(h)

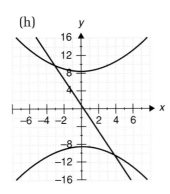

In Problems 21–28, graph the solution of each system of inequalities.

21. $\begin{cases} -x^2 + y \geq 0 \\ x + y < 1 \end{cases}$

22. $\begin{cases} \sqrt{x} - y > 0 \\ x - 9y \leq 0 \end{cases}$

23. $\begin{cases} x^2 + y \leq 0 \\ x + y > -2 \end{cases}$

24. $\begin{cases} y > x - 3 \\ y \leq \sqrt{x} - 1 \\ y \geq 0 \end{cases}$

25. $\begin{cases} x \geq 0, y \geq 0 \\ y \leq 4 - x^2 \end{cases}$

26. $\begin{cases} x^2 + y^2 \leq 1 \\ x + y > 1 \end{cases}$

27. $\begin{cases} 4x^2 - y^2 \geq 0 \\ x + y < 9 \end{cases}$

28. $\begin{cases} y \geq (x - 1)^2 + 2 \\ 2x - 3y \leq -9 \end{cases}$

Applying the Concepts

29. Computer Monitor Design: An electronics engineer designs a rectangular computer monitor screen with a 10 inch diagonal and a viewing area of 48 square inches. Find the dimensions of the screen (Figure 5).

Figure 5

10 in

30. Salt Storage: A certain northern state stores road salt, used to melt ice and snow in winter, in buildings made of galvanized sheet iron in the shape of right circular cones (Figure 6). Each building has a height h equal to 1.5 times the radius r of its base, and each building has a volume of 12,000 cubic feet. Find h and r rounded off to two decimal places.

Figure 6

h

r

$$\left(V = \frac{\pi r^2 h}{3}\right)$$

31. Construction: Figure 7 shows a child's sandbox made of galvanized sheet iron with a square base of side x and of height y. If the volume of the sandbox is 50 cubic feet, and the combined area of the sides and the bottom is 65 square feet, find the dimensions x and y of the box. Round off to one decimal place.

Figure 7

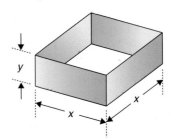

y

x

x

32. Investment Portfolio: The annual simple interest earned on an investment is \$170. If the interest had been 1% higher, the interest earned from this investment would have been \$238. What was the amount of the investment and what was the lower interest rate?

33. Engineering Design: An engineer designs a rectangular solar collector with a surface area of 750 square feet. If the length of the collector is 30 times its width, find its length and width.

34. Template Design: An artist designs a plastic template in the shape of a right triangle with one base length of 60 centimeters and an area of 1500 square centimeters. Find the lengths of the sides of the triangular template. Round off to one decimal place.

35. Fencing: A farmer uses 30 meters of fence to enclose a rectangular pen next to a barn, with the barn wall forming one side of the pen (Figure 8). What are the dimensions of the pen if it encloses an area of 108 square meters?

Figure 8

36. Fencing: A rancher wishes to fence a rectangular corral with a total of 110 meters of fence. By adding a fence lengthwise inside the original corral, there are two corrals. What are the dimensions of these corrals, if one corral is a square region and is 144 square meters more than the other?

37. Puzzle: Find two positive numbers whose sum is 15 and the difference of their squares is also 15.

38. Number Problem: The sum of the squares of the digits of a certain two-digit number is 61. The product of the number and the number with the digits reversed is 3640. What is the number? Hint: Write the number in the form $10y + x$

Developing and Extending the Concepts

In Problems 39–42, solve each system algebraically.

39. $\begin{cases} -4x^2 + 3xy = -133 \\ 2x^2 - xy = 77 \end{cases}$

40. $\begin{cases} -2uv + 3v^2 = 8 \\ uv + v^2 = 6 \end{cases}$

41. $\begin{cases} (p - 2)^2 + (q - 2)^2 = 2 \\ pq = 1 \end{cases}$

42. $\begin{cases} (u-4)^2 + (v + 4)^2 = 25 \\ u + v = 7 \end{cases}$

In Problems 43–46, graph each system to approximate the solutions of each system. Round off the answer to two decimal places.

43. $\begin{cases} x^2 - 3y = 1 \\ x - y = 1 \end{cases}$

44. $\begin{cases} x - y = 0 \\ x^2 + y^2 = 8 \end{cases}$

45. $\begin{cases} x - y = 21 \\ \sqrt{x} + \sqrt{y} = 7 \end{cases}$

46. $\begin{cases} \dfrac{3}{x^2} - \dfrac{2}{y^2} = 1 \\ \dfrac{5}{x^2} - \dfrac{6}{y^2} = 1 \end{cases}$

In Problems 47–50, solve each system of inequalities graphically.

47. $\begin{cases} -x^2 + y^2 \le 64 \\ x^2 - y^2 \ge 4 \end{cases}$

48. $\begin{cases} \dfrac{x^2}{36} + \dfrac{y^2}{16} \ge 1 \\ \dfrac{x^2}{16} + \dfrac{y^2}{36} \le 1 \end{cases}$

49. $\begin{cases} x^2 + y^2 \ge 4 \\ \dfrac{x^2}{4} + \dfrac{y^2}{25} \le 1 \end{cases}$

50. $\begin{cases} x^2 - y^2 \ge 9 \\ x^2 + y^2 \le 25 \end{cases}$

Objectives

1. Describe Sequences
2. Defining a Sequence Recursively
3. Find the General Term of a Sequence
4. Graph Sequences
5. Solve Applied Problems

9.4 Sequences

In this section we give a mathematical definition of a *sequence*. Then we concentrate on finding the sum of the terms of a sequence called a *series*.

Describing Sequences

Intuitively, a **sequence** is a collection or list of numbers arranged in a definite order. The individual numbers of a sequence are called **terms**. Examples of sequences are:

$$2, 4, 6, 8, 10 \quad \text{and} \quad 2, 4, 6, 8, 10, \ldots$$

The sequence in the first example contains five terms, a finite number; it is called a **finite sequence.** The second sequence contains infinitely many terms (indicated by the three periods called an **ellipsis**), and is called an **infinite sequence** that lists the positive multiples of 2. The definition of a sequence can be clearly stated by using the terminology of functions. A **sequence** is a function with the positive integers as its set of inputs (domain). When the domain is apparent, we refer to either a finite or an infinite sequence as a sequence.

Following tradition, we write the function values in the form

$$a_1, a_2, a_3, \ldots, a_n, \ldots$$

These elements in the range of a sequence are called its **terms**. The terms are named with a letter, such as a, and a subscript corresponding to the position of the term. In the above sequence, a_1 is referred to as the **first term**, a_2 is the **second term,** and a_n is the **nth term** or the **general term.**

The notation $\{a_n\}$, in which the general term is enclosed in braces, is also used to denote a sequence. To specify a particular sequence, we give a rule by which the nth term, a_n, is determined. This is often done by means of a formula such as

$$a_n = (n + 1)^2.$$

In that case, we generate the terms of a sequence by substituting values for n in the formula, where

$$n = 1, 2, 3, \ldots$$

EXAMPLE 1   **Determining the Term of a Sequence**

Find the first five terms of each sequence.

(a) $a_n = (-1)^n$ (b) $a_n = 3 - \dfrac{1}{n}$

Solution   To find the first five terms of each sequence we substitute the positive integers 1, 2, 3, 4, and 5, in turn, for n in the formula for the general term.

(a) We have

$$a_1 = (-1)^1 = -1, \quad a_2 = (-1)^2 = 1, \quad a_3 = (-1)^3 = -1$$
$$a_4 = (-1)^4 = 1, \quad \text{and} \quad a_5 = (-1)^5 = -1.$$

Thus, the first five terms of the sequence $\{a_n\}$ are

$$-1, 1, -1, 1, \text{ and } -1.$$

(b) We have

$$a_1 = 3 - \frac{1}{1} = 2, \quad a_2 = 3 - \frac{1}{2} = \frac{5}{2}, \quad a_3 = 3 - \frac{1}{3} = \frac{8}{3},$$

$$a_4 = 3 - \frac{1}{4} = \frac{11}{4}, \quad \text{and} \quad a_5 = 3 - \frac{1}{5} = \frac{14}{5}.$$

Therefore, the first five terms of the sequence $\{a_n\}$ are

$$2, \frac{5}{2}, \frac{8}{3}, \frac{11}{4}, \text{ and } \frac{14}{5}.$$

Defining a Sequence Recursively

So far, the formulas used for the nth term, a_n, of a sequence have expressed the nth term as a function of n. In another approach, a formula that relates the general term a_n of a sequence to one or more of the terms that come before it is called a **recursion formula.** A sequence that is specified by giving the first term (or the first few terms) together with a recursion formula is said to be defined **recursively.**

EXAMPLE 2   **Using a Recursion Formula**

Find the first five terms of the sequence defined recursively by

$$a_1 = 1 \text{ and } a_n = na_{n-1}.$$

Solution   We are given $a_1 = 1$. By the recursion formula $a_n = na_{n-1}$ we have

$$a_2 = 2a_{2-1} = 2a_1$$
$$= 2(1) = 2.$$

Now, using the fact that $a_2 = 2$, we find that

$$a_3 = 3a_{3-1} = 3a_2$$
$$= 3(2) = 6.$$

Continuing in this way, we have

$$a_4 = 4a_{4-1} = 4a_3$$
$$= 4(6) = 24$$

and

$$a_5 = 5a_{5-1}$$
$$= 5a_4 = 5(24) = 120.$$

Finding the General Term of a Sequence

Now we look at the problem in reverse. We often have the terms of a sequence and want to write a formula that defines the sequence. That is, if the first few terms of a sequence are known and the sequence continues in some indicated pattern, then we can predict a formula for a_n in terms of n.

EXAMPLE 3   **Finding the General Term of a Sequence**

Predict the general term of each sequence.

(a) 4, 7, 10, 13, . . .   (b) 1, −2, 4, −8, 16, . . .

Solution   (a) Notice that $7 - 4 = 3$, $10 - 7 = 3$, $13 - 10 = 3$, and so on. Since the domain of a sequence consists of the positive integers, then the expression $3n + 1$ defines the general term. That is, the general term

$$a_n = 3n + 1, \text{ where } n = 1, 2, 3, 4, . . .$$

(b) These terms are the powers of 2 with alternating signs, so the general term a_n is given by

$$a_n = (-1)^{n+1}2^{n-1}, \qquad n = 1, 2, 3, 4, \ldots$$

Graphing Sequences

In graphing a sequence, we designate the sequence numbers a_n as outputs and positive integers n as inputs. As usual, we use the horizontal axis for the inputs n and the vertical axis for the outputs $a_n = f(n)$. Then we plot

(n, a_n) as discrete points in a coordinate system.

EXAMPLE 4   **Graphing a Sequence**

(a) Graph the first five terms of the sequence whose general term is

$$a_n = 2n - 1 \quad \text{for } 1 \le n \le 5.$$

(b) Graph the function $f(x) = 2x - 1$ for $x \ge 1$.

Solution   (a) To find the first five terms of the sequence, we substitute

$n = 1, 2, 3, 4, 5$ into the general term $a_n = 2n - 1$ to get

$$a_1 = 2(1) - 1 = 1, \qquad a_2 = 2(2) - 1 = 3 \qquad a_3 = 2(3) - 1 = 5$$
$$a_4 = 2(4) - 1 = 7 \qquad a_5 = 2(5) - 1 = 9.$$

The graph of this sequence consists of the following discrete points

$$(1,1), (2,3), (3,5), (4,7), \text{ and } (5,9) \text{ (Figure 1a)}.$$

(b) The graph of $f(x) = 2x - 1$ for $x \ge 1$ is a straight line (Figure 1b).

Figure 1

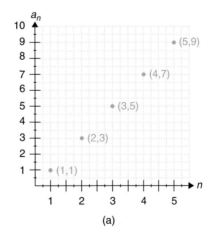

(a)

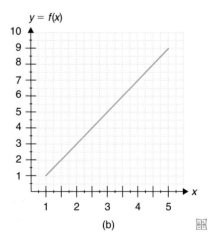

(b)

Solving Applied Problems

We use a recursion formula to investigate one of the most famous sequences, which was introduced by the Italian mathematician, Leonardo Fibonacci in 1202 AD and is defined as follows:

$$a_1 = 1, a_2 = 1 \text{ and } a_n = a_{n-1} + a_{n-2}, \text{ for } n \ge 3.$$

EXAMPLE 5   **Using a Fibonacci Sequence**

Suppose that we have a pair of rabbits that grow to maturity in one month. How many pairs of rabbits can be produced from the single pair in a year if every month each pair begets a new pair that from the second month onward becomes productive?

Solution   Assume that the rabbits in the initial pair are newborn, so that they produce the first new pair after two months. The first month we have one pair of rabbits, the second month we still have one pair of rabbits, but they are mature enough to mate. By the third month, we have our first new pair of rabbits as well as the original pair. The number of pairs present after 12 months is given by:

$$a_1 = 1 \qquad\qquad a_2 = 1$$
$$a_3 = a_2 + a_1 = 1 + 1 = 2 \qquad a_4 = a_3 + a_2 = 2 + 1 = 3$$
$$a_5 = a_4 + a_3 = 3 + 2 = 5 \qquad a_6 = a_5 + a_4 = 5 + 3 = 8$$
$$a_7 = a_6 + a_5 = 8 + 5 = 13 \qquad a_8 = a_7 + a_6 = 13 + 8 = 21$$
$$a_9 = a_8 + a_7 = 21 + 13 = 34 \qquad a_{10} = a_9 + a_8 = 34 + 21 = 55$$
$$a_{11} = a_{10} + a_9 = 55 + 34 = 89 \qquad a_{12} = a_{11} + a_{10} = 89 + 55 = 144.$$

Therefore, the first twelve terms of the sequence are:

1, 1, 2, 3, 5, 8, 13, 21, 34, 55, 89, 144.

This pattern of numbers is called a Fibonacci sequence.

PROBLEM SET 9.4

Mastering the Concepts

In Problems 1–12, find the first five terms in each sequence.

1. $a_n = n - 4$

2. $a_n = \dfrac{n + 4}{n}$

3. $a_n = \dfrac{n(n + 2)}{2}$

4. $a_n = \dfrac{n + 1}{n + 2}$

5. $a_n = \dfrac{n(n - 3)}{2}$

6. $a_n = \dfrac{n(n + 3)}{2}$

7. $a_n = \dfrac{1}{n^2}$

8. $a_n = \dfrac{n^2 - 2}{2}$

9. $a_n = 2^{n-3}$

10. $a_n = 3^{n-4}$

11. $a_n = (-1)^n + 3$

12. $a_n = \dfrac{3}{n(n + 1)}$

In Problems 13–20, find the general term a_n for each sequence.

13. 2, 4, 6, 8, 10, . . .

14. 3, 6, 9, 12, 15, . . .

15. 1, 3, 5, 7, 9, . . .

16. $1, \dfrac{1}{8}, \dfrac{1}{27}, \dfrac{1}{64}, \dfrac{1}{125}, \ldots$

17. 1, 8, 27, 64, 125, . . .

18. 1, 16, 81, 256, 625, . . .

19. −3, 6, −9, 12, −15, . . .

20. 2, −4, 6, −8, 10, . . .

In Problems 21–28, find the first five terms of each sequence defined recursively as indicated.

21. $a_1 = 1, a_{n+1} = \dfrac{1}{2} a_n$

22. $a_1 = 2, a_{n+1} = 2a_n$

23. $a_1 = 1, a_{n+1} = a_n + \dfrac{1}{2}$

24. $a_1 = 1, a_{n+1} = a_n - \dfrac{1}{2}$

25. $a_1 = -10, a_n = a_1 + (n - 1)(10)$

26. $a_1 = 10, a_n = a_1 + (n - 1)(-10)$

27. $a_1 = -2, a_n = 3a_{n-1}, n \geq 2$

28. $a_1 = 3, a_n = 4a_{n-1}, n \geq 2$

Applying the Concepts

29. Pay Raise: A person is offered a job with an annual salary of $20,000, and a raise of 4% at the end of each year. Write a sequence showing the salary at the end of 2 years, 3 years, 4 years, and 5 years.

30. Drug Measurement: A patient takes 15 milligrams of a drug each morning. Suppose that 80% of the drug is eliminated each day. Write out the first six terms of the sequence where a_n is the amount of the drug present in the patient's body immediately after the nth dose.

31. Investment: One dollar is deposited in a savings account that pays an annual interest rate 6%. If no money is withdrawn, what is the amount accrued in the account after the first, second, and third years?

32. Bacteria Growth: A colony of bacteria doubles every day. If the colony has 100 bacteria at the beginning of an experiment, find the number of bacteria after 6 days.

Developing and Extending the Concepts

In Problems 33–36, find the first five terms in each sequence.

33. $a_n = (-1)^n(n + 2)$

34. $a_n = \dfrac{(-1)^n}{2n + 1}$

35. $a_n = \sqrt{n + 1} - \sqrt{n}$

36. $a_n = n \log 10^{n - 1}$

In Problems 37–40, find the indicated term in each sequence.

37. The 10th term of $a_n = 3n + 2$

38. The 15th term of $a_n = 2n + 3$

39. The 20th term of $a_n = \dfrac{n - 2}{n + 2}$

40. The 7th term of $a_n = 5(2)^{3 - n}$

41. Write the first four terms of the sequence.
$$a_n = \sqrt{n} + \sqrt{n + 9} - 6\sqrt{n}$$

42. Write the first four terms of the sequence.
$$a_n = \sqrt{n} + \sqrt{|\sqrt{n} - 3|}$$

Round off each answer to two decimal places.

43. Predict the general term of the sequence.
$$\sqrt{5}, \sqrt{8}, \sqrt{11}, \sqrt{14}, \ldots$$

44. Predict the general term of the sequence.
$$-3, 3, -3, 3, -3, 3, \ldots$$

In Problems 45–48, write the first five terms of each sequence defined recursively.

45. $a_1 = 1, a_{n+1} = 3 + 2a_n$

46. $a_1 = 2, a_{n + 1} = 2a_{n-1} - 3$

47. $a_1 = 1, a_2 = 1, a_{n+2} = a_{n+1} + a_n$

48. $a_1 = 5, a_2 = 7, a_{n+2} = a_{n+1} + a_n$

In Problems 49–52, graph the continuous function f and the first five ordered pairs of each sequence a_n.

49. $f(x) = 3x - 4$; $a_n = 3n - 4$

50. $f(x) = 1 - 3x$; $a_n = 1 - 3n$

51. $f(x) = 2^x$; $a_n = 2^n$

52. $f(x) = \dfrac{2x + 1}{x}$; $a_n = \dfrac{2n + 1}{n}$

Objectives

1. Describe a Series
2. Use Summation Notation
3. Using Properties of Summation
4. Solving Applied Problems

9.5 Series and Summation

In many applied problems, it is sometimes necessary to add the terms of a sequence. Special terminology, notation, and techniques have been developed for that purpose.

Describing a Series

The sum of the terms of a sequence is called a **series.** The sum of the first n terms of a sequence $\{a_n\}$ is written in the form

$$a_1 + a_2 + a_3 + \ldots + a_n.$$

Because sequences can be finite or infinite, series can also be *finite* or *infinite*. A series of the form

$$a_1 + a_2 + a_3 + \ldots + a_n + \ldots$$

is called an **infinite series.** The term a_n in the above series is called the **general term.**

EXAMPLE 1 **Find the Sum of a Series**

Find the sum of the terms of the sequence.

$$5, 8, 11, 14, 17, 20, 23, 26, 29, 32, 35, 38$$

Solution Of course, we can add these terms by using a calculator.

Without using a calculator, we pair these numbers as follows.

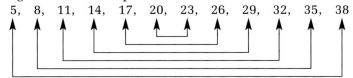

There are 12 numbers, and $\dfrac{12}{2} = 6$ pairs, each adding up to 43.

The sum is $6(43) = 258$.

Using Summation Notation

The sum of the first n terms of a sequence $a_1 + a_2 + a_3 + \ldots + a_n$ may be written in the form

$$a_1 + a_2 + a_3 + \ldots + a_n.$$

This sum can be written using a compact notation denoted by the Greek capital letter *sigma,* Σ, as follows.

$$\sum_{k=1}^{n} a_k = a_1 + a_2 + a_3 + \cdots + a_n$$

The integer k is called the **index** of the summation. The notation $k = 1$ under Σ indicates that the summation starts at 1, which is called the **lower limit** of the summation. The value n above Σ indicates the **upper limit** of the summation. Letters such as i, j, k, and n are commonly used to represent indices of the summation. For instance, we may write

$$\sum_{k=1}^{n} 3^k = \sum_{i=1}^{n} 3^i = \sum_{j=1}^{n} 3^j = 3^1 + 3^2 + 3^3 + \cdots + 3^n.$$

EXAMPLE 2 **Using Summation Notation**

Evaluate each sum.

(a) $\displaystyle\sum_{k=1}^{3} (4k^2 - 3k)$ (b) $\displaystyle\sum_{j=2}^{5} \frac{j-1}{j+1}$ (c) $\displaystyle\sum_{k=1}^{4} 5$

Solution   (a) Since $k = 1$ and $n = 3$, we substitute 1, 2, and 3 for k, then add the resulting terms.

$$\sum_{k=1}^{3}(4k^2 - 3k) = [4(1)^2 - 3(1)] + [4(2)^2 - 3(2)] + [4(3)^2 - 3(3)]$$

$$= 1 + 10 + 27 = 38$$

(b) Since $j = 2$ and $n = 5$, we substitute 2, 3, 4, and 5 for j, then add the resulting terms.

$$\sum_{j=2}^{5}\frac{j-1}{j+1} = \frac{2-1}{2+1} + \frac{3-1}{3+1} + \frac{4-1}{4+1} + \frac{5-1}{5+1}$$

$$= \frac{1}{3} + \frac{2}{4} + \frac{3}{5} + \frac{4}{6} = \frac{21}{10}$$

(c) This expression represents the sum of the first 4 terms of the sequence $a_n = 5$ specifically,

$$\sum_{k=1}^{4}5 = 5 + 5 + 5 + 5 = 20.$$

EXAMPLE 3   **Writing a Series Using Summation Notation**

Express each series using summation notation.

(a) $3 + 5 + 7 + 9 + 11$ (b) $\dfrac{1}{2} - \dfrac{1}{4} + \dfrac{1}{8} - \dfrac{1}{16} + \dfrac{1}{32} - \dfrac{1}{64}$

Solution   (a) The general term of this sequence is given by

$$a_k = 2k + 1, \text{ where } 1 \le k \le 5.$$

Thus, this sum is written in summation notation as

$$\sum_{k=1}^{5}(2k + 1).$$

(b) The terms are those of the sequence $\dfrac{1}{2}, \dfrac{1}{4}, \dfrac{1}{8}, \dfrac{1}{16}, \dfrac{1}{32}, \dfrac{1}{64}$.

There may be more than one way to express a sum in summation notation.

Using the factor $(-1)^{k+1}$ to make the signs alternate between the positive and negative, we write the sum as

$$\sum_{k=1}^{6}(-1)^{k+1}\left(\frac{1}{2}\right)^{k}.$$

Using Properties of Summation

In calculations involving sums in sigma notation, the following properties may be used.

Summation Properties

Let $\{a_n\}$ and $\{b_n\}$ be sequences. For every positive integer n and any constant c, we have

1. $\displaystyle\sum_{k=1}^{n} c = nc$

   Sum of a constant.

2. $\displaystyle\sum_{k=1}^{n} ca_k = c\sum_{k=1}^{n} a_k$

   Constant times a sum.

3. $\displaystyle\sum_{k=1}^{n} (a_k + b_k) = \sum_{k=1}^{n} a_k + \sum_{k=1}^{n} b_k$

   Sum of sums.

4. $\displaystyle\sum_{k=1}^{n} (a_k - b_k) = \sum_{k=1}^{n} a_k - \sum_{k=1}^{n} b_k$

   Difference of sums.

EXAMPLE 4 **Using Properties of Summation**

Suppose that $\displaystyle\sum_{k=1}^{10} a_k = 30$ and $\displaystyle\sum_{k=1}^{10} b_k = 45$, evaluate each summation.

(a) $\displaystyle\sum_{k=1}^{10} (2a_k + 3b_k)$

(b) $\displaystyle\sum_{k=1}^{10} (a_k + 5)$

Solution

(a) $\displaystyle\sum_{k=1}^{10} (2a_k + 3b_k) = \sum_{k=1}^{10} 2a_k + \sum_{k=1}^{10} 3b_k$ Use property 3

$= 2\displaystyle\sum_{k=1}^{10} a_k + 3\sum_{k=1}^{10} b_k$ Use property 2

$= 2(30) + 3(45)$

$= 60 + 135$

$= 195$

(b) $\displaystyle\sum_{k=1}^{10} (a_k + 5) = \sum_{k=1}^{10} a_k + \sum_{k=1}^{10} 5$ Use property 3

$= 30 + 5(10)$

$= 30 + 50$

$= 80$

Solving Applied Problems

A story is told about the great German mathematician Carl Friedrich Gauss (1777–1858). When he was 10 years old, his teacher wanted to keep his young students busy for a considerable period. He asked the class to add the numbers from 1 to 100. Within only a few moments, young Gauss raised his hand and said that he had the answer. The teacher was amazed at the boy's speed. How did he find the sum so quickly? His technique was based on adding the first and last terms and multiplying by half the number of terms. That is

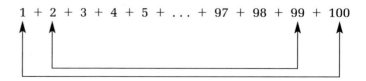

Because $1 + 100 = 101$, $2 + 99 = 101$, $3 + 98 = 101$, and so on, all he needed to finish the problem is to find the number of such pairs, which is 50. The sum is $50(101) = 5050$. Gauss's pairing method was used in finding the sum of even terms of the sequence in Example 1. His method also works with an odd number of terms of a sequence as the next example shows.

EXAMPLE 5   **Finding a Sum of Salary**

Suppose that a starting salary for an employee is \$15,000 during the first year and the employee receives an \$8,000 pay increase at the end of each subsequent year. What are the employee's earnings over the first 7 years?

Solution   The employee's salary over the first 7 years is given by the sum of the terms of the sequence.

$$\$15,000, \$23,000, \$31,000, \$39,000, \$47,000, \$55,000, \$63,000$$

That is

$$\$15,000 + \$23,000 + \$31,000 + \$39,000 + \$47,000 + \$55,000 + \$63,000.$$

Using Gauss's pairing method, there are $3\frac{1}{2}$ pairs each adding to \$78,000. The sum is $3\frac{1}{2}(78,000) = \$273,000$.

PROBLEM SET 9.5

Mastering the Concepts

In Problems 1 and 2, use Gauss's system of pairing numbers to find the sum of the terms of each sequence.

1. 4, 9, 14, 19, 24, 29, 34, 39, 44, 49, 54, 59, 64, 69

2. 11, 17, 23, 29, 35, 41, 47, 53, 59, 65, 71, 77, 83

In Problems 3–14 evaluate each sum.

3. $\displaystyle\sum_{k=1}^{5} k$

4. $\displaystyle\sum_{k=1}^{4} \frac{2^4}{k+1}$

5. $\displaystyle\sum_{i=1}^{10} 2i(i-1)$

6. $\displaystyle\sum_{k=0}^{4} 3^{2k}$

7. $\displaystyle\sum_{k=2}^{5} 2^{k-2}$

8. $\displaystyle\sum_{i=2}^{6} \frac{1}{i}$

9. $\displaystyle\sum_{k=1}^{3} (2k+1)$

10. $\displaystyle\sum_{k=1}^{5} (3k^2 - 5k + 1)$

11. $\displaystyle\sum_{i=2}^{4} \frac{i}{i+1}$

12. $\displaystyle\sum_{k=1}^{4} k^k$

13. $\displaystyle\sum_{k=1}^{100} 5$

14. $\displaystyle\sum_{i=3}^{7} (i+2)$

In Problems 15–22, express each series using summation notation.

15. $4 + 8 + 12 + 16 + 20 + 24$

16. $5 + 10 + 15 + 20 + 25 + 30$

17. $4 - 1 - 6 - 11 - 16 - 21$

18. $3 + 8 + 13 + 18 + 23 + 28$

19. $-1 - 4 - 7 - 10 - 13 - 16$

20. $-8 - 1 + 6 + 13 + 20 + 27$

21. $\dfrac{1}{1.2} + \dfrac{1}{2.3} + \dfrac{1}{3.4} + \dfrac{1}{4.5} + \dfrac{1}{5.6}$

22. $\dfrac{1}{1.2^3} + \dfrac{1}{2.3^3} + \dfrac{1}{3.4^3} + \dfrac{1}{4.5^3}$

In Problems 23–28, suppose $\sum_{k=1}^{30} a_k = 25$ and $\sum_{k=1}^{30} b_k = 15$. Use the properties of summation to evaluate each sum.

23. $\sum_{k=1}^{30} (4a_k + b_k)$ **24.** $\sum_{k=1}^{30} (a_k + 7b_k)$

25. $\sum_{k=1}^{30} (5a_k - 4)$ **26.** $\sum_{k=1}^{30} (4b_k + 5)$

27. $\sum_{k=1}^{30} (2a_k + 3b_k + 2)$ **28.** $\sum_{k=1}^{30} (3a_k - 4b_k + 8)$

In Problems 29–32, the formula

$$\bar{x} = \frac{x_1 + x_2 + x_3 + \cdots + x_n}{n}$$

determines the *arithmetic mean* of the sequence $x_1, x_2, x_3, \ldots, x_n$. Find the arithmetic mean for the numbers in each sequence.

29. 1, 5, 9, 13, 17, 21, 25, 29
30. 3, 6, 9, 12, 15, 18, 21, 24
31. 1.3, 3.3, 5.3, 7.3, 9.3, 11.3
32. 1.1, 3.3, 5.5, 7.7, 9.9, 11.11

Applying the Concepts

33. Investment: Each year the amount of money accumulated on a $1,000 investment from 1998 to 2002 is approximated by the model $a_k = 1,000(1 + 0.04)^k$, where $k = 1, 2, 3, 4, 5$ and a_k is the amount accumulated after the kth year, with $k = 1$ representing 1998. Use this model to determine the value of investment at the end of each year.

34. Distance Traveled: A skier travels 4 feet down a mountain slope during the first second of descent. During the next second the skier travels 9 feet, and during the third second the skier travels 14 feet, and so on. During the kth second the skier distance traveled is given by

$$a_k = 4 + 5(k - 1), k = 1, 2, 3, \ldots, n$$

Find the total distance traveled by the skier during the first 10 seconds of descent.

Developing and Extending the Concepts

In Problems 35–38, compare the following sums. Round off to 4 decimal places.

35. (a) $\sum_{n=1}^{20} 1$ **36.** (a) $\sum_{n=1}^{20} (n + 1)$

(b) $\sum_{n=1}^{20} 2$ (b) $\sum_{n=1}^{20} n + \sum_{n=1}^{20} 1$

37. (a) $\sum_{n=1}^{10} \frac{1}{n}$ **38.** (a) $\sum_{n=1}^{10} \frac{1}{n^2}$

(b) $\sum_{n=1}^{10} (-1)^n \frac{1}{n}$ (b) $\sum_{n=1}^{10} \frac{1}{n^3}$

In Problems 39–44, use the summation formulas

$$\sum_{k=1}^{n} k = \frac{n(n + 1)}{2} \text{ and } \sum_{k=1}^{n} k^2 = \frac{n(n + 1)(2n + 1)}{6}$$

and the properties to evaluate each sum.

39. $\sum_{k=1}^{20} (3k + 4)$ **40.** $\sum_{k=1}^{20} (4k - 3)$

41. $\sum_{k=1}^{10} (2k^2 + 3k)$ **42.** $\sum_{k=1}^{10} (-3k^2 + 5k)$

43. $\sum_{k=1}^{15} \frac{k(k - 1)}{2}$ **44.** $\sum_{k=1}^{15} \frac{k(2k + 1)}{3}$

In Problems 45 and 46, solve each equation for x.

45. $\sum_{n=4}^{6} nx = 30$ **46.** $\sum_{n=3}^{5} (nx + 3) = 45$

For Problems 47–50, consider the data $x_1, x_2, x_3, \ldots, x_n$. In statistics, the *standard deviation* of this data is given by

$$s = \sqrt{\frac{1}{n - 1} \sum_{k=1}^{n} (x_k - \bar{x})^2}, \text{ where } \bar{x} = \frac{1}{n} \sum_{k=1}^{n} x_k.$$

Use the formula to find the standard deviation of the given data.

47. 5, 9, 11, 2, 7, 1   **48.** 3, 5, 5, 7, 10
49. 1, 3, 4, 6, 7   **50.** 4, 2, 2, 8, 5, 3

Objectives

1. Define an Arithmetic Sequence
2. Find the Sum of an Arithmetic Sequence
3. Solve Applied Problems

9.6 Arithmetic Sequences and Series

Many sequences are generated by starting with an initial term, then repeatedly adding a fixed number. Such a sequence is known as the *arithmetic sequence.* The general term of this type of a sequence corresponds to a linear function.

Defining an Arithmetic Sequence

Consider the following sequence

$$3, 7, 11, 15, 19, \ldots$$

Note that each term of this sequence is 4 more than the previous one. That is, consecutive terms differ by 4 since

$$a_2 - a_1 = 7 - 3 = 4$$
$$a_3 - a_2 = 11 - 7 = 4$$
$$a_4 - a_3 = 15 - 11 = 4$$

and so on.

This sequence is an example of an *arithmetic sequence,* which has a feature that consecutive terms differ by a constant value.

More precisely, we have the following definition.

Arithmetic Sequences

A sequence of the form

$$a_1, a_2, a_3, \ldots, a_n, \ldots$$

is called an **arithmetic sequence** if there is a constant d such that any two consecutive terms differ by d. That is, for every positive integer $n \geq 1$

$$a_{n+1} - a_n = d.$$

The constant d is called the **common difference.**

In the above sequence, each term is created from the previous one (except for the first) by adding the common difference d. That is, every term of the sequence can be written in terms of the first term a_1 and common difference d as follows.

First term: $a_1 = a_1$

Second term: $a_2 = a_1 + d$

Third term: $a_3 = a_2 + d$
$$= (a_1 + d) + d$$
$$= a_1 + 2d$$

Fourth term: $a_4 = a_3 + d$
$$= (a_1 + 2d) + d$$
$$= a_1 + 3d$$

and so on. This pattern leads to a formula for the nth term which is given by

$$a_n = a_1 + (n - 1)d, n \geq 1.$$

EXAMPLE 1

Finding the *n*th Term of a Sequence

Find the *n*th term of the arithmetic sequence 7, 10, 13, 16, ... then find a_{10} and a_{20}.

Solution

The first term is 7, and the common difference is

$$d = a_2 - a_1 = 10 - 7 = 3.$$

Using the formula for the nth term a_n, we have

$$a_n = a_1 + (n - 1)d$$
$$= 7 + (n - 1)3 \qquad \text{Replace } a_1 \text{ by 7, } d \text{ by 3}$$
$$= 7 + 3n - 3$$
$$= 4 + 3n. \qquad \text{Simplify}$$

Thus, the *n*th term is $a_n = 4 + 3n$, so that

$$a_{10} = 4 + 3(10) = 4 + 30 = 34$$

and

$$a_{20} = 4 + 3(20) = 4 + 60 = 64.$$

EXAMPLE 2

Finding the Number of Terms of a Sequence

Find the number of terms of the finite arithmetic sequence

$$-9, -2, 5, 12, 19, \ldots, 96.$$

Solution

Here $a_1 = -9$, $d = -2 - (-9) = 7$, and $a_n = 96$.

Substituting these values into the formula for a_n, we have

$$a_n = a_1 + (n - 1)d$$
$$96 = -9 + (n - 1)7$$
$$96 = -9 + 7n - 7$$
$$96 = -16 + 7n$$
$$112 = 7n$$
$$16 = n.$$

Therefore, there are 16 terms of this sequence.

EXAMPLE 3

Finding a Specific Term of a Sequence

Find the 50th term of the arithmetic sequence whose common difference is 4 and whose 10th term is 37.

Solution

Using $a_n = a_1 + (n - 1)d$, with $d = 4$, $a_{10} = 37$, and $n = 10$, we have

$$37 = a_1 + (10 - 1)4$$
$$37 = a_1 + 36$$
$$1 = a_1.$$

Therefore,

$$a_{50} = a_1 + (n-1)d$$
$$= 1 + (50-1)4$$
$$= 1 + 196 = 197.$$

Finding the Sum of an Arithmetic Sequence

Associated with any arithmetic sequence, with common difference d,

$$a_1, a_2, a_3, \ldots, a_n$$

is an **arithmetic series** of the form

$$S_n = \sum_{k=1}^{n} a_k = a_1 + a_2 + a_3 + \cdots + a_n.$$

Suppose we are to find the value of S_n, the sum of the first n terms of this sequence. Using Gauss's pairing method in section 9.5 leads us to a general formula for the sum of the first n terms of an arithmetic sequence. This is accomplished by adding the first and last terms and multiply by half the number of terms. That is, the sum S_n of the first n terms of an arithmetic sequence is given by

$$S_n = \sum_{k=1}^{n} a_k = \frac{n}{2}(a_1 + a_n).$$

Using the formula $a_n = a_1 + (n-1)d$ for the nth term, we find the following alternative form for the sum S_n.

$$S_n = \frac{n}{2}\left[2a_1 + (n-1)d\right]$$

EXAMPLE 4

Find the Sum of an Arithmetic Sequence

Find the sum of the first 20 terms of the arithmetic sequence

$$2, 6, 10, 14, 18, \ldots$$

Solution

In this sequence, the first term is $a_1 = 2$ and the common difference is

$$d = 6 - 2 = 4.$$

Since we want to find the sum of the first 20 terms, we have $n = 20$. Substituting these values into the alternative form for the sum, we get

$$S_n = \frac{n}{2}\left[2a_1 + (n-1)d\right]$$

$$= \frac{20}{2}\left[2(2) + (20-1)4\right]$$

$$= 10(4 + 76)$$

$$= 800.$$

EXAMPLE 5   **Finding the Sum of an Arithmetic Sequence**

Find the sum of the terms of the finite arithmetic sequence

$$2, 5, 8, 11, 14, 17, 20, \ldots, 86.$$

Solution   The first term of this sequence is $a_1 = 2$, and the common difference is

$$d = 5 - 2 = 3.$$

We need to find the number of terms in the sequence such that $a_n = 86$. We substitute these values into the formula for a_n to get

$$a_n = a_1 + (n - 1)d$$
$$86 = 2 + (n - 1)3$$
$$84 = (n - 1)3$$
$$28 = n - 1$$
$$29 = n.$$

Thus the sequence contains 29 terms. Using the formula for the sum of an arithmetic sequence, we have

$$S_n = \frac{n}{2}(a_1 + a_n)$$

$$= \frac{29}{2}(2 + 86)$$

$$= 1,276.$$

Thus, the sum of this sequence is 1,276.

Solving Applied Problems

Many applied problems have solutions that lead to various arithmetic sequences and series as the next example shows.

EXAMPLE 6   **Finding the Number of Seats in an Auditorium**

The first 12 rows of the center section of the Macomb Center for the Performing Arts are staggered. The first row has 11 seats, the second row has 12 seats, and the third row has 13 seats, and so on in an arithmetic sequence. Use a series to find the total number of seats in this section.

Solution   Here $a_1 = 11$, $d = 1$ and $n = 12$. Substituting these values in the formula for a_n we get

$$a_n = a_1 + (n - 1)d$$
$$a_{12} = 11 + (12 - 1)(1)$$
$$= 11 + 11$$
$$= 22.$$

The total number of seats is obtained from the formula

$$S_n = \frac{n}{2}(a_1 + a_n)$$

$$= \frac{12}{2}(11 + 22) \qquad \text{Substitute 22 for } a_n$$

$$= 198.$$

Therefore, there are 198 seats in this section.

The graph of an arithmetic sequence can give us additional information about its characteristics and properties as the next example shows.

EXAMPLE 7 **Using a Graph to Show the Properties of an Arithmetic Sequence**

Graph the first five ordered pairs of the sequence

$$11, 8, 5, 2, -1, \ldots$$

Compare the graph of this sequence to the graph of a linear function.

Solution The first term is $a_1 = 11$ and $d = 8 - 11 = -3$, so the nth term of this sequence is given by

$$a_n = a_1 + (n - 1)d$$

$$= 11 + (n - 1)(-3) = 11 - 3n + 3$$

$$= 14 - 3n.$$

The first five ordered pairs are

$$(1, 11), (2, 8), (3, 5), (4, 2) \text{ and } (5, -1).$$

Figure 1a shows the graph of this sequence. The graph suggests that the graph of an arithmetic sequence is a set of collinear points which corresponds to the graph of the linear function in Figure 1b,

$$f(x) = 14 - 3x.$$

Figure 1

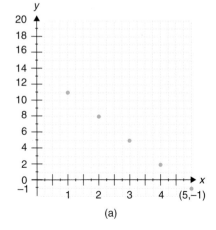

(a)

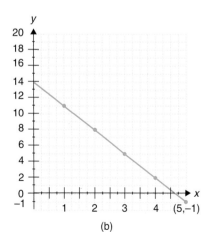

(b)

PROBLEM SET 9.6

Mastering the Concepts

In Problems 1–10, find (a) a_n, (b) a_{10}, and (c) a_{20} in each arithmetic sequence.

1. $-2, 1, 4, 7, \ldots$
2. $-9, -5, -1, 3, \ldots$
3. $11, 8, 5, 2, \ldots$
4. $25, 16, 7, -2, \ldots$
5. $2, 5, 8, 11, \ldots$
6. $7, 12, 17, 22, \ldots$
7. $5, -1, -7, -13, \ldots$
8. $3, -1, -5, -9, \ldots$
9. $-9, -6, -3, 0, \ldots$
10. $1, -1, -3, -5, \ldots$

In Problems 11–16, find the number of terms of each finite sequence.

11. $5, 10, 15, 20, \ldots, 125$
12. $5, 2, -1, -4, \ldots, -55$
13. $-4, -1, 2, 5, 8, \ldots, 83$
14. $2, 6, 10, 14, \ldots, 118$
15. $-7, -2, 3, 8, 13, \ldots, 143$
16. $-6, -3, 0, 3, \ldots, 114$

In Problems 17–32, find the indicated term of each arithmetic sequence with the given information.

17. a_{10} if $a_4 = 8$ and $d = 7$
18. a_{30} if $a_3 = 15$ and $d = -2$
19. a_6 if $a_5 = 16$ and $d = 14$
20. a_{15} if $a_2 = 8$ and $d = -6$
21. a_{12} if $a_1 = 15$ and $d = -8$
22. a_{30} if $a_2 = 13$ and $d = 4$
23. a_{10} if $a_1 = 9$ and $d = -4$
24. a_{25} if $a_1 = 3$ and $d = -0.5$
25. a_1 and a_{20} if $a_5 = 8$ and $a_{10} = 23$
26. a_1 and a_{20} if $a_{10} = -26$ and $a_{14} = -6$
27. a_1 and a_{15} if $a_5 = 2$ and $a_{11} = -10$
28. a_1 and a_{20} if $a_{10} = 3$ and $a_{16} = 21$
29. a_1 and a_{20} if $a_{15} = 16$ and $a_{19} = -12$
30. a_1 and a_{50} if $a_{10} = -11$ and $a_{40} = -71$
31. a_1 and a_{40} if $a_5 = 3$ and $a_{50} = 30$
32. a_1 and a_{15} if $a_{10} = -37$ and $a_{12} = -45$
33. Find the sum of the first 10 terms of an arithmetic sequence whose first term is 1 and whose common difference is 3.
34. Find the sum of the first 15 terms of an arithmetic sequence whose first term is 1/2 and whose common difference is 1/2.
35. Find the sum of the first 8 terms of an arithmetic sequence whose first term is -5 and whose common difference is 3/7.
36. Find the sum of the first 12 terms of an arithmetic sequence whose first term is 11 and whose common difference is -2.

In Problems 37–44, find the sum of the terms of each finite arithmetic sequence.

37. $4, 10, 16, 22, \ldots, 184$
38. $7, 19, 31, 43, \ldots, 247$
39. $-12, -3, 6, 15, \ldots, 123$
40. $-15, -7, 1, 9, \ldots, 225$
41. $100, 95, 90, 85, \ldots, 10$
42. $4, 8, 12, 16, \ldots, 92$
43. $7, 11, 15, 19, \ldots, 127$
44. $120, 116, 112, 108, \ldots, 12$

Applying the Concepts

45. **Bricklayer's Pattern:** A bricklayer wishes to arrange 376 bricks in a pile. The bricks were stacked so that there will be 1 brick in the top row, 4 bricks in the second row, 7 in the third row, and so on in an arithmetic sequence. Can this be done? If so,
    (a) How many rows will it take?
    (b) And how many bricks will be in the bottom row?

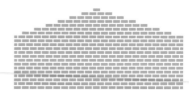

46. **Advertising:** During a special sales promotion, an automobile salesperson received a bonus of $100 for the first car, $105 for the second car, $110 for the third car, and so on in an arithmetic sequence. Find
    (a) The bonus received for the 30th car.
    (b) The total bonus received.
47. **Accumulated earnings:** A college professor started working in 1978 in a teaching job that started at an annual salary of $25,600 and each year received a $1400 pay raise.
    (a) How much would she be paid during the 1998 school year?
    (b) What would be the total salary paid from 1978 through 2002?
48. **Accumulated Savings:** If a newspaper carrier saved $1 on September 1, another $2 on

September 2, another $3 on September 3, and so on, how much money was saved during the month of September (30 days)?

49. **Drilling a Well:** A well-drilling contractor charges $1.50 to drill the first foot, $1.65 for the second foot, $1.80 for the third foot, and so on in an arithmetic sequence. At this rate find
   (a) The cost to drill the last foot of a well 200 feet deep.
   (b) The total cost to drill the well.

50. **Pollution Standards:** A company is informed that if it fails to meet EPA pollution standards by a fixed date, it is to be fined $2,000 the first day, and each day thereafter the fine would be increased by $400. What is the fine for a company that fails to meet the standards for 10 days after the deadline?

Developing and Extending the Concepts

51. Find the sixth term of the arithmetic sequence

$$a + 24b, 4a + 20b, 7a + 16b, \ldots$$

52. Find the sixteenth term of the arithmetic sequence

$$7a^2 - 4b, 2a^2 + 7b, -3a^2 + 18b, \ldots$$

53. Given an arithmetic sequence with $a_1 = -2$, $d = 7$ and $a_n = 138$, find n.

54. Given an arithmetic sequence with $a_3 = 25$, $d = -14$ and $a_n = -507$, find n.

For Problems 55–58,
   (a) Graph the first five ordered pairs of each arithmetic sequence.
   (b) Find and graph the corresponding linear function.
   (c) Compare the graph of the sequence to the graph of the corresponding linear function.

55. $23, 17, 11, 5, -1, \ldots$
56. $10, 13, 16, 19, 22, \ldots$
57. $-3, 0, 3, 6, 9, \ldots$
58. $-10, -6, -2, 2, \ldots$
59. (a) Find the general terms a_n and b_n of each arithmetic sequence

$$2, 5, 8, 11, \ldots \text{ and } 7, 9, 11, 13, \ldots$$

   (b) Find the general term of a sequence, c_n, whose terms are the sum of the corresponding terms of the sequences in part (a), i.e. find $c_n = a_n + b_n$.
   (c) Find $p_n = (a_n)(b_n)$. That is, find the sequence whose general term corresponds to the product of the corresponding terms of the sequences in part (a).

60. Consider the sequence a_n defined by $a_n = n^2$. Suppose that that b_n is another sequence defined by

$$b_n = a_n - a_{n-1};$$

   show that $b_1, b_2, b_3, \ldots$ is an arithmetic sequence.

Objectives

1. Define a Geometric Sequence
2. Find a Sum of a Geometric Series
3. Evaluate an Infinite Geometric Series
4. Solve Applied Problems

9.7 Geometric Sequences and Series

In this section, we focus on another important sequence called a *geometric sequence,* where each term except the first is determined by multiplying the previous term, by a fixed number.

Defining a Geometric Sequence

Consider the sequence

$$3, 6, 12, 24, 48, \ldots$$

Each term of this sequence, except the first, is 2 times the previous one. That is, the ratio of consecutive terms is 2.

$$\frac{a_2}{a_1} = \frac{6}{3} = 2, \quad \frac{a_3}{a_2} = \frac{12}{6} = 2, \quad \frac{a_4}{a_3} = \frac{24}{12} = 2, \quad \text{and so on.}$$

This sequence is an example of a *geometric sequence,* which is characterized by the feature that the ratio of consecutive terms is constant.

More precisely, we have the following definition.

Geometric Sequence

A sequence of the form

$$a_1, a_2, a_3, \ldots, a_n, \ldots$$

is called a **geometric sequence** if there is a constant r, called the **common ratio** such that

$$\frac{a_{n+1}}{a_n} = r,$$

for n positive integer.

EXAMPLE 1   **Finding the Common Ratio of a Geometric Sequence**

Find the common ratio of the sequence

$$-4, 16, -64, 256, \ldots$$

Solution   To find the common ratio r of this sequence choose any two adjacent terms and divide the second by the first. Choosing the second and the third terms of this sequence, we have

$$r = -\frac{64}{16} = -4.$$

Therefore, the common ratio is -4.

Each term of a geometric sequence can be expressed in terms of the first term a_1 and the common ratio r as follows.

First term: $a_1 = a_1$

Second term: $a_2 = a_1 r$

Third term: $a_3 = a_2 r = (a_1 r)r$
$$= a_1 r^2$$

Fourth term: $a_4 = a_3 r = (a_1 r^2)r$
$$= a_1 r^3$$

Fifth term: $a_5 = a_4 r = (a_1 r^3)r$
$$= a_1 r^4$$

nth term: $a_n = a_{n-1} r = (a_1 r^{n-2})r$
$$= a_1 r^{n-1}$$

which suggests the following formula for the general term a_n:

$$a_n = a_1 r^{n-1}$$

EXAMPLE 2   **Finding the *n*th Term of a Geometric Sequence**

Find the *n*th term of each geometric sequence.

(a) $3, 6, 9, 12, \ldots$ (b) $-32, 16, -8, 4, \ldots$

Solution   (a) The first term is $a_1 = 3$ and the common ratio $r = \dfrac{6}{3} = 2$.

Using the formula for the nth term of a geometric sequence, we get

$$a_n = a_1 r^{n-1}$$
$$a_n = 3(2)^{n-1}. \qquad \text{Replace } a_1 \text{ by 3 and } r \text{ by 2}$$

(b) Here $a_1 = -32$ and $r = -\dfrac{16}{32} = -\dfrac{1}{2}$. Using the formula for the *n*th term, we have

$$a_n = a_1 r^{n-1}$$
$$a_n = -32\left(-\dfrac{1}{2}\right)^{n-1}. \qquad \text{Replace } a_1 \text{ by } -32 \text{ and } r \text{ by } -\dfrac{1}{2}.$$

EXAMPLE 3   **Finding a Specific Term of a Geometric Sequence**

Suppose that the first two terms of a geometric sequence are 2 and 4, respectively. Find (a) the term of the sequence that is equal to 512 (b) the fifteen term a_{15}.

Solution   (a) Since the first two terms of the sequence are: $a_1 = 2$ and $a_2 = 4$, we conclude that $r = 2$.

If $a_n = 512$, then $512 = 2(2^{n-1})$

or $2^9 = 2^{1+n-1} = 2^n,$

so that $n = 9.$

Therefore, the ninth term of the sequence is 512.

(b) The general term of this sequence is given by

$$a_n = 2(2^{n-1}) \qquad \text{Replace } n \text{ by 15}$$
$$= 2^n$$
$$a_{15} = 2^{15}$$
$$= 32{,}768.$$

Finding a Sum of a Geometric Series

Consider the geometric sequence

$$a_1, a_1 r, a_1 r^2, a_1 r^3, \ldots, a_1 r^{n-1}$$

with first term a_1 and common ratio r.

The associated sum

$$S_n = a_1 + a_1 r + a_1 r^2 + a_1 r^3 + \ldots + a_1 r^{n-1}$$

is called a **geometric series.** The value of S_n is the sum of n terms of the sequence.

For example, consider the sequence

$$5, 15, 45, 135, 405, \ldots, 5(3^{n-1}).$$

The sum of the first five terms of this sequence is given by

$$S_1 = 5$$
$$S_2 = 5 + 15 = 20$$
$$S_3 = 5 + 15 + 45 = 65$$
$$S_4 = 5 + 15 + 45 + 135 = 200$$
$$S_5 = 5 + 15 + 45 + 135 + 405 = 605.$$

Following this pattern, the sum S_n is given by

$$S_n = \frac{5}{2}(3^n - 1).$$

Notice that the sequence and its sum were based on powers of 3. This suggests that there is a formula for finding the sum of the terms of a geometric sequence based on the powers of r.

Sum of a Geometric Sequence

The **sum** S_n of the first n terms of a geometric sequence with first term a_1 and common ratio r is given by

$$S_n = \frac{a_1(1 - r^n)}{1 - r}, r \neq 1.$$

EXAMPLE 4   **Finding the Sum of a Geometric Sequence**

Find the sum of the first 10 terms of the geometric sequence whose first term is 1/2 and whose common ratio is 2.

Solution   We substitute $n = 10$, $a_1 = 1/2$, and $r = 2$ in the formula

$$S_n = \frac{a_1(1 - r^n)}{1 - r}, r \neq 1$$

and obtain

$$S_{10} = \frac{\frac{1}{2}(1 - 2^{10})}{1 - 2}$$

$$= \frac{\frac{1}{2}(-1{,}023)}{-1}$$

$$= 511.5.$$

EXAMPLE 5   **Finding the First Five Terms of a Geometric Series**

If the sum of the first five terms of a geometric series is 61/27, find the first five terms of this sequence that has a common ratio $r = -1/3$.

Solution   Substituting $S_n = 61/27$ and $r = -1/3$ in the formula for the sum, we get

$$S_n = \frac{a_1(1 - r^n)}{1 - r}$$

$$\frac{61}{27} = \frac{a_1\left[1 - \left(-\frac{1}{3}\right)^5\right]}{1 - \left(-\frac{1}{3}\right)}$$ Replace r by $-\dfrac{1}{3}$ and n by 5.

$$\frac{61}{27} = \frac{\frac{244}{243}}{\frac{4}{3}}a_1$$

$$\frac{61}{27} = \frac{61}{81}a_1$$

$$a_1 = 3$$

The first five terms are $3, -1, \frac{1}{3}, -\frac{1}{9}, \frac{1}{27}$.

Evaluating an Infinite Geometric Series

Consider the geometric series

$$1 + \frac{1}{2} + \frac{1}{2^2} + \frac{1}{2^3} + \cdots \frac{1}{2^{n-1}} + \cdots$$

Here $a_1 = 1$ and $r = 1/2$, so the sum S_n of the n terms is given by

$$S_n = \frac{a_1(1 - r^n)}{1 - r}$$

$$= \frac{1\left(1 - \left(\frac{1}{2}\right)^n\right)}{1 - \frac{1}{2}}$$

$$= 2\left[1 - \left(\frac{1}{2}\right)^n\right].$$

Now, as n gets larger and larger, the expression $\left(\dfrac{1}{2}\right)^n$ gets closer and closer to zero. In fact, by taking n large enough, $\left(\dfrac{1}{2}\right)^n$ can be made as close to zero as we please. This leads to the conclusion that

$$\frac{1 - \left(\frac{1}{2}\right)^n}{1 - \frac{1}{2}} = 2 \text{ as } n \text{ gets large enough.}$$

Similar reasoning works for obtaining the sum of any infinite geometric series with common ratio r for which $|r| < 1$.

Evaluating an Infinite Geometric Sequence

If the common ratio r is between -1 and 1, the **sum** S of an infinite number of terms of a geometric sequence with first term a_1 is given by

$$S = \frac{a_1}{1 - r}.$$

EXAMPLE 6   **Finding the Sum of an Infinite Geometric Sequence**

Find the indicated sum of infinite number of terms for each geometric sequence.

(a) $\dfrac{3}{10} + \dfrac{3}{100} + \dfrac{3}{1,000} + \dfrac{3}{10,000} + \cdots$ (b) $10 - 4 + 1.6 - 0.64 + \ldots$

Solution   (a) The common ratio $r = \dfrac{\frac{3}{100}}{\frac{3}{10}} = \dfrac{1}{10}$ and $a_1 = \dfrac{3}{10}$. So that

$$S = \frac{a_1}{1-r} = \frac{\frac{3}{10}}{1 - \frac{1}{10}} = \frac{\frac{3}{10}}{\frac{9}{10}} = \frac{3}{9} = \frac{1}{3}.$$

(b) Here $r = \dfrac{-4}{10} = \dfrac{-2}{5}$ and $a_1 = 10$. So

$$S = \frac{a_1}{1-r} = \frac{10}{1 - \left(-\frac{2}{5}\right)} = \frac{10}{\frac{7}{5}} = \frac{50}{7}.$$

Solving Applied Problems

Geometric sequences and series are often used to solve applied problems from different fields as the next example shows.

EXAMPLE 7   **Finding the Distance of a Child Swing**

A child swing starts with an initial distance of 18 feet and goes 7/9 of the prior distance on each subsequent swing.

(a) How far does the swing travel through the nth swing?

(b) What is the total distance traveled through the sixth swing?

(c) How far does the swing travel before eventually coming to rest?

Solution   (a) The child swing travels 18 feet through the first swing, $18\left(\dfrac{7}{9}\right)$ feet through the second swing, $18\left(\dfrac{7}{9}\right)^2$ feet through the third swing, and so on.

The distance the swing travels through the nth swing is given by

$$a_n = 18\left(\frac{7}{9}\right)^{n-1}.$$

(b) The total distance traveled through the sixth swing is given by

$$S_n = \frac{a_1(1 - r^n)}{1 - r}$$

$$S_6 = \frac{18\left(1 - \left(\frac{7}{9}\right)^6\right)}{1 - \frac{7}{9}} \qquad \text{Replace } n \text{ by 6}$$

$$= \frac{18\left[\frac{413,792}{531,441}\right]}{\frac{2}{9}} = 63.07.$$

(c) The total distance traveled before eventually coming to rest is given by

$$S = \frac{a_1}{1 - r}$$

$$= \frac{18}{1 - \frac{7}{9}} \qquad\qquad \text{Replace } a_1 \text{ by 18 and } r \text{ by } \frac{7}{9}$$

$$= \frac{18}{\frac{2}{9}} = 81$$

The swing travels 81 feet before it comes to rest.

In section 9.6 we noted that an arithmetic sequence corresponds to a linear function. By graphing a geometric sequence, we can see that geometric sequences correspond to exponential functions as the next example shows.

EXAMPLE 8

Compare Graphs of a Geometric Sequence and an Exponential Function

Graph the first five ordered pairs and the corresponding exponential function on different planes.

$$a_n = \left(\frac{2}{3}\right)^n \ \text{ and } \ f(x) = \left(\frac{2}{3}\right)^x$$

Solution

Substituting 1, 2, 3, 4, and 5 for n into $a_n = \left(\frac{2}{3}\right)^n$ we get

$$a_1 = \left(\frac{2}{3}\right)^1 = \frac{2}{3} \qquad a_2 = \left(\frac{2}{3}\right)^2 = \frac{4}{9} \qquad a_3 = \left(\frac{2}{3}\right)^3 = \frac{8}{27}$$

$$a_4 = \left(\frac{2}{3}\right)^4 = \frac{16}{81} \qquad a_5 = \left(\frac{2}{3}\right)^5 = \frac{32}{243}.$$

The first five ordered pairs of this sequence are given by

$$\left(1, \frac{2}{3}\right), \left(2, \frac{4}{9}\right), \left(3, \frac{8}{27}\right), \left(4, \frac{16}{81}\right), \text{ and } \left(5, \frac{32}{243}\right).$$

Figure 1a shows the plot of these points and Figure 1b shows the graph of

$$f(x) = \left(\frac{2}{3}\right)^x.$$

Figure 1

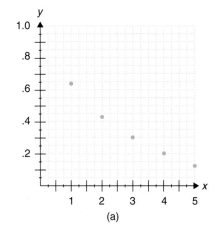

(a)

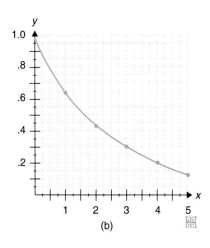

(b)

◈ PROBLEM SET 9.7

Mastering the Concepts

In Problems 1–14, determine whether each sequence is arithmetic, geometric, or neither. If it is geometric find the common ratio, if arithmetic, find the common difference.

1. $2, 6, 18, 54, \ldots$
2. $1, -2, 4, -8, \ldots$
3. $81, 54, 36, 24, \ldots$
4. $2, 4, 8, 16, \ldots$
5. $9, -6, 4, -\frac{8}{3}, \ldots$
6. $3, 9, 27, 81, \ldots$
7. $64, -32, 16, -8, \ldots$
8. $18, -6, 2, -\frac{2}{3}, \ldots$
9. $1, \frac{1}{5}, \frac{1}{25}, \frac{1}{125}, \ldots$
10. $\frac{4}{9}, \frac{1}{6}, \frac{1}{16}, \frac{1}{32}, \ldots$
11. $9, 3, -3, -9, \ldots$
12. $1, 4, 7, 10, \ldots$
13. $7, 19, 31, 43, \ldots$
14. $-15, -7, 1, 9, \ldots$

In Problems 15–26, find (a) the general term a_n, and (b) the eighth term a_8 for each geometric sequence.

15. $-4, 2, -1, \frac{1}{2}, \ldots$
16. $\frac{1}{8}, \frac{1}{4}, \frac{1}{2}, 1, \ldots$
17. $32, 16, 8, 4, \ldots$
18. $1, 1.03, (1.03)^2, (1.03)^3, \ldots$
19. $6, 12, 24, 48, \ldots$
20. $3, 6, 12, 24, \ldots$
21. $5, 10, 20, 40, \ldots$
22. $\frac{1}{4}, -8, 256, -8192, \ldots$
23. $\frac{1}{2}, -2, 8, -32, \ldots$
24. $\frac{1}{3}, -1, 3, -9, \ldots$
25. $8, 4, 2, 1, \ldots$
26. $-1, -\frac{3}{2}, -\frac{9}{4}, -\frac{27}{8},$

27. If the first two terms of a geometric sequence are 2 and 1 respectively, which term of the sequence is 1/16?
28. If the first two terms of a geometric sequence are $3\sqrt{3}$ and 9, respectively, which term of the sequence is $243\sqrt{3}$?
29. If the third term of a geometric sequence is 9/4 and the sixth term is 243/256, what is its first term?
30. The sixth term of a geometric sequence is 27, and the common ratio is $-1/3$. Find the third term.
31. Find the sum of the first 6 terms of the geometric sequence whose first term is 3/2 and whose common ratio is 2.
32. Find the sum of the first 10 terms of the geometric sequence whose first term is 6 and whose common ratio is 1/2.
33. Find the sum of the first 12 terms of the geometric sequence whose first term is -4 and whose common ratio is -2.
34. Find the sum of the first 8 terms of the geometric sequence whose first term is 5 and whose common ratio is $-1/2$.

35. Find the sum of the first 8 terms of the geometric sequence
$$\frac{1}{2}, \frac{3}{2}, \frac{9}{2}, \frac{27}{2}, \ldots$$
36. Find the sum of the first 5 terms of the geometric sequence
$$-\frac{1}{24}, \frac{1}{12}, -\frac{1}{6}, \frac{1}{3}, \ldots$$
37. Find the sum of the first 7 terms of the geometric sequence
$$16, 4, 1, \frac{1}{4}, \ldots$$
38. Find the sum of the first 5 terms of the geometric sequence
$$3, 12, 48, 192, \ldots$$
39. The sum of the first 5 terms of a geometric sequence is 31 and the common ratio is 2. Find the first four terms of the sequence.
40. The sum of the first 4 terms of a geometric sequence is 20 and the common ratio is -2. Find the first five terms of the sequence.

In Problems 41–48, find the sum of each infinite geometric series.

41. $1 + \frac{1}{2} + \frac{1}{4} + \frac{1}{8} + \ldots$
42. $3 + 1 + \frac{1}{3} + \frac{1}{9} + \ldots$
43. $8 + 4 + 2 + 1 + \ldots$
44. $1 + \frac{2}{3} + \frac{4}{9} + \frac{8}{27} + \ldots$
45. $54 - 18 + 6 - 2 + \ldots$
46. $4 - 2 + 1 - \frac{1}{2} + \ldots$
47. $\frac{4}{5} + \frac{8}{15} + \frac{16}{45} + \frac{32}{135} + \ldots$
48. $\frac{4}{3} + \frac{2}{9} + \frac{1}{27} + \ldots$

Applying the Concepts

49. **Monthly Earning:** A firm offers an employee a starting salary of $24,000 with 5% annual increases for 6 years.
   (a) List the annual salaries during each of the 6 years as a sequence.
   (b) Does the annual salary during the fourth year of employment represent a 15% raise over the original salary? Explain.
   (c) Suppose that 5% annual raise continued for 10 years. Determine the annual salary during the tenth year of employment.

50. Depreciation: For tax purposes, a company depreciates the value of a company car by 20% every year. The car was originally purchased for $32,000.

(a) List the depreciated values of the car at the end of each of the first 4 years as a sequence.

(b) Does the depreciated value of the car at the end of the fourth year represent a 60% decrease? Explain.

(c) Determine the depreciated value of the car at the end of the seventh year.

51. Rebound Distance: A Super Ball dropped from the top of the Eiffel Tower (300 meters high) rebounds two thirds of the distance falling (Figure 2).

(a) How far (up and down) will the ball have traveled when it hits the ground for the fifth time?

(b) How far will it travel before it comes to rest?

Figure 2

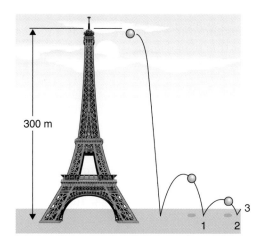

300 m

1   2   3

52. Investment: At the beginning of each month $1,000 is deposited into an annuity for a college professor that pays 7% annual interest compounded monthly. Determine the balance of this investment after

(a) 2 years   (b) 5 years   (c) 10 years.

53. Orange Grower: An orange grower receives $10,000 for the oranges he sold for the first year of his young trees. If each succeeding year his income increases 10%, what would be the grower's total income over the first 8 years?

54. Pendulum Swings: The tip of a pendulum swings through an arc of 8 inches, and each arc thereafter is 11/13 of the length of the preceding one (Figure 3). How far does the tip move before the pendulum comes to rest?

Figure 3

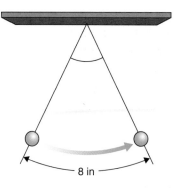

8 in

Developing and Extending the Concepts

55. If the first term of a geometric sequence is 1/16, the nth term is 32 and the common ratio is 2, find the number of terms n.

56. If the first term of a geometric sequence is 250, the nth term is 32 3/5, and the common ratio is 3/5, find the sum of the first n terms.

57. If the sum of the terms of an infinite geometric sequence is 35 and the first term a_1 is 7, find the common ratio r.

58. If the sum of the terms of an infinite geometric sequence is 20, and the common ratio r is $-1/2$, find the first term a_1.

59. The repeating decimal $0.\overline{7}$ can be written as the infinite geometric series

$$0.7 + 0.07 + 0.007 + 0.0007 + \ldots,$$

with $a_1 = 0.7$ and $r = 0.1$. Use the sum formula to show that the infinite sum is 7/9. Note that this is another way to convert $0.\overline{7}$ to a fraction in the form a/b.

60. The repeating decimal $0.\overline{14}$ can be written as the infinite geometric series as $0.14 + 0.0014 + 0.000014 + 0.00000014 + \ldots$ with $a_1 = 0.14$ and $r = 0.01$. Use the formula for the sum of an infinite series to convert $0.\overline{14}$ to a fraction.

61. If 16, x, 9/4 is a geometric sequence, how can we find the value of x?

62. Does there exist an infinite geometric series for which $a_1 = 2$ and its sum $S = 1$? Justify your answer.

In Problems 63–66, graph the first five ordered pairs of the sequence and the corresponding exponential function.

63. $a_n = 2^n, f(x) = 2^x$

64. $a_n = \dfrac{1}{2^n}, f(x) = \dfrac{1}{2^x}$

65. $a_n = \left(\dfrac{3}{2}\right)^n, f(x) = \left(\dfrac{3}{2}\right)^x$

66. $a_n = (1.1)^n, f(x) = (1.1)^x$

Objectives

1. Compute Binomial Coefficients by Pascal's Triangle
2. Compute Binomial Coefficients Using Factorials
3. Expand Binomials Using the Binomial Theorem
4. Find a Specific Term in a Binomial Expansion
5. Solve Applied Problems

9.8 The Binomial Theorem

In many applied problems in statistics, calculus, and other branches of mathematics, we are required to write expressions such as $(a + b)^n$ as a sum, which we call the **expanded form** of $(a + b)^n$. In this section we will develop a formula that enables us to accomplish this objective.

Computing Binomial Coefficients by Pascal's Triangle

In section 4.2, we considered the special products $(a + b)^2$ and $(a + b)^3$. We often work with expressions of the form $(a + b)^n$, where n is a positive integer. Since the expression $a + b$ is a binomial, the formula for expanding $(a + b)^n$ is called the **binomial theorem.**

We can expand $(a + b)^n$, for small values of n, by using direct calculations. For instance,

$$(a + b)^1 = a + b$$

$$(a + b)^2 = a^2 + 2ab + b^2$$

$$(a + b)^3 = a^3 + 3a^2b + 3ab^2 + b^3$$

$$(a + b)^4 = a^4 + 4a^3b + 6a^2b^2 + 4ab^3 + b^4$$

$$(a + b)^5 = a^5 + 5a^4b + 10a^3b^2 + 10a^2b^3 + 5ab^4 + b^5$$

The pattern holds for the expansion of $(a + b)^n$, where n is any positive integer. The following rules are used for this expansion:

1. There are $n + 1$ terms.
2. The first term is a^n and the last term is b^n.
3. The powers of a decrease by 1 and the powers of b increase by 1 for each term.
4. The sum of the exponents of a and b is n for each term.
5. Each term has a numerical coefficient that must be determined.

One way to display the coefficients in the expansion of $(a + b)^n$ for $n = 1, 2, 3, \ldots$ is the following array of numbers (Figure 1), known as **Pascal's triangle** named in honor of the French Mathematician Blaise Pascal (1623–1662):

Figure 1

Row 0 → $(a + b)^0$	1
Row 1 → $(a + b)^1$	1 1
Row 2 → $(a + b)^2$	1 2 1
Row 3 → $(a + b)^3$	1 3 3 1
Row 4 → $(a + b)^4$	1 4 6 4 1
Row 5 → $(a + b)^5$	1 5 10 10 5 1

The first and last numbers in each row are always 1. The other numbers can be found by adding the pair of numbers from the preceding row as indicated by the V's. For example,

4 6
10

indicates that 10 was obtained by adding 4 and 6.

The advantage of Pascal's triangle is that it allows us to calculate the binomial coefficients using addition rather than multiplication. The disadvantage, however, is that it is impractical for large powers, since it requires us to calculate all rows preceding the one in which we are interested.

EXAMPLE 1   **Using Pascal's Triangle to Expand a Binomial**

Use Pascal's triangle to expand $(3x + 2y)^5$.

Solution   The coefficients are given in Row 5 of Figure 1, and the sum of the exponents of each term is 5. So that

$$(3x + 2y)^5 = (3x)^5 + 5(3x)^4(2y) + 10(3x)^3(2y)^2 + 10(3x)^2(2y)^3 + 5(3x)(2y)^4 + (2y)^5$$

$$= 243x^5 + 810x^4y + 1080x^3y^2 + 720x^2y^3 + 240xy^4 + 32y^5.$$

Computing Binomial Coefficients Using Factorials

We indicated that the disadvantage in using Pascal's triangle is that it becomes cumbersome when a binomial contains a larger power. However, by using the idea of *factorial notation,* we can determine the binomial coefficient more efficiently. The symbol $n!$ (read "n factorial") is defined as follows.

$$n! = 1 \cdot 2 \cdot 3 \cdot 4 \ldots (n - 1)n$$

or

$$n! = n(n - 1)(n - 2) \ldots 2 \cdot 1$$

Here are some examples of the value of factorials.

$$1! = 1$$
$$2! = 2 \cdot 1 = 2$$
$$3! = 3 \cdot 2 \cdot 1 = 6$$
$$4! = 4 \cdot 3 \cdot 2 \cdot 1 = 24$$
$$5! = 5 \cdot 4 \cdot 3 \cdot 2 \cdot 1 = 120$$

We also define

$$0! = 1$$

Notice that $(n - 1)! = (n - 1)(n - 2) \ldots 3 \cdot 2 \cdot 1$.

Therefore,

$$n! = n(n - 1)!$$

where n is a positive integer.

The following example shows how to simplify expressions involving factorial notation.

EXAMPLE 2   **Simplifying Expressions Involving Factorials**

Find the value of each expression.

(a) $\dfrac{7!}{5!}$ (b) $\dfrac{8!}{3!5!}$ (c) $\dfrac{(n+1)!}{(n-1)!}$

Solution   (a) $\dfrac{7!}{5!} = \dfrac{7 \cdot 6 \cdot 5 \cdot 4 \cdot 3 \cdot 2 \cdot 1}{5 \cdot 4 \cdot 3 \cdot 2 \cdot 1} = 7 \cdot 6 = 42$

$\dfrac{8!}{3! \cdot 5!} = \dfrac{8 \cdot 7 \cdot 6 \cdot 5 \cdot 4 \cdot 3 \cdot 2 \cdot 1}{(3 \cdot 2 \cdot 1)(5 \cdot 4 \cdot 3 \cdot 2 \cdot 1)} = 8 \cdot 7 = 56$

$\dfrac{(n+1)!}{(n-1)!} = \dfrac{(n+1)(n)[(n-1)!]}{(n-1)!} = (n+1)n = n^2 + n$

Now, we can express the coefficients in a binomial expansion in terms of factorials. The following notation is used in its formulation.

The Binomial Coefficient

The expression $\dbinom{n}{k}$ is called a **binomial coefficient,** where

$$\binom{n}{k} = \frac{n!}{k!(n-k)!}, \quad n \geq k.$$

EXAMPLE 3   **Evaluating Binomial Coefficients**

Evaluate each binomial coefficient.

(a) $\dbinom{5}{3}$ (b) $\dbinom{2}{2}$ (c) $\dbinom{3}{0}$

Solution   (a) $\dbinom{5}{3} = \dfrac{5!}{3! \,(5-3)!} = \dfrac{5!}{3!2!} = \dfrac{3! \cdot 4 \cdot 5}{3!2!} = \dfrac{20}{2!} = 10$

(b) $\dbinom{2}{2} = \dfrac{2!}{2! \,(2-2)!} = \dfrac{2!}{2!0!} = \dfrac{1}{0!} = \dfrac{1}{1} = 1$

(c) $\dbinom{3}{0} = \dfrac{3!}{0! \,(3-0)!} = \dfrac{3!}{0!3!} = \dfrac{1}{0!} = \dfrac{1}{1} = 1$

Expanding Binomials Using the Binomial Theorem

If we use factorials and the above pattern for the coefficients, we can use a formula called the *Binomial theorem* to generalize the expansion of a binomial for any positive integer power.

The Binomial Theorem

Let a and b be real numbers or expressions and n any positive integer. Then the **binomial expansion** of $(a + b)^n$ is given by

$$(a+b)^n = \binom{n}{0}a^n + \binom{n}{1}a^{n-1}b + \binom{n}{2}a^{n-2}b^2 + \ldots + \binom{n}{n-1}ab^{n-1} + \binom{n}{n}b^n.$$

EXAMPLE 4   **Using the Binomial Theorem**

Expand the binomial $(x + 2y)^4$.

Solution   Using the Binomial theorem with $a = x$, $b = 2y$, and $n = 4$, we have,

$$(x + 2y)^4 = \binom{4}{0}x^4 + \binom{4}{1}x^3(2y) + \binom{4}{2}x^2(2y)^2 + \binom{4}{3}x(2y)^3 + \binom{4}{4}(2y)^4$$

where

$$\binom{4}{0} = 1 \qquad \binom{4}{3} = 4$$

$$\binom{4}{1} = 4 \qquad \binom{4}{4} = 1$$

$$\binom{4}{2} = 6$$

so that

$$(x + 2y)^4 = 1x^4 + 4(x^3)(2y) + 6x^2(4y^2) + 4x(8y^3) + 1(16y^4)$$

$$= x^4 + 8x^3y + 24x^2y^2 + 32xy^3 + 16y^4.$$

Finding a Specific Term in a Binomial Expansion

Suppose we want to find a specific term, say the $(k + 1)^{\text{th}}$ term, in the binomial expansion of $(a + b)^n$. If we examine each term in the expansion of $(a + b)^n$, we can write

The first term as $\binom{n}{0}a^n b^0$

The second term as $\binom{n}{1}a^{n-1}b^1$

The third term as $\binom{n}{2}a^{n-2}b^2$

and so on. This can be generalized by using a formula for finding a *specific term* without writing the entire expansion.

Finding a Specific Term

The $(k + 1)^{\text{th}}$ term of the binomial expansion of $(a + b)^n$ is

$$\binom{n}{k}a^{n-k}b^k.$$

EXAMPLE 5 **Finding a Specific Term of a Binomial**

Find and simplify the indicated term in each expansion:

(a) the fifth term of $(x - y)^8$

(b) the term involving x^7 in the expansion of $(2 - x)^{12}$.

Solution (a) First we note that $5 = 4 + 1$. Thus, $k = 4$, $a = x$, $b = -y$, and $n = 8$. So the fifth term of the expansion is:

$$\binom{8}{4}x^{8-4}(-y)^4 = \binom{8}{4}x^4y^4$$

$$= \frac{8!}{4!(8-4)!}x^4y^4$$

$$= 70x^4y^4.$$

(b) For the term x^7, $k = 7$. If $n = 12$, $a = 2$, and $b = -x$, then

$$\binom{n}{k}a^{n-k}b^k = \binom{12}{7}2^{12-7}(-x)^7$$

$$= -\frac{12!}{7!(12-7)!}2^5x^7 = -25{,}344x^7.$$

Solving Applied Problems

Many important applications from different fields involve activities that require the use of binomial expansion.

EXAMPLE 6 **Solving a Compound Interest Problem**

Suppose that \$1000 is invested in an account paying 4% annual interest compounded annually. Use the binomial expansion $1000(1 + 0.04)^6$ to compute the balance at the end of the sixth year.

Solution Substituting $P = 1000$, $r = 4\% = 0.04$, $n = 1$, and $t = 6$ in the compound interest formula to get

$$S = P\left(1 + \frac{r}{n}\right)^{nt}$$

$$= 1000(1 + 0.04)^6$$

$$= 1000[1 + 6(0.04) + 15(0.04)^2 + 20(0.04)^3 + 15(0.04)^4 + 6(0.04)^5 + (0.04)^6]$$

$$= \$1265.32.$$

Notice that the direct calculation of the expression $1000(1.04)^6$ will also result in \$1265.32.

PROBLEM SET 9.8

Mastering the Concepts

In Problems 1–6, use Pascal's triangle to expand each binomial expression.

1. $(x + 2y)^4$
2. $(2x - y)^5$
3. $(y + 1)^6$
4. $(x^2 + 3y)^5$
5. $(y^2 - 2x)^4$
6. $(2 + x/y)^6$

In Problems 7–16, write each expression in expanded form and simplify.

7. $\dfrac{4!}{6!}$
8. $\dfrac{10!}{5!7!}$

9. $\dfrac{2!}{4! - 3!}$
10. $\dfrac{1}{4!} + \dfrac{1}{3!}$

11. $\dfrac{3!8!}{4!7!}$
12. $\dfrac{4!6!}{8! - 5!}$

13. $\dfrac{0!}{0!}$
14. $\dfrac{(n - 2)!}{(n - 1)!}$

15. $\dfrac{(n + 1)!}{(n - 3)!}$
16. $\dfrac{(n + k)!}{(n + k - 2)!}$

In Problems 17–24, evaluate each binomial coefficient.

17. $\dbinom{8}{3}$
18. $\dbinom{7}{2}$
19. $\dbinom{10}{4}$
20. $\dbinom{5}{4}$

21. $\dbinom{8}{8}$
22. $\dbinom{7}{6}$
23. $\dbinom{12}{10}$
24. $\dbinom{16}{14}$

In Problems 25–32, expand each expression by using the binomial theorem and simplify each term.

25. $(x + 2)^5$
26. $(a - 2b)^4$
27. $(x^2 + 4y^2)^3$
28. $(1 - a^2)^4$
29. $(a^3 - a^{-1})^6$
30. $\left(1 - \dfrac{x}{y^2}\right)^5$
31. $\left(2 + \dfrac{x}{y}\right)^5$
32. $\left(x + \dfrac{y}{2}\right)^6$

In Problems 33–36, find the first four terms of each expansion and simplify.

33. $(x^2 - 2a)^{10}$
34. $\left(2a - \dfrac{1}{b}\right)^6$
35. $\left(\sqrt{\dfrac{x}{2}} + 2y\right)^7$
36. $\left(\dfrac{x}{2} + \dfrac{1}{a}\right)^{11}$

In Problems 37–44, find the first five terms in each expansion and simplify.

37. $(x + y)^{16}$
38. $(a^2 + b^2)^{12}$
39. $(a - 2b^2)^{11}$
40. $(a + 2y^2)^8$
41. $(x - 2y)^7$
42. $\left(1 - \dfrac{x^2}{y}\right)^8$
43. $(a^3 - a^2)^9$
44. $\left(x + \dfrac{1}{2y}\right)^{15}$

In Problems 45–52, find the specified term of each binomial expansion.

45. $(x + y)^6$; fourth term
46. $(x + 2y)^8$; sixth term
47. $(t - 2)^{10}$; fifth term
48. $(2p + 3q)^4$; middle term
49. $(x - 3y)^{12}$; middle term
50. $(x + 5)^{36}$; fourth term
51. $(2x + 5y)^{19}$; term containing y^7
52. $(3p - 2q)^{21}$; term containing q^9

In Problems 53–56, find the indicated term of each expression.

53. $\left(2x^2 - \dfrac{a^2}{3}\right)^9$, seventh term
54. $(x + \sqrt{a})^{12}$, middle term
55. $\left(a + \dfrac{x^3}{3}\right)^9$, term containing x^{12}
56. $\left(2\sqrt{y} - \dfrac{x}{2}\right)^{10}$, term containing y^4

Applying the Concepts

57. **Depreciation:** A small business depreciates a company car that costs $25,000 at an annual rate of 15% for the first 5 years. Use the binomial expansion of $25{,}000(1 - 0.15)^5$ to determine the value of the car after 5 years.

58. **Population Growth:** The population of a small town grows at the rate of 3% per year, with an initial population of 20,000. Use the binomial expansion of

$$20{,}000(1 + 0.03)^4$$

to determine the size of the city after 4 years. Round off to the nearest person.

Developing and Extending the Concepts

In Problems 59 and 60, find the binomial expansion of each expression.

59. $(1 + \sqrt{2})^5 + (1 - \sqrt{2})^5$
60. $(1 - \sqrt{3})^4 - (1 + \sqrt{3})^4$

61. Find a formula for $(a + b + c)^n$.

62. Find the expansion of $(x^2 + y^2 + z^2)^4$.

63. Expand and simplify the expression

$$\frac{(x + h)^7 - x^7}{h}.$$

64. Solve each equation for x.

(a) $\binom{4}{x} = \binom{4}{1}$ (b) $\binom{6}{x} = \binom{6}{2}$

 CHAPTER 9 REVIEW PROBLEM SET

In Problems 1–6, find the value of each determinant.

1. (a) $\begin{vmatrix} 5 & 2 \\ 2 & -3 \end{vmatrix}$

(b) $\begin{vmatrix} 2 & 8 \\ 3 & 7 \end{vmatrix}$

2. (a) $\begin{vmatrix} 2 & -5 \\ 1 & 6 \end{vmatrix}$

(b) $\begin{vmatrix} 3 & 1 \\ 2 & -4 \end{vmatrix}$

3. $\begin{vmatrix} 1 & 1 & 2 \\ 2 & -3 & -1 \\ 3 & 7 & -2 \end{vmatrix}$

4. $\begin{vmatrix} 2 & 1 & 2 \\ 1 & 3 & -1 \\ 3 & -1 & -2 \end{vmatrix}$

5. $\begin{vmatrix} 3 & 2 & -1 \\ 1 & -1 & 2 \\ 5 & 3 & -4 \end{vmatrix}$

6. $\begin{vmatrix} 1 & -1 & 2 \\ 3 & 1 & 1 \\ 2 & -1 & 5 \end{vmatrix}$

In Problems 7 and 8, solve each equation for x.

7. $\begin{vmatrix} x + 1 & x \\ x & x - 2 \end{vmatrix} = -6$

8. $\begin{vmatrix} x & 4 & 5 \\ 0 & 1 & x \\ 5 & 2 & 0 \end{vmatrix} = 7$

In Problems 9–12, use Cramer's rule to solve each system.

9. $\begin{cases} x - y = 3 \\ 2x + y = 3 \end{cases}$

10. $\begin{cases} 5x + 2y = 3 \\ 2x + 3y = -1 \end{cases}$

11. $\begin{cases} x - y + 2z = 0 \\ 3x + y + z = 2 \\ 2x - y + 5z = 5 \end{cases}$

12. $\begin{cases} 3x + 2y - z = -4 \\ x - y + 2z = 13 \\ 5x + 3y - 4z = -15 \end{cases}$

In Problems 13–16, identify the intercepts and the foci and sketch the graph of each conic.

13. $4x^2 + 9y^2 = 36$

14. $25x^2 + 4y^2 = 100$

15. $7x^2 - 16y^2 = 112$

16. $16y^2 - 4x^2 = 64$

In Problems 17 and 18, find the standard equation of each conic with the given information. Also sketch the graph.

17. The hyperbola with vertices at $(-3,0)$ and $(3,0)$, center at $(0,0)$, and one focus at $(5,0)$.

18. The ellipse with vertices at $(-5,0)$, $(5,0)$, $(0,3)$, and $(0,-3)$.

In Problems 19–22, solve each system algebraically. Check the solutions by graphing each equation and determining the points of intersection.

19. $\begin{cases} 3x - 4y = 25 \\ x^2 + y^2 = 25 \end{cases}$

20. $\begin{cases} 2x - y = 2 \\ x^2 + 2y^2 = 12 \end{cases}$

21. $\begin{cases} x^2 - y^2 = 21 \\ x^2 + y^2 = 29 \end{cases}$

22. $\begin{cases} 3x^2 - 2y^2 = 27 \\ 7x^2 + 5y^2 = 63 \end{cases}$

In Problems 23 and 24, find the first five terms of the sequence with the specified general term a_n.

23. (a) $a_n = (-1)^n n$

(b) $a_n = \dfrac{n}{2^{n+1}}$

(c) $a_1 = 3$, $a_2 = 1$, $a_{n+1} = 2a_n$

24. (a) $a_n = \dfrac{(-1)^n}{n^2}$

(b) $a_n = \dfrac{1 + (-1)^n}{2 + 3n}$

(c) $a_1 = -2$, $a_{n+1} = -a_n$, $n \geq 1$

In Problems 25 and 26, find the general term a_n for each sequence.

25. $3, 12, 27, 48, 75, \ldots$

26. $1, 8, 27, 64, 125, \ldots$

In Problems 27 and 28, write out and evaluate each sum.

27. (a) $\sum_{k=1}^{4} (3k + 2)$

(b) $\sum_{k=1}^{5} (-1)^k 3^k$

28. (a) $\sum_{k=0}^{4} 2^{-k}$

(b) $\sum_{k=1}^{4} (3 + (-1)^k)$

29. Express the series $2 + 4 + 6 + 8 + 10 + 12$ using summation notation.

30. Suppose that $\sum_{k=1}^{13} a_k = 4$ and $\sum_{k=1}^{13} b_k = -12$,

evaluate $\sum_{k=1}^{13} (3a_k - 2b_k)$.

In Problems 31–34, find the indicated term in each arithmetic sequence.

31. (a) a_{15} if $5, 9, 13, 17, \ldots$
　　(b) a_{15} if $2, -2, -6, -10, \ldots$
32. (a) a_{30} if $27, 33, 39, 45, \ldots$
　　(b) a_{20} if $5, 1, -3, -7, \ldots$
33. (a) a_{20} if $a_1 = 2$ and $d = 5$
　　(b) a_{30} if $a_2 = -2$ and $a_{15} = -210$
34. (a) a_{40} if $a_1 = 26$ and $d = -10$
　　(b) a_1 if $a_3 = -18$ and $a_{10} = -53$

In Problems 35–38, find the sum of the first n terms, S_n, for the given value of n of each arithmetic sequence.

35. S_{15}; $4, 10, 16, 22, \ldots$
36. S_{20}; $-12, -3, 6, 15, \ldots$
37. S_{10} if $a_1 = 4, d = 3$
38. S_{11} if $a_1 = 5$, $d = 2$

In Problems 39 and 40, find the number of terms n of each finite arithmetic sequence.

39. $-2, 1, 4, 7, \ldots, 43$
40. $51, 45, 39, 33, \ldots, -123$

In Problems 41–44, find the indicated term in each geometric sequence.

41. a_{10} if $1, 2, 4, 8, \ldots$
42. a_{15} if $2, -\dfrac{2}{3}, \dfrac{2}{9}, \ldots$
43. a_1 if $a_2 = 15$, $a_5 = -1875$
44. a_6 if $a_3 = -\dfrac{1}{4}$ and $r = \dfrac{1}{2}$

In Problems 45 and 46, find the sum of the first n terms S_n of a geometric sequence for a given value of n.

45. S_{10} if $a_1 = -\dfrac{1}{9}$ and $r = 3$
46. S_{10} if $a_1 = -4$ and $r = -2$

In Problems 47–52, find the sum of each infinite geometric series.

47. $10 + 4 + 1.6 + 0.64 + \ldots$
48. $10 + 6 + 3.6 + 2.16 + \ldots$
49. $\displaystyle\sum_{k=1}^{\infty} 5(0.3)^{k-1}$
50. $\displaystyle\sum_{k=1}^{\infty} 40\left(\tfrac{1}{4}\right)^{k-1}$
51. $\displaystyle\sum_{k=1}^{\infty} \dfrac{3}{5^{k-1}}$
52. $\displaystyle\sum_{k=1}^{\infty} \dfrac{0.25}{4^{k-1}}$

In Problems 53–56, find the value of each expression.

53. $\dfrac{5! \cdot 2!}{3! \cdot 6!}$
54. $\dfrac{8!}{4! \cdot 6!}$
55. $\dbinom{8}{3}$
56. $\dbinom{15}{12}$

In Problems 57 and 58, use Pascal's triangle to expand each binomial expression and simplify.

57. $(x + 3y)^4$
58. $(2x - 3y)^5$

In Problems 59 and 60, use the binomial theorem to expand each binomial expression and simplify.

59. $(3x + 2)^5$
60. $(3x - 4y)^4$

In Problems 61 and 62, find the specified term of each binomial expansion and simplify.

61. The fourth term of $(3x + 4y)^{11}$
62. The seventh term of $(2x + 5y)^{10}$
63. Health Food: A health food store has two different kinds of granola–cashew nut selling at $4.20 per pound, and golden granola selling at $3.40 per pound. How much of each kind should be mixed to produce a 200 pound mixture selling at $3.72 per pound?
64. Geometry: The perimeter of a rectangular waterbed is 270 inches. If 5 times its width equals 4 times its length, what are the length and width of the waterbed?
65. Loan: Suppose that at the end of 8 months, the balance on a business loan including simple annual interest is $2563, and at the end of 18 months, the balance is $2754. How much money was borrowed and what is the annual rate of interest?
66. Trucking: An owner of a 5000 gallon fuel truck loads his truck with gasoline and kerosene. For a shipment of fuel, the owner charges $0.15 for delivering each gallon of gasoline and $0.18 for delivering each gallon of kerosene. How many gallons of each fuel should the trucker load in order to charge $780 for the load of fuel?
67. Investment: A businesswoman invested a total of $62,000 in three mutual funds. At the end of one year, one fund had returned 15%, the second returned 8%, and the third returned −3% respectively on their parts of the investment. If the total return on the entire investment for a year was $6230, and if the businesswoman invested $8000 more in the

15% fund than she did in the 8% fund, how much did she invest in each fund? Round to the nearest dollar.

68. Carpeting: A homeowner purchases carpets for two square rooms with a combined area of 52 square yards. The price of the first carpet is $13 per square yard, and the price of the second is $18 per square yard. If the total cost of the carpets is $856, find the dimensions of each square room.

69. Investment: One investor receives $500 annual income from an investment. His brother invested $4000 more, but at an annual interest rate of 1% less, and receives an annual income of $560. Find the amount and the interest rate of each investment.

70. Travel Time and Rate: Linda drove her car for a trip of 350 miles. Her boyfriend made the same trip at a speed 5 miles per hour faster than Linda and required 20 minutes less time. Find the time and the speed for Linda to make the trip.

71. Investment: Suppose that $2000 is deposited into an account that pays 4% annual interest compounded quarterly, the balance A_n in the account after n quarters is given by

$$A_n = 2000\left(1 + \frac{0.04}{4}\right)^n.$$

Determine (a) the first five terms of this sequence (b) the balance in the account after 4 years.

72. Depreciation: A company depreciates a machine that costs $40,000 by $2500 each year. Determine

(a) The value of the machine after 1 year, 2 years, 3 years, and 4 years.

(b) The value of the machine after the nth year.

73. Graph the first five ordered pairs of each sequence. Compare the graph of each sequence to the graph of the corresponding linear function.

(a) $f(x) = 3 - 2x$, $a_n = 3 - 2n$

(b) $f(x) = 4x + 1$, $a_n = 4n + 1$

74. Graph the first five ordered pairs of each sequence and the corresponding exponential function.

(a) $f(x) = 3^x$, $a_n = 3^n$ (b) $f(x) = 2^{-x}$, $a_n = 2^{-n}$

CHAPTER 9 PRACTICE TEST

1. Evaluate each determinant.

(a) $\begin{vmatrix} -4 & 3 \\ 3 & -2 \end{vmatrix}$ (b) $\begin{vmatrix} 3 & 0 & -1 \\ 3 & 1 & -1 \\ 1 & 1 & 2 \end{vmatrix}$

2. Use Cramer's rule to solve each system.

(a) $\begin{cases} 3x + 4y = 7 \\ -2x + 3y = 1 \end{cases}$ (b) $\begin{cases} 3x + y + z = 5 \\ 5x - y + z = 5 \\ 3x + 2y - z = 4 \end{cases}$

3. Identify the intercepts, the vertices, and the foci; sketch the graph for each conic.

(a) $4x^2 + y^2 = 4$ (b) $4x^2 - 9y^2 = 36$

4. Find the standard equation of the hyperbola whose vertices are $(-5,0)$ and $(5,0)$, center at $(0,0)$, and one focus at $(7,0)$.

5. Find the general term a_n for each sequence.

(a) $19, 13, 7, 1, -5, \ldots$

(b) $1, -\dfrac{1}{7}, \dfrac{1}{49}, -\dfrac{1}{343}, \ldots$

6. Suppose that $\displaystyle\sum_{k=1}^{20} a_k = 7$ and $\displaystyle\sum_{k=1}^{20} b_k = 3$, evaluate each sum.

(a) $\displaystyle\sum_{k=1}^{20} (4a_k + 3b_k)$ (b) $\displaystyle\sum_{k=1}^{20} (8a_k - 4b_k)$

7. Find the sum of the infinite geometric series

$$8 + 4 + 2 + 1 + \ldots$$

8. Simplify each expression.

(a) $\dfrac{4! \cdot 6!}{8! - 5!}$ (b) $\dbinom{7}{4}$

9. Find the first five terms of the expansion of $(2x + 3y)^{11}$ and simplify.

10. A gold membership for a fitness center includes an initial fee for joining as well as monthly dues. The total cost after 9 months' membership is $385, and after 14 months' membership is $560. Find both the initial fee and the monthly dues.

ANSWERS TO ODD NUMBERED PROBLEMS AND REVIEWS
ALL ANSWERS TO PRACTICE TESTS

—CHAPTER R—

PROBLEM SET R1 (on page 5)

1a. $4 + 5 = 9$
1b. $5 + 4 = 9$
1c. $4 + 5 = 5 + 4$
3. $x \cdot 5 > x/5$
5a. The product of 5 and a number is 20.
5b. The product of 5 and a number less 4 is 20.
7a. The sum of a number and 2 is less than 10.
7b. 10 is greater than the sum of a number and 2.
9a. $6 \cdot 6 = 36$
9b. $3 \cdot 3 \cdot 3 \cdot 3 \cdot 3 \cdot 3 = 729$
11a. $2^2 x y^3$
11b. $2^2 3^2 y^2 z$
13a. 8
13b. 8
15a. 33

15b. 33
17. 65
19. 45
21a. 6
21b. 3
23. 42
25a. 60
25b. 60
27a. 9
27b. 9
29a. $5(3 + 4) = 35$
29b. $5 \cdot 3 + 5 \cdot 4 = 35$
31a. $x = 10(20 + y)$
31b. $x = 20 + 10y$
33a. Three times a number less one is not equal to 5.
33b. Three times the difference of a number and 1 is equal to 5.
35a. $(4 + 5) \cdot 2 + 3 = 21$

35b. $(4 + 5) \cdot (2 + 3) = 45$
35c. $4 + 5 \cdot (2 + 3) = 29$
37a. $4 + 3 \cdot (8 - 1) + 6 = 31$
37b. $4 + 3 \cdot 8 - 1 + 6 = 33$
37c. $(4 + 3) \cdot (8 - (1 + 6)) = 7$
39a. 0
39b. undefined
41. 173
43. 0
45a. 3, +, 4, ÷, 5
45b. (, 3, +, 4,), ÷, 5
45c. 4, +, 5, *, (, 3, −, 2,), ÷, 10
47a. $2(95,000) - y$
47b. $185,000, $183,000
49a. $\dfrac{w - 300}{5}$
49b. $760

PROBLEM SET R2 (on page 12)

1a. $\varnothing, \{0\}, \{1\}, \{0,1\}$
1b. $\varnothing, \{a\}, \{b\}, \{c\}, \{a,b\}, \{a,c\}, \{b,c\}, \{a,b,c\}$
3a. $\{x|x \leq 0\}$
3b. $\{x|x < 0\}$
3c. $\{x|x \geq 0\}$
5a. $\{-5,-4,-3,-2,\ldots,3,4,5\}$
5b. $\{9,10,11\}$
5c. $\{x| x \neq 5\}$
5d. $\{x| x \, \varepsilon \, I \text{ and } x \neq 5\}$
7a.

$$3.1 \quad \tfrac{223}{71} \quad \pi \quad \tfrac{22}{7} \quad 3.2$$

7b.

$$1.4 \quad 1.4141 \quad 1.4142 \quad \sqrt{2} \quad 1.5$$

9a. 0.9
9b. 0.09
9c. 0.009
11a. $0.\overline{2}$
11b. $0.\overline{5}$
11c. $0.\overline{7}$
13a. integer, rational
13b. natural number, integer, rational
15a. irrational

15b. natural, integer, rational
17a. 3.14159
17b. 3.14286
17c. 3.14110
19a. 2.99982
19b. 3.00000
19c. 3.00329
21a. $-\tfrac{3}{5}$

$$-1 \quad -\tfrac{3}{5} \quad 0 \quad \tfrac{3}{5} \quad 1$$

21b. $\tfrac{2}{3}$

$$-1 \quad -\tfrac{2}{3} \quad 0 \quad \tfrac{2}{3} \quad 1$$

23a. distributive
23b. commutative for multiplication
23c. commutative for addition
25a. commutative for addition
25b. associative for addition
25c. commutative for addition
27. symmetric
29. substitution
31.

$$-4 \quad -\sqrt{10} \quad -\sqrt{4} \quad 0 \quad \tfrac{1}{2} \quad \sqrt{2} \quad 2$$

33a. $19 \cdot 7 + 82 \cdot 7$
33b. $(82 + 19) \cdot 7$
33c. $7(19 + 82)$
35a. 25/5
35b. $-\sqrt{9}$, 0, 25/5
35c. $-\sqrt{9}$, 0, 25/5, 25/4
35d. $-\sqrt{5}$
35e. all are real
37a. $3.60
37b. $2.40 + $1.20
37c. $a = 0.10, b = 0.05, c = 24$
39. $8 \div 4 = 2$ but $4 \div 8 = 0.5$
    $(12 \div 6) \div 2 = 1$ but
    $12 \div (6 \div 2) = 4$
41. 2^n
43a. distributive
43b. additive inverse
45a. associative
45b. multiplicative identity
45c. distributive
45d. additive inverse
45e. additive identity

PROBLEM SET R3 (on page 20)

1a. 25	9a. 2	17a. −225	27. 0	41a. 36
1b. 25	9b. 2	17b. 4500	29. undefined	41b. 36
1c. −25	11a. −18	19a. 2	31a. 25	43a. −25/7
3a. −2	11b. 4	19b. −2	31b. −25	43b. 1
3b. −3	13a. −20	19c. −2	33a. −64	45a. $\lvert 18° - (-2°) \rvert$
3c. −4	13b. 20	21a. 4	33b. −64	45b. 20°
5a. 5 − (−3)	15a. 5000	21b. 32	35. −135	47a. 20 − 4 + 7 − 5
5b. −3 − 1	15b. 5000	23a. 25	37. −12	47b. 18 yard line
7a. 10	15c. −5000	23b. −20	39a. −2	
7b. −10	15d. −5000	25. −15	39b. −4	

PROBLEM SET R4 (on page 26)

1a. 1/3	9a. 37.5%	17a. 30	27. −1/3	37a. 5/8
1b. 9/14	9b. 87.5%	17b. 8.4	29. 1/50	37b. 62.5%
3a. −13/19	11a. 7/100	19. 20%	31a. 7/4	37c. 5/3
3b. −5/17	11b. 1/4	21a. 3/5	31b. 10/11	39. 16/99
5a. 2/3	13a. 1/200	21b. 2/3	33a. 3/5	41. 1/5
5b. 3/5	13b. 1/160	23a. 5/8	33b. 1	43. 46,000,000 acres
7a. 80%	15a. 7/200	23b. 2/9	35a. −7/3	
7b. 35%	15b. 53/500	25. 7/5	35b. −47/10	

PROBLEM SET R5 (on page 32)

1a. 243	11a. 12^6	21b. 3.85×10^8	29. $-\dfrac{27a^9 b^{21}}{8}$
1b. x^5	11b. x^{600}	21c. 3.8×10^4	
3a. 4^{12}	13a. y^{65}	23a. 9.6×10^{14}	31a. 3.584×10^8 km
3b. x^{12}	13b. $-y^{65}$	23b. 2.4×10^6	31b. 3×10^8 inches
5. y^{12}	15a. $-4x^2 y^4$	25. 7.8×10^4	33. 50 nsec/operation
7a. 13^3	15b. $-16x^2 y^6$		35. 1.8×10^{27} tons
7b. x^{30}	17. $x^4 y^7$	27. $\dfrac{a^4 + b^4}{a^2 b^2}$	37. 2.955×10^{51} kg.
9a. $-x^5$	19. $x^2 y^4$		
9b. $-x^5$	21a. 3.85×10^5		

PROBLEM SET R6 (on page 38)

1. 0.0588 m^2, 58,800 m^2	19b. S = 150 in^2	35. 4%	47a. cube
3. 0.01 m^2, 10,000 mm^2	21a. V = 57.76 m^3	37. 2200 mi	47b. cube
5. 14,256 ft	21b. S = 103.36 m^2	39. 312.5 ft/sec	49a. 57.6 acre
7. A = 24 cm^2 , P = 24 cm	23a. V = 15.70 ft^3	41a. 77°F	49b. $18,000/acre
9. A = 112 in^2, P = 44 in	23b. S = 37.68 ft^2	41b. −5°C	51. 0.35 hours
11. A = 38.47 cm^2,	25a. V = 1149.76 yd^3	43a. 62.6°F	53a. $1400
P = 24.99 cm.	25b. S = 530.66 yd^2	43b. −10°C	53b. $2800
13. A = 15 in^2, P = 18 in	27. yes	45a. $241.78	55a. $1469.33
15. 3.57cm^2	29. no	45b. $53.21	55b. $2938.66
17. 3.14 cm^2	31. yes	45c. $294.99	
19a. V = 125 in^3	33. $360		

CHAPTER R REVIEW PROBLEM SET (on page 40)

1a. 5 + (−5 + 3)	5g. associative & commutative	9b. −3
1b. (−2)(−6) − 3	5h. distributive	11a. 1/2
1c. −30/5 − 2	7a. $0.\overline{3}$	11b. −1
3a. 3·3·3·3·3	7b. $0.2\overline{7}$	11c. 9/2
3b. $3^3 \cdot x^4$	7c. $0.\overline{285714}$	11d. 1/3
5a. associative	7d. $0.1\overline{8}$	11e. 1/24
5b. identity for multiplication	7e. $-0.07\overline{0}$	11f. 23/60
5c. commutative	7f. 0.3333, 0.2778, 0.2857,	11g. 55/48
5d. additive inverse	0.1818, −0.0700	11h. 31/120
5e. identity for addition	9a. 1	11i. 5/12
5f. distributive		11j. 3/40

13a. 7.3×10^8

13b. 235,000

13c. 10^5

15a. 51.8°F

15b. 8 °C

17a. $2(100 + 70) - 123$

17b. $217

19. $225

21. $750

23. 1/5

25. 43.9 ft

27. 152.85 units^2

—CHAPTER 1—

PROBLEM SET 1.1 (on page 48)

1a. 8

1b. 0

3a. 10

3b. 9

5a. 4

5b. −2

7a. −6

7b. 1/2

9. 1

11. 5

13. 1

15. 3

17. 3

19. −11/2

21. −15

23. 210

25. 2

27. 59

29. 1

31. 1

33. 5

35. 2

37. −5

39. 6

41. 800

43. 0.25

45a. identity

45b. identity

47. inconsistent

49a. $135

49b. 600 mi

51a. $130,200

51b. $160,000

53. −1.5

55. −2

57. 2

59. 9°

PROBLEM SET 1.2 (on page 54)

1. $5c$

3. $\dfrac{c-b}{a}$

5. $-\dfrac{b}{2}$

7. $-\dfrac{5a}{3}$

9. $-\dfrac{c}{2}$

11. $\dfrac{7d}{2m}$

13. $-7b$

15. $\dfrac{12-4x}{3}$

17. $\dfrac{99-11x}{9}$

19. $\dfrac{63-27x}{14}$

21. $100x + 3$

23. $\dfrac{4x-90}{23}$

25. $\dfrac{A}{rt+1}$

27. $\dfrac{Ty}{x}$

29. $\dfrac{P}{0.7}$

31. $\dfrac{y-b}{m}$

33. $\dfrac{C}{2\pi}$

35. $\dfrac{P-2w}{2}$

37a. $\dfrac{3V}{h}$

37b. $\dfrac{3V}{B}$

39a. $V + 32t$

39b. $\dfrac{V_o - V}{32}$

41a. $\dfrac{bx}{b-y}$

41b. $\dfrac{ay}{a-x}$

43. 168.2 ft

45a. 339.1 inches

45b. $22.51

47a. 113.04 ft^3

47b. 94.2 min

47c. $\dfrac{V}{\pi r^2}$

47d. 3 ft

49a. $114

49b. $\dfrac{T}{0.15} + 1000$

49c. $2680

51a. $\dfrac{V}{\pi r^2} - \dfrac{4}{3}r$

51b. 15.6 m

53. $\dfrac{7}{a}$

55. $-\dfrac{17}{a^2}$

57a. $0.5N(N-3)$

57b. 9 diagonals

PROBLEM SET 1.3 (on page 63)

1. 18, 20, 22

3. 4, 6, 8

5. 2, 4, 6

7. 32 ft by 15 ft

9. 199m by 399m

11. 8 in by 10 in

13. 11 ft from one end

15. 29 ft

17. first: 96 cm
    second: 32 cm
    third: 52 cm

19. 350 mi @ 35 mph;
    850 mi @ 85 mph

21. 21 and 23.25 mi

23. 20 mph

25. bus: 52.5 mph;
    train: 83.5 mph

27. 450 and 520 mph

29. 30 tens, 90 fives

31. son 11, father 36

33. $1029

35. $50

37. $23.54

39. 135

41. 9%

43. 22 weeks

45. 2500 m

47. 2.66 minutes

49. $9.00

51. 12°

PROBLEM SET 1.4 (on page 70)

1. $20,000 at 9%,
   $12,000 at 8.5%

3. $45,000

5. $1068 @ 4.25%;
   $2382 @ 7.75%

7. 25 gal

9. 300 lbs @ 30%,
   100lbs @ 10%

11. 15 tons sand; 30
    tons gravel

13. $\dfrac{20}{9}$ days

15. 3.6 hours

17. 4500 copies per hr.

19. 14 hrs.

21. $8576 @ 3%
    $6424 @5%

23. 640 ml. of 12.5%
    and 320ml of 20%.

25a. 59, 60, 61

25b. 58, 60, 62

25c. not possible, only answers are even

27. 14 m by 26 m
29. $8.26

PROBLEM SET 1.5 (on page 80)

1a. (−1, 1]
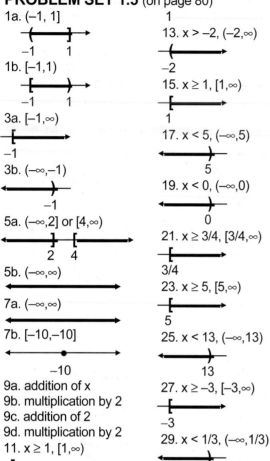
−1   1

1b. [−1,1)

−1   1

3a. [−1,∞)

−1

3b. (−∞,−1)

−1

5a. (−∞,2] or [4,∞)

2   4

5b. (−∞,∞)

7a. (−∞,∞)

7b. [−10,−10]

−10

9a. addition of x
9b. multiplication by 2
9c. addition of 2
9d. multiplication by 2
11. x ≥ 1, [1,∞)

1

13. x > −2, (−2,∞)

−2

15. x ≥ 1, [1,∞)

1

17. x < 5, (−∞,5)

5

19. x < 0, (−∞,0)

0

21. x ≥ 3/4, [3/4,∞)

3/4

23. x ≥ 5, [5,∞)

5

25. x < 13, (−∞,13)

13

27. x ≥ −3, [−3,∞)

−3

29. x < 1/3, (−∞,1/3)

1/3

31. x < 12, (−∞, 12)

12

33. x ≤ 24, (−∞,24]

24

35. x > 3, (3,∞)

3

37. x ≥ −2, [−2,∞)

−2

39. x > 1, (1,∞)

1

41. 2 ≤ x ≤ 4, [2,4]

2   4

43. −3 < x < 2, (−3,2)

−3   2

45. −1/2 ≤ x ≤ 3/2, [−1/2,3/2]

−1/2   3/2

47. x < 1/2 or x > 2, (−∞,1/2) or (2,∞)

1/2   2

49. x < −2 or x > −1/2, (−∞,−2) or (−1/2,∞)

−2   −1/2

51. x < −3 or x > 4, (−∞,−3) or (4,∞)

−3   4

53. 15 or more
55. 15 mins or less
57. 82 or more
59a. 3 ≤ x ≤ 4
59b. 21 ≤ x ≤ 28 per week

59c. 1095 ≤ x ≤ 1460 per year
61. 33 cm
63 x ≥ 6, [6,∞)

6

65. x − 1 is negative, so must reverse the inequality when dividing.

PROBLEM SET 1.6 (on page 86)

1a. 2
1b. 5 − π ≈ 1.858
1c. √5 − 2 ≈ 0.236
3a. 9
3b. 2
3c. 3
5a. 11, −1
5b. 11, −1
7a. 3,1
7b. −1, −3
9a. 5
9b. no solution
11a. 0
11b. 0
13a. 6, −6
13b. 6, −10
15. no solution
17. −2/3
19. −3, −11
21. 13, −7

23 −1, −3
25a. x ≤ −1 or x ≥ 9 (−∞,−1] or [9,∞)

−1   9

25b. −1 < x < 9 (−1,9)

−1   9

27a. x ≤ −39 or x ≥ 11 (−∞,−39] or [11,∞)

−39   11

27b. −39 < x < 11 (−39,11)

−39   11

29. −5 < x < 5, (−5,5)

−5   5

31a. −1 ≤ x ≤ 2, [−1,2]

−1   2

31b. x ≤ −1 or x ≥ 2, (−∞,−1] or [2,∞)

−1   2

33a. all real numbers
33b. no solution
35. x < −1 or x > 1 (−∞,−1) or (1,∞)

−1   1

37. −2 < x < 3, (−2,3)

−2   3

39. x ≤ −2 or x ≥ 4/3 (−∞,−2] or [4/3,∞)

−2   4/3

41. 1 ≤ x ≤ 2, [1,2]

1   2

43. x ≤ −1 or x ≥ 5 (−∞,−1] or [5,∞)

−1   5

45a. no solution
45b. x = −7

−7

47a. x ≠ −6 (−∞,−6) or (−6,∞)

−6

47b. (−∞,∞)

49. $|T - 98.6| \leq 2.3$
    $96.3° \leq T \leq 100.9°$
51. $|220 - V| \leq 25$
    $195\ v \leq x \leq 245\ v$

53.
$|x - 4 \times 10^6| \leq 0.5 \times 10^6$
$3.5 \times 10^6 \leq x \leq 4.5 \times 10^6$
55. $|x - 2.5| \leq 0.2$

$2.3\ cc \leq x \leq 2.7\ cc$
57a. $|x - 3| = 8$;
   $x = 11, -5$
57b. $|x - 3| < 8$,

$-5 < x < 11$
59a. $|x| < 5$
59b. $|x| \leq 4$

CHAPTER 1 REVIEW PROBLEM SET (on page 88)

1. 3
3. 1/2
5. 100
7. −8
9. −15
11. 11/2
13. 3
15. 1, 2
17. 0, 1
19. 3
21. 3
23. $\dfrac{5 - y}{2}$
25. $\dfrac{2A - ah}{h}$
27. $\dfrac{S - 2\pi r^2}{2\pi r}$

29. $\dfrac{ac}{a - c}$
31. $x \leq 2,\ (-\infty, 2]$

2

33. $x > 1,\ (1, \infty)$

1

35. $-3 \leq x \leq 3,\ [-3, 3]$

−3   3

37. $2 < x < 3,\ (2, 3)$

2   3

39. $-2 \leq x \leq 5,$
   $[-2, 5]$

−2   5

41. $-4 < x < 0,$
   $(-4, 0)$

−4   0

43. $x \leq -12$ or $x \geq 20,$
   $(-\infty, -12]$ or $[20, \infty)$

−12   20

45. $x < -14$ or $x > 10,$
   $(-\infty, -14)$ or $(10, \infty)$

−14   10

47. $-11/3 \leq x \leq 1,$
   $[-11/3, 1]$

−11/3   1

49a. −2/3

49b. x is any real
    number
51. $1 \leq x \leq 4,\ [1, 4]$

1   4

53. 35 years old
55. 10 ft by 24 ft
57. 24 tons, 36 tons
59. 112, 224 seats
61. $4\frac{4}{9}$ hours
63. 8.25 minutes
65. $-4 \leq x \leq 13/2$

−4   13/2

67. $x \leq 5/2$

5/2

CHAPTER 1 PRACTICE TEST (on page 90)

1a. −85
1b. −1
1c. $12.\overline{6} = \dfrac{38}{3}$
2a. 1
2b. no solution
3a. 8, −8
3b. 3, −5/3
3c. −5/3
4a. −3a/7
4b. $\dfrac{A}{rt + 1}$

5a. $x \leq -3$

−3

5b. $-6 \leq x \leq 0$

−6   0

5c. $x \leq 5$ or $x > 8$

5   8

6a. $x \leq -3,\ (-\infty, -3]$

6b. $x < -3,\ (-\infty, -3)$

−3

6c. $-4 < x \leq 1,$
   $(-4, 1]$

−4   1

6d. $x \geq 1,\ [1, \infty)$

1

6e. $-3 \leq x \leq 3,\ [-3, 3]$

−3   3

6f. $x \leq 3$ or $x \geq 5,$
   $(-\infty, 3]$ or $[5, \infty)$

3   5

7. $1,500
8. 4 inches
9. 3 mph
10. $x > -5$

— CHAPTER 2 —

PROBLEM SET 2.1 (on page 98)

1.

(0,7)
(−2,4)
(3,4)
(−7,0)
(5,−1)
(0,−6)

Quadrant I: (3,4)

Quadrant II:(−2,4)
Quadrant IV: (5,−1)
y axis:(0,−6) and (0,7)
x axis: (−7,0)
3. (−3,4)
5. (−8,0)
7. (−5,−5)
9.a no   b. yes
11. a no   b. no
13.

15.
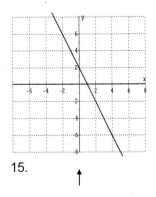

17.
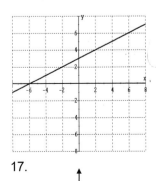

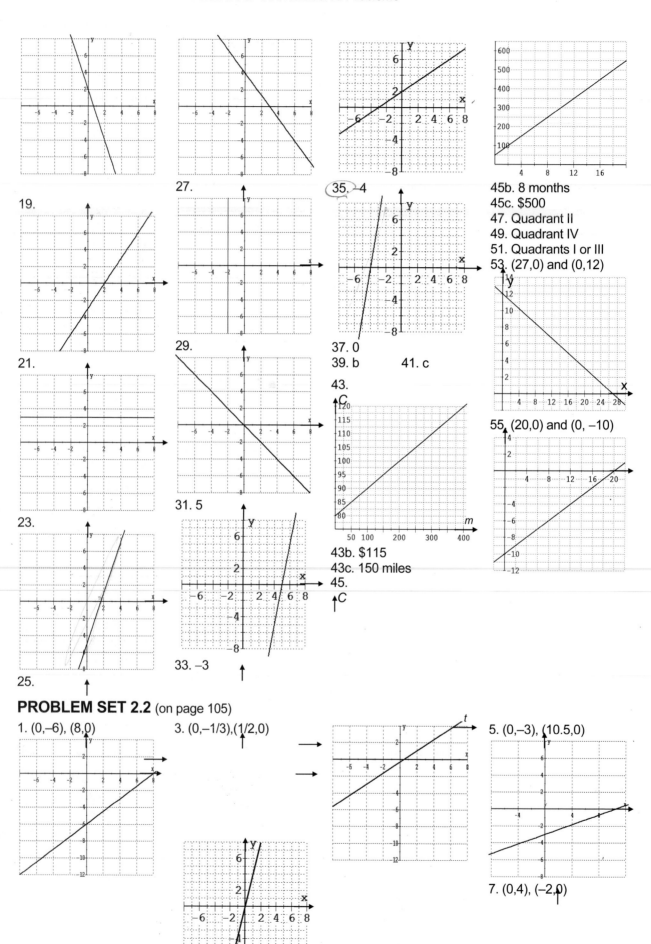

27.

19.

21.

23.

25.

29.

31. 5

33. –3

35. –4

37. 0

39. b   41. c

43.

43b. $115
43c. 150 miles

45.

45b. 8 months
45c. $500
47. Quadrant II
49. Quadrant IV
51. Quadrants I or III
53. (27,0) and (0,12)

55. (20,0) and (0, –10)

PROBLEM SET 2.2 (on page 105)

1. (0,–6), (8,0)

3. (0,–1/3),(1/2,0)

5. (0,–3), (10.5,0)

7. (0,4), (–2,0)

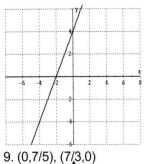

9. (0,7/5), (7/3,0)

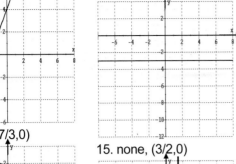

11. (0,0), (0,0)

13. (0,–3), none

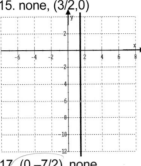

15. none, (3/2,0)

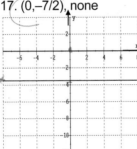

17. (0,–7/2), none

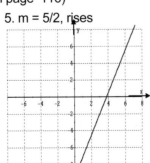

19. none, (1/3,0)

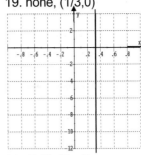

21. (2,0), (0,–1)
23. (2,0), (0,–4)
25. (–3.5,0), (–0.5,0), (0,1)

27.

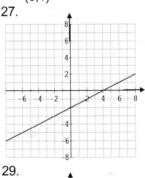

29.

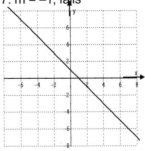

31a. 2, 6

31b. (2,6)
33a. –2, 1
33b. (–∞,–2) or (1,∞)
35a. a loss of $400,000
35b. 100 service calls
37a. 4 ft.
37b. 10 ft.
37c. 1 hour
39a. –2, 4
39b. –8
39c. (–2,4)
41a. $5000
41b. $2000
41c. 4 years
41d. There is no data since the graph stops at 6 years.

PROBLEM SET 2.3 (on page 116)

1. (1,3), (–3,–3)
   m = 3/2

3. (2,3), (–1,1)
   m = 2/3

11. m = 3/4, rises

5. m = 5/2, rises

7. m = –1, falls

9. m = 2/3, rises

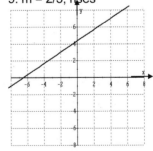

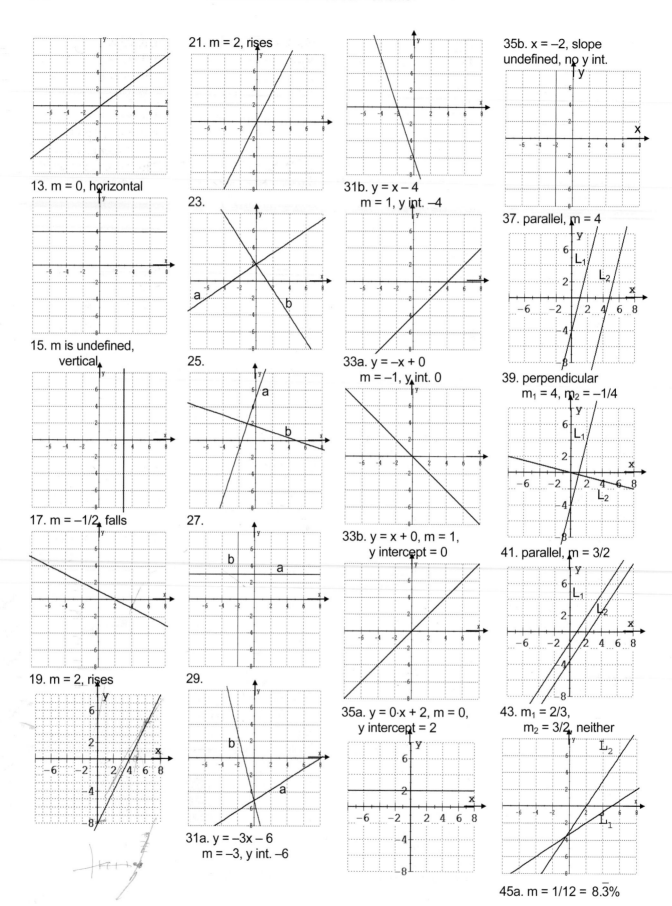

21. m = 2, rises

35b. x = −2, slope
undefined, no y int.

13. m = 0, horizontal

23.

31b. y = x − 4
    m = 1, y int. −4

37. parallel, m = 4

15. m is undefined,
    vertical

25.

33a. y = −x + 0
    m = −1, y int. 0

39. perpendicular
    m₁ = 4, m₂ = −1/4

17. m = −1/2, falls

27.

33b. y = x + 0, m = 1,
    y intercept = 0

41. parallel, m = 3/2

19. m = 2, rises

29.

31a. y = −3x − 6
    m = −3, y int. −6

35a. y = 0·x + 2, m = 0,
    y intercept = 2

43. m₁ = 2/3,
    m₂ = 3/2, neither

45a. m = 1/12 = 8.3%

45b. 31.2 ft.
47a. 5%
47b. 0.9 ft.
49a. 25/8° per hour
49b. 8.75°F
51a. M = (−1,7)
51b. M = (−7/2,7/2)
53. M = (6.02,6.92)
55a. y = 6
55b. slope
57a. W = 6w

57b.

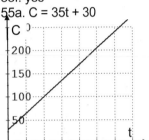

57c. earth weight is 6 times moon weight

59a. k = 12.5
59b.

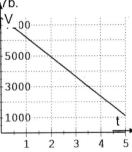

59c. k = m = 12.5

59d. 75 cm.
61a. 4, −2, 6, 0, 8, −10
61b. 2, 4, 6
61c. 5
61d. 3, 7
61e. 7
63a. It is a right triangle with angle B = 90°
63b. It is a right triangle with angle B = 90°

PROBLEM SET 2.4 (on page 125)

1. $y = \dfrac{4}{5}x + \dfrac{2}{5}$

3. $y = -3x + 1$

5. $y = 1$

7. $y = -\dfrac{1}{6}x + \dfrac{2}{3}$

9. $y = 2x - 3$

11. $y = -\dfrac{2}{3}x + 2$

13. $y = 3x + 2$

15. $y = -\dfrac{5}{8}x + 5$

17a. $y - 2 = 5(x + 1)$
17b. $y = 5x + 7$
19a.

$y + 1 = -\dfrac{3}{7}(x - 5)$

19b. $y = -\dfrac{3}{7}x + \dfrac{8}{7}$

21a. $y - 0 = \dfrac{3}{8}(x - 0)$

21b. $y = \dfrac{3}{8}x$

23a. $y + 4 = 0(x - 3)$
23b. $y = -4$

25a. $y + 1 = \dfrac{5}{3}(x + 1)$

25b. $y = \dfrac{5}{3}x + \dfrac{2}{3}$

27a. $y - 3 = -4(x - 2)$
27b. $y = -4x + 11$

29a. $y + 2 = \dfrac{3}{8}(x - 4)$

29b. $y = \dfrac{3}{8}x - \dfrac{7}{2}$

31a.

$y + 8 = -\dfrac{1}{4}(x + 5)$

31b. $y = -\dfrac{1}{4}x - \dfrac{37}{4}$

33. $2x - y = -7$
35. $2x + 3y = 8$
37. $4x - 5y = 32$
39. $4x + 3y = -26$
41. $8x + y = 37$
43. $37x + 70y = -127$
45. $x + 0 \cdot y = -1$

47a. $y = -\dfrac{1}{2}x + \dfrac{1}{2}$

47b. $x + 2y = 1$

49a. $y = \dfrac{3}{2}x + \dfrac{7}{2}$

49b. $3x - 2y = -7$

51a. $y = \dfrac{1}{2}x + 3$

51b. $x - 2y = -6$
53a. (12,1.80)
    (20,2.80)

53b. $C = \dfrac{1}{8}t + 0.3$

53c.

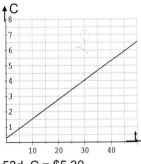

53d. C = $5.30
53e. t = 46.8 min.

53f. yes
55a. C = 35t + 30

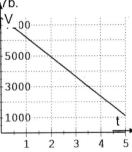

55c. $240
55d. 4 months
55e. yes
57a.
V = −1280t + 7500
57b.

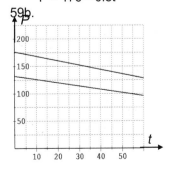

57c. $4940
59a. P = 132 − 0.6t

    P = 176 − 0.8t

59b.

59c. 108 ≤ P ≤ 144
59d. t = 50
61a. m = −b/a

$y - 0 = -\dfrac{b}{a}(x - a)$

$ay = -bx + ab$

$bx + ay = ab$

$\dfrac{x}{a} + \dfrac{y}{b} = 1$

61b. i. $\dfrac{x}{5} + \dfrac{y}{6} = 1$

61b. ii. $-\dfrac{x}{2} + \dfrac{y}{7} = 1$

63a. M = (2,2)
63b.

$y - 2 = -\dfrac{3}{5}(x - 2)$

PROBLEM SET 2.5 (on page 136)

1a. is a function
    D = {1, 2, 3, 5}
    R = {5, 7, 9, 13}
1b. is a function
    D = {−4, −3, 3}
    R = {16,9}
3a. yes
    D = {each citizen}
    R = {all social
         security numbers}
3b. yes
    D = {person's age}
    R = {person's height}
5a. −6
5b. 0
5c. 6
7a. −4
7b. −2
7c. 0
9a. 3
9b. 0
9c. 5
11a. 1/5
11b. 0
11c. 7/5
13a. −14
13b. 0
13c. 22
15a. 7t + 3
15b. 7t + 17
15c. 7t + 5
15d. 14
17a. 5 − 8t
17b. 5 − 4t − 4h
17c. 5 − 4t + h
17d. 4t − 4h
19. 3
21. −9
23 −5/2

25a.
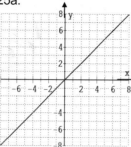

25b.
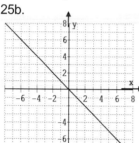

27.
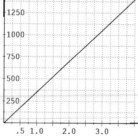

29.
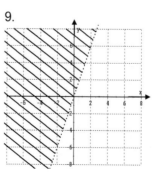

31.
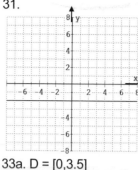

33a. D = [0,3.5]
33b. R = [0,210]
33c. 87 miles
33d. 3 hours
35a. yes
35b. no
37a. D = (−∞,2]
     R = [0,∞)
37b. D = [−3,3]
     R = [0,9]

39a. $C(m) = \frac{1}{5}m + 24$

39b. $50
39c. 10 miles
39d. $84
41a. C(t) = 350t
41b.
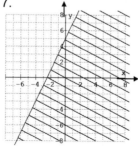

41c. 875 calories
43a. $23,500
43b. 2000
43c. no, 0 ≤ t ≤ 10

45a. S(l) = 3l − 22
45b. 9.5
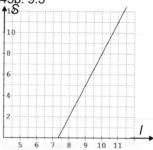

45d. l > 22/3
47a. 5x + 5
47b. x − 9
49a. $6x^2 + 17x − 14$
49b. $3x^2 − 2x$
51. f(x) = 2x + 1
53. any vertical line
55a.
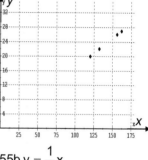

55b. $y = \frac{1}{6}x$

55c. ⅙ , anything that
    weights 6 pounds
    on earth will weight
    1 pound on the
    moon.
55d. 192 lbs.

PROBLEM SET 2.6 (on page 144)

1a. iv
1b. i
1c. vi

3a. $y > \frac{2}{3}x + 2$

3b. y > −x + 3

3c. $y > \frac{1}{2}x + 1$

5a. line and below
5b. line and above

7.
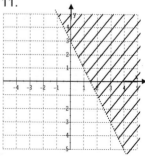

9.

11.

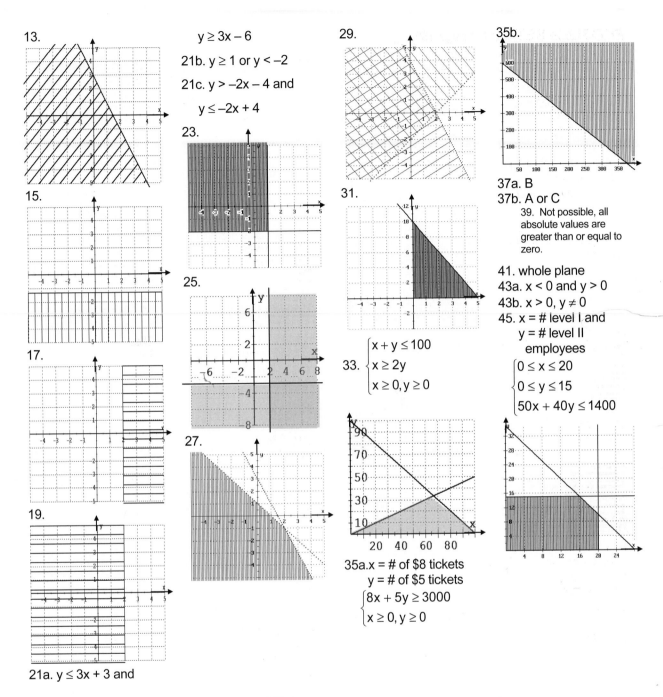

13.

21b. $y \geq 1$ or $y < -2$

21c. $y > -2x - 4$ and

$\qquad y \leq -2x + 4$

$y \geq 3x - 6$

23.

15.

25.

17.

$$33. \begin{cases} x + y \leq 100 \\ x \geq 2y \\ x \geq 0, y \geq 0 \end{cases}$$

27.

19.

35a. $x = $ # of $8 tickets

$y = $ # of $5 tickets

$$\begin{cases} 8x + 5y \geq 3000 \\ x \geq 0, y \geq 0 \end{cases}$$

21a. $y \leq 3x + 3$ and

29.

31.

35b.

37a. B

37b. A or C

39. Not possible, all
absolute values are
greater than or equal to
zero.

41. whole plane

43a. $x < 0$ and $y > 0$

43b. $x > 0$, $y \neq 0$

45. $x = $ # level I and
$y = $ # level II
employees

$$\begin{cases} 0 \leq x \leq 20 \\ 0 \leq y \leq 15 \\ 50x + 40y \leq 1400 \end{cases}$$

CHAPTER 2 REVIEW PROBLEM SET (on page 147)

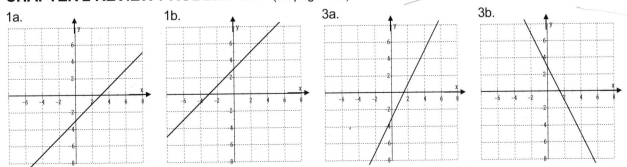

1a.

1b.

3a.

3b.

5.

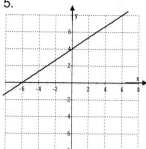

15. slope = 0, horizontal

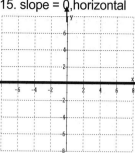

23a. perpendicular

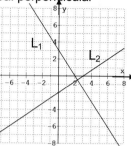

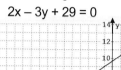

7. x int = 5, y int = −2

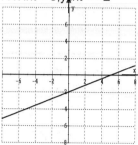

17.

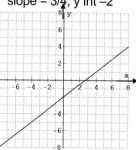

25a. $y = 3x - 2$

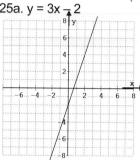

31. $y - 5 = \dfrac{2}{3}(x + 7)$

$2x - 3y + 29 = 0$

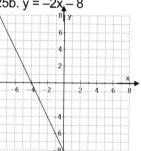

9. x int = −3, no y int

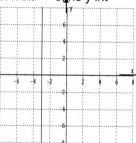

19a. $y = \dfrac{3}{4}x - 2$

slope = 3/4; y int −2

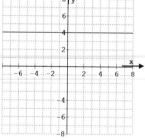

25b. $y = -2x - 8$

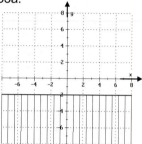

33a.

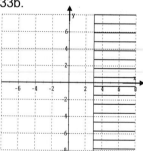

11. slope = 1, rises

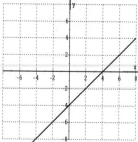

19b. $y = -\dfrac{3}{4}x - 2$

slope : −3/4, y int −2

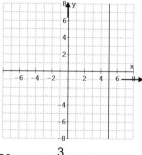

27a. $y = 4$

33b.

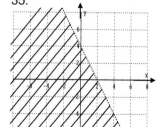

13. slope = −1/2, falls

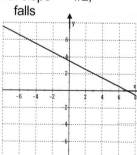

21. parallel

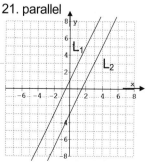

27b. $x = 5$

29. $y - 3 = \dfrac{3}{2}(x - 2)$

$3x - 2y = 0$

35.

37.
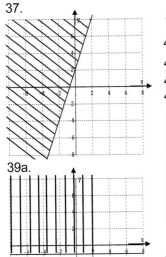

39a.
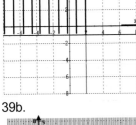

39b.
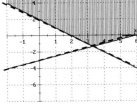

41a. 9
41b. 2t + 5
41c. −1
41d. 2
43a. 2

43b. $\frac{1}{2}$

43c. 0

43d. $\sqrt{3}$

45. −5

47a.
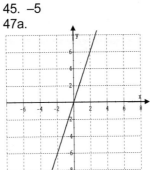

47b.
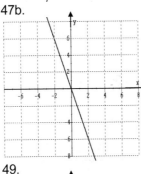

49.
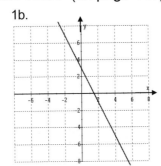

51a. yes
51a. Domain = [−4,4]
      Range = [−4,0]
51b. not a function
53a.
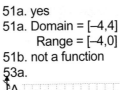
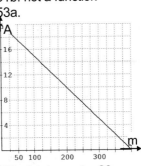

53b. A intercept = 20
      means number of
      gallons at the
      start. The m
      intercept is 400,
      the number of
      miles traveled
      when the tank is
      empty.
53c. 10 gallons
53d. 300 miles

55a. $C = \frac{1}{20}n + 5$

55b.

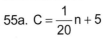

55c. $42.50
57a. t = 10n + 490
57b. slope = 10, each
      month the
      number of tons
      increases by 10.
57c. 8 months
59. y = 2x

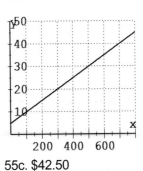
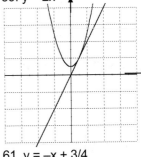
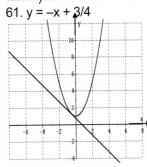

61. y = −x + 3/4
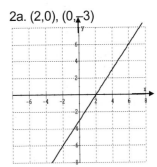

CHAPTER 2 PRACTICE TEST (on page 149)

1a.

1b.

1c.

2a. (2,0), (0,−3)

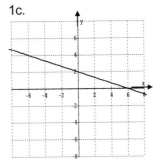

2b. (3,0), no y int

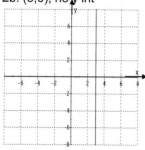

2c. no x int, (0,–1)

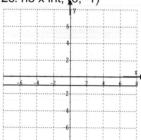

3a. slope = 5/2, rises

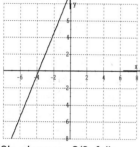

3b. slope = –2/3, falls

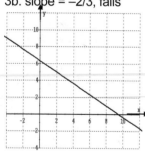

4. $y = \dfrac{5}{3}x - 5$

slope = 5/3, y int –5

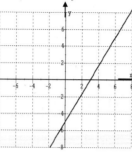

5. perpendicular

6a. $y = -x + 5$
$x + y - 5 = 0$

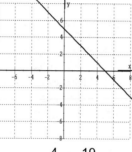

6b. $y = -\dfrac{4}{3}x - \dfrac{19}{3}$

$4x + 3y + 19 = 0$

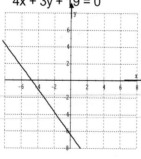

6c. $y = -\dfrac{3}{7}x - \dfrac{18}{7}$

$3x + 7y + 18 = 0$

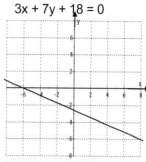

7a.

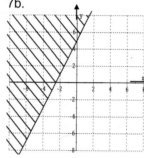

7b.

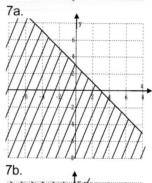

7c.

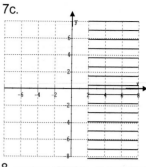

8.

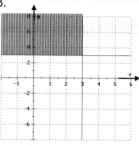

9a. 4
9b. 0
9c. 12
9d. 3
10. D = (–∞,∞)
    R = (–∞,2]
11a.
y = 0.45x + 4500
11b. $4725
11c. 1900 cones
12a.
V(t)= –1000x + 6000
12b. $6000
12c. $1000

— CHAPTER 3 —

PROBLEM SET 3.1 (on page 157)

1. (–1,9)

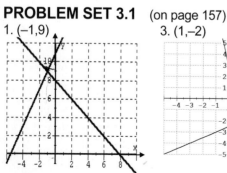

3. (1,–2)

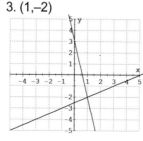

5. (2,–2)

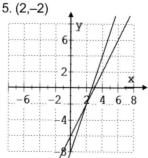

7. no solution

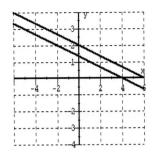

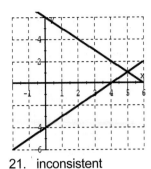

$$\begin{cases} y = -x + 0.5 \\ y = x + 3 \end{cases}$$

29. infinite number of
     solutions

$$\begin{cases} y = x - 6 \\ y = x - 6 \end{cases}$$

31b. (68,874)

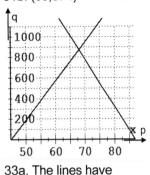

33a. The lines have
      different slopes.
33b. The lines have the
      same slope but
      different x and y
      intercepts.
33c. The lines have the
      same slope and
      the same
      intercepts.

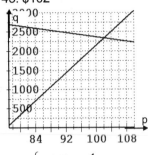

45. $102

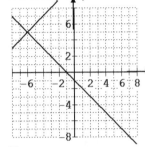

47a. $$\begin{cases} x + y = -1 \\ y - x = 11 \end{cases}$$

47b & c. (−6,5)

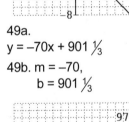

9. no solution

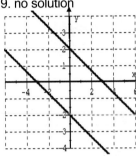

21. inconsistent

$$\begin{cases} y = -\frac{2}{3}x + \frac{5}{3} \\ y = -\frac{2}{3}x + \frac{7}{6} \end{cases}$$

21.

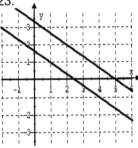

23. inconsistent

$$\begin{cases} y = -\frac{2}{3}x + \frac{5}{3} \\ y = -\frac{2}{3}x + \frac{10}{3} \end{cases}$$

23.

35a. $$\begin{cases} y = mx + b_1 \\ y = mx + b_2 \\ \text{where } b_1 \neq b_2 \end{cases}$$

35b.

$$\begin{cases} y = ax + b \\ y = cx + d \\ \text{where } a = c \text{ and } b = d \end{cases}$$

37. k = −2
39. k = −2
41. $m_1 = m_2, b_1 \neq b_2$
43. $9

11. infinite number of
    solutions

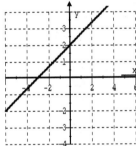

13. One
15. None
17. Infinite
19.

$$\begin{cases} y = -x + 6 \\ y = x - 4 \end{cases}$$

19. consistent and
    independent

25. no solution

$$\begin{cases} y = \frac{3}{4}x + \frac{1}{2} \\ y = \frac{3}{4}x - \frac{7}{4} \end{cases}$$

27. one solution

49a.
y = −70x + 901 ⅓

49b. m = −70,
     b = 901 ⅓

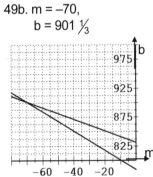

PROBLEM SET 3.2 (on page 165)

1. (4,3)
3. (7/3,−4/3)
5. (7/11,−10/11)
7. (1,5)
9. (159,−186)
11. (−3,0)
13. (3,1)
15. (7/3,−4/3)
17. (0,6)

19. (−6,9)
21. (−10,0)
23. (1,1)
25. (−5,3)
27. (−3/2,27/2)
29. (22/25, 21/25)
31. (3,2)
33. (2,−1)
35. (2/5,21/5)
37. (9,−5/3)

39. (10,24)
41. consistent &
    dependent

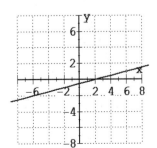

43. inconsistent

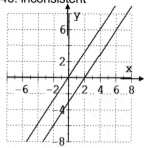

45. (12,15) consistent & independent

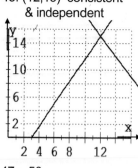

47a. 50
47c. A loss occurs when 0 ≤ x < 50.

47d. A profit occurs when x > 50
47b.

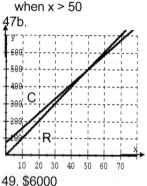

49. $6000
51. (−3,4)

53. (10,13)

55. (1/13,1/17)

57. (1/4,−1/5)

59a. a = 4/7,
59a. b = 40/7
59b. c = 4
61. $y = \frac{3}{4}x + \frac{11}{4}$

PROBLEM SET 3.3 (on page 171)

1. yes
3. (1,1,1)
5. (−5,5,8)
7. (2,2,1)
9. (5,2,1)
11. (−3,4,1)
13. (1,−3,−2)
15. (3,1,0)
17. no solution, inconsistent, the three planes have no common intersection.
19. infinite number of solutions, consistent and dependent, the planes intersect along a line

21. no solution, inconsistent, the three planes have no common intersection.
23. infinite number of solutions, consistent and dependent, there is only one plane
25. (6,−3,12) consistent and independent. The three planes intersect at a point
27. (2,0,2) same as #25
29. a = −2, b = 144, c = −80
31. a = 1/12 ≈ 0.08
    b = −1.5
    c = 23/3 ≈ 7.67

33. $\begin{cases} x + y - z = 6 \\ -x + 2y + z = 0 \\ -2x + z = -7 \end{cases}$
The number of systems is infinite.
35. (2,3,−1)
37. (4,3,2)
39. (2,−1,3)
41. (3,4,6)
43. a = 5, b = 3, c = 7
45a. (0.3, −0.2, 0.5)
45b. (0.1, 0.3, 2)

PROBLEM SET 3.4 (on page 180)

1. l = 250 ft w = 200 ft
   Cost = $70,833.33
3. W = 18 in.
5. A = 6.5 oz, B = 4.5 oz
7. airliner = 502.23 mph
   wind = 33.49 mph
9. 22 kph and 17 kph
11. $25,000 @ 6% and $15,000 @ 7%
13. $12,000 @ 10.5%, $18,000 @ 6%
15. 4 lbs @ $9
    16 lbs @ $4
17. 30 oz @ 30%,

20 oz @ 90%
19. $6000 in CD, $9000 in stocks, $5000 in bonds
21. 3000 @ $20, 6000 @ $15, 11,000 @ $10
23. 15 model A, 10 model B, 20 model C
25. 33, 21
27. 17 dimes 28 quarters
29. Tower is 437 m
31. More than 58 visits

33. 185 pairs of tennis shoes, 75 pairs of running shoes
35. Bride has $1000, groom has $3000
37. $23,055 @ 6%, $15,041 @ 8.5%, tax included
39. $0.90 for juice; $0.80 for bagel; $0.25 for coffee
41. 13 dryers, 9 washers, 26 microwaves
43. 380 lbs A, 60 lbs B 160 lbs C.
45. 32 nickels, 8 quarters

PROBLEM SET 3.5 (on page 187)

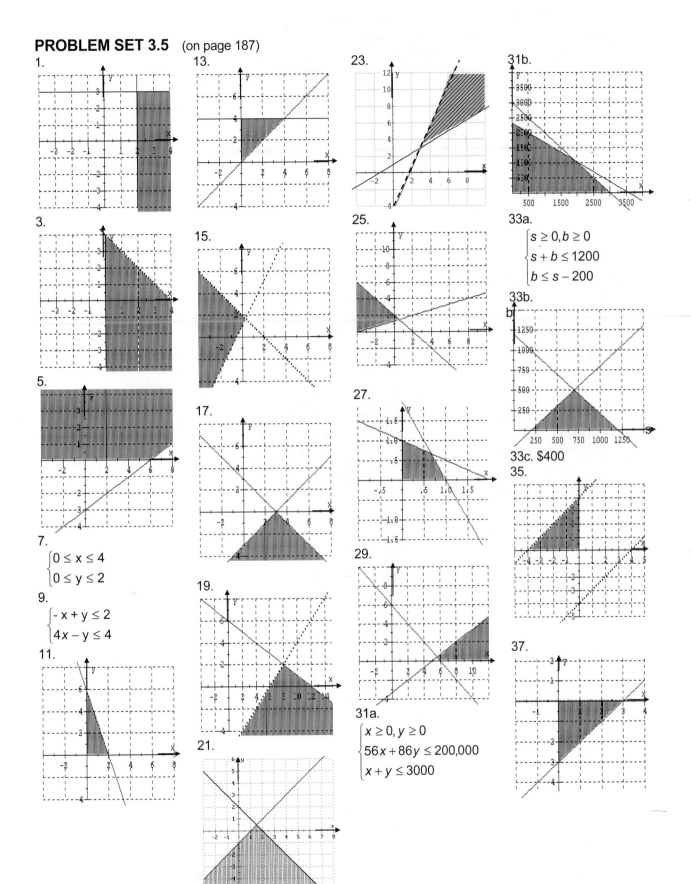

7.
$$\begin{cases} 0 \le x \le 4 \\ 0 \le y \le 2 \end{cases}$$

9.
$$\begin{cases} -x + y \le 2 \\ 4x - y \le 4 \end{cases}$$

31a.
$$\begin{cases} x \ge 0, y \ge 0 \\ 56x + 86y \le 200,000 \\ x + y \le 3000 \end{cases}$$

33a.
$$\begin{cases} s \ge 0, b \ge 0 \\ s + b \le 1200 \\ b \le s - 200 \end{cases}$$

33c. $400

39.
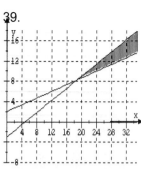

41.
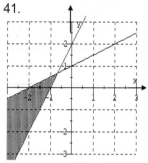

CHAPTER 3 REVIEW PROBLEM SET (on page 189)

1a. (3,2)
1b. y = –x + 5
    y = x – 1
1c. consistent and
    independent
3a. no intersection
3b. y = –2x + 6
    y = –2x + 4
3c. inconsistent
5. (7/3,1/3)
7. (12,6)
9. (2,0) consistent and
    independent
11. (–1,–2) consistent and
    independent
13. (–1,13/4)
15. (4,3,2)
17. (1,0,0)
19. (–2,0,1)
21. (13,–15,–2)
23. (1/2,1/2,1/2)

25.
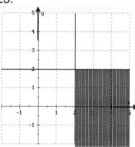

27.
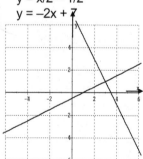

29.
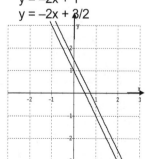

31.
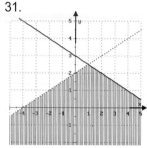

33. 106°, 74°
35. TV:$570, VCR $320
37. 4000 of A, 1000 of B
39. 440 mph, 560 mph
41.
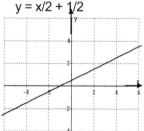

all the combinations in
the shaded area will
satisfy the conditions.

43. (–1.00,3.25)
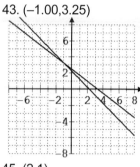

45. (2,1)
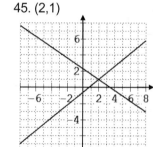

CHAPTER 3 PRACTICE TEST (on page 190)

1a. (3,1)
    y = x/2 – 1/2
    y = –2x + 7

1b. no intersection
    y = –2x + 1
    y = –2x + 3/2

1c. identical lines
    y = x/2 + 1/2
    y = x/2 + 1/2

2a. consistent and
    independent
2b. inconsistent
2c. consistent and
    dependent
3a. (–1,–5),
    consistent
3b. (5,10), consistent
3c. no solution;
    inconsistent
4. (–2,5,3) is a
    solution

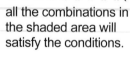

5a. 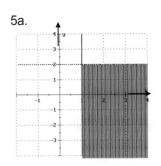   5b.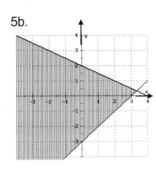

6. 30 dimes,
   25 nickels
7. $8,460 in mutual
   funds, $4,230 in
   bond fund

-- CHAPTER 4 --

PROBLEM SET 4.1   (on page 198)

	Degree	Coefficients
1.	2	$-7, 3, -9$
3.	3	$1, 0, -8, 5$
5.	4	$1, 0, 5, 0, -2$
7.	5	$1, 0, 0, 4, -3, -1$

Descending powers	Leading term	Leading coefficient
9. $-4x^5 + x^3 + x^2 - 11x + 2$	$-4x^5$	-4
11. $-8x^4 + 4x^2 - 7x + 5$	$-8x^4$	-8
13. $-10x^4 + 2x^3 - x + 9$	$-10x^4$	-10
15. $4a^4 + a^2 + 7a - 8$	$4a^4$	4

17a. 23   17b. 137

19a. 67   19b. -209

21a. 3   21b. $-\dfrac{t^3}{4} - \dfrac{5t^2}{4} - t + 3$

23a. 0   23b. 0

23c. 15   23d. 48

25a. 29   25b. $-3t^3 + 2t^2 - 3$

25c. $-24u^3 + 8u^2 - 3$ 25d. $81v^3 + 18v^2 - 3$

	x intercepts	y intercept	Domain
27.	$-3, -1, 2$	-6	Reals
29.	$-2, 0, 1, 3$	0	Reals

31. $7x^2 - x - 1$
33. $t^2 + 10t + 3$
35. $4t^3 + 2t^2 - 11t - 6$
37. $x^6 - x^5 - 2x^4 + 2x^3 - 3x^2 + 3x + 4$

39a. -6 39b. 198
39c. $3x^2 + 5x - 8$ 39d. $13x^2 + x - 6$
41a. 22   41b. 462
41c. $-x^3 + x^2 + x + 12$ 41d. $-7x^3 + x^2 - x - 6$
43a. 64 feet   43b. 60 feet
43c. 55 feet
45a. $4x + 10$ 45b. 26 feet
47a. $-6x^2 + 420x - 500$ 47b. $6250, $6850, $5500
49. $5x^2 + 2x - 2$
51. $2t^3 + 2t^2 - t - 2$
53. $w^2 + w + 9$
55. $6x + 10$
57. $10y^4 - 5y^3 + 3y^2 + 3y - 8$

59. $-\dfrac{4x^3}{3} - \dfrac{7x^2}{2} + \dfrac{7x}{3} - \dfrac{7}{4}$

61. $20x^2 - 39x + 15$
63a. any number except 0
63b. degree 2

PROBLEM SET 4.2 (on page 207)

1a. $6x^6$ 1b. $35x^5y^3$
3a. $-30y^7$ 3b. $-6x^4y^4z^3$
5a. $-12y^3 + 16y^2$ 5b. $15x^6 + 10x^2$
7a. $-8x^3y^7 + 28x^3y^5 + 12x^2y^3$
7b. $10x^4y^4 - 4x^2y^3 + 10xy^2$
9a. $2.7x^6y^4 - 1.2x^4y^6 + 1.5x^3y^4$
9b. $5.98m^7n^3 - 11.96m^4n^3 + 23m^3n^5$
11. $9x^3 + 21x^2 - 2x - 8$ 13. $3y^3 - 10y^2 - 4y + 35$
15. $4u^4 - 4u^3 + 2u^2 + 6u - 12$
17. $9x^3 - 21x^2 + 22x - 8$
19. $m^5 + 2m^4 + 10m^2 - 9m + 12$
21. $y^4 - 3y^3 - 4y^2 - 9y + 9$
23. $x^5 - x^4 - 8x^3 + 9x^2 + 6x - 6$

25a. -35 25b. -35
25c. $-6x^4 + 19x^2 - 15$ 25d. $-6x^4 + 19x^2 - 15$
27a. $x^2 + 3x + 2$ 27b. $x^2 - 3x + 2$
29a. $x^2 + 2x - 15$ 29b. $x^2 - 2x - 15$
31a. $5x^2 - x - 4$ 31b. $5x^2 + xy - 4y^2$
33. $10x^4 + 13x^2 - 3$

35. $0.8x^2 + 5.8xy - 1.5y^2$ 37. $x^2 + \dfrac{143}{12}xy - y^2$

39a. $w^2 - 49$ 39b. $49 - w^2$
41a. $x^2 + 2x + 1$ 41b. $x^2 - 2x + 1$
43a. $16y^4 + 40y^2z + 25z^2$ 43b. $16y^4 - 40y^2z + 25z^2$
45a. $x^3 + y^3$ 45b. $x^3 - y^3$

47. $9x^2 - \dfrac{1}{16}$

49a. $x^2 - 5x$ 49b. $24\ ft^2$

51a. $-2x^2 - 180x + 7200$ 51b. 6250 oranges

53. $-100x^4 + 400x^3 - 400x^2$

55. $9x^3 - 12x^2 - 80x - 64$ 57. $25x^2 - 10x - y^2 + 1$

59. $x^2 + 2xy + 10x + y^2 + 10y + 25$

61. $x^2 - 9y^2 - 24y - 16$

63. $x^4 + 8x^3y + 24x^2y^2 + 32xy^3 + 16y^4$

65. $x^2 - 2xy + 8x + y^2 - 8y + 16$

PROBLEM SET 4.3 (on page 214)

1a. $-18xy$ 1b. $-3xy^2$

3a. $-8x^3 - 5x - 4$ 3b. $-9x^5 + 4x^3 + 3x$

	Quotient	Remainder
5.	$x - 2$	0
7.	$x - 2$	-8
9.	$4x - 1$	2
11.	$x^2 + 5x + 14$	35
13.	$x^3 + x^2 - 4x + 9$	-8
15.	$x^2 + 3x + 4$	0
17.	$x + 2$	0
19.	$xy^2 - 3y^3 + 10y$	0
21.	$3x + 1$	$-12 = P(-2)$

23.	$2x + 4$	$5 = P(-2)$
25.	$4x^2 + 8x + 6$	$3 = P(1)$
27.	$4x^2 + 7x + 4$	$3 = P(1)$
29.	$5b^2 + 4b + 11$	$33 = P(2)$
31.	$x^4 + 2x^3 + 5x^2 + 10x + 20$	$42 = P(2)$

33a. $2x + 7$ feet   33b. 4 feet

35a. $2x + 5$ inches   35b. 9 inches

37a. 1   37b. x

37c. x^2 37d. x = 1

37e. all positive or negative values of x

37f. no value of x

39. $x^2 - xy + y^2$; multiply to verify

PROBLEM SET 4.4 (on page 218)

1a. 1, 2, 3, 6, 9, 18, 1x, 2x, 3x, 6x, 9x, 18x, x^2, $2x^2$, $3x^2$, $6x^2$, $9x^2$, $18x^2$

1b. 1, 2, 13, 26, 1x, 2x, 13x, 26x, x^2, $2x^2$, $13x^2$, $26x^2$

3a. 1, 2, 31, 62, y, 2y, 31y, 62y

3b. 1, 2, 4, 5, 8, 10, 16, 20, 40, 80, y, 2y, 4y, 5y, 8y, 10y, 16y, 20y, 40y, 80y, y^2, $2y^2$, $4y^2$, $5y^2$, $8y^2$, $10y^2$, $16y^2$, $20y^2$, $40y^2$, $80y^2$

5. 7   7. 5

9. x^2 11. $6x^2$

13. $7x^3$ 15. $2x$

17a. $5(x + 2)$ 17b. $7(t + 3)$

19a. $3(x - 21)$ 19b. $8(2y - 3)$

21a. $4(2x - 3)$ 21b. $6(5 - 2y)$

23a. $3x^2(1 + 8x)$ 23b. $x(9 - 4x)$

25a. $x(14 + 11x)$

25b. $y^2(9 - 13y)$

27a. $12(u^2 + 2)$

27b. $2y^2(5y - 6)$

29a. $a(a^2 - 4a - 1)$

29b. $y(y^2 - 7y - 3)$

31a. $4y^2(4y + 3 - 2y^2)$

31b. $y^3(y^3 + 7y^2 - 11)$

33a. $12a^2b^2(a + 3b)$

33b. $6ab(b + 5a)$

35a. $(a + b)(x + y)$

35b. $(x^2 + 1)(a + b)$

37a. $(b^2 - d)(a - c)$

37b. $(x^2 + z^2)(2 - y)$

39a. $(x - 5)(x + 2)$

39b. $(x + 3)(x + 4)$

41. $A = 2\pi r(r + h)$

43. $4x + 4 = 4(x + 1)$

45a. $xy^2(4z + xz - x^2y)$

45b. $9mn(m + 2n - 3)$

47a. $3x^2(y^2 + 2z^2 - 3)$

47b. $(2a + b)(3x + 5y)$

49a. $(x + 1)(y - z)$

49b. $2x^{-5}(2x^2 + 3)$

51. $(2x + 5z)(4x - 3y)$

53. $(x^2y^2 + 4t^3)(a - 5)$

55. $(x^2 + 1)(3 + 2y + z)$

PROBLEM SET 4.5 (on page 228)

1a. $(x + 3)(x + 4)$

1b. $(x - 3)(x - 1)$

3a. $(x + 7)(x + 2)$

3b. $(x - 5)(x - 4)$

5a. $(d - 8)(d - 7)$

5b. $(t - 11)^2$

7a. $(x - 6)(x + 1)$

7b. $(t + 4)(t - 3)$

9a. $(y - 7)(y + 4)$

9b. $(d + 11)(d - 3)$

11a. $(y + 6)(y - 5)$

11b. $(u - 13)(u + 1)$

13a. $(u + 9)(u - 3)$

13b. $(t - 12)(t + 8)$

15a. $(x - 10y)(x + 3y)$

15b. $(b - 12d)(b + 3d)$

17a. $(x - 22y)(x + 2y)$

17b. $(y + 11z)(y - 10z)$

19a. $(3m - 4)(m - 2)$

19b. $(3x + 4)(2x - 3)$

21a. $(2y + 3)(y + 4)$

21b. $(4x - 3)(x + 1)$

23a. $(2x + 5)(7x - 3)$

23b. $(2x - 3)(3x - 2)$

25a. $-(y - 9)(y + 3)$

25b. $-(t - 5)(t + 3)$

27. c, $(x + 1)(x - 4)$

29. b, $(2x - 5)(x + 2)$

31a. D = 76, not factorable

31b. D = 25, $(2x + 1)(x - 2)$

33a. D = 25, $(x - 3y)(x + 2y)$

33b. D = -23, not factorable

35. $(2x - 1)(x - 1)$

37. $16(t - 15)(t - 1)$

39. $x^2(x + 5y)(3x - 2y)$

41. $-(x + 6y)(3x - 2y)$

43. $(x + 2)(2x + 7)$

45. $x(5x + 8)$

47. $(x - 1)(2x - 5)$

49. $(x^2 + 3)(x^2 + 2)$

51. $(2y^3 - 1)(y^3 - 3)$

53. $(2x^2 - 3y)(x^2 + 2y)$

PROBLEM SET 4.6 (on page 234)

1a. $(x-10)(x + 10)$
1b. $(10 - x)(10 + x)$
3a. $(4x - 7)(4x + 7)$
3b. $(4x - 9y^2)(4x + 9y^2)$
5a. $(4u - 5v)(4u + 5v)$
5b. $(5m - 7n)(5m + 7n)$
7a. $(7x - 4y^2)(7x + 4y^2)$
7b. $(2xy -1)(2xy + 1)(4x^2y^2 + 1)$
9. $(x - 3)(x + 3)(x^2 + 9)$
11a. $(x - 1)(x^2 + x + 1)$
11b. $(y + 5)(y^2 - 5y + 25)$
13a. $(w - 2yz)(w^2 + 2wyz + 4y^2z^2)$
13b. $(4z + 3b)(16z^2 - 12zb + 9b^2)$
15a. $(x - 1)(x^2 + x + 1)(x^6 + x^3 + 1)$
15b. $(y + 2)(y^2 - 2y + 4)(y^6 - 8y^3 + 64)$
17a. $(x + y - 3)(x^2 - xy - 6x + y^2 + 3y + 9)$
17b. $(2x + 3)(4x^2 + 3)$
19. $(y - x)(x^2 + xy + y^2)(x^6 + x^3y^3 + y^6)$
21a. $(x + 3)^2$ 21b. $(y - 4)^2$

23a. $4(x - 1)^2$ 23b. $3(2x -1)^2$
25a. $(2x^2 -1)^2$ 25b. $(3x^2 -2)^2$
27a. $(x^4 + 1)^2$ 27b. $(x^3 - 2)^2$
29. $(2x - y^2)^2$ 31a. $(x^2 + y^2)^2$
31b. $(2x + y)(2x - y)(4x^2 + y^2)$
33a. $5x(x - 7)(x - 4)$
33b. $yz^2(x - 4)(x + 3)$
35a. $(x + 15)^2$ 35b. $(x - 15y)^2$
37. $3x(3x + 1)(9x^2 - 3x + 1)$
39. $2(b - 2c)(x + 3y)$
41a. $(3x - 1)(5x + 6)$
41b. $A(5) = 434$ m^2, length 31m, width 14m.
43a. $3x(x - 2y)(x + 2y)(x^2 + 4y^2)$
43b. $3t(t - 2)(t^2 + 2t + 4)$
45a. $(2u - 3)(2u + 3)(4u^2 - 6u + 9)(4u^2 + 6u + 9)$
45b. $3t^2(t^2 + 3)(t^4 - 3t^2 + 9)$
47. $(2x + y - 1)(x^2 + y)$
49. $(x + y + z + 1)(x - y - z + 1)$
51. $(5x - 3y)(5x + 3y)(25x^2 + 9y^2)$

PROBLEM SET 4.7   (on page 243)

1a. $-1, -6$
1b. $1, 11$
3a. $4, 2$
3b. $1, 36$
5a. 3
5b. $\dfrac{2}{3}$
7a. $8, -1$
7b. $6, -11$
9a. $7, -5$
9b. $-7, 5$
11a. $\dfrac{2}{3}, -\dfrac{5}{2}$

11b. $-\dfrac{2}{3}, \dfrac{5}{2}$
13a. $\dfrac{4}{3}, -\dfrac{5}{2}$
13b. $\dfrac{7}{2}, -\dfrac{2}{5}$
15a. $0, 7$
15b. $0, \dfrac{9}{2}$
17. $11, -\dfrac{3}{2}$
19. ± 3
21. $-\dfrac{4}{3}, \dfrac{3}{2}$

23. $-1, \dfrac{1}{3}$
25. $-\dfrac{7}{2}, \dfrac{1}{9}$
27. $1, -2, \dfrac{3}{2}$
29. $0, 2, -3, -\dfrac{7}{2}$
31. $11, 2, -2$
33. $0, -12, -\dfrac{1}{2}$
35a. $32, 12, 0, 0$
35b. each lies on the graph

35c. $0, 4$
35d. same values
37a. $20, 0, -7, -15, 0$
37b. each lies on the graph
37c. $-3, 5$
37d. same values
39. $c, -1, 4$
41. $b, 5$
43a. $5, 6$
43b. $-7, -8$
45. 19 by 22 feet or 11 by 38 feet
47. 3 feet

49. 5 feet
51. 7 cm, 24 cm, 25 cm
53. $3, \dfrac{3}{2}$
55. ± 1
57. $5a, -3a$
59. $0, \dfrac{1}{b - a}, a \neq b$

CHAPTER 4 REVIEW PROBLEM SET (on page 246)

1a. 4
1b. 5, 1, 2, -1, -13
1c. 81
3. $4x^3 - 2x^2 + 2x$
5. $2y^3 + y^2 + 8y - 9$
7. $a^3 - 3a^2b + 3ab^2 - b^3$
9. $12y^2 - 9xy - 81x^2$
11. $9x^2 + 42x + 49$
13. $5x - 1$
15. $x - 2$, Remainder 12
17. $7xy(x - 3y^2)$
19. $(5a - b)(m^2 + n)$
21. $9u(u - 3v)(u + 3v)$
23. $(m - 9)(m + 4)$

25. $2ab(a - 7b)(a + 2b)$
27. $(m - 5)(m + 2)$
29. $(2y + 1)(y + 3)$
31. $(5x + 2)(x + 3)$
33. $(3t - 2)^2$
35. $(6c - d)(6c + d)$
37. $4(5c - 2d)(5c + 2d)$
39. $(3w - 2)(3w + 2)(9w^2 + 4)$
41. 8, 3
43. $-5, \dfrac{1}{4}$
45. $-4, -7$
47. $\pm\dfrac{5}{9}$

49. b, $(9,0), (-1,0), (0,-9)$
51. c, $(-2,0), (7,0), (0,14)$
53a. $-\$600$
53b. $\$400$
53c. 60 or 110 units
55a. $\dfrac{x(x + 2)}{2}$
55b. 24 square inches
55c. 12 inches
57a. $4r^2 + \pi r^2$
57b. $r^2(4 + \pi)$
57c. 28.6m^2

CHAPTER 4 PRACTICE TEST (on page 248)

	Degree	Coefficients	Values		1a.	2	7, -3, 5	$G(-2) = 39,$

			$G(3) = 59$
1b.	3	4, 0, 1, –1	$h(0) = -1, h(-1) = -6$

2a. $5x^2 + 2x + 3$

2b. $5x^2 - 8x + 7$

2c. $4x^3 + 4x^2 - 5x - 3$

2d. $2x^2 + 7x + 4, R = 0$

2e. $7x^2 + x + 8, R = 19$

3a. $49x^2 + 14x + 1$

3b. $1 - 10x + 25x^2$

3c. $9x^2 - 4$

3d. $x^3 + 8$

4a. $(3m - 10n)(3m + 10n)$

4b. $(a - b + c)(a + b - c)$

4c. $(y^2 + 2)(x - 4)$

4d. $(x - 1)(x + 17)$

4e. $(y + 1)(6y - 7)$

4f. $2(w - 1)(w + 3)$

5a. 2, 15

5b. $\pm\dfrac{7}{5}$

5c. $\pm 3, \pm 2$

5d. 6, –9

6a. $(12 - 2x)(16 - 2x)$

6b. 60 square inches

6c. 4 inches

— CHAPTER 5 —

PROBLEM SET 5.1 (on page 256)

1a. 1

1b. –8

1c. undefined

1d. 0

3a. $x \neq 3$, asy: $x = 3$

3b. $x \neq -6$, asy: $x = -6$

5a. $x \neq 6, -7$

    asy: $x = 6, x = -7$

5b. $x \neq -3, 2$

    asy: $x = -3, x = 2$

7a. $\dfrac{2y}{3x}$

7b. $2x - 1$

9a. y

9b. x

11a. $\dfrac{x+1}{x-1}$

11b. $\dfrac{x-3}{x}$

13a. $\dfrac{y+2}{4y}$

13b. $\dfrac{x+2}{3x}$

15a. $\dfrac{x+4}{x+7}$

15b. $\dfrac{x+4}{x+2}$

17a. $\dfrac{x-y}{x+y}$

17b. $\dfrac{3x-6y}{x+2y}$

19a. $\dfrac{x+2y}{3x+4y}$

19b. $\dfrac{x-y}{x+y}$

21a. 6xy

21b. 39xy

23a. $20m^7y^4$

23b. 18x

25a. $9a^3(a - b)^2$

25b. $m^2n(m + n)$

27. $2x^2 - 9x + 9$

29. $2x^2 + x - 6$

31a. yes

31b. yes

33. no

35. yes

37a. $\dfrac{7}{x-4}$

37b. $\dfrac{-4}{x-4}$

39. $\dfrac{46}{3}$ ft

41. $-\dfrac{x+y}{2x+y}$

43. $\dfrac{w+z}{y+z}$

45. $x^n - 3$

47. d

49. b

51. D = reals, $x \neq 3$

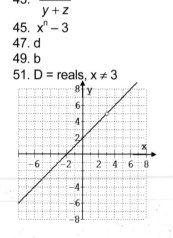

PROBLEM SET 5.2 (on page 263)

1a. $\dfrac{8}{9y}$

1b. $\dfrac{2y}{9x}$

3. $\dfrac{2x}{3y^3}$

5. $-10x^2y^2z$

7. $\dfrac{1}{2}$

9. $\dfrac{5}{3x}$

11a. $\dfrac{5}{2}$

11b. $\dfrac{15}{16}$

13. 75yz

15. $\dfrac{1}{64b^2c}$

17. $-\dfrac{7z}{4y}$

19. $\dfrac{x}{3}$

21. $\dfrac{7}{m-n}$

23. $\dfrac{x^2}{5(x-1)}$

25. $x^2 - 16x + 48$

27. $\dfrac{1}{(x-y)(2x+y)}$

29. $1 - 3a$

31. $\dfrac{(y-1)^2}{(y+1)^2}$

33. $\dfrac{x-3}{3x+1}$

35. $-\dfrac{t+2}{t+5}$

37. $\dfrac{x-1}{x+2}$

39. $\dfrac{a^2-b^2}{a^3}$

41. $\dfrac{2x}{3}$

43. $\dfrac{2x+5y}{x+5y}$

45. $3x-1$

47. $\dfrac{x+1}{7(x+3)}$, $x\ne$ $-1,-3,3$

49. $\dfrac{x(x-3)}{2x+1}$, $x\ne \pm\frac{1}{2}$

51a. $\dfrac{x}{y}$

51b. 20 mpg

53. $\dfrac{25y^2}{24x^2}$

55. $\dfrac{1}{(x+1)^2}$

57. x^2+4x+3

59. $\dfrac{(x-1)^2(x+1)}{x^2+1}$

61. $\dfrac{a-1}{b(3-a)(b+1)}$

63. $\dfrac{2x}{x+1}$

65a. 1

65b. x

65c. x^2

65d. 1

65e. all values, $x\ne 0$

65f. no values

PROBLEM SET 5.3 (on page 272)

1. $\dfrac{3}{4x}$

3. $\dfrac{-8}{5x}$

5. $\dfrac{8}{7x}$

7. $\dfrac{1}{2t}$

9. $\dfrac{t}{2k}$

11. $\dfrac{2a+3}{a+3}$

13. $\dfrac{15}{y-5}$

15. $\dfrac{2}{x+2}$

17. $\dfrac{1}{4-x}$

19. $\dfrac{6}{y+1}$

21. $\dfrac{21x+5y}{35y}$

23. $\dfrac{y+12}{4}$

25. $\dfrac{ab+a^3-b^2}{b^3}$

27. $\dfrac{30+9t+4t^2}{6t^2}$

29. $\dfrac{8x+7}{(x+2)(x-1)}$

31. $\dfrac{15y-57}{(y-5)(y-3)}$

33. $\dfrac{x^2-x+2}{(3x+2)(x-4)}$

35. $\dfrac{5x-3}{x(x^2-1)}$

37. $\dfrac{2m^2+4}{m(m-6)(m+1)}$

39. $\dfrac{8x^2+15}{(x-4)(2x+3)(3x-1)}$

41. $\dfrac{3-4c}{(c-3)(c-2)(c+3)}$

43. $\dfrac{4x^2-5x}{(x-4)(2x+3)(5x+2)}$

45. $\dfrac{5x^2-12xy^2+21y}{12x^3y^3}$

47. $\dfrac{2}{a+1}$

49. $\dfrac{x^2-6x}{x^2-9}$

51. $-\dfrac{17t^2+t-9}{3t^2-27}$

53a. $\dfrac{x}{60}+\dfrac{y}{60}=\dfrac{x+y}{60}$

53b. $28.\overline{3}$ hours

55. $\dfrac{3}{x(x+3)}$

57a. $\dfrac{4x^3+79x^2+256x}{4x+16}$

57b. $11,575.96

59a. $\dfrac{x^2+x-9}{(x-3)(x-2)(x+3)}$

59b. $\dfrac{x^2-5x+9}{(x-3)(x-2)(x+3)}$

61a. $\dfrac{2x^2-2x+6}{(x+2)(x-2)(x+3)}$

61b. $\dfrac{8x-6}{(x+2)(x-2)(x+3)}$

63a. $\dfrac{16}{15}$

63b. -6.9908

65. $\dfrac{x^4-y^4}{x^2y^2}$

67. 2

69a. $\dfrac{65}{24}=2.708\overline{3}$

69b. $\dfrac{x^4+4x^3+12x^2+24x+24}{24}=\dfrac{65}{24}$

69c. $e\approx 2.71828$

71a. $\dfrac{y+x}{xy}$

71b. 5/6

71c. 2/5, not equal

PROBLEM SET 5.4 (on page 282)

1a. $\dfrac{4y}{3x^2}$

1b. $\dfrac{x}{z}$

3a. $-\dfrac{3}{10a}$

3b. $-\dfrac{2x}{y}$

5a. $-xy$

5b. $-6xy$

7. $-\dfrac{1}{3a+3}$

9. $\dfrac{3x+1}{2x-3}$

11a. $\dfrac{5}{6}$

11b. $\dfrac{3}{4}$

13a. $-\dfrac{y}{15ab^2}$

13b. $\dfrac{1}{3b}$

15. $\dfrac{y(x^2-y)}{x^2(y-1)}$

17. $\dfrac{1}{5x}$

19. $\dfrac{2y+7x}{8y-6x}$

21. $\dfrac{m}{m-2}$

23. $\dfrac{-c-d}{4}$

25. $\dfrac{-1-2t}{3t+2}$

27. $\dfrac{1+x}{1-x}$

29. $\dfrac{1}{u-v}$

31. $x-3$

33. $\dfrac{3t+2}{2t-1}$

35. $\dfrac{m^2-mn-n}{m+mn-n^2}$

37a. $\dfrac{1}{y^2}$

37b. x^6

39. $\dfrac{128}{3}$

41. $\dfrac{1}{a^2b^2+ab^3}$

43. $\dfrac{x+y}{xy}$

45. $\dfrac{x+2}{3}$

47. $-\dfrac{3}{a(a+h)}$

49. $\dfrac{3}{(a-1)(a+h-1)}$

51a. 2 hrs
51b. 10 hrs
51c. 20 weeks
51d. 0.3 mpg
53a. $411.98
53b. $3775.04

55. $\dfrac{-2x}{x^2+1}$

57. $\dfrac{y-1}{2y-1}$

59a. $\dfrac{7}{5}$

59b. $\dfrac{17}{12}$

59c. $\dfrac{41}{29}$

59d. $\dfrac{x^4+x^3+3x^2+2x+1}{x^4+3x^2+1}$

61. $\dfrac{14a+7h-14}{(a-1)^2(a+h-1)^2}$

PROBLEM SET 5.5 (on page 290)

1. $x \neq 0$, $x = -1/2$
3. $t \neq 0$, $t = 5$
5. $x \neq 0$, $x = 80$
7. $x \neq 0$, $x = 1$
9. $y \neq 0$, $y = 6/5$
11. $x \neq -4$, $x = 1$
13. $x \neq -5/4$, $x = 15/8$
15. $x \neq -7$, 0 $x = -35/2$
17. $x \neq \pm 3$, $x = 15$
19. $y \neq 2,3$, $y = 4$
21. $t \neq 3$, $t = 3/4$
23. $y \neq -1$, $y = 1/4$
25. $x \neq -1$, $x = 4$
27. $x \neq 1$, 1 extraneous solution
29. $x \neq -2, -4$, $x = -1$

31. $x \neq 2$, 2 extraneous solution
33. $x \neq 4$, 4 extraneous solution
35. $y \neq 0$, 0 extraneous solution
37. $a, b \neq 0$, $y = \dfrac{bc+d}{a}$
39. $x = \dfrac{6a-7b}{4}$
41. $-1, -7$
43. 20
45. 60 mph
47. 55 mph, 70 mph
49a. $y \neq 1, -2, -3$
49b. ½, –6
51a. $y \neq 0, 2$
51b. –3

53a. $-\dfrac{fq}{f-q}$
53b. $\dfrac{RI+2r}{2}$
55a. $\dfrac{x+6}{2} = \dfrac{3x+26}{8}$ or
$\dfrac{5}{2} = \dfrac{5x+10}{8}$ or
$\dfrac{x+6}{5} = \dfrac{3x+26}{5x+10}$
55b. 2
55c. 2, 5, 8; 8, 20, 32
57a. s wave: 3 km/sec;
     p wave: 5 km/sec
57b. 18.75 km

PROBLEM SET 5.6 (on page 297)

1. 7 and 28
3. 7
5. 2 and 6
7. $1110
9. 1875 women
11. 1
13. 15 hours
15. 15 hours
17. 15/4 hours
19. $22.50 midsize,
    $32.50 mini vans
21. $500
23. 2 assistants
25. 9 hits
27. bus: 60 mph
    plane: 150 mph
29. 25/6 mph and
    25/3 mph
31. 5 ft/sec or 3.4
    mph
33. 6 hours
35. 5 hours

CHAPTER 5 REVIEW PROBLEM SET (on page 299)

1a. $\dfrac{1}{3(m-n)}$

1b. $\dfrac{2y+3}{y}$

3. $\dfrac{x-4}{x+6}$

5. $\dfrac{x-2}{x}$

7. $3a^2 + 3ab$

9. $x \neq 3, -2$, $f(x) = \dfrac{x+3}{x-3}$

11. $x \neq 2, -5$, $f(x) = \dfrac{x+2}{x+5}$

13. $\dfrac{4t^2-2}{(2t-1)(2t+1)}$

15. $\dfrac{x^2+3x}{4x^3+8x^2-9x-18}$

17. $\dfrac{3m+1}{2m^3+3m^2-2m-3}$

19. $\dfrac{2(x-4)^2}{x(x^2-8)}$

21. 1

23. $\dfrac{m+1}{m-3}$

25. $\dfrac{(x-2)(x+2)}{(x-4)(x+1)}$

27. y

29. $-\dfrac{1}{x+y}$

31. −15

33. 11/2

35. 7

37. $\dfrac{bc}{b-c}$

39. 2

41a. $\dfrac{2x^2+8}{x^2-4}$

41b. $\dfrac{8x}{x^2-4}$

41c. 1

41d.

$\dfrac{x^2+4x+4}{x^2-4x+4}$

43. 2

45. b

47. a

49a. $21 thousand

49b. $13 thousand

51a. $1,246.67

51b. $1,668.26

53. 6 mph

PRACTICE TEST CHAPTER 5 (on page 301)

1a. $\dfrac{3x+1}{a-b}$

1b. $\dfrac{2x-3}{3x+2}$

2a. $\dfrac{2(x+2)}{3}$

2b. 1

2c. $\dfrac{7x+3}{4(x-1)(x+1)}$

2d. $\dfrac{b^2+2ab-a^2}{b(a+b)(a-b)}$

3a. $\dfrac{3(x-1)}{3x-7}$

3b. $\dfrac{y}{y+x}$

4a. $\dfrac{2x^2+32}{(x-4)(x+4)}$

4b. $\dfrac{-16x}{(x-4)(x+4)}$

4c. 1

4d. $\dfrac{(x-4)^2}{(x+4)^2}$

5a. x ≠ 1, b

5b. x ≠ 0, a

6. 6 hours

7. 68 mph

— CHAPTER 6 —

PROBLEM SET 6.1 (on page 309)

1a. 1

1b. 1

1c. 1

3a. 1

3b. 1

3c. 7

5a. 1/25

5b. 1/16

5c. 1/49

7a. 9/16

7b. 9

7c. x

9a. 25

9b. x^4

9c. y^3

11a. 9/8

11b. $\dfrac{y^2}{x^4}$

11c. $\dfrac{b^5}{a^2}$

13a. 49

13b. $\dfrac{1}{5^5}=\dfrac{1}{3125}$

15a. $\dfrac{1}{x^5}$

15b. $\dfrac{1}{y^9}$

17a. $\dfrac{1}{256}$

17b. p^8

19a. $\dfrac{25}{p^8}$

19b. $\dfrac{1}{p^5 q^{10}}$

21a. $\dfrac{q^4}{p^6}$

21b. $\dfrac{x^2}{y^2}$

23. 1/9

25. $\dfrac{x^2}{y^{10}}$

27a. x = 2

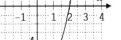
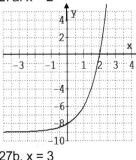
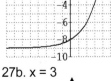

27b. x = 3

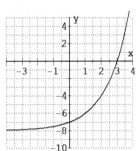

29a. x = −1

29b. x = 1

31a. x = −3

31b. x = −2

33. 1.77×10^{11}

35. 1.67×10^4

37. V = 9.2×10^{-15}cm^3

39. 3×10^2 hrs

41. S = $10.08 per hr

43. S = $2,254.54

45a. $1,268.24

45b. $5,636.36

47a.

x	f(x) = 3^x
−2	1/9
−1	1/3
0	1
1	3
2	9
3	27

47b.

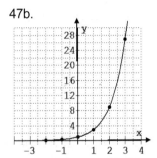

47c. Domain = Reals
Range = (0,∞)

49a.

x	h(x) = 4^x
−2	1/16
−1	¼
0	1
1	4
2	16
3	64

49b.

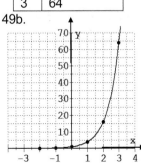

49c. Domain = Reals
　　　Range = $(0,\infty)$
51. Domain = Reals
　　　Range = $(0,\infty)$
　　　graph c
53. Domain = Reals
　　　Range = $(0,\infty)$
　　　graph a
55a. x = 4

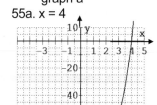

55b. x = 5

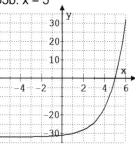

57a. x = –1

57b. x = 1

59a. $a + b$
59b. x^6

61a. $\dfrac{x^3 y^6}{27}$

61b. $\dfrac{u^{28} z^4}{v^{20}}$

63a. $x^{-2} y^{-4}$
63b. $5(2 + a)^{-5}$

PROBLEM SET 6.2 (on page 318)

1a. 5
1b. 4
3a. 4
3b. –2
5a. 0.2
5b. –0.6
7a. 3
7b. not real, the square
　　of any real number
　　is nonnegative.
9a. –2/3
9b. not real, the square
　　of any real number
　　is non negative.
11a. –1/2
11b. 3/4
13a. –11
13b. 11
15a. –3
15b 3
17a. –2x
17b. 6y
19a. $2x^3 y^2$
19b. $2xy^2$
21a. 5x
21b. –3x
23a. y^7
23b. –3x
25a. –2x
25b. not real
27a. 2|x|
27b. –5|x|
29a. 3|x|
29b. y^2

31a. x
31b. |x – 3|
33. x ≥ 7/3
35. |x – 5|
37a. x ≥ 3
37b. x ≥ –3
39a. 1
39b. 1.73
39c. 2.65
41a. –2.08
41b. –2
41c. –1.71
43a. –5.39
43b. –2
43c. –5.39
45a.

x	f(x) = $\sqrt{x}$ + 2
0	2
1	3
4	4
9	5
16	6

45b.

45c. Domain = $[0,\infty)$,

Range = $[2,\infty)$
47. 5%
49a. 3,000
49b. 4,300
51a. 1.7 sec
51b. 10.5 sec
53. 7.78 km/sec
55a. n odd
55b. n even
57a. $\sqrt{2}$
57b. 1
57c. 0
59a. 1
59b. 0
59c. not real
61a. Domain = $[4,\infty)$

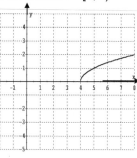

61b. Domain: $(-\infty,4]$

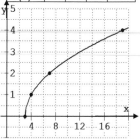

63a.

x	3	4	7	19
f(x)	0	1	2	4

63c. Domain = $[3,\infty)$
　　　Range = $[0,\infty)$
63d. x = 7
65a. 8 times larger
65b. 27 times larger
65c. 64 times larger
65d. The number is
increased by the cube
of the factor.

PROBLEM SET 6.3 (on page 326)

1a. 6
1b. 1/7
1c. 3
3a. −3
3b. 3
3c. −2
5a. −3/2
5b. 5/9
5c. 3/4
7a. 4
7b. −16
9a. −8
9b. 4
11a. not real
11b. not real
13a. 1/16
13b. 1/9
15a. $\sqrt[3]{5}$
15b. $\dfrac{1}{\sqrt[4]{6}}$
17a. $\sqrt[7]{x^2}$
17b. $\sqrt[4]{y^3}$
19a. $\sqrt[3]{(8m)^2}$
19b. $\sqrt[5]{\left(4xy^2\right)^2}$
21a. $7^{1/2}$
21b. $3^{1/2}$

23a. $m^{1/3}$
23b. $x^{3/5}$
25a. $(7x^2)^{2/5}$
25b. $(3xy)^{2/4}$
27a. 2
27b. x
29a. 25
29b. $\dfrac{1}{x^5}$
31a. $16p^{12}$
31b. $\dfrac{y}{5}$
33a. x
33b. $x^{1/2}y^{1/3}$
35. $\dfrac{1}{xy^6}$
37. $\dfrac{2x^3}{5y^2}$
39a. $x^2 + 1$
39b. $x^{1/2} + 1$
41a. f(−27) = 9,
f(−8) = 4, f(0) = 0,
f(1) = 1, f(27) = 9

41b.
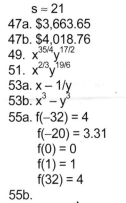

41c. Domain = Reals
Range = $[0,\infty)$

43a. h(−27) = 1/9,
h(−1/8) = 4,
h(0) = undefined,
h(1) = 1, h(27) = 1/9

43b.
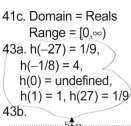

43c. D: reals $x \neq 0$
R: $(0,\infty)$

45. $s \approx 11.5$

$s \approx 21$
47a. $3,663.65
47b. $4,018.76
49. $x^{35/4}y^{17/2}$
51. $x^{2/3}y^{19/6}$
53a. $x − 1/y$
53b. $x^3 − y^3$
55a. f(−32) = 4
f(−20) = 3.31
f(0) = 0
f(1) = 1
f(32) = 4

55b.
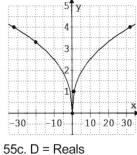

55c. D = Reals
R = $[0,\infty)$
57a. D of g = $[0,\infty)$
R of g = $[0,\infty)$
57a. D of h = $[0,\infty)$
R of h = $[−8,\infty)$
57b. x = 4

PROBLEM SET 6.4 (on page 333)

1a. $3\sqrt{3}$
1b. $-3\sqrt[3]{2}$
3a. $\dfrac{\sqrt{5}}{2}$
3b. $-\dfrac{1}{2}\sqrt[3]{5}$
5a. $4\sqrt{3x}$
5b. $2y\sqrt[3]{3}$
7a. $4c^2\sqrt{2c}$
7b. $-2p^2\sqrt[3]{4}$

9a. $\dfrac{\sqrt{7}}{2x}$
9b. $-2p\sqrt[5]{p}$
11a. $\dfrac{\sqrt{3}}{5x}$
11b. $\dfrac{\sqrt{5}}{3x^2}$
13a. 2a
13b. $x\sqrt{2}$
15a. $x^2\sqrt[3]{5}$
15b. x^2
17. $5ab^2c^3$

19. $3xy^3z^4$
21a. $2\sqrt[3]{4}$
21b. x
23a. $\sqrt[6]{432}$
23b. $\sqrt[6]{5488}$
25a. $\sqrt[6]{x^5}$
25b. $\sqrt[4]{x^3}$
27a. $\sqrt[6]{(5x)^5}$
27b. $2\sqrt[6]{2x^2y^3}$
29a. $2yz\sqrt[5]{2x^3y}$

29b. $18,664,000
31a. 1.626577
31b. 1.626577
31c. $\sqrt[4]{7} = \sqrt{\sqrt{7}}$
33a. 1.732, 1.316
1.147, 1.071
33b. seems to
approach 1
35. $x^2\sqrt[8]{x^3y}$
37. $\dfrac{54x^2\sqrt{x}}{5}$
39. $\dfrac{5}{2}\sqrt[3]{p^2q}$

PROBLEM SET 6.5 (on page 340)

1a. $8\sqrt{7}$
1b. $14\sqrt{11}$
3a. $8\sqrt{5}$
3b. $7\sqrt[5]{2}$
5. $4\sqrt{x} + 5\sqrt{y}$

7. $5\sqrt{2}$
9. $-10\sqrt{2}$
11. $18\sqrt{5}$
13. $18\sqrt[3]{3}$
15. $3\sqrt{3x} - \sqrt{3y}$

17a. $\sqrt{6} - 2$
17b. $2\sqrt{3} - 15$
19a. $\sqrt{3x} - 3\sqrt{2}$
19b. $4\sqrt{n} - 3n$
21a. $2\sqrt{33} - 44$

21b. $7 + \sqrt{5}$

23a. $34 + 3\sqrt{6}$

23b. $7 - 4\sqrt{3}$

25a. $4x - 12\sqrt{x} + 9$

25b. -4

27a. 42

27b. $9x - 121$

29a. $\dfrac{2\sqrt{3}}{3}$

29b. $\dfrac{3\sqrt{21}}{7}$

31a. $\sqrt{2}$

31b. $\dfrac{\sqrt{10}}{4}$

33a. $\dfrac{8\sqrt{11}}{77}$

33b. $\dfrac{2\sqrt{5x}}{3x}$

35. $\dfrac{5\sqrt[3]{3}}{3}$

37. $3(\sqrt{5} + 2)$

39. $7(\sqrt{10} - 3)$

41. $\sqrt{5} - \sqrt{2}$

43. $\dfrac{8(2\sqrt{7} + \sqrt{5})}{23}$

45. $\dfrac{\sqrt{35} - 2\sqrt{7}}{7} \approx 0.0892$

47. $\dfrac{\sqrt{21} + 3}{6} \approx 1.2638$

49a. $\dfrac{4 + 3\sqrt{3}}{6}$

49b. $\dfrac{\sqrt{x}(3 + x)}{x}$

51a. 54 in^2

51b. $13\sqrt{6}$ in

51c. 54 in^2, 31.84 in

53a. 34.55

53b. 34.55

53c. $(\sqrt{63} + \sqrt{2})(\sqrt{63} - 3\sqrt{2})$

$\quad = 63 - 3\sqrt{126} + \sqrt{126} - 6$

$\quad = 57 - 2\sqrt{126}$

$\quad = 57 - 6\sqrt{14}$

55a. $\dfrac{17\sqrt{3}}{6} \approx 4.91$

55b. $\dfrac{38\sqrt{7}}{7} \approx 14.36$

55c. $\dfrac{5\sqrt{15}}{3} \approx 6.45$

57a. $1.41, 2.73, 3.41, 4.45$

57b. (b)

57c. Domain $= [1, \infty)$

$\quad$ Range $= [\sqrt{2}, \infty)$

59a. $8.12, 3.73, 2.41, 1$

59b. d

59c. Domain $= (-\infty, 0]$

$\quad$ Range $= [1, \infty)$

61a. $\sqrt{m}\,(m - 2m^3 + 3m^4)$

61b.

$2y^2 \sqrt[5]{xy^3} - 8y\sqrt[5]{x^2 y} + x^2\sqrt[5]{xy^3}$

63a. $\dfrac{5x\sqrt{2x}}{y}$

63b. $28x\sqrt[3]{x}$

65a.

$\dfrac{2}{\sqrt{x+2} - \sqrt{x}} \cdot \dfrac{\sqrt{x+2} + \sqrt{x}}{\sqrt{x+2} + \sqrt{x}}$

$= \dfrac{2(\sqrt{x+2} + \sqrt{x})}{2}$

$= \sqrt{x+2} + \sqrt{x}$

65b. $1.41, 3.41, 4.45, 5.99,$
$\quad 6.63$

67. $\dfrac{3}{\sqrt{10} + 2}$

69. $15 + \sqrt{2} + \sqrt{3} +$
$\quad \sqrt{4} + \ldots + \sqrt{15} \approx 54.47$ in.

PROBLEM SET 6.6 (on page 347)

1a. 9

1b. 10

3a. 25

3b. -4

5. 25

7. 3

9. $11/2$

11. 5

13. no solution

15. 8

17. $7/2$

19. 0

21. -10

23. 7

25. 1

27. 1

29. ± 8

31. -243

33. ± 64

35. ± 128

37. $1/81$

39. $2, 4$

41. $-7, 9$

43a. 301.72 in^2

43b. $14,657,415$ mi^2

45a. 18.0 ft

45b. $\left(\sqrt[4]{x - 2}\right)^2 = 10^2 + 15^2$

$\quad x = 105,627$

47. g earth $= 32.08$ ft/sec^2

$\quad$ g moon $= 5.35$ ft/sec^2

49. 2144 lbs.

51. 41

53. 8

55. $-2/3$

57a. 16

57b. -16

57c. graph b

59a. 0

59b. 0

59c. graph d

PROBLEM SET 6.7 (on page 355)

1a. $4i$

1b. $9i$

3a. $-2i\sqrt{2}$

3b. $6i\sqrt{2}$

5a. $-3i/4$

5b. $-5i/2$

7a. 2

7b. 4

9. $8\sqrt{2}\,i$

11. -10

13. $-2(\sqrt{5} + \sqrt{6})$

15. $x = 5, y = 4$

17a. $-1 + 9i$

17b. $13 - 17i$

19a. $1 - i$

19b. $0 + 4i$

21. $0 + 6i$

23. $3 - 6i$

25. $28 + 14i$

27. $14 - 8i$

29. $21 - i$

31. $27 - 5i$

33. $45 + 28i$

35. $41 + 0i$

37. $169 + 0i$

39a. $\dfrac{4}{25} - \dfrac{3}{25}i$

39b. $-1 + 0i$

41a. $\dfrac{35}{41} + \dfrac{28}{41}i$

41b. $-\dfrac{3}{26} + \dfrac{37}{26}i$

43a. $\dfrac{7}{13} - \dfrac{4}{13}i$

43b. $\dfrac{3}{5} - \dfrac{11}{5}i$

45a. i
45b. −i
45c. −1
47a. 1
47b. i
47c. i
49. 4 − 2i
51. 0 + 0i

53. −3 − 7i
55. 0 + 0i
57. $\dfrac{10}{841} + \dfrac{21}{1682}i$

59a.
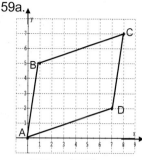

59b. $\overline{AC} = \sqrt{113}$

59c. 8 + 7i

CHAPTER 6 REVIEW PROBLEM SET (on page 356)

1a. 3
1b. 40/9
1c. 1/500
1d. $\dfrac{y^6}{x^4}$
3a. 81
3b. 1/8
3c. b^5
5a. $x^{13/20}$
5b. $\dfrac{1}{x^3}$
5c. $\dfrac{x^9 y^{12}}{8}$
7a. $\dfrac{y^5}{x}$
7b. $\dfrac{a^7}{b^6}$
7c. $\dfrac{4}{x^8 y^4}$
9a. −2
9b. 1
11a. 9
11b. −4
11c. not real, even
  index and
  negative radicand
13a. 1/8
13b. 2
13c. 1/7776
15a. $(-7)^{1/3}$
15b. $x^{4/5}$
17a. $\dfrac{1}{\sqrt[3]{7^2}}$

17b. $\sqrt[5]{(x+1)^2}$
19a. $5\sqrt{5}$
19b. −5
19c. $2t\sqrt[4]{2t}$
21a. 2/5
21b. $-\dfrac{\sqrt[3]{7}}{3}$
21c. $\dfrac{\sqrt[5]{3}}{2}$
23a. t/2
23b. $2x^2/3$
23c. $-\dfrac{\sqrt[3]{5}}{x^3}$
25a. $5x^3$
25b. $2x^2$
27a. $4x^2 y^3$
27b. 2x
29a. $\dfrac{y^6}{8x^3}$
29b. $\dfrac{8}{x^3 y^6}$
31a. $8\sqrt{2}$
31b. $10\sqrt{2}$
33a. $5\sqrt{7x}$
33b. $4\sqrt[3]{2y}$
35a. $12 + 4\sqrt{5}$
35b. $4x - 12\sqrt{xy} + 9y$
37a. 6
37b. $a - b^2$
39a. $2x - \sqrt{xy} - y$

39b. $3x + 5y\sqrt{x} - 2y^2$
41a. $\dfrac{2\sqrt{7}}{7} \approx 0.756$
41b. $\dfrac{3\sqrt{2}}{10} \approx 0.424$
43a. $2 - \sqrt{3} \approx 0.268$
43b. $-\sqrt{14} - \sqrt{21}$
  ≈ -8.324
45. $\dfrac{x + 2\sqrt{x} + 1}{x - 1}$
47a. $-\dfrac{2}{\sqrt{5} - 3}$
47b. $\dfrac{4}{\sqrt{21} - 3}$
49a. $-8\sqrt{3}$
49b. $-3xy^2$
51a. 64
51b. 6
53a. 11
53b. 1, −3
55a. 0 + 2i
55b. −12 + 0i
57a. 35 + 5i
57b. 5 + 0i
59a. $\dfrac{14}{17} + \dfrac{5}{17}i$
59b. $\dfrac{5}{13} + \dfrac{1}{13}i$
61a. 4
61b. 1
61c. 0.12
61d. 5.28
63a. 2.65
63b. 1

63c. 2.86
63d. 0
65a. 1/4
65b. undefined
65c. 0.47
65d. 0.38
67. $1235.02
69. 5.92 ft
71a.
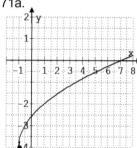
71b. D = [−1,∞)
  R = [−4,∞)
71c. x = 7
73a.
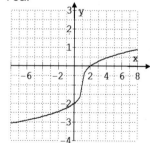
73b. D = reals
  R = reals
73c. x = 2

CHAPTER 6 PRACTICE TEST (on page 359)

1a. 1
1b. 49/9

1c. 81

1d. $\dfrac{1}{x^4}$

1e. $\dfrac{1}{xy^5}$

2a. 4

2b. −1

2c. 3/2

3a. 2

3b. not real, negative radicand, even index

3c. −5

3d. −3

4a. −3

4b. 5x

4c. 3x

4d. not real

5. $|x - 9|$

6a. $f(0) = -2$

6b. $f(-19) = -3$

7a. 1/2

7b. 8

7c. 9

7d. 1/16

8a. $x^{2/7}$

8b. $\sqrt[11]{x^3}$

9a. 1/5

9b. 9

9c. 2

10a. $f(-8) = 4$

10b. $f(0) = 0$

10c. $f(27) = 9$

11a. $3\sqrt[3]{2}$

11b. $\dfrac{\sqrt{3}}{5x}$

11c. $-3x\sqrt[3]{3x}$

11d. $x\sqrt{15}$

12. $\sqrt[6]{432}$

13a. $6\sqrt{3}$

13b. $3\sqrt{5} + 3\sqrt{2}$

13c. $19 + 4\sqrt{21}$

13d. 5

14a. $\dfrac{\sqrt{21}}{3}$

14b. $\sqrt{3} - 1$

15a. ± 8

15b. 4

15c. 10

15d. $\pm \dfrac{5}{2}$

16a. $3 + 3i$

16b. $-5 + 2i$

16c. $12 - i$

16d. $\dfrac{1}{13} + \dfrac{5}{13} i$

16e. 1

17. $6.38

— CHAPTER 7 —

PROBLEM SET 7.1   (on page 367)

1a. $1, -\dfrac{2}{3}$

1b. 0, 5

1c. 6, −1

3a. 6, −4

3b. −7

3c. $\dfrac{4}{3}, -\dfrac{2}{5}$

5a. ± 8

5b. $\pm \dfrac{2}{3}$

5c. $\pm 3i$

5d. $\pm 3i$

7a. $2 \pm 2i$

7b. $\dfrac{2 \pm 7i}{3}$

9. 6, −12

11a. 9

11b. 16

13a. $\dfrac{49}{4}$

13b. $\dfrac{121}{4}$

15a. 1, −3

15b. $-2 \pm \sqrt{11} \approx 1.32, -5.32$

17a. 6, −1

17b. 1, −4

19. $2 \pm \sqrt{5} \approx 4.24, -0.24$

21. $\dfrac{2 \pm \sqrt{7}}{2} \approx 2.32, -0.32$

23. $\dfrac{4 \pm \sqrt{101}}{5} \approx 2.81, -1.21$

25. $\dfrac{-7 \pm \sqrt{129}}{8} \approx 0.54, -2.29$

27. $\dfrac{-5 \pm \sqrt{53}}{14} \approx -0.88, 0.16$

29. $\dfrac{3 \pm \sqrt{14}}{4} \approx -0.19, 1.69$

31a. a. −5, −6

31a. b. $(x - 1)^2 - 6$

31a. c. (1,−6) lowest point

31a. d. $1 \pm \sqrt{6} \approx -1.45, 3.45$

31b. a. 7, −2

31b. b. $f(x) = (x - 3)^2 - 2$

31b. c. lowest point = (3,−2)

31b. d. $3 \pm \sqrt{2} \approx 4.41, 1.59$

33a. 5, −2

33b. $2, -\dfrac{5}{6}$

35a. $0.0375(x - 20)^2 + 15$

35b. (20,15) lowest point

37a. 25

37b. ± 30

39a. $\sqrt{A}$

39b. $\dfrac{\sqrt{S\pi}}{2\pi}$

41a. $\sqrt{A} - 1$

41b. $\dfrac{\sqrt{2hg}}{g}$

43. $\dfrac{v \pm \sqrt{v^2 - 4gs}}{2g}$

45. $\dfrac{3 \pm \sqrt{6}\, i}{3}$

47. $\dfrac{-5 \pm i}{3}$

49. $\dfrac{-2 \pm \sqrt{3}\, i}{2}$

51a. 0

51b. 0

53a. −3

53b. 1/3

55a. $f(x) = (x - 3)^2 - 1$

55b. lowest point (3,−1)

55c. 2, 4

55d. c

57a. $4(x - 5/2)^2 - 5$

57b. lowest point (5/2,−5)

57c. $\dfrac{5 \pm \sqrt{5}}{2} \approx 1.38, 3.62$

57d. d

PROBLEM SET 7.2   (on page 375)

1a. 1, 4

1b. 3, −1

3a. −2, −2/3

3b. 5, 4/3

5. 3/2, ½

7. 5/3, −1/2

9. $\dfrac{7 \pm \sqrt{97}}{12} \approx 1.40, -0.24$

11. $\dfrac{1 \pm \sqrt{34}\,i}{5}$

13. $\dfrac{1 \pm 2\sqrt{2}\,i}{6}$

15. $\dfrac{1 \pm \sqrt{83}\,i}{6}$

17. $-8 \pm 8\sqrt{2} \approx 3.31, -19.31$

19. $3 \pm \sqrt{13} \approx 6.61, -0.61$

21. two rational

23. two complex

25. one rational

27. two irrational

29a. 8.2 and 3.8 sec.

29b. 6 sec

29c. 580 ft highest, doesn't reach 600 feet.

31. 1.06 and 0.19 sec

33. 640 phones

35a. (a) $a = 1$, b = –9, c = 8

35a. (b) 8, 1

35a. (c) $x^2 - 9x + 8 = 0$

35b. (a) $a = 1$, b = –4, c = 3

35b. (b) 1, 3

35b. (c) $x^2 - 4x + 3 = 0$

37. 6.66, –3.42

39. 1.81, –3.91

41. $\dfrac{-mn \pm m\sqrt{n^2 + 4n}}{2n}$

43. $\dfrac{-RC \pm \sqrt{R^2C^2 - 4LC}}{2LC}$

45a. k = –9/8

45b. k > –9/8

45c. k < –9/8

47a. k = 4

47b. k < 4

47c. k > 4

49a. $x_1 + x_2 = 5/3$, $-\dfrac{b}{a} = -\dfrac{5}{3}$

49b. $x_1 \cdot x_2 = -2/3$, $\dfrac{c}{a} = \dfrac{-2}{3}$

51a. pick any non zero value for a, then b = $3a$, c = $-18a$.

51b. pick any non zero value for a, then b = $5a$, c = $4a$.

53a. pick any non zero value for a, then b = 0, c = $-5a$.

53b. pick any non zero value for a, then b = 0, c = $4a$.

55a. $3 \pm \sqrt{17} \approx 7.12, -1.12$

55b. $f(x) = (x - 3)^2 - 17$, (3,–17)

55c. b

57a. $\dfrac{5 \pm \sqrt{41}}{2} \approx -0.70, 5.70$

57b. $f(x) = (x - 2.5)^2 - 10.25$, (2.5,–10.25)

57c. a

59a. 3.5, 0.5

59b. 3.41, 0.59

61a –0.5, 0.7

61b. –0.43, 0.77

PROBLEM SET 7.3 (on page 381)

1. 3/7, 4

3. 1/4, 1/2

5. –5/2, 1

7. 9

9. 4

11. 4

13. 0, 3

15a. $\pm 2, \pm \sqrt{3}$

15b. 1/4, 1/3

15c. 9, 16

17a. $\pm 1, \pm 4$

17b. 1, 4096

17c. 256, 1

19a. ± 1

19b. 1

19c. 4

21. 2, 3

23. $\pm 1, -3$

25. 4, 36

27. 1/3, –2, 2/3, –1

29. 15 years

31. 16

33. $-9, 3, -3 \pm 2\sqrt{3}$

35. 1/2, 5/4

37. –1/4

39. 1

41. 16

43. $\pm 2, \pm 5$

45. 1, 5

47. 27, 125

49. 1

51a. (–3,0), (3,0)

51b. ± 3

53a. (–3,0), (3,0)

53b. ± 3

55a. (2,0)

55b. 2

PROBLEM SET 7.4 (on page 389)

1a. 23, 24

1b. –18, –19

3. 20 in by 37 in

5. 3 ft

7. 5 ft

9. 8 ft

11. 6% and 8%

13. 20, 30 mph

15. 12, 13 years

17. 13th and 29th

19a. 18.2 in

19b. 1.52 ft

21. 6, 8, 10

23. 2 ft

25a. $\dfrac{1 + \sqrt{5}}{2}$

25b. 3.1 in

27. 10%

29. 24 min

31. 300, 400 mph

33. 9 mph

35. 15 kph

37. 6, 12 hrs

39. 78.8 in

41. 7.16 hrs

43. 350 rackets

45. 13.1, 8.1 hrs

47. $1.20 per gal

PROBLEM SET 7.5 (on page 399)

1. (–3,4)

−3   4

3. [–5,2]

−5   2

5. (−∞,−3] or [1,∞)

−3   1

7. (–1,2)

−1   2

9. (−∞,1] or [2,∞)

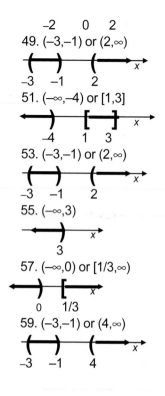

$$11. \left(-\infty, -\frac{\sqrt{15}}{3}\right] \text{ or } \left[\frac{\sqrt{15}}{3}, \infty\right)$$

13. $(-\infty, \infty)$

15. $[3/4, 3/4]$

17. $(-\infty, -3]$ or $[1, \infty)$

19. $(-1, 2)$

21. $(-\infty, 1]$ or $[2, \infty)$

23. $(3, \infty)$

25. $(-4, 1]$

27. $[-1, 2)$

29. $(-\infty, -2)$ or $(1/2, \infty)$

31. $(-\infty, -4)$ or $[0, \infty)$

33. $(3, \infty)$

35. $(-4, 1]$

37. $[-1, 2)$

39. $(-\infty, -3)$ or $(0, 3)$

41. $[-3, 0]$ or $[3, \infty)$

43a. $(0, 180)$

43b. $(180, \infty)$

43c. $[80, 100]$

45a. $0.25 < t < 1.75$ sec

45b. $2.75 < t < 3.45$ sec

47. $(-\infty, -2]$ or $[0, 2]$

49. $(-3, -1)$ or $(2, \infty)$

51. $(-\infty, -4)$ or $[1, 3]$

53. $(-3, -1)$ or $(2, \infty)$

55. $(-\infty, 3)$

57. $(-\infty, 0)$ or $[1/3, \infty)$

59. $(-3, -1)$ or $(4, \infty)$

PROBLEM SET 7.6 (on page 410)

1.
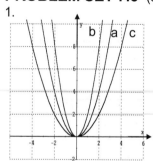

3.
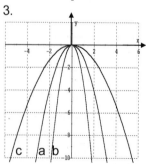

5.
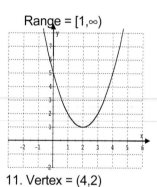

7.
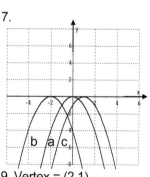

9. Vertex = (2,1)
   Axis x = 2
   Domain = Reals

Range = $[1, \infty)$
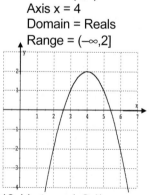

11. Vertex = (4,2)
    Axis x = 4
    Domain = Reals
    Range = $(-\infty, 2]$

13. Vertex = (−2,1)

Axis x = –2
Domain = Reals
Range = (–∞,1]

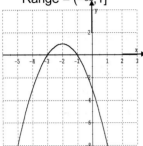

15. Vertex = (–3,2)
   Axis x = –3
   Domain = Reals
   Range = [2,∞)

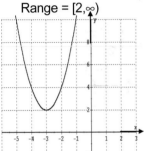

17. Vertex = (1,1)
   Axis x = 1
   Domain = Reals
   Range = (–∞,1]

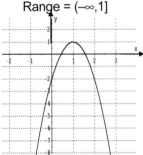

19. Vertex = (–3,1)
   Axis x = –3
   Domain = Reals
   Range = (–∞,1]

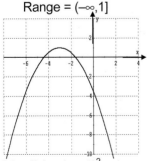

21a. $f(x) = (x – 1)^2 – 1$
   Vertex = (1,–1)
   Axis x = 1

Range = [–1,∞)

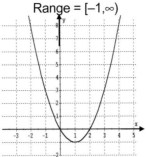

21b. $f(x) = –(x – 1)^2 + 1$
   Vertex = (1,1)
   Axis x = 1
   Range = (–∞,1]

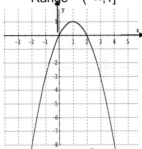

23. $f(x) = (x – 3)^2 + 1$
   Vertex = (3,1)
   Axis x = 3
   Range = [1,∞)

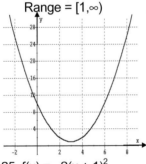

25. $f(x) = –2(x + 1)^2$
   Vertex = (–1,0)
   Axis x = –1
   Range = (–∞,0]

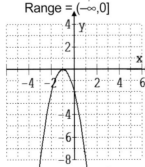

27. $f(x) = 2(x + 1.5)^2 + 4.5$
   Vertex = (–3/2,9/2)
   Axis x = –3/2
   Range = [9/2,∞)

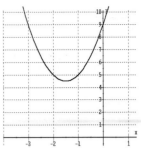

29a. x int = ± 2, y int = 4
   Vertex = (0,4)
   Axis x = 0
   Range = (–∞,4]

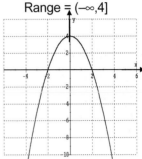

29b. x int = ± 2, y int = –4
   Vertex = (0,–4)
   Axis x = 0
   Range = [–4,∞)

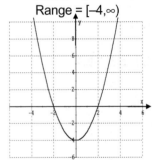

31. x int = 5, y int = 25
   Vertex = (5,0)
   Axis x = 5
   Range = [0,∞)

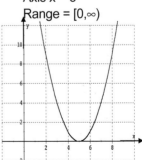

33. x int = –1,3; y int = –3
   Vertex = (1,–4)
   Axis x = 1
   Range = [–4,∞)

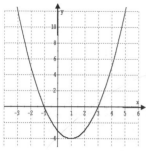

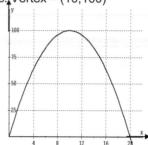

43a. 0, $91.13, $118.13, $121.13

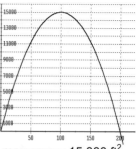

35. Vertex = (7,5) min = 5
37. Vertex = (–6,1) max = 1
39. Vertex = (1,2) min = 2
41a. d(5) =560 ft, d(10) = 320 ft,

    d(15) is less than 0 so the rocket is on the ground
41b. $-16(t-6)^2 + 576$
41c. Vertex = (6,576) max = 576 ft

41d. 6 sec
41e. 12 sec

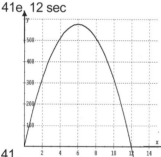

41.

43b. $0.80 or $1.00
43c. $R(p) = -150(p-0.9)^2 + 121.50$
43d. Vertex = (0.9,121.5)
    max $121.50
45a. P = 40 = 2x + 2w
    width = 20 – x
    Area = length X width
45b. 0 < x < 20
45c. Vertex = (10,100)

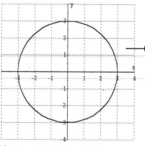

45d. 10 m by 10 m
45e. 100 m²
47a. P = 600 = 3x + 2 l

    length $\dfrac{600 - 3x}{2}$

47b. 0 < x < 200
47c. Vertex = (100, 15,000)

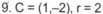

47d. max area = 15,000 ft²
47e. w = 100 ft   l = 150 ft
49a. $f(x) = -(x-2)^2 + 4$
49b. $f(x) = x^2 - 4$
51. $F(x) = 2(x-3)^2 + 2$
53. $f(x) = -16x^2 + 80x + 20$

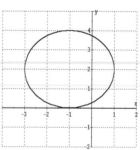

55a. 10 and 10
55b. –3 and 3

PROBLEM SET 7.7 (on page 418)

1a. 10
1b. $\sqrt{13} \approx 3.61$
3a. 5
3b. $\sqrt{241} \approx 15.52$
5a. $\sqrt{103.61} \approx 10.18$
5b. $\sqrt{165.64} \approx 12.87$
7a. C = (0,0) r = 1

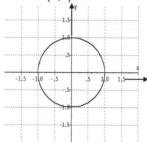

7b. C = (0,0) r = 3

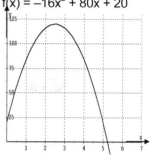

9. C = (1,–2), r = 2

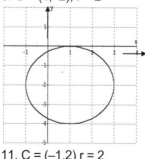

11. C = (–1,2) r = 2

13. C = (2,–4) r = 4

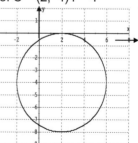

15. C = (3,–4) r = 5

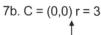

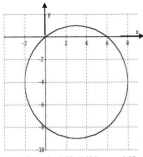

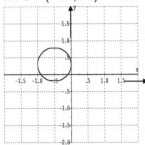

17. $C = (-1/2, 1/3)$ $r = 1/2$

19a. $x^2 + y^2 = 16$
19b. $x^2 + y^2 = 9$
21a. $(x + 4)^2 + (y - 3)^2 = 48$
21b. $(x + 5)^2 + (y + 2)^2 = 20$
23. $(x - 2)^2 + (y - 3)^2 = 9$
   $C = (2,3)$ $r = 3$
25. $(x - 3)^2 + (y + 2)^2 = 9$
   $C = (3,-2)$ $r = 3$
27. $(x - 3)^2 + (y + 4)^2 = 16$
   $C = (3,-4)$ $r = 4$
29. $|\overline{AB}| = 3\sqrt{13}$ yes
   $|\overline{AC}| = 6$, $|\overline{BC}| = 9$

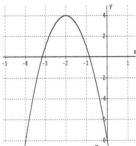

31. $(x + 5)^2 + (y - 12)^2 = 25$
33a. $\sqrt{x^2 + y^2} = 25$
33b. 24 m
35. $(x - 3)^2 + (y - 6)^2 = 36$
37. $(x + 2)^2 + (y - 5)^2 = 16$
39a. $(x + 2)^2 + (y - 1)^2 = 25$
39b. $(x - 2)^2 + (y + 3)^2 = 16$

CHAPTER 7 REVIEW PROBLEM SET (on page 419)

1a. -1, 15
1b. 0, 5/2
1c. -2, 12/5
3a. $1 \pm 3i$
3b. $\dfrac{1 \pm 5i}{3}$
5a. 16
5b. 121/4
7a. $3 \pm \sqrt{2} \approx 1.59$, 4.41
7b. $3/2 \pm i$
9a. $-5 \pm \sqrt{38} \approx -11.16$, 1.16
9b. $\dfrac{-3 \pm \sqrt{65}}{4} \approx -2.77$, 1.27
11a. $x^2 + 7x + 12 = 0$
11b. $x^2 - 10x + 18 = 0$
11c. $x^2 - 6x + 25 = 0$
13. 1/2, 1/4
15. 4
17. 4
19a. ± 1, ± 2
19b. 1, ¼
21a. $-3 \pm i$, $-3 \pm \sqrt{3}/3$
21b. $\pm \sqrt{3}$, $\pm i$
23. 1, 2, -2, -3
25. 1, -3, $-1 \pm \sqrt{2}$
27. 8, 27
29. 1
31a. -1, 2/3
31b. -1, 9
33a. $-3/8$, 0, 5/2
33b. -3, $-2/3$, 3

35. $\dfrac{-9 \pm \sqrt{73}}{2} \approx -8.77$, -0.23
37. $-3 \pm \sqrt{21} \approx -7.58$, 1.58
39. -1, 2
41. 25, 64
43. ± 2, and $\pm \sqrt{3} \approx \pm 1.73$
45. 2.01
47a. $(-5,3)$

(-5 to 3)

47b. $(-1,1)$ or $(2,\infty)$

(-1 to 1, 2 onward)

49a. $[1/3, 7/5)$

(1/3 to 7/5)

49b. $(-\infty, -1]$ or $[2,5)$

(-1, 2 to 5)

51a. $f(x) = -3(x + 2)^2 + 4$
51a. $(-2,4)$
51a. $x = \dfrac{-6 \pm 2\sqrt{3}}{3}$
   ≈ -3.15, -0.84 y int = -8
51a. $(-\infty, 4]$
51a. max 4
51a.

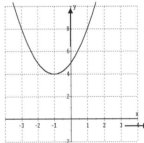

51b. $f(x) = (x + 1)^2 + 4$
51b. $(-1,4)$
51b. x int = none y int = 5
51b. $[4,\infty)$
51b. min 4
51b.

51c. $f(x) = -(x - 2)^2 - 4$
51c. $(2,-4)$
51c. x int = none y int = -8
51c. $(-\infty, -4]$
51c. max -4
51c.

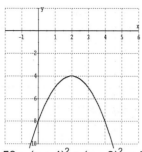

53a. $(x + 4)^2 + (y - 3)^2 = 4$
    $C = (-4,3), r = 2$
53b. $(x - 3)^2 + (y - 2)^2 = 9$
    $C = (3,2), r = 3$
55a. $\sqrt{10} \approx 3.16$
55b. $\sqrt{181} \approx 13.45$
57a. $\dfrac{-V \pm \sqrt{V^2 + 2gs}}{g}$

57b. $\dfrac{-\pi h \pm \sqrt{\pi^2 h^2 + \pi s}}{\pi}$

59. ± 2
61. h = 10.1 ft, b = 5.55 ft
63. 30 mph
65. 9%
67. 10 in by 16 in or
    24 in by 6.67 in
69a. 64 ft
69b. 4 sec
71. $1 \le t \le 3$
73. –0.25, 1
73. $(-\infty, -0.25)$ or $(1, \infty)$

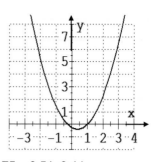

75. –0.51, 3.11
75. [–0.51, 3.11]

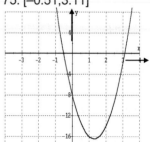

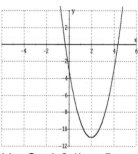

11a. C = (–3,1) r = 5
11b.

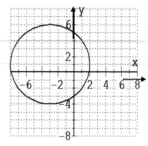

12. 8 ft by 13 ft
13a. 124 ft
13b. 2 sec

CHAPTER 7 PRACTICE TEST (on page 422)

1a. –3, 5
1b. –1,2
1c. –3, –7/2, 0
2a. $-1 \pm 2i$
2b. $2 \pm \sqrt{2} \approx 0.59, 3.41$
2c. $\dfrac{9 \pm 4\sqrt{6}}{3} \approx 6.27, -0.27$
3a. $\dfrac{5 \pm \sqrt{23}i}{6}$
3b. 2/3
3c. $\dfrac{-1 \pm \sqrt{10}}{3} \approx 0.72, -1.39$
4. 0
5. 1, 4096
6. $x^2 - 3x - 18 = 0$
7. $(-\infty, -1)$ or $(1/2, \infty)$

    D = x < –1 or x > 1/2
8. $(-\infty, -3/2]$ or $(5, \infty)$

    D = x ≤ –3/2 or x > 5
9. V = (1,3)
    axis x = 1
    Domain = Reals
    Range = $(-\infty, 3]$

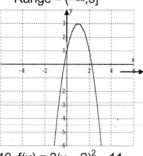

10. $f(x) = 2(x - 2)^2 - 11$
    Range = $[-11, \infty)$

— CHAPTER 8 —

PROBLEM SET 8.1 (on page 431)

1a. 0.086
1b. 1.903
1c. 0.624
1d. 0.011
3a. 3.669
3b. 0.819
3c. 0.107
3d. 29.964
5a. 0.808

5b. 0.798
5c. 1294.789
7. C
9. D
11. B
13. A
15. Domain = (−∞,∞)
    Range = (0,∞)

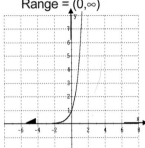

17. Domain = (−∞,∞)
    Range = (0,∞)

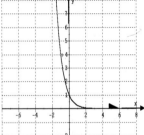

19. Domain = (−∞,∞)
    Range = (0,∞)

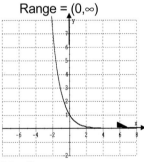

21a. 3
21b. 2
23a. 2
23b. 5/2
25a. 3/2
25b. 5/4
27a. Domain = (−∞,∞)
     Range = (0,∞)

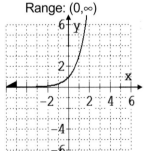

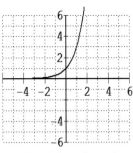

27b. Domain = (−∞,∞)
     Range = (0,∞)

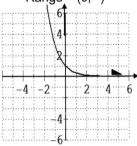

27c. Domain: (−∞,∞)
     Range: (0,∞)

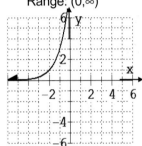

29a. Domain: (−∞,∞)
     Range: (0,∞)

29b. Domain = (−∞,∞)
     Range = (−∞,0)

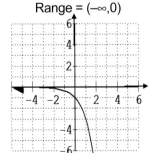

29c. Domain = (−∞,∞)
     Range = (−1,∞)

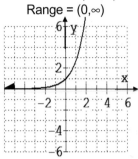

31a. Domain = (−∞,∞)
     Range = (0,∞)

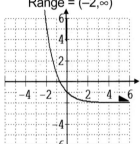

31b. Domain = (−∞,∞)
     Range = (−2,∞)

31c. Domain: (−∞,∞)
     Range: (1,∞)

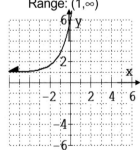

33a. $16,473.43
33b. $16,507.99
33c. $16,525.53
35a. 16.7 million
35b. 17.3 million
37a. $20,844.40
37b. $14,015.77
37c. $11,492.94

2	894	900
3	917	917
4	934	934
5	952	952
6	970	971

39. $422,240

41a. 84%

41b. 77%

43a. 32.0 gm

43b. 25.6 gm

43c. 13.1 gm

45a. 127.0 & 333.6 million

45b. 2011

47a. −5/4

47b. 0

49a. (0,1)

49b. Domain = (−∞,∞)

Range = (0,∞) for both

49c. g is steeper than f

51a. $277,224.54

51b. $144,000

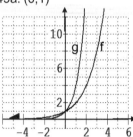

53a.
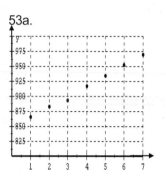

53b.

T	table	$866e^{0.019t}$
0	866	866
1	883	883

55a.
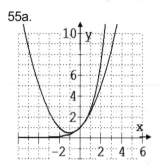

55b. When −1 ≤ x ≤ 1, the graphs are almost the same.

PROBLEM SET 8.2 (on page 442)

1a. $9^2 = 81$

1b. $6^2 = 36$

3a. $(1/3)^{-2} = 9$

3b. $10^{-1} = 1/10$

5a. $5^{\frac{1}{2}} = \sqrt{5}$

5b. $7^{\frac{1}{4}} = \sqrt[4]{7}$

7a. $\log_5 125 = 3$

7b. $\log_4 1/16 = -2$

9a. $\log_9 3 = \frac{1}{2}$

9b. $\log_{1/8} 4 = -2/3$

11a. $\log_5 \sqrt{5} = \dfrac{1}{2}$

11b. $\log_4 \sqrt[3]{4} = \dfrac{1}{3}$

13a. 4

13b. 27

15a. 3

15b. 5

17a. 2

17b. 3

19a. −2

19b. −2

21a. 2

21b. 2

21c. −2/3

23a. 1

23b. 0

23c. 0

25a. −1

25b. 5/2

25c. 1

27a. −3

27b. −3

27c. −1

29a. 2.5011

29b. 3.5926

31a. 0.8954

31b. −0.1046

33a. −1.4559

33b. −2.1549

35a. 3.9890

35b. 5.0370

37a. −2.6451

37b. −4.9618

39a. −5.7764

39b. −2.5744

41a. 3

41b. −2

43a. ¼

43b. −3

45a. 3

45b. −1

47a. ¼

47b. −4

49a. 0.5518

49b. 2.2731

49c. −0.2688

49d. −1.2238

51a. 2.59

51b. 3.07

51c. 19.70

53a. 19.65

53b. 0.20

53c. 0.16

55a. −1.66

55b. 0.73

57a. 1.2091

57b. 1.3383

59a. −2.5850

59b. −0.6826

61. 8.4

63a. 110 db

63b. 70.4 db

63c. 84.2 db

65a. 7.9×10^{-4}

65b. 5.0×10^{-5}

65c. 5.0×10^{-3}

65d. 1.3×10^{-8}

67a. 9.9 days

67b. 27.8 days

67c. 47.1 days

69a. 1/3

69b. −1/3

69c. −1/4

71a. −3

71b. −3

71c. −3

73a. 1.8484

73b. 1.7205

75a. −0.0903

75b. 0.4236

77a. no, ln 1 = 0 and log 0 not defined

77b. no log 0.5 < 0, ln not defined for negative number

79. A = 5, B = 2

PROBLEM SET 8.3 (on page 449)

1a. $\log_3 5 + \log_3 y$

1b. $\log_2 7 + \log_2 x$

3a. $\log_5 x - \log_5 3$

3b. $\log_3 t - \log_3 11$

5a. $12\log_4 x$

5b. $\dfrac{2}{3}\log_2 p$

7a. $\frac{1}{2}\ln Q$

7b. $\frac{1}{3}\ln P$

9a. $\log_3 x + 2\log_3 y - \log_3 4$

9b. $3\log_3 x + \log_3 y - \log_3 4$

11a. $4\log x + \log y$

11b. $\frac{1}{5}\log x + \frac{1}{5}\log y$

13a. $2\log_5 x - 4\log_5 y$

13b. $7\log_3 x - 2\log_3 y$

15a. $\log_3 4x$

15b. $\log_5 3y$

17a. $\ln\frac{4}{x}$

17b. $\log\frac{x}{3}$

19a. $\log x$

19b. $\log_3 y$

21a. $\log_2 x^5$

21b. $\log_3 y^5$

23a. $\log\dfrac{z^3}{y^2}$

23b. $\log_2\dfrac{64}{v^2}$

25. $\log(x^2 - 9)$

27a. 9

27b. y^{-2}

29a. $4x$

29b. $-2y$

31a. $\dfrac{e}{x}$

31b. $\dfrac{e^2}{x^3}$

33. $-\log[H^+]$

35a. 1

35b. -0.40

37a. -1

37b. -0.80

39a. 2.02

39b. -0.05

39c. 0.41

41a. $\frac{4}{7}\log_4 x + \frac{5}{7}\log_4 y$

41b. $\frac{1}{5}\log_5 x + \frac{4}{5}\log_5 y$

43. $7\log c + \frac{2}{9}\log d - \log 5 - \frac{1}{2}\log f$

45. $\ln 4 + \frac{1}{3}\ln x + \frac{1}{6}\ln y$

47a. $\log_a\frac{y}{3}$

47b. $\log_2\dfrac{y}{x^2}$

49. $\ln(a + 1)$

51. $\log_3\dfrac{x+5}{x+1}$

53. $\log_5(5^7 x^3)$

55. $\ln(e^{uv}w)$

57. product: $\log_3 uv = \log_3 u + \log_3 v$

change of base: $= \dfrac{\ln u + \ln v}{\ln 3}$

PROBLEM SET 8.4 (on page 454)

1a.

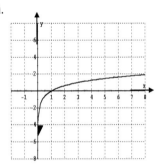

1b.

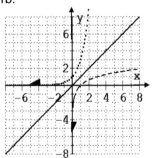

1c. Domain: $(0, \infty)$

  Range: $(-\infty, \infty)$

3. 4

5. 1/81

7. 4

9. 2

11. 3

13. Domain: $(0, \infty)$

  x intercept: $(1,0)$

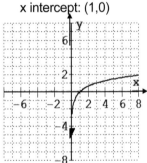

15. Domain: $(0, \infty)$

  x intercept: $(1,0)$

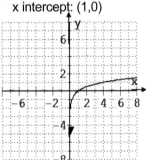

17. Domain: $(0, \infty)$

  x intercept: $(1,0)$

19. Domain: $(0, \infty)$

  x intercept: $(1,0)$

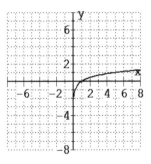

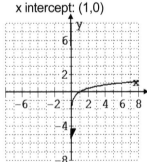

21. c; Domain: $(-\infty, 0)$

23. d; Domain: $(1, \infty)$

25. $g(x) = -\log_3 x$

27. $g(x) = \log_3(x + 1)$

29. Domain: $(-1, \infty)$

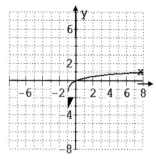

31. Domain: (−4,∞)

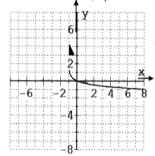

33. Domain: (−1,∞)

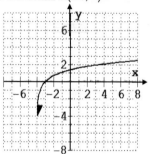

35. Domain: (−∞,0)

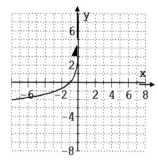

37a. 3300 units
37b. 3000 units
37c. $4,000

39a. $\dfrac{\log x}{\log 3}$

39b. $\dfrac{\ln x}{\ln 3}$

41a. $-\dfrac{\log x}{\log 3}$

41b. $-\dfrac{\ln x}{\ln 3}$

43a. $-2\dfrac{\log x}{\log 7}$

43b. $-2\dfrac{\ln x}{\ln 7}$

45a. $-3\dfrac{\log x}{\log \frac{2}{3}}$

45b. $-3\dfrac{\ln x}{\ln \frac{2}{3}}$

47a. $2 + \dfrac{\log x}{\log 3}$

47b. $2 + \dfrac{\ln x}{\ln 3}$

49.

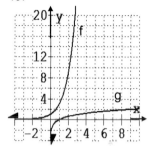

51.

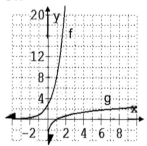

53a. 60
53b. 12 months
55. $f^{-1}(x) = \log_3 x$

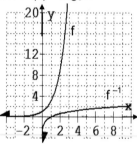

PROBLEM SET 8.5   (on page 462)

1. 1.46	19. 0.19	37a. 42	51b. 27.7 years	67. ±1
3. 0.63	21. 16	37b. 8/15	53. 2000	69. 1/3
5. 1.32	23. −3,1	39. 2	55. 14 years	71. 0, −2/3
7. 3.70	25. 10, −4	41. 3	57. 26.0 years	73. 1, e^2
9. −0.88	27. 2	43. 5	59a. 4222 years	75a. 19 m/s
11. 1.33	29. 5, −2	45. 98	59b. 5728 years	75b. 7.98 sec
13. −0.33	31. 3/2	47. 2/3	61. 3 years	77a. $2,356 thousand
15. −1.93	33. 0.49	49. 14 years	63. 5080ft	77b. $3,486 thousand
17. 13.42	35. 1.70	51a. 3.4 years	65. ¾	77c. $4,049 thousand

CHAPTER 8 REVIEW PROBLEM SET (on page 465)

1a. $\dfrac{1}{7^3} \approx 0.00$

1b. $\dfrac{1}{7^2} \approx 0.02$

1c. 1

1d. 87.85

1e. 15.67

3a. Domain = (−∞,∞)

Range = (0,∞)

↑

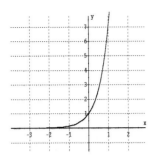

3b. Domain = $(-\infty,\infty)$

　　Range = $(0,\infty)$

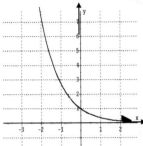

5. g is moved up 2 units from f

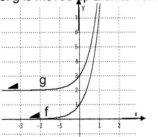

7. g is moved right 3 units from f

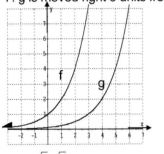

9a. $3^{\sqrt{2}+\sqrt{5}} \approx 55.16$

9b. $7^{\sqrt{6}} \approx 117.51$

11. 420.78

13a. $\log_4 \frac{1}{16} = -2$

13b. $\log_x 5 = y$

15a. $\pi^1 = \pi$

15b. $1/49 = 7^{-2}$

17a. 5/2

17b. -1

19a. 3

19b. ½

19c. 3

19d. 3

21a. 3.8924

21b. 2.0283

21c. -2.9208

21d. -3.0748

23a. 0.6297

23b. 0.1079

23c. not real

23d. not real

25a. 2.59

25b. 1.63

25c. 0.02

27a. 2

27b. 2.5

27c. 2

29a. $\log_6 7 + \log_6 x$

29b. $\ln 4 + 3\ln x$

29c. $\log 4 + \log x + 5\log y$

29d. $\frac{1}{2} \ln x + \frac{3}{4} \ln y$

31a. $\log \frac{2}{9}$

31b. $\ln \frac{x+1}{x}$

31c. $\ln \frac{y}{z}$

33a. x

33b. $\dfrac{1}{3e^2}$

33c. $x^2 - 4$

35. Domain of f = $(0,\infty)$
　　Domain of g = $(0,\infty)$
　　Domain of h = $(-\infty,0)$

h　　　　　　　　f

g

37. Domain of f = $(0,\infty)$
　　Domain of g = $(1,\infty)$
　　Domain of h = $x \neq -1$

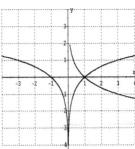

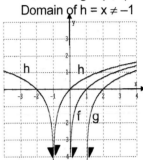

39a. 1

39b. 2.73, -0.73

39c. 0, 4/3

41a. 2.32

41b. 0.23

41c. 0

43a. 50

43b. 1

45. $830.61

47a. 4,848 bacteria

47b. 47 hrs

49. $122,200

51. 107 mph

53. Domain = $(-\infty,\infty)$
　　Range = $[2,\infty)$

55. Domain = $(0,\infty)$
　　Range = $(-\infty,\infty)$

57. Domain = $x \neq 0$
　　Range = $(-\infty,\infty)$

CHAPTER 8 PRACTICE TEST (on page 468)

1a. 3.32, 0.30, 1.50
1b. 0.54, 1.85, 0.81
1c. 4.71, 26.61, 6.23
2. g is f reflected through the x axis. h is f moved up 3 units.

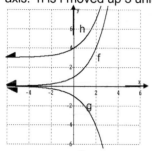

3a. $\log_8 4 = \dfrac{2}{3}$

3b. $\log_{49} \dfrac{1}{7} = -\dfrac{1}{2}$

4a. $25^{-\frac{1}{2}} = \dfrac{1}{5}$

4b. $\left(\dfrac{1}{2}\right)^{-4} = 16$

5a. 3
5b. −3
5c. 3.5

5d. −2
6a. 9
6b. 2.5
6c. 2/7
6d. 1.49
6e. ¼
6f. 0.46
7a. (i) 1.2455, (ii) −2.4169, (iii) 32
7b. (i) 25.25, (ii) 11.20, (iii) 0.05, (iv) 0.10
8a. Domain = $(0,\infty)$
8b. Domain = $(-\infty,0)$
8c. Domain = $(-1,\infty)$
8. g is f reflected through the y axis. h is f moved to the left 1 unit.

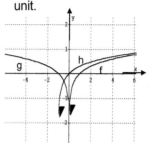

9a. $\log_3 8 + \log_3 x$
9b. $\ln x - \ln 7$

9c. $3\ln x + 4\ln y$
9d. $\frac{2}{3}\log x + \frac{1}{3}\log y$
10a. $\log 2/5$
10b. $\ln \dfrac{x^3}{y^2}$
10c. $\log[x^3(x+2)]$
11a. 0.00
11b. 1.61
11c. 0.49
12. $\dfrac{4B-2A-7}{3}$
13a. \$8,239.16
13b. 4.16 years
14a. 113 mph
14b. 10 miles

—CHAPTER 9—

PROBLEM SET 9.1 (on page 477)

1a. 3
1b. −13
3a. 22
3b. −17
5. 49
7. 4
9. (1/3,2/3)
11. (−2,3)
13. (5,7)

15. (1,−3/2)
17. (2/5, 21/5)
19. $\left(\frac{19}{4},-\frac{3}{4},-\frac{29}{4}\right)$
21. $\left(-5,-\frac{14}{3},-\frac{16}{3}\right)$
23. (3,−1,−1)
25. (−1,2,1)

27. $\left(-5,\frac{5}{3},\frac{28}{3}\right)$
29. 30°, 60°
31. \$4000, \$6500
33. A: 10 oz
     B: 5 oz
     C: 8 oz
35. 25%: 500 lbs

35%: 2000 lbs
40%: 1500 lbs
37. a = 10,000
     b = 1000
     c = 6000
39. 72
41. −1
43. 4
45. 2,8

47a. 6
47b. −6
47c. opposite signs
49a. −4
49b. −40
49c. 10 times larger

PROBLEM SET 9.2 (on page 487)

1. B
3. D
5. x int ± 4, y int ± 3
   vertices (± 4,0)
   foci (± $\sqrt{7}$,0)

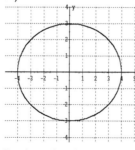

7. x int ± 2, y int ± 4

vertices (0,± 4)
foci (0,± 2$\sqrt{3}$)

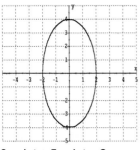

9. x int ± 5, y int ± 2
    vertices (± 5,0)
    foci (±√21 ,0)

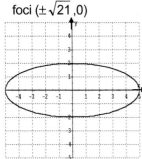

11. x int ± √3
    y int ± √15
    vertices (0, ± √15)
    foci (0, ± 2√3)

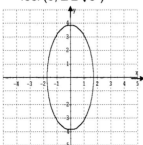

13. x int ± 3
    vertices (± 3,0)
    foci (± √34 ,0)
    asymptotes y = ± 5x/3

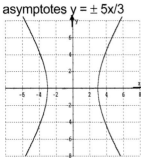

15. x int ±2

vertices (± 2,0)
foci (± √13 ,0)
asymptotes y = ± 3x/2

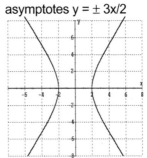

17. x int ± 3
    vertices (± 3,0)
    foci (± √10 ,0)
    asymptotes y = ± x/3

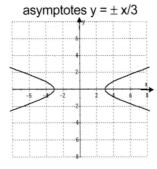

19. y int ± √3
    vertices (0,± √3)
    foci (0,± √15)
    asymptote y = ± x/2

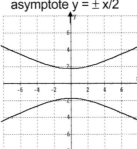

21. y int ± √5
    vertices (0,± √5)
    foci (0,± 3)
    asymptote y = ± √5 x/2
    ↑

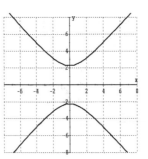

23. $\dfrac{x^2}{4} + \dfrac{y^2}{2} = 1$

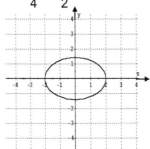

25. $y^2 - x^2 = 1$

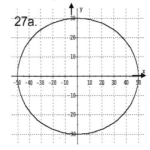

27a.

27b. $\dfrac{x^2}{2500} + \dfrac{y^2}{900} = 1$

27c. foci (± 40,0)

29b. $\dfrac{x^2}{130^2} - \dfrac{y^2}{560.1^2} = 1$
    for x > 0, y ≥ 0

29c. although $d_2 - d_1$ is 260 m, the
    exact distances cannot be
    found.

29a.
↑

        →

                                    →

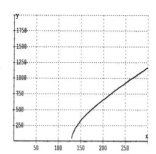

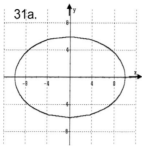

31a.

31b. foci $(\pm 8,0)$
31c. 20
31d. $36\cdot 0 + 100(-6)^2 = 3600$
     $36(10)^2 + 100\,(0) = 3600$

31e. $d_1 = \sqrt{8^2 + 6^2} = 10$

$d_2 = \sqrt{8^2 + 6^2} = 10$

so $d_1 + d_2 = 20$

33. $\dfrac{(x + 2)^2}{9} + \dfrac{(y + 1)^2}{4} = 1$

35. $\dfrac{(x - 3)^2}{9} + \dfrac{(y - 4)^2}{16} = 1$

37. $\dfrac{(x + 2)^2}{4} - \dfrac{(y - 1)^2}{1} = 1$

39. $\dfrac{(y + 3)^2}{16} - \dfrac{(x - 3)^2}{4} = 1$

PROBLEM SET 9.3 (on page 492)

1. $(-3,-4)$, $(2,1)$
3. $(-4/5,-22/5)$, $(2,4)$
5. $(3,-4)$, $(-5,8)$
7. $(\pm 3,2)$, $(\pm 3,-2)$
9. $(\pm 2\sqrt{6},1)$, $(\pm 2\sqrt{6},-1)$
11. $(\pm 2\sqrt{3},2)$, $(\pm 2\sqrt{3},-2)$
13. A $(\pm 6,4/5)$, $(\pm 6, -4/5)$
15. C $(\pm 3,2)$, $(\pm 3,-2)$
17. H, $(-3,10)$, $(4,-11)$
19. D, $(0,6)$, $(4,2)$
21.

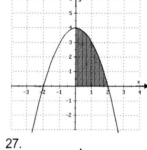

23.

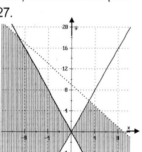

25.

27.

29. 6 in by 8 in

31. x = 5 ft, y = 2 ft or

  x = 4.3 ft, y = 2.7 ft

33. 150 ft by 5 ft
35. 9 m by 12 m or

  6 m by 18 m

37. 8 and 7
39. $(7,3)$, $(-7,-3)$

41. $(1,1)$

43. $(1,0)$, $(2,1)$

45. $(25,4)$
47.

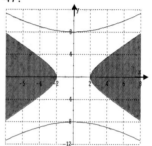

49.

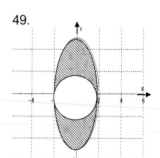

PROBLEM SET 9.4 (on page 498)

1. $-3, -2, -1, 0, 1$
3. $3/2, 4, 15/2, 12, 35/2$
5. $-1, -1, 0, 2, 5$
7. $1, 1/4, 1/9, 1/16, 1/25$
9. $\frac{1}{4}, \frac{1}{2}, 1, 2, 4$
11. $2, 4, 2, 4, 2$
13. $a_n = 2n$

15. $a_n = 2n - 1$
17. $a_n = n^3$
19. $a_n = (-1)^n 3n$
21. $1, \frac{1}{2}, \frac{1}{4}, 1/8, 1/16$
23. $1, 3/2, 2, 5/2, 3$
25. $-10, 0, 10, 20, 30$

27. $-2, -6, -18, -54, -$
162
29. $20,800, $21,632,
  $22,497.28,
  $23,397.17,
  $24,333.06
31. $1.06, $1.12, $1.19

33. –3, 4, –5, 6, –7

35. $\sqrt{2}-1$, $\sqrt{3}-\sqrt{2}$, $2-\sqrt{3}$, $\sqrt{5}-2$, $\sqrt{6}-\sqrt{5}$

37. 32

39. 9/11

41. 3, 3, 3, 3

43. $a_n = \sqrt{3n+2}$

45. 1, 5, 13, 29, 61

47. 1, 1, 2, 3, 5

49.

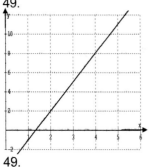

49.

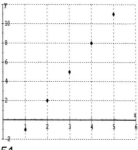

51.

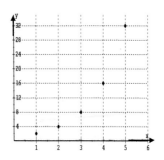

51.

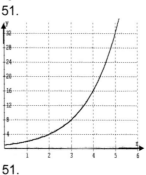

51.

PROBLEM SET 9.5 (on page 503)

1. 511
3. 15
5. 660
7. 15
9. 15
11. 133/60
13. 500

15. $\sum\limits_{k=1}^{6} 4k$

17. $\sum\limits_{k=1}^{6} (9-5k)$

19. $\sum\limits_{k=1}^{6} (2-3k)$

21. $\sum\limits_{k=1}^{5} \dfrac{1}{1.1k+0.1}$

23. 115

25. 5
27. 155
29. 15
31. 6.3
33. end of 1998 = $1040
　　end of 1999 = $1081.60
　　end of 2000 = $1124.86
　　end of 2001 = $1169.86
　　end of 2002 = $1216.65
35a. 20

35b. 40, twice as large
37a. 2.9290
37b. –0.6456
39. 710
41. 935
43. 560
45. 2
47. 3.92
49. 2.39

PROBLEM SET 9.6 (on page 510)

1a. $a_n = 3n-5$
1b. $a_{10} = 25$
1c. $a_{20} = 55$
3a. $a_n = 14-3n$
3b. $a_{10} = -16$
3c. $a_{20} = -46$
5a. $a_n = 3n-1$
5b. $a_{10} = 29$
5c. $a_{20} = 59$
7a. $a_n = 11-6n$
7b. $a_{10} = -49$
7c. $a_{20} = -109$
9a. $a_n = -12+3n$
9b. $a_{10} = 18$
9c. $a_{20} = 48$
11. n = 25
13. n = 30
15. n = 31
17. $a_{10} = 50$

19. $a_6 = 30$
21. $a_{12} = -73$
23. $a_{10} = -27$
25. $a_1 = -4$
　　$a_{20} = 53$
27. $a_1 = 10$
　　$a_{15} = -18$
29. $a_1 = 114$
　　$a_{20} = -19$
31. $a_1 = 3/5$
　　$a_{40} = 24$
33. $S_{10} = 145$
35. $S_8 = -28$
37. $S_{31} = 2914$
39. $S_{16} = 888$
41. $S_{19} = 1045$
43. $S_{31} = 2077$
45a. 16 rows
45b. 46 bricks

47a. $53,600
47b. $1,060,000
49a. $31.35
49b. $3,285
51. 16a + 4b
53. 21
55. $a_n = 23-6(n-1)$

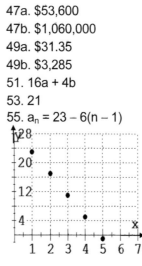

55b. $f(x) = 29-6x$

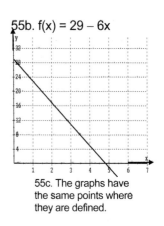

55c. The graphs have the same points where they are defined.

57. $a_n = -3+3(n-1)$

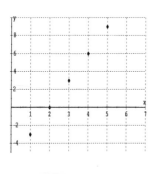

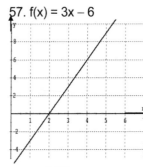

57. $f(x) = 3x - 6$

57c. The graphs have the same points where they are defined.

59a. $a_n = 3n - 1$
$b_n = 2n + 5$
59b. $c_n = 5n + 4$
59c. $p_n = 6n^2 + 13n - 5$

PROBLEM SET 9.7 (on page 518)

1. geometric, $r = 3$
3. geometric, $r = 2/3$
5. geometric, $r = -2/3$
7. geometric, $r = -1/2$
9. geometric, $r = 1/5$
11. arithmetic, $d = -6$
13. arithmetic, $d = 12$
15a. $(-4)\left(-\frac{1}{2}\right)^{n-1}$
15b. $-1/32$
17a. $32\left(\frac{1}{2}\right)^{n-1}$
17b. $1/4$
19a. $6(2)^{n-1}$
19b. 768
21a. $5(2)^{n-1}$
21b. 640
23a. $\frac{1}{2}(-4)^{n-1}$
23b. -8192
25a. $8\left(\frac{1}{2}\right)^{n-1}$
25b. $1/16$
27. 6th
29. 4
31. 94.5
33. 5460
35. 1640

37. $\dfrac{5461}{256}$

39. $1, 2, 4, 8$
41. 2
43. 16
45. $81/2$
47. $12/5$
49a. $\$24{,}000$, $\$25{,}200$, $\$26{,}460$, $\$27{,}783$, $\$29{,}172.15$, $\$30{,}630.76$
49b. no, the raises are geometric not arithmetic
49c. $\$37{,}231.88$
51a. 1262.96 m
51b. 1500 m
53. $\$114{,}358.88$
55. 10
57. $4/5$
59. $S =$
$$\frac{0.7}{1-0.1} = \frac{0.7}{0.9} = \frac{7}{9}$$
61. $x = 16r$ and $9/4 = 16r^2$, so $x = \pm 6$

63a. $a_n = 2^n$

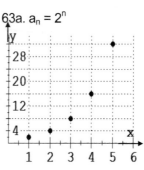

63b. $f(x) = 2^x$

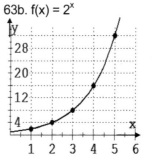

65a. $a_n = \left(\dfrac{3}{2}\right)^n$

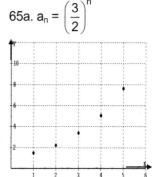

65b. $f(x) = \left(\dfrac{3}{2}\right)^x$

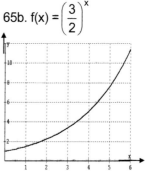

PROBLEMS SET 9.8 (on page 525)

1. $x^4 + 8x^3y + 24x^2y^2 + 32xy^3 + 16y^4$
3. $y^6 + 6y^5 + 15y^4 + 20y^3 + 15y^2 + 6y + 1$
5. $y^8 - 8xy^6 + 24x^2y^4 - 32x^3y^2 + 16x^4$
7. $1/30$
9. $1/9$
11. 2
13. 1
15. $(n + 1)(n)(n - 1)(n - 2)$
17. 56
19. 210

21. 1
23. 66
25. $x^5 + 10x^4 + 40x^3 + 80x^2 + 80x + 32$
27. $x^6 + 12x^4y^2 + 48x^2y^4 + 64y^6$
29. $a^{18} - 6a^{14} + 15a^{10} - 20a^6 + 15a^2 - 6a^{-2} + a^{-6}$
31. $32 + 80\dfrac{x}{y} + 80\dfrac{x^2}{y^2} + 40\dfrac{x^3}{y^3} + 10\dfrac{x^4}{y^4} + \dfrac{x^5}{y^5}$
33. $x^{20} - 20ax^{18} + 180a^2x^{16} - 960a^3x^{14}$

35. $\dfrac{x^3}{16}\sqrt{2x} + \dfrac{7x^3 y}{4} + \dfrac{21x^2 y^2}{2}\sqrt{2x} + 70x^2 y^3$

37. $x^{16} + 16x^{15}y + 120x^{14}y^2 + 560x^{13}y^3 + 1820x^{12}y^4$

39. $a^{11} - 22a^{10}b^2 + 220a^9 b^4 - 1320a^8 b^6 + 5280a^7 b^8$

41. $x^7 - 14x^6 y + 84x^5 y^2 - 280x^4 y^3 + 560x^3 y^4$

43. $a^{27} - 9a^{26} + 36a^{25} - 84a^{24} + 126a^{23}$

45. $20x^3 y^3$

47. $3360\, t^6$

49. $673{,}596x^6 y^6$

51. $16{,}124{,}160{,}000{,}000x^{12}y^7$

53. $\dfrac{224}{243}x^6 a^{12}$

55. $\dfrac{14}{9}a^5 x^{12}$

57. \$11,092.63

59. 82

61. $\dbinom{n}{0}(a+b)^n + \dbinom{n}{1}(a+b)^{n-1}c + \dbinom{n}{2}(a+b)^{n-2}c^2 +.$

$\quad\ldots + \dbinom{n}{n-1}(a+b)^1 c^{n-1} + \dbinom{n}{n}c^n$

63. $7x^6 + 21x^5 h + 35x^4 h^2 + 35x^3 h^3 + 21x^2 h^4 + 7xh^5 + h^6$

CHAPTER 9 REVIEW PROBLEM SET (on page 526)

1a. −19

1b. −10

3. 60

5. 14

7. 4

9. (2,−1)

11. (−1,3,2)

13. x int ± 3; y int ± 2
   foci $(\pm\sqrt{5}, 0)$

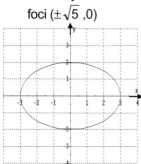

15. x int ± 4
   y int none
   foci $(\pm\sqrt{23}, 0)$

17. $\dfrac{x^2}{9} - \dfrac{y^2}{16} = 1$

19. (3,−4)

21. $(\pm 5, 2),\ (\pm 5, -2)$

23a. −1, 2, −3, 4, −5

23b. 1/4, 1/4, 3/16, 1/8, 5/64

23c. 3, 1, 2, 4, 8

25. $3n^2$

27a. 38

27b. −183

29. $\displaystyle\sum_{k=1}^{6} 2k$

31a. 61

31b. −54

33a. 97

33b. −450

35. 690

37. 175

39. 16th term

41. 512

43. −3

45. −29524/9

47. 50/3

49. 50/7

51. 15/4

53. 1/18

55. 56

57. $x^4 + 12x^3 y + 54x^2 y^2 + 108xy^3 + 81y^4$

59. $243x^5 + 810x^4 + 1080x^3 + 720x^2 + 240x + 32$

61. $69{,}284{,}160x^8 y^3$

63. 80 lbs at \$4.20, 120 lbs at \$3.40

65. \$2410.20 at 9.5%

67. \$30,931 at 15%
   \$22,931 at 8%
   \$ 8,138 at −3%

69. \$10,000 at 5%
   \$14,000 at 4%

71a. \$2020,
   \$2040.20,
   \$2060.60,
   \$2081.21,
   \$2102.02

71b. \$2345.16

73. f(x) = 3 − 2x

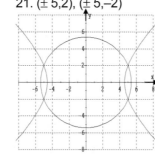

73. $a_n = 3 - 2n$

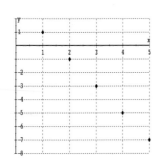

73b. $f(x) = 4x + 1$

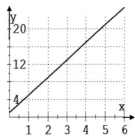

$a_n = 4n + 1$

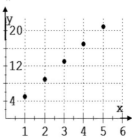

CHAPTER 9 PRACTICE TEST (on page 528)

1a. -1

1b. 7

2a. $(1,1)$

2b. $(1,1,1)$

3a. x int ± 1, y int ± 2
   vertices $(0, \pm 2)$
   foci $(0, \pm \sqrt{3})$

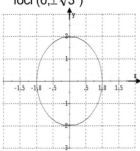

3b. x int ± 3,
   y int none
   vertices $(\pm 3, 0)$
   foci $(\pm \sqrt{13}, 0)$

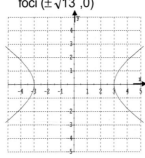

4. $\dfrac{x^2}{25} - \dfrac{y^2}{24} = 1$

5a. $25 - 6n$

5b. $\left(-\dfrac{1}{7}\right)^{n-1}$

6a. 37

6b. 44

7. 16

8a. 144/335

8b. 35

9. $2048x^{11} + 33{,}792x^{10}y + 253{,}440x^9y^2 + 1{,}140{,}480x^8y^3 + 3{,}421{,}440x^7y^4$

10. initial fee $70

   monthly fee $35

Index